# ACI SP-17(14)

# THE REINFORCED CONCRETE DESIGN HANDBOOK

## A Companion to *ACI 318-14*

### VOLUME 1

BUILDING EXAMPLE

STRUCTURAL SYSTEMS

STRUCTURAL ANALYSIS

DURABILITY

ONE-WAY SLABS

TWO-WAY SLABS

BEAMS

DIAPHRAGMS

COLUMNS

STRUCTURAL REINFORCED CONCRETE WALLS

FOUNDATIONS

### VOLUME 2

RETAINING WALLS

SERVICEABILITY

STRUT-AND-TIE MODEL

ANCHORING TO CONCRETE

# ACI SP-17(14)
# Volume 1

# THE REINFORCED CONCRETE DESIGN HANDBOOK

## A Companion to *ACI 318-14*

**Editors:**
**Andrew Taylor**
**Trey Hamilton III**
**Antonio Nanni**

Second Printing
November 2016
ISBN: 978-1-942727-37-8
Errata as of September 2, 2016

# THE REINFORCED CONCRETE DESIGN HANDBOOK
## Volume 1 ~ Ninth Edition

Copyright by the American Concrete Institute, Farmington Hills, MI. All rights reserved. This material may not be reproduced or copied, in whole or part, in any printed, mechanical, electronic, film, or other distribution and storage media, without the written consent of ACI.

The technical committees responsible for ACI committee reports and standards strive to avoid ambiguities, omissions, and errors in these documents. In spite of these efforts, the users of ACI documents occasionally find information or requirements that may be subject to more than one interpretation or may be incomplete or incorrect. Users who have suggestions for the improvement of ACI documents are requested to contact ACI via the errata website at http://concrete.org/Publications/DocumentErrata.aspx. Proper use of this document includes periodically checking for errata for the most up-to-date revisions.

ACI committee documents are intended for the use of individuals who are competent to evaluate the significance and limitations of its content and recommendations and who will accept responsibility for the application of the material it contains. Individuals who use this publication in any way assume all risk and accept total responsibility for the application and use of this information.

All information in this publication is provided "as is" without warranty of any kind, either express or implied, including but not limited to, the implied warranties of merchantability, fitness for a particular purpose or non-infringement.

ACI and its members disclaim liability for damages of any kind, including any special, indirect, incidental, or consequential damages, including without limitation, lost revenues or lost profits, which may result from the use of this publication.

It is the responsibility of the user of this document to establish health and safety practices appropriate to the specific circumstances involved with its use. ACI does not make any representations with regard to health and safety issues and the use of this document. The user must determine the applicability of all regulatory limitations before applying the document and must comply with all applicable laws and regulations, including but not limited to, United States Occupational Safety and Health Administration (OSHA) health and safety standards.

Participation by governmental representatives in the work of the American Concrete Institute and in the development of Institute standards does not constitute governmental endorsement of ACI or the standards that it develops.

Order information: ACI documents are available in print, by download, on CD-ROM, through electronic subscription, or reprint and may be obtained by contacting ACI. Most ACI standards and committee reports are gathered together in the annually revised ACI Manual of Concrete Practice (MCP).

**American Concrete Institute**
**38800 Country Club Drive**
**Farmington Hills, MI 48331 USA**
**+1.248.848.3700**

**Managing Editor:** Khaled Nahlawi
**Staff Engineers:** Daniel W. Falconer, Matthew R. Senecal, Gregory M. Zeisler, and Jerzy Z. Zemajtis
**Technical Editors:** Shannon B. Banchero, Emily H. Bush, and Cherrie L. Fergusson
**Manager, Publishing Services:** Barry Bergin
**Lead Production Editor:** Carl Bischof
**Production Editors:** Kelli Slayden, Kaitlyn Hinman, Tiesha Elam
**Graphic Designers:** Ryan Jay, Aimee Kahaian
**Manufacturing:** Marie Fuller

www.concrete.org

# DEDICATION

This edition of *The Reinforced Concrete Design Handbook, SP-17(14)*, is dedicated to the memory of Daniel W. Falconer and his many contributions to the concrete industry. He was Managing Director of Engineering for the American Concrete Institute from 1998 until his death in July 2015.

Dan was instrumental in the reorganization of *Building Code Requirements for Structural Concrete (ACI 318-14) and Commentary (ACI 318R-14)* as he served as ACI staff liaison to ACI Committee 318, Structural Concrete Building Code; and ACI Subcommittee 318-SC, Steering Committee. His vision was to simplify the use of the Code for practitioners and to illustrate the benefits of the reorganization with this major revision of *SP-17*. His oversight and review comments were instrumental in the development of this Handbook.

An ACI member since 1982, Dan served on ACI Committees 344, Circular Prestressed Concrete Structures, and 373, Circular Concrete Structures Prestressed with Circumferential Tendons. He was also a member of the American Society of Civil Engineers. Prior to joining ACI, Dan held several engineering and marketing positions with VSL Corp. Before that, he was Project Engineer for Skidmore, Owings, and Merrill in Washington, DC. He received his BS in civil engineering from the University of Buffalo, Buffalo, NY and his MS in civil and structural engineering from Lehigh University, Bethlehem, PA. He was a licensed professional engineer in several states.

In his personal life, Dan was an avid golfer, enjoying outings with his three brothers whenever possible. He was also an active member of Our Savior Lutheran Church in Hartland, MI, and a dedicated supporter and follower of the Michigan State Spartans basketball and football programs. Above all, Dan was known as a devoted family man dedicated to his wife of 33 years, Barbara, his children Mark, Elizabeth, Kathryn, and Jonathan, and two grandsons Samuel and Jacob.

In his memory, the ACI Foundation has established an educational memorial. For more information visit http://www.scholarshipcouncil.org/Student-Awards. Dan will be sorely missed for many years to come.

# FOREWORD

*The Reinforced Concrete Design Handbook* provides assistance to professionals engaged in the design of reinforced concrete buildings and related structures. This edition is a major revision that brings it up-to-date with the approach and provisions of *Building Code Requirements for Structural Concrete (ACI 318-14)*. The layout and look of the Handbook have also been updated.

*The Reinforced Concrete Design Handbook* now provides dozens of design examples of various reinforced concrete members, such as one- and two-way slabs, beams, columns, walls, diaphragms, footings, and retaining walls. For consistency, many of the numerical examples are based on a fictitious seven-story reinforced concrete building. There are also many additional design examples not related to the design of the members in the seven story building that illustrate various *ACI 318-14* requirements.

Each example starts with a problem statement, then provides a design solution in a three column format—code provision reference, short discussion, and design calculations— followed by a drawing of reinforcing details, and finally a conclusion elaborating on a certain condition or comparing results of similar problem solutions.

In addition to examples, almost all chapters in the *Reinforced Concrete Design Handbook* contain a general discussion of the related *ACI 318-14* chapter.

All chapters were developed by ACI staff engineers under the auspices of the ACI Technical Activities Committee (TAC). To provide immediate oversight and guidance for this project, TAC appointed three content editors: Andrew Taylor, Trey Hamilton III, and Antonio Nanni. Their reviews and suggestions improved this publication and are appreciated. TAC also appreciates the support of Dirk Bondy and Kenneth Bondy who provided free software to analyze and design the post-tensioned beam example, in addition to valuable comments and suggestions. Thanks also go to JoAnn Browning, David DeValve, Anindya Dutta, Charles Dolan, Matthew Huslig, Ronald Klemencic, James Lai, Steven McCabe, Mike Mota, Hani Nassif, Jose Pincheira, David Rogowski, and Siamak Sattar, who reviewed one or more of the chapters.

Special thanks go to StructurePoint and Computers and Structures, Inc. (SAP 2000 and Etabs) for providing a free copy of their software to perform analyses of structure and members.

Special thanks also go to Stuart Nielsen, who provided the cover art using SketchUp.

*The Reinforced Concrete Design Handbook* is published in two volumes: Chapters 1 through 11 are published in Volume 1 and Chapters 12 through 15 are published in Volume 2. Design aids and a moment interaction diagram Excel spreadsheet are available for free download from the following ACI webpage links:

https://www.concrete.org/store/productdetail.aspx?ItemID=SP1714DAE
https://www.concrete.org/store/productdetail.aspx?ItemID=SP1714DA

**Keywords:** anchoring to concrete; beams; columns; cracking; deflection; diaphragm; durability; flexural strength; footings; frames; piles; pile caps; post-tensioning; punching shear; retaining wall; shear strength; seismic; slabs; splicing; stiffness; structural analysis; structural systems; strut-and-tie; walls.

Khaled Nahlawi
*Managing Editor*

# VOLUME 1: CONTENTS

## CHAPTER 1—BUILDING EXAMPLE
1.1—Introduction, p. 9
1.2—Building plans and elevation, p. 9
1.3—Loads, p. 12
1.4—Material properties, p. 12

## CHAPTER 2—STRUCTURAL SYSTEMS
2.1—Introduction, p. 13
2.2—Materials, p. 13
2.3—Design loads, p. 13
2.4—Structural systems, p. 14
2.5—Floor subassemblies, p. 20
2.6—Foundation design considerations for lateral forces, p. 22
2.7—Structural analysis, p. 23
2.8—Durability, p. 23
2.9—Sustainability, p. 23
2.10—Structural integrity, p. 23
2.11—Fire resistance, p. 23
2.12—Post-tensioned/prestressed construction, p. 23
2.13—Quality assurance, construction, and inspection, p. 23

## CHAPTER 3—STRUCTURAL ANALYSIS
3.1—Introduction, p. 25
3.2—Overview of structural analysis, p. 25
3.3—Hand calculations, p. 26
3.4—Computer programs, p. 26
3.5—Structural analysis in ACI 318, p. 27
3.6—Seismic analysis, p. 29

## CHAPTER 4—DURABILITY
4.1—Introduction, p. 31
4.2—Background, p. 33
4.3—Requirements for concrete in various exposure categories, p. 33
4.4—Concrete evaluation, acceptance, and inspection, p. 35
4.5—Examples, p. 35

## CHAPTER 5—ONE-WAY SLABS
5.1—Introduction, p. 39
5.2—Analysis, p. 39
5.3—Service limits, p. 39
5.4—Required strength, p. 40
5.5—Design strength, p. 40
5.6—Flexure reinforcement detailing, p. 40
5.7—Examples, p. 42

## CHAPTER 6—TWO-WAY SLABS
6.1—Introduction, p. 81
6.2—Analysis, p. 81
6.3—Service limits, p. 81
6.4—Shear strength, p. 82
6.5—Calculation of required shear strength, p. 83
6.6—Calculation of shear reinforcement, p. 84
6.7—Flexural strength, p. 84
6.8—Shear reinforcement detailing, p. 84
6.9—Flexure reinforcement detailing, p. 85
6.10—Examples, p. 88

## CHAPTER 7—BEAMS
7.1—Introduction, p. 133
7.2—Service limits, p. 133
7.3—Analysis, p. 134
7.4—Design strength, p. 134
7.5—Temperature and shrinkage reinforcement, p. 140
7.6—Detailing, p. 140
7.7—Examples, p. 143

## CHAPTER 8—DIAPHRAGMS
8.1—Introduction, p. 281
8.2—Material, p. 281
8.3—Service limits, p. 281
8.4—Analysis, p. 281
8.5—Design strength, p. 283
8.6—Reinforcement detailing, p. 284
8.7—Summary steps, p. 286
8.8—Examples, p. 289

## CHAPTER 9—COLUMNS
9.1—Introduction, p. 353
9.2—General, p. 353
9.3—Design limits, p. 353
9.4—Required strength, p. 354
9.5—Design strength, p. 356
9.6—Reinforcement limits, p. 357
9.7—Reinforcement detailing, p. 357
9.8—Design steps, p. 359
9.9—Examples, p. 362

## CHAPTER 10—STRUCTURAL REINFORCED CONCRETE WALLS
10.1—Introduction, p. 391
10.2—General, p. 391
10.3—Required strength, p. 393
10.4—Design strength, p. 394
10.5—Detailing, p. 398
10.6—Summary, p. 399
10.7—Examples, p. 400

## CHAPTER 11—FOUNDATIONS
11.1—Introduction, p. 419
11.2—Footing design, p. 419
11.3—Design steps, p. 420
11.4—Footings subject to eccentric loading, p. 422
11.5—Combined footing, p. 423
11.6—Examples, p. 425

# CHAPTER 1—BUILDING EXAMPLE

## 1.1—Introduction

The building depicted in this chapter was developed to show how, by various examples in this Handbook, to design and detail a common concrete building according to ACI 318-14. This example building is seven stories above ground and has a one story basement. The building has evenly spaced columns along the grid lines. One column has been removed along Grid C on the second level so that there is open space for the lobby. The building dimensions are:
- Width (north/south) = 72 ft (5 bays @ 14 ft)
- Length (east/west) = 218 ft (6 bays @ 36 ft)
- Height (above ground) = 92 ft
- Basement height = 10 ft

The basement is used for storage, building services and mechanical equipment. It is ten feet high and has an extra column added in every bay along Grids A through F to support a two-way slab at the second level. There are basement walls at the perimeter.

The structural system is an ordinary concrete shear wall in the north/south direction and an ordinary concrete moment frame in east/west direction. These basic systems were chosen as a starting point for the examples. Member examples may be expanded to show how they may be designed in intermediate or special systems but a new structural analysis is not done. The following analysis results provide the moments, shears, and axial loads given in the examples in other chapters in the manual. Those examples may modify this initial data to demonstrate some specific code requirement.

## 1.2—Building plans and elevation

The following building plans and elevation provide the illustration of the example building.

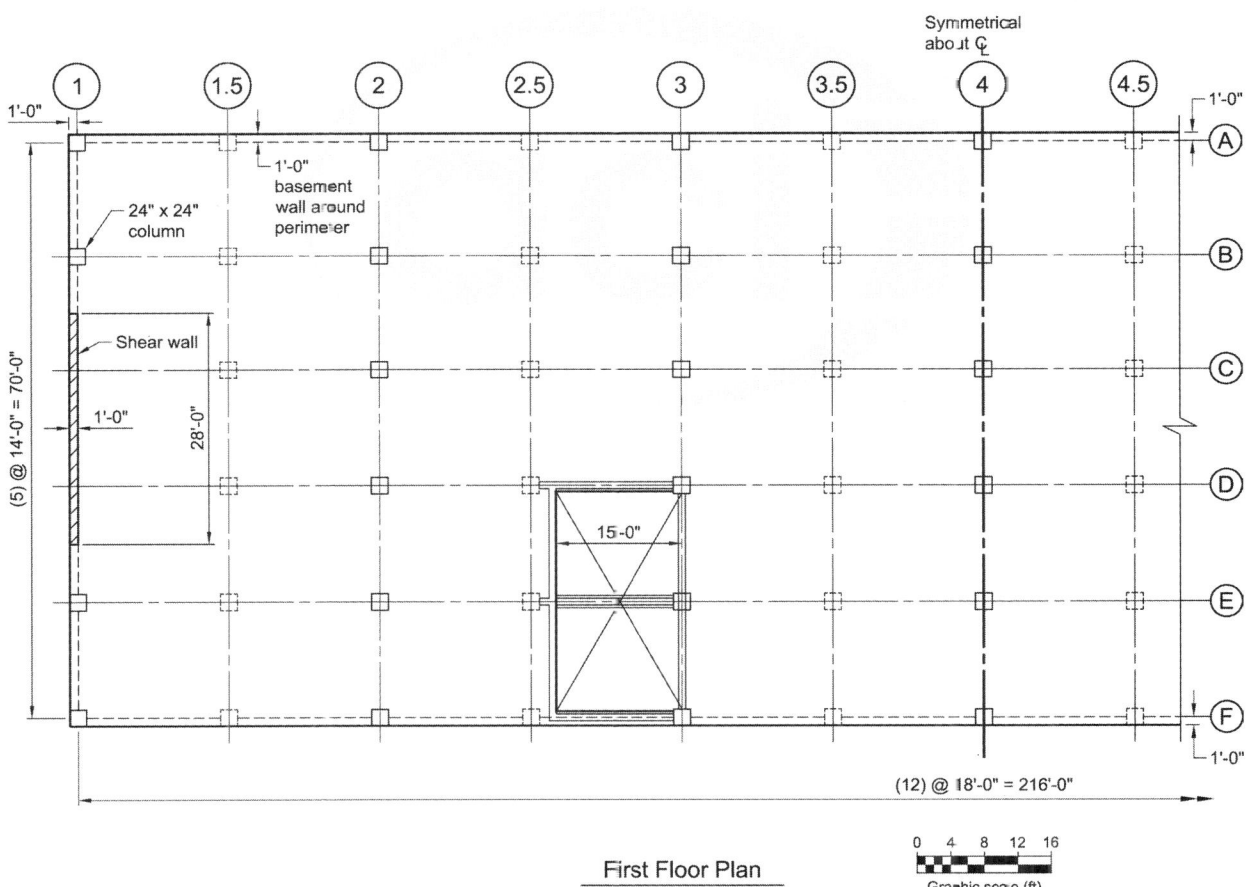

First Floor Plan

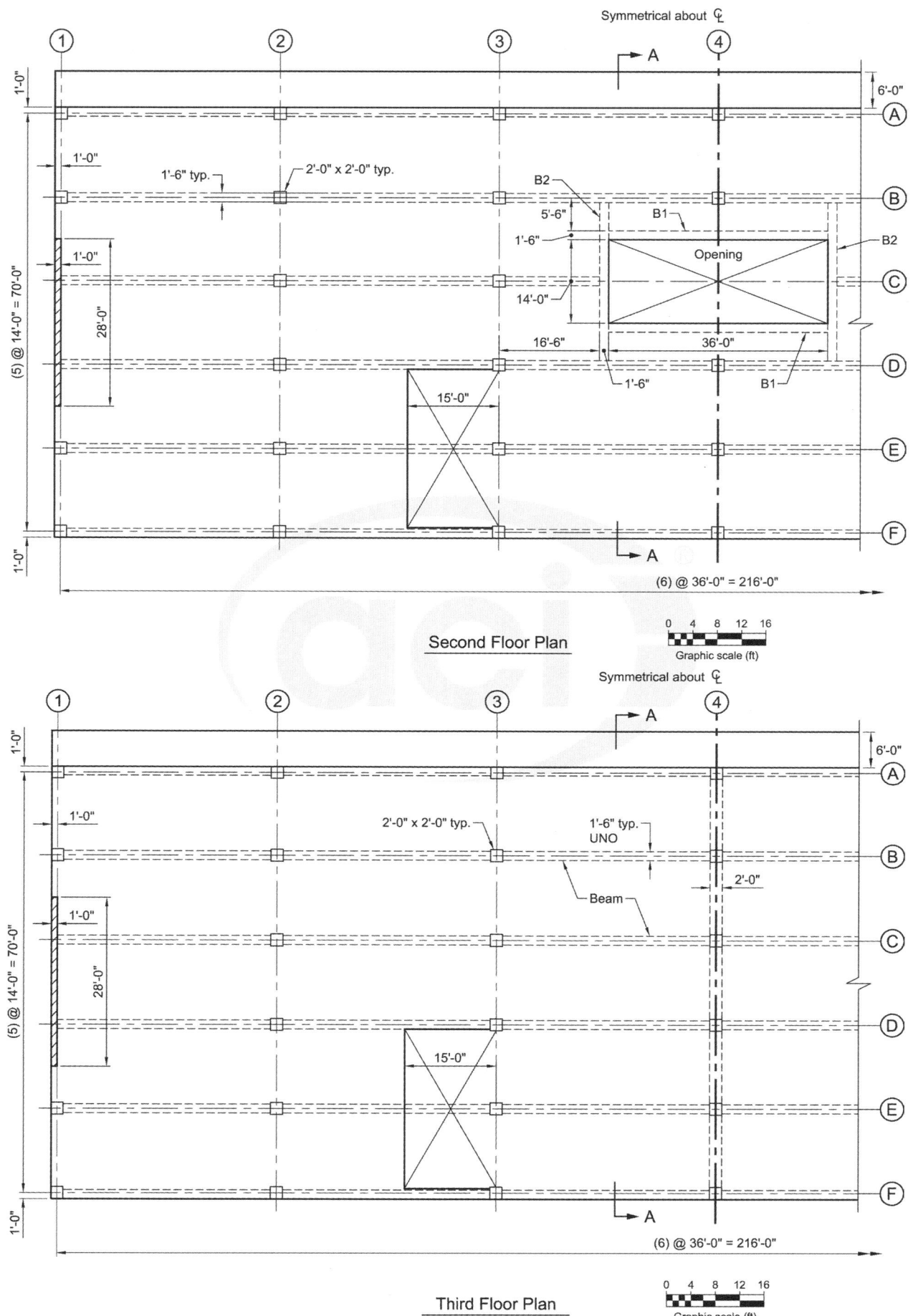

Second Floor Plan

Third Floor Plan

# CHAPTER 1—BUILDING EXAMPLE

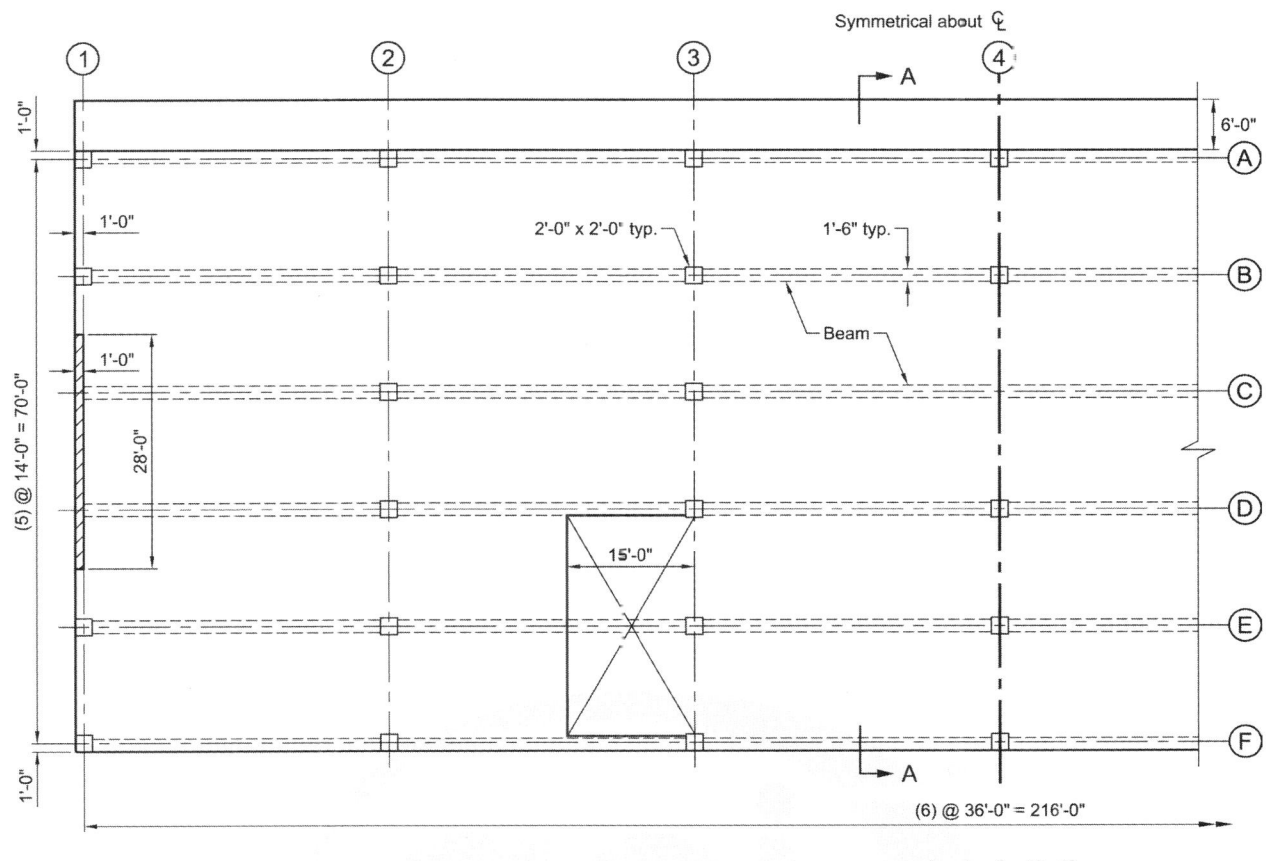

Typical Floor Plan 4th-7th

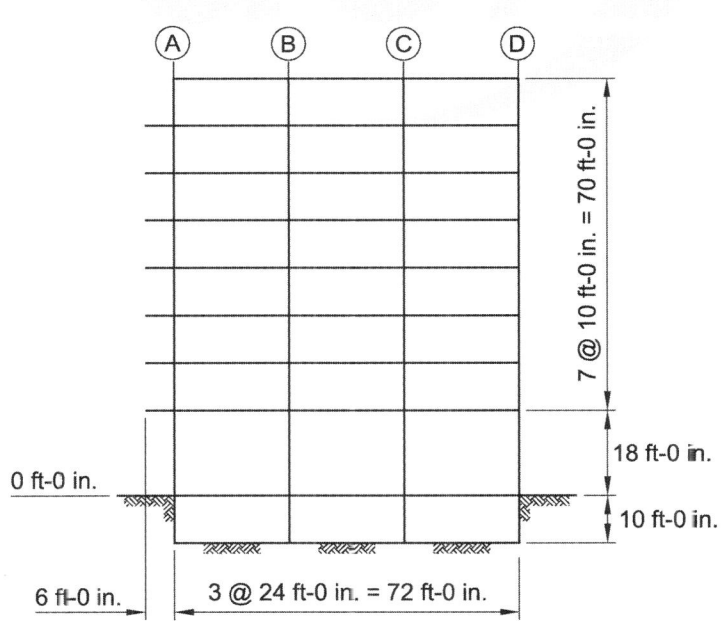

Elevation A-A

## 1.3—Loads
The following loads for the example building are generated in accordance with ASCE7-10. The Risk Category is II.
**Gravity Loads**
Dead Load, $D$:
- Self weight
- Additional D = 15 lb/ft²
- Perimeter walls = 15 lb/ft²

Live Load:
- 1st and 2nd Floors: Lobbies, public rooms, and corridors serving them = 100 lb/ft²
- Typical Floor: Private rooms and corridors serving them = 65 lb/ft²

Roof Live Load:
- Unoccupied = 20 lb/ft²

Snow Load:
- Ground load, $P_g$ = 20 lb/ft²
- Thermal, $C_t$ = 1.0
- Exposure, $C_e$ = 1.0
- Importance, $I_s$ = 1.0
- Flat roof load, $P_f$ = 20 lb/ft²

**Lateral Loads**
Wind Load:
- Basic (ultimate) wind speed = 115 mph
- Exposure category = C
- Wind directionality factor, $K_d$ = 0.85
- Topographic factor, $K_{st}$ = 1.0
- Gust-effect factor, $G_f$ = 0.85 (rigid)
- Internal pressure coefficient, $GC_{pi}$ = +/-0.18
- Directional Procedure

Seismic Load:
- Importance, $I_e$ = 1.0
- Site class = $D$
- $S_S$ = 0.15, $S_{DS}$ = 0.16
- $S_1$ = 0.08, $S_{D1}$ = 0.13
- Seismic design category = $B$
- Equivalent lateral force procedure
- Building frame system; ordinary reinforced concrete shear walls in the north-south direction
  - $R$ = 5
  - $C_s$ = 0.046
- Moment-resting frame system; ordinary reinforced concrete moment frame in the east-west direction
  - $R$ = 3
  - $C_s$ = 0.032

## 1.4—Material properties
The material properties for any building should have a reasonable knowledge of locally available concrete and steel materials. As a preliminary value for this example, a specified concrete compressive strength, $f_c'$, of 4000 psi usually provides for a satisfactory floor design. In the US, reinforcing steel for floor design is usually specified as 60,000 psi.

The $f_c'$ for columns and walls in multi-story buildings may be different than the $f_c'$ used for the floor system. Concrete placement usually proceeds in two stages for each story; first, the vertical members, such as columns, and second, the floor members, such as beams and slabs. It is desirable to keep the concrete strengths of the vertical members within a ratio of 1.4 of the floor concrete strength. Section 15.3.1 in ACI 318-14 states that if this ratio is exceeded, the floor concrete in the area immediately around the vertical members must be "puddled" with higher strength concrete. Usually this situation only becomes an issue for taller buildings.

For this example, the building height is moderate and the loads are typical. The locally available aggregate is a durable dolomitic limestone. Thus, the concrete can readily have a higher $f_c'$ than the initial assumption of 4000 psi. A check of the durability requirements of Table 19.3.2.1 in ACI 318-14 shows that 5000 psi will satisfy the minimum $f_c'$ for all exposure classes. For this concrete, a check of Table 19.2.1.1 in ACI 318-14 shows that all the code minimum limits are satisfied. The following concrete material properties are chosen:
- $f_c'$ = 5000 psi
- Normalweight, $w_c$ = 150 lb/ft³
- $E_c$ = 4,030,000 psi
- $\nu$ = 0.2
- $e_{th}$ = 5.5 × 10⁻⁶/F

The use of lightweight concrete can reduce seismic forces and foundation loads. Based on local experience, however, this type of building won't greatly benefit from the use of lightweight. The modulus of elastic for concrete, $E_c$, is calculated according to 19.2.2 in ACI 318. For normalweight concrete, Eq. 19.2.2.1.b in ACI 318 is applicable. Software programs using finite element analysis can account for the Poisson effect. The Poisson ratio can vary due to material properties, but an average value for concrete is 0.2. Recommendations for the thermal coefficient of expansion, $e_{th}$, of concrete can be found in ACI 209R.

The most common and most available nonprestressed reinforcement is Grade 60. Higher grades are available but 20.2.2.4 in ACI 318-14 limits many uses of reinforcing steel to 60 ksi. The modulus of elastic for reinforcement, Es, is given in 20.2.2.2 in ACI 318.

Reinforcement Material Properties
- $f_y$ = 60,000 psi
- $f_{yt}$ = 60,000 psi
- $E_s$ = 29,000,000 psi

## REFERENCES
*American Concrete Institute*
ACI 209R-92—Prediction of Creep, Shrinkage, and Temperature Effects in Concrete Structures

# CHAPTER 2—STRUCTURAL SYSTEMS

## 2.1—Introduction

A chapter on structural systems of reinforced concrete buildings has been introduced into the ACI Code (ACI 318-14). This chapter gives guidance on the relationships among the different chapters and their applicability to structural systems.

A structural engineer's primary concern is to design buildings that are structurally safe and serviceable under design vertical and lateral loads. Prior to the 1970s, reinforced concrete buildings that were of moderate height (less than 20 stories), not in seismically active areas, or constructed with nonstructural masonry walls and partitions, were seldom explicitly designed for lateral forces (ACI Committee 442 1971). Continuing research, advancement in materials science, and improvements in analysis tools have allowed structural engineers to develop economical building designs with more predictable structural performance.

Structural systems and their component members must provide sufficient stability, strength, and stiffness so that overall structural integrity is maintained, design loads are resisted, and serviceability limits are met. The individual members of a building's structural system are generally assumed to be oriented either vertically or horizontally, with the common exception of parking structure ramps. Chapter 4 of ACI 318-14 identifies the structural members and connection types that are common to reinforced concrete building structural systems with design and detailing code provisions (ACI 318-14):

(a) Horizontal floor and roof members (one-way and two-way slabs, Chapters 7 and 8)

(b) Horizontal support members (beams and joists, Chapter 9)

(c) Vertical members (columns and structural walls, Chapters 10 and 11)

(d) Diaphragms and collectors (Chapter 12)

(e) Foundations—isolated footings, mats, pile caps, and piles (Chapter 13)

(f) Plain concrete—unreinforced foundations, walls, and piers (Chapter 14)

(g) Joints and connections (Chapters 15 and 16)

In Table 2.1, code chapters are correlated with the chapters in Volumes 1 and 2 of this Handbook.

## 2.2—Materials

The concrete mixture proportion needs to satisfy the design properties and limits in ACI 318-14, Chapter 19, and the reinforcing steel needs to satisfy the design properties and limits in Chapter 20 (ACI 318-14).

## 2.3—Design loads

ACI 318-14 assumes that ASCE 7-10 design loads are applied to the building's structural system and to individual members, as applicable. Loads are assumed to be applied vertically and horizontally. Horizontal loads are assumed to act in orthogonal directions. Two types of lateral loads are discussed in this chapter:

1. Wind loading (elastic analysis, ACI 318-14, Chapter 6)
2. Earthquake loading (ACI 318-14, Chapter 18)

Wind and earthquake loads are dynamic in nature; however, they differ in the manner in which these loads are induced in a structure. Wind loads are externally applied loads and, hence, are related to the structure's exposed surface. Earthquake loads are inertial forces related to the magnitude and distribution of the mass in the structure.

**2.3.1** *Wind loading*—Wind kinetic energy is transformed into potential energy when it is resisted by an obstruction. Wind pressure is related to the wind velocity, building height, building surface, the surrounding terrain, and the location and size of other local structures. The structural response to a turbulent wind environment is predominantly in the first mode of vibration.

The quasi-static approach to wind load design has generally proved sufficient. It may not be satisfactory, however, for very tall buildings, especially with respect to the comfort of the occupants and the permissible horizontal movement, "or drift," which can cause the distress of partitions and glass. Therefore, to determine design wind loads for very tall buildings, wind tunnel testing is not unusual.

**2.3.2** *Earthquake loading*—The main objective of structural design is life safety; that is, preserving the lives of occupants and passersby. Serviceability and minimizing economical loss, however, are also important objectives. By studying the results of previous earthquakes on various structural systems, improvements to code provisions and design practices have been achieved. These improvements have led to a reduction in damage of reinforced concrete structures that experience an earthquake. Some code improvements for members that resist significant seismic accelerations are:

### Table 2.1—Member chapters

| Volume no. ACI SP-17(14) | Chapter name ACI SP-17(14) | Chapter no. ACI 318-14 | Chapter no. ACI SP-17(14) |
|---|---|---|---|
| I | Building system | — | 1 |
| I | Structural systems | 4 and 5 | 2 |
| I | Structural analysis | 6 | 3 |
| I | Durability | 19 | 4 |
| I | One-way slab | 7 | 5 |
| I | Two-way slab | 8 | 6 |
| I | Beams | 9 | 7 |
| I | Diaphragm | 12 | 8 |
| I | Columns | 10 | 9 |
| I | Walls | 11 | 10 |
| I | Foundations | 13 | 11 |
| II | Retaining walls | 7 and 11 | 12 |
| II | Serviceability | 24 | 13 |
| II | Strut-and-tie | 23 | 14 |
| II | Anchoring to concrete | 17 | 15 |

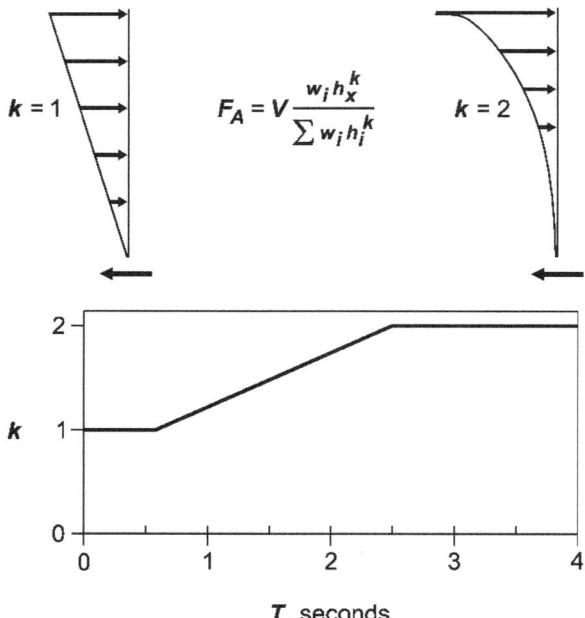

*Fig. 2.3.2—Typical distribution of equivalent static lateral forces representing seismic forces "ASCE 7-10."*

1. The requirement that columns in a frame are flexurally stronger than beams—the so-called "strong column-weak beam" concept
2. Improve detailing to increase ductility and large energy dissipation capacity (with less deterioration in stiffness and strength)
3. Designing and detailing members to ensure flexural yielding before reaching nominal shear strength
4. Designing and detailing the connections to be stronger than the members framing into them
5. Limiting structural system irregularities.

For most structures, the equivalent lateral force procedure given in ASCE 7-10 is used.

Based on this procedure, the distribution of design forces along the height of a building roughly approximates the building's fundamental mode of vibration (Fig. 2.3.2).

Applying recorded earthquake motions to a structure through elastic dynamic analyses usually result in greater force demands than from the earthquake design forces specified by most codes. This is because codes generally account for force reductions due to inelastic response. For example, ASCE 7-10 applies an $R$ factor (response modification factor), which accounts for the ductility of a building, system over-strength, and energy dissipation through the soil-foundation system "ASCE 7-10." It simplifies the seismic design process such that linear static elastic analysis can be used for building designs. The $R$ factor reduces the calculated lateral loads and assumes a building may be damaged during an earthquake event, but will not collapse. The higher the R-value, the lower the lateral design load on a structure. R-values range from 1-1/2 for structures with stiff systems, having low deformation capacity, to 8 for ductile systems, having significant deformation capacity. In a design-level earthquake, it is expected that some building members will yield. To promote appropriate inelastic behavior, ACI 318 contains provisions meant to ensure inelastic deformation capacity in regions where yielding is likely, which then protects the overall integrity and stability of the building.

Dynamic (modal) analysis is commonly used for larger structures, important structures, or for structures with an irregular vertical or horizontal distribution of stiffness or mass. For very important and potentially critical structures—for example, nuclear power plants—inelastic dynamic analysis may be used (ACI Committee 442 1988).

## 2.4—Structural systems

All structures must have a continuous load path that can be traced from all load sources or load application to the foundation. The joints between the vertical members (columns and walls) and the horizontal members (beams, slabs, diaphragms, and foundations) are crucial to this concept. Properly detailed cast-in-place (CIP) reinforced concrete joints transfer moments and shears from the floor into columns and walls, thus creating a continuous load path. The joint design strength (ACI 318-14, Chapter 15) must, of course, adequately resist the factored forces applied to the joint. Refer also to ACI 352R-02, for joint design and detailing information.

Engineers commonly refer to a structure's gravity-load-resisting system (GLRS) and lateral-force-resisting system (LFRS). All members of a CIP reinforced concrete structure contribute to the GLRS and most contribute to both systems. For low-rise structures, the inherent lateral stiffness of the GLRS is often sufficient to resist the design lateral forces without any changes to the design or detailing of the GLRS members. As the building increases in height, the importance of designing and detailing the LRFS to resist lateral loads increases. At some point, stiffness rather than strength will govern the design of the LFRS. In the design process, the type of LFRS is usually influenced by architectural considerations and construction requirements.

There are several types of structural systems or a combination thereof to resist gravity, lateral, and other loads, with deformation behavior as follows:

1. **Frames**—Lateral deformations are primarily due to story shear. The relative story deflections therefore depend on the horizontal shear applied at each story level.
2. **Walls**—Lateral deformations are due to both shear and bending. The behavior predominate mode depends on the wall's height-to-width aspect ratio.
3. **Dual systems**—Dual systems are a combination of moment-resisting frames and structural walls. The moment-resisting frames support gravity loads, and up to 25 percent of the lateral load. The structural walls resist the majority of the lateral loading.
4. **Frames with closely spaced columns, known as cantilevered column system or a tube system**—Lateral deformations are due to both shear and bending, similar to a wall. Wider openings in a tube, however, can produce a behavior intermediate between that of a frame and a wall.

Regardless of the system, a height is reached at which the resistance to lateral sway will govern the design of the

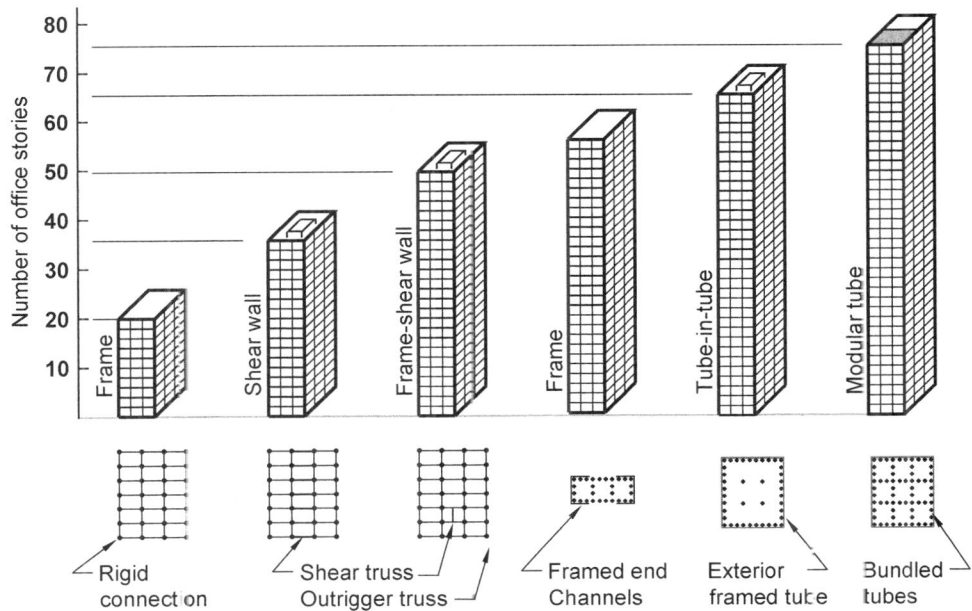

*Fig. 2.4—Structural systems and optimum height limitations (Ali and Moon 2007).*

structural system. At such a height, stiffness, not strength, controls the building design.

ASCE 7 (2010) provides provisions that determine the Seismic Design Categories (SDC A through F). As a building's Seismic Design Category increases, from A through F, ASCE 7 (2010) requires a progressively more rigorous seismic design and a more ductile system to maintain an acceptable level of seismic performance.

ACI 318 provides three categories of earthquake detailing; ordinary, intermediate, and special. These categories provide an increasing level of system toughness.

Building height limits in ASCE 7 (2010) are related to the LFRS.

For buildings in SDC A and B, wind load will usually control the design of the LFRS.

For buildings in SDC C, seismic loads are likely to control design forces, and seismic detailing is required. LFRSs are not limited in height for most systems for this SDC, but interstory drift limits from ASCE 7 (2010) must be met. Again, stiffness, not strength, will likely control the lateral-force-resisting system design.

For buildings in SDC D, E, and F, seismic loads almost always control design forces, and increased seismic detailing is required. LFRS often have maximum height limitations based on assumed structural performance level behavior. Figure 2.4 shows approximate height limits for different structural systems.

Table 2.4 provides ASCE limits for choosing a structural system for a particular building. The ranges of applicability shown are influenced by occupancy requirements, architectural considerations, internal traffic flow (particularly in the lower floors), the structure's height and aspect ratio, and load intensity and types (live, wind, and earthquake).

**2.4.1** *Gravity-resisting systems*—A gravity-load-resisting system (GLRS) is composed of horizontal floor members and vertical members that support the horizontal members. Gravity loads are resisted by reinforced concrete members through axial, flexural, shear, and torsional stiffness and strength. The related deformations are exaggerated and shown in Fig. 2.4.1.

**2.4.2** *Lateral-load-resisting system*—A lateral-force-resisting system (LFRS) must have an adequate toughness to maintain integrity during high wind loading and design earthquake accelerations. Buildings are basically cantilevered members designed for strength (axial, shear, torsion, and moment) and serviceability (deflection and creep must be considered for tall buildings).

ACI 318-14, Section 18.2.1, lists the relevant code sections for each SDC as it applies to a specific seismic-force-resisting system. The following lateral-force-resisting systems are addressed as follows.

**2.4.2.1** *Moment-resisting frames*—Cast-in-place moment-resisting frames derive their load resistance from member strengths and connection rigidity. In a moment-resisting frame structure, the lateral displacement (drift) is the sum of three parts: 1) deformation due to bending in columns, beams, slabs, and joint deformations; 2) deformation due to shear in columns and joints; and 3) deformations due to axial force in columns.

Yielding in the frame members or the foundation can significantly increase the lateral displacement. The effect of secondary moments caused by column axial forces multiplied by lateral deflections ($P$-$\Delta$ effect) will further increase the lateral deflection.

In buildings, moment-resisting frames are usually arranged parallel to the principal orthogonal axes of the structure and the frames are interconnected by floor diaphragms (ACI 318, Chapter 12). Moment-resisting frames usually allow the maximum flexibility in space planning, and are an economical solution up to a certain height.

## Table 2.4—Approximate building height limits for various LFRS

| | Practical limit of system (ASCE 7-10 limit according to SDC) | | | | |
|---|---|---|---|---|---|
| | SDC | | | | |
| Type of LFRS | A and B | C | D | E | F |
| Moment-resisting frames (only): | | | | | |
| Ordinary moment frame (OMF) | NL | NP | NP | NP | NP |
| Intermediate moment frame (IMF) | NL | NL | NP | NP | NP |
| Special Moment frame (SMF) | NL | NL | NL | NL | NL |
| Structural walls (only): | | | | | |
| Building frame systems (structural walls are the primary LFRS and frames are the primary GLRS): | | | | | |
| Ordinary structural wall (OSW) | NL | NL | NP | NP | NP |
| Special structural wall (SSW)* | NL | NL | 160 ft | 160 ft | 100 ft |
| Bearing wall systems (structural walls are the primary lateral- and gravity-load-resisting system): | | | | | |
| OSW | NL | NL | NP | NP | NP |
| SSW* | NL | NL | 160 ft | 160 ft | 100 ft |
| Dual systems (structural walls are the primary LRFS, and the moment-resisting frames carry at least 25% of the lateral load): | | | | | |
| OSW with OMF | NL | NP | NP | NP | NP |
| OSW with IMF | NL | NL | NP | NP | NP |
| OSW with SMF | NL | NL | NP | NP | NP |
| SSW with OMF | NP | NP | NP | NP | NP |
| SSW with IMF | NL | NL | 160 ft | 100 ft | 100 ft |
| SSW with SMF | NL | NL | NL | NL | NL |

*Height limits can be increased per ASCE 7 (2010), Section 12.2.5.4.

Notes: NL = no limit; NP = not permitted.

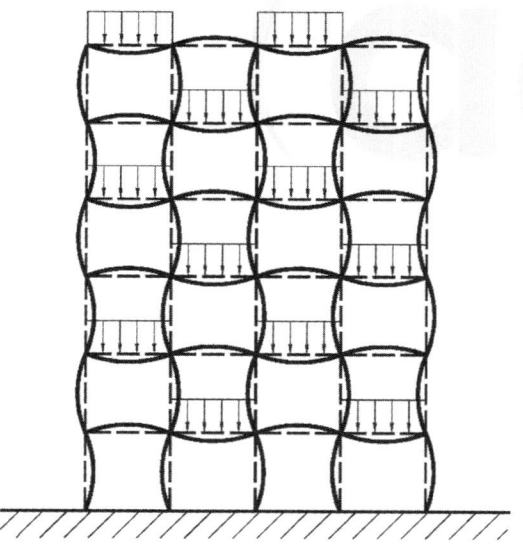

Fig. 2.4.1—Deflections due to gravity load.

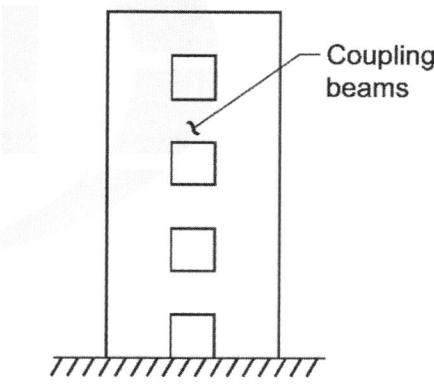

Fig. 2.4.2.2a—Coupled shear walls.

**2.4.2.2** *Shear walls*—Reinforced concrete shear walls are often introduced into multistory buildings because of their high in-plane stiffness and strength to resist lateral forces or when the building program is conducive to layout of structural walls. For buildings without a significant moment frame, shear walls behave as vertical cantilevers. Walls can be designed with or without openings (Fig. 2.4.2.2a). Separate walls can be coupled to act together by beams/slabs or deep beams, depending on design forces and architectural requirements. Coupling of shear walls introduces frame action to the LFRS and thus reduces lateral deflection of the system. Reinforced concrete walls are often used around elevator and stair shafts to achieve the required fire rating. For shear wall types and functions, refer to Table 2.4.2.2.

A shear wall building usually consists of a series of parallel shear walls in orthogonal directions that resists lateral loads and supports vertical loads.

In multistory bearing wall buildings, significant discontinuities in mass, stiffness, and geometry should be avoided. Bearing walls should be located close to the plan perimeter if possible and should preferably be symmetric in plan to reduce torsional effects from lateral loading (refer to Fig. 2.4.2.2c).

**2.4.2.3** *Staggered wall-beam system*—This system uses story-high solid or pierced walls extending across the entire width of the building and supported on two lines of columns placed along exterior faces (Fig. 2.4.2.3). By staggering the locations of these wall beams on alternate floors, large clear areas are created on each floor.

## Table 2.4.2.2—Shear wall types and functions

| Structural walls | Behavior | Reinforcement | Remarks |
|---|---|---|---|
| Short—height-to-length ratio does not exceed 2 | Lateral design is usually concerned only with shear strength. | Bars evenly distributed horizontally and vertically. | Wall foundation must be capable of resisting the actions generated in the wall. Consider sliding resistance provided by foundation. |
| Height-to-length ratio is greater than 2 | Lateral design must consider both the wall's shear and moment strength. | Evenly distributed vertical and horizontal reinforcement. Part of the vertical reinforcement may be concentrated at wall ends—boundary elements. Vertical reinforcement in the web contributes to the flexural strength of the wall. | Wall foundation must be capable of resisting the actions generated in the wall. Consider overturning resistance provided by foundation. |
| Ductile structural wall | Lateral design is heavily influenced by flexure stiffness and strength. | Flexural bar spacing and size should be small enough so that flexural cracking is limited if yielding occurs. Over-reinforcing for flexure is discouraged because flexural yielding is preferred over shear failure. | Acceptable ductility can be obtained with proper attention to axial load level, confinement of concrete, splicing of reinforcement, treatment of construction joints, and prevention of out-of-plane buckling. |
| Coupled walls with shallow coupling beams or slabs (Fig. 2.4.2.2b(a)) | Link slab flexural stiffness deteriorates quickly during inelastic reversed loading. | Place coupling slab bars to limit slab cracking at the stress concentrations at the wall ends. | Punching shear stress around the wall ends in the slab needs to be checked. |
| Coupled walls with coupling beams (Fig. 2.4.2.2b(b)) | Depending on span-to-depth ratio, link beams may be designed as deep beams. | Main reinforcement is placed horizontally or diagonally. For coupling beams with main reinforcement placed diagonally from the deep beam's corner to corner may be confined by spiral or closed ties and designed to resist flexure and shear directly. | Properly detailed coupling beams can achieve ductility. Coupling beams should maintain their load-carrying capacity under reverse inelastic deformation. |
| Infilled frames (structural or nonstructural) (Fig. 2.4.2.2b(c)) | Frames behave as braced frames, increasing the lateral strength and stiffness. The infilling acts as a strut between diagonally opposite frame corners, and creates high shear forces in the columns. | Infill walls should either be sufficiently separated from the moment frame (making them nonstructural), or detailed to be connected structurally with the moment frame. | Uneven infilling can cause irregularities of the moment frame. If there are no infills at a given story level, that story acts as a weak or soft story that is vulnerable to concentrated damage and instability. |

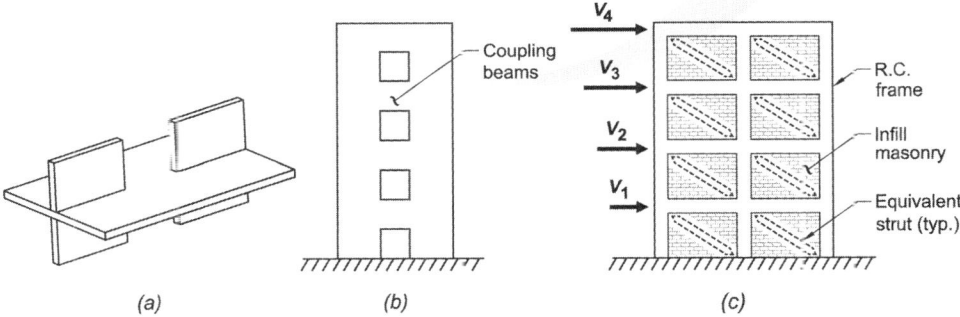

*Fig. 2.4.2.2b—Coupled and infill walls: (a) shallow coupling beams or slabs; (b) coupling beams; and (c) infill walls.*

The staggered wall-beam building is suitable for multi-story construction having permanent interior partitions such as apartments, hotels, and student residences.

An advantage of the wall-beam building is the large open area that can be created in the lower floors when needed for parking, commercial use, or even to allow a highway to pass under the building. This system should be considered in low seismic areas because of the stiffness discontinuity at each floor.

**2.4.2.4** *Tubes*—A tube structure consists of closely spaced columns in a moment frame, generally located around the perimeter of the building (Fig. 2.4.2.4(a)).

Because tube structures generally consist of girders and columns with low span-to-depth ratios (in the range of 2 to 4), shearing deformations often contribute to lateral drift and should be included in analytical models. Tubes are often thought of as behaving like a perforated diaphragm.

Frames parallel to direction of force act like webs to carry the shear from lateral loads, while frames perpendicular to the direction of force act as flanges to carry the moment from lateral loads. Gravity loads are resisted by the exterior frames and interior columns.

A reinforced concrete braced tube is a system in which a tube is stiffened and strengthened by infilling in a diagonal pattern over the faces of the building (Fig. 2.4.2.4(b)). This bracing increases the structure's lateral stiffness, reduces the

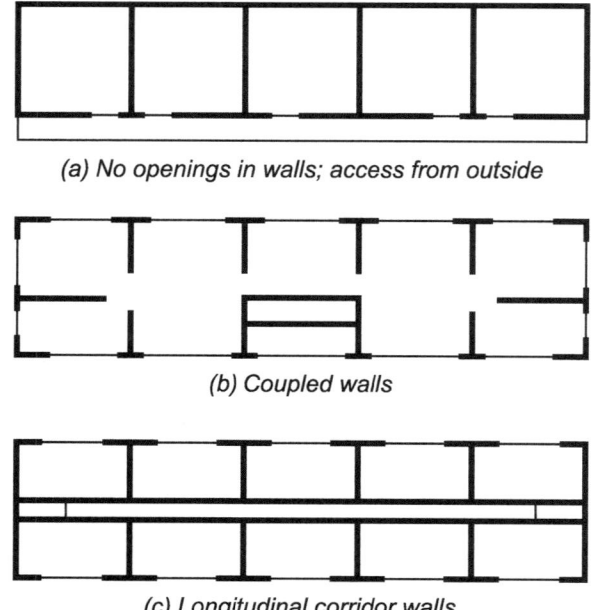

*Fig. 2.4.2.2c—Example of shear wall layouts.*

(a) No openings in walls; access from outside

(b) Coupled walls

(c) Longitudinal corridor walls

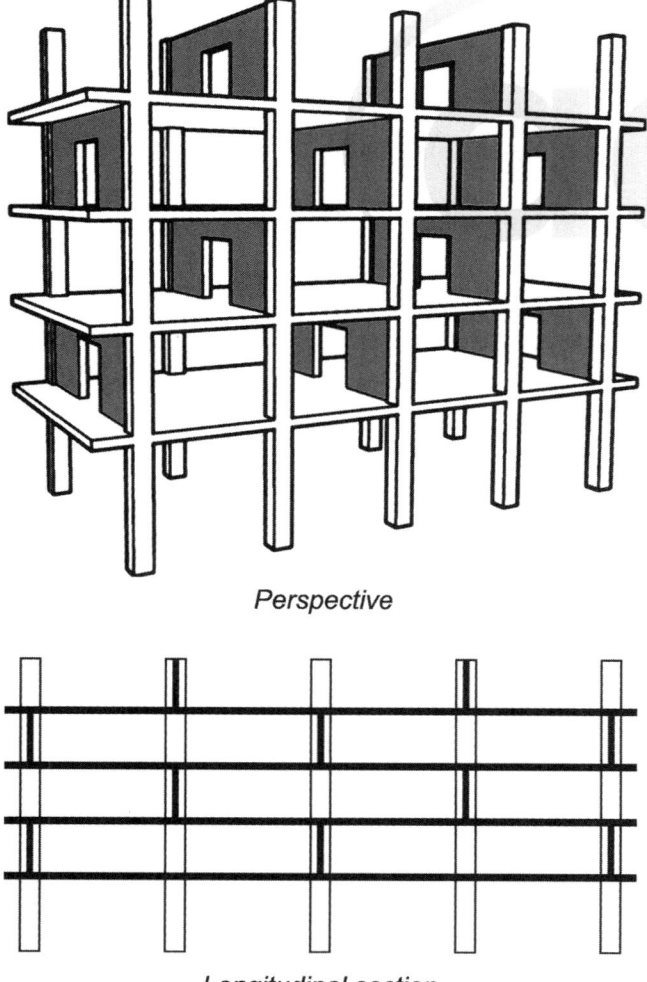

Perspective

Longitudinal section

*Fig. 2.4.2.3—Staggered wall-beam system.*

moments in the columns and girders, and reduces the effects of shear lag.

**2.4.3** *Dual systems*—Dual systems consist of combining two of the structural systems discussed in the previous section. They are used to achieve specific response characteristics, particularly with respect to seismic behavior. Some of the more common dual systems are discussed in 2.4.3.1 through 2.4.3.6.

**2.4.3.1** *Wall-frame systems*—Rigid-jointed frames and isolated or coupled structural walls can be combined to produce an efficient lateral-force-resisting system. Because of the different shear and flexural lateral deflection characteristics of moment frames and structural walls, careful attention to the interaction between the two systems can improve the structure's lateral response to loads by reducing lateral deflections (Fig. 2.4.3.1).

The wall's overturning moment is greatly reduced by interaction with the frame. Because drift compatibility is forced on both the frame and the wall, and the frame-alone and wall-alone drift modes are different, the building's overall lateral stiffness is increased. Design of the frame columns for gravity loads is also simplified in such cases, as the frame columns are assumed to be braced against sidesway by the walls.

The wall-frame dual system permits the structure to be designed for a desired yielding sequence under strong ground motion. Beams can be designed to experience significant yielding before inelastic action occurs at the bases of the walls. By creating a hinge sequence, and considering the relative economy with which yielded beams can be repaired, wall-frame structures are appropriate for use in higher seismic zones. However, note that the variation of shears and overturning moments over the height of the wall and frame is very different under inelastic versus elastic response conditions.

**2.4.3.2** *Outrigger systems*—An outrigger system uses orthogonal walls, girders, or trusses, one or two stories in height, to connect the perimeter columns to central core walls, thus enhancing the structure lateral stiffness (Fig. 2.4.3.2)

In addition to the outrigger girders that extend out from the core, girders or trusses are placed around the perimeter of the structure at the outrigger levels to help distribute lateral forces between the perimeter columns and the core walls. These perimeter girders or trusses are called "hat" or "top-hat" bracing if located at the top, and "belt" bracing if located at intermediate levels. Some further reductions in total drift and core bending moments can be achieved by increasing the cross section of the columns and, therefore, the axial stiffness, and by adding outriggers at more levels. Outriggers are effective in increasing overall building stiffness and, thus, resist wind loads with less drift. Design of outrigger-type systems for SDC D through F must consider the effect of the high local stiffness of the outriggers on the inelastic response of the entire system. Members framing into the outriggers should be detailed for ductile response.

**2.4.3.3** *Tube-in-tube*—For tall buildings with a reasonably large service core, it is generally advantageous to use

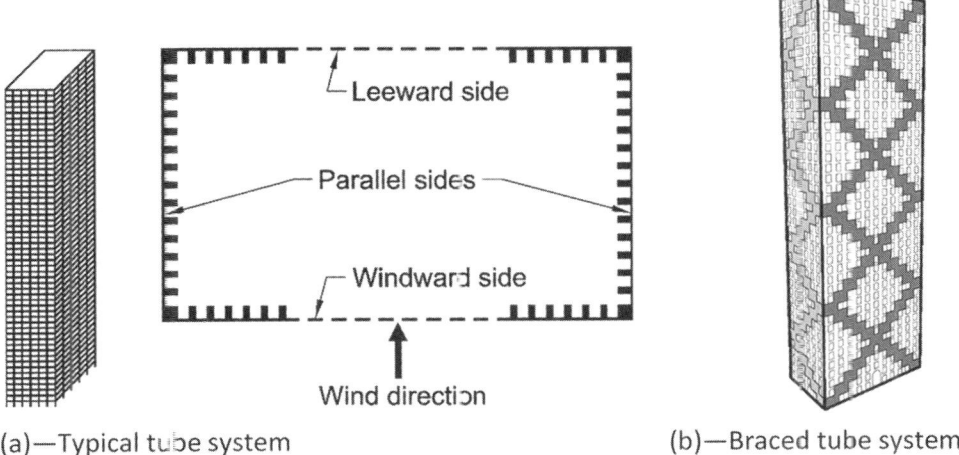

(a)—Typical tube system

(b)—Braced tube system

*Fig. 2.4.2.4—Tube systems.*

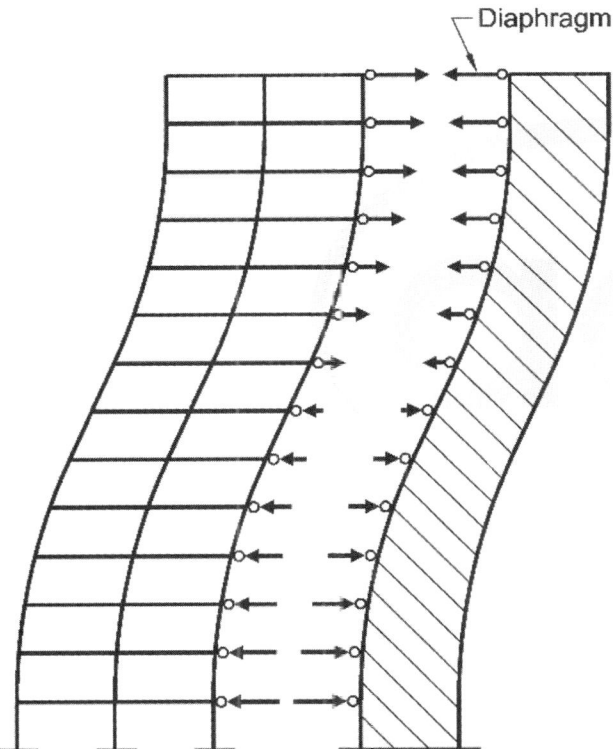

Frame and shear wall connected by floor diaphragm (equal lateral deflection at each level)

*Fig. 2.4.3.1—Shear wall and moment frame system.*

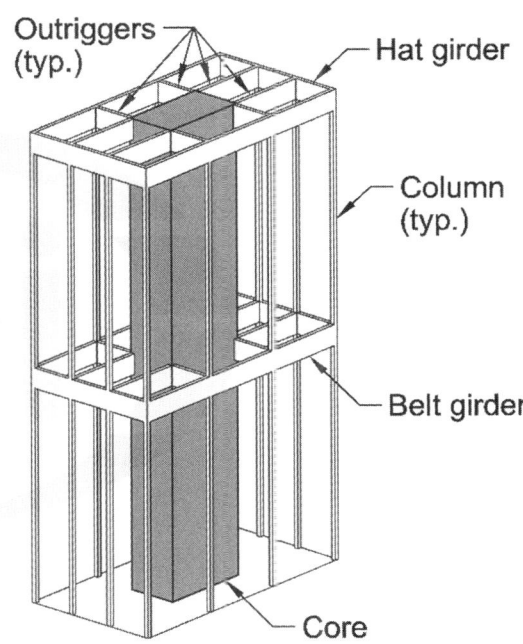

*Fig. 2.4.3.2—Outrigger system.*

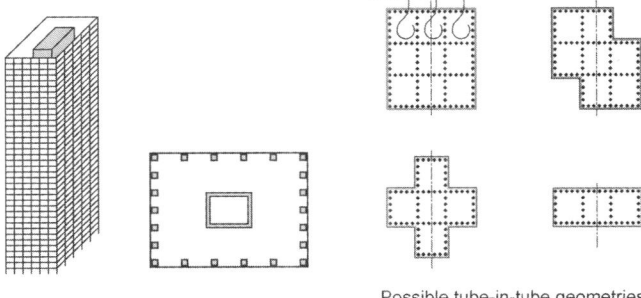

Possible tube-in-tube geometries (Bundled tube structure)

*Fig. 2.4.3.3—Tube-in-tube and bundled tube systems.*

shear walls enclosing the entire service core (inner tube) as part of the lateral-force-resisting system. The outer tube is formed by the closely-spaced column-spandrel beam frame. A bundled tube system consists of several framed tubes bundled into one larger structure that behaves as a multicell perforated box (Fig. 2.4.3.3).

The tube-in-tube system combines the advantages of both the perimeter framed tube and the inner shear walls. The inner shear walls enhance the structural characteristics of the perimeter framed tube by reducing the shear deformation of the columns in the framed tube. The tube-in-tube system can be considered a refined version of the shear wall-frame interaction type structure.

**2.4.3.4** *Bundled tubes*—A bundled tube system consists of several framed tubes bundled into one larger structure that behaves as a multicell perforated box. Individual tubes can be terminated at different heights. The bundled tube system offers considerable flexibility in layout and possesses large torsional and flexural stiffness.

**2.4.3.5** *Mixed concrete-steel structures*—Mixed concrete-steel systems consist of interacting concrete and steel assemblies. The resulting composite structure displays most or all of the advantages of steel structures (large spans and lightweight construction) as well as the favorable characteristics of concrete structures (high lateral stiffness of shear walls and cores, and high damping). Engineers must address the differential vertical creep and shrinkage between steel and concrete to prevent uneven displacement. Because the erection of steel and concrete structures involves different building trades and equipment, engineers who design mixed construction should consider scheduling issues.

**2.4.3.6** *Precast structures*—Precast concrete members are widely used as components in frame, wall, and wall-frame systems. Mixed construction, consisting of precast concrete assemblies connected to a cast-in-place concrete core, is also used. The efficiency of such systems depends on the extent of standardization, the ease of manufacture, the simplicity of assembly, and the speed of erection.

Precast floor systems include large standardized reinforced (and usually prestressed) concrete slabs, with or without interior cylindrical voids (hollow core), as well as prefabricated rib slabs. Rigid-jointed frames are usually assembled from H- or T-units, and shear walls and cores are assembled from prefabricated single-story panels. Planning and designing appropriate connection details for panels, frame members, and floor assemblies is the single most important operation related to prefabricated structures.

Three main types of connections are described as follows:

1. Steel reinforcement bars protruding from adjacent precast members are made continuous by mechanical connectors, welding or lap splices, and the joint between the members is filled with cast-in-place concrete. If welding is used, the engineer should specify appropriate welding procedures to avoid brittle connections.

2. Steel inserts (plates and angles) provided in the precast members are bolted or welded together and the gaps are grouted.

3. The individual precast units are post-tensioned together across the joint, with or without a mortar bed.

The behavior of a precast system subjected to seismic loading depends to a considerable degree on the characteristics of the connections. Connection details can be developed that ensure satisfactory performance under seismic loadings, provided that the engineer pays particular attention to steel ductility and positive confinement of concrete in the joint area.

## 2.5—Floor subassemblies

Selection of the floor system significantly affects a structure's cost as well as the performance of its lateral-force-resisting system. The primary function of a floor system is to resist gravity load. Additional important functions in most buildings are:

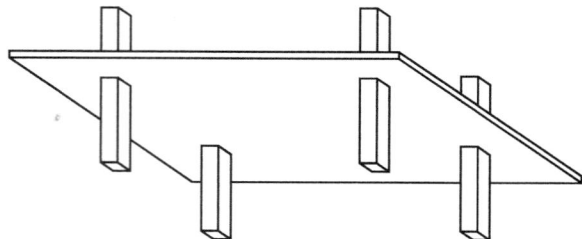

*Fig. 2.5.1—Two-way flat plate system.*

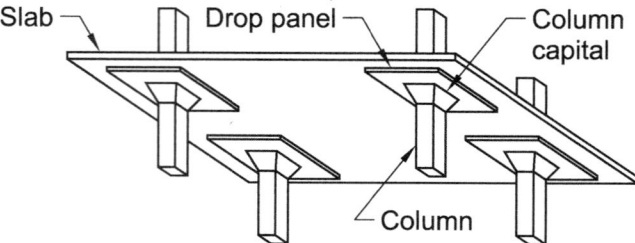

*Fig. 2.5.2—Flat slab with drop panels and capitals.*

(a) Diaphragm action: The slab's in-plane stiffness maintains the plan shape of the structure, and distributes horizontal forces to the lateral-force-resisting system.

(b) Moment resistance: The flexural stiffness of the floors may be an integral and necessary part of the lateral-force-resisting system.

Concrete structures are commonly analyzed for lateral loads assuming the floor system acts as a diaphragm, infinitely stiff in its plane. This assumption is not valid for all configurations and geometries of floor systems. Factors affecting diaphragm stiffness are: span-to-depth ratio of the slab's plan dimensions relative to the location of the lateral-load-resisting members, slab thickness, locations of slab openings and discontinuities, and type of floor system used.

The floor system flexural stiffness can add to the lateral stiffness of the structure. If the slab is assumed to act as part of a frame to resist lateral moments, engineers usually limit the effective slab width (acting as a beam within the frame) to between 25 and 50 percent of the bay width.

**2.5.1** *Flat plates*—A flat plate is a two-way slab supported by columns, without column capitals or drop panels (Fig. 2.5.1).

The flat plate system is a very cost-effective floor for commercial and residential buildings. Simple formwork and reinforcing patterns, as well as lower overall building height, are advantages of this system. In designing and detailing plate-column connections, particular attention must be paid to the transfer of shear and unbalanced moment between the slab and the columns (ACI 318, Chapters 8 and 15). This is achieved by using a sufficient slab thickness or shear reinforcement (stirrups or headed shear studs) at the slab-column joint, and by concentrating slab flexural reinforcement over the column area.

**2.5.2** *Flat slabs with drop panels, column capitals, or both*—The shear strength of flat slabs can be improved by thickening the slab around columns with drop panels, column capitals (either constant thickness or tapered), shear caps, or a combination (Fig. 2.5.2). Like flat plates, flat slab

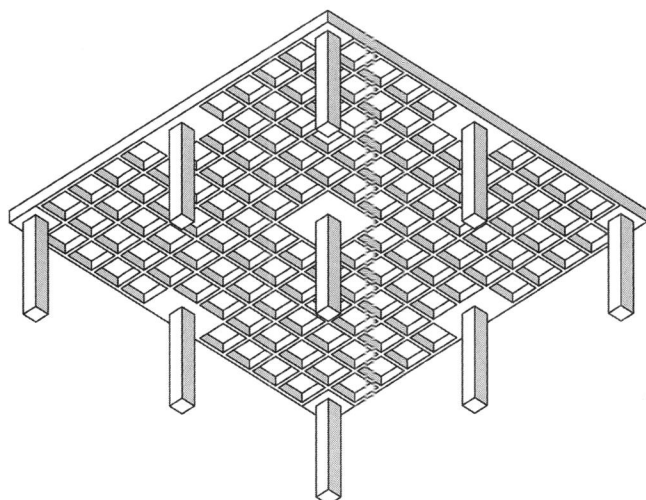

Fig. 2.5.3—Two-way grid (waffle) slab.

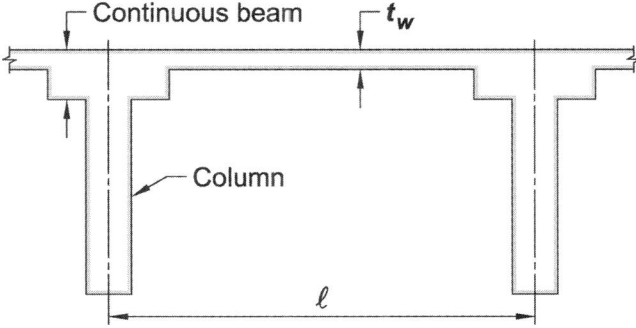

Fig. 2.5.6—One-way banded one-way slab.

systems normally act as diaphragms transmitting lateral forces to columns and walls.

Drop panels increase a slab's flexural and shear strength at the column, and thus improve the ability of the flat slab to participate in the LFRS. Shear caps and column capitals improve the slab shear strength by increasing the slab thickness around the column. To improve the slab shear strength without increasing the slab thickness, engineers can provide closely spaced stirrups or shear studs radiating out from the column.

Lateral-force-resisting systems consisting only of flat slab or flat plate frames, without ductile frames, structural walls, or other bracing members, are unsuitable in high seismic areas (SDC D through F).

**2.5.3** *Two-way grid (waffle) slabs*—For longer spans, a slab system consisting of a grid of ribs intersecting at a constant spacing can be used to achieve an appropriate slab depth for the longer span with much less dead load than a solid slab (Fig. 2.5.3).

The ribs are formed by standardized dome or pan forms that are closely spaced. The slab thickness between the ribs is thin and normally governed by fire rating requirements. Some pans adjacent to the columns are omitted to form a solid concrete drop panel, to satisfy requirements for transfer of shear and unbalanced moment between the slab and columns.

A waffle slab provides an adequate shear diaphragm. The solid slab adjacent to the column provides significant two-way shear strength. Slab flexural and punching shear strength can be increased by the addition of closely spaced stirrups radiating out from the column face in two directions. Stirrups may also be used in the ribs. Because a waffle slab behaves similarly to a flat slab, LRFSs consisting only of waffle slab frames are unsuitable in high seismic design areas (SDC D through F).

**2.5.4** *One-way slabs on beams and girders*—One-way slabs on beams and girders consist of girders that span between columns and beams that span between the girders. One-way slabs span between the beams. This system provides a satisfactory diaphragm, and uses the girder-column frames and beam-column frames to resist lateral loads. Adequate flexural ductility can be obtained by proper detailing of the beam and girder reinforcement.

The beams and slabs can be placed in a composite fashion (with precast elements). If composite, shear connectors are placed at the beam-slab interface to ensure composite action. This system can provide good lateral force resistance, provided that the shear connectors are detailed with sufficient strength and ductility. Some examples of this type of slab system include:

(a) Precast concrete joists with steel shear connectors between the top of the beam and a cast-in-place concrete slab. The concrete joists are usually fabricated to readily support the formwork for the cast-in-place slab. In this system, the joists are supported on walls or cast-in-place concrete beams framing directly into columns.

(b) Steel joists with the top chord embedded in a cast-in-place concrete slab. The slab formwork is supported from the joists, which supports the fresh slab concrete.

(c) Steel beams supporting a noncomposite steel deck with a cast-in-place concrete slab. Note that ACI 318 does not govern the structural design of concrete slabs for composite steel decks.

**2.5.5** *One-way ribbed slabs (joists)*—One-way ribbed slab (joist) systems consist of concrete ribs in one direction, spanning between beams, which span between columns. The size of pan forms available usually determines rib depth and spacing. As with a two-way ribbed system, the thickness of the thin slab between ribs is often determined by the building's fire rating requirements.

This system provides an adequate shear diaphragm and is used in a structure whose lateral resistance comes from a moment-resisting frame or shear walls. One row of pans can be eliminated at column lines, giving a wide, flat beam that may be used as part of the LFRS. Even if the slab system does not form part of the designated LRFS, the engineer should investigate the actions induced in the ribs by building drift.

**2.5.6** *One-way banded slabs*—A one-way banded slab is a continuous drop panel (shallow beam) spanning between columns, usually in the long-span direction, and a one-way slab spanning in the perpendicular direction (Fig. 2.5.6). The shallow beam can be reinforced with closely spaced stirrups near the support to increase the slab's shear strength. This

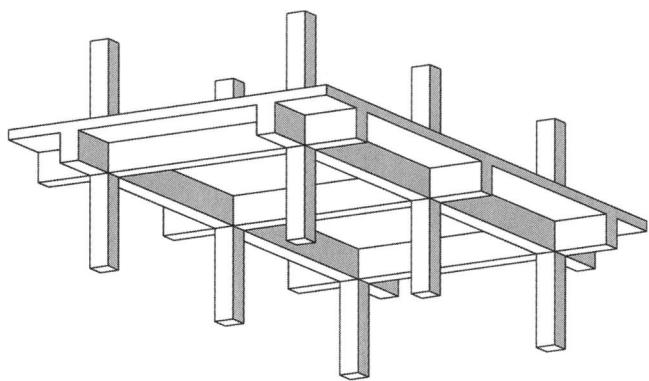

*Fig. 2.5.7—Two-way slab with edge beams around perimeter.*

system is also sometimes referred to as wide-shallow beams with one-way slabs.

A structure using this type of floor system is less stiff laterally than a structure using a ductile moment frame with beams of normal depth. Lateral-force-resisting systems consisting only of flat slab or flat plate frames, without ductile frames, structural walls, or other bracing members, are not suitable in SDC D through F.

**2.5.7** *Two-way slabs with edge beams*—As shown in Fig. 2.5.7, the slab is supported by beams in two directions on the perimeter column lines. This system is useful where a beam-column frame is required as part of the LFRS. The slab provides high diaphragm stiffness, and the perimeter beams can provide sufficient lateral stiffness and strength though frame action for use in SDC D through F.

For longer spans, a two-way grid (Section 2.5.3) slab may be used instead of a flat plate.

**2.5.8** *Precast slabs*—Precast, one-way slabs are usually supported by bearing walls, precast beams, or cast-in-place beams. Precast slabs may be solid, hollow-core slabs, or single- or double-T-sections. They are sometimes topped by a thin cast-in-place concrete layer, referred to as a "topping slab."

Welded connections are normally used to transfer in-plane shear forces between precast slabs and their supports. Because precast slabs are individual units interconnected mechanically, the ability of the assembled floor system to act as a shear diaphragm must be examined by the engineer. Boundary reinforcement may be required, particularly where the lateral-force-resisting members are far apart. In areas of high seismicity, the connections between the precast slab system and the lateral-load-resisting system must be carefully detailed. A concrete topping bonded to the precast slab improves the ability of the slab system to act as a shear diaphragm, and can be used in SDC D through F.

## 2.6—Foundation design considerations for lateral forces

A foundation design must consider the weight of the building, live loads, and the transmission of lateral forces to the ground. A distinction should be drawn between external forces, such as wind, and inertia forces that result from the building's response to ground motions during an earthquake.

External lateral forces can include static pressures due to water, earth or fill, and equivalent static forces representing the effects of wind pressures, where a gust factor or impact factor is included to account for their dynamic nature.

The soil type and strata usually dictate whether the foundation is deep or shallow. A soils report from a licensed geotechnical engineer provides the detailed information and foundation recommendations that the licensed design professional (LDP) needs to design the foundation. For shallow footings, the geotechnical engineer provides an allowable soil-bearing pressure for the soil at the foundation elevation. That pressure limit targets a certain amount of soil deflection, and includes consideration of the anticipated use of the building. If allowable soil pressure is less than 2500 lb/ft$^2$, the soil is very soft and deep foundation options are usually considered. Other soil situations, such as expansive clay or nonstructural fill, may preclude the use of shallow foundations. If the building is below grade, concrete walls can be part of the foundation system.

The two types of deep foundations are caissons (also known as piers) and piles. If hard rock is not far below existing grade, caissons can transfer a column load directly to the bedrock. Bearing values for solid rock can be more than 10 kip/ft$^2$. Caissons are large in diameter, usually starting at approximately 30 in. Piles are generally smaller in diameter, starting at around 12 in., and can be cast-in-place in augered holes or precast piles that are driven into place. Piles are usually designed for lighter loads than caissons are. Groups of piles may be used where bedrock is too deep for a caisson. Tops of piles or caissons are bridged by pile caps and grade beams to distribute column loads as needed.

Shallow foundations are referred to as footings. Types of footings are isolated, combined, and mat. Isolated rectangular or square footings are the most common types. Combined footings are often needed if columns are too close together for two isolated footings, if an exterior column is too close to the boundary line, or if columns are transmitting moments to the footing, such as if the column is part of a lateral-force-resisting system. If the column loads are uniformly large, such as in multistory buildings, or if column spacing is small, mat foundations are considered.

**2.6.1** *Resistance to lateral loads*—The vertical foundation pressures resulting from lateral loads are usually of short duration and constitute a small percentage of the total vertical load effects that govern long-term soil settlements. Allowing a temporary peak in vertical bearing pressures under the influence of short-term lateral loads is usually preferred to making the footing areas larger.

The geotechnical engineer should report the likelihood of liquefaction of sands or granular soils in areas with a high groundwater table, or the possibility of sudden consolidation of loose soils when subjected to jarring. The capacity of friction piles founded in soils susceptible to liquefaction or consolidation should be checked.

**2.6.2** *Resistance to overturning*—The engineer should investigate the safety factor of the foundation against overturning and ensure it is within the limits of the local building code. Overturning calculations should be made with removable soil fill or live load completely removed and should be based on a safe (low) estimate of the building's actual dead load.

## 2.7—Structural analysis

The analysis of concrete structures "shall satisfy compatibility of deformations and equilibrium of forces," as stated in Section 4.5.1 of ACI 318. The LDP may choose any method of analysis as long as these conditions are met. This discussion is intended to be a brief overview of the analysis process as it relates to structural concrete design. For more detailed information on structural analysis, refer to Chapter 3 of this Handbook.

## 2.8—Durability

Reinforced concrete structures are expected to be durable. The design of the concrete mixture proportions should consider exposure to temperature extremes, snow and ice, and ice-removing chemicals. Chapter 19 of ACI 318-14 provides mixture requirements to protect concrete and reinforcement against various exposures and deterioration. Chapter 20 of ACI 318-14 provides concrete cover requirements to protect reinforcement against steel corrosion. For more information, refer to Chapter 4 of this Handbook.

## 2.9—Sustainability

Reinforced concrete structures are expected to be as sustainable as practical. ACI 318 allows sustainability requirements to be incorporated in the design, but they must not override strength and serviceability requirements.

## 2.10—Structural integrity

The ACI Code concept of structural integrity is to "improve the redundancy and ductility in structures so that in the event of damage to a major supporting element or an abnormal loading event, the resulting damage may be confined to a relatively small area and the structure will have a better chance to maintain overall stability" (ACI Committee 442 1971). The Code addresses this concept by providing system continuity through design and detailing rules within the beam and two-way slab chapters.

## 2.11—Fire resistance

Minimum cover specified in Chapter 20 of ACI 318-14 is intended to protect reinforcement against fire; however, the Code does not provide a method to determine the fire rating of a member. The International Building Code (IBC) 2015 Section 722 permits calculations that determine fire ratings to be performed in accordance with ACI 216.1 for concrete, concrete masonry, and clay masonry members.

## 2.12—Post-tensioned/prestressed construction

The introduction of post-tensioning/prestressing to concrete floor, beams, and wall elements imparts an active, permanent force within the structural system. Because cast-in-place structural systems are monolithic, this force affects the behavior of the entire system. The engineer should consider how elastic and plastic deformations, deflections, changes in length, and rotations due to post-tensioning/prestressing affect the entire system. Special attention must be given to the connection of post-tensioned/prestressed members to other members to ensure the proper transfer of forces between, and maintain a continuous load path. Because the post-tensioning/prestressing force is permanent, the system creep and shrinkage effects require attention.

## 2.13—Quality assurance, construction, and inspection

The International Standardization Organization (ISO) defines "quality" as the degree to which a set of inherent characteristics fulfills a set of requirements. The goal of quality assurance is to establish confidence that projects are built in compliance with project construction documents. Chapter 26 of ACI 318-14 contains requirements to facilitate the implementation of competent construction documents, material, construction, and inspection.

## REFERENCES

*American Concrete Institute*
ACI 216.1-14—Code Requirements for Determining Fire Resistance of Concrete and Masonry Construction Assemblies
ACI 352R-02—Recommendations for Design of Beam-Column Connections in Monolithic Reinforced Concrete Structures
ACI 442R-71—Response of Buildings to Lateral Forces
ACI 442R-88—Response of Concrete Buildings to Lateral Forces

*American Society of Civil Engineers*
ASCE 7-10—Minimum Design Loads for Buildings and other Structures

*International Code Council*
IBC 2015 International Building Code

## Authored documents

Ali, M. M., and Moon, K. S., 2007, "Structural Development in Tall Buildings: Current Trends and Future Prospects," *Architectural Science Review*, V. 50, No. 3, pp. 205-223.

# CHAPTER 3—STRUCTURAL ANALYSIS

## 3.1—Introduction

Structural engineers mathematically model reinforced concrete structures, in part or in whole, to calculate member moments, forces, and displacements under the design loads that are specified by a standard such as ASCE 7-10. In all conditions, equilibrium of forces and compatibility of deformations must be maintained. The stiffnesses values of individual members for input into the model, under both service loads and factored loads, are discussed in detail in ACI 318-14, Chapter 6. The factored moments and forces resulting from the analysis are used to determine the required strengths for individual members. The calculated displacements and drift are also checked against commonly accepted serviceability limits.

## 3.2—Overview of structural analysis

**3.2.1** *General*—The analysis of concrete structures "shall satisfy compatibility of deformations and equilibrium of forces," as stated in Section 4.5.1 of ACI 318-14. The licensed design professional (LDP) may choose any method of analysis as long as these conditions are met. ACI 318-14, Chapter 6, is divided into three levels of analysis: 1) elastic first-order; 2) elastic second-order; and 3) inelastic second-order. In addition, ACI 318 permits the use of strut-and-tie modeling for the analysis of discontinuous regions.

Except as noted in Chapter 18, ACI 318 provisions state that the designer may assume that reinforced concrete members behave elastically under design loads. It is also generally acceptable to model concrete members with constant sectional properties along the member length. These assumptions simplify analysis models but they may differ from the actual behavior of the concrete member.

**3.2.2** *Elastic analysis*—Most concrete structures are modeled using an elastic analysis. The stability of columns and walls must be considered by both first-order and second-order analysis. For first-order analysis, end moments of columns and walls are conservatively amplified to account for second-order, or $P$-$\Delta$, effects. For second-order analysis, the second-order effects are calculated directly considering the loads applied on the laterally deformed structure. A series of analyses are made where the secondary moment from each analysis is added to the subsequent analysis until equilibrium is achieved.

**3.2.3** *Inelastic analysis*—Inelastic second-order analysis determines the ultimate capacity of the deformed structure. The analysis may take into account material nonlinearity, member curvature and lateral deformation (second-order effects), duration of loads, shrinkage and creep, and interaction with the supporting foundation. The resulting strength must be compatible with results of published tests. This analysis is used for seismic retrofit of existing buildings; design of materials and systems not covered by the code; and evaluation of building performance above code minimum requirements (Deierlein et al. 2010). This handbook does not include discussion of the inelastic analysis approach.

**3.2.4** *Strut-and-tie*—The strut-and-tie method in Chapter 23, ACI 318-14, is another analysis method that is permitted by ACI 318. This method does not assume that plane sections of unloaded members remain plane under loading. Because this method also provides design provisions, it is considered both an analysis and design method. This method is applicable where the sectional strength assumptions in ACI 318, Chapter 22, do not apply for a discontinuity region of a member or a local area.

**3.2.5** *Analysis types and tools*—ACI 318 identifies three general types of analysis see Section 3.2.1: 1) first-order

### Table 3.2.5—Common analysis types and tools

| Analysis type | Applicable member or assembly | Analysis tool |
|---|---|---|
| First-order Linear elastic Static load Hand calculations | One-way slab | Analysis tables* |
| | Continuous one-way slab | Simplified method in Section 6.5 of ACI 318-14 |
| | Two-way slab | Direct design method in Section 8.10 of ACI 318-14 |
| | | Equivalent frame method in Section 8.11 of ACI 318-14 |
| | Beam | Analysis tables* |
| | Continuous beam | Simplified method in Section 6.5 of ACI 318-14 |
| | Column | Interaction diagrams* |
| | Wall | Interaction diagrams* |
| | | Alternative method for out-of-plane slender wall analysis in Section 11.8 of ACI 318-14 |
| | Two-dimensional frame | Portal method in Section 3.3.3 of this chapter |
| First-order Linear elastic Static load Computer programs | Gravity-only systems | Spreadsheet program based on the analysis tools for hand calculations above |
| | | Program based on matrix methods but only analyze floor assemblies for gravity loads |
| | Two-dimensional frames and walls | Program based on matrix methods without iterative capability |
| Second-order Linear elastic Static or dynamic load Computer programs | Two-dimensional frames and walls | Programs based on matrix methods with iterative capability |
| | Three-dimensional structure | Programs based on finite element methods with iterative capability |
| Second-order Inelastic | Three-dimensional structure | Beyond the scope of this Handbook |

*Information can downloaded from the ACI website; refer to Table of Contents.

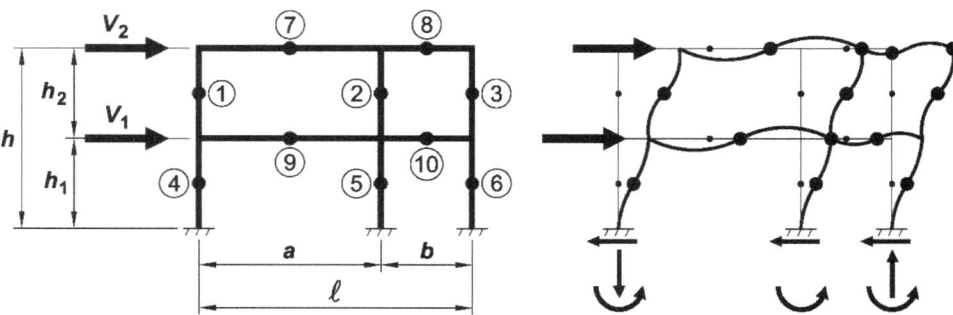

*Fig. 3.3.3—Frame analyzed by portal method.*

linear elastic; 2) second-order linear elastic; and 3) second-order inelastic. Table 3.2.5 shows some common analysis tools used for different analysis methods, loads, and systems.

## 3.3—Hand calculations

**3.3.1** *General*—Before computers became widely available, designers used simplified code equations to calculate gravity design moments and shears (Section 6.5, 8.10, and 8.11 in ACI 318) along with a simplified frame analysis, such as the portal method, to calculate frame moments and shears due to lateral forces. In very limited applications, the design of an entire building using hand calculations is still possible with today's set of building codes. For the large majority of building designs, however, a hand calculation design approach is not practical due to the large number and complexity of design load combinations necessary to fully meet ASCE 7-10 requirements.

**3.3.2** *Code design equations for moment and shear*—The simplified code equations are useful for purposes of preliminary estimating or member sizing, for designing isolated members or subassemblies, and to complete rough checks of computer program output. Because these equations and expressions are easy to incorporate into electronic spreadsheets and equation solvers, they continue to be helpful. In the member chapters of this handbook, examples of hand calculations are provided.

**3.3.3** *Portal method*—The portal method was commonly used before computers were readily available to calculate a frame's moments, shears, and axial forces due to lateral forces (Hibbeler 2015). This method has been virtually abandoned as a design tool with the widespread use of commercial design software programs. The portal method has limitations as stated in the assumptions and considerations that follow, but is still a useful tool for the designer. With complex, three-dimensional modeling becoming commonplace, there is always a chance of modeling error. The portal method allows the designer to independently and quickly find approximate moments and shears in a frame. This can be very useful for spot-checking the program results (Fig. 3.3.3).

The basics assumptions to the portal method are:
(a) Apply only the lateral load to the frame
(b) Exterior columns resist the overturning from lateral loads
(c) Shear at each column is based on plan tributary area
(d) Inflection points are assumed to be located at midheight of column and midspan of beams
(e) Shear in the beam is the difference between column axial forces at a joint
(f) Beam axial force is to be zero

These assumptions reduce a statically indeterminate problem to a statically determinate one.

The following should be considered when using this method:
(a) Discontinuity in geometry or stiffness, such as setbacks, changes in story height, and large changes in member sizes, can cause member moments to differ significantly from those calculated by a computer analysis.
(b) The lateral deformation will be larger than the lateral deformation calculated by a computer analysis.
(c) Axial column deformation is ignored.

## 3.4—Computer programs

**3.4.1** *General*—Computing power and structural software have advanced significantly from the time computers were introduced to the designer. Numerous complex computer programs and specialized analysis tools have been developed, taking advantage of increasing computer speeds. Currently, designers commonly use finite element analysis to design structures. A multistory building only takes minutes of computing time on a personal computer compared to the past, when it would take several hours on a large mainframe computer. Computer software has also greatly improved: user interfaces have become more intuitive; members can be automatically meshed; and input and output data can be reviewed graphically and tabular in a variety of preprogrammed or user-defined menus.

Although three-dimensional models are becoming commonplace, many engineers still analyze the building as a series of two-dimensional frames. Matrix methods mentioned in Table 3.2.5 are programs based on the direct stiffness method. Simpler programs model the structure as discrete members connected at joints. The members can be divided into multiple elements to account for changes in member properties along its length. The two-dimensional stiffness method is relatively easy to program and evaluate. A more sophisticated use of the direct stiffness method is the finite element method. The structure is modeled as discrete elements connected at nodes. Each member of the structure consists of multiple rectangular elements, which more accurately determine the behavior of the member. The act of modeling members as an assembly of these discrete elements is called "meshing" and can be a time-consuming task. A large amount of data is generated from this type of analysis, which may be tedious for the designer to review and process. For straightforward designs, a designer may more efficiently analyze the structure by dividing it into parts and using less complicated programs to analyze each part.

**3.4.2** *Two-dimensional frame modeling*—A building can be divided into parts that are analyzed separately. For example, structures are often symmetrical with regularly spaced columns in both directions. There may be a few isolated areas of the structure where columns are irregularly spaced. These columns can be designed separately for gravity load and checked for deformation compatibility when subjected to the expected overall lateral deflection of the structure.

Buildings designed as moment-resisting frames can often be effectively modeled as a series of parallel planar frames. The complete structure is modeled using orthogonal sets of crossing frames. Compatibility of vertical deflections at crossing points is not required. The geometry of beams can vary depending on the floor system. For slab-column moment frames, it may be possible to model according to the equivalent frame method in Section 8.11 in ACI 318. For beam-column moment frames, it is permitted to model T-beams, with the limits on geometry given in Section 6.3.2 of ACI 318; however, it is often simpler to ignore the slab and model the beams as rectangles. For beams in intermediate or special moment frames, the assumption of a rectangular section may not be conservative; refer to Sections 18.4.2.3, 18.6.5.1, and 18.7.3.2 in ACI 318.

The stiffness of the beam-column joint is underestimated if the beam spans are assumed to extend between column centerlines and the beam is modeled as prismatic along the entire span. Many computer programs thus allow for the beam to be modeled as spanning between faces of columns. To do this, a program may add a rigid zone that extends from the face of the column to the column centerline. If the program does not provide this option, the designer can increase the beam stiffness in the column region 10 to 20 percent to account for this change of rigidity (ACI 442-71).

Walls with aspect ratios of total height to width greater than 2 can sometimes be modeled as column elements. A thin wall may be too slender for a conventional column analysis, and a more detailed evaluation of the boundary elements and panels may be necessary. Where a beam frames into a wall that is modeled as a column element, a rigid link should be provided between the edge of the wall and centerline of the wall. All walls can be modeled using finite element analysis. Where a finite beam element frames into a finite wall element, rotational compatibility should be assured.

Walls with openings can be more difficult to analyze. A wall with openings can be modeled as a frame, but the rigidity of the joints needs to be carefully considered. Similar to the beam-column joint modeling discussed previously, a rigid link should be modeled from the centerline of the wall to the edge of the opening (Fig. 3.4.2a). Finite element analysis and the strut-and-tie method, however, are more commonly used to analyze walls with openings.

For lateral load analysis, all of the parallel plane frames in a building are linked into one plane frame to enforce lateral deformation compatibility. Alternately, two identical frames can be modeled as one frame with doubled stiffness, obtained by doubling the modulus of elasticity. Structural walls, if present, should be linked to the frames at each floor level (Fig. 3.4.2b). Note that torsional effects need to be considered after the lateral deformation compatibility analysis is run. For seismic loads, rigid diaphragms are required to account for accidental torsion according to Section 12.8.4.2 in ASCE 7-10. For wind loads, a torsional moment should be applied according to Fig. 27.4-8 in ASCE 7-10.

**3.4.3** *Three-dimensional modeling*—A three-dimensional model allows the designer to observe structural behavior that a two-dimensional model would not reveal. The effects of structural irregularities and torsional response can be directly analyzed. Current computer software that provides three-dimensional modeling often use finite element analysis with automatic meshing. These high-end programs are capable of running a modal response spectrum analysis, seismic response history procedures, and can perform a host of other time-consuming mathematical tasks.

To reduce computation time, concrete floors are sometimes modeled as rigid diaphragms, reducing the number of dynamic degrees of freedom to only three per floor (two horizontal translations and one rotation about a vertical axis). ASCE 7-10 allows for diaphragms to be modeled as rigid if the following conditions are met:

(a) For seismic loading, no structural irregularities and the span-to-depth ratios are 3 or less (Section 12.3.1.2 in ASCE 7-10)

(b) For wind loading, the span-to-depth ratios are 2 or less (Section 27.5.4 in ASCE 7-10)

If a rigid diaphragm is assumed, the stresses in the diaphragm are not calculated and need to be derived from the reactions in the walls above and below the floor. A semi-rigid diaphragm requires more computation power but provides a distribution of lateral forces and calculates slab stresses. A semi-rigid diaphragm can also be helpful in analyzing torsion effects. For more information on torsion, refer to the code and commentary in Section 12.8 of ASCE 7.

## 3.5—Structural analysis in ACI 318

**3.5.1** *Arrangement of live loads*—Section 4.3.3 in ASCE 7 states that the "full intensity of the appropriately reduced live load applied only to a portion of a structure or member shall be accounted for if it produces a more unfavorable load effect than the same intensity applied over the full structure or member." This is a general requirement that acknowledges greater moments and shears may occur with a pattern load than with a uniform load. There have been a variety of methods used to meet this requirement. Cast-in-place concrete is inherently continuous, and Section 6.4 in ACI 318 provides acceptable arrangements of pattern live load for continuous one-way and two-way floor systems.

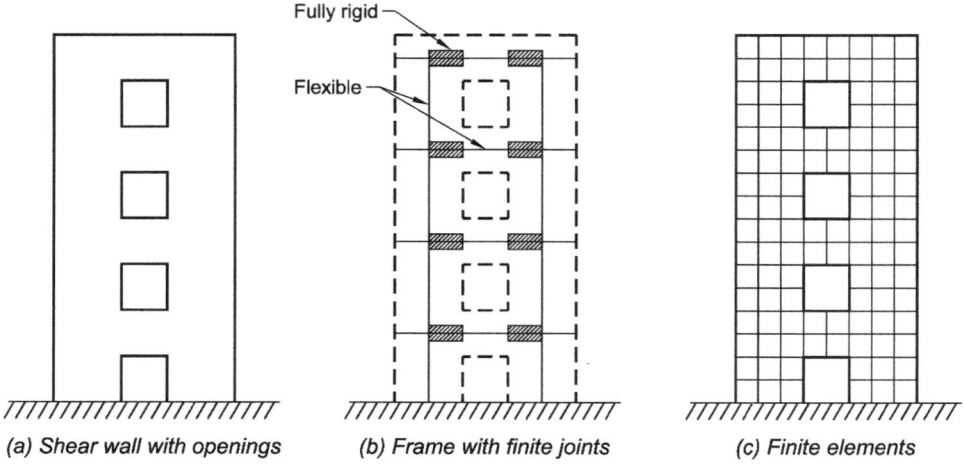

Fig. 3.4.2a—Element and frame analogies.

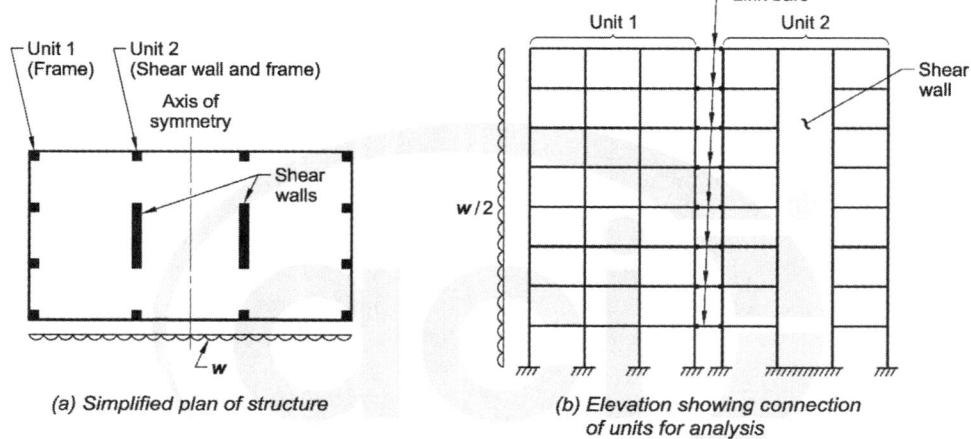

Fig. 3.4.2b—Idealization for plane frame analysis.

**3.5.2** *Simplified method of analysis for nonprestressed continuous beams and one-way slabs*—Section 6.5 in ACI 318 provides approximate equations for conservative design moments and shears, which greatly simplifies the design of continuous floor members. This method is probably used more often to estimate initial member sizes for computer input, or for initial cost estimates, than for final design.

**3.5.3** *First-order analysis*—Code requirements for first-order analysis are provided in Section 6.6 in ACI 318.

**3.5.3.1** *Section properties*—Section properties for elastic analysis are given in Table 6.6.3.1.1(a) of ACI 318. The moment of inertia values have a stiffness reduction $\phi_k$ of 0.875 already applied. These properties are acceptable for the analysis of the structure for strength design. For service level load analysis, the moment of inertia values in Table 6.6.3.1.1(a) can be multiplied by 1.4. Table 6.6.3.1.1(b) offers a more accurate estimation of stiffness by including the effects of axial load, eccentricity, reinforcement ratio, and concrete compressive strength. These equations can also be used to calculate member stiffness at factored load levels by using the factored axial load and moment, as presented, but the equations can be used to calculate member stiffness for any given axial load and moment. These moment-of-inertia equations also have the 0.875 stiffness reduction $\phi_k$ already applied. Section 6.6.3.1.2 of ACI 318 also allows a simplification of using $0.5I_g$ for all members in a lateral load analysis. This is helpful for hand-calculation methods such as the portal method.

It is important to note that the stiffness reduction factor used for moment of inertia discussed previously is for global building behavior. The moment of inertia for second-order effects related to an individual column or wall should have a stiffness reduction $\phi_k$ of 0.75, as discussed in R6.6.4.5.2 of ACI 318.

**3.5.3.2** *Slenderness effects*—A first-order analysis in ACI 318 assumes that only primary stresses are calculated. Secondary stresses caused by the lateral deflection caused by the design loads are not calculated. First-order analysis is typical when hand-calculation methods are used or basic matrix analysis computer programs are used that are not programmed for iterative analysis.

This method ignores the $P$-$\Delta$ effects, which are the second-order moments caused by vertical loads acting on the building's laterally deformed configuration (Fig. 3.5.3.2). To approximately account for these secondary effects, a moment magnifier is applied to first-order column design moments.

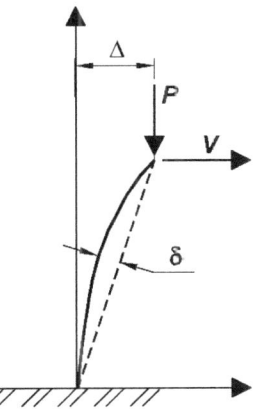

*Fig. 3.5.3.2—P-Δ effects.*

The designer must account for slenderness in a first-order analysis. Figure R6.2.6 in ACI 318 provides a flow chart that illustrates the options to account for slenderness. In summary, slenderness can be neglected if the column or wall meets the requirement of Section 6.2.5 in ACI 318. If slenderness cannot be neglected, the next step is to determine if the building story being analyzed is sway or nonsway. If the story is nonsway, the column or wall end moments are only magnified for $P$-$\delta$ effects along the member. If the story is sway, the column or wall end moments are magnified for $P$-$\Delta$ effects along the member ($P$-$\delta$) and at the ends due to story drift ($P$-$\Delta$).

**3.5.3.3** *Superposition*—Linear analysis allows for superposition to be used when combining loads. This is very helpful when performing hand calculations. The designer calculates the member moment, shear, and axial load for each load. The reactions are then superimposed according to the applicable load combination. Many hand-calculation tools, such as the moment magnification method, assume that the designer is performing a linear analysis with superposition of multiple load effects.

**3.5.3.4** *Redistribution of moments*—ACI 318 allows for the designer to adjust design slab and beam moments and shears by taking advantage of the ductility provided through the code detailing requirements. Ductile detailing is required for continuous fibers at supports and midspan. Moment redistribution can be very helpful in creating economical designs. For example, in a final design, moment redistribution may permit the designer to specify uniform beam sizes over multiple beam spans. If column spacing is not uniform, some beam design moments may be slightly lower than the beam required moments. Once steel yielding has developed at factored loads, however, the moments will redistribute to regions that have not yet yielded, and the beam design moments will satisfy the beam required moments throughout the multiple beam spans.

**3.5.4** *Second-order analysis*—In Section 6.7 of ACI 318, a second-order analysis assumes that the effect of loads on the laterally deformed structure is included in the computer analysis. The initial $P$-$\Delta$ effects on the member due to story drift are computed. A computer algorithm then automatically carries out a series of iterative analyses using the new deflection values until the solution converges to the final secondary moments. Note that linear material properties are used with this method, but the results of a second-order analysis is a nonlinear solution. This is referred to as "geometric nonlinearity." This means that the load cases cannot be computed separately and then combined for the calculation of the secondary moments.

Software should be checked to determine how it accounts for $P$-$\Delta$ effects. Software can easily calculate the additional moment due to building lateral deformation but some software does not calculate the secondary moments along the member length. The designer may have to model the column as at least two segments to capture this effect. Even though a column member is modeled by two elements, the designer must account for the smaller stiffness reduction factor for the moment of inertia (refer to Section 3.5.3.1), because deflection along the member is a local effect. Because of the difficulty of appropriately capturing the secondary moment along the column length, many programs calculate the secondary moments due to lateral deformation and use 6.6.4.5 of ACI 318 in a post-processing program to account for the secondary moment along the length of the column.

**3.5.5** *Inelastic second-order analysis*—The consideration of the nonlinear behavior of structures arising from nonlinearity of the stress-strain curve for concrete and steel reinforcement, particularly under large deformations, can become important in seismic analysis. Nonlinearities in structural response, whether arising from material properties as for concrete or steel, loading conditions (for example, axial load effects on bending stiffness) or geometry (for example, second-order moments) are best handled by numerical iterative or step-by-step procedures. For inelastic second-order analysis, the principle of superposition should not be used. Nonlinear analysis is beyond the scope of this Handbook. Several references that provide further information on nonlinear analysis are ASCE 41-13, FEMA report, and Deierlein et al. (2010).

**3.5.6** *Finite element analysis*—The finite element analysis of concrete structures is permitted by ACI 318 and can be used to satisfy each the first-order, elastic second-order, and inelastic second-order analyses as long as the element types are compatible with the response required.

Section 6.9 in ACI 318 was added to acknowledge that finite element analysis is a widely used and acceptable tool for analysis. Many programs are based on finite element analysis and have sophisticated auto-mesh capabilities. Finite element analysis is a tool that may be used for either linear or nonlinear analyses, but care should be exercised in selecting element types, numerical solver methods, and nonlinear element properties.

## 3.6—Seismic analysis

For seismic loads, the structure may go through multiple cycles of significant inelastic deformations. ASCE 7 provides the equivalent lateral force (ELF) analysis procedure (Section 12.8 of ASCE 7) to allow a linear elastic analysis even though the structure will actually behave inelastically. The ELF procedure is a commonly used design method and is adequate for most structures. The ELF analysis assumes

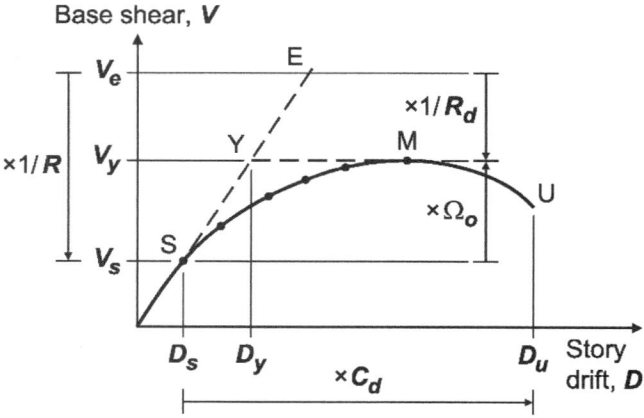

*Fig. 3.6—Inelastic force-deformation curve (SEAOC Seismology Committee 2008).*

an approximately uniform distribution of mass and stiffness along the building height with minor torsional effects. Structures analyzed using the ELF procedure for seismic loads must comply with several limitations. Depending on the Seismic Design Category (SDC), the building height and type, and the type of structural irregularities, a Modal Response Spectrum analysis (Section 12.9 of ASCE 7) or Seismic Response History procedure, either linear or nonlinear (Ch. 16 of ASCE 7), may be required. Regardless of irregularities, the ELF procedure is acceptable for all buildings in SDC A and B up to 160 ft in height.

Section 12.3 of ASCE 7 provides limitations related to types of structural irregularities. ASCE 7 describes five horizontal irregularities: 1) torsion; 2) reentrant corner; 3) diaphragm discontinuity; 4) out-of-plane offset; and 5) nonparallel systems. ASCE 7 also describes five vertical irregularities: 1) stiffness-soft story; 2) weight; 3) vertical geometry; 4) in-plane discontinuity in lateral force-resisting systems (LFRSs); and 5) discontinuity in lateral strength.

Because the actual structure will undergo greater deflections and stresses than predicted by an ELF analysis, ASCE 7 and ACI 318 have additional requirements to account for the anticipated behavior. Lateral-force-resisting systems for concrete structures are defined in ASCE 7 and Chapter 18 of ACI 318. Each LFRS has a response modification coefficient $R$, overstrength factor $\Omega$, and deflection amplification factor $C_d$ (Fig 3.6) used in analysis. These factors account for the difference between the actual expected design forces and displacements and the estimated behavior. Detailed explanations of these factors and their application can be found in ASCE 7 and FEMA P-750.

For reinforced concrete structures, ACI 318 provides structures with the ability to deform inelastically by enforcing special seismic detailing requirements. The seismic detailing requirements in Chapter 18 of ACI 318 are additive to the detailing requirements in the member chapters, or the seismic detailing requirements supersede the member chapter requirements. The detailing requirements for a particular LFRS need to be applied even if seismic loads do not govern the required strength of the structure.

## REFERENCES
*American Concrete Institute*
ACI Committee 442-77—Response of Buildings to Lateral Forces

*American Society of Civil Engineers*
ASCE 7-10—Minimum Design Loads for Buildings and Other Structures
ASCE 41-13—Seismic Evaluation and Retrofit Rehabilitation of Existing Buildings

*Federal Emergency Management Agency*
FEMA 440-05—Improvement of Nonlinear Static Seismic Analysis Procedures
FEMA P-750—NEHRP Recommended Seismic Provisions for New Buildings and Other Structures

## Authored references
Deierlein, G. G.; Reinhorn, A. M.; and Willford, M. R., 2010, "Nonlinear Structural Analysis for Seismic Design," *NEHRP Seismic Design Technical Brief No. 4 (NIST GCR 10-917-5)*, National Institute of Standards and Technology, Gaithersburg, MD, 36 pp.

Hibbeler, R., 2015, *Structural Analysis*, ninth edition, Prentice Hall, New York, 720 pp.

SEAOC Seismology Committee, 2008, "A Brief Guide to Seismic Design Factors," *Structure Magazine*, Sept., pp. 30-32. http://www.structurearchives.org/article.aspx?articleID=756

# CHAPTER 4—DURABILITY

## 4.1—Introduction

Durability of structural concrete is its ability, while in service, to resist possible deterioration due to the surrounding environment, and to maintain its engineering properties. This can be accomplished by proper proportioning and selection of materials for the concrete mixture design. Other aspects influencing durability include reinforcing bar selection, detailing, and construction practices. The ACI 318 Code provides minimum requirements to protect the structure against early serviceability deterioration. Depending on exposure conditions, structural concrete may be required to resist chemical or physical attack, or both. The attack mechanisms the Code covers include exposure to freezing and thawing, soil and water sulfates, wetting and drying, and reinforcement corrosion due to chlorides. All these failure mechanisms depend on transport of water through concrete. For this reason, it is essential to understand the mechanisms themselves and how different concrete-making materials, including admixtures and their proportions, influence concrete's resistance to these mechanisms.

**4.1.1** *Permeability*—Permeability can be defined as "the ease with which a fluid can flow through a solid" or as "the ability of concrete to resist penetration by water or other substances (liquid, gas, or ions)" (ACI 365.1R; Kosmatka and Wilson 2011). Low-permeability concretes are more resistant to resaturation, freezing and thawing, sulfate and chloride ion penetration, and other forms of chemical attack (Kosmatka and Wilson 2011). Concrete permeability is related to porosity (volume of voids/pores in concrete) and connectivity of these pores. Out of the pores present in concrete, capillary pores in cement paste are relevant to concrete durability, as they are responsible for the transport properties of concrete (ACI 201.2R; Kosmatka and Wilson 2011). The influence of capillary porosity in cement paste on permeability was reported by Powers (Fig. 4.1.1a) (Powers 1958).

Concrete permeability, diffusivity, and electrical conductivity can be reduced with lower water-cement ratios ($w/c$), the use of SCMs, and extended moist curing (Kosmatka and Wilson 2011). Effects of the $w/c$ and duration of the moist curing on permeability is presented in Fig. 4.1.1b.

**4.1.2** *Freezing and thawing*—When water freezes in concrete, it causes cement paste to dilate destructively by generating hydraulic and osmotic pressure. While hydraulic pressure forces water away from the freezing water-filled capillary cavities, osmotic pressure is produced by water

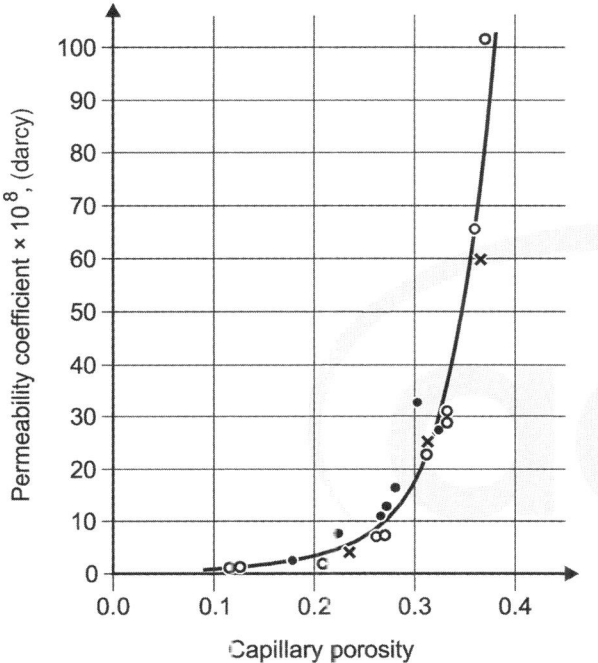

*Fig. 4.1.1a—Permeability versus capillary porosity for cement paste. Different symbols designate different cement pastes (Powers 1958).*

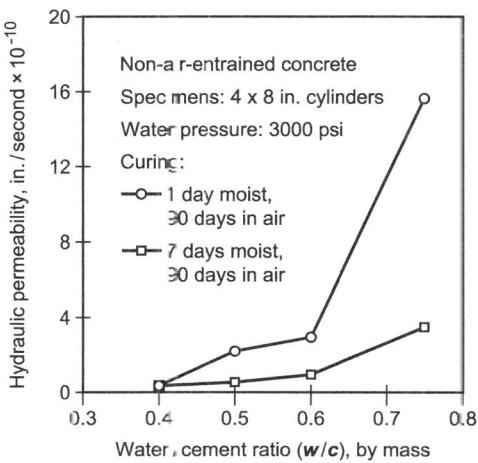 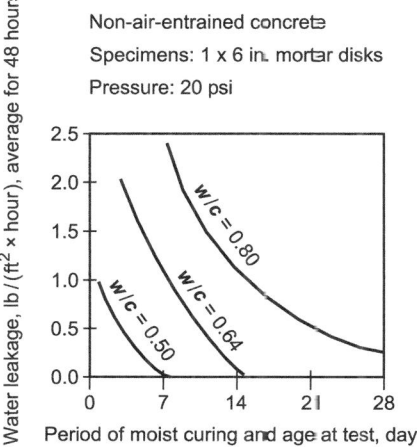

*Fig. 4.1.1b—(left) Effect of $w/c$ and initial curing on hydraulic (water) permeability; and (right) effect of $w/c$ and curing duration on permeability (leakage) of mortar (Kosmatka and Wilson 2011).*

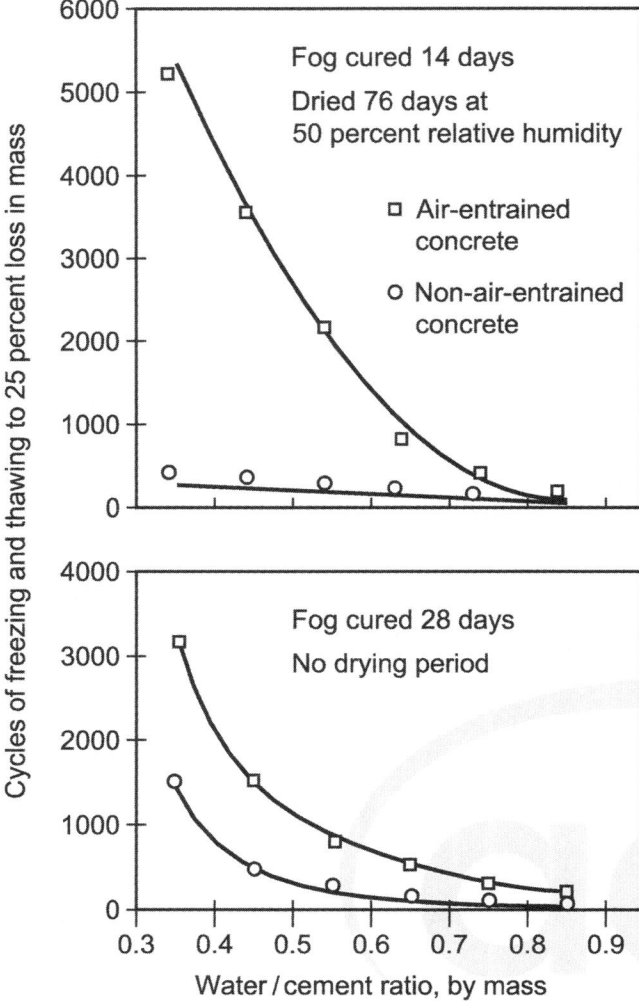

*Fig. 4.1.2—Effect of w/c, air-entrainment, and curing/drying on resistance to freezing and thawing of concrete (Kosmatka et al. 2008).*

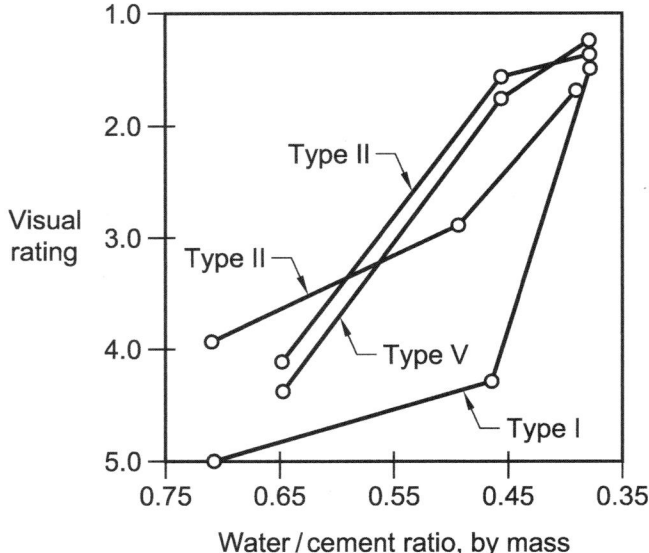

*Fig. 4.1.3—Effect of w/c ratio on sulfate resistance for different ASTM C150/C150M types of cement, (lower visual rating indicates better resistance) (Stark 1989).*

entering partly frozen capillary cavities (Powers 1958). Hydraulic pressures in cement paste are generated by the 9 percent expansion of water when it freezes and changes to ice. For the freezing to take place, a capillary has to reach its critical saturation: 91.7 percent filled with water (Kosmatka and Wilson 2011). Osmotic pressures develop due to various concentrations of alkali solutions in the paste.

When pressure in concrete due to freezing exceeds the tensile strength of concrete, some damage occurs, especially if concrete is saturated with water and exposed to repeated cycles of freezing and thawing. The resulting damage is cumulative as it increases with additional repetitions of freezing and thawing cycles. Deterioration due to freezing and thawing can appear in the form of cracking, scaling, disintegration, or all three of these (Kosmatka and Wilson 2011).

Low permeability and low absorption are main characteristics needed for concrete to be frost resistant, while air-entraining admixtures are used to control the pressure generated in concrete paste during freezing-and-thawing cycles. In other words, high resistance to freezing and thawing is associated with entrained air, low w/c, and a drying period prior to freezing-and-thawing exposure, which is demonstrated by data presented in Fig. 4.1.2.

**4.1.3** *Sulfates*—Sulfates present in soil and water can react with hydrated compounds in the hardened cement paste and induce sufficient pressure to disintegrate the concrete. However, formation of new crystalline substances due to those reactions is partly responsible for the expansion. If water can freely diffuse out of capillaries in the cement paste, the volume of growing crystals cannot exceed the space available to them. At the same time, however, the swelling pressure can also arise from the diffusion of the sulfate salts into the gel pores, which disturbs the equilibrium between the gel and its surrounding liquid phase, resulting in the movement of more water from the outside into the gel pores (Hewlett 1998).

Although ordinary portland cements are most susceptible to sulfate attack, the use of sulfate-resistant cements will not stop the sulfate attack, either (Fig. 4.1.3). Resistance to sulfate attack can be greatly increased by decreasing the permeability of concrete through reduction of the water-cementitious material ratio (*w/cm*) (Stark 1989).

**4.1.4** *Corrosion*—Alkaline nature of concrete (pH greater than 13) will induce formation of a passive, noncorroding layer on reinforcing steel. If, however, chloride ions are present in concrete, they can reach and disrupt that layer and lead to corrosion of steel in the presence of water and oxygen. Once corrosion initiates, corrosion products form and may cause cracking, spalling, or delamination of concrete. This allows for easier access of aggressive agents to the steel surface and increases the rate of corrosion. Cross-sectional area of the corroding steel will decrease and the load-carrying capacity of the member will be reduced (Neville 2003).

Chlorides can be introduced to concrete with materials used to produce the mixture (contaminated aggregate or water, or some admixtures), with deicing chemicals, or through marine exposure (seawater or brackish water). To reduce likelihood of corrosion initiation, total chloride-ion content should not exceed a certain concentration value, referred to as the chloride threshold. A literature review of reported chloride threshold values revealed "that there is no single threshold value, but a range based on the conditions and materials in use" and were found to vary from 0.1 to 1 percent by mass of cement (Taylor et al. 1999). ACI 318-14 limits water-soluble chlorides to 0.15 percent by mass of cement for concrete exposed to external chlorides or seawater. Value of 0.40 percent total chloride by mass of cement is given in British and European Standards (Neville 2003; Whiting et al. 2002).

Corrosion of the reinforcing steel in concrete can be reduced or prevented by minimizing the $w/cm$ ratio (permeability), ensuring maximum cover depth of concrete over steel (Stark 2001), and use of corrosion inhibitors or corrosion resistant steel.

## 4.2—Background

To produce durable structural concrete, concrete materials and mixture proportions are selected based on design strength requirements, anticipated exposure conditions, and required service life of the structure. The selection of materials and mixture proportions has to be accompanied by appropriate field practices, such as quality control, testing, inspection; and proper placement, finishing, and curing practices.

As stated in Section 1.3.1 of ACI 318-14, "*The purpose of this Code is to provide for public health and safety by establishing minimum requirements for strength, stability, serviceability, and integrity of concrete structures.*" Section 4.8 of ACI 318-14 addresses global durability requirements related to material selection for concrete mixtures and corrosion protection of reinforcement. ACI 318-14, Chapters 19 and 20, provide detailed durability requirements for concrete and reinforcing steel, respectively. ACI 318-14, Chapter 26, discusses what durability requirements must be specified in a project's construction documents.

The Code's durability focus is mainly on concrete resistance to fluid penetration, which is primarily affected by the $w/cm$ and the composition of those materials. The use of SCMs, such as Type F and Type C fly ashes, slag cement, silica fume, calcined shale, calcined clay or metakaolin, or their combinations, can result in a significant improvement in concrete durability. The SCMs affect concrete properties in many ways, depending on their type, dosage, and other mixture proportions and composition. In general, SCMs have the following impacts on hardened concrete properties (Kosmatka and Wilson 2011):
- Increase long-term strength
- Have varied effect on early age strength gain (Type F fly ash, calcined shales, and clays lower early strength; silica fume and metakaolin increase early strength gain)
- Reduce permeability and absorption
- Improve resistance to corrosion
- Increase sulfate resistance (with the exception of Type C fly ash, which may have either a positive or negative effect)
- Have no significant impact on abrasion resistance, drying creep and shrinkage, and freezing and thawing
- May reduce resistance to deicer scaling

The Code does not cover all topics related to concrete durability. It does not include recommendations for extreme exposure conditions (that is, acids, high temperature, or exposure to fire), alkali-aggregate reaction, or abrasion. The Code commentary (R4.8) identifies the importance of preventive maintenance; however, the topic is not explicitly addressed in the Code. Additionally, the Code does not cover waterproofing, routine inspections, condition assessment, or service life prediction. Information related to these topics are found in other ACI documents, including:
- ACI 201.2R—Guide to Durable Concrete
- ACI/TMS 216.1—Code Requirements for Determining Fire Resistance of Concrete and Masonry Construction Assemblies
- ACI 221.1R—Report on Alkali-Aggregate Reactivity
- ACI 362.1R—Guide for the Design and Construction of Durable Concrete Parking Structures
- ACI 222R—Protection of Metals in Concrete Against Corrosion
- ACI 222.2R—Report on Corrosion of Prestressing Steels
- ACI 222.3R—Guide to Design and Construction Practices to Mitigate Corrosion of Reinforcement in Concrete Structures
- ACI 224.1R—Causes, Evaluation, and Repair of Cracks in Concrete Structures
- ACI 311.4R—Guide for Concrete Inspection
- ACI 365.1R—Service-Life Prediction
- ACI 515.2R—Guide to Selecting Protective Treatments for Concrete
- ACI 562—Code Requirements for Evaluation, Repair, and Rehabilitation of Concrete Buildings and Commentary

## 4.3—Requirements for concrete in various exposure categories

The Code addresses durability by requiring that four exposure categories be assigned to each concrete member. The four exposure categories are:
1. F: concrete exposed to moisture and cycles of freezing and thawing (with or without deicing chemicals);
2. S: concrete in contact with soil or water containing deleterious amounts of water-soluble sulfate ions;
3. W: concrete in contact with water but not exposed to freezing and thawing, chlorides, or sulfates;
4. C: concrete exposed to conditions that require additional protection against corrosion of reinforcement.

Each exposure category is divided into exposure classes that define severity of the exposure, starting with 0 for a negligible exposure. Once all structural members are assigned exposure classes and the concrete mixtures for these members satisfy those various requirements, the Code's minimum durability requirements are met.

**4.3.1** *Freezing and thawing (F)*—The volume of ice is 9 percent larger than water. As water freezes in saturated concrete, cement phase and aggregates are subject to internal pressure, which then causes concrete tensile stresses. If those stresses are greater than the tensile strength of concrete, cracking will occur. The cumulative expansion after many cycles of freezing and thawing may lead to significant concrete damage. One method to protect concrete from freezing-and-thawing damage is to reduce moisture penetration so it does not become critically saturated; however, this is not always possible. The other method is to generate small air bubbles in fresh concrete by addition of an air-entraining admixture, which creates voids for the freezing water to expand into without creating internal stress.

The Code requires concrete in structural members exposed to cycles of freezing and thawing to be protected by using air-entrained concrete. Air entrainment significantly improves resistance of saturated concrete to freezing and thawing. ACI 212.3R provides an in-depth discussion on these materials, their applications, dosage rates, effects on fresh and hardened concrete, and other factors they influence.

The specified amount of air entrainment depends primarily on frequency of exposure to water (exposure class), but also on nominal maximum aggregate size and concrete compressive strength. To achieve similar freezing-and-thawing protection, higher air content is generally required for concrete mixtures with smaller nominal maximum aggregate size. For example, concrete with 3/8 in. aggregate requires 50 percent higher air content than concrete with 2 in. aggregate (ACI 318-14, Table 19.3.3.1). The Code requires that the licensed design professional (LDP) specify the nominal maximum aggregate size for each concrete mixture in the construction documents. Nominal maximum aggregate size depends on locally available aggregates, as well as construction issues such as size of formwork, member depth, and clear bar spacing. The criteria for maximum size selection are given in Section 26.4.2.1 of ACI 318-14. Table 19.3.3.1 lists target air content for Classes F1, F2 and F3, depending on the nominal maximum aggregate size.

Another factor affecting selection of target air content is compressive strength. An air content reduction of 1 percent is allowed for concrete with specified compressive strength exceeding 5000 psi (ACI 318-14, Section 19.3.3.3). The reason for air content reduction is that concretes with higher strengths are characterized by lower $w/cm$ and reduced porosity, which improve resistance to freezing-and-thawing cycles.

For example, a structural member in Exposure Class F2 with 1/2 in. nominal maximum aggregate size requires concrete with a target air content of 7 percent (or 6 percent for concrete with compressive strength exceeding 5000 psi). Because exact air content is difficult to achieve, the Code allows tolerance for air content in as-delivered concrete of ±1.5 percentage points. This is consistent with the tolerances in ASTM C94/C94M and ASTM C685/C685M (Section R26.4.2.1(a)(5)). The required air content range, therefore, is from 5.5 to 8.5 percent (or 4.5 to 7.5 percent for concrete with compressive strength exceeding 5000 psi).

Additional requirements or limitations, such as minimum compressive strength, minimum $w/cm$, or limits on cementitious materials, depend on the exposure class assigned to a particular member. Interior members, foundations below the frost line, or structures in climates where freezing temperatures are not anticipated are assigned Exposure Class F0. These conditions, therefore, do not require air entrainment and there is no limit on maximum $w/cm$ or on the use of cementitious materials. The minimum compressive strength for concrete in Exposure Class F0 is the Code minimum: 2500 psi.

Freezing-and-thawing cycles have little effect on concrete that is not critically saturated. Structural members exposed to freezing and thawing cycles, but with low likelihood of being saturated, are assigned exposure class F1. Concrete for this exposure must be air entrained (Table 19.3.3.1 of ACI 318) in case there is occasional saturation during freezing. In addition, the concrete should have maximum $w/cm$ of 0.55 and at least 3500 psi compressive strength.

Exposure Classes F2 and F3 are assigned to concrete in structural members with a high likelihood of water saturation during freezing. The distinction between the two classes is that Class F2 anticipates no exposure to deicing chemicals or seawater, while Class F3 does. Concrete in F2 and F3 exposure classes must be air entrained (Table 19.3.3.1 of ACI 318) and have a maximum $w/cm$ of 0.45 and 0.40, respectively. The minimum concrete compressive strengths for F2 and F3 classes are 4500 and 5000 psi, respectively. The most severe class of exposure, F3, also has a limit on cementitious materials in concrete mixtures, given in ACI 318, Table 26.4.2.2(b).

The summary of requirements for concrete in Exposure Category F is listed in Table 19.3.2.1 of ACI 318.

**4.3.2** *Sulfate (S)*—All soluble forms of sulfate, sodium, calcium, potassium, or magnesium have a detrimental effect on concrete. Depending on the sulfate form, they react with hydrated cement phases and result in formation of ettringite or gypsum. Depending on the reaction product, the concrete either expands and cracks (ettringite), or softens and loses strength (gypsum). The most effective measure to reduce the effects of sulfate reactions, apart from reducing moisture ingress, is to use cements with a low content of tricalcium aluminate ($C_3A$). A more detailed discussion on sulfate's effect on concrete can be found in ACI 201.2R.

Exposure Category S applies to structural members that will likely be affected by external source of sulfates, which predominantly come from exposure to soil, groundwater, or seawater. The exposure classification (class) is selected based on the concentration of sulfate ions ($SO_4^{2-}$), which should be determined in accordance with ASTM C1580 for soil samples and with ASTM D516 or ASTM D4130 for water samples. The Code requires the LDP to specify the exposure class by comparing field test results with concentration ranges in Table 19.3.1.1 of ACI 318. Note that seawater exposure is classified as S1 even though the sulfate concentration (in seawater) is usually higher than 1500 ppm. The reason for lower class for seawater is the presence of chloride ions, which inhibit expansive reaction due to sulfate attack.

Class S0 is assigned to concrete in members not exposed to sulfates and there is no restriction on $w/cm$, or type or limit on cementitious materials. The only requirement for concrete classified as S0 is the minimum compressive strength be at least 2500 psi. Greater minimum compressive strength and maximum w/cm limits are imposed on concrete in Exposure Classes S1 through S3. For these exposure classes, the type of cement is the major requirement.

A summary of all requirements for concrete in Exposure Category S is listed in Table 19.3.2.1 of ACI 318-14.

**4.3.3** *In contact with water (W)*—The durability of structural members in direct contact with water, such as foundation walls below the groundwater table, may be affected by water penetration into or through concrete. Apart from external systems, such as drainage systems or waterproofing membranes for foundations, the most effective way to reduce concrete permeability is to keep the w/cm low.

Concrete for members assigned to Exposure Class W0 has no unique requirements except that it has a minimum compressive strength of 2500 psi. Concrete in structural members assigned to Exposure Class W1 requires low permeability. Table 19.3.2.1 of ACI 318-14 requires w/cm not to exceed 0.50 and compressive strength to be at least 4000 psi. Note that additional requirements are imposed if the member's durability is to be affected by reinforcement corrosion, sulfate exposure, or exposure to cycles of freezing and thawing. Recommendations for the design and construction of water tanks and reservoirs are provided in ACI 350.4R, ACI 334.1R, and ACI 372R.

Requirements for concrete in Exposure Class W are listed in Table 19.3.2.1 of ACI 318-14.

**4.3.4** *Corrosion (C)*—Corrosion of reinforcement may significantly affect durability and structural capacity of a member. Reinforcement corrosion usually occurs as a result of the presence of chlorides or steel depassivation due to carbonation. Corrosion products (rust) are larger in volume than the original steel and therefore exert internal pressure on the surrounding concrete, causing it to crack or delaminate. A significant loss of reinforcing bar cross section leads to increased steel stresses under service load and reduced member nominal strength. Because moisture and oxygen must be present at the steel surface for corrosion to occur, the quality of concrete and the reinforcing bar cover are of great importance. Corrosion can be mitigated by proper mixture design and construction practices; application of sealers, coatings, or membranes that protect concrete from moisture and chloride penetration; use of corrosion resistant reinforcement; or inclusion of corrosion inhibitors in the mixture to elevate the corrosion threshold concentration. Refer to ACI 222R, ACI 222.3R, and ACI 212.3R for additional information.

Each exposure class within the corrosion exposure category has a limit on water-soluble chloride-ion content in concrete. The chloride-ion content is measured in accordance with ASTM C1218/C1218M, which requires the sample be representative of all concrete-making ingredients—that is, cementitious materials, fine and coarse aggregate, water, and admixtures. Because chloride limits are imposed even on concrete in Exposure Class C0, all structural concrete must comply with the Code's maximum chloride ion limits. Chloride limits for nonprestressed concrete, expressed as percent of cement weight, are 1 percent for Class C0, 0.30 percent for Class C1, and 0.15 percent for Class C2. Chloride limit for prestressed concrete is 0.06 percent by cement weight, regardless of exposure class. Apart from chloride limits, Exposure Classes C0 and C1 have no additional requirements, as there is no limit on w/cm and the minimum compressive strength is 2500 psi.

Class C2 requires concrete strength of at least 5000 psi, a maximum w/cm of 0.40, and reinforcing steel specified cover to satisfy the Code's minimum concrete cover provisions. The minimum concrete cover depends on exposure to weather, contact with ground, type of member, type of reinforcement, diameter and arrangement (bundling) of reinforcement, method of construction (cast-in-place or precast), and if the member is prestressed. Tables 20.6.1.3.1, 20.6.1.3.2, and 20.6.1.3.3 of ACI 318-14 provide cover provisions for cast-in-place nonprestressed, cast-in-place prestressed, and precast nonprestressed or prestressed concrete members, respectively. If the design requires bundled bars, check Section 20.6.1.3.4 of ACI 318-14 for specific requirements. Concrete cover requirements in corrosive environments or other severe exposure conditions are more stringent and are provided in Section 20.6.1.4 of ACI 318-14.

Requirements for concrete in Exposure Class C are listed in Table 19.3.2.1 of ACI 318-14

## 4.4—Concrete evaluation, acceptance, and inspection

Durability requirements are met once concrete proportions and properties satisfy the minimums set by the Code. To assure that the delivered concrete achieves the desired durability, the LDP should specify concrete evaluation and acceptance criteria consistent with ACI 318-14, Section 26.12 and field inspection consistent with ACI 318-14, Section 26.13.

## 4.5—Examples

The following examples illustrate one approach of implementing minimum durability requirements of the Code. In some cases, durability requirements for material properties may exceed those of the structural design. This is more likely for severe exposure conditions, which require a minimum compressive strength of 5000 psi. In some cases, SCMs may be required, which may extend setting time and early-age strength, and result in modifications to construction schedule. For these reasons, cooperation with engineers experienced with concrete materials and mixture proportioning, and with concrete suppliers, is recommended.

**4.5.1** *Example 1: Interior suspended slab not exposed to moisture or freezing and thawing*—Consider the design of a cast-in-place, nonprestressed slab in a multistory office building. It is located in a climate zone with frequent freezing-and-thawing cycles; however, the slab will be constructed during summer and the temperatures at night during construction are expected to remain above 40 to 45°F. It is desirable for the slab to quickly gain strength to meet the construction schedule. For this reason, calcium chloride was proposed as an accelerating admixture. The required minimum compressive strength, from structural analysis, is 4000 psi. The slab is 7 in. thick with top and bottom mats of No. 5 bars spaced at 8 in. What additional information should be specified for the slab concrete to meet durability requirements?

**Answer:** The first step is to assign exposure classes within every exposure category to each structural member or group of members. Once exposure classes are assigned, the Code guides the LDP to satisfy the durability requirements. The step-by-step instructions are as follows:

| Step description/ action item | Selection and discussion | Code reference |
|---|---|---|
| Assign exposure classes within each exposure category | **F0** (concrete not exposed to freezing-and-thawing cycles) **S0** (soil not in contact with concrete; low and injurious sulfate attack is not a concern) **W0** (there are no specific requirements for low permeability) **C0** (concrete dry or protected from moisture) | Table 19.3.1.1 |
| Assign required minimum compressive strength | **2500 psi** (based on F0) | Table 19.3.2.1 |
| Assign maximum *w/cm* | **Not limited** (based on all exposure classes) | Table 19.3.2.1 |
| Assign minimum concrete cover | **0.75 in.** (not exposed to weather, slabs..., No. 11 bars and smaller) | Table 20.6.1.3.1 |
| Assign nominal maximum size of aggregate | **2 in.** (3/4 x 3 in. clear bar spacing – top and bottom mat, or 1/3 x 7 in. – slab thickness); **use 1 in.** as readily available | 26.4.2.1(a)(4) |
| Assign required air content | **Not air entrained** | Table 19.3.2.1 |
| Assign limits on cementitious materials | **No limits** | Table 19.3.2.1 |
| Assign limits on calcium chloride admixture | **No restriction** (based on S0) [Note: chloride ions from the admixture will significantly affect measured chloride ion content in concrete.] | Table 19.3.2.1 |
| Assign maximum water-soluble chloride ion (Cl–) content in concrete, percent by weight of cement | **1.00** (based on C0, water-soluble chloride-ion content from all concrete ingredients determined by ASTM C1218/C1218M at age between 28 and 42 days) | Table 19.3.2.1 |

**4.5.2** *Example 2: Balcony slab exposed to moisture and freezing and thawing*—An LDP designs a cast-in-place, nonprestressed balcony slab in a multistory office building, located in a climate zone with frequent freezing-and-thawing cycles. It is anticipated that the balconies will be exposed to moisture, but not deicing salts. The required minimum compressive strength, from structural analysis, is 4000 psi. The balcony slabs are 6 in. thick with top mat of No. 4 bars spaced at 6 in. What additional information is needed for balcony concrete to meet durability requirements?

**Answer:** Durability requirements are met once the most rigorous requirements of the Code are satisfied. The first step is to assign exposure classes within every exposure category to each structural member or group of members. Once exposure classes are assigned, the code guides the LDP to set the minimum durability requirements. The step-by-step instructions are as follows:

| Step description/ action item | Selection and discussion | Code reference |
|---|---|---|
| Assign exposure classes within each exposure category | **F2** (concrete exposed to freezing-and-thawing cycles with frequent exposure to water) **S0** (soil not in contact with concrete; low and injurious sulfate attack is not a concern) **W0** (there are no specific requirements for low permeability) **C1** (concrete exposed to moisture but not to an external source of chlorides) | Table 19.3.1.1 |
| Assign required minimum compressive strength | **4500 psi** (based on F2); because 4500 psi is greater than design strength of 4000 psi, the 4500 psi governs | Table 19.3.2.1 |
| Assign maximum *w/cm* | **0.45** (based on F2) | Table 19.3.2.1 |
| Assign minimum concrete cover | **1.5 in.** (exposed to weather, No. 5 bar and smaller) | Table 20.6.1.3.1 |
| Assign nominal maximum size of aggregate | **2 in.** (1/3 x 6-in. – slab thickness, 3/4 x 6-in. clear bar spacing – top mat bars); **use 1 in.** as readily available | 26.4.2.1(a)(4) |
| Assign required air content | **6% ± 1.5%** (for 1 in. aggregate and F2 class) [Note: 1 in. aggregate can be substituted with 3/4 in. aggregate with no air content change] | Table 19.3.3.1 and Section R26.4.2.1(a)(5) |
| Assign limits on cementitious materials | **No limits** | Table 19.3.2.1 |
| Assign maximum water-soluble chloride ion (Cl–) content in concrete, percent by weight of cement | **0.30** (water-soluble chloride ion content from all concrete ingredients determined by ASTM C1218/1218M at age between 28 and 42 days) | Table 19.3.2.1 |
| Provide guidance on cold weather construction | Consult ASTM C94/C94M, ACI 306R, and ACI 301 for guidance on temperature limits for concrete delivered in cold weather. | Section 26.5.4.1 |

**4.5.3** *Example 3: Wall foundation exposed to sulfate soil and deicing salts while in service*—An LDP designs a cast-in-place, nonprestressed foundation wall of a partially underground parking structure. The structure is located in a northern climate zone with frequent freezing-and-thawing cycles, high sulfate soil content (6 percent $SO_4^{2-}$ by mass) and exposure to deicing salts are anticipated as a runoff from the nearby streets and a sidewalk. It is desirable for the foundation wall to quickly gain strength to reduce possible frost damage and to meet the construction schedule. The required minimum compressive strength, from structural analysis, is 4000 psi. The foundation wall is 8 in. thick with inside face and outside face mats of No. 4 bars spaced at 12 in. What additional information should be specified for foundation wall concrete to meet durability requirements?

**Answer:** The first step is to assign exposure classes within every exposure category to each structural member or group of members. Once exposure classes are assigned, the code guides the LDP to set the minimum durability requirements. The step-by-step instructions are as follows:

| Step description/ Action item | Selection and discussion | Code reference |
|---|---|---|
| Assign exposure classes within each exposure category | F3 (concrete exposed to freezing-and-thawing cycles with frequent exposure to water and exposure to deicing chemicals) S3 (structural concrete members in direct contact with soluble sulfates in soil or water) W1 (concrete in contact with water and low permeability is required) C2 (concrete exposed to moisture and an external source of chlorides from deicing chemicals) | Table 19.3.1.1 |
| Assign minimum compressive strength | 5000 psi (based on F3 and C2); because 5000 psi is greater than design strength of 4000 psi, 5000 psi governs | Table 19.3.2.1 |
| Assign maximum $w/cm$ | 0.40 (based on F3 and C2) | Table 19.3.2.1 |
| Assign minimum concrete cover | 2.0 in. – outside face of wall (1.5 in. cover is listed in Table 20.6.1.3.1 for exposure to weather or in contact with ground for No. 5 bar and smaller; cover increased to 2.0 in. based on 20.6.1.4.1) 3/4 in. – inside face of wall (side of the wall not exposed to weather or in contact with ground) | Table 20.6.1.3.1 20.6.1.4.1 |
| Assign nominal maximum size of aggregate | 1.5 in. (1/5 x 8 in. – wall thickness, 3/4 x 3-1/4 in. clear bar spacing – between interior and exterior mats of reinforcing steel); use 1.5 in. | Section 26.4.2.1(a)(4) |
| Assign required air content | 5.5% ± 1.5% (for 1.5 in. aggregate and F3 class) [Notes: 1. Changing to lower nominal maximum aggregate size will require higher air content; 2. Air content reduction of 1% (to 4.5% ± 1.5%) is allowable if concrete compressive strength exceeds 5000 psi; refer to 19.3.3.3] | Table 19.3.3.1 |
| Assign limits on cementitious materials | Limits in accordance with Table 26.4.2.2(b) Cement combinations (for Class S3 in Table 19.3.2.1) must be tested in accordance with ASTM C1012/C1012M and meet the maximum expansion requirement of 0.10% (Class S3); check Table 26.4.2.2(c) | Table 19.3.2.1 Table 26.4.2.2(b) Table 26.4.2.2(c) |
| Assign limits on calcium chloride admixture | Not permitted (based on S2 and C2) | Table 19.3.2.1 |
| Assign maximum water-soluble chloride-ion (Cl⁻) content in concrete, percent by weight of cement | 0.15 (based on C2, water-soluble chloride ion content from all concrete ingredients determined by ASTM C1218/C1218M at age between 28 and 42 days) | Table 19.3.2.1 |

# REFERENCES

*American Concrete Institute*
ACI 201.2R-08—Guide to Durable Concrete
ACI 212.3R-10—Report on Chemical Admixtures for Concrete
ACI 222R-01—Protection of Metals in Concrete Against Corrosion
ACI 222.3R-11—Guide to Design and Construction Practices to Mitigate Corrosion of Reinforcement in Concrete Structures
ACI 301-10—Specification for Structural Concrete
ACI 306R-10—Guide to Cold Weather Concreting
ACI 334.1R-92—Concrete Shell Structures-Practice and Commentary
ACI 350—Please correct document number and add title here and correct reference in body of text (Section 4.3.3)
ACI 372R-13—Guide to Design and Construction of Circular Wire- and Strand-Wrapped Prestressed Concrete Structures

*ASTM International*
ASTM C94/C94M-15—Standard Specification for Ready-Mixed Concrete
ASTM C1012/C1012M-13—Standard Test Method for Length Change of Hydraulic-Cement Mortars Exposed to a Sulfate Solution
ASTM C1218/C1218M-99—Standard Test Method for Water-Soluble Chloride in Mortar and Concrete
ASTM C150/C150M-12—Standard Specification for Portland Cement
ASTM D516-11—Standard Test Method for Sulfate Ion in Water
ASTM C685/C685M-14—Standard Specification for Concrete Made by Volumetric Batching and Continuous Mixing
ASTM C1580-09-Standard Test Method for Water-Soluble Sulfate in Soil
ASTM D4130-15—Standard Test Method for Sulfate Ion in Brackish Water, Seawater, and Brines

## Authored documents

Hewlett, P. C., ed., 1998, *Lea's Chemistry of Cement and Concrete*, fourth edition, John Wiley & Sons, New York, Toronto, 1057 pp.

Kosmatka, S. H.; Kerkhoff, B.; and Panarese, W. C., 2002, *Design and Control of Concrete Mixtures (EB001)*, 14th edition, fourth printing (rev.), Portland Cement Association, Skokie, IL, Feb., 358 pp.

Kosmatka, S. H., and Wilson, M. L., 2011, *Design and Control of Concrete Mixtures (EB001)*, 15th edition, Portland Cement Association, Skokie, IL, 444 pp.

Neville, A., ed., 2003, *Neville on Concrete: An Examination of Issues in Concrete Practice*, American Concrete Institute, publisher, Farmington Hills, MI, 510 pp.

Powers, T. C., 1958, "Structure and Physical Properties of Hardened Portland Cement Paste," *Journal of the American Ceramic Society*, V. 41, No. 1, pp. 1-6.

Stark, D., 1989, "Durability of Concrete in Sulfate-Rich Soils (RD097)," Portland Cement Association, Skokie, IL, 14 pp.

Stark, D., 2001, "Long-Term Performance of Plain and Reinforced Concrete in Seawater Environments (RD119)," Portland Cement Association, Skokie, IL, 14 pp.

Taylor, P. C.; Nagi, M. A.; and Whiting, D. A., 1999, "Threshold Chloride Content for Corrosion of Steel in Concrete: A Literature Review (RD2169)," Portland Cement Association, Skokie, IL, 32 pp.

Whiting, D. A.; Taylor, P. C.; and Nagi, M. A., 2002, "Chloride Limits in Reinforced Concrete (RD2438)," Portland Cement Association, Skokie, IL, 72 pp.

# CHAPTER 5—ONE-WAY SLABS

## 5.1—Introduction

A one-way slab is generally used in buildings with vertical supports (columns or walls) that are unevenly spaced, creating a long span in one direction and a short span in the perpendicular direction. One-way slabs typically span in the short direction and are supported by beams in the long direction. During preliminary design, the designer determines the loads and spans, reinforcement type (post-tensioned [PT] or nonprestressed), and slab thickness. The designer determines the concrete strength based on experience and the Code's exposure and durability provisions.

This chapter discusses cast-in-place, nonprestressed, and PT slabs. The Code allows for either bonded or unbonded tendons in a PT slab. Because bonded tendons are not usually used in slabs in the United States, this chapter will address PT slabs with unbonded tendons.

At times, the design of a one-way slab will require point load considerations, such as wheel loads in parking garages. These result in local shear forces on the slab, requiring verification of the slab's punching shear strength. Punching shear is addressed in Chapter 6 for two-way slabs in this Handbook.

For relatively small slab openings, trim bars can be used to limit crack widths caused by geometric stress concentrations. For larger openings, a local increase in slab thickness, as well as additional reinforcement, may be necessary to provide adequate serviceability and strength.

## 5.2—Analysis

ACI 318 allows for the designer to use any analysis procedure that satisfies equilibrium and geometric compatibility, as long as design strength and serviceability requirements are met. The Code includes a simplified method of analysis for one-way slabs that relies on coefficients to calculate moments and shears.

## 5.3—Service limits

**5.3.1** *Minimum thickness*—For nonprestressed slabs, the Code allows the designer for slabs not supporting or attached to partitions or other construction likely to be damaged by large deflection to either calculate deflections or simply satisfy a minimum slab thickness (Section 7.3.1, ACI 318-14). In the case where loads are heavy, nonuniform, or deflection is a concern, calculations should verify that short- and long-term deflections are within the Code limits (Section 24.2.2, ACI 318-14).

The Code does not provide a minimum thickness-to-span ratio for PT two-way slabs, but Table 9.3 of *The Post-Tensioning Manual* (Post-Tensioning Institute (PTI) 2006), lists span-to-depth ratios for different members that have been found from experience to provide satisfactory structural performance.

**5.3.2** *Deflections*—For nonprestressed slabs that are thinner than the ACI 318 minimum, or if the slab resists a

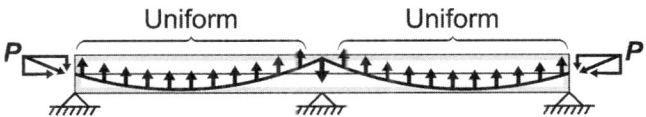

Fig. 5.3.3—Load balancing concept.

very heavy live load, superimposed dead load, or both, and for PT slabs as well, the designer calculates deflections. The calculated deflections must not exceed the limits given in Section 24.2, ACI 318-14. Deflections can be calculated by any appropriate method, such as classical equations or software results.

Note that the spacing of slab reinforcing bar to limit crack width, timing of form removal, concrete quality, timing of construction loads, and other construction variables all could affect the actual deflection. These variables should be considered when assessing the accuracy of deflection calculations. In addition, creep over time will increase the immediate deflections.

Typically, with a PT slab, slab deflections are usually small. If the designer limits the maximum net concrete tensile stress to below cracking stress under service loads, deflection calculations can consider the gross slab properties.

**5.3.3** *Concrete service stress*—Nonprestressed slabs are designed for strength without reference to a concrete pseudo-service flexural stress limit.

For PT slabs, the analysis of concrete flexural tension stresses is a critical part of the design. In Section 8.3.4.1 of ACI 318-14, the concrete flexural tensile stress in negative moment areas at columns in PT slab is limited to $6\sqrt{f_c'}$. At positive moment sections, Section 8.6.2.3 of ACI 318-14 requires slab reinforcing bar if the concrete tensile stress exceeds $2\sqrt{f_c'}$. These service tensile flexural stress limits are below the concrete cracking stress of $7.5\sqrt{f_c'}$, thus having the effect of limiting deflections. In addition, Section 8.6.2.1 of ACI 318-14 requires a PT slab's axial compressive stress in both directions, due to PT, to be at least 125 psi.

Before the slab flexural stresses in a design strip can be calculated, the tendon profile should be defined. Both profile and tendon force are directly related to slab forces and moments created by PT. A common approach to calculate PT slab moments is the use of the "load balancing" concept. Tendons are typically placed in a parabolic profile such that the tendon is at the minimum cover requirements at midspan and over supports; this maximizes the parabolic drape. Anchors are typically placed at mid-depth at the slab edge (Fig. 5.3.3).

The tendon exerts a uniform upward force along its length that counteracts a portion of the gravity loads, usually 60 to 80 percent of the slab self-weight according to Libby (1990); hence, the term "load balancing." The load effect from the prestressing force in the tendon is then combined with the load effect of the gravity loads to determine net concrete stresses.

To achieve Code stress limits, the designer can use an iterative or direct approach. In the iterative approach, the tendon profile is defined and the tendon force is assumed. The analysis is executed, flexural stresses are calculated, and the designer then adjusts the profile or force or both, depending on results and design constraints.

In the direct approach, the designer determines the highest tensile stress permitted, then rearranges equations so that the analysis calculates the tendon force required to achieve the stress limit.

The Code does not impose a minimum concrete compressive stress due to PT, but majority of engineers use 125 psi as a general minimum for cast-in-place slabs. For slabs exposed to aggressive environments, engineers usually design slabs with an increased minimum concrete compressive stress.

### 5.4—Required strength

The design of one-way slabs are typically controlled by moment strength, not concrete stress or shear strength. Assuming a uniform load, the designer calculates the unit (usually a 1 ft width) factored slab moments. The required area of flexural reinforcement over a unit slab width is calculated with the same assumptions as a beam.

**5.4.1** *Calculation of required moment strength*—For nonprestressed reinforced slabs, a quick way to calculate a slab's gravity design moments (if the slab meets the specified geometric and load conditions) is by the moment coefficients in Section 6.5 of ACI 318-14. Chapter 6 of ACI 318-14 permits other, more exact analysis methods.

For PT slabs, effects of reactions induced by prestressing (secondary moments) should be added to the factored gravity moments per Section 5.3.11 of ACI 318-14 to calculate $M_u$. The slab's secondary moments are a result of the beam's vertical restraint of the slab against the PT "load" at each support. Because the PT force and drape are determined during the service stress checks, secondary moments can be quickly calculated by the "load-balancing" analysis concept.

A simple method for calculating the secondary moment is to subtract the tendon force multiplied by the tendon eccentricity (distance from the neutral axis) from the total balance moment, expressed mathematically as $M_2 = M_{bal} - Pe$.

The critical locations to calculate $M_u$ along the span are usually at the support and midspan. Section 7.4.2.1 of ACI 318-14 allows $M_u$ to be calculated at the face of support rather than the support centerline.

**5.4.2** *Calculation of required shear strength*—Assuming a uniform load, the designer calculates the unit (usually a 1 ft width) factored slab shear force by either the coefficient method or more exact calculations.

### 5.5—Design strength

One-way slabs must have adequate one-way shear strength and moment strength in all design strips.

**5.5.1** *Calculation of design moment strength*—The required area of flexural reinforcement for a nonprestressed and PT unit slab width are calculated with the same assumptions as for a beam, Chapter 7, of this Handbook.

**Table 5.6.1a—$A_{s,min}$ for nonprestressed one-way slabs (Table 7.6.1.1, ACI 318-14)**

| Reinforcement type | $f_y$, psi | $A_{s,min}$, in.² | |
|---|---|---|---|
| Deformed bars | < 60,000 | $0.0020A_g$ | |
| Deformed bars or welded wire reinforcement | ≥ 60,000 | Greater of: | $\dfrac{0.0018 \times 60,000}{f_y} A_g$ |
| | | | $0.0014A_g$ |

**Table 5.6.1b—Maximum spacing of bonded reinforcement in nonprestressed and Class C prestressed (Table 24.3.2 one-way slabs and beams, ACI 318-14, partial)**

| Lesser of: | $15\left(\dfrac{40,000}{f_s}\right) - 2.5c_c$ |
|---|---|
| | $12(40,000/f_s)$ |

**5.5.2** *Calculation of design shear strength*—Discussion for nominal one-way shear strength is the same as for a beam, Chapter 7, of this Handbook.

### 5.6—Flexure reinforcement detailing

The Code requires a minimum area of flexural reinforcement be placed in tension regions to ensure that the slab deformation and crack widths are limited when the slab's cracking strength is exceeded. If more than the minimum area is required by analysis, that reinforcement area must be provided. Reinforcement in one way slabs is usually uniformly spaced, unless there is a significant point load or opening.

**5.6.1** *Nonprestressed reinforced slab – Flexural reinforcement area and placing*—For nonprestressed slabs, the minimum flexural bar area, $A_{s,min}$, is given in Table 7.6.1.1 of ACI 318-14 (Table 5.6.1a of this Handbook).

The maximum spacing of flexural bars is given in Table 24.3.2 of ACI 318-14 (Table 5.6.1b of this Handbook). Because $f_s$ is usually taken as 40,000 psi, the maximum spacing will not exceed 12 in.

The bar termination rules in Section 7.7.3 of ACI 318-14 cover general conditions that apply to beams, but because one-way slab bars are usually spaced close to the maximum, bars generally cannot be terminated without violating the maximum spacing. This usually results in all bottom bars extending full length into the beams.

**5.6.2** *Post-tensioned slab – Flexural tendon area and placing*—For PT one-way slabs, the Code does not have a limit for minimum tendon area or a minimum compressive stress due to PT. This is consistent with the flexible approach on service stresses. The Code limits the maximum tendon spacing to $8h$ or 5 ft.

**5.6.3** *Post-tensioned slab – Flexural reinforcing bar area and placing*—The Code requires the reinforcing bar area, $A_{s,min}$, to be placed close to the slab face at the bottom at midspan and the top at the support. For one-way PT slabs $A_{s,min} = 0.004A_{ct}$. Because the one-way slab strip is rectangular, $A_{ct} = 0.5A_g$. This minimum is independent of service stress level.

The maximum spacing of reinforcing bar in a PT one-way slab is the lesser of $3h$ and 18 in.

If the slab design moment strength is fully satisfied by the tendons alone, the termination length of $A_{s,min}$ bars for bottom bars is (a) and for top bars is (b):

(a) At least $\ell_n/3$ in positive moment areas and be centered in those areas

(b) At least $\ell_n/6$ on each side of the face of support

The termination length for bars that are required for strength are the same as for nonprestressed slabs.

**5.6.4** *Temperature and shrinkage reinforcement and placing*—Shrinkage and temperature (S&T) slab reinforcement is required and could be either reinforcing bar or tendons placed perpendicular to flexural reinforcement.

If the designer uses reinforcing bar, the minimum area of temperature and shrinkage Grade 60 bar is $0.0018A_g$.

If the designer uses tendons, the minimum slab effective compression force due to temperature and shrinkage tendons is 100 psi.

The purpose of this reinforcement is to restrain the size and spacing of slab cracks, which can occur due to volume variations caused by temperature changes and slab shrinkage over time. In addition, if the slab is restrained against movement, the Code requires the designer to provide reinforcement that accounts for the resulting tension stress in the slab.

## Authored references

Libby, J., 1990, *Modern Prestressed Concrete: Design Principles and Construction Methods*, fourth edition, Springer, 871 pp.

Post-Tensioning Institute (PTI), 2006, *Post-Tensioning Manual*, sixth edition, PTI TAB.1-06, 354 pp.

## 5.7—Examples

**One-way Slab Example 1**: *Nonprestressed one-way slab—*

Design and detail the second story of the seven story building. The one-way slab consists of five spans of 14 ft each. The slab is supported by 18 in. beams. A 6 ft cantilever extends at the left end of the slab (Fig. E1.1).

Given:
*Load—*
Service live load $L = 100$ psf

*Concrete—*
$f_c' = 5000$ psi (normalweight concrete)
$f_y = 60,000$ psi

*Geometry—*
Span length: 14 ft
Beam width: 18 in.
Column dimensions: 24 in. x 24 in.

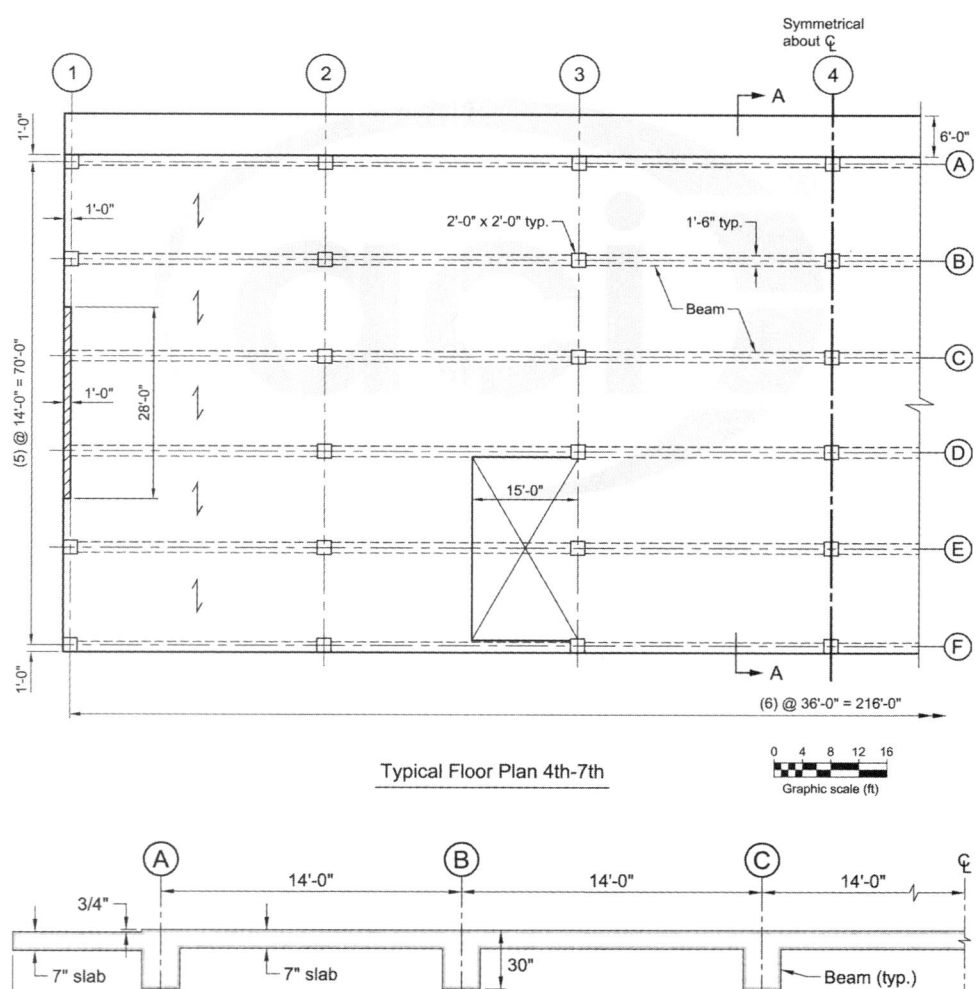

*Fig. E1.1—Plan of five-span one-way slab.*

# CHAPTER 5—ONE-WAY SLABS

| ACI 318-14 | Discussion | Calculation |
|---|---|---|
| **Step 1: Geometry** | | |
| 7.3.1.1 | Determine slab thickness using ratios from Table 7.3.1.1. | $h \geq \dfrac{\ell}{24} = \dfrac{(14 \text{ ft})(12 \text{ in./ft})}{24} = 7$ in. |
| | Determine cantilever thickness: | $h \geq \dfrac{\ell}{10} = \dfrac{(6 \text{ ft})(12 \text{ in./ft})}{10} = 7.2$ in., say, 7 in. |
| | | Because the slab and cantilever satisfy ACI 318-14 span-to-depth ratios (Table 7.3.1.1), the designer does not need to check deflections unless the slab is supporting breakable partitions. |
| | Note: Architectural requirements specify a 3/4 in. step at the cantilever. Detail to maintain 7 in. slab thickness. | |
| | Self-weight<br>Slab: | $w_s = (7 \text{ in.}/12 \text{ in./ft})(150 \text{ lb/ft}^3) = 87.5$ psf |
| **Step 2: Loads and load patterns** | | |
| 5.3.1 | The service live load is 100 psf in assembly areas and corridors per Table 4-1 in ASCE 7-10. For cantilever use 100 psf. To account for weights from ceilings, partitions, HVAC systems, etc., add 15 psf as miscellaneous dead load.<br><br>$U = 1.4D$             (5.3.1a)<br>$U = 1.2D + 1.6L$   (5.3.1b)<br><br>The slab resists gravity only and is not part of a lateral force-resisting system, except to act as a diaphragm. | The required strength equations to be considered are:<br>$U = 1.4 (87.5 \text{ psf} + 15 \text{ psf}) = 143.5$ psf<br>$U = 1.2 (102.5 \text{ psf}) + 1.6 (100 \text{ psf})$<br>   $= 123 \text{ psf} + 160 \text{ psf} = 283$ psf   **Controls** |
| 6.4.2 | Both ASCE 7 and ACI provide guidance for addressing live load patterns. Either approach is acceptable.<br>ACI 318 allows the use of the following two patterns, Fig. E1.2:<br>Factored dead load is applied on all spans and factored live load is applied as follows:<br>(a) Maximum positive $M_u$ near midspan occurs with factored live load on the span and on alternate spans.<br>(b) Maximum negative $M_u$ at a support occurs with factored live load on adjacent spans only. | |

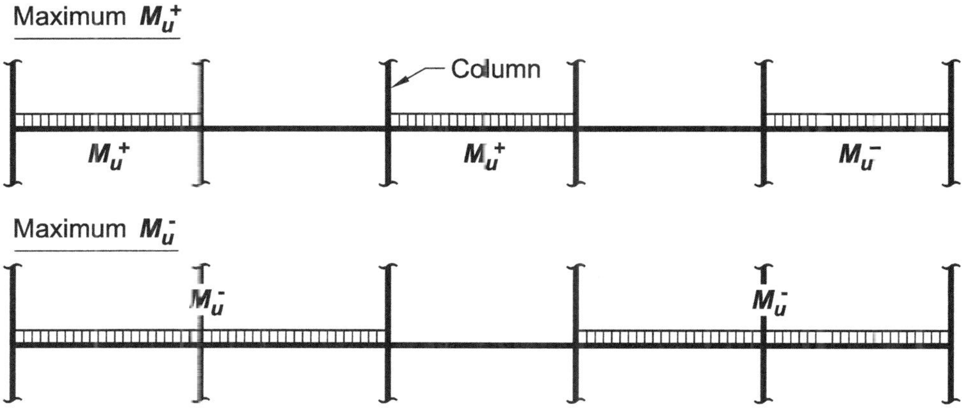

Fig. E1.2—Live load loading pattern.

| | | |
|---|---|---|
| Step 3: Concrete and steel material requirements | | |
| 7.2.2.1 | The mixture proportion must satisfy the durability requirements of Chapter 19 and structural strength requirements of ACI 318-14. The designer determines the durability classes. Please refer to Chapter 4 of this design Handbook for an in-depth discussion of the categories and classes.<br><br>ACI 301 is a reference specification that is coordinated with ACI 318. ACI encourages referencing ACI 301 into job specifications.<br><br>There are several mixture options within ACI 301, such as admixtures and pozzolans, which the designer can require, permit, or review if suggested by the contractor. | By specifying that the concrete mixture must be in accordance with ACI 301-10 and providing the exposure classes, Chapter 19 (ACI 318-14) requirements are satisfied.<br><br>Based on durability and strength requirements, and experience with local mixtures, the compressive strength of concrete is specified at 28 days to be at least 5000 psi. |
| 7.2.2.2 | The reinforcement must satisfy Chapter 20 of ACI 318-14.<br><br>The designer determines the grade of bar and if the reinforcing bar should be coated by epoxy or galvanized, or both. | By specifying the reinforcing bar grade and any coatings, and that the reinforcing bar must be in accordance with ACI 301-10, Chapter 20 requirements are satisfied. In this case, assume Grade 60 bar and no coatings. |
| Step 4: Slab analysis | | |
| 6.3 | Because the building relies on the building's other members to resist lateral loads, the slab qualifies for braced frame assumptions, as discussed in the commentary. | Modeling assumptions:<br>Assume an effective moment of inertia for the entire length of the slab.<br>Ignore torsional stiffness of beams.<br>Only the slab at this level is considered. |
| 6.6 | The analysis should be consistent with the overall assumptions about the role of the slab within the building system. Because the lateral force-resisting-system only relies on the slab to transmit axial forces, a first-order analysis is adequate. | Analysis approach:<br>The connection to the beams is monolithic; however, when the slab is fully loaded, flexural cracking will soften the joint. |
| Step 5: Required moment strength | | |
| 7.4.2 | The slab's negative design moments are taken at the face of support as is permitted by the Code (Fig. E1.3). | |

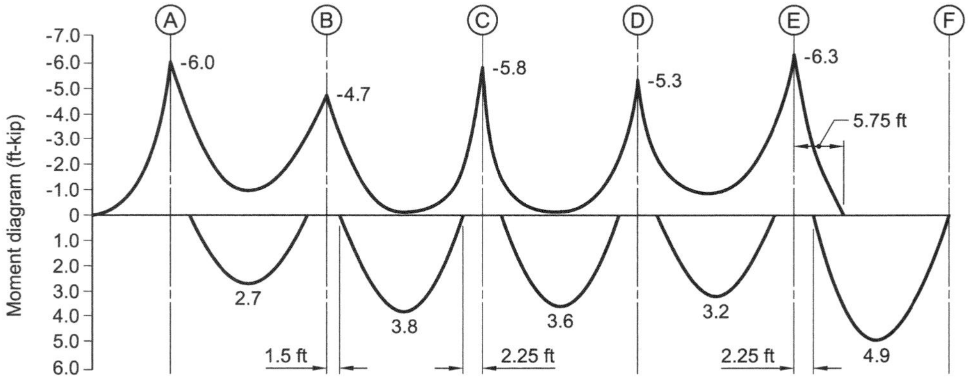

*Fig. E1.3—Moment envelope.*

The negative moment at the centerline of the end right support is 0.0 ft-kip.

The maximum positive moment in the end span, EF, is 4.9 ft-kip. The inflection points for positive moments are 0.0 ft from the exterior support centerline and 2.25 ft from the first interior support centerline, column line (CL) E.

The maximum negative moment at the face of the first interior support from the right end (CL E) is 6.3 ft-kip. The negative moment's right inflection point is 5.75 ft from the support centerline. On the left side and for the full length of the slab, there is no inflection point.

The maximum positive moment in the interior span, CL BC, is 3.8 ft-kip. The inflection points for positive moments are 1.5 ft from the first interior support centerline and 2.25 ft from the second interior support centerline. Because of pattern loading, a small negative moment can exist across all spans with the exception of the last span.

The maximum negative moment at the exterior left support, CL A, is 6 ft-kip because of the cantilevered slab.

### Table 1.1—Maximum moments at supports and midspans

| Required strength | Location from left to right along the span | | | | | |
|---|---|---|---|---|---|---|
| | Exterior support | First midspan | Second support | Second midspan | Third support | Third midspan |
| $M_u$, ft-kip | −6.0 | +2.7 | −4.7 | +3.8 | −5.8 | +3.6 |

Continue:

| Required strength | Location from left to right along the span | | | | |
|---|---|---|---|---|---|
| | Fourth support | Fourth midspan | Fifth support | Fifth midspan | End support |
| $M_u$, ft-kip | −5.3 | +3.3 | −6.3 | +4.4 | 0 |

### Step 6: Required shear strength

| 7.4.3.1 | The slab's maximum shear is taken at the support centerline for simplicity. The maximum shear under all conditions is 2.4 kip (Fig E1.4). |
|---|---|

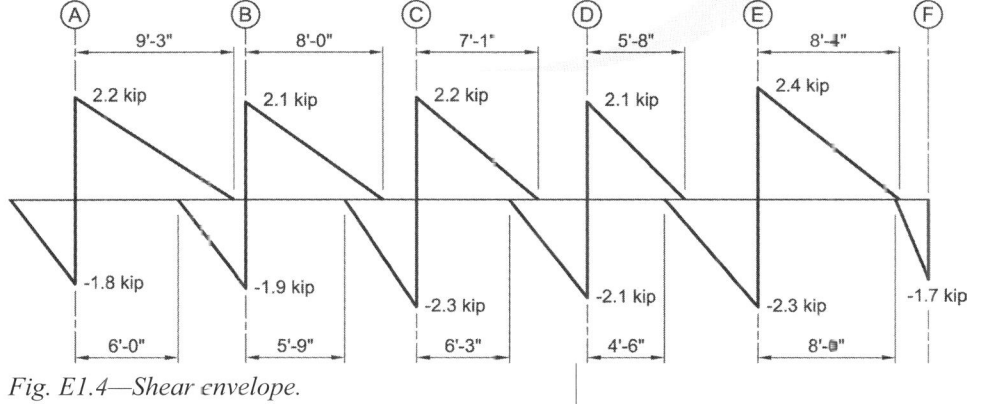

Fig. E1.4—Shear envelope.

| | Step 7: Design moment strength | |
|---|---|---|
| 7.5.1 | The two common strength inequalities for one-way slabs, moment and shear, are noted in Section 7.5.1.1. | |
| 7.5.2 | The one-way slab chapter refers to Section 22.3 for calculation of flexural strength. | |
| 7.3.3.1 | To ensure a ductile failure usually the Code requires slabs to be designed such that the steel strain at ultimate strength exceeds 0.004 in./in.<br><br>In most reinforced slabs, such as this example, reinforcing bar strain is not a controlling issue. | |
| 21.2.1(a) | The design assumption is that slabs will be tensioned controlled, $\phi = 0.9$. This assumption will be checked later. | |
| 22.2.2.1 | Determine the effective depth assuming No.5 bars and 0.75 in. cover (Fig. E1.5): | $d_{estimated} = t - \text{cover} - d_b/2$<br>$d_{estimated} = 7 \text{ in.} - 0.75 \text{ in.} - 0.625 \text{ in.}/2$<br><br>*Fig. E1.5—Effective depth.* |

| | | |
|---|---|---|
| 7.7.1.1<br>20.6.1.3.1 | One row of reinforcement<br>$d = t - \text{cover} - d_b/2$ | $d = 7 \text{ in.} - 0.75 \text{ in.} - 0.625 \text{ in.}/2 = 6.18 \text{ in.}$, say, 6.0 in. |
| 22.2.2.1 | The concrete compressive strain at nominal moment strength is calculated at: $\varepsilon_{cu} = 0.003$ | |
| 22.2.2.2 | The tensile strength of concrete in flexure is a variable property and is approximately 10 to 15 percent of the concrete compressive strength. ACI 318 neglects the concrete tensile strength to calculate nominal strength. | |
| 22.2.2.3 | Determine the equivalent concrete compressive stress at nominal strength:<br><br>The concrete compressive stress distribution is inelastic at high stress. The Code permits any stress distribution to be assumed in design if shown to result in predictions of ultimate strength in reasonable agreement with the results of comprehensive tests. Rather than tests, the Code allows the use of an equivalent rectangular compressive stress distribution of $0.85f_c'$ with a depth of: | |
| 22.2.2.4.1 | $a = \beta_1 c$, where $\beta_1$ is a function of concrete compressive strength and is obtained from Table 22.2.2.4.3. | |
| 22.2.2.4.3 | For $f_c' \leq 5000$ psi: | $\beta_1 = 0.85 - \dfrac{0.05(5000 \text{ psi} - 4000 \text{ psi})}{1000 \text{ psi}} = 0.8$ |
| 22.2.1.1 | Find the equivalent concrete compressive depth, $a$, by equating the compression force to the tension force within the beam cross section:<br><br>$C = T$<br><br>$0.85f_c'ba = A_s f_y$<br>Effective width: 12 in. | $0.85(5000 \text{ psi})(b)(a) = A_s(60{,}000 \text{ psi})$<br><br>$a = \dfrac{A_s(60{,}000 \text{ psi})}{0.85(5000 \text{ psi})(12 \text{ in.})} = 1.176 A_s$ |

| | | |
|---|---|---|
| 7.5.1.1 | The slab is designed for the maximum flexural moments obtained from the approximate method above.<br><br>The first interior support will be designed for the larger of the two moments.<br><br>The beam's design strength must be at least the required strength at each section along its length:<br><br>$\phi M_n \geq M_u$<br>$\phi V_n \geq V_u$<br><br>Calculate the required reinforcement area:<br><br>$\phi M_n \geq M_u = \phi A_s f_y \left( d - \dfrac{a}{2} \right)$<br><br>A No. 5 bar has a $d_b = 0.625$ in. and an $A_s = 0.31$ in.$^2$ | Maximum positive moment:<br><br>$4.9 \text{ ft-kip} \leq (0.9)(60 \text{ ksi}) A_s \left( 6.0 \text{ in.} - \dfrac{1.176 A_s}{2} \right)$<br><br>$A^+_{s,req'd} = 0.18$ in.$^2$/ft<br><br>Use No. 4 at 12 in. on center or No. 5 at 18 in. center bottom. Try No.5 at 18 in. on center<br>$A_{s,provd.} = (0.31 \text{ in.}^2/\text{ft})(12 \text{ in.}/18 \text{ in.}) = 0.21$ in.$^2$/ft<br><br>$A_{s,provd.} = 21$ in.$^2$/ft $> A^+_{s,req'd} = 0.18$ in.$^2$/ft   **OK**<br><br>Maximum negative moment:<br><br>$6.3 \text{ ft-kip} \leq (0.9)(60 \text{ ksi}) A_s \left( 6.0 \text{ in.} - \dfrac{1.176 A_s}{2} \right)$<br><br>$A^-_{s,req'd} = 0.24$ in.$^2$/ft<br><br>Use No. 5 at 12 in. on center top<br><br>$A_{s,provd.} = 31$ in.$^2$/ft $> A^-_{s,req'd} = 0.24$ in.$^2$/ft   **OK** |
| | Check if the calculated strain exceeds 0.005 in./in. (tension controlled). Form similar triangles (Fig. E1.6).<br><br>$a = \dfrac{A_s f_y}{0.85 f'_c b}$ and $c = a/\beta_1$<br><br>where $\beta_1 = 0.8$ for $f'_c = 5000$ psi<br><br>$\varepsilon_t = \dfrac{\varepsilon_{cu}}{c}(d-c)$ | Top reinforcement<br>$a = 1.176 A_s = (1.176)(0.31 \text{ in.}^2) = 0.36$ in.<br><br>$c = 0.36/0.8 = 0.46$ in.<br><br>$\varepsilon_t = \dfrac{0.003}{0.46 \text{ in.}}(6 \text{ in.} - 0.46 \text{ in.}) = 0.036 \geq 0.005$   **OK** |

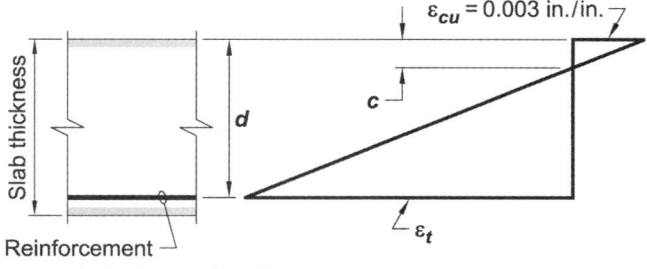

Fig. E1.6—Strain distribution

## CHAPTER 5—ONE-WAY SLABS

| | Step 8: Design shear strength | |
|---|---|---|
| 7.5.3 | Assuming a one-way slab won't contain shear reinforcement, $V_n$ is equal to $V_c$. Assuming negligible axial force, the Code provides the following expression, | |
| 22.5.5.1 21.2.1 | $V_c = 2\sqrt{f_c'}bd$ <br> Shear strength reduction factor: | $V_c = 2\sqrt{5000 \text{ psi}}(12 \text{ in.})(6 \text{ in.}) = 10{,}180 \text{ lb/ft} \cong 10 \text{ kip/ft}$ <br> $\phi = 0.75$ <br><br> $\phi V_c = (0.85)(10{,}000 \text{ lb}) = 8500 \text{ lb} > 2400 \text{ lb}$ **OK** <br> This exceeds the maximum $V_u$ (2.4 kip/ft); therefore, no shear reinforcement is required. |
| | **Step 9: Minimum flexural reinforcement** | |
| 7.6.1 | Check if design reinforcement exceeds the minimum required by the Code. | $A_{s,min} = 0.0018 \times 6 \times 12 = 0.13 \text{ in.}^2/\text{ft}$ <br><br> At all critical sections, the required $A_s$ is greater than the minimum. |
| | **Step 10: Shrinkage and temperature reinforcement** | |
| 7.6.4 24.4.3.2 | For one-way with Grade 60 bars, the minimum area of shrinkage and temperature (S+T) bars is $0.0018A_g$. The maximum spacing of S+T reinforcing bar is the lesser of $3h$ and 18 in. | S+T steel area = $0.0018 \times 12 \times 7 = 0.15 \text{ in.}^2$ <br><br> Based on S+T steel area, solutions are No. 4 at 16 in. or No. 5 at 18 in.; use No. 4 at 16 in. placed atop and perpendicular to the bottom flexure reinforcement. |
| | **Step 11: Minimum and maximum spacing of flexural reinforcement** | |
| 7.7.2.1 25.2.1 | The minimum spacing between bars must not be less than the greatest of: <br> (a) 1 in. <br> (b) $d_b$ <br> (c) $4/3 d_{agg}$ <br><br> Assume 1 in. maximum aggregate size. | (a) 1 in. <br> (b) 0.625 in. <br> (c) $(4/3)(1 \text{ in.}) = 1.33 \text{ in.}$ **Controls** |
| 7.7.2.2 24.3.2 | For reinforcement closest to the tension face, the spacing between reinforcement is the lesser of (a) and (b): <br> (a) $12(40{,}000/f_s)$ <br> (b) $15(40{,}000/f_s) - 2.5c_c$ | (a) $12(40{,}000/40{,}000) = 12 \text{ in.}$ **Controls** <br> (b) $15(40{,}000/40{,}000) - 2.5(0.75 \text{ in.}) = 13.1 \text{ in.}$ |
| 24.3.2.1 | $f_s = 2/3 f_y = 40{,}000 \text{ psi}$ | |
| 7.7.2.3 | The maximum spacing of deformed reinforcement is the lesser of $3h$ and 18 in. | $3(7 \text{ in.}) = 21 \text{ in.} > 18 \text{ in.}$ <br> Therefore, Section 24.3.2 controls; 12 in. |

| | | |
|---|---|---|
| **Step 12: Select reinforcing bar size and spacing** | | |
| | Based on the above requirement, use No. 5 bars. Spacing on top and bottom bars is 12 in.<br><br>Note that there is no point of zero negative moment along all spans except the last bay, so continue the top bars across all spans.<br><br>Also, No. 4 bars can be used instead of No. 5. While this solution is slightly conservative (No. 5 versus No. 4 bars), the engineer may desire consistent spacing and reinforcing bar use for easier installation and inspection. | |
| **Step 13: Top reinforcing bar length at the exterior support** | | |
| | The top bars have to satisfy the following provisions: | Inflection points<br>The inflection point for negative moment at end span is 5.0 ft from support centerline. |
| 7.7.3.3 | Reinforcement shall extend beyond the point at which it is no longer required to resist flexure for a distance equal to the greater of $d$ and $12d_b$, except at supports of simply-supported spans and at free ends of cantilevers. | Bar cutoffs<br>Extend bars beyond the inflection point at least:<br>$d = 6$ in. or $(12)(0.625$ in.$) = 7.5$ in.<br>Therefore, use 7.5 in. |
| 7.7.3.8.4 | At least one-third the negative moment reinforcement at a support shall have an embedment length beyond the point of inflection at least the greatest of $d$, $12d_b$, and $\ell_n/16$. | 33 percent of the bars to extend beyond the inflection point at least<br>$(14 \text{ ft} - 1.5 \text{ ft})(12)/16 = 9.4$ in. $> 12d_b = 7.5$ in. $> d = 6$ in.<br>Because the reinforcing bar is already at maximum spacing, no percentage of bars (as permitted by Section 7.7.3.3 check) can be cut off in the tension zone. |
| **Step 14: Development and splice lengths** | | |
| 7.7.1.2<br>25.4.2.3 | ACI provides two equations for calculating development length; simplified and detailed. In this example, the detailed equation is used:<br><br>$$\ell_d = \left( \frac{3}{40} \frac{f_y}{\lambda \sqrt{f_c'}} \frac{\psi_t \psi_e \psi_s}{\left( \frac{c_b + K_{tr}}{d_b} \right)} \right) d_b$$<br><br>where<br>$\psi_t$ = bar location; not more than 12 in. of fresh concrete below horizontal reinforcement<br>$\psi_e$ = coating factor; uncoated<br>$\psi_s$ = bar size factor; No. 7 and larger<br><br>However, the expression: $\frac{c_b + K_{tr}}{d_b}$ must not be taken greater than 2.5. | The development length of a No. 5 black bar in a 7 in. slab with 0.75 in. cover is:<br><br>$$\ell_d = \left( \frac{3}{40} \frac{60{,}000 \text{ psi}}{(1.0)\sqrt{5000 \text{ psi}}} \frac{(1.0)(1.0)(0.8)}{1.7 \text{ in.}} \right)(0.625 \text{ in.})$$<br>$= 19$ in.<br><br>$\psi_t = 1.0$, because not more than 12 in. of concrete is placed below bars.<br>$\psi_e = 1.0$, because bars are uncoated<br>$\psi_s = 0.8$, because bars are smaller than No. 7<br><br>$\frac{1.06 \text{ in.} + 0}{0.625 \text{ in.}} = 1.7$ in. |

# CHAPTER 5—ONE-WAY SLABS

| | | |
|---|---|---|
| 7.7.1.3<br>25.5<br>25.5.1.1 | Splice<br>The maximum bar size is No. 5, therefore, splicing is permitted. | |
| 25.5.2.1 | Tension lap splice length, $\ell_{st}$, for deformed bars in tension must be the greater of:<br><br>$1.3\ell_d$ and 12 in. | $\ell_{st} = (1.3)(19 \text{ in.}) = 24.7 \text{ in.}$; use 36 in. |

### Step 15: Bottom reinforcing bar length along first span

| | | |
|---|---|---|
| 7.7.3.3 | The bottom bars have to satisfy the following provisions:<br>Reinforcement must extend beyond the point at which it is no longer required to resist flexure for a distance equal to the greater of $d$ and $12d_b$, except at supports of simply-supported spans and at free ends of cantilevers. | Inflection points<br>The inflection points for positive moments are 0 ft from exterior support centerline at CL F and 2.1 ft from the first interior support centerline CL E. |
| 7.7.3.4 | Continuing flexural tensile reinforcement must have an embedment length not less than $\ell_d$ beyond the point where bent or terminated tensile reinforcement is no longer required to resist flexure. | This condition is satisfied along at any section along the beam span. |
| 7.7.3.5 | Flexural tensile reinforcement must not be terminated in a tensile zone unless (a), (b), or (c) is satisfied.<br>$V_u \leq (2/3)\phi V_n$ at the cutoff point.<br><br>Note that (b) and (c) do not apply. | 2400 lb < (2/3)(10,800 lb) = 7200 lb  **OK** |
| 7.7.3.8.2 | At least one-fourth the maximum positive moment reinforcement must extend along the slab bottom into the continuous support a minimum of 6 in. | |
| 7.7.3.8.3 | At points of inflection, $d_b$ for positive moment tensile reinforcement shall be limited such that $\ell_d$ for that reinforcement satisfies condition (b), because end reinforcement is not confined by a compressive reaction.<br><br>$\ell_d \leq M_n/V_u + \ell_a$<br>$\ell_a$ is the greater of $d$ and $12d_b = 7.5$ in.<br>$$M_n = A_s f_y \left( d - \frac{a}{2} \right)$$<br>The elastic analysis indicates that $V_u$ at inflection point is 1800 lb. | Check if bar size is adequate<br>$M_n$ for an 7 in. slab with No. 5 at 12 in., 0.75 in. cover is:<br><br>$M_n = (0.31 \text{ in.}^2/\text{ft})(60,000 \text{ psi})(6 \text{ in.} - 18 \text{ in.}) =$<br>108,252 in.-lb $\cong$ 108,000 in.-lb<br><br>$\ell_d = 19 \text{ in.} \leq \dfrac{108,000 \text{ in.-lb}}{1800 \text{ lb}} + 7.5 \text{ in.} = 67.5 \text{ in.}$<br><br>Therefore, No. 5 bar is **OK** |

### Step 16: Bottom bar length

| | | |
|---|---|---|
| | The bar cut offs that are implicitly permitted from the Code provisions because of reduced required strength along the span do not apply before the inflection points for this slab, because if any bars were cut off, the maximum reinforcing bar spacing would be violated. Because all bottom bars extend past the tensile zone, Section 7.7.3.5 does not apply. All bottom bars need to extend at least 7 in. (refer to Section 7.7.3.3) beyond the positive moment inflection points. | The Code requires that at least 25 percent of bottom bars be full length, extending 6 in. into the support. Because the cut off locations are close to the supports and for field placing simplicity, extend all bars 6 in. into both supports. |

### Step 17: Slab detailing

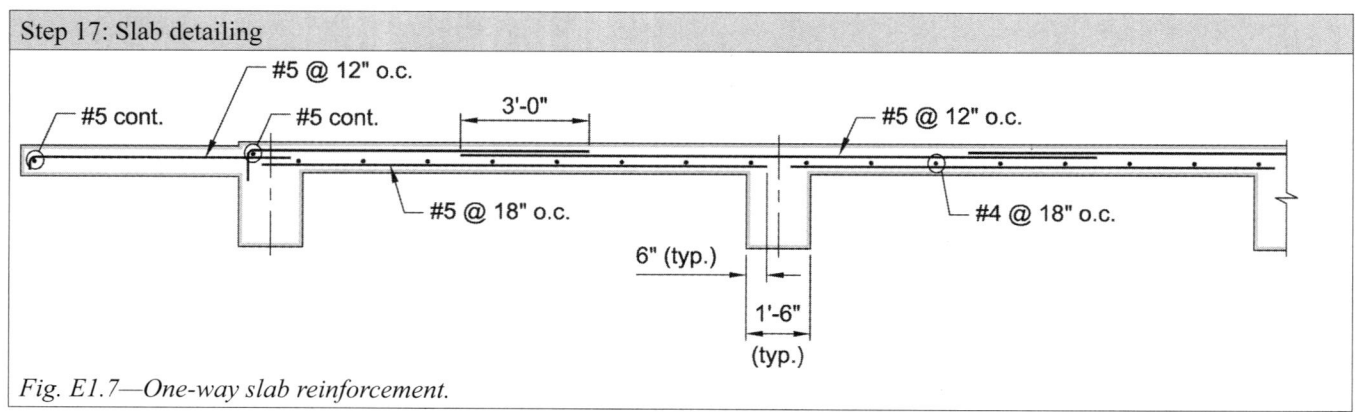

*Fig. E1.7—One-way slab reinforcement.*

**One-way Slab Example 2**: *Assembly loading*—
Design and detail a one-way nonprestressed reinforced concrete slab both for service conditions and factored loads. The one-way slab spans 20 ft-0 in. and is supported by 12 in. thick walls on the exterior, and 12 in. wide beams on the interior.

Given:
*Load*—
Live load $L$ = 100 psf
Concrete unit weight $\gamma_s$ = 150 lb/ft³

*Geometry*—
Span = 20 ft
Slab thickness $t$ = 9 in.

*Material properties*—
$f_c'$ = 5000 psi (normalweight concrete)
$f_y$ = 60,000 psi

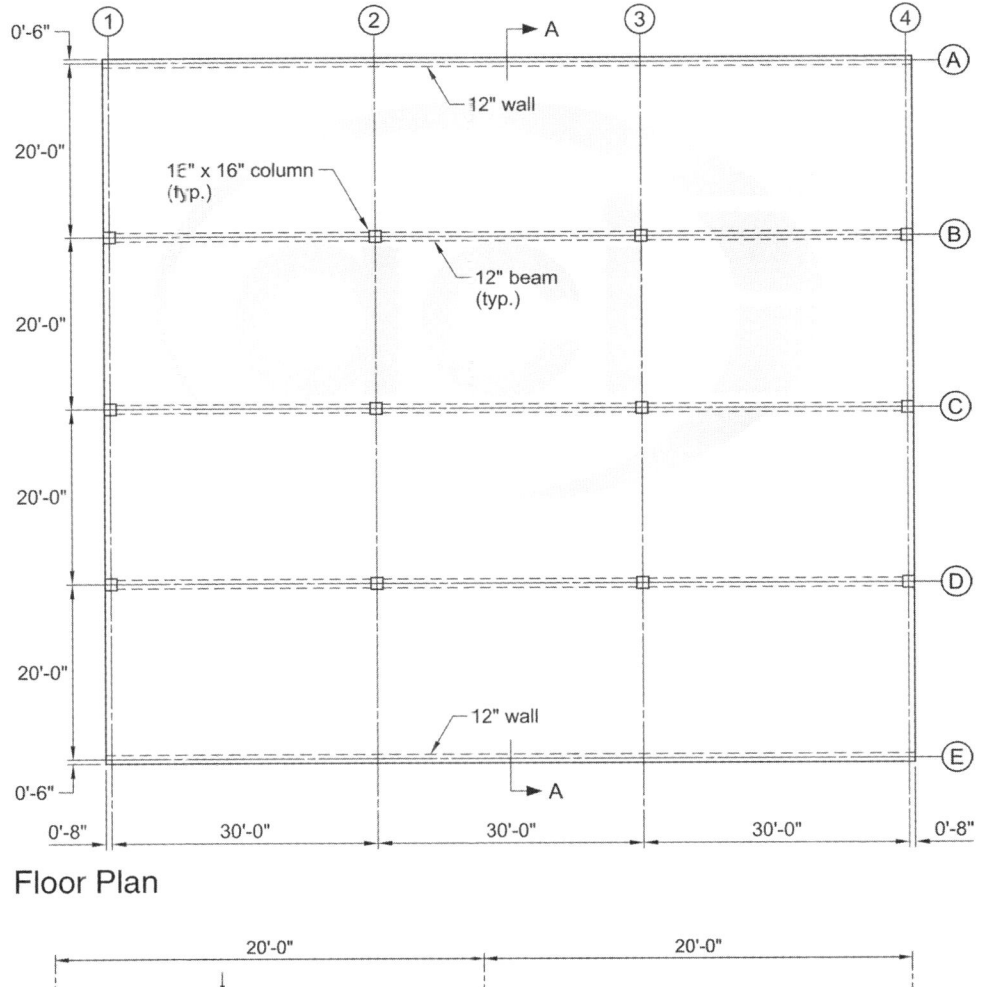

Floor Plan

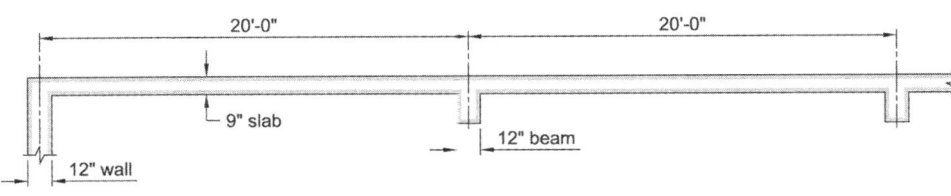

Section A-A

*Fig. E2.1—One-way slab framing plan.*

| ACI 318-14 | Discussion | Calculation |
|---|---|---|
| **Step 1: Geometry** | | |
| 7.3.1.1 | The specified slab thickness is 9 in. Since the slab satisfies the ACI 318-14 span-to-depth ratios (Table 7.3.1.1), the designer does not need to check deflections unless supporting or attached to partitions or other construction likely to be damaged by large deflections. | $h \geq \dfrac{\ell}{27} = \dfrac{(20 \text{ ft})(12 \text{ in./ft})}{27} = 8.89$ in., say, 9 in.<br><br>This ratio is less than the table value for "both ends continuous," so deflections are not required to be checked. |
| **Step 2: Loads and load patterns** | | |
| 5.3.1 | For hotel lobbies, the live load is assembly occupancy; the design live load is 100 psf per Table 4-1 in ASCE 7-10. A 9 in. slab is a 112 psf dead load. To account for loads due to ceilings, partitions, HVAC systems, etc., add 10 psf as miscellaneous dead load.<br><br>$U = 1.4D$            (5.3.1a)<br>$U = 1.2D + 1.6L$    (5.3.1b)<br><br>The slab resists gravity only and is not part of a lateral force-resisting system, except to act as a diaphragm.<br><br>Both ASCE 7 and ACI provide guidance for addressing live load patterns. Either approach is acceptable.<br><br>ACI 318 allows the use of the following two patterns, Fig. E2.2: | The required strength equations to be considered are:<br>$U = 1.4(122) = 171$ psf<br>$U = 1.2(122) + 1.6(100) = 146 + 160 = 306$ psf<br>**Controls** |
| 6.4.2 | Factored dead load is applied on all spans and factored live load is applied as follows:<br>(a) Maximum positive $M_u$ near midspan occurs with factored live load on the span and on alternate spans.<br>(b) Maximum negative $M_u$ at a support occurs with factored live load on adjacent spans only. | |

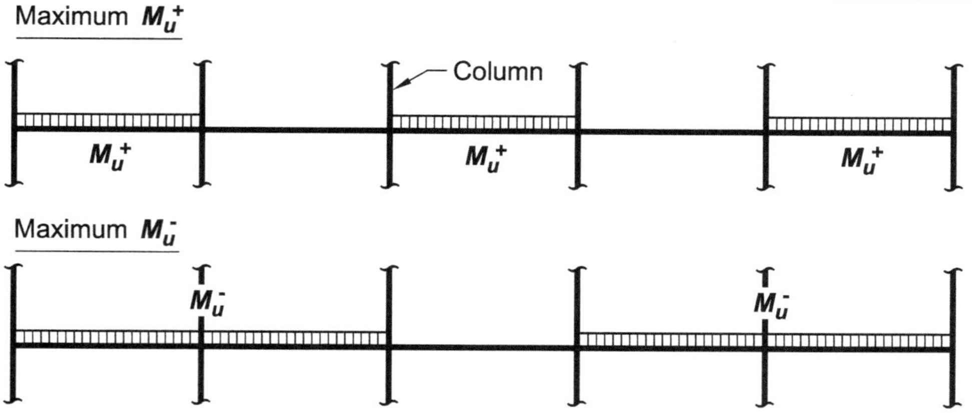

Fig. E2.2—Live load loading pattern.

## CHAPTER 5—ONE-WAY SLABS

| Step 3: Concrete and steel material requirements | | |
|---|---|---|
| 7.2.2.1 | The mixture proportion must satisfy the durability requirements of Chapter 19 and structural strength requirements of ACI 318-14. The designer determines the durability classes. Please refer to Chapter 4 of this Handbook for an in-depth discussion of the categories and classes.<br><br>ACI 301 is a reference specification that is coordinated with ACI 318. ACI encourages referencing ACI 301 into job specifications.<br><br>There are several mixture options within ACI 301, such as admixtures and pozzolans, which the designer can require, permit, or review if suggested by the contractor. | By specifying that the concrete mixture must be in accordance with ACI 301-10 and providing the exposure classes, Chapter 19 requirements are satisfied.<br><br>Based on durability and strength requirements, and experience with local mixtures, the compressive strength of concrete is specified at 28 days to be at least 5000 psi. |
| 7.2.2.2 | The reinforcement must satisfy Chapter 20 of ACI 318-14.<br>The designer determines the grade of bar and if the reinforcing bar should be coated by epoxy or galvanized, or both. | By specifying the reinforcing bar grade and any coatings, and that the reinforcing bar must be in accordance with ACI 301-10, Chapter 20 requirements are satisfied. In this case, assume Grade 60 bar and no coatings. |
| Step 4: Slab analysis | | |
| 6.3 | Because the building relies on the building's other members to resist lateral loads, the slab qualifies for braced frame assumptions, as discussed in the commentary. | Modeling assumptions<br>Apply the effective moment of inertia for the entire length of the slab.<br><br>Ignore torsional stiffness of beams.<br><br>Only the slab at this level is considered. |
| 6.6 | The analysis should be consistent with the overall assumptions about the role of the slab within the building system. Because the lateral force-resisting system only relies on the slab to transmit axial forces, a first order analysis is adequate.<br><br>Analysis approach:<br>The connection to the wall is monolithic; however, when the slab is fully loaded, flexural cracking will soften the joint. Rather than attempting to estimate an appropriate level of softening, the slab is simply modeled twice:<br><br>To simulate effect of cracking, reduce flexural stiffness by increasing the length of support. In this example, support lengths are increased to 100 ft long columns.<br><br>Assume fully connected slab to exterior wall with uncracked moment of inertia, modelled by a 10 ft, 12 in. x 12. in. columns, and the middle three supports are slender, 100 ft columns.<br><br>Note: The moments resulting from analysis maximize the moments and may be used to design the exterior wall. | |
| Step 5: Required moment strength | | |
| 7.4.2 | The slab's negative design moments are taken at the face of support, as is permitted by the Code. | |

Analysis (a):
Note: The moment at the exterior support is near zero (refer to Fig. E2.3):

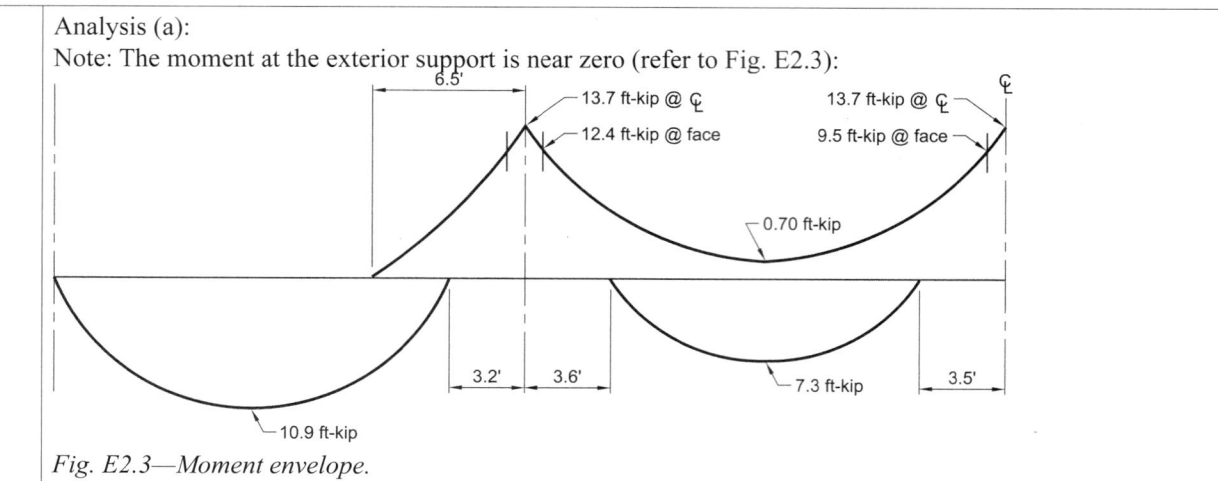

*Fig. E2.3—Moment envelope.*

The negative moment at the centerline of the exterior support is 0.0 ft-kip

The maximum positive moment in the end span is 10.9 ft-kip. The inflection points for positive moments are 0.0 ft from the exterior support centerline and 3.2 ft from the first interior support centerline.

The maximum negative moment at the face of the first interior support is 12.4 ft-kip. The negative moment's left inflection point is 6.5 ft from the support centerline. On the right side, under the pinned-at-wall assumption, there is no inflection point. Because of pattern loading, a small negative moment can exist across the span.

The maximum positive moment in the interior span is 7.3 ft-kip. The inflection points for positive moments are 3.6 ft from the first interior support centerline and 3.6 ft from the second interior support centerline.

The maximum negative moment at the face of the second interior support is 9.5 ft-kip. On the left side, under the pinned-at-wall assumption, there is no inflection point.

Because of pattern loading, a small negative moment can exist across the span.

## Table 2.1—maximum moment for hinged end condition (Approach (a))

| Required strength | Location from left to right along the span | | | | |
|---|---|---|---|---|---|
| | Exterior support | First midspan | Second support | Second midspan | Middle support |
| $M_u$, ft-kip | 0.0 | +10.9 | −12.4 | +7.3 | −9.5 |

Analysis (b) (end support is monolithic with wall)

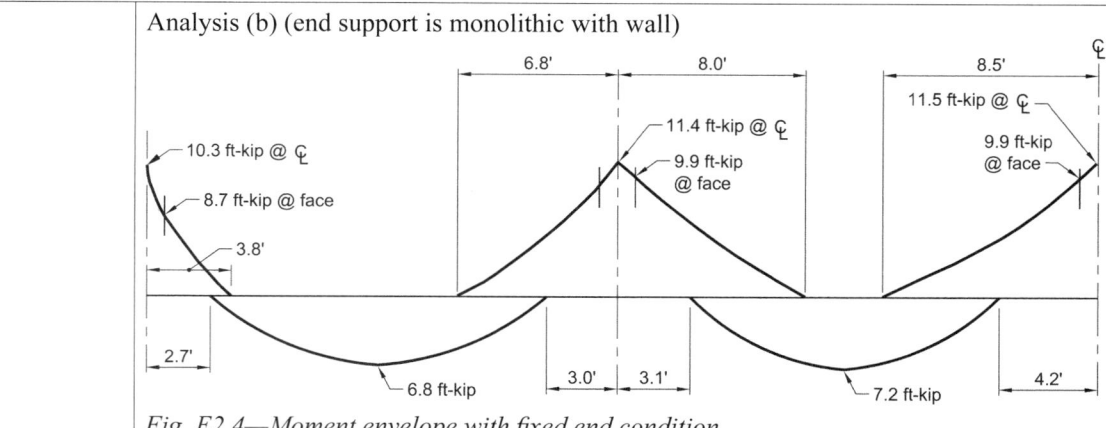

*Fig. E2.4—Moment envelope with fixed end condition.*

The negative moment at the face of the exterior support is 8.7 ft-kip. The negative moment's inflection point is 3.8 ft from the support centerline.

The maximum positive moment in the end span is 6.8 ft-kip. The inflection points for positive moments are 2.7 ft from the exterior support centerline and 3.0 ft from the first interior support centerline.

The maximum negative moment at the face of the first interior support is 9.9 ft-kip. The negative moment's left inflection point is 6.8 ft from the support centerline and the right inflection point is 8.0 ft from the support centerline.

The maximum positive moment in the interior span is 7.2 ft-kip. The inflection points for positive moments are 3.1 ft from the first interior support centerline and 4.2 ft from the second interior support centerline. The maximum negative moment at the face of the second interior support is 9.9 ft-kip. The negative moment's left inflection point is 8.5 ft from the support centerline and the right inflection point is 8.5 ft from the support centerline.

Following are the maximum moments from a combination of Analysis (a) and (b) (conservative approach).

### Table 2.2—Maximum moment for continuity condition between slab on wall (Approach (b))

| Required strength | Location from left to right along the span | | | | |
|---|---|---|---|---|---|
| | Exterior support | First midspan | Second support | Second midspan | Middle support |
| $M_u$, ft-kip | –8.7 | +6.8 | –9.9 | +7.2 | –9.9 |

| | Step 6: Required shear strength |
|---|---|
| 7.4.3.1 | The slab's maximum shear is taken at the support centerline for simplicity. The maximum shear under any condition is 3.4 kips. |

| | Step 7: Design moment strength |
|---|---|
| 7.5.1<br><br>21.2.1 | The two common strength inequalities for one-way slabs, moment and shear, are noted in Section 7.5.1.1 (ACI 318-14). The flexural strength reduction factor in Section 7.5.1.2 is assumed to be 0.9, which will be checked later. |
| 7.5.2 | The one-way slab chapter refers to Section 22.3 (ACI 318-14) for calculation of flexural strength. |

| | | |
|---|---|---|
| 22.2.1<br>22.2.2.2<br><br><br><br><br>22.2.2.4.1 | Chapter 22 (ACI 318-14) provides the design assumptions for reinforced concrete members.<br><br>Generate the minimum area of steel for the required moment.<br><br>$A_s f_y = 0.85 f_c'(b)a$<br><br>The effective depth, $d$, is the overall slab height minus the cover (3/4 in.) minus half the bar diameter (for a single layer of reinforcing bar). Assuming a No. 5 bar, therefore,<br><br>$d = 9 \text{ in.} - 0.75 \text{ in.} - 0.625 \text{ in.}/2$<br>$\quad = 7.9 \text{ in.}$ | To calculate $A_s$ in terms of the depth of the compression block, $a$, set the section's concrete compressive strength equal to steel tensile strength:<br><br>$T = C$<br><br>$A_s = \dfrac{0.85(5000 \text{ psi})(12 \text{ in.})a}{60,000 \text{ psi}} = 0.85a$<br><br>To calculate the minimum required $A_s$, set the design strength moment equal to the required strength moment.<br><br>$\phi A_s f_y (d - a/2) = M_u$<br><br>$0.9 A_s (60,000 \text{ psi}) \left( 7.9 \text{ in.} - \dfrac{A_s}{2(0.85)} \right) = M_u$ |
| | Table 2.3 following shows the required area of steel corresponding to the maximum moments from a conservative combination of Analysis (a) and (b). | |

### Table 2.3—Summary of maximum moment of Approaches (a) and (b)

| | Location from left to right along the span | | | | |
|---|---|---|---|---|---|
| | Exterior support | First midspan | Second support | Second midspan | Middle support |
| $M_u$, ft-kip | –8.7 | +10.9 | –12.4 | +7.3 | –9.9 |
| Req'd $A_s$, in.² per foot | 0.26 | 0.32 | 0.37 | 0.22 | 0.28 |

| | | |
|---|---|---|
| <br><br><br><br><br><br><br><br><br><br>22.2.2.1 | To ensure a ductile failure mode, the steel strain at ultimate strength must be at least 0.004 in./in. For usual reinforced slabs, such as this example, bar strain does not usually control the design.<br><br>A strain diagram is drawn (Fig. E2.5).<br><br><br><br><br><br>Maximum strain at the extreme concrete compression fiber is assumed equal to:<br>$\varepsilon_{cu} = 0.003$ in./in. | To calculate reinforcing bar strain, begin with force equilibrium within the section:<br>$T = C$<br>$A_s f_y = 0.85 f_c' b a$<br><br>where $b = 12$ in./ft; $f_y = 60,000$ psi; and the slab's maximum reinforcement is $A_s = 0.37$ in.²<br><br>From above calculations:<br>$A_s = 0.85a$ or $a = A_s/0.85$<br><br>Therefore, $a = 0.44$ in.<br>where $a = \beta_1 c$ and $\beta_1 = 0.80$ for $f_c'$ of 5,000 psi;<br>so $c = 0.55$ in.<br>From similar triangles (Fig. E2.5):<br>$\varepsilon_t = \dfrac{0.003(7.9 \text{ in.} - 0.55 \text{ in.})}{(0.55 \text{ in.})} = 0.040 \geq 0.004$<br>Therefore, the assumption of $\phi = 0.9$ is correct. |

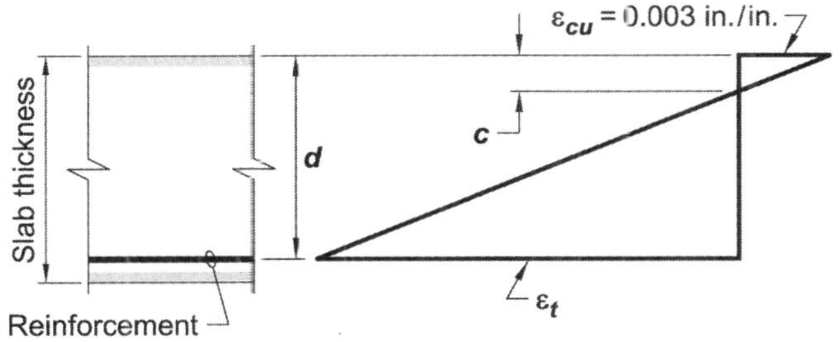

# CHAPTER 5—ONE-WAY SLABS

Fig. E2.5—Strain distribution.

| Step 8: Design shear strength | | |
|---|---|---|
| 7.5.3 | Assuming a one way slab won't contain shear reinforcement, $V_n$ is equal to $V_c$. Assuming negligible axial force, the Code provides a simple expression: | |
| 22.5.5.1 | $V_c = 2\sqrt{f'_c}bd$ | $V_c = 2\sqrt{5000\text{ psi}}(12\text{ in.})(7.9\text{ in.}) = 13{,}400\text{ lb}$ |
| 21.2.1 | Shear strength reduction factor: | $\phi = 0.75$<br>$\phi V_c = (0.75)(13{,}400\text{ lb}) = 10{,}055\text{ lb} > 3400\text{ lb}$  OK<br>This exceeds the maximum $V_u$ (3.4 kip/ft); therefore, no shear reinforcement is required. |
| Step 9: Minimum flexural reinforcement | | |
| 7.6.1 | Check if design reinforcement exceeds the minimum required reinforcement by Code. | $A_{s,\,min} = 0.0018 \times 9\text{ in.} \times 12\text{ in.} = 0.20\text{ in.}^2/\text{ft.}$<br>At all critical sections, the required $A_s$ is greater than the minimum. |
| Step 10: Shrinkage and temperature reinforcement | | |
| 7.6.4<br>24.4.3.2 | For one-way slabs with grade 60 bars, the minimum area of shrinkage and temperature (S+T) bars is $0.0018 A_g$. The maximum spacing of S+T reinforcing bar is the lesser of $3h$ and 18 in. | S+T steel area = $0.0018 \times 12\text{ in.} \times 9\text{ in.} = 0.20\text{ in.}^2$<br><br>Based on S+T steel area, solutions are No. 4 at 12 in. or No. 5 at 18 in.; use No. 5 at 18 in. placed atop and perpendicular to the bottom flexure reinforcement. |
| Step 11: Minimum and maximum spacing of flexural reinforcement | | |
| 7.7.2.1<br>25.2.1 | The minimum spacing between bars must not be less than the greatest of:<br>(a) 1 in.<br>(b) $d_b$<br>(c) $4/3 d_{agg}$<br>Assume 1 in. maximum aggregate size. | (a) 1 in.<br>(b) 0.625 in.<br>(c) $(4/3)(1\text{ in.}) = 1.33\text{ in.}$  **Controls** |

| | | |
|---|---|---|
| 7.7.2.2<br>24.3.2 | For reinforcement closest to the tension face, the spacing between reinforcement is the lesser of (a) and (b):<br><br>(a) $12(40,000/f_s)$<br>(b) $15(40,000/f_s) - 2.5c_c$ | (a) $12(40,000/40,000) = 12$ in.  **Controls**<br>(b) $15(40,000/40,000) - 2.5(0.75$ in.$) = 13.1$ in. |
| 24.3.2.1 | $f_s = 2/3 f_y = 40,000$ psi | |
| 7.7.2.3 | The maximum spacing of deformed reinforcement is the lesser of 3h and 18 in. | $3(7$ in.$) = 21$ in. $> 18$ in.<br>Therefore, Section 24.3.2 controls; 12 in. |

## Step 12: Select reinforcing bar size and spacing

### Table 2.4—Bar spacing

| | Location from left to right along the span | | | | |
|---|---|---|---|---|---|
| Bar size | Exterior support | First midspan | Second support | Second midspan | Middle support |
| No. 4 at spacing, in. | 9 | 7 | 6 | 10 | 8 |
| No. 5 at spacing, in. | 12 | 11 | 10 | 12 | 12 |
| No. 6 at spacing, in. | 12 | 12 | 12 | 12 | 12 |

Refer to Fig. E2.6 for provided bar spacing.

| | |
|---|---|
| | Based on the above, use No. 5 bars. Spacing of top bars at exterior support is 12 in., interior supports is 10 in. Note that there is no point of zero negative moment along the second and third span, so continue the top bars across both spans. While this solution is slightly conservative (10 in. versus 12 in. spacing), the engineer may desire consistent spacing for easier installation and inspection. |

# CHAPTER 5—ONE-WAY SLABS

| Step 13: Top reinforcing bar length at the exterior support | | |
|---|---|---|
| | The top bars have to satisfy the following provisions: | **Inflection points**<br>The inflection point for negative moment is 3.8 ft from support centerline. |
| 7.7.3.3 | Reinforcement shall extend beyond the point at which it is no longer required to resist flexure for a distance equal to the greater of $d$ and $12d_b$, except at supports of simply-supported spans and at free ends of cantilevers. | **Bar cutoffs**<br>Extend bars beyond the inflection point at least:<br>$d = 7.9$ in. or $(12)(0.625$ in.$) = 7.5$ in.;<br>Therefore, use 8 in. ~ 7.9 in. |
| 7.7.3.8.4 | At least one-third the negative moment reinforcement at a support shall have an embedment length beyond the point of inflection at least the greatest of $d$, $12d_b$, and $\ell_n/16$. | 33 percent of the bars to extend beyond the inflection point at least<br>$(19 \text{ ft} \times 12 \text{ in./ft})/16 = 15$ in. $> d = 8$ in. $> 12d_b = 7.5$ in.<br>Because the reinforcing bar is already at maximum spacing, no percentage of bars (as permitted by Section 7.7.3.3 of ACI 318-14) can be cut off in the tension zone. |
| | **Bars at the wall connection**<br>It is assumed that the wall is placed several days before the first floor slab. Because the wall and the slab will be firmly connected, the wall will tend to restrain the slab from shrinking as it cures. Many designers place extra reinforcement along the slab edge, parallel to the wall, to limit widths of possible cracks due to this restraint. | **Solution**<br>The top bar length is 6 in. (wall beyond centerline) plus 3.8 ft (inflection) plus an extension of either 8 in. or 15 in. Because the two cut off locations are close together, use a 15 in. extension for all bars. A practical length for top bars is:<br><br>3.8 ft + 0.5 ft + 1.25 ft = 5.55 ft, say, 6 ft. |
| **Step 14: Development and splice lengths** | | |
| 7.7.1.2<br>25.4.2.3 | ACI provides two equations for calculating development length; simplified and detailed. In this example, the detailed equation is used:<br><br>$$\ell_d = \left( \frac{3}{40} \frac{f_y}{\lambda \sqrt{f_c'}} \frac{\psi_t \psi_e \psi_s}{\left( \frac{c_b + K_{tr}}{d_b} \right)} \right) d_b$$<br><br>where<br>$\psi_t$ = bar location; not more than 12 in. of fresh concrete below horizontal reinforcement<br>$\psi_e$ = coating factor; uncoated<br>$\psi_s$ = bar size factor; No. 7 and larger<br><br>But the expression: $\frac{c_b + K_{tr}}{d_b}$ must not be taken greater than 2.5. | The development length of a No. 5 black bar in an 7 in. slab with 0.75 in. cover is:<br><br>$$\ell_d = \left( \frac{3}{40} \frac{60,000 \text{ psi}}{(1.0)\sqrt{5000 \text{ psi}}} \frac{(1.0)(1.0)(0.8)}{1.7 \text{ in.}} \right)(0.625 \text{ in.})$$<br>$= 19$ in.<br><br>$\psi_t = 1.0$, because not more than 12 in. of concrete is placed below bars.<br>$\psi_e = 1.0$, because bars are uncoated<br>$\psi_s = 0.8$, because bras are smaller than No. 7<br><br>$\frac{1.06 \text{ in.} + 0}{0.625 \text{ in.}} = 1.7$ in. |
| 7.7.1.3<br>25.5<br>25.5.1.1 | **Splice**<br>The maximum bar size is No. 5, therefore, splicing is permitted. | |
| 25.5.2.1 | Tension lap splice length, $\ell_{st}$, for deformed bars in tension must be the greater of:<br><br>$1.3\ell_d$ and 12 in. | $\ell_{st} = (1.3)(19 \text{ in.}) = 24.7$ in.; use 36 in. |

## Step 15: Bottom reinforcing bar length along first span

| | | |
|---|---|---|
| | The bottom bars have to satisfy the following provisions: | |
| 7.7.3.3 | Reinforcement must extend beyond the point at which it is no longer required to resist flexure for a distance equal to the greater of $d$ and $12d_b$, except at supports of simply-supported spans and at free ends of cantilevers. | Inflection points<br>The inflection points for positive moments are 0.0 ft from the exterior support centerline (analysis (a)) and 3.0 ft from the first interior support centerline (analysis (b)). |
| 7.7.3.4 | Continuing flexural tensile reinforcement must have an embedment length not less than $\ell_d$ beyond the point where bent or terminated tensile reinforcement is no longer required to resist flexure. | This condition is satisfied along at any section along the beam span. |
| 7.7.3.5 | Flexural tensile reinforcement shall not be terminated in a tensile zone unless (a), (b), or (c) is satisfied.<br>(a) $V_u \leq (2/3)\phi V_n$ at the cutoff point<br><br>Note that (b) and (c) do not apply. | |
| 7.7.3.8.2 | At least one-fourth the maximum positive moment reinforcement must extend along the slab bottom into the continuous support a minimum of 6 in. | |
| 7.7.3.8.3 | At points of inflection, $d_b$ for positive moment tensile reinforcement must be limited such that $\ell_d$ for that reinforcement satisfies condition (b), because end reinforcement is not confined by a compressive reaction.<br><br>$\ell_d \leq M_n/V_u + \ell_a$<br><br>$$M_n = A_s f_y \left(d - \frac{a}{2}\right)$$<br><br>The elastic analysis indicates that $V_u$ at inflection point is 2800 lb. The term $\ell_a$ is 8 in. | Check if bar size is adequate<br>$M_n$ for an 9 in. slab with No. 5 at 12 in., 0.75 in. cover is:<br><br>$M_n = (0.31$ in.$^2)(60{,}000$ psi$)(7.9$ in. $- 0.4$ in.$)$<br>$\quad = 140{,}000$ in.-lb<br><br>$\ell_d \leq \dfrac{140{,}000 \text{ lb}}{2800 \text{ lb}} + 8$ in. $= 58$ in. $> 19$ in.<br><br>Therefore, No. 5 bar is **OK** |

## Step 16: First span bottom bar length

| | | |
|---|---|---|
| | The bar cut offs that are implicitly permitted from prior Code provisions because of reduced required strength along the span do not apply before the inflection points for this slab, because if any bars were cut off, the maximum reinforcing bar spacing would be violated. Because all bottom bars extend past the tensile zone, Section 7.7.3.5 does not apply. All bottom bars need to extend at least 7 in. (refer to Section 7.7.3.3) beyond the positive moment inflection points. | The Code requires that at least 25 percent of bottom bars be full length, extending 6 in. into the support. Because the cut off location is close to the right support and for field placing simplicity, extend all bars 6 in. into both supports. |

| Step 17: Interior span bottom reinforcing bar lengths | | |
|---|---|---|
| | Inflection points<br>The inflection points for positive moments are 3.6 ft from the left support centerline and 4.2 ft from the right support centerline. | Create a partial length bar that is symmetrical within the span, so assume both inflection points are 3.6'. The minimum length is<br>20 ft – 3.6 ft – 3.6 ft + (2 ft)(0.5) = 13.8 ft, say, 14 ft 0 in. |
| | Bar cutoffs<br>Similar to the first interior span, all bottom bars must extend at least 8 in. past inflection points. The Code requires at least 25 percent of bottom bars be full length, extending 6 in. into the support. | In a repeating pattern, use 3 No. 5 at 14 ft long and 1 No. 5 at 21 ft long. |
| Step 18: Top reinforcing bar length at the middle support | | |
| | Inflection points<br>There are no inflection points over either span that frame into the middle support. | The required reinforcing bar is No. 5 at 12 in. in the middle support. Because the top bar from the first support is No. 5 at 10 in., extend the No. 5 at 10 in. top over the middle support for simplicity. |
| | Bar cutoffs<br>Because the top bars will be continuous, no bars are cut off. | |
| Step 19: Detailing | | |

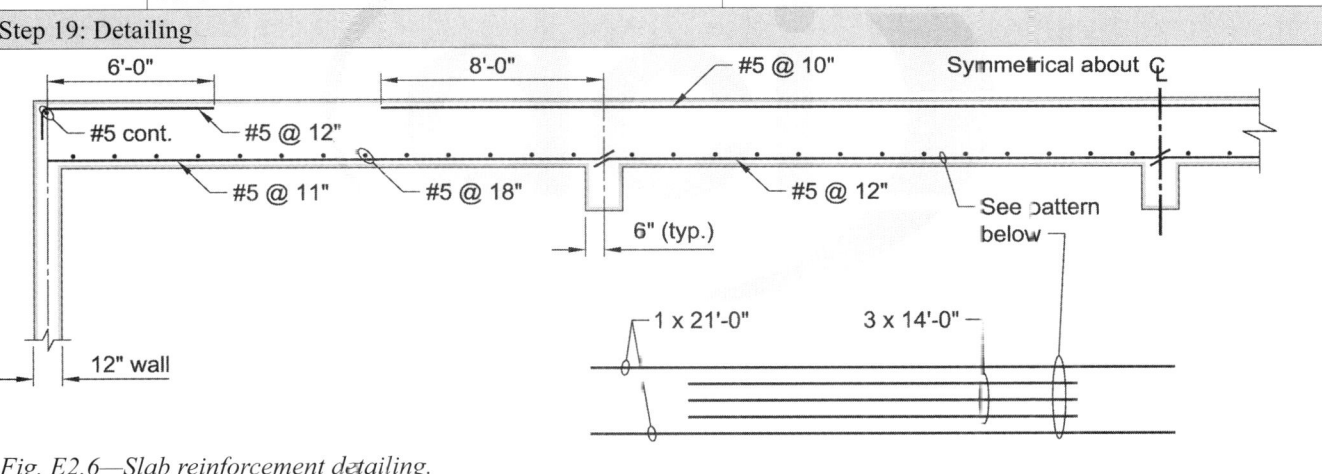

*Fig. E2.6—Slab reinforcement detailing.*

**One-way Slab Example 3**: *One-way slab post-tensioned – Hotel loading*

There are four spans of 20 ft-0 in. each, with a 3 ft-0 in. cantilever balcony at each end. The slab is supported by 12 in. walls on the exterior, and 12 in. wide beams on the interior (Fig. E3.1). This example will illustrate the design and detailing of a one-way post-tensioned (PT) slab, both for service conditions and factored loads.

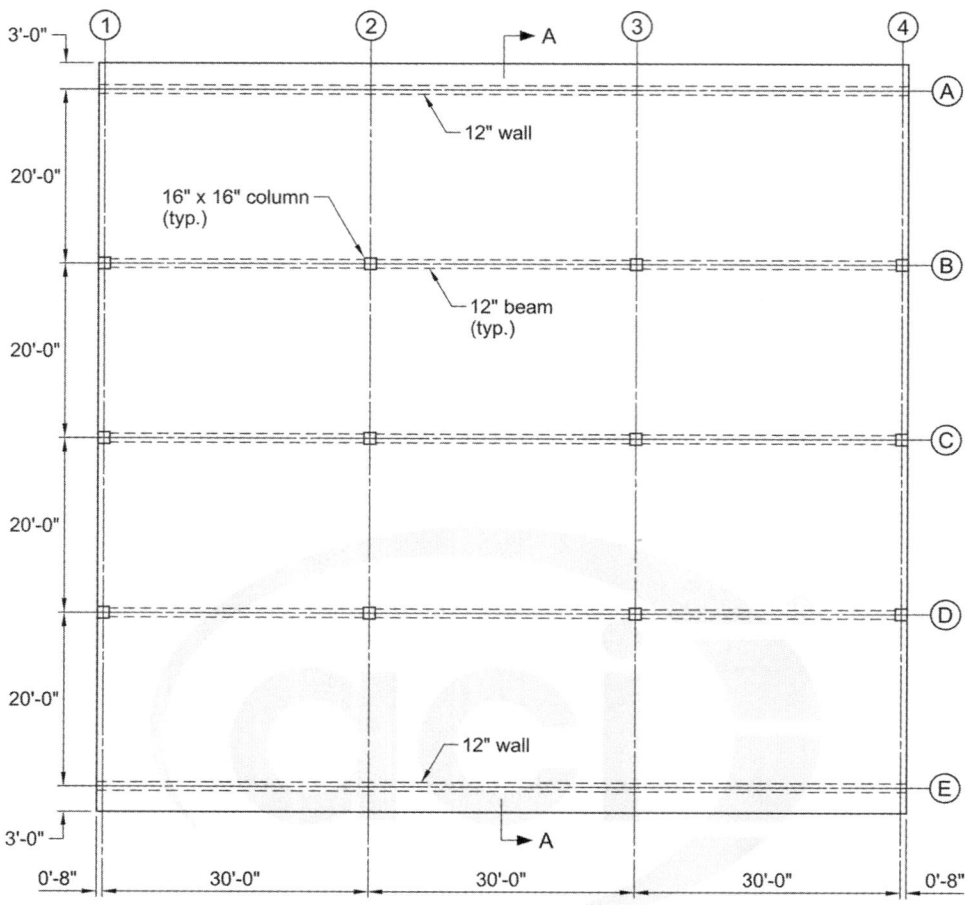

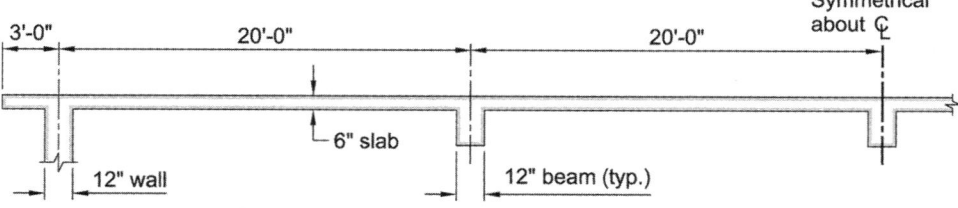

*Fig. E3.1—One-way slab.*

# CHAPTER 5—ONE-WAY SLABS

| ACI 318-14 | Discussion | Calculation |
|---|---|---|
| **Step 1: Geometry** | | |
| 7.3.2 | The ACI 318-14 span-to-depth ratios do not apply to PT slabs. The *Post-Tensioning Manual*, 2006, sixth edition Chapter 9, Table 9.3, suggests a ratio limit of $\ell/48$. | |
| | For this example, this ratio gives a slab thickness of | (20 ft)(12 in./ft)/48 = 5.0 in<br>This example uses a 6 in. thick slab. |
| **Step 2: Loads and load patterns** | | |
| 7.4.1.1 | For hotel occupancy, the design live load is 40 psf per Table 4-1 in ASCE 7-10. A 6 in. slab is a 75 psf dead load. To account for weights from ceilings, partitions, HVAC systems, etc., add 10 psf as miscellaneous dead load. | $D$ = 75 psf + 10 psf = 85 psf |
| | The slab resists gravity only and is not part of a lateral-force-resisting system, except to act as a diaphragm. | The required strength equations to be considered are: |
| 5.3.1 | $U = 1.4D$<br>$U = 1.2D+1.6L$ | $U = 1.4(85) = 119$ psf<br>$U = 1.2(85) + 1.6(40) = 102 + 64 = 166$ psf  **Controls** |
| 7.4.1.2 | Both ASCE 7 and ACI provide guidance for addressing live load patterns. Either approach is acceptable.<br><br>ACI 318 allows the design to use the following two patterns (Fig. E3.2): | |
| 6.4.2 | Factored dead load is applied on all spans and factored live load is applied as follows:<br><br>(a) Maximum positive $M_u^+$ near midspan occurs with factored L on the span and on alternate spans.<br>(b) Maximum negative $M_u^-$ at a support occurs with factored L on adjacent spans only. | |

One-Way Slabs

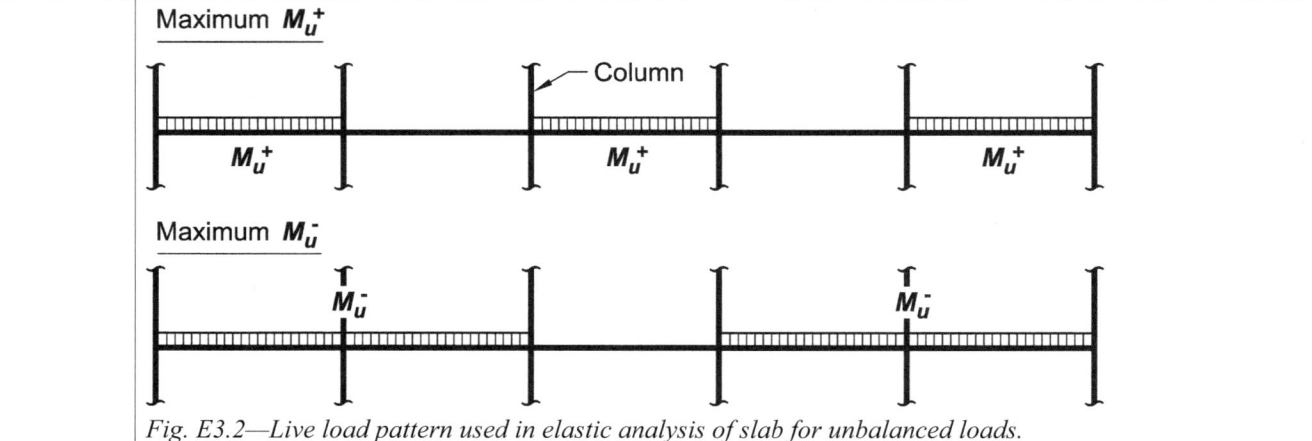

*Fig. E3.2—Live load pattern used in elastic analysis of slab for unbalanced loads.*

| | Step 3: Concrete and steel material requirements | |
|---|---|---|
| 7.2.2.1 | The mixture proportion must satisfy the durability requirements of Chapter 19 and structural strength requirements. The designer determines the durability classes. Please refer to Chapter 4 of this design Handbook for an in-depth discussion of the Categories and Classes. | |
| | ACI 301 is a reference specification that is coordinated with ACI 318. ACI encourages referencing ACI 301 into job specifications. | By specifying that the concrete mixture shall be in accordance with ACI 301-10 and providing the exposure classes, Chapter 19 requirements are satisfied. |
| | There are several mixture options within ACI 301, such as admixtures and pozzolans, which the designer can require, permit, or review if suggested by the contractor. | Based on durability and strength requirements, and experience with local mixtures, the compressive strength of concrete is specified at 28 days to be at least 5,000 psi. |
| 7.2.2.2 | The reinforcement must satisfy Chapter 20. | |
| | In this example, unbonded, 1/2 in. single-strand tendons are assumed. | By specifying the reinforcement shall be in accordance with ACI 301-10, the PT type and strength, and reinforcing bar grade (and any coatings), Chapter 20 requirements are satisfied.<br>In this example, assume grade 60 bar and no coatings. |
| | The designer determines the grade of bar and if the reinforcement should be coated by epoxy or galvanized, or both. | |
| 20.3 | The Code requires strand material to be 270 ksi, low relaxation (ASTM A416a). | |
| 20.3.2.5.1 | The U.S. industry usually stresses, or jacks, monostrand to impart a force equal to the least of $0.80f_{pu}$ and $0.9f_y$. immediate and long-term losses will reduce this force. | The jacking force per individual strand is:<br>(270 ksi)(0.8)(0.153 in.$^2$) = 33 kip.<br>This is immediately reduced by seating and friction losses, and elastic shortening of the slab. Long term losses will further reduce the force per strand. Refer to Commentary R20.3.2.6 of ACI 318-14. |
| 20.3.2.5.1 | where<br>$f_y = 0.94f_u$ (Table 20.3.2.5.1)<br>$0.80f_{pu}$ controls, the maximum allowed by the Code. | A PT force design value of 26.5 kips per strand is common. |

## Step 4: Slab analysis

| | | |
|---|---|---|
| 6.3 | Because the building relies on the building's other members to resist lateral loads, the slab qualifies for braced frame assumptions, as discussed in the commentary. | Modeling assumptions:<br>Slab will be designed as Class U. Consequently, use the gross moment of inertia of the slab in the analysis.<br><br>Assume the supporting beams have no torsional resistance and act as a knife edge support.<br><br>Only the slab at this level is considered. |
| 6.6<br><br><br><br><br><br><br><br><br><br><br><br>20.6.1.3.2 | The analysis performed should be consistent with the overall assumptions about the role of the slab within the building system. Because the lateral force-resisting system only relies on the slab to transmit axial forces, a first order analysis is adequate.<br><br>Although gravity moments are calculated independent of PT moments, the same model is used for both.<br><br>The strands profile is chosen to provide maximum resistance to dead load and live load. At the exterior support an eccentricity of 0.25 in. is chosen to balance the cantilever load. At the interior supports and midspans, the maximum possible eccentricity is chosen (1 in. cover) | Analysis approach:<br>To analyze the flexural effects of post-tensioning on the concrete slab under service loads, the tendon drape is assumed to be parabolic with a discontinuity at the support centerline shown as follows, which imparts a uniform uplift over each span when tensioned. The magnitude of the uplift, $w_p$, or "balanced load," in each span of a prismatic member is calculated as:<br><br>$w_p = 8Fa/\ell^2$<br><br>where $F$ is the effective PT force and $a$ is the tendon drape (average of the two high points minus the low point). In this example the PT force is assumed constant for all spans, but the uplift force varies due to different tendon drapes. |

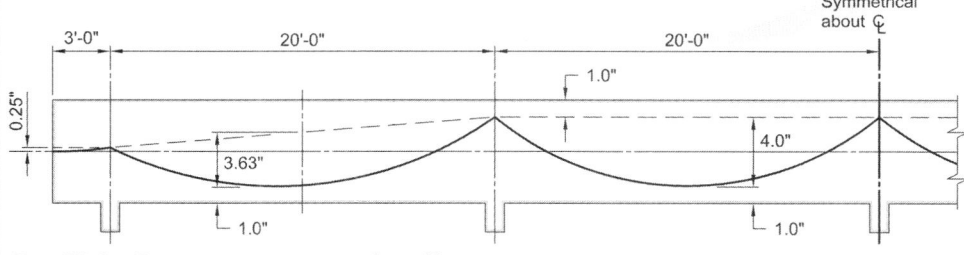

*Fig. E3.3—Post-tensioning strand profile.*

| | | |
|---|---|---|
| **Step 5: Slab stress limits** | | |
| 7.3.4.1<br>19.2.3.1 | This example assumes a Class U slab; that is, a slab under full service load with a concrete tension stress not exceeding $7.5\sqrt{f_c'}$. The slab analysis model for the service condition is the same as for the nominal condition. | $7.5\sqrt{5000 \text{ psi}} = 530 \text{ psi}$ |
| 19.2.3.1 | To verify that the concrete tensile stresses are less than $7.5\sqrt{5000 \text{ psi}}$, the net service moments and tensile stresses at the face of supports are needed.<br><br>This example assumes two parameters:<br>(a) The PT force provides a $F/A$ slab compressive stress of at least 125 psi (9 kip/ft), and<br>(b) The combination of PT force and strand profile provides an uplift force $w_p$ to balance at least 75 percent of the slab weight, or 56 psf (Fig. E3.4).<br><br>The basic equation for concrete tensile stress is:<br><br>$f_t = M/S - F/A$, where $M$ is the net service moment. At the exterior support, the drape is 3.25 in. (Fig. E3.3). The required force is calculated from: $w_p = 8Fa/\ell^2$<br>where $w_p = 0.056$ kip/ft | $S = (12 \text{ in.})(6 \text{ in.})^2/6 = 72 \text{ in.}^3$ (section modulus),<br>$A = (12 \text{ in.})(6 \text{ in.}) = 72 \text{ in.}^2$ (gross slab area per foot).<br><br>$F = \dfrac{(0.056 \text{ kip/ft})(20 \text{ ft})^2}{8(3.25 \text{ in.})/(12 \text{ in./ft})} = 10.3 \text{ kip/ft}$ |

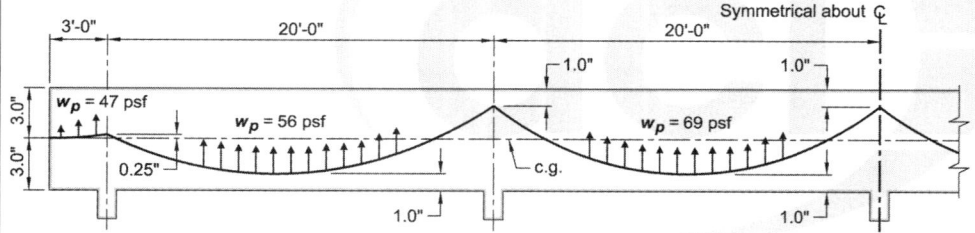

Fig. E3.4—Uplift force due to tendon layout.

| | | | | |
|---|---|---|---|---|
| 7.3.4.2 | The service gravity load is 125 psf. The PT uplift is subtracted from the gravity load. | | | |

| | Location from left to right along the span | | |
|---|---|---|---|
| **Service loads** | **Cantilever** | **First span** | **Second midspan** |
| Gravity uniform load, psf | 125 | 125 | 125 |
| PT uniform uplift, psf | 47 | 56 | 69 |
| Net "unbalanced load," psf | 78 | 69 | 56 |

| | | |
|---|---|---|
| 7.3.4.2 | The slab stresses are determined from an elastic analysis using the "net" load, minus the slab's axial compression.<br><br>Slab axial compression force: | $F/A = (10.3 \text{ kip/ft})(1000 \text{ lb/kip}) (6 \text{ in.})(12 \text{ in./ft}) = 143$ psi. |

| | | Location from left to right along the span | | | |
|---|---|---|---|---|---|
| | | Exterior support | First midspan | Interior support | Second midspan |
| 7.3.4.2 | Net service unbalanced moments, ft-kip/ft | 0.2 | 1.5 | 2.2 | 1.1 |
| | Net tensile stress, psi | 33 − 143 = −110 | 250 − 143 = 107 | 367 − 143 = 224 | 183 − 143 = 40 |
| 7.4.2.1 | The cantilever moment at the support centerline is: $w_{net}\ell^2/2 = (0.078\ psf)(3\ ft)2/2 = 0.351$ kip-ft/ft. ACI 318-14 permits to calculate the design slab moment at the face of the support: 0.2 kip-ft/ft. The moments for the interior supports are calculated at the faces of interior supports. | | | | |
| 7.3.4.2 19.2.3.1 | The aforementioned results show the maximum slab tensile stress (224 psi) calculated for an average PT force of 10.3 kip/ft is less than $7.5\sqrt{f'_c} = 530$ psi Therefore, the slab is uncracked and the aforementioned assumption is correct. | | | | |

| Step 6: Deflections | | |
|---|---|---|
| 7.3.2.1 | This chapter refers to Section 24.2 of ACI 318-14, "Deflections due to service-level gravity loads," for allowable stiffness approximations to calculate immediate and time-dependent (long term) deflections.<br><br>Section 24.2.2 provides maximum allowed span-to-deflection ratios.<br><br>Because slab is a Class U, use $I_g$: | $I_g = \dfrac{(12 \text{ in.})(6 \text{ in.})^3}{12} = 216 \text{ in.}^4$ |
| 24.2.3.8 | The balanced portion of the total load is offset by the camber from the prestressing, which results in a zero net deflection. Unbalanced load, however, will result in short- and long-term deflections that must be checked. The following equation, which can be downloaded from the Reinforced Concrete Design Handbook Design Aid – Analysis Tables: https://www.concrete.org/store/productdetail.aspx?ItemID=SP1714DA, will be used to calculate an approximate maximum deflection of the slab span with the largest unbalanced load (69 psf). In general, deflections do not control the design of PT slabs.<br><br>$\Delta_{max} = 0.0065 w \ell^4 / EI$<br><br>The additional time-dependent deflection is the immediate deflection due to sustained load multiplied by two (refer to Scanlon and Suprenant, 2011, "Estimating Two-Way Slab Deflections," *Concrete International*, V. 33, No. 7, July, pp. 24-29). | $\Delta_{max} = \dfrac{(0.0065)(69 \text{ psf})(240 \text{ in.})^4}{(4{,}030{,}000 \text{ psi})(216 \text{ in.}^4)/(12 \text{ in./ft})} = 0.14 \text{ in.}$<br><br>Expressed as a ratio,<br>$\ell/\Delta = 240 \text{ in.}/0.14 \text{ in.} = 1700$ |
| 24.2.2 | Assume that no portion of the live load is sustained. Calculate the immediate deflection based on a total sustained load of 85 psf reduced by the unbalanced load of 56 psf.<br><br>The long-term multiplier on the immediate deflection is 2, so the ratio is: | $(0.14 \text{ in.})(85 \text{ psf} - 56 \text{ psf})/(69 \text{ psf}) = 0.06 \text{ in.}$<br><br>$\ell/\Delta = 240/0.26 \text{ in} = \ell/920$<br><br>The deflection ratios are much less than the limit of $\ell/480$, so deflections are satisfied without more detailed calculations. |
| Step 7: Required moment strength | | |
| 7.4.2 | The gravity design moments, including pattern loading, are shown in Fig. E3.5: 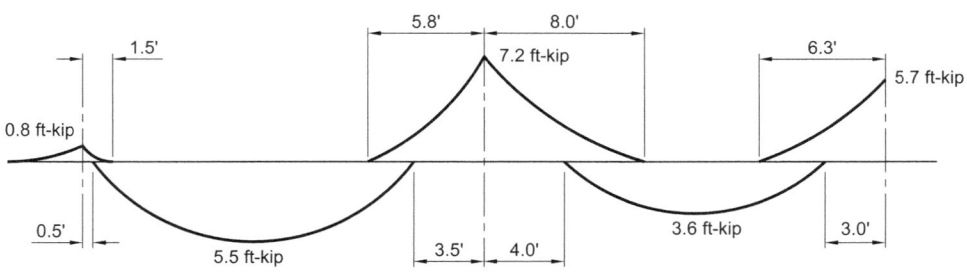*Fig. E3.5—Moment envelope due to factored loads.* | |

| | | |
|---|---|---|
| 7.4.1.3 | The Code requires that moments due to reactions induced by prestressing (secondary moments) be included with a load factor of 1.0. Secondary moments are calculated at each support as:<br>$M_2 = M_{pt} - Fe$.<br>The secondary moment diagram is linear between supports (Fig. E3.6). | |
| 7.4.1.3 | The PT secondary moments are:<br>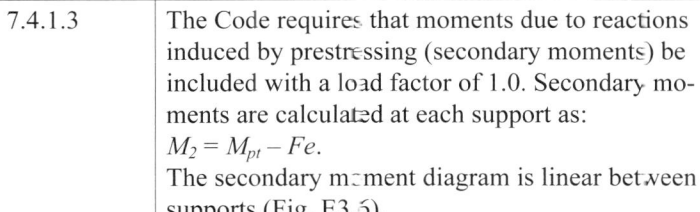<br>*Fig. E3.6—Secondary moments due to post-tensioning.* | |
| 7.4.2.1 | The combined factored and secondary moments, which are the required moment strengths, are shown in Fig. E3.7:<br>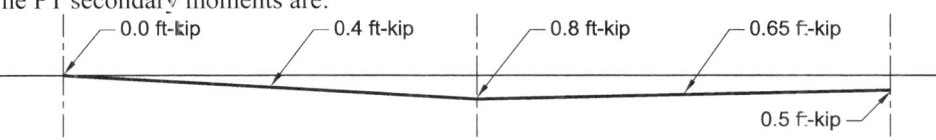<br>*Fig. E3.7—Moment envelope including factored load effect and secondary moments.* | |

| Required strength (Gravity only) | Location from left to right along the span ||||| 
|---|---|---|---|---|---|
| | Face of exterior support | First midspan | Face of second support | Second midspan | Face of middle support |
| Factored mom. at face, ft-kip | −0.8 | 5.5 | −7.2 | 3.6 | −5.7 |
| $M_2$, ft-kip | 0.0 | 0.4 | 0.8 | 0.65 | 0.5 |
| $M_u$, ft-kip | −0.8 | 5.9 | −6.4 | 4.3 | −5.2 |

## Step 8: Calculate required $A_s$

| | | |
|---|---|---|
| 7.7.4.2 | <u>Check flexural strength considering only PT tendons</u><br>If the PT tendons alone provide the design strength, $\geq \phi M_n$, then the Code permits the reinforcing bar to be a reduced length. If the PT tendons alone do not provide the design strength, then the reinforcing bar length is required to conform to standard lengths. | The depth of the equivalent stress block, $a$, is calculated by<br>$$a = \frac{A_{ps} f_{ps}}{0.85 f_c'(12 \text{ in.-ft})}$$<br>where $A_{ps}$ is the tendon area perfoot of slab. |
| | 7.5.2.1 refers to 22.3 for the calculation of $\phi M_n$. Section 22.3 refers to 22.2 for calculation of $M_n$. Section 22.2.4 refers to 20.3.2.4 to calculate $f_{ps}$. The span-to-depth ratio is 240/6 = 40, so the following equation applies:<br><br>The reinforcing bar and tendons are usually at the same height at the support and at midspan. | $$f_{ps} = f_{se} + 10{,}000 + \frac{f_c'}{300\rho_p}$$ |

| | | |
|---|---|---|
| 20.3.2.4 | Each single unbonded tendon is stressed to the value prescribed by the supplier. Friction losses cause a variation of $f_{se}$ along the tension length, but for design purposes, $f_{se}$ is usually taken as the average value. | The tendon supplier usually calculates $f_{se}$, and 175,000 psi is a common value. The force per strand is therefore: 175,000 psi × 0.153 in.$^2$ = 26,800 lb |
| | The effective force per foot of slab is 10.3 kip/ft, so the spacing of tendons is: | 26.8 kip/10.3 kip × 12 in./ft = 31 in., or 2 ft-7 in. |
| | The value of $A_{ps}$ is therefore,<br>The value of $\rho_p$ in $A_{ps}/(b \times d_p)$ | $A_{ps}$ = (0.153 in.$^2$)(12 in./ft)/(31 in.) = 0.059 in.$^2$/ft.<br>$\rho_p$ = 0.059 in.$^2$/ft / 60 in$^2$ = 0.00098<br>$f_{ps} = 175{,}000 \text{ psi} + 10{,}000 \text{ psi} + \dfrac{5000 \text{ psi}}{0.294} = 202{,}000 \text{ psi}$ |
| 20.3.2.4.1 | $f_{ps}$ limit as follows:<br>(a) $f_{se}$ + 30,000 and<br>(b) $f_{py} = 0.9 f_{pu}$ | (a) 175,000 psi + 30,000 psi = 205,000 psi and<br>(b) (0.9)(270,000 psi) = 243,000 psi<br><br>So the design value of $f_{ps}$ = 202,000 psi |
| 22.2.2.4.1 | The compression block depth is therefore:<br>$a = \dfrac{A_{ps} f_{ps}}{0.85 f'_c (12 \text{ in/ft.})}$<br><br>Note that the effective depth is 5 in. at critical locations, except at the exterior joint. Therefore, the Code permits a minimum $d$ of 0.8h, or 4.8 in. For 3/4 in. cover,<br>$\phi M_n = \phi A_{ps} f_{ps}(d - a/2)$ | $a = \dfrac{(0.059 \text{ in.}^2)(202{,}000 \text{ psi})}{(0.85)(5000 \text{ psi})(12 \text{ in./ft})} = 0.23$ in.<br><br><br>$\phi M_n$ = (0.9)(0.059 in.2)(202,000 psi)(5 in. − 0.12 in.)<br>= 52,129 in.-lb<br>= 4.34 ft-kip/ft. |

| | Location from left to right along the span | | | | |
|---|---|---|---|---|---|
| | Face of exterior support | First midspan | Face of second support | Second midspan | Face of middle support |
| $\phi M_n$, only tendons, ft-kip | 4.16 | 4.34 | 4.34 | 4.34 | 4.34 |
| $M_u$, ft-kip | −0.8 | 5.9 | −5.5 | 4.3 | −4.6 |

Because almost all the design moments are greater than $\phi M_n$ when considering the tendons alone, standard reinforcement lengths must be used.

### Step 9: Minimum flexural reinforcement

| | | |
|---|---|---|
| 7.6.2.3 | The minimum area of flexural reinforcing bar per foot is a function of the slab's cross-sectional area. | $A_{s,\ min}$ = 0.004 × 12 in. × 3 in. = 0.15 in.$^2$/ft. |

### Step 10: Design moment strength

| | | |
|---|---|---|
| 7.5.1 | The two common strength inequalities for one-way slabs, moment and shear, are noted in Section 7.5.1.1 of ACI 318-11. The strength reduction factor in Section 7.5.1.2 is assumed to be 0.9. | |

| | | |
|---|---|---|
| 7.5.2 | Determine if supplying the minimum area of reinforcing bar is sufficient to achieve a design strength that exceeds the required strength.<br><br>Comparing this value with the required moment strength $M_u$ indicates that the minimum reinforcement plus the tendons supply enough tensile reinforcement for slab to resist the factored loads at all locations.<br><br>$$a = \frac{A_{ps}f_{ps} + A_s f_y}{0.85 f_c'(12 \text{ in./ft})}$$ | Set the section's concrete compressive strength equal to steel tensile strength, and rearrange for compression block depth $a$:<br><br>$$a = \frac{(0.059 \text{ in.}^2)(202,000 \text{ psi}) + (0.15 \text{ in.}^2)(60,000 \text{ psi})}{(0.85)(5,000 \text{ psi})(12)}$$<br><br>$a = 0.41$ in.<br><br>For 3/4 in. cover:<br><br>$$M_n = \phi \left[ A_{ps}f_{ps} + A_s f_y \right]\left(d - \frac{a}{2}\right)$$<br><br>$= 0.9[0.059 \times 202,000 + 0.15 \times 60,000](5 - 0.21)$<br>$= 90,200$ in.-lb $= 7.5$ ft-kip<br><br>Therefore, minimum reinforcement provides adequate strength to resist the applied moment. **OK** |
| **Step 11: Design shear strength** | | |
| 7.5.3.1 | The slab's maximum shear is taken at the support centerline for simplicity. | $V_c = 2\left(\sqrt{5000 \text{ psi}}\right)(12 \text{ in.})(5 \text{ in.}) = 8500$ lb |
| 21.2.1 | Shear strength reduction factor:<br><br>Note: Shear does not typically control the thickness of one-way post-tensioning slab system. | $\phi = 0.75$<br><br>$\phi V_c = (0.75)(2)\left(\sqrt{5000 \text{ psi}}\right)(12 \text{ in.})(5 \text{ in.}) = 6364$ lb<br><br>The maximum $V_u = 2.0$ kips at the first interior support. Therefore, **OK**. |

| | | |
|---|---|---|
| **Step 12: Shrinkage and temperature reinforcement** | | |
| 7.6.4.2 | To control shrinkage and temperature stresses in the direction to the span, it is typical to use tendons rather than mild reinforcement in one-way post tensioning slabs. To calculate the number and spacing of temperature tendons, the Code allows the designer to consider the effect of beam tendons on the slab. | |
| | Assuming the beam is 12 in. x 30 in., the concrete cross-sectional area in the beam influence area is: | $A_{\text{infl area}}$ = (6 in.)(20 ft)(12 in./ft) + (12 in.)(24 in.) = 1726 in.$^2$ |
| | Assume the beam has an effective post-tensioning force of 189 kips, which results in an average compression of:<br>This amount is, therefore, sufficient to meet the Code minimum of 100 psi. The Code also has three spacing requirements which apply: | $\sigma$ = (189,000 lb)/(1726 in.$^2$) = 109 psi. |
| | Provide at least one tendon on each side of the beam.<br>If temperature tendon spacing does not exceed 4.5 ft, additional reinforcing bar is not needed; but if temperature tendon spacing exceeds 4.5 ft, supplemental reinforcement is required along the edge of the slab adjacent to tendon anchors. Spacing above 6 ft is prohibited. | |
| | In this example, temperature tendons, starting at 4 ft from the beam, are specified at 4 ft on center. No supplemental edge reinforcement is needed. | |
| **Step 13: Maximum spacing of flexural reinforcement** | | |
| 7.7.2.3 | The maximum spacing of flexural reinforcing bar in a PT slab is the lesser of $3h$ and 18 in. | The area of flexural reinforcing bar must be at least 0.15 in.$^2$/ft, use No. 4 bar at 16 in. on center, which also satisfies the maximum spacing requirement. |

| Step 14: Top reinforcing bar length at the exterior support | | |
|---|---|---|
| 7.7.3 | Reinforcing bar length and details at the exterior support | Inflection points<br>The inflection point for negative moment is 1.5 ft from exterior support centerline. |
| 7.7.3.3 | The top bars have to satisfy:<br>Reinforcement must extend beyond the point at which it is no longer required to resist flexure for a distance equal to the greater of $d$ and $12d_b$, except at supports of simply-supported spans and at free ends of cantilevers. | Bar cutoffs<br>Section 7.7.3.3 requires all bars to extend beyond the inflection point at least $d$ (5 in.) or 12 × 0.5 in; therefore 6 in. |
| 7.7.3.8.4 | At least one-third the negative moment reinforcement at a support shall have an embedment length beyond the point of inflection at least the greatest of $d$, $12d_b$, and $\ell_n/16$. | In addition, Section 7.7.3.8.4 requires 33 percent of the bars to extend beyond the inflection point at least (19 ft)(12 in./ft)/16 = 15 in. The top bars at the exterior joint will extend from the end of the cantilever, past the support and into the span. Because the reinforcing bar is at wide spacing, no percentage of bars (as permitted by Section 7.7.3.8) can be cut off in the tension zone. |
| | Balcony considerations<br>The architect usually specifies a slab recess of about 0.75 in. at the exterior wall of residential units to guard against water intrusion. In addition, the architect usually specifies a balcony slope of about 1/4 in./ft. These two details result in a slab thickness at the edge of 4.5 in. Balcony considerations are discussed in detail by Suprenant in, "Understanding Balcony Drainage," 2004, *Concrete International*, Jan | Solution<br>The top bar length is 3 ft. (balcony) plus 1.6 ft (inflection) plus an extension of 15 in. A practical length for top bars is 6 ft.<br><br>Trim bar at the outside edge<br>The PT suppliers usually require two No. 4 continuous "back-up" bars behind the anchorages, about 2 to 3 in. from the edge. These bars can also limit widths of possible cracks due to unexpected restraint, drying shrinkage, or other local issues. At the edge of the balcony, it is recommended to hook the top flexure bars around the continuous edge bars. |

| Step 15: Bottom reinforcing bar length along first span | | |
|---|---|---|
| 7.7.3 | The bottom bars have to satisfy the following provisions: | |
| 7.7.3.3 | Reinforcement must extend beyond the point at which it is no longer required to resist flexure for a distance equal to the greater of $d$ and $12d_b$, except at supports of simply-supported spans and at free ends of cantilevers. | |
| 7.7.3.4 | Continuing flexural tensile reinforcement must have an embedment length not less than $\ell_d$ beyond the point where bent or terminated tensile reinforcement is no longer required to resist flexure. $$\ell_d = \left( \frac{3}{40} \frac{f_y}{\lambda\sqrt{f'_c}} \frac{\psi_t \psi_e \psi_s}{\left(\frac{c_b + K_{tr}}{d_b}\right)} \right) d_b$$ | Note: the development length of a No. 4 black bar in an 6 inch slab with 0.75 in cover is: $$\ell_d = \left( \frac{3}{40} \frac{60{,}000 \text{ psi}}{1.0\sqrt{5000 \text{ psi}}} \frac{1.0 \times 1.0 \times 0.8}{\left(\frac{1.0 + 0}{0.5}\right)} \right) 0.5 = 13 \text{ in.}$$ |
| 7.7.3.5 | Flexural tensile reinforcement must not be terminated in a tensile zone unless (a), (b), or (c) is satisfied. | $V_u \leq (2/3)\,\phi V_n$ at the cutoff point.<br>$V_u = 2000$ lb and $\phi V_n = 6364$ lb (refer to Step 11)<br><br>$2/3\,\phi V_n = 4243$ lb $> V_u = 2000$ lb therefore, **OK**<br>Note that (b) and (c) do not apply. |
| 7.7.3.8.2 | At least one-fourth the maximum positive moment reinforcement must extend along the slab bottom into the continuous support a minimum of 6 in. | Extend bottom reinforcement minimum 6 in. into the supports. |
| 7.7.8.3 | At points of inflection, $d_b$ for positive moment, tensile reinforcement must be limited such that $\ell_d$ for that reinforcement satisfies Eq. (11.7.3.3.2) of ACI 318-14.<br><br>$\ell_d \leq M_n/V_u + \ell_a$<br><br>Inflection points<br>The inflection points for positive moments are 0.5 ft from the exterior support centerline and 3.0 ft from the first interior support centerline. | |

## Bar cutoffs

The bar cutoffs that are implicitly permitted in the aforementioned code provisions do not apply before the inflection points for this slab, because if any bars were cut off, the maximum reinforcing bar spacing would be violated. Because all bottom bars extend past the tensile zone, Section 7.7.3.5 does not apply. All bottom bars need to extend at least 5 in. (Refer to Section 7.7.3.3) beyond the positive moment inflection points.

The Code requires that at least 25 percent of bottom bars be full length, extending 6 in. into the support.

## Check bar size

$M_n$ is:

$$M_n = [A_s f_y]\left(d - \frac{a}{2}\right) + [A_{ps} f_{ps}]\left(d_p - \frac{a}{2}\right)$$

$M_n = (0.15 \text{ in.}^2)(60,000 \text{ psi})(5 \text{ in.} - 0.21 \text{ in.})$
$\quad + [(0.059 \text{ in.}^2)(202,000 \text{ psi})(4.8 \text{ in.} - 0.21 \text{ in.})]$
$= 97.8$ in.-kip

The elastic analysis indicates that $V_u$ at inflection point is 2.0 kips. The term $\ell_a$ is 5 in.

$\ell_d \leq M_n/V_u + \ell_a$

$\ell_d \leq \dfrac{M_n}{V_u} + 5 \text{ in.} = \dfrac{97.8 \text{ in.-kip}}{2.0 \text{ kip}} + 5 = 54$ in.

> 13 in.; therefore, No. 4 bar is OK.

Because the cut off location is within a foot of the left support, extend all bottom bars 6 in. into left support and then to the edge of the cantilever. These bottom bars will support shrinkage and temperature bars in the balcony.

For field placing simplicity, specify all bottom bars in this span also extend 6 in. into the right support.

### Step 16: Top reinforcing bar length at the first interior support

| | | |
|---|---|---|
| | | Inflection points<br>The inflection points for negative moments are 5.3 ft from the support centerline and 7.5 ft from the first interior support centerline. |
| 7.7.3.3 | Bar cutoffs<br>All bars must extend beyond the inflection point at least $d$ (5 in.) or<br>$12 \times 0.5$ in.; | Therefore, 6 in. |
| 7.7.3.8.4 | In addition, 33 percent of the bars must extend beyond the inflection point at least: | (19 ft × 12 in./ft)/16 = 15 in. |
| | However, the top bars at the exterior joint will extend from the end of the cantilever, past the support and into the span. Because the reinforcing bar is at wide spacing, no percentage of bars (as permitted by Section 7.7.3.8) can be cut off in the tension zone. | Therefore, the top bar length is 5.3 ft (inflection) plus an extension of 15 in. plus 7.5 ft plus 15 in. A practical length for top bars is 16 ft. |

| Step 17: Second span bottom reinforcing bar lengths | | |
|---|---|---|
| | | Inflection points<br>The inflection points for positive moments are 3.5 ft from the left support centerline and 2.5 ft from the right interior support centerline.<br><br>Bar cutoffs<br>The bar cut offs that are implicitly permitted by the Code do not apply before the inflection points for this slab, because if any bars were cut off, the maximum reinforcing bar spacing would be violated. Because all bottom bars extend past the tensile zone, Section 7.7.3.5 doesn't apply. All bottom bars need to extend at least 5 in. (Refer to Section 7.7.3.3) beyond the positive moment inflection points.<br>The Code requires that at least 25 percent of bottom bars be full length, extending 6 in. into the support.<br><br>Solution<br>The minimum bottom bar length is (20 ft minus 3.5 ft (inflection) plus an extension of 5 in. minus 2.5 ft (inflection) plus 5 in. A practical length for bottom bars is one at 20 ft and three at 16 ft. |
| Step 18: Top reinforcing bar length at the middle support, CL C | | |
| 7.7.3 | The top bars have to satisfy the following provisions: | Inflection points<br>The inflection points for negative moments are 5.8 ft from the support centerline on both sides. |
| 7.7.3.3 | Reinforcement must extend beyond the point at which it is no longer required to resist flexure for a distance equal to the greater of $d$ and $12d_b$, except at supports of simply-supported spans and at free ends of cantilevers. | |
| 7.7.3.8.4 | At least one-third the negative moment reinforcement at a support must have an embedment length beyond the point of inflection at least the greatest of $d$, $12d_b$, and $\ell_n/16$. | Bar cutoffs<br>Section 7.7.3.3 requires all bars to extend beyond the inflection point at least $d$ (5 in.) or 12 x 0.625 in; therefore, 8 in. In addition, Section 7.7.3.8.4 requires 33 percent of the bars to extend beyond the inflection point at least $(19 \times 12)/16 = 15$ in.<br>Because the reinforcing bar is already at maximum spacing, no percentage of bars (as permitted by Section 11.7.3.5 or 7.7.3.8 of ACI 318-14) can be cut off in the tension can be cut off in the tension zone.<br><br>Solution<br>The top bar length is two times (5.8 ft (inflection) plus an extension of 15 in. at each end) A practical length for top bars is 15 ft. |

## Step 19: Tendon termination

| | | |
|---|---|---|
| | There are requirements both for anchorage zones (the reinforced concrete around the anchorage) and for the anchorages themselves. | |
| 7.7.4.3.1 | Post-tensioned anchorage zones must be designed and detailed in accordance with Section 25.9 of ACI 318-14. | The concrete around the anchorage is divided into a local zone and a general zone. For monostrand anchorages, the local zone reinforcement, according to Code, "shall meet the bearing resistance requirements of ACI 423.7." ACI 423.7 limits the bearing stresses an anchorage can impose on the concrete, unless the monostrand anchorage is tested to perform, as well as those meeting those stresses. All U.S. manufacturers' supply tested anchorages.<br>For the general zone, Section 25.9.3.1 (a) of ACI 318-14 requires two "back up" bars for monostrand anchorages at the edge of the slab, and Section 25.9.3.2 (b) is not applicable to this example. |
| 7.7.4.3.2 | Post-tensioning anchorages and couplers must be designed and detailed in accordance with 25.7. | The information in Section 25.7 of ACI 318-14 provides performance requirements for the design of PT anchorages. These only apply to the anchorage design, so the engineer rarely (if ever) is concerned about Section 25.7. |

## Step 20: Shrinkage and temperature tendons

| | | |
|---|---|---|
| 7.7.6.3 | There are 4 temperature tendons evenly spaced per span (at 4 ft-0 in. O/C). Because the spacing is less than 4 ft 6 in., no additional edge reinforcement is required by the Code. The commentary recommends placing temperature tendons so that the resultant is within the kern of the slab (middle 2 inches). Anchors are usually attached to the outside forms at mid-height of the slab and longitudinally supported directly by the flexural tendons in such a manner as to meet this recommendation. | |

## Step 21: Detailing

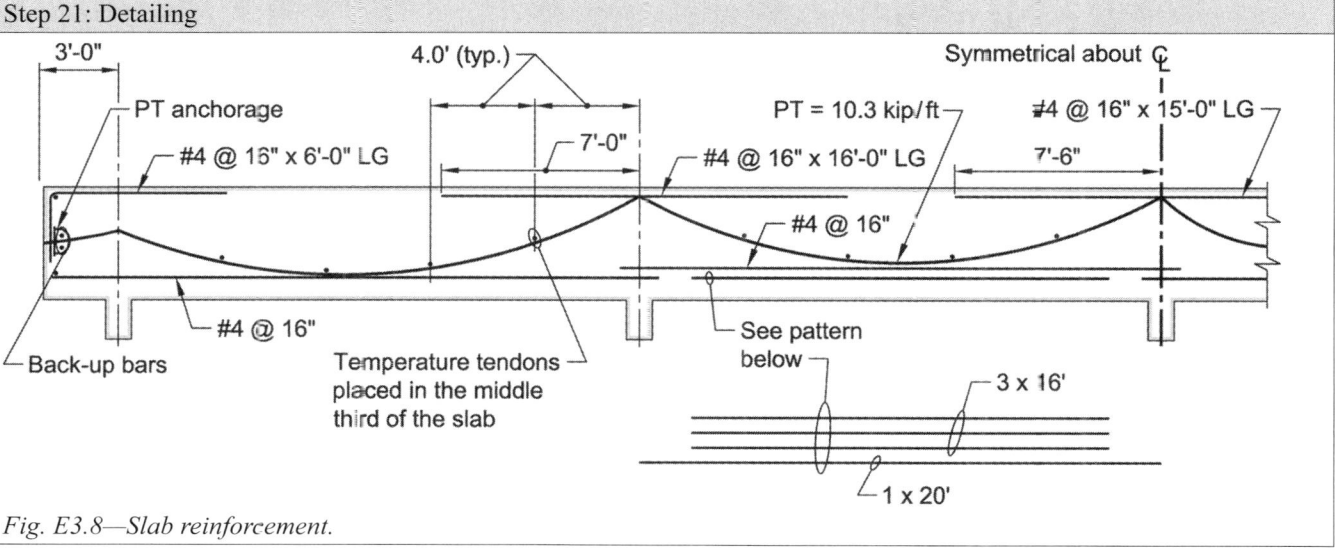

*Fig. E3.8—Slab reinforcement.*

# CHAPTER 6—TWO-WAY SLABS

## 6.1—Introduction

A two-way slab is usually used in buildings with columns that are approximately evenly spaced, creating a span length in one direction that is within a factor of 2 to the perpendicular direction. Structural concrete two-way slabs, which have been constructed for over 100 years, have taken many forms. The basic premise for these forms is that the slab system transmits the applied loads directly to the supporting columns through internal flexural and shear resistance.

This chapter discusses cast-in-place, nonprestressed, and post-tensioned (PT) slabs. The Code (ACI 318-14) allows for either bonded or unbonded tendons in a PT slab. Because bonded tendons are not usually placed in two-way slabs in the U.S., this chapter only discusses PT slabs with unbonded tendons.

At the preliminary design level, with spans given by the architect, the designer determines the loads, reinforcement type (prestressed or nonprestressed), and slab thickness. The preliminary concrete strength is based on experience and the Code's exposure and durability provisions.

## 6.2—Analysis

ACI 318-14 allows the designer to use any analysis procedure that satisfies equilibrium and geometric compatibility, as long as design strength and serviceability requirements are met. The Code includes detailed provisions for the Direct Design Method (DDM) and the Equivalent Frame Method (EFM), as well as general provisions for Finite Element Analysis (FEA). The commentary notes that while the analysis of a slab system is important, the design results should not deviate far from common practice, unless it is justified based on the reliability of the calculations used in the analysis.

**6.2.1** *Direct Design Method*—The DDM (ACI 318-14, Section 8.10) is a simplified method of analysis that has several geometric and loading limitations. Nonprestressed reinforced flat plates, flat slabs, and waffle slabs can all be designed by this method. The Code does not permit PT slabs to be designed by DDM. The results of the DDM are the approximate magnitude and distribution of slab moments, both along the span and transverse to it. The coefficients that distribute the total static moment in the design panel to the column and middle strips are based on papers by Corley, Jirsa, Sozen, and Siess (Corley et al. 1961; Jirsa et al. 1963, 1969; Corley and Jirsa 1970). The total static moment is determined assuming that the reactions are along the faces of the support perpendicular to the span considered. Once the total static moment is determined, it is then distributed to negative and positive moment areas of the slab. From there, it is further distributed to the column strip and middle strips. The designer uses these moments to calculate the flexural reinforcement area in the direction being designed. The designer needs to perform calculations in both directions to determine two-way slab reinforcement. The DDM also provides the design shear at each column.

**6.2.2** *Equivalent Frame Method*—The EFM (ACI 318-14, Section 8.11) can be used for a broader range of slab geometries than are allowed for DDM use, as well as PT slabs. Flat plates, flat slabs, and waffle slabs can all be designed by this method. The EFM assumptions used to calculate the effective stiffness of the slab, torsional beams, and columns at each joint are based on papers by Corley, Jirsa, Sozen, and Siess (Corley et al. 1961; Jirsa et al. 1963, 1969; Corley and Jirsa 1970). The EFM models a three-dimensional slab system by a series of two-dimensional frames that are then analyzed for loads acting in the plane of the frames. The original analysis method used with the EFM was the moment distribution method; however, any linear elastic analysis will work. The analysis calculates design moments and shears along the length of the model. For nonprestressed slabs, the EFM uses DDM coefficients to distribute the total moments into column strips and middle strips. For prestressed slabs, the slab strip is from the middle of one bay to the middle of the next bay, and is designed in flexure as a wide, shallow beam.

**6.2.3** *Finite Element Method*—A great variety of FEA computer software programs are available, including those that perform static, dynamic, elastic, and inelastic analysis. Any two-way slab geometry can be accommodated. Finite element models could have beam-column elements that model structural framing members along with plane stress elements; plate elements; and shell elements, brick elements, or both, that are used to model the floor slabs, mat foundations, diaphragms, walls, and connections. The model mesh size selected should be capable of determining the structural response in sufficient detail. Any set of reasonable assumptions for member stiffness is allowed.

## 6.3—Service limits

**6.3.1** *Minimum thickness*—For nonprestressed flat plates and flat slabs, the Code allows the designer to either calculate slab deflections or simply satisfy a minimum slab thickness (ACI 318-14, Section 8.3.1). Most flat slab and flat plate designs simply conform to the minimum thickness criteria and, therefore, designers do not usually calculate deflections for nonprestressed reinforced two-way slabs. The Code does not provide a minimum thickness-to-span ratio for PT two-way slabs, but the ratio for usual conditions is in the range of 37 to 45.

**6.3.2** *Deflections*—For nonprestressed two-way flat plate or flat slabs that are thinner than the ACI 318 minimum, for slabs that resists a heavy live load, for waffle slabs, and for PT slabs, the designer calculates deflections. Deflections can be calculated by EFM, the FEM, or classical methods. For EFM, the slab system is modelled in both directions, and the calculated deflection at midspan of a panel is the sum of the column strip deflection and the perpendicular middle strip deflection (refer to the crossing beam method (ACI 435R)).

The calculated deflections must not exceed the limits in Section 24.2 of ACI 318-14. For most buildings, the limit of $\ell/480$ for long term deflections usually controls.

Note that the spacing of slab reinforcing bar to limit crack width, timing of form removal, concrete quality, timing of construction loads, and other construction variables all can affect the actual measured deflection. These variables should be considered when assessing the accuracy of deflection calculations. In addition, creep over time will increase the immediate deflections.

Typically, with a PT slab thickness-to-span ratios in the range of 37 to 45, slab deflections are usually within the Code allowable limits. The Code limits the maximum service concrete tensile stress to below cracking stress, so deflection calculations use the gross slab properties.

**6.3.3** *Concrete service stress*—Nonprestressed slabs are designed for strength without reference to a pseudo-concrete service flexural stress limit.

For PT slabs, the analysis of concrete flexural tension stresses is a critical part of the design. In ACI 318-14, Section 8.3.4.1, the concrete tensile flexural stress in negative moment areas at columns PT slab is limited to $6\sqrt{f'_c}$. At positive moment sections, Section 8.6.2.3 requires slab reinforcing bar if the concrete tensile stress exceeds $2\sqrt{f'_c}$. This bottom reinforcing bar is often not required. These service tensile flexural stress limits are below the concrete cracking stress of $7.5\sqrt{f'_c}$, thus having the effect of reducing deflections. In addition, Section 8.6.2.1 requires a PT slab's axial compressive stress in both directions due to post-tensioning to be at least 125 psi.

Before the slab flexural stresses in a design strip can be calculated, the tendon profile needs to be defined. The profile and the tendon force are directly related to the slab forces and moments created by the PT. A common approach to calculate PT slab moments is to use the "load balancing" concept, where the profile is usually the maximum practical considering cover requirements, the tendon profile is parabolic, the parabola has an angular "break" at the column centerlines, and that the tendon terminates at middepth at the exterior (refer to Fig 6.3.3).

The load balancing concept assumes the tendon exerts a uniform upward "load" in the parabolic length, and a point load down at the support. These loads are then combined with the gravity loads, and the analysis is performed with a net load. Fig. 6.3.3 shows the commonly used simplification of the tendon profile. The real tendon profile is smooth with reverse parabolas over the interior supports rather than cusps.

To conform to the Code stress limits, the designer can use an iterative approach or a direct approach. In the iterative approach, the tendon profile is defined and the tendon force is assumed. The analysis is executed, flexural stresses are calculated, and the designer then adjusts the profile or force or both, depending on results and design constraints.

In the direct approach, the designer determines the highest tensile stress permitted, then rearranges equations so that the analysis calculates the tendon force needed to achieve the stress limit.

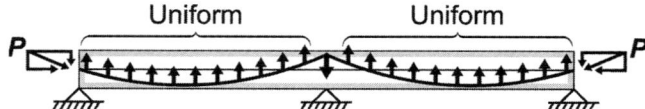

*Fig. 6.3.3—Load balancing concept.*

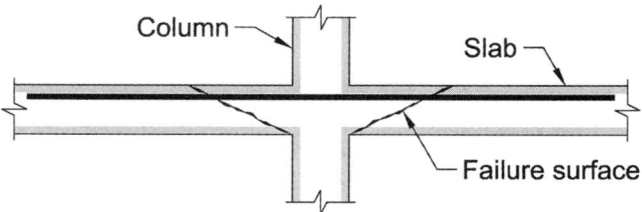

*Fig. 6.4.1—Punching shear failure.*

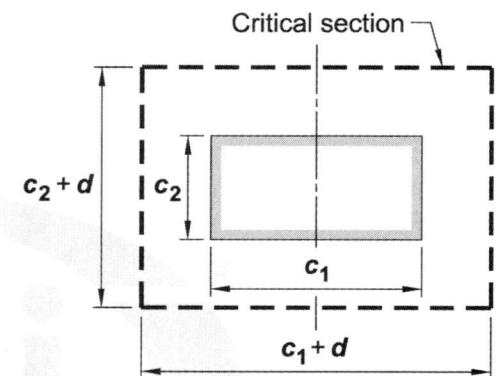

*Fig. 6.4.2—Critical section geometry.*

## 6.4—Shear strength

Two-way slabs must have adequate one-way shear strength in each design strip (assuming the slab is a wide, shallow beam) and adequate two-way shear strength at each column. The discussion for the nominal one-way shear strength are the same as provided in Chapter 7 (Beams) of this Handbook and is not reproduced here.

**6.4.1** *Punching shear strength*—Two-way shear strength, also called punching shear strength, is considered a critical strength for two-way slabs. ACI 318 calculates nominal punching shear strength based on the slab's concrete strength and shear reinforcement when provided. The effect of the slab's flexural reinforcement on punching shear strength is ignored.

The assumed punching shear failure shape (Fig. 6.4.1) is usually a truncated cone or pyramid-shape surface around the column.

**6.4.2** *Critical section*—For geometric simplicity, ACI 318 assumes a critical section. This is a vertical section extended from the column at a distance $d/2$, where $d$ is the slab's effective depth. In Fig. 6.4.2, the critical perimeter is $b_o = 2[(c_1 + d) + (c_2 + d)]$. The critical section to calculate concrete shear stress is $b_o d$.

**6.4.3** *Calculation of nominal shear strength*—In ACI 318, punching shear strength limits are given in terms of stress. As shown below, the shear stress limit for a nonprestressed reinforced slab is the least of three expressions:

The Code punching shear strength limit for PT slabs are usually slightly higher than those for nonprestressed rein-

## Table 6.4.3—Calculation of $v_c$ for two-way shear (ACI 318-14, Table 22.6.5.2)

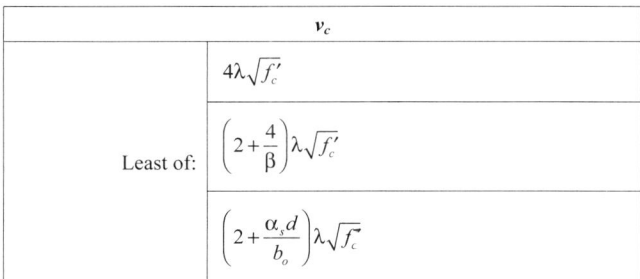

| | $v_c$ |
|---|---|
| Least of: | $4\lambda\sqrt{f_c'}$ |
| | $\left(2+\dfrac{4}{\beta}\right)\lambda\sqrt{f_c'}$ |
| | $\left(2+\dfrac{\alpha_s d}{b_o}\right)\lambda\sqrt{f_c'}$ |

Note: $\beta$ is the ratio of long side to short side of the column and $\alpha_s$ is 40 for interior columns, 30 for edge columns, and 20 for corner columns.

forced slabs as shown in the following Eq. (a) and (b). For PT two-way slabs, the designer can use Eq. (a) and (b) unless the column is closer to a discontinuous edge than four times the slab thickness $h$. For many edge columns, this requires shear strength to be calculated by Table 6.4.3.

For prestressed, two-way members, $v_c$ is permitted to be the lesser of (a) and (b) (ACI 318-14, Eq. 22.6.5.5 (a) and (b):

(a) $v_c = \left(3.5\lambda\sqrt{f_c'} + 0.3f_{pc}\right) + \dfrac{V_p}{b_o d}$

(b) $v_c = \left(1.5 + \dfrac{\alpha_s d}{b_o}\right)\lambda\sqrt{f_c'} + 0.3f_{pc} + \dfrac{V_p}{b_o d}$

where $\alpha_s$ is the same as in Table 6.4.3, the value of $f_{pc}$ is the average of $f_{pc}$ in the two directions, limited to 500 psi, $V_p$ is the vertical component of the effective prestress force crossing the critical section, and the value of $\sqrt{f_c'}$ is limited to 70 psi.

Because of the shallow depth of most PT slabs, many engineers conservatively ignore the $V_p/b_o d$ component when calculating $v_c$.

The Code also requires the engineer to consider slab openings close to the column. Such openings, which are commonly used for heating, ventilating, and air conditioning (HVAC), and plumbing chases, will reduce the shear strength. The Code requires a portion of $b_o$ enclosed by straight lines projecting from the centroid of the column and tangent to the boundaries of the opening to be considered ineffective (ACI 318-14, Section 22.6.4.3).

## 6.5—Calculation of required shear strength

The factored punching shear stress at a column, $v_u$, is the total of two components: 1) direct shear stress, $v_{ug}$; and 2) shear stresses due to moments transferred from the slab to the column. The two stress diagrams are added and the total is the required shear stress diagram at the critical section.

Direct shear stress $v_{ug}$ is calculated by $v_{ug} = V_u/b_o d$. To calculate the shear stresses due to slab bending, ACI 318 first stipulates that a percentage of the unbalanced slab moment at the column, $M_{sc}$, is resisted by slab flexure within a limited width over the column. The remaining percentage of $M_{sc}$ is transferred to the slab by eccentricity of shear. The follow two sections of Chapter 8 (ACI 318-14) state that:

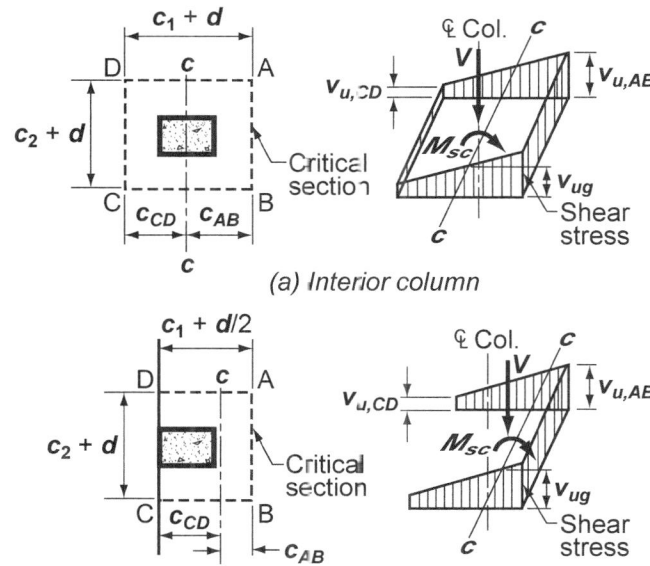

Fig. 6.5—Assumed distribution of shear stress (ACI 318-14, Commentary Section R8.4.4.2.3).

**8.4.2.3.2** The fraction of factored slab moment resisted by the column, $\gamma_f M_{sc}$, shall be assumed to be transferred by flexure, where $\gamma_f$ shall be calculated by:

$$\gamma_f = \dfrac{1}{1 + \left(\dfrac{2}{3}\right)\sqrt{\dfrac{b_1}{b_2}}}$$

**8.4.4.2.2** The fraction of $M_{sc}$ transferred by eccentricity of shear, $\gamma_v M_{sc}$, shall be applied at the centroid of the critical section in accordance with Section 8.4.4.1 (ACI 318-14), where

$$\gamma_v = 1 - \gamma_f$$

Under certain circumstances given in Table 8.4.2.3.4 of ACI 318-14, the value of $\gamma_f$ can be increased, which then decreases the fraction of $M_{sc}$ required to be transferred by eccentricity of shear.

Note that these modified values do not apply for PT slabs.

The slab shear stresses due to the unbalanced moment transferred to the column by eccentricity of shear is calculated by $\gamma_v M_{sc} c/J_c$, where $c$ is the distance from $b_0$ to the critical section centroid, and $J_c$ is the polar moment of inertia of the critical section about its centroidal axis. When $v_{ug}$ is added, the total shear stress diagram is shown by Fig. 6.5.

If the maximum total factored shear stress does not exceed the design shear stress, the slab's concrete shear strength is adequate. If the maximum total factored shear stress exceeds the design shear stress, the slab thickness near the column can be increased by using, for example, shear capitals (ACI 318-14, Section 8.2.5) or shear reinforcement can be added.

At times, the design of a two-way slab requires point loads to be considered, such as wheel loads in parking garages.

### Table 6.6a—Maximum $v_c$ for two-way members with shear reinforcement (ACI 318-14, Table 22.6.6.1)

| Type of shear reinforcement | Maximum $v_c$ at critical sections defined ACI 318-14, Section 22.6.4.1 | Maximum $v_c$ at critical section defined in ACI 318-14, Section 22.6.4.2 |
|---|---|---|
| Stirrups | $2\lambda\sqrt{f_c'}$ | $2\lambda\sqrt{f_c'}$ |
| Headed shear stud reinforcement | $3\lambda\sqrt{f_c'}$ | $2\lambda\sqrt{f_c'}$ |

The contribution of shear reinforcement is calculated by: $v_s = A_v f_{yt}/b_o s$ (ACI 318-14, Eq. (22.6.8.2)).

### Table 6.6b—Maximum $v_u$ for two-way members with shear reinforcement (ACI 318-14, Table 22.6.6.2)

| Type of shear reinforcement | Maximum $v_u$ at critical sections defined in 22.6.4.1 |
|---|---|
| Stirrups | $\phi 6\sqrt{f_c'}$ |
| Headed shear stud reinforcement | $\phi 8\sqrt{f_c'}$ |

These result in local shear slab stresses, and the slab's punching shear strength in that area needs to be verified.

## 6.6—Calculation of shear reinforcement

Shear reinforcement can be provided to increase the slab's nominal shear strength close to a column. Assuming the shear reinforcement is uniformly spaced, shear strength is first checked at the first critical section at $d/2$ beyond the column face including the contribution of shear reinforcement. The shear strength is then checked at $d/2$ beyond the outermost peripheral line of shear reinforcement, without the contribution of shear reinforcement. In slabs without shear reinforcement, $v_c$ is usually $4\sqrt{f_c'}$; however, for slab sections with shear reinforcement, the concrete contribution to shear strength is limited to the values in Table 6.6a.

There is an upper limit to a slab's nominal shear strength even with shear reinforcement, as shown in Table 6.6b. The Code states this limit in terms of the maximum factored two-way shear stress, $v_u$, calculated at a critical section.

Note that the use of stirrups as slab shear reinforcement is limited to slabs with an effective depth $d$ that satisfy (a) and (b):

(a) $d$ is at least 6 in.

(b) $d$ is at least $16d_b$, where $d_b$ is the diameter of the stirrups

The use of shear studs is not limited by the slab thickness, but the studs must fit within the geometric envelop. The overall height of the shear stud assembly needs to be at least the thickness of the slab minus the sum of (a) through (c):

(a) Concrete cover on the top flexural reinforcement

(b) Concrete cover on the base rail

(c) One-half the bar diameter of the flexural tension reinforcement

## 6.7—Flexural strength

After the designer calculates the factored slab moments, the required area of flexural reinforcement over a slab width is calculated with the same behavior assumptions as a beam.

**6.7.1** *Calculation of required moment strength*—There are two calculations for required moment strength for two-way slabs. The first calculation is to determine factored moments over the entire panel in the positive and negative moment areas.

For nonprestressed reinforced slabs, the slab analysis should provide the distribution of panel factored moments to the column strip and middle strip.

For PT slabs, effects of reactions induced by prestressing (secondary moments) need to be included. The slab's secondary moments are a result of the column's vertical restraint of the slab against the PT load at each support. Because the PT force and drape are determined during the service stress checks, secondary moments can be quickly calculated by the load-balancing analysis concept.

A simple way to calculate the secondary moment is to subtract the tendon force times the tendon eccentricity (distance from the NA) from the total balance moment, expressed mathematically as $M_2 = M_{bal} - P \times e$.

The second calculation is to determine $\gamma_f M_{sc}$ at each slab-column joint. The value of $M_{sc}$ is the difference between the design moments on either side of the column.

**6.7.2** *Calculation of design moment strength*—In a nonprestressed slab, the required reinforcement area $A_s$ resisting the column and middle strip's negative and positive $M_u$ is usually placed uniformly across each strip. The required reinforcement area $A_s$ resisting $\gamma_f M_{sc}$ must be placed within a width $b_{slab}$.

For PT slabs, the tendons are banded, which is where all tendons are placed together in a line that follows the column lines in one direction and uniformly distributed in the other. The slab flexural strength calculations for tendons (with $f_{ps}$ determined from Section 20.3.2.4 of ACI 318-14 substituted for $f_y$ in the $M_n$ equation) in the banded direction and in the uniform direction are the same, regardless of the tendon's horizontal location within the slab.

For PT slabs, the reinforcement area $A_s$ resisting the panel's negative $M_u$ is usually placed only at the column region. The $A_s$ plus $A_{pt}$ resisting $\gamma_f M_{sc}$ must be placed within a width $b_{slab}$ per ACI 318-14, Section 8.4.2.3.3. If the panel reinforcement already within $b_{slab}$ is not sufficient, designers usually add only $A_s$ to increase the flexural strength.

For PT slabs, the $A_{pt}$ provided to limit concrete service tensile stresses will usually be sufficient to also resist the panel's positive $M_u$.

## 6.8—Shear reinforcement detailing

**6.8.1** *Stirrups*—If stirrups are provided to increase shear strength, ACI 318 provides limits on their location and spacing in Table 6.8.1.

The related ACI 318-14 Commentary Fig. R8.7.6d as shown in the following Fig. 6.8.1 of this Handbook, also includes the two critical section locations:

**6.8.2** *Shear studs*—If shear studs are provided to increase shear strength, ACI 318-14 provides limits on shear stud locations and spacing in Table 6.8.2.

The related ACI 318-14 Commentary Fig. R8.7.7 as shown as the following Fig. 6.8.2, which also includes the two critical section locations.

## 6.9—Flexure reinforcement detailing

**6.9.1** *Nonprestressed reinforced slab reinforcement area and placing*—The Code requires a minimum area of flexural reinforcement in tension regions, with the area as shown in Table 6.9.1 (ACI 318-14, Table 8.6.1.1). If more than the minimum area is required by analysis, that reinforcement area must be provided.

Two-way slab flexural reinforcement is placed in top and bottom layers. For nonprestressed reinforced two-way slabs without beams, Fig. 6.9.1 (ACI 318-14, Commentary Fig. R8.6.1.1) provides a typical layout of column strip and middle strip top and bottom bars. ACI 318-14, Fig. 8.7.4.1.3(a) provides the minimum reinforcing bar extensions, lap locations, and the minimum $A_s$ at various sections. If the panel geometry is rectangular rather than square, the outer layer is usually placed parallel to the longer span.

**6.9.2** *Corners*—Corner restraint, created by walls or stiff beams, induces slab moments in the diagonal direction and perpendicular to the diagonal. These moments are in addition to the calculated flexural moments. Additional reinforcement per ACI 318-14, Section 8.7.3, is required for this condition.

**6.9.3** *Post-tensioned slab – Reinforcing bar area and placing*—Over each column region, the Code requires an area of flexural reinforcing bar of at least $0.0075A_{cf}$ in each direction, placed within $1.5h$ of the outside of the column. The Code also requires reinforcing bar in positive moment areas if the calculated service tensile flexural stress in other areas (usually midspan at the bottom) exceeds $2\sqrt{f_c'}$ or if required for strength. Bottom bar placement is at the discretion of the designer.

The Code allows for a reduced top and bottom minimum bar lengths if the design strength, calculated with only the PT tendons, is at least the required design strength. The top bars must extend at least $\ell/6$ on each side of the column. The bottom bars (if needed) must be at least $\ell/3$ and be centered at the maximum moment. The shorter lengths often control under typical spans and loadings. If the sectional strength using only the area of PT is insufficient to satisfy design strength, then the minimum top and bottom bar lengths are the same as a nonprestressed reinforced slab.

**6.9.4** *Post-tensioned slab – Tendon area and placing*—A minimum of 125 psi axial compression in each direction is required in a PT two-way slab. Post-tensioned tendons are usually placed in two orthogonal directions. In this configuration, the Code allows banding of tendons in one direction and in the other direction the tendon spacing is uniform across the design panel, within the spacing limits of $8h$ and 5 ft. This layout is predominant in the U.S. The Code also requires at least two tendons to be placed within the column

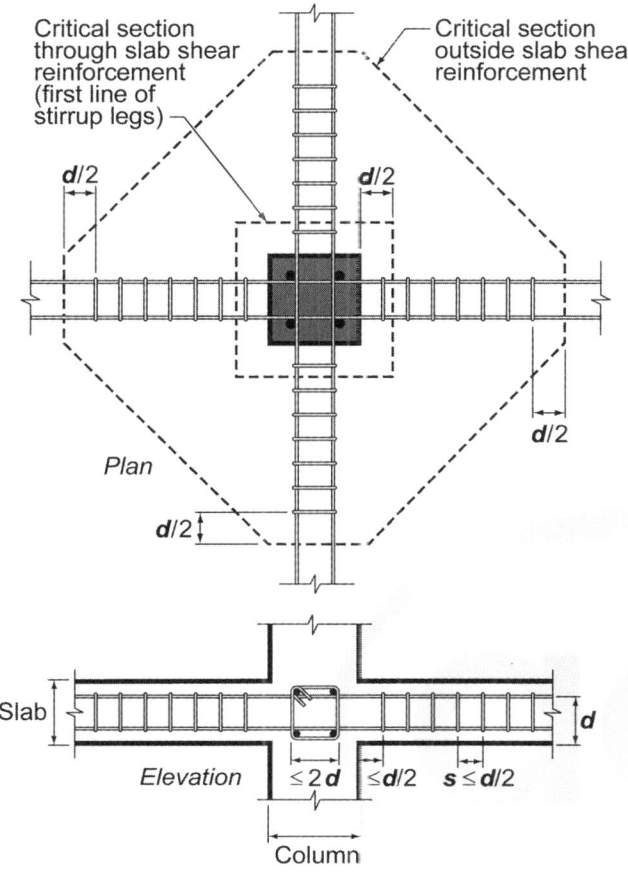

*Fig. 6.8.1—Arrangement of stirrup shear reinforcement, interior column (ACI 318-14, Commentary Fig. R8.7.6d).*

### Table 6.8.1—First stirrup location and spacing limits (ACI 318, Table 8.7.6.3)

| Direction of measurement | Description of measurement | Maximum distance or spacing, in. |
|---|---|---|
| Perpendicular to column face | Distance from column face to first stirrup | $d/2$ |
| | Spacing between stirrups | $d/2$ |
| Parallel to column face | Spacing between vertical legs of stirrups | $2d$ |

### Table 6.8.2—Shear stud location and spacing limits (ACI 318-14, Table 8.7.7.1.2)

| Direction of measurement | Description of measurement | Condition | | Maximum distance or spacing, in. |
|---|---|---|---|---|
| Perpendicular to column face | Distance from column face to first peripheral line of shear studs | All | | $d/2$ |
| | Constant spacing between peripheral lines of shear studs | Nonprestressed slab with | $v_u \leq \phi 6\sqrt{f_c'}$ | $3d/4$ |
| | | Nonprestressed slab with | $v_u > \phi 6\sqrt{f_c'}$ | $d/2$ |
| | | Prestressed slabs conforming to Section 22.6.5.4 of ACI 318-14. | | $3d/4$ |
| Parallel to column face | Spacing between adjacent shear studs on peripheral line nearest to column face | All | | $2d$ |

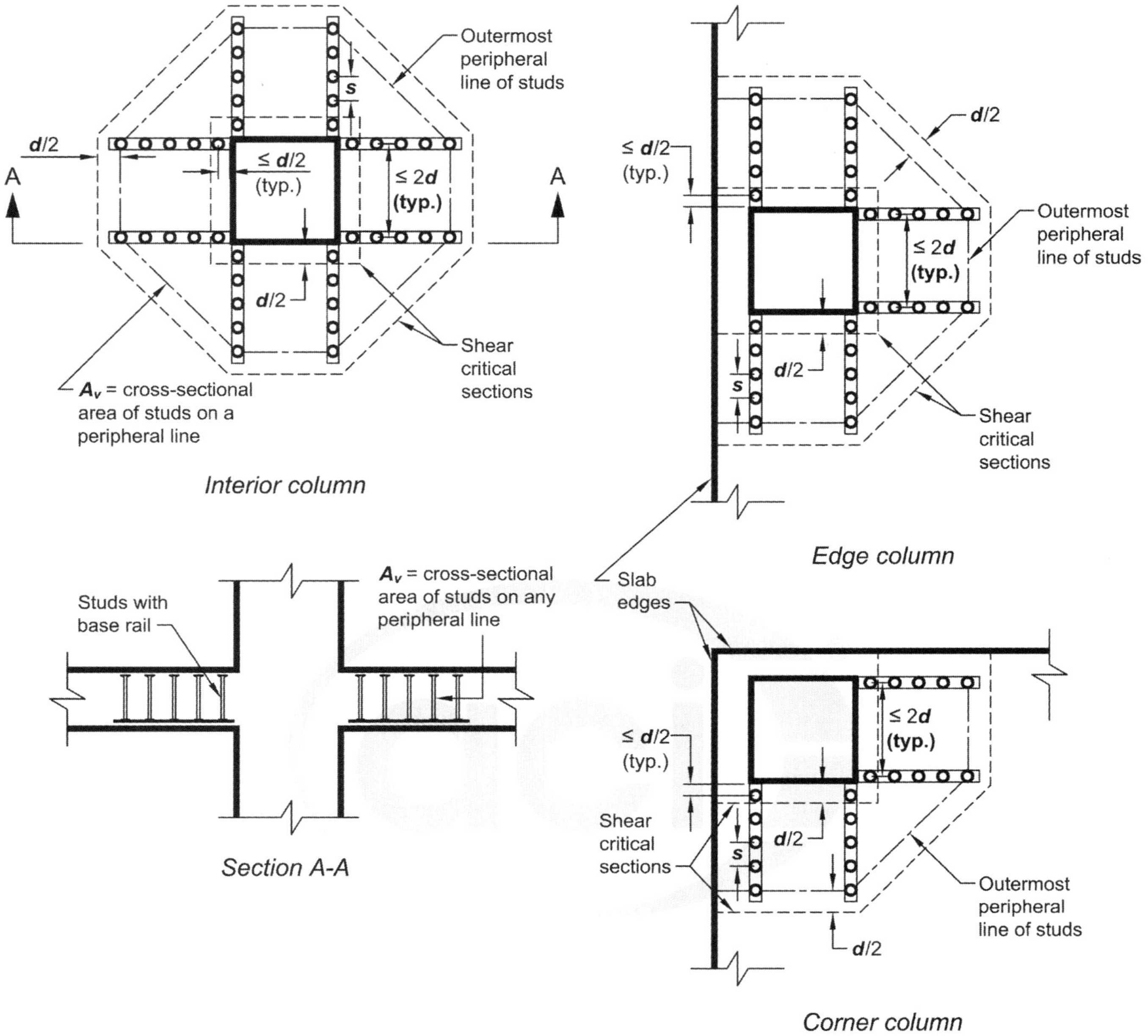

Fig. 6.8.2—*Typical arrangements of headed shear stud reinforcement and critical sections (ACI 318-14, Commentary Fig. R8.7.7)*

### Table 6.9.1—$A_{s,min}$ for nonprestressed two-way slabs

| Reinforcement type | $f_y$, psi | $A_{s,min}$, in.² | |
|---|---|---|---|
| Deformed bars | < 60,000 | | $0.0020A_g$ |
| Deformed bars or welded wire reinforcement | ≥ 60,000 | Greater of: | $\dfrac{0.0018 \times 60,000}{f_y} A_g$ |
| | | | $0.0014A_g$ |

reinforcement cage in either direction for overall building integrity.

**6.9.5** *Slab openings*—For relatively small slab openings, trim reinforcing bar usually limits crack widths that can be caused by geometric stress concentrations and provides adequate strength. For larger openings, a local increase in slab thickness as well as additional reinforcement may be necessary to provide adequate serviceability and strength.

## REFERENCES
*American Concrete Institute (ACI)*
ACI 435R-95—Control of Deflection in Concrete Structures (Reapproved 2000)

### Authored references
Corley, W. G.; Sozen, M. A.; and Siess, C. P., 1961,"Equivalent-Frame Analysis for Reinforced Concrete Slabs," *Structural Research Series* No. 218, Civil Engineering Studies, University of Illinois, June, 166 pp.

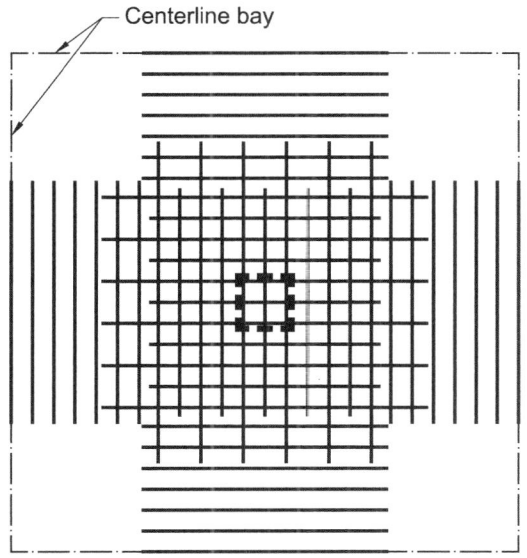

*Fig. 6.9.1—Arrangement of minimum reinforcement near the top of a two-way slab (ACI 318-14, Commentary Fig. R8.6.1.1).*

Corley, W. G., and Jirsa. J. O., 1970, "Equivalent Frame Analysis for Slab Design," *ACI Journal Proceedings*, V. 67, No. 11, Nov., pp 875-884.

Jirsa, J. O.; Sozen, M. A.; and Siess, C. P., 1963, "Effects of Pattern Loadings on Reinforced Concrete Floor Slabs," *Structural Research Series* No. 269, Civil Engineering Studies, University of Illinois, Urbana, IL, July.

Jirsa, J. O.; Sozen, M. A.; and Siess, C. P., 1969, "Pattern Loadings on Reinforced Concrete Floor Slabs," *Proceedings*, ASCE, V. 95, No. ST6 June, pp 1117-1137.

## 6.10—Examples

**Two-way Slab Example 1:** *Two-way slab design using direct design method (DDM) – Internal Frame*

This two-way slab is nonprestressed without interior beams between supports. This example designs the internal strip along grid line B. Material properties are selected based on the code requirements of Chapters 5 and 6, engineering judgment, and locally available materials. Lateral loads are resisted by shear walls; therefore, the design is for gravity loads only. Diaphragm design is not considered in this example.

### Given:

*Uniform loads:*

Self-weight dead load is based on concrete density including reinforcement at 150 lb/ft³
Superimposed dead load $D = 0.015$ kip/ft²
Live load $L = 0.100$ kip/ft²

*Material properties*:

$f_c' = 5000$ psi
$f_y = 60,000$ psi

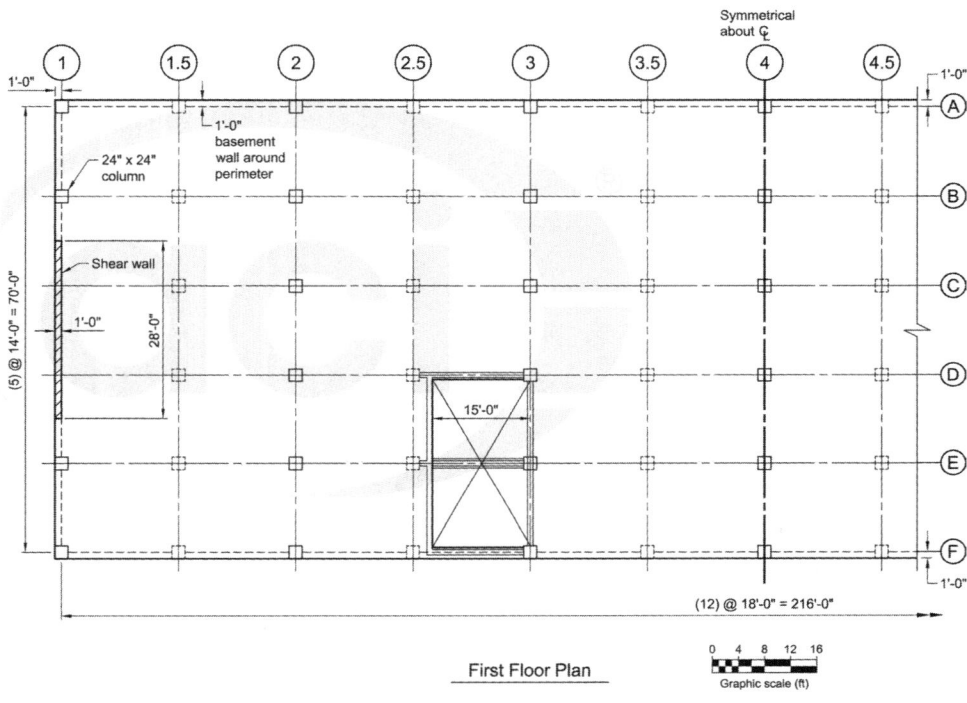

*Fig. E1.1—First floor plan*

# CHAPTER 6—TWO-WAY SLABS

| ACI 318-14 | Discussion | Calculation |
|---|---|---|
| **Step 1: Geometry** | | |
| 8.2.1<br>8.10<br>8.10.1.1<br>8.10.2.1<br>8.10.2.2<br>8.10.2.3<br>8.10.2.4<br>8.10.2.5<br>8.10.2.6<br>8.10.2.7<br>8.2.4<br>8.2.5 | This slab is designed using the Direct Design Method (DDM) in Section 8.10.<br><br>The slab geometry satisfies the limits of Sections 8.10.2.1 through 8.10.2.4, which allows the use of DDM.<br><br>The uniform design loads satisfy the limits of Sections 8.10.2.5 and 8.10.2.6 to allow use of DDM. | There are at least three continuous spans in each direction so Section 8.10.2.1 is satisfied.<br><br>The successive spans are the same lengths so Section 8.10.2.2 is satisfied.<br><br>The ratio of the longer to the shorter panel dimension is 1.29 so Section 8.10.2.3 is satisfied.<br><br>Columns are not offset through the slab so Section 8.10.2.4 is satisfied.<br><br>All design loads are distributed uniformly and due to gravity only so Section 8.10.2.5 is satisfied.<br><br>Ratio of unfactored live load to unfactored dead load is approximately $100/102.5 = 0.98$. This ratio is less than 2, so Section 8.10.2.6 is satisfied.<br><br>There are no supporting beams so Section 8.10.2.7 is not applicable.<br><br>This example does not include drop panels or shear caps so Sections 8.2.4 and 8.2.5 are not applicable. |
| 8.3.1.1 | Check the slab thickness for deflection control. | Using Table 8.3.1.1 with $f_y = 60{,}000$ psi, without drop panels, and assuming the wall performs as a stiff edge beam, the minimum thickness for the external panel is:<br><br>$$\frac{\ell_n}{33} = \frac{192 \text{ in.}}{33} = 5.8 \text{ in.}$$<br><br>The minimum thickness for the internal panels is calculated using the same table and the result is the same as that for the external panel<br>The remainder of the building is using a slab thickness of 7 in.; therefore, use 7 in. The slightly thicker than necessary slab aids with both deflections and shear strength. |
| 8.3.1.3 | No concrete floor finish is placed monolithically with the slab or composite with the floor slab. | |
| 8.3.2 | Calculated deflections are not required because the slab thickness-to-span ratio satisfies Section 8.3.2.1 of ACI 318-14. | |
| **Step 2: Load and load patterns** | | |
| 8.4.1.1 | The load factors are provided in Table 5.3.1 of ACI 318-14. | The load combination that controls is $1.2D + 1.6L$. Because Section 8.10.2.5 is satisfied in Step 1, pattern loading need not be checked. |

| Step 3: Initial two-way shear check | | |
|---|---|---|
| | Before performing detailed calculations, it is often beneficial to perform an approximate punching shear check. This check should reduce the probability of having to repeat the calculations shown in this example.<br><br>This check uses the following limits on the ratio of the design shear strength to the effects of shear stress based on direct shear stress alone ($\phi v_n/v_{ug}$):<br>For interior columns:<br>$\phi v_n/v_{ug} \geq 1.2$<br>For edge columns:<br>$\phi v_n/v_{ug} \geq 1.6$<br>For corner columns:<br>$\phi v_n/v_{ug} \geq 2.0$<br>If these ratios are not exceeded, it is possible that the slab will not satisfy two-way shear strength requirements. The design slab could be thickened, drop panels added, or other options for adding two-way shear strength may be considered. | For $\phi v_n$, the calculations here are discussed in Step 10 more fully:<br><br>$v_n = 4\sqrt{f'_c} = \dfrac{4\sqrt{5000}}{1000}$ ksi = 0.283 ksi<br><br>$v_n = \left(2 + \dfrac{4}{\beta}\right)\sqrt{f'_c} = \dfrac{6\sqrt{5000}}{1000}$ ksi = 0.424 ksi<br><br>$v_n = \left(2 + \dfrac{\alpha_s d}{b_o}\right)\sqrt{f'_c} = \dfrac{3.89\sqrt{5000}}{1000}$ ksi = 0.275 ksi*<br><br>*controls<br><br>$\phi v_n = 0.75 \times 0.275$ ksi = 0.206 ksi<br><br>For $v_{ug}$, the calculations here are discussed in Step 7 more fully:<br><br>$v_{ug} = \dfrac{V_u}{b_o d}$<br><br>$V_u = \left(14 \text{ ft} \times 18 \text{ ft} - \dfrac{29.6 \text{ in.} \times 29.6 \text{ in.}}{144}\right) \times \dfrac{283}{1000} \dfrac{\text{kip}}{\text{ft}^2}$<br><br>$V_u = 70$ kip<br><br>$v_{ug} = \dfrac{70 \text{ kip}}{118.4 \text{ in.} \times 5.6 \text{ in.}} = 0.106$ ksi<br><br>$\phi v_n/v_{ug} = 0.206/0.106 = 1.94 \geq 1.2 \therefore$ proceed.<br><br>Note that due to the basement wall supporting the exterior perimeter of the slab, the punching shear for the edge and corner columns will not need to be checked in this example. |
| Step 4: Analysis – Direct design method moment determination | | |
| 8.4.1.3<br>8.4.1.4<br>8.4.1.5<br>8.4.1.6<br>8.10.3.1 | The geometry of the design is shown in Fig. E1.2. | The design strip is bounded by the panel center line on each side of the column line and consists of a column strip and two half-middle strips. |

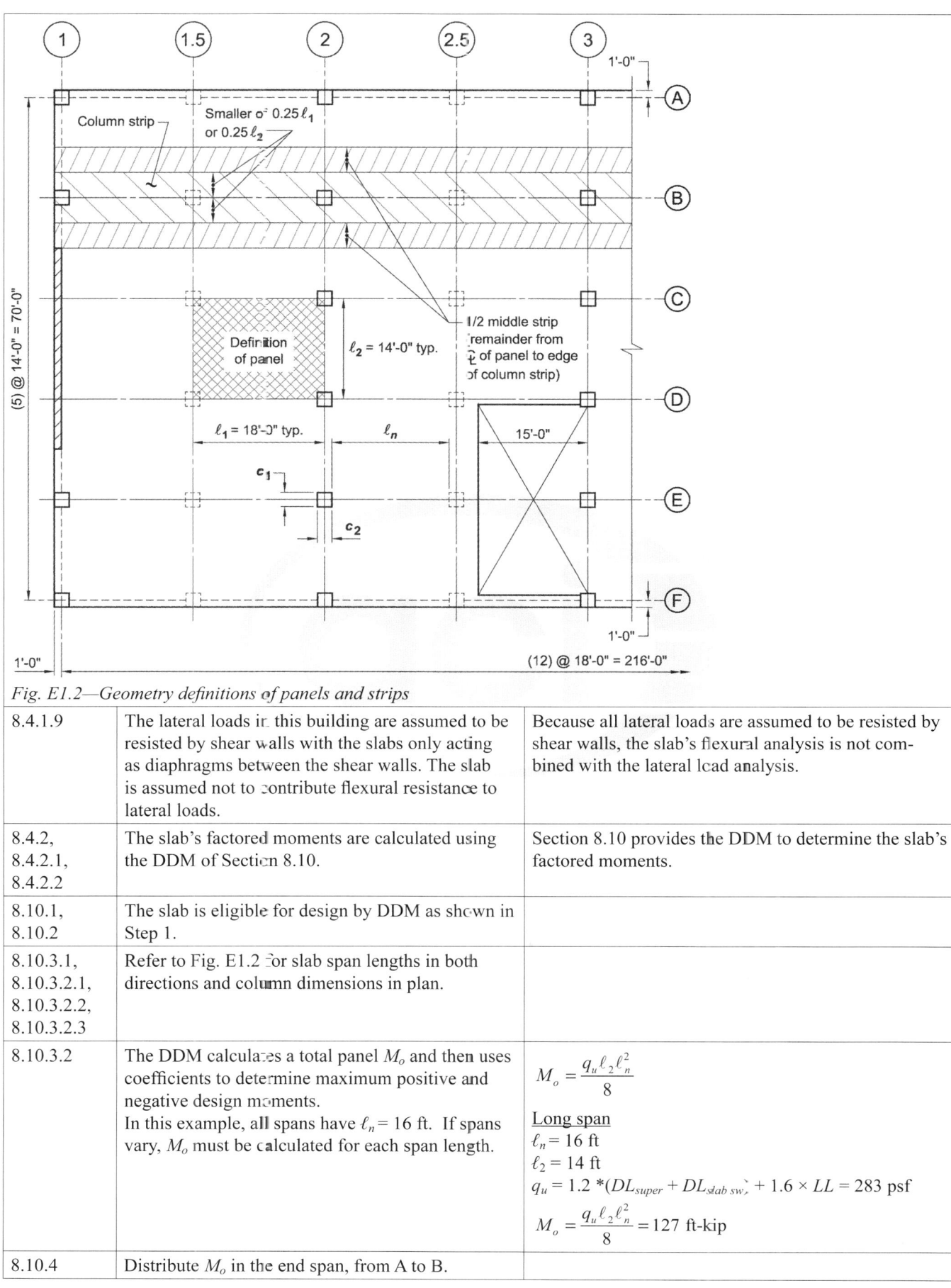

Fig. E1.2—Geometry definitions of panels and strips

| | | | |
|---|---|---|---|
| 8.4.1.9 | The lateral loads in this building are assumed to be resisted by shear walls with the slabs only acting as diaphragms between the shear walls. The slab is assumed not to contribute flexural resistance to lateral loads. | Because all lateral loads are assumed to be resisted by shear walls, the slab's flexural analysis is not combined with the lateral load analysis. | |
| 8.4.2, 8.4.2.1, 8.4.2.2 | The slab's factored moments are calculated using the DDM of Section 8.10. | Section 8.10 provides the DDM to determine the slab's factored moments. | |
| 8.10.1, 8.10.2 | The slab is eligible for design by DDM as shown in Step 1. | | |
| 8.10.3.1, 8.10.3.2.1, 8.10.3.2.2, 8.10.3.2.3 | Refer to Fig. E1.2 for slab span lengths in both directions and column dimensions in plan. | | |
| 8.10.3.2 | The DDM calculates a total panel $M_o$ and then uses coefficients to determine maximum positive and negative design moments.<br>In this example, all spans have $\ell_n$ = 16 ft. If spans vary, $M_o$ must be calculated for each span length. | $M_o = \dfrac{q_u \ell_2 \ell_n^2}{8}$<br><br>Long span<br>$\ell_n$ = 16 ft<br>$\ell_2$ = 14 ft<br>$q_u = 1.2 * (DL_{super} + DL_{slab\,sw}) + 1.6 \times LL = 283$ psf<br>$M_o = \dfrac{q_u \ell_2 \ell_n^2}{8} = 127$ ft-kip | |
| 8.10.4 | Distribute $M_o$ in the end span, from A to B. | | |

| | | |
|---|---|---|
| 8.10.4.1<br>8.10.4.2<br>8.10.4.3<br>8.10.4.4<br>8.10.4.5 | Table 8.10.4.2 gives the $M_o$ distribution coefficients for the slab panel. In Table 8.10.4.2, this example uses the fully restrained column of the table. The reason for this is that the combined member of the wall and column is much stiffer than the slab and little rotation is expected at the slab-to-wall connection.<br><br>Section 8.10.4.3 gives the option of modifying the factored moments by up to 10 percent, but that allowance is not used in this example. Section 8.10.4.4 indicates the negative moments are at the face of the supporting columns. Section 8.10.4.5 requires that the greater value of the two interior negative moments at the first interior column controls the design of the slab. | In Table 8.10.4.2, for the exterior edge being fully restrained:<br>Negative $M_u$ at face of exterior column = $0.65M_o$<br>= 83 ft-kip<br>Maximum positive $M_u$ = $0.35M_o$ = 45 ft-kip<br>Negative $M_u$ at face of first interior column = $0.65M_o$<br>= 83 ft-kip |
| 8.10.5 | Proportion the total panel factored moments from 8.10.4 to the column and middle strips for the end span, from A to B. | |
| 8.10.5.1 | After distributing the total panel negative and positive $M_u$ as described earlier in Section 8.10.4, Table 8.10.5.1 then proportions the interior negative $M_u$ assumed to be resisted by the column strip. | In Table 8.10.5.1, $\ell_2/\ell_1 = 14/18 = 0.778$ and $\alpha_{f1} = 0$. Therefore, the top line of the table controls:<br>$M_{u,int.neg,\,cs} = 0.75 \times 83$ ft-kip = 63 ft-kip |
| 8.10.5.2 | After distributing the total panel $M_u$ as described earlier in Section 8.10.4, Table 8.10.5.2 then proportions the exterior negative $M_u$ assumed to be resisted by the column strip. | In Table 8.10.5.2, $\ell_2/\ell_1 = 14/18 = 0.778$ and $\alpha_{f1} = 0$. Assuming the wall behaves as a beam, C is calculated to determine $\beta_t$ using Eq. (8.10.5.2(a) and (b)).<br>$$\beta_t = \frac{E_{cb}C}{2E_{cs}I_s}$$<br>$$C = \left(1 - 0.63\frac{x}{y}\right)\frac{x^3 y}{3}$$<br>$x = 10$ in.<br>$y = 120$ in.<br>$C = 37,900$ in.$^4$<br>$E_{cb} = E_{cs}$<br>$$I_s = \frac{bh^3}{12} = \frac{168 \text{ in.} \times (7 \text{ in.})^3}{12} = 4802 \text{ in.}^4$$<br>$$\beta_t = \frac{37,900}{2 \times 4802} = 3.9$$ |
| 8.10.5.5 | After distributing the total panel $M_u$ as described earlier in Section 8.10.4, Table 8.10.5.5 then proportions the positive $M_u$ assumed to be resisted by the column strip. | In Table 8.10.5.5, $\ell_2/\ell_1 = 14/18 = 0.778$ and $\alpha_{f1} = 0$. Therefore, the top line of the table controls.<br>$M_{u,\,pos.\,cs} = 0.60 \times 45$ ft-kip = 27 ft-kip |
| 8.10.6 | The total panel $M_u$ from 8.10.4 is distributed into column strip moments and middle strip moments. The middle strip $M_u$ is the portion of the total panel $M_u$ not resisted by the column strip. | Determine the amounts distributed to the middle strips. Subtract the amounts distributed to the column strips in Section 8.10.5 from the panel $M_u$ calculated in Section 8.10.4.<br>$M_{u,\,int.\,neg,\,ms}$ = 83 ft-kip – 63 ft-kip = 20 ft-kip<br>$M_{u,\,ext.\,neg,\,ms}$ = 83 ft-kip – 63 ft-kip = 20 ft-kip<br>$M_{u,\,pos.,\,ms}$ = 45 ft-kip – 27 ft-kip = 18 ft-kip |

| | | |
|---|---|---|
| 8.10.7.3 | The gravity load moment transferred between slab and edge column by eccentricity of shear is $0.3M_o$. | If there was no wall supporting the exterior edge of the slab, this moment would be used to calculate the two-way shear in the slab at the exterior column in 8.5. However, because of the wall, two-way shear does not apply to the design at the exterior column. |
| 8.10 | Repeat the $M_u$ calculations for the interior span. | The results are shown for interior panels, with the same negative $M_u$ at either end of the panel.<br>$M_{u,\,neg,\,cs}$ = 63 ft-kip<br>$M_{u,\,neg,\,ms}$ = 20 ft-kip<br>$M_{u,\,pos,\,cs}$ = 27 ft-kip<br>$M_{u,\,pos,\,ms}$ = 18 ft-kip<br><br>Refer to Fig. E1.3 for final distribution along this column line. The middle strip moments are split into two half-middle strips, one on either side of the column strip. |

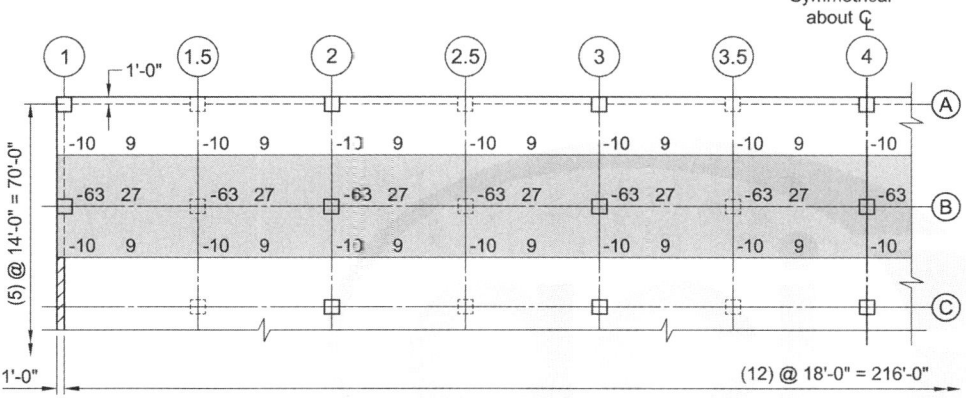

Fig. E1.3—Final moment distribution

### Step 5: Required strength – Factored slab moment resisted by the column

| | | |
|---|---|---|
| 8.4.2.3<br>8.4.2.3.1<br>8.4.2.3.2<br>8.4.2.3.3<br>8.4.2.3.4<br>8.4.2.3.5<br>8.4.2.3.6 | Slab negative moments at a column can be unbalanced; that is, different on either side of the column. This difference in slab moments, $M_{sc}$, must be transferred into the column, usually by a combination of flexure or shear. Eq. (8.4.2.3.2) calculates a factor that determines the fraction of $M_{sc}$ transferred by flexure. In this example, the permitted modifications to this factor are not used. | The columns are square, so $b_1/b_2 = 1$.<br><br>$$\gamma_f = \frac{1}{1+\frac{2}{3}\sqrt{\frac{b_1}{b_2}}} = 0.6$$ |
| 8.4.2.3.3,<br>8.4.2.3.5 | The effective slab width to resist $\gamma_f M_{sc}$ is the width of the column plus $1.5h$ of the slab on either side of the column. Section 8.4.2.3.5 requires sufficient reinforcement within the effective slab width to resist $\gamma_f M_{sc}$. | This concentration of reinforcement within the effective slab width is considered during the detailing of the column slab joint in Section 8.5.<br><br>Figure E1.4 shows undistributed total panel moments. The moment diagram is symmetric about the axis of the column in the center of the building (108 ft). Note that using this moment diagram will result in a net zero $M_{sc}$. The DDM uses an artificial unbalanced load condition in Section 8.10.7 to avoid an unconservative design for two-way shear. |

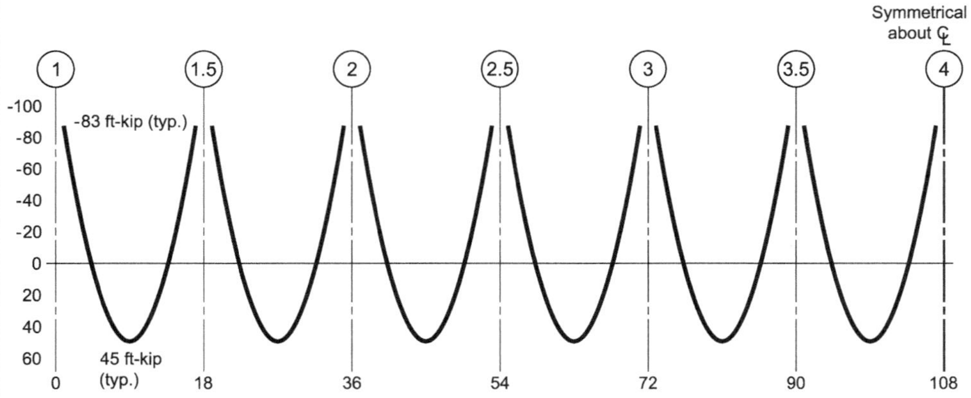

Fig. E1.4—Total panel moments

| | | |
|---|---|---|
| 8.10.7<br>8.10.7.1<br>8.10.7.2 | $M_{sc}$ to satisfy the DDM provisions at an interior column is calculated by Eq. (8.10.7.2):<br>$M_{sc} = 0.07[(q_{Du} + 0.5q_{Lu})\ell_2\ell_n^2 - q_{Du}'\ell_2'(\ell_n')^2]$<br>where the ' indicates the shorter span. When the spans are the same as in our example, the ' simply indicates the next span. | At an interior column:<br>$M_{sc} = 0.07[(q_{Du} + 0.5q_{Lu})\ell_2\ell_n^2 - q_{Du}'\ell_2'(\ell_n')^2]$<br>$M_{sc} = 0.07[(0.123 \text{ kip/ft}^2 + 0.5(0.160 \text{ kip/ft}^2))$<br>$14 \text{ ft}(16 \text{ ft})^2 - 0.123 \text{ kip/ft}^2(14 \text{ ft})(16 \text{ ft})^2] = 20.1 \text{ ft-kip}$ |
| 8.10.7.3 | $M_{sc}$ to satisfy the DDM provisions at an exterior column is calculated by Section 8.10.7.3:<br>$M_{sc} = 0.3M_o$ | At an exterior column:<br>$M_{sc} = 0.3M_o$<br>$M_{sc} = 0.3(127 \text{ ft-kip}) = 38.1 \text{ ft-kip}$ |
| 8.4.2.3.2,<br>8.4.4.2.2 | $M_{sc}$ is required to be transferred through both flexure and two-way shear into the column. The two-way shear calculations are discussed in Step 7. The amount of steel required to transfer $\gamma_f M_{sc}$ into the column via flexure is determined.<br>The flexural reinforcement determined in later steps is allowed to be used to meet the required $A_s$ in this step. Therefore, at step 13, the required reinforcement from this step will be checked. | At interior columns:<br>$\gamma_f = 0.6$<br>$M_{sc} = 20.1$ ft-kip<br>$\gamma_f M_{sc} = (0.6)(20.1 \text{ ft-kip})$<br>$\gamma_f M_{sc} = 12.1$ ft-kip<br><br>Using the method described in Step 8, the amount of flexural steel required within $1.5h$ of the column is:<br>$A_s = 0.49$ in.$^2$/45 in. or 0.13 in.$^2$/ft<br><br>At exterior columns:<br>$\gamma_f = 0.6$<br>$M_{sc} = 38.1$ ft-kip<br>$\gamma_f M_{sc} = (0.6)(38.1 \text{ ft-kip})$<br>$\gamma_f M_{sc} = 22.9$ ft-kip<br><br>Using the method described in Step 8, the amount of flexural steel required within $1.5h$ of the column is:<br>$A_s = 0.94$ in.$^2$/45 in. or 0.25 in.$^2$/ft |
| **Step 6: Required strength — Factored one-way shear** | | |
| 8.4.3<br>8.4.3.1<br>8.4.3.2 | One-way shear slab rarely controls over two way shear in the design of a two-way slab, but it must be checked. In this section, $V_u$ is determined. In 8.5, it is verified that the slab shear strength, $\phi V_n$, is sufficient to resist $V_u$. | Figure E1.5 shows one-way shears. The shear diagram is symmetric about the axis of the column at the center of the building (108 ft).<br><br>$V_u = 32$ kip |

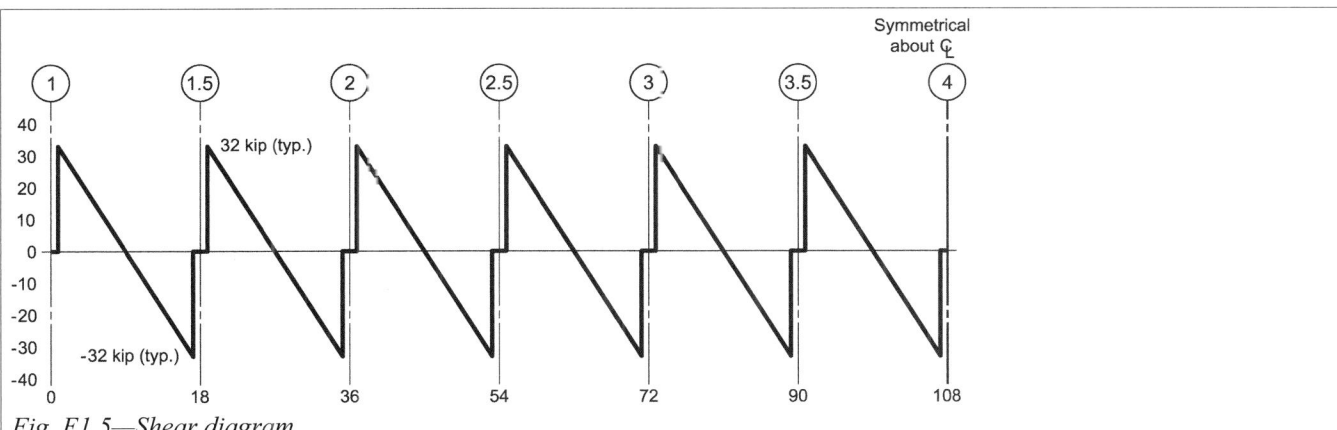

Fig. E1.5—Shear diagram

## Step 7: Required strength — Factored two-way shear

| | | |
|---|---|---|
| 8.3.1.4 | Stirrups are not used as shear reinforcement in this example. | |
| 8.4.4, 8.4.4.1, 22.6.4, 20.6.1.3.1 | Determine the critical section for two-way shear without shear reinforcement.<br>Calculate $b_o$ at an interior column:<br><br>$b_o = 2 \times (c_1 + d) + 2 \times (c_2 + d)$<br><br>where $d$ is the average effective depth (Fig. E1.6) and this example assumes No. 5 bars when determining $d$.<br>Cover is assumed to be 0.75 in. per Table 20.6.1.3.1 | $b_o = 2 \times (24 \text{ in.} + 5.6 \text{ in.}) + 2 \times (24 \text{ in.} + 5.6 \text{ in.})$<br>$b_o = 118.4$ in.<br><br>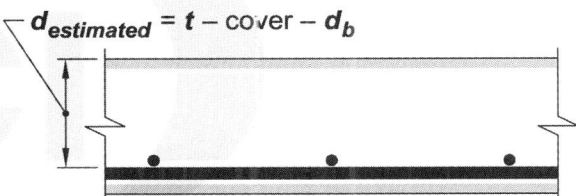<br><br>Fig. E1.6—Average slab effective depth<br><br>Figure E1.7 shows two-way critical sections, $b_o$, at an interior column. |

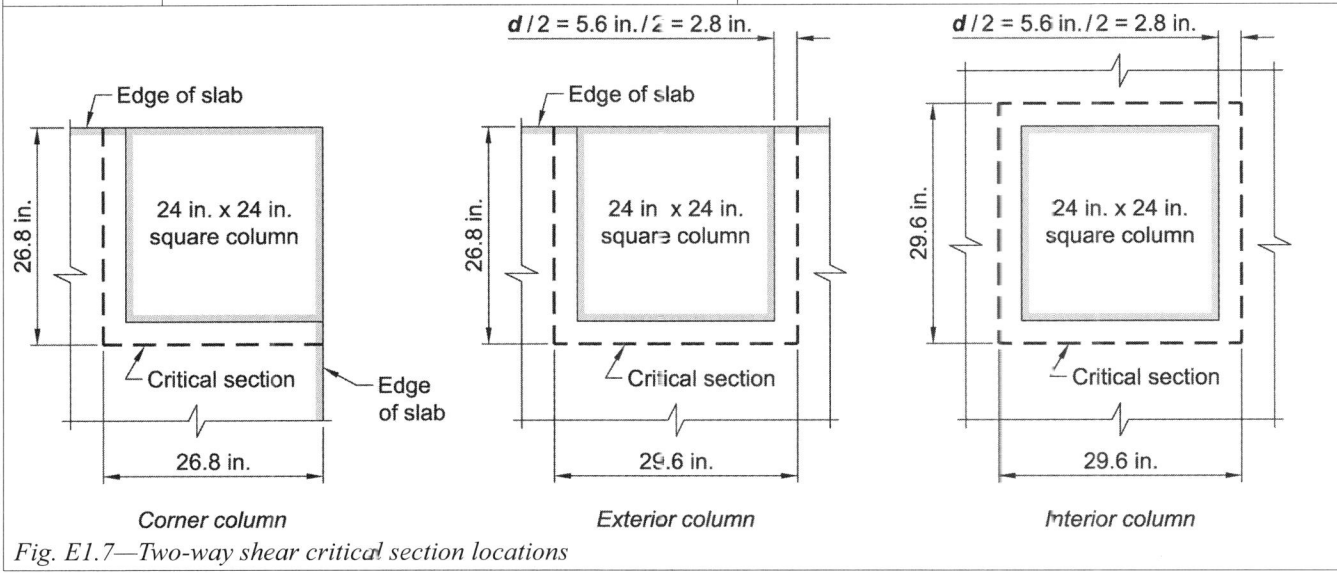

Fig. E1.7—Two-way shear critical section locations

| | | |
|---|---|---|
| 8.4.4.2<br>8.4.4.2.1 | Determine $v_{ug}$ due to direct slab shear stress. | Calculate the direct shear stress at the interior column with full factored load on all spans:<br>$$v_{ug} = \frac{V_u}{b_o d}$$<br>$$V_u = \left(14 \text{ ft} \times 18 \text{ ft} - \frac{29.6 \text{ in.} \times 29.6 \text{ in.}}{144}\right) \times \frac{283}{1000} \frac{\text{kip}}{\text{ft}^2}$$<br>$$V_u = 70 \text{ kip}$$<br>$$v_{ug} = \frac{70 \text{ kip}}{118.4 \text{ in.} \times 5.6 \text{ in.}} = 0.106 \text{ ksi}$$ |
| 8.4.4.2.1<br>8.4.4.2.2 | Determine the slab shear stress due to moment. | Calculate the shear stress due to moments at an interior column:<br>$$\gamma_v = 0.4$$<br>$$M_{sc} = 20.1 \text{ ft-kip}$$<br>$$c_{AB} = 14.8 \text{ in.}$$<br>$$J_c = 97688 \text{ in.}^4$$<br>$$\frac{\gamma_v M_{sc} c_{AB}}{J_c} = 0.015 \text{ ksi}$$ |
| 8.4.4.2.3 | Calculate $v_u$ by combining the two-way direct shear stress and the stress due to moment transferred to the column via eccentricity of shear. | $$v_u = v_{ug} + \frac{\gamma_v M_{sc} c_{AB}}{J_c}$$<br>Calculate the design shear stress at an interior column:<br>$v_u = 0.106$ ksi + 0.015 ksi = 0.121 ksi<br>Note that these calculations are conservative. $M_{sc}$ assumes that some live load is not present to produce unbalanced moments, but $v_{ug}$ assumes that full live load and dead load are present. |

## Step 8: Design strength—Reinforcement required to resist factored moments

**8.5.1, 8.5.1.1, 8.5.1.2, 8.5.2, 8.5.2.1, 8.4.2.3.5**

There are many methods available to determine the flexural reinforcement required at all sections within the span in each direction.

To determine the amount of flexural steel required, this example solves the following quadratic equation:

$$\phi M_n = \phi \left( A_s f_y \left( d - \frac{a}{2} \right) \right)$$

$$\phi M_n = \phi \left( A_s f_y \left( d - \frac{A_s f_y}{2 \times 0.85 b f_c'} \right) \right)$$

$$\omega = \frac{A_s f_y}{b d f_c'}$$

$$\phi M_n = \phi (b d f_c' \omega (d - 0.59 \omega d))$$

$$\phi M_n = \phi (b d^2 f_c' \omega (1 - 0.59 \omega))$$

Set $\phi M_n = M_u$ and solve for $\omega$.

$\phi$ is assumed to be 0.9 for flexure as the slab is lightly reinforced. Using the moments shown in Fig. E1.8 and E1.9 for the column strip and middle strip, respectively, to determine the reinforcement required at each location.

Reinforcement in an exterior panel

Column strip at the columns
$M_u = 63$ ft-kip

Solving the quadratic equation gives
$\omega = 0.0664$ ∴

$$A_s = \frac{\omega b d f_c'}{f_y}$$

$$A_s = \frac{(0.0664)(84 \text{ in.})(5.6 \text{ in.})(5000 \text{ psi})}{60,000 \text{ psi}}$$

$A_s = 2.61$ in.$^2$

Column strip at midspan:
$M_u = 27$ ft-kip

Solving the quadratic equation gives:
$\omega = 0.0278$ ∴

$$A_s = \frac{\omega b d f_c'}{f_y}$$

$$A_s = \frac{(0.0278)(84 \text{ in.})(5.6 \text{ in.})(5000 \text{ psi})}{60,000 \text{ psi}}$$

$A_s = 1.09$ in.$^2$

Using the same method, the following can be found:

<u>Exterior Panels:</u>

Column strip at column line:
$A_s = 2.61$ in.$^2$

Middle strip at column line:
$A_s = 0.81$ in.$^2$

Column strip at midspan:
$A_s = 1.09$ in.$^2$

Middle strip at midspan:
$A_s = 0.73$ in.$^2$

| | | |
|---|---|---|
| | | Interior Panels:

Column strip at column lines:
$A_s = 2.61$ in.$^2$

Middle strip at column lines:
$A_s = 0.81$ in.$^2$

Column strip at midspan:
$A_s = 1.09$ in.$^2$

Middle strip at midspan:
$A_s = 0.73$ in.$^2$ |

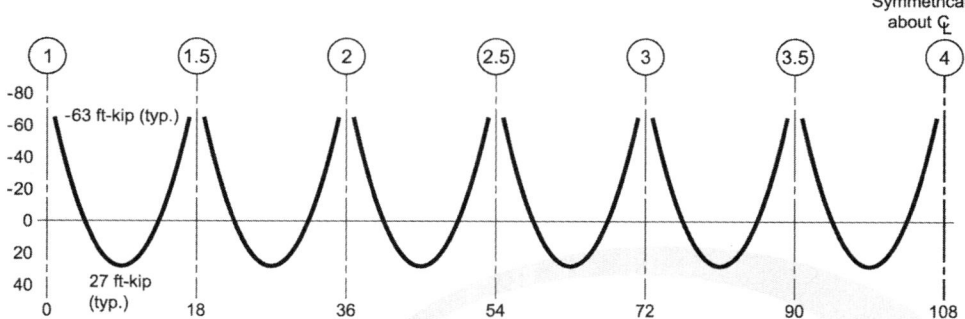

Fig. E1.8—Column strip moment diagram

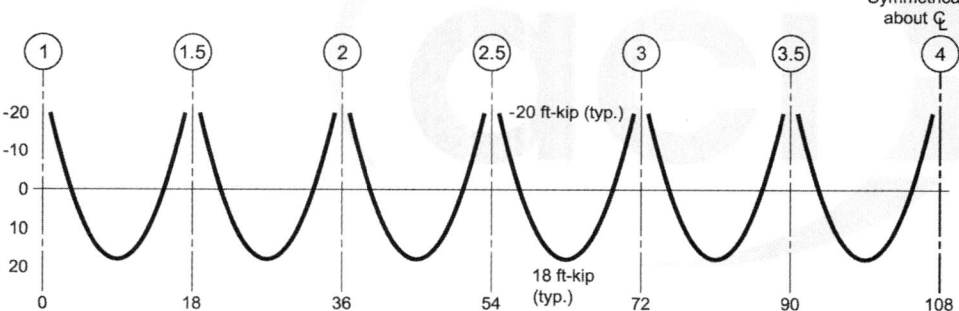

Fig. E1.9—Middle strip moment diagram

| Step 9: Design strength – One-way shear |||
|---|---|---|
| 8.5.3.1.1<br>22.5 | One-way design shear strength is calculated in accordance with 22.5. | $\phi = 0.75$<br><br>$V_n = 2\sqrt{f_c'}bd$<br><br>$\phi V_n = (0.75)(2)\left(\sqrt{5000 \text{ psi}}\right)(168 \text{ in.})(5.6 \text{ in.})\left(\dfrac{1 \text{ kip}}{1000 \text{ lb}}\right)$<br><br>$\phi V_n = 100$ kip<br><br>This is greater than the required strength of 32 kip from Step 6; therefore, one-way shear is okay. |

## Step 10: Design strength – Two-way shear

| | | |
|---|---|---|
| 8.5.3.1.2<br>22.6 | Two-way design shear strength is calculated in accordance with Section 22.6. Shear reinforcement is assumed to not be required. | Interior column:<br>$v_n$ = the least of the three equations from Table 22.6.5.2 of ACI 318-14<br><br>$v_n = 4\sqrt{f_c'} = \dfrac{4\sqrt{5000}}{1000}$ ksi = 0.283 ksi<br><br>$v_n = \left(2 + \dfrac{4}{\beta}\right)\sqrt{f_c'} = \dfrac{6\sqrt{5000}}{1000}$ ksi = 0.424 ksi<br><br>$v_n = \left(2 + \dfrac{\alpha_s d}{b_o}\right)\sqrt{f_c'} = \dfrac{3.89\sqrt{5000}}{1000}$ ksi = 0.275 ksi *<br><br>*controls<br><br>$\phi v_n = 0.75 \times 0.275$ ksi = 0.206 ksi<br><br>This is greater than the required strength for interior columns of 0.121 ksi from Step 7; therefore, two-way shear at interior columns is okay.<br><br>The assumption that two-way shear reinforcement is not required at these locations is confirmed. |

## Step 11: Reinforcement limits — Minimum flexural reinforcement in nonprestressed slabs

| | | |
|---|---|---|
| 8.6<br>8.6.1<br>8.6.1.1 | At least a minimum area of flexural reinforcement are provided at locations where tension is calculated in the slab. | From Table 8.6.1.1,<br><br>$f_y$ = 60,000 psi<br>$A_{s,min} = 0.0018 \times A_g$<br>$A_{s,min} = 0.0018 \times 7$ in. $\times 14$ ft $\times (12$ in./ft.$)$<br>$A_{s,min} = 2.12$ in.$^2$<br><br>This minimum area of reinforcement is split evenly between the column and middle strips; therefore 1.06 in.$^2$ per strip.<br><br>Minimum flexural reinforcement controls at the middle strip of all panels. |

## Step 12: Reinforcement detailing – General requirements

| | | |
|---|---|---|
| 8.7.1<br>8.7.1.1<br>20.6.1 | Concrete cover, development lengths, and splice lengths are determined in these sections. | Concrete cover requirements are provided in Table 20.6.1.3.1. The slab is not exposed to weather or in contact with the ground. Assuming No. 5 bars for reinforcement, the specified cover is 0.75 in. |

| | | |
|---|---|---|
| 8.7.1.2<br>25.4 | Development length is needed to determine splice length. | Development length is calculated by Eq. (25.4.2.3(a)). Using normalweight concrete with No. 6 and smaller uncoated bars and the casting position with less than 12 in. of fresh concrete placed below the horizontal reinforcement, modification factors from Table 25.4.2.4 are as follows:<br><br>$\psi_s = 0.8$<br>$\psi_e = 1.0$<br>$\psi_t = 1.0$<br>$\lambda = 1.0$<br><br>The bar spacing is larger than the distance from the center of the bottom bar to the concrete surface.<br><br>$c_b = 0.75$ in. $+ (0.625$ in./$2) = 0.94$ in.<br>$d_b = 0.625$ in.<br><br>$K_{tr}$ is assumed 0 as permitted by 25.4.2.3.<br><br>$$\ell_d = \left( \frac{3}{40} \frac{f_y}{\lambda\sqrt{f'_c}} \frac{\psi_t \psi_e \psi_s}{\left( \frac{c_b + K_{tr}}{d_b} \right)} \right) d_b$$<br><br>$\ell_d = 21.2$ in.; use 22 in. |
| 8.7.1.3, 25.5 | It is likely that splices will be required during construction. Allowable locations for splices are shown in ACI 318-14 Fig. 8.7.4.1.3. | Lap splice lengths are determined in accordance with Table 25.5.2.1. The provided $A_s$ does not exceed the required $A_s$ by a substantive amount. Therefore, class B splices are required.<br><br>$\ell_{st} = 1.3 \times 21.2$ in. $= 27.5$ in.<br><br>use $\ell_{st} = 28$ in. |

# CHAPTER 6—TWO-WAY SLABS

| | Step 13: Reinforcement detailing – Spacing requirements | |
|---|---|---|
| 8.7.2<br>8.7.2.1<br>25.2.1<br>8.7.2.2 | Minimum and maximum spacing limits are determined. The bar spacing for design strength is also reviewed. | Minimum spacing is determined in accordance with Section 25.2.1. Minimum spacing is 1 in., $d_b$, and $(4/3)d_{agg}$. Assuming that the maximum nominal aggregate size is 1 in., than minimum clear spacing is 1.33 in. With a No. 5 bar, this equates to a minimum spacing of approximately 2 in.<br><br>Maximum spacing is limited by Section 8.7.2.2:<br>At critical sections, the maximum spacing is<br>the lesser of $2h$ (2 × 7 in.) and 18 in., so 14 in. controls.<br><br>All other sections, the critical spacing is<br>the lesser of $3h$ (3 × 7 in.) and 18 in., so 18 in. However, because all of the bars cross a critical section, use a maximum spacing of 14 in. for all sections.<br><br>Assuming No. 5 bars are used, the spacing for the different areas of the slab are as follows:<br><br>All spans:<br>Column strip at column line:<br>2.61 in.$^2$/0.31in.$^2$ = nine No. 5 bars over 7 ft – spacing is 9 in.<br><br>Middle strip at column line:<br>0.81 in.$^2$/0.31 in.$^2$ = six No. 5 bars over 7 ft – spacing is 14 in. (maximum spacing controls over minimum area in the middle strip)<br><br>Column strip at midspan:<br>1.09 in.$^2$/0.31 in.$^2$ = six No. 5 bars over 7 ft – spacing is 14 in. (maximum spacing controls over strength requirements at this location)<br><br>Middle strip at midspan:<br>0.73 in.$^2$/0.31in.$^2$ = six No. 5 bars over 7 ft – spacing is 14 in. (maximum spacing controls over minimum area in the middle strip) |
| 8.4.2.3.2 | This is a check to verify that the reinforcement amounts required to transfer the fraction of factored slab moment via flexure are satisfied using the design slab reinforcement. | The minimum requirements for all column strips is a 14 in. spacing of No. 5 bars. This equates to 0.26 in.$^2$/ft. This meets or exceeds the 0.13in.$^2$/ft and 0.26 in.$^2$/ft required from Step 5, therefore, Section 8.4.2.3.2 is satisfied. Note that if this had not been met, additional steel would have been required to be placed within the effective slab width as defined in Section 8.4.2.3.3. |
| | Step 14: Reinforcement detailing – Reinforcement termination | |
| 8.7.4<br>8.7.4.1<br>8.7.4.1.1<br>8.7.4.1.2<br>8.7.4.1.3 | Reinforcement lengths and extensions are at least that required by Fig. 8.7.4.1.3 of ACI 318-14. | Use ACI 318-14, Fig. 8.7.4.1.3 to determine reinforcement lengths. The figure for final layout of reinforcement in these panels shows the design lengths. |

| | | |
|---|---|---|
| **Step 15: Reinforcement detailing – Structural integrity** | | |
| 8.7.4.2<br>8.7.4.2.1<br>8.7.4.2.2 | Structural integrity for a two-way slab is met by satisfying ACI 318-14 detailing provisions. | Section 8.7.4.2.1 is met when reinforcement is detailed in accordance with Fig. 8.7.4.1.3 (ACI 318-14).<br><br>Section 8.7.4.2.2 requires at least two of the column strip bottom bars pass through the column inside the column reinforcement cage. |
| **Step 16: Slab-column joints** | | |
| 8.2.7<br>15.2.1<br>15.2.2<br>15.2.3<br>15.2.5<br>15.3.1<br>15.4.2 | Joints are designed to satisfy Chapter 15 of ACI 318-14. | The specified concrete strength of the slab and columns are identical and therefore, 15.2.1 and 15.3 are met.<br><br>Section 15.2.2 is met in Steps 7 and 10 of this example.<br><br>Section 15.2.3 is met by satisfying the detailing Sections in 15.4.<br><br>Section 15.2.5 states that interior columns are restrained because they are laterally supported on four sides by the slab.<br><br>Section 15.4.2 applies to columns along the exterior of the building.<br><br>Assuming that No. 4 bars are used as column ties, Eq. (15.4.2 (a)) and (b) are satisfied if the spacing of the column ties in the slab-column joint satisfies (a) and (b):<br><br>(a) (15.4.2(a))<br>$$s = 0.4 \text{ in.}^2 \times \frac{60,000 \text{ psi}}{0.75\sqrt{5000 \text{ psi}} \times 24 \text{ in.}}$$<br>$s = 18.9$ in.<br><br>(b) (15.4.2(b))<br>$$s = 0.4 \text{ in.}^2 \times \frac{60,000 \text{ psi}}{50 \times 24 \text{ in.}}$$<br>$s = 20.0$ in.<br><br>Because the spacing is larger than the joint depth of 7 in., only one tie is required within the slab-column joint in each exterior and corner column. |

## Step 17: Summary tables of required $A_s$

| $A_s$ required, column strip, in.² | | | | |
|---|---|---|---|---|
| | External bays | | Internal bays | |
| | Column lines | Midspan | Column lines | Midspan |
| Strength | 2.61 | 1.09 | 2.61 | 1.09 |
| Minimum | 1.06 | 1.06 | 1.06 | 1.06 |
| Maximum spacing, assuming No. 5 bars | 1.86 | 1.86 | 1.86 | 1.86 |

| $A_s$ required, middle strip, in.² | | | | |
|---|---|---|---|---|
| | External bays | | Internal Bays | |
| | Column lines | Midspan | Column lines | Midspan |
| Strength | 0.81 | 0.73 | 0.81 | 0.73 |
| Minimum | 1.06 | 1.06 | 1.06 | 1.06 |
| Maximum spacing, assuming No. 5 bars | 1.86 | 1.86 | 1.86 | 1.86 |

Note: The highlighted cells indicate the required reinforcement that controls design.

## Step 18: Summary sketches of required bars

Figures E1.10 and E1.11 are summary sketches of required bars for this example.

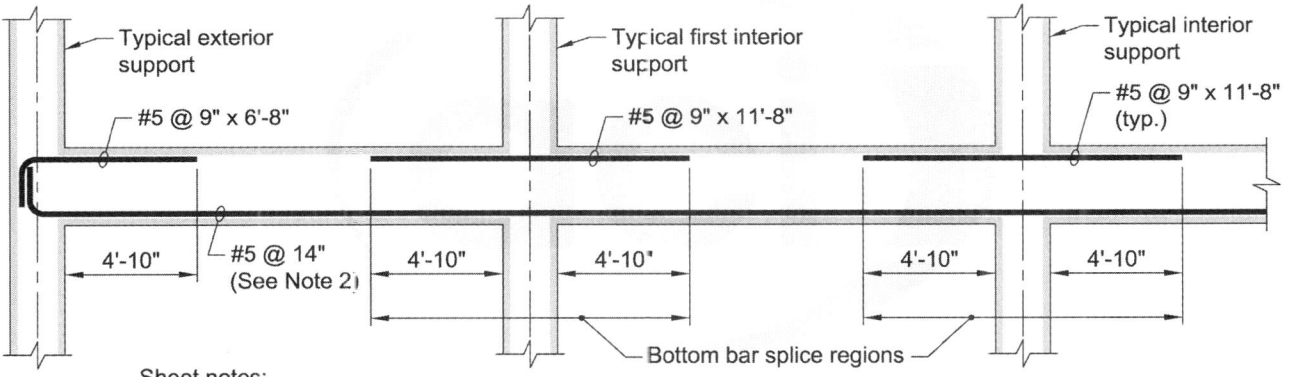

Sheet notes:

1. ⌐ – Indicates standard hook. Standard hooks may be 90 or 180 degree hooks. Standard hook bars may be inclined up to 45 degrees to allow hook to fit in slab.

2. Minimum of two bottom bars must pass within longitudinal column bars.

*Fig. E1.10—Summary sketch of required bars, column strip*

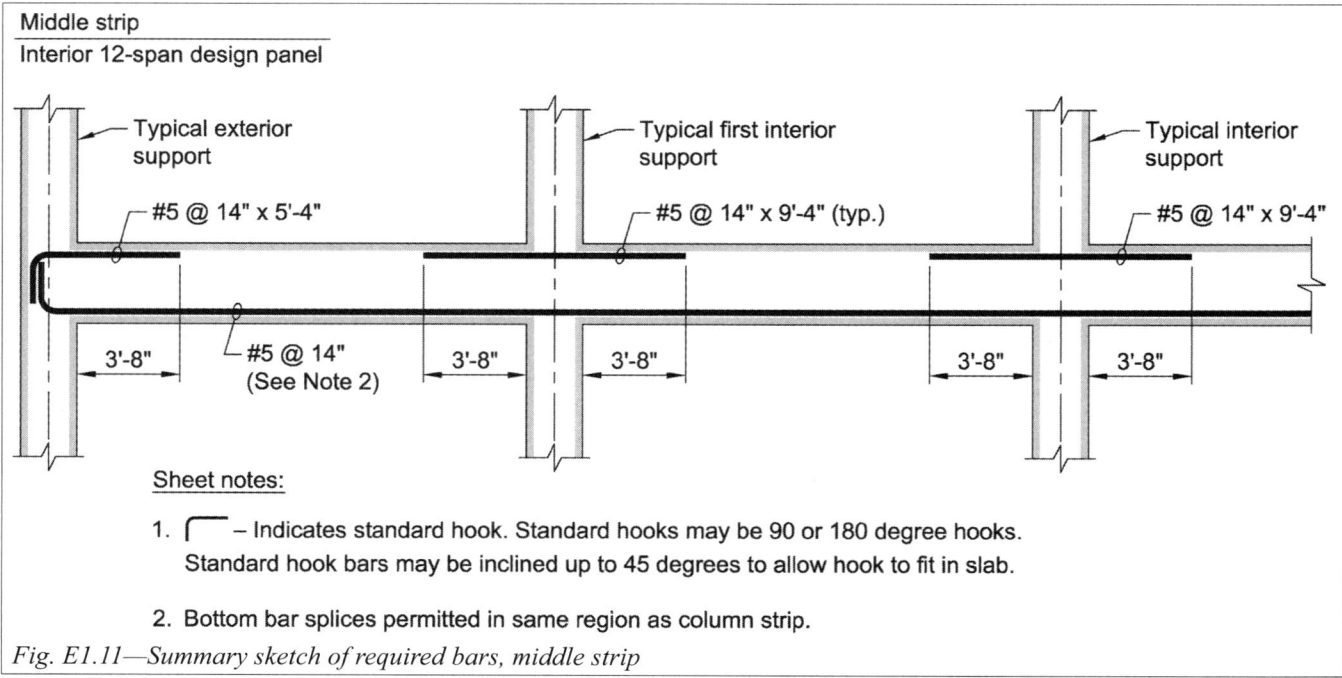

Fig. E1.11—Summary sketch of required bars, middle strip

# CHAPTER 6—TWO-WAY SLABS

**Two-way Slab Example 2:** *Equivalent Frame Method (EFM) – Internal column line design*

This example is an interior column strip along grid line B in a nonprestressed two-way slab without beams between supports. This example uses the moment distribution method to determine design moments, but any method for analyzing a statically indeterminate structure can be used. This example uses the Hardy column analogy to determine the structural stiffness for the members analyzed.

### Given:
*Uniform loads:*
Self-weight dead load is based on concrete density including reinforcement at 150 lb/ft$^3$
Superimposed dead load $D$ = 0.015 kip/ft$^2$
Live load $L$ = 0.100 kip/ft$^2$

Material properties:
$f_c'$ = 5000 psi
$f_y$ = 60,000 psi

Thickness of slab
$t$ = 7 in.

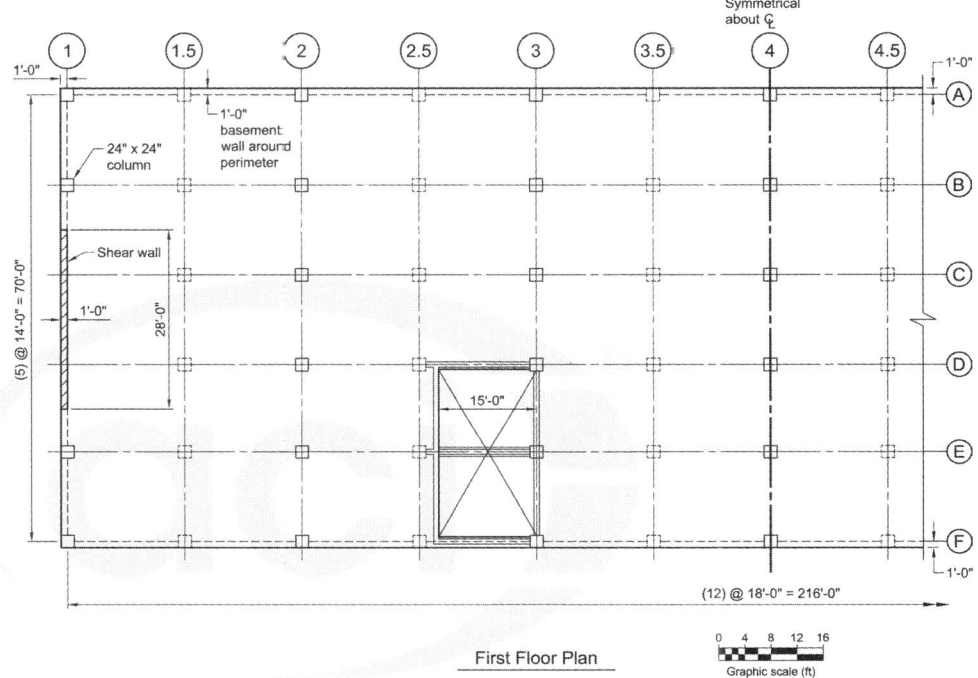

*Fig. E2.1—First-floor plan.*

| ACI 318-14 | Discussion | Calculation |
|---|---|---|
| **Step 1: Geometry** | | |
| 8.4.2<br>8.4.2.1<br>8.4.2.2 | The moments in the slab are calculated using the Equivalent Frame Method (EFM) in accordance with Section 8.11 of ACI 318-14. | |

| | | |
|---|---|---|
| 8.11.1<br>8.11.1.1<br>8.11.1.2<br>8.11.1.3<br>8.11.1.4<br>8.11.2<br>8.11.2.1<br>8.11.2.2<br>8.11.2.3<br>8.11.2.4<br>8.11.2.5 | Fig. E2.2 shows an isometric of the slab-beam strip and the attached torsional members of the equivalent frame model.<br><br>A key element of the EFM is that, unlike a beam and column frame, in a slab and column frame some of the unbalanced moments can redistribute around the column into the next span regardless of the stiffness of the columns. The EFM softens the columns to simulate this effect on slab moments by incorporating the flexibility of the slab torsional member in the equivalent column stiffness. | *Fig. E2.2—Equivalent frame strip* |
| **Step 2: Analysis – Equivalent column stiffness determination** | | |
| 8.11.3<br>8.11.4<br>8.11.5 | Determine the geometry coefficients necessary to use the design aids to determine the equivalent column stiffness, moment coefficient, and carry-over factor for use in the moment distribution method. (Corley and Jirsa, 1970, "Equivalent Frame Analysis for Slab Design," *ACI Journal Proceedings*, V. 67, No. 11, Nov. pp. 875-884). | |

| | | |
|---|---|---|
| 8.11.3, 8.11.5 | To determine the equivalent column stiffness, $K_{ec}$, the stiffness of the torsional member intersecting with the column, $K_t$, is needed at each intersection. $K_t$ is determined using an equation given in Commentary Section R8.11.5. The effects of cracking on $K_t$ are neglected in ACI 318-14. | $K_t = \sum \dfrac{9 E_{cs} C}{\ell_2 \left(1 - \dfrac{c_2}{\ell_2}\right)^3}$ <br><br> $\ell_2 = 14$ ft $= 168$ in. <br> $E_{cs} = E_{cc}$ <br> $c_2 = 24$ in. |
| | This example uses the same concrete strength throughout the structure, so the modulus of elasticity is also considered equal. This simplifies the calculations. | Interior column torsional members <br> For the torsional member at the interior columns and the slab portion of the torsional member at the exterior columns, <br> $x = 7$ in. <br> $y = 24$ in. <br> and <br><br> $C = \sum \left(1 - 0.63 \times \dfrac{x}{y}\right) \dfrac{x^3 y}{3}$ <br><br> $C = \left(1 - 0.63 \times \dfrac{7 \text{ in.}}{24 \text{ in.}}\right) \dfrac{(7 \text{ in.})^3 \times 24 \text{ in.}}{3}$ <br> $C = 2240$ in.$^4$ <br> $K_t = 191$ <br><br> for each side of the column. Therefore, the total torsional member stiffness at an internal column is <br> $K_t = 2 \times 191 = 382$ |
| | The basement wall is monolithic with the column and provides substantial stiffness to the exterior equivalent column, but the wall rotation will be greater than the column rotation so the torsional stiffness of the wall will be considered along with its flexural stiffness. The basement wall dimension, $y = 113$ in., in the calculations here is the distance from the bottom of the slab being designed to the top of the mat foundation. | External column torsional members <br> For the wall portion of the torsional member at the exterior columns, <br> $x = 12$ in. <br> $y = 113$ in. <br> And the total $C$ for the exterior column torsional members is <br><br> $C = \sum \left(1 - 0.63 \times \dfrac{x}{y}\right) \dfrac{x^3 y}{3}$ <br><br> $C = \left(1 - 0.63 \times \dfrac{12 \text{ in.}}{113 \text{ in.}}\right) \dfrac{(12 \text{ in.})^3 \times 113 \text{ in.}}{3} + 2240$ in.$^4$ <br> $C = 60{,}733$ in.$^4$ <br> $K_t = 5357$ <br><br> for each side of the column. Therefore, the total torsional member stiffness at an external column is <br> $K_t = 2 \times 5357 = 10{,}714$ |

| 8.11.4 | To determine the equivalent column stiffness, $K_{ec}$, the stiffness coefficients for the columns above and below the slab are needed at each intersection. Because the slab thickness, column heights, and foundation thickness geometry is uniform, $K_{ctop}$ and $K_{cbot}$ are consistent at each interior joint in this design strip.<br><br>$K_{ctop}$ and $K_{cbot}$ are determined using the Hardy column analogy. (K. Wang, *Intermediate Structural Analysis*, McGraw-Hill, New York, 1983). Note that if Fig. E2.3 and E2.4 were combined, it provides a section cut through the basement slab being designed in this example. The bottom-most slab in Fig. E2.4 is the mat foundation while the upper-most beam and slab in Fig. E2.3 is the first floor above the entrance/lobby level of the structure.<br><br>Please refer to the short discussion at the end of this example regarding an alternate method for determining the stiffness for the columns, beams, and slabs. | $K_c = \dfrac{k_c \times E_{cc} \times I_c}{\ell_c}$<br><br>The following values are used in the calculations for $K_{ctop}$ and correspond to Fig. E2.3:<br>$t_{top} = 7$ in.<br>$h_{beam} = 2.5$ ft<br>$\ell_{col} = 15.5$ ft<br>$h = 18$ ft<br>$t_{bottom} = 7$ in.<br><br>$K_{ctop}$ is determined using the geometry from Fig. E2.3.<br>$k_{ctop} = \ell_c \left( \dfrac{1}{A_a} + \dfrac{Mc}{I_a} \right)$<br>$A_a = \ell_{col} = 15.5$<br>$I_a = \dfrac{A_a^3}{12} = \dfrac{15.5^3}{12} = 310.3$<br>$M_{bot} = 1.0 c_{bottom} = 8.04$<br>$c = c_{bottom} = 8.04$<br>$k_{ctop} = 18 \times \left( \dfrac{1}{15.5} + \dfrac{8.04^2}{310.3} \right) = 4.91$<br>$K_{ctop} = \dfrac{k_{ctop} \times I_{col}}{\ell_c} = \dfrac{4.91 \times 24^4}{12 \times 18(12)} = 629$ |

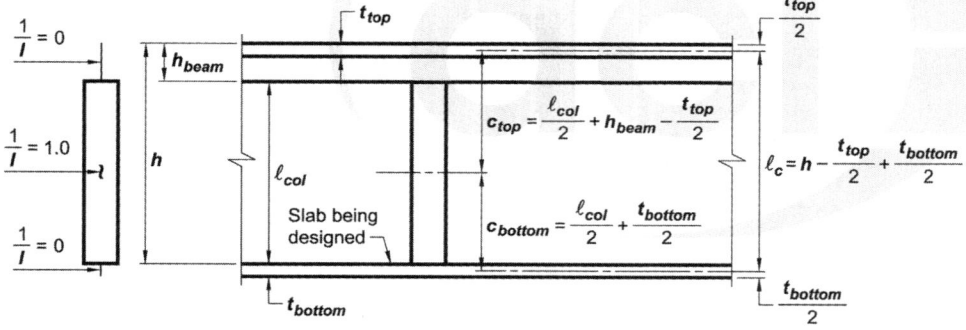

$c_{bottom}$ in this figure is used to determine $K_{ctop}$ for the slab being designed

Figure for Hardy column analogy to determine $K_{ctop}$

*Fig. E2.3—Hardy column analogy for the columns above the slab being designed*

| 8.11.4 | | The following values are used in the calculations for $K_{cbot}$ and correspond to Fig. E2.4: |
|---|---|---|
| | | $t_{top}$ = 7 in. |
| | | $\ell_{col}$ = 9.42 ft |
| | | $h$ = 10 ft |
| | | $t_{bottom}$ = 3.5 ft (assumed mat foundation thickness) |
| | | $K_{cbot}$ is determined using the geometry from Fig. E2.4. |
| | | $k_{cbot} = \ell_c \left( \dfrac{1}{A_a} + \dfrac{Mc}{I_a} \right)$ |
| | | $A_a = \ell_{col} = 9.42$ |
| | | $I_a = \dfrac{A_a^3}{12} = \dfrac{9.42^3}{12} = 69.6$ |
| | | $M_{bot} = 1.0 c_{top} = 5$ |
| | | $c = c_{top} = 5$ |
| | | $k_{cbot} = 11.46 \times \left( \dfrac{1}{9.42} + \dfrac{5^2}{69.6} \right) = 5.33$ |
| | | $K_{cbot} = \dfrac{k_{cbot} \times I_{col}}{\ell_c} = \dfrac{5.33 \times 24^4}{12 \times 11.46(12)} = 1072$ |

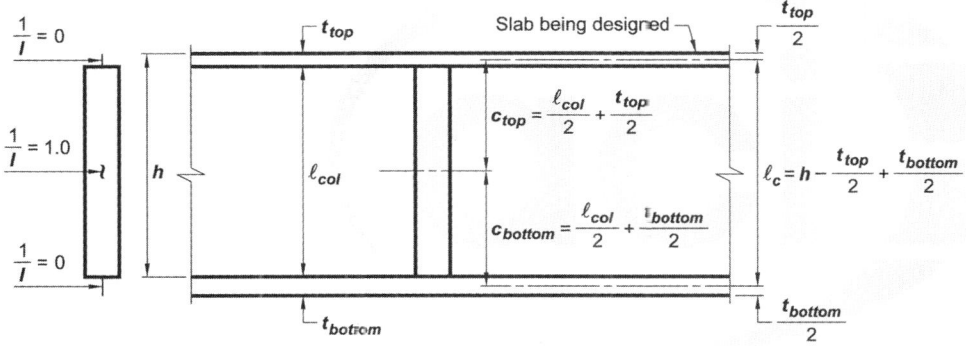

$c_{top}$ in this figure is used to determine $K_{cbot}$ for the slab being designed

Figure for Hardy column analogy to determine $K_{cbot}$

*Fig. E2.4—Hardy column analogy for the columns below the slab being designed*

| 8.11.4 | To determine the equivalent column stiffness, $K_{ec}$, combine the torsional beam stiffness with the column stiffness's determined above. | External column $$K_{ec} = \frac{1}{\frac{1}{\Sigma K_c} + \frac{1}{K_t}}$$ $$K_{ec} = \frac{1}{\frac{1}{629+1072} + \frac{1}{10714}}$$ $$K_{ec} = 1468$$ Internal column $$K_{ec} = \frac{1}{\frac{1}{\Sigma K_c} + \frac{1}{K_t}}$$ $$K_{ec} = \frac{1}{\frac{1}{629+1072} + \frac{1}{382}}$$ $$K_{ec} = 312$$ |
|---|---|---|

| Step 3: Analysis – Slab stiffness | | |
|---|---|---|
| 8.11.2 | The slab stiffness is determined using the Hardy column analogy. | Slab panel (refer to Fig. E2.5)<br>$c_1 = c_2 = 2$ ft = 24 in.<br>$\ell_s = 18$ ft = 218 in.<br>$\ell_2 = 14$ ft = 168 in.<br>$\ell_n = 16$ ft = 192 in.<br>$M = 18$ ft/2 = 9 ft<br>$c = 18$ ft/2 = 9 ft<br>$$I_{col} = \left(1 - \frac{c_2}{\ell_2}\right)^2 = \left(1 - \frac{2}{14}\right)^2 = 0.7347$$<br>$$I_s = \frac{bh^3}{12} = \frac{168 \times 7^3}{12} = 4802$$<br>$$I_a = \frac{\ell_n^3}{12} + \frac{\frac{c_1+c_2}{2}}{1/I_{col}}\left(\frac{\ell_n}{2} + \frac{c_1/2}{2}\right)^2$$<br>$$I_a = \frac{16^3}{12} + \frac{\frac{2+2}{2}}{1.361}\left(\frac{16}{2} + \frac{2/2}{2}\right)^2$$<br>$I_a = 448$<br>$$A_a = \ell_n + \frac{c_1}{2}(I_{col}) + \frac{c_2}{2}(I_{col})$$<br>$$A_a = 16 + \frac{2}{2}(0.7347) + \frac{2}{2}(0.7347)$$<br>$A_a = 17.53$<br>$$k_s = \ell_s\left(\frac{1}{A_a} + \frac{Mc}{I_a}\right)$$<br>$$k_s = 18\left(\frac{1}{17.53} + \frac{9^2}{448}\right)$$<br>$k_s = 4.281$<br>$$K_s = \frac{k_s \times I_s \times E_{cs}}{\ell_s} = \frac{4.281 \times 4802}{18 \times 12} = 95$$ |

$$C.O.F. = \frac{\ell_s \left( \frac{1}{A_a} - \frac{Mc}{I_a} \right)}{k_s}$$

$$C.O.F. = \frac{-18 \left( \frac{1}{17.53} - \frac{9^2}{448} \right)}{k_s}$$

$$C.O.F. = \frac{2.228}{4.255} = 0.524$$

$$A_m = A_1 + A_2 + 2 \times A_3$$

$$A_1 = \frac{2}{3} \ell_n \times \left( \frac{\ell_s^2}{8} - \frac{c_1/2}{2} \times \left( \ell_s - \frac{c_1}{2} \right) \right)$$

$$A_1 = \frac{2}{3} 16 \times \left( \frac{18^2}{8} - \frac{1}{2} \times (18-1) \right)$$

$$A_1 = 341.3$$

$$A_2 = \frac{c_1/2}{2} \times \left( \ell_s - \frac{c_1}{2} \right) \times \ell_n$$

$$A_2 = \frac{1}{2} \times (18-1) \times 16 = 136$$

$$A_3 = \frac{c_1/2}{2} \times \left( \ell_s - \frac{c_1}{2} \right) \times \frac{c_1}{2} \times \frac{1}{2}$$

$$A_3 = \frac{1}{2} \times (18-1) \times 1 \times \frac{1}{2} = 4.25$$

$$A_m = 341.33 + 136 + 2 \times 4.25$$

$$A_m = 481.6$$

$$FEM = \frac{A_m}{A_a \ell_s^2} = \frac{481.6}{17.53 \times 18^2} = 0.085$$

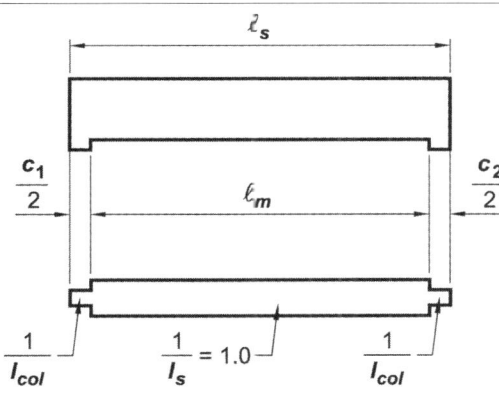

$$c_1 = \text{width of near column in direction being checked}$$

$$c_2 = \text{width of far column in direction being checked}$$

$$\ell_2 = \text{span width in other direction (into the page)}$$

$$I_{col} = \left(1 - \frac{c_2}{\ell_c}\right)^2$$

$$M = \frac{\ell_s}{2}$$

$$C = \frac{\ell_s}{2}$$

$$I_a = \frac{\ell_n^3}{12} + \frac{2 \times \frac{c_2}{2}}{I_{col}}\left(\frac{\ell_n}{2} + \frac{\frac{c_2}{2}}{2}\right)^2$$

$$A_a = \ell_n + \frac{c_1}{2}\left(\frac{1}{I_{col}}\right) + \frac{c_2}{2}\left(\frac{1}{I_{col}}\right)$$

$$k_s = \left(\frac{1}{A_a} + \frac{MC}{I_a}\right)\ell_s$$

$$\text{C.O.F.} = \frac{\left(\frac{1}{A_a} - \frac{MC}{I_a}\right)\ell_s}{k_s}$$

$$\text{FEM} = \frac{A_m}{A_a \ell_s^2}$$

$A_m$ = Area under unit moment curve

*Fig. E2.5—Section properties*

### Step 4: Analysis – Moment distribution

| 8.11.1 | This example uses moment distribution with pattern live load in accordance with Section 6.4.3 of ACI 318-14. Loading all spans simultaneously does not necessarily produce the maximum flexural stresses in the slab. Therefore, in Section 6.4.3, live load patterns are defined for use with two-way slab systems. Fig. E2.6 shows examples of the different live load patterns considered in the code.<br><br>When reduced to face of support, these results are comparable to the DDM analysis in Example 1. | The moment distribution in Fig. E2.7 shows the first four column lines when full live load is applied to all spans. The structure is symmetrical and repeats from column line 2.5 through column line 5.5. The moment distributions for the different live load patterns are not included here but have been incorporated into the example.<br>The moment diagram (Fig. E2.8) and shear diagram (Fig. 2.9) show the final results considering the live load patterns per Section 6.4.3.3 of ACI 318-14. The shear and moment diagrams in Fig. E2.8 and E2.9 are determined using known moments at the end of the slab along with known loads on the slab. The numerical values shown on these diagrams are the maximums determined at each location from the live load patterns discussed in Section 6.4.3.3 of ACI 318-14.<br>Note that the numerical values shown on these diagrams are the moments and shears reduced to the moments at the face of the columns, not at the midline of the columns as shown in the moment distribution (Fig. E2.7). |

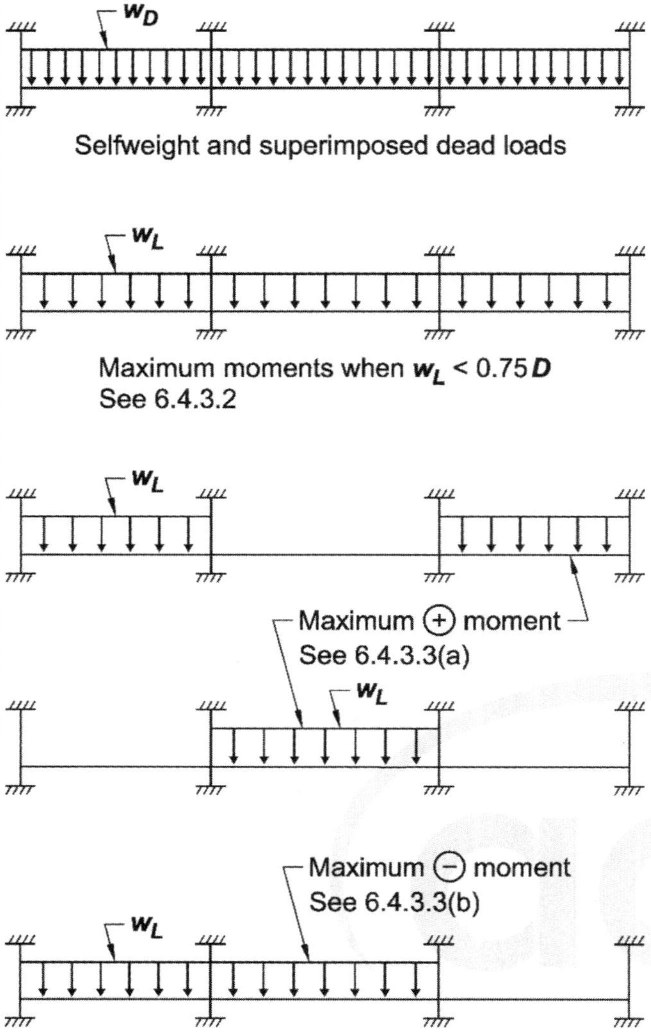

Fig. E2.6—Code live load patterns, example uses 6.4.3.3 (a) and (b).

# CHAPTER 6—TWO-WAY SLABS

**Moment distribution example**

Given:
- $w_u$ = 0.283 k/ft²
- $\ell_{1\,external}$ = 216 in
- $\ell_{1\,internal}$ = 216 in
- $\ell_2$ = 168 in

| Column line | 1 | | 1.5 | | 2 | | 2.5 |
|---|---|---|---|---|---|---|---|
| $K_{sc}$ | 1468 | | 312 | | 312 | | 312 |
| $K_{s,left}$ | 0 | | 95 | | 95 | | 95 |
| $K_{s,right}$ | 95 | | 95 | | 95 | | 95 |
| $\Sigma K$ | 1563 | | 502 | | 502 | | 502 |
| Moment Coefficient, M | 0.085 | | 0.085 | | 0.085 | | 0.085 |
| C.O.F., C | 0.52 >> | << | 0.52 >> | << | 0.52 >> | << | 0.52 |
| Slab Distribution Factor | | 0.06 | 0.19 | 0.19 | 0.19 | 0.19 | 0.19 |
| Column Distribution Factor | 0.94 | | 0.62 | | 0.62 | | 0.62 |
| FEM | | -109 | 109 | -109 | 109 | -109 | 109 |
| bal | 102 | 7 | 0 | 0 | 0 | 0 | 0 | 0 |
| carryover | | 0 | 4 | 0 | 0 | 0 | 0 |
| bal | 0 | 0 | -1 | -2 | -1 | 0 | 0 | 0 | 0 |
| carryover | | -1 | 0 | 0 | -1 | 0 | 0 |
| bal | 1 | 0 | 0 | 0 | 0 | 0 | 1 | 0 | 0 | 0 |
| carryover | | 0 | 0 | 0 | 0 | 0 | 0 |
| bal | 0 | 0 | 0 | 0 | 0 | 0 | 0 | 0 | 0 |
| Balanced moment at Column Centerline | 103 | -103 | 112 | -2 | -110 | 108 | 1 | -109 | 109 | 0 |

Fig. E2.7—Moment distribution: example partial distribution with all spans with full live load.

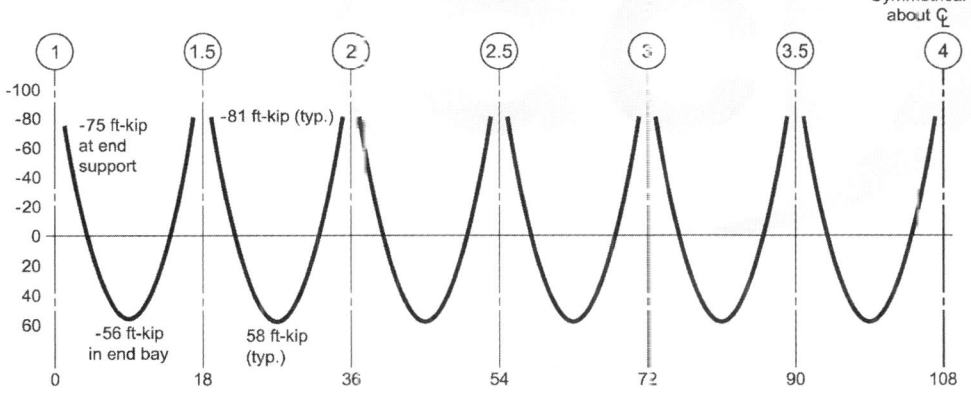

Fig. E2.8—Moment diagram maximum values at face of support and midspan

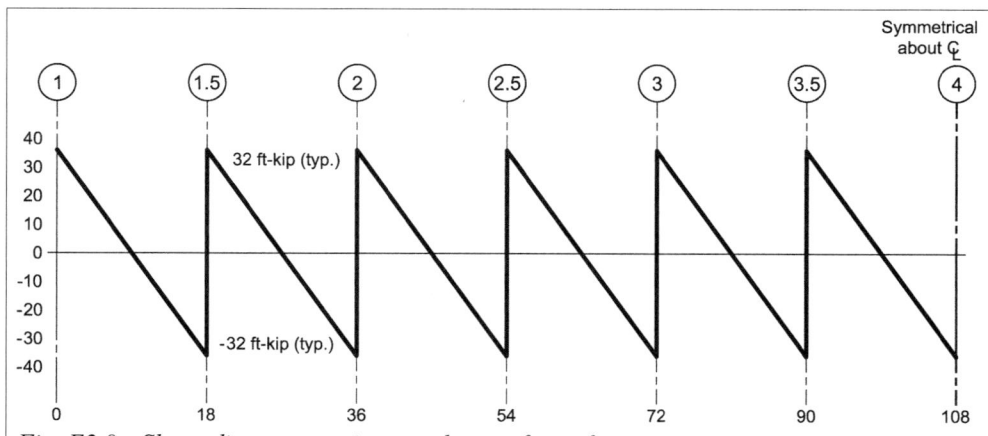

*Fig. E2.9—Shear diagram maximum values at face of support*

| Step 5: Design | | |
|---|---|---|
| 8.11.6.6 | The Code permits the moments determined from the EFM to be distributed to the column and middle strips in accordance with the Direct Design Method (DDM) in Section 8.10 of ACI 318-14. Continuing on, the design solution follows a similar method as the direct design method. | Refer to the DDM in the Two-way Slab Example 1 of this Handbook for this procedure. |

Alternative method for determining stiffness coefficients for use in the moment distribution calculations:
ACI 318-14, Section 6.3.1.1 states that:
"Relative stiffnesses of members within structural systems shall be based on reasonable and consistent assumptions."
This provision allows the designer to use any set of reasonable assumptions for determining the stiffnesses of the members in a two-way slab system in the EFM. In this example, the Hardy column analogy was used. An alternative method is suggested in the following discussion.

Given that Table 6.6.3.1.1(a) of ACI 318-14 will be used to account for the effects of cracking and the approximations in Table 6.6.3.1.1, detailed calculations for $k_c$ to include the effects of rigid ends on the column stiffness are not warranted (The effects of rigid ends are small compared to the effects of cracking). Therefore, take $k_c = 4.0$ and :

Columns: $I_c = 0.7 I_g = 0.7 \dfrac{24(24)^3}{12} = 19,353$ in.$^4$

Walls: $I_w = 0.35 I_g = 0.35 \dfrac{168(12)^3}{12} = 8467$ in.$^4$

Slabs: $I_s = 0.25 I_g = 0.25 \dfrac{168(7)^3}{12} = 1200$ in.$^4$

Using these stiffness values to determine $K_c$ to use for moment distribution calculations. Note that because all of the concrete strengths are the same, the modulus of elasticity is assumed equal to 1 ksi in this example.

Upper column:
$$K_c = \dfrac{k_c E_{cc} I_c}{\ell_c}$$
$$K_c = \dfrac{(4)(1)(19,353)}{216} = 358$$

Lower column:
$$K_c = \dfrac{k_c E_{cc} I_c}{\ell_c}$$
$$K_c = \dfrac{(4)(1)(19,353)}{137.5} = 563$$

Walls (neglecting torsion action with the column):
$$K_c = \dfrac{k_c E_{cc} I_w}{\ell_c}$$
$$K_c = \dfrac{(4)(1)(8467)}{137.5} = 246$$

Slabs:
$$K_s = \dfrac{k_s E_{sc} I_s}{\ell_1}$$
$$K_s = \dfrac{(4)(1)(1200)}{216} = 22$$

Combining these values with $K_t$ (Step 2 torsional members from this Example), and using the resulting stiffness values in the moment distribution along with the fixed end moments (without modification of the fixed end moment factor—that is, FEM = $(1/12)(w\ell^2)$), gives results that are approximately 5 percent different from the values shown in the example above.

**Two-way Slab Example 3:** *Post-tensioning*

This two-way slab is a prestressed solid slab roof without beams between supports. The strength of the slab is checked and two-way shear reinforcement at the external columns is designed. Material properties were selected based on the code requirements of Chapters 5 and 6, engineering judgment, and known available materials.

Given:
Load:

Superimposed dead load $D = 0.015$ kip/ft$^2$
Roof live load $L = 0.040$ kip/ft$^2$

*Material properties—*
$f_c' = 5000$ psi
$f_y = 60,000$ psi

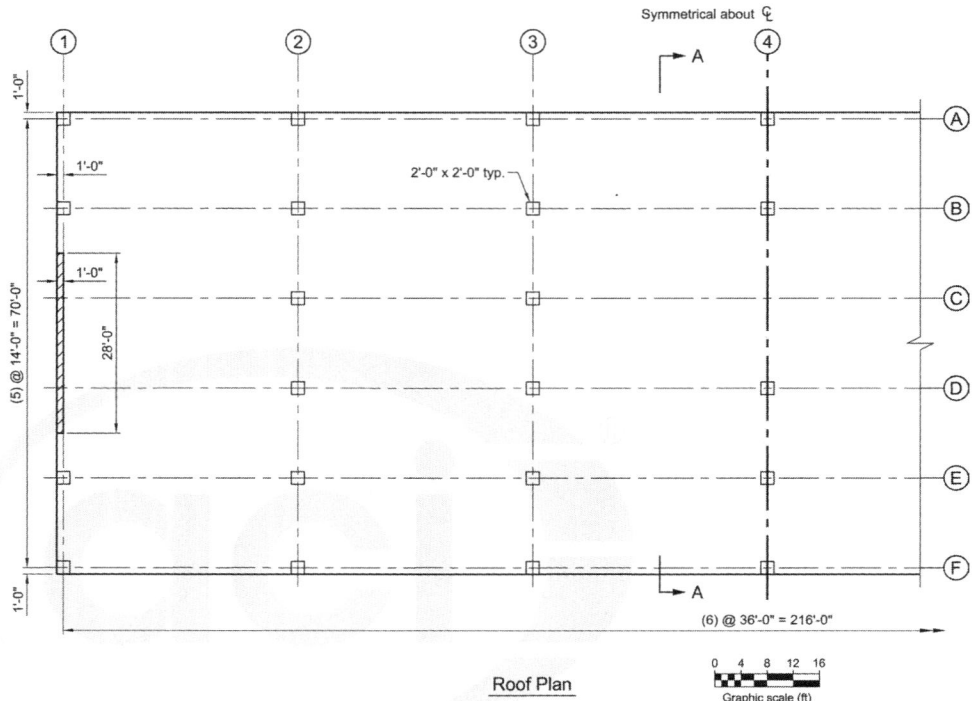

*Fig. E3.1—Roof plan.*

| ACI 318-14 | Discussion | Calculation |
|---|---|---|
| Step 1: Geometry | | |
| 8.3.2 | In the direction taken, there are six spans of 36 ft. The slab is supported by 24 in. square columns.<br><br>The ACI 318-14 span-to-depth ratios do not apply to post-tensioned (PT) slabs. Span/depth ratios between 40 and 50 are typically reasonable for two-way slab designs (Nawy G., 2006, *Prestressed Concrete: A Fundamental Approach, Fifth edition*, Pearson Prentice Hall, New Jersey, 945 pp).<br><br>Use a ratio of 45 to set the initial thickness of the slab. | $45 \leq \dfrac{\ell}{t}$<br><br>$\ell = 432$ in.<br><br>$\therefore$<br><br>$t = 9.6$ in.<br><br>Use a thickness of 10 in. |

| Step 2: Load and load patterns | | |
|---|---|---|
| 8.4.1.2 | Loading all spans simultaneously does not necessarily produce the maximum flexural stresses in the slab. Therefore, in Section 6.4.3 of ACI 318-14, live load patterns are defined for use with two-way slab systems. Fig. E3.2 shows examples of the different live load patterns considered in the code, using Section 6.4.3.2 of ACI 318-14.<br><br>Section 6.4.3.2 is applicable for this example because the roof live load is less than 75 percent of the combined dead loads. | |

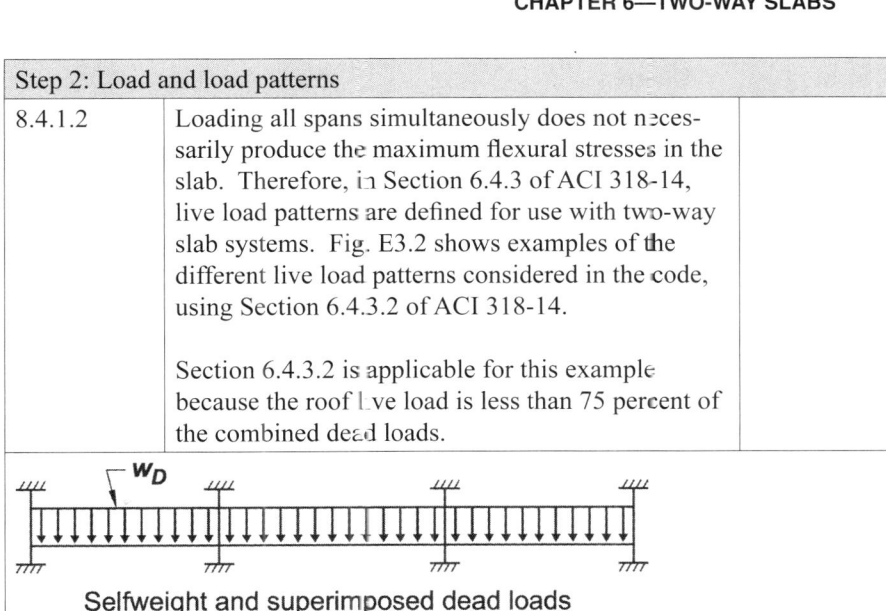

Selfweight and superimposed dead loads

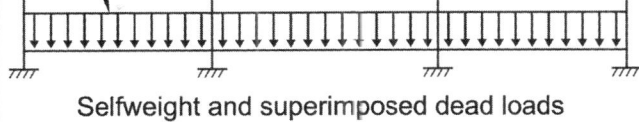

Maximum moments when $w_L < 0.75D$
See 6.4.3.2

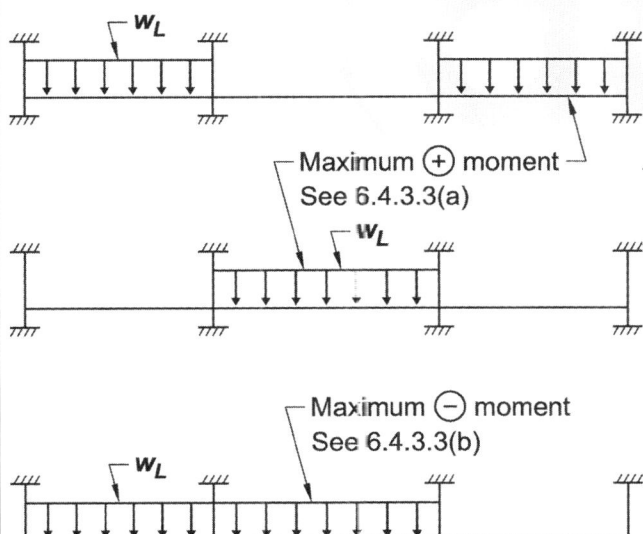

Fig. E3.2—Code live load patterns, example uses 6.4.3.2

| Step 3: Concrete and steel material requirements | | |
|---|---|---|
| 8.2.6.1 | The mixture proportion must satisfy the durability requirements of Chapter 19 (ACI 318-14) and structural strength requirements. The designer determines the durability classes. Please refer to Chapter 4 of this Handbook for an in-depth discussion of the categories and classes.<br><br>ACI 301 is a reference specification that correlates with ACI 318. ACI encourages referencing ACI 301 into job specifications.<br><br>There are several mixture options within ACI 301, such as admixtures and pozzolans, which the designer can require, permit, or review if suggested by the contractor. | By specifying that the concrete mixture shall be in accordance with ACI 301-10 and providing the exposure classes, Chapter 19 (ACI 318-14) requirements are satisfied.<br>Based on durability and strength requirements, and experience with local mixtures, the compressive strength of concrete is specified at 28 days to be at least 5000 psi. |
| 8.2.6.2 | The reinforcement must satisfy Chapter 20 of ACI 318-14.<br><br>In this example, unbonded, 1/2 in. single strand tendons are assumed.<br><br>The designer determines the grade of bar and if the reinforcing bar should be coated by epoxy or galvanized, or both. In this case, assume grade 60 bar and no coatings. | Chapter 20 (ACI 318-14) requirements are satisfied by specifying that the reinforcement shall be in accordance with ACI 301-10. This includes the PT type and strength, and reinforcing bar grade and any coatings for the reinforcing bar. |
| 20.3 | The code requires strand material to be 270 ksi, low relaxation (ASTM A416a). The U.S. industry usually stresses, or jacks, monostrand to impart a force equal to $0.80 f_{pu}$, which is the maximum allowed by the Code.<br><br>The final stress after all losses is usually between 60 to 64 percent of the specified tensile strength of low relaxation strands. | The jacking force per individual strand is:<br>270 ksi × 0.8 × 0.153 in.² = 33 kip<br><br>This is immediately reduced by seating and friction losses, and elastic shortening of the slab. Long term losses will further reduce the force per strand. Refer to R20.3.2.6 of the Code.<br><br>An effective PT force design value of around 26.5 k (64%$f_{pu}$) per strand is common. |

## Step 4: Analysis

**6.6, 8.11**

The analysis performed should be consistent with the overall assumptions about the role of the slab within the building system. Because the lateral force resisting system relies on the slab to transmit axial forces, a first order analysis is adequate.

Although gravity moments are calculated independent of PT moments, the same model is used for both.

Modeling assumptions:
Assume a single moment of inertia for the entire length of the slab. Refer to Section 24.5.2.2 of the Code regarding cracked vs uncracked. Prestressed two-way slabs are required to be designed with service load limits of:

$$f_t \leq 6\sqrt{f'_c}$$

The direct design method is not permitted for use with a prestressed slab, therefore, the equivalent frame method is used. In practice, computer analysis software is typical.

Analysis approach:
To analyze the flexural effects of post-tensioning on the concrete slab under service loads, the tendon drape is assumed to be parabolic with a discontinuity at the support centerline as shown below, which imparts a uniform uplift over each span when tensioned. The magnitude of the uplift, $w_p$, or "balanced load," in each span of a prismatic member is calculated as:

$$w_p = \frac{8Fa}{\ell^2}$$

where $F$ is effective PT force and $a$ is tendon drape (average of the two high points minus the low point). In this example the PT force is assumed constant for all spans, but the uplift force varies due to different tendon drapes. Figure E3.3 shows the tendon profile assumed in this example.

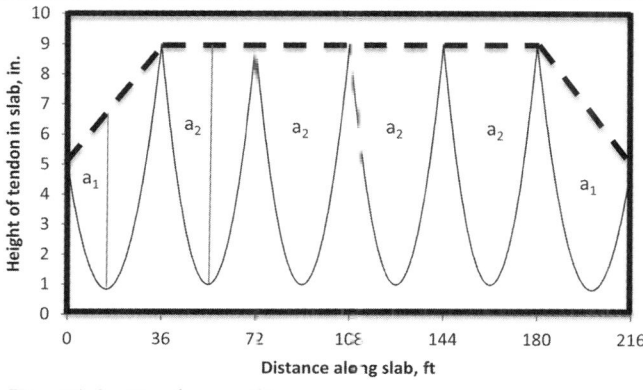

Fig. E3.3—Tendon profile

$$a_1 = \frac{(5 \text{ in.} - 1 \text{ in.}) + (9 \text{ in.} - 1 \text{ in.})}{2} = 6 \text{ in.}$$

$$a_2 = \frac{(9 \text{ in.} - 1 \text{ in.}) + (9 \text{ in.} - 1 \text{ in.})}{2} = 8 \text{ in.}$$

## Step 5: Analysis – Slab stress limits

| | | |
|---|---|---|
| 8.3.4.1 | The code requires a Class U slab assumption; that is, a slab under full service load with a concrete tension stress not exceeding $6\sqrt{f_c'}$. The slab analysis model for the service condition is the same as for the nominal condition. | For 5000 psi concrete, this limit is $6\sqrt{5000}$ psi = 424 psi |
| 8.6.2.1 | To verify that the concrete tensile stresses are less than $6\sqrt{5000}$ psi, the net service moments and tensile stresses at the face of supports are needed. This example assumes two parameters:<br><br>(a) the PT force provides a $F/A$ slab compressive stress of at least 125 psi (15 kip/ft)<br>(b) the combination of PT force and profile provides a uplift force $w_p$ of at least 7 percent of the slab weight, or 94 psf.<br><br>Solve for $F$: | The basic equation for concrete tensile stress is:<br>$f_t = M/S - F/A$, where $M$ is the net service moment,<br><br>$$S = \frac{bt^2}{6} = \frac{(12 \text{ in.})(10 \text{ in.})^2}{6} = 200 \text{ in.}^3$$<br>(section modulus),<br><br>and $A = bt = 12$ in.$*10$ in. $= 120$ in.$^2$ (gross slab area per foot).<br><br>At the exterior support, the drape is $a = 6$ in. The equation for<br><br>$$0.094 \text{ kip/ft} = \frac{8Fa}{l^2} = \frac{8F(6 \text{ in.})}{(36 \text{ ft})^2(12 \text{ in./ft})^2}$$<br><br>$\therefore$<br>$F = 30.5$ kip/ft |

| | Location from left to right along the span | | | | | |
|---|---|---|---|---|---|---|
| Service loads | First midspan | Second midspan | Third midspan | Fourth midspan | Fifth midspan | Sixth midspan |
| Gravity uniform load, psf | 180 | 180 | 180 | 180 | 180 | 180 |
| PT Uniform uplift, psf | 94 | 126 | 126 | 126 | 126 | 94 |
| Net load, psf | 86 | 54 | 54 | 54 | 54 | 86 |

| | | |
|---|---|---|
| | Using the above information and performing an equivalent frame analysis (refer to Two-way Slab Example 2), the following maximum service moments in the slab are determined:<br>Negative moment is maximum at the face of the first interior support and is 8.1 ft-kip/ft<br>Positive moment is maximum at midspan of the first and sixth spans and is 5.3 ft-kip/ft. | Use these moments to determine the stresses at service load:<br>At the face of the first interior support:<br>$$f_t = -\frac{P}{A} + \frac{M}{S}$$<br>$$f_t = \left(-\frac{30.5 \text{ kip}}{120 \text{ in.}^2} + \frac{8.1 \text{ ft-kip}}{200 \text{ in.}^3} \times \frac{12 \text{ in.}}{\text{ft}}\right)\left(\frac{1000 \text{ lb}}{1 \text{ kip}}\right)$$<br>$f_t = 232$ psi $\leq 424$ psi $\therefore$ OK<br><br>For the positive moment at midspan, it is usually desirable to avoid additional reinforcement required by Section 8.6.2.3. To avoid this, the tensile stresses in the slab should not exceed $2\sqrt{5000}$ psi $= 141$ psi.<br>$$f_t = -\frac{P}{A} + \frac{M}{S}$$<br>$$f_t = \left(-\frac{30.5 \text{ kip}}{120 \text{ in.}^2} + \frac{5.3 \text{ ft-kip}}{200 \text{ in.}^3} \times \frac{12 \text{ in.}}{\text{ft}}\right) \times \left(\frac{1000 \text{ lb}}{1 \text{ kip}}\right)$$<br>$f_t = 64$ psi $\leq 141$ psi $\therefore$ OK |

## Step 6: Analysis – Deflections

**8.3.2**

The two-way slab chapter refers the user to Section 24.2.2 (ACI 318-14) that states, "Deflections due to service-level gravity loads…" for allowable stiffness approximations to calculate immediate and time-dependent (long term) deflections. Section 24.2.2 provides maximum allowed span-to-deflection ratios. Section 24.2.3.8 permits using $I_g$ to calculate deflections for Class U slabs. Commentary Section R24.2.3.3 of the Code alerts the designer that calculations for deflections of two-way slabs is challenging. This example determines the deflections in one direction and doubles it for the effect from the other direction. This is not an accurate assumption, but it should give conservative and reasonable results. Note that excessive deflections are generally not experienced in PT slabs.

The example assumes the deflections in each direction are identical and combines them to give the maximum deflection at the midpoint of the slab. Deflections are checked in the long direction of the slab. Deflections due to the uniform live load are checked in Section 24.2.2, therefore, the uniform live load only is applied in this deflection calculation. The analysis is approximate due to several simplifying assumptions, but it provides a reasonable result.

$$\Delta_{max} = \frac{0.0065 w \ell^4}{EI} = \frac{0.0065(0.040/12)(432)^4}{4030(1000)} = 0.19 \text{ in.}$$

Assuming twice this to account for the two-way action of the slab,
$\Delta_{max} = 2 \times 0.19 \text{ in.} = 0.38 \text{ in.}$
Expressed as a ratio, $\ell/\Delta = 432/0.38 = \ell/2400$. This is much less than the limit of $\ell/180$, so deflection limits are satisfied.

## Step 7: Analysis – Balanced, secondary, factored, and design moments

Balanced moments are determined using the PT Uniform uplift load from the table of service loads in Step 5 and performing an equivalent frame analysis (refer to Two-way Slab Example 2). Secondary moments are determined using balanced moments and primary moments.
Factored moments are determined using the factored load combinations required by code and performing an equivalent frame analysis (refer to Two-way Slab Example 2).
Design moments are determined by subtracting the secondary moments from the factored moments.
The following table gives the balanced, secondary, factored, and design moments at the face of supports across the slab section.
Fig. E3.4 shows the design moments as determined by the equivalent frame analysis.
The slab is symmetrical about the third column.

The moment curve below the table is the full design moment curve with critical section moments shown on the curve. The design moments shown are at the face of supports and at the point of maximum positive moment in the span.

|  | Location from left to right along the span (sym about Col 4) | | | | | |
|---|---|---|---|---|---|---|
|  | Col 1 (ext) | Col 2 (ext) | Col 2 (int) | Col 3 (int) | Col 3 (int) | Col 4 (int) |
| Balanced moment, $M_{bal}$, ft-kip/ft | 6.3 | 10 | 11.1 | 11.6 | 11.6 | 11.4 |
| Eccentricity, $e$, in. | 0 | 4 | 4 | 4 | 4 | 4 |
| Primary moment, $M_1$, ft-kip/ft | 0 | 10.2 | 10.2 | 10.2 | 10.2 | 10.2 |
| Secondary moment, $M_s = M_{bal} - M_1$, ft-kip/ft | 6.3 | -0.2 | 0.9 | 1.4 | 1.4 | 1.2 |
| Factored load moment, $M_u'$, ft-kip/ft | 16.1 | 23.4 | 21.6 | 21 | 21.2 | 21.3 |
| Design moment, $M_t = M_u' - M_s$, ft-kip/ft | 9.8 | 23.6 | 20.7 | 19.6 | 19.8 | 20.1 |

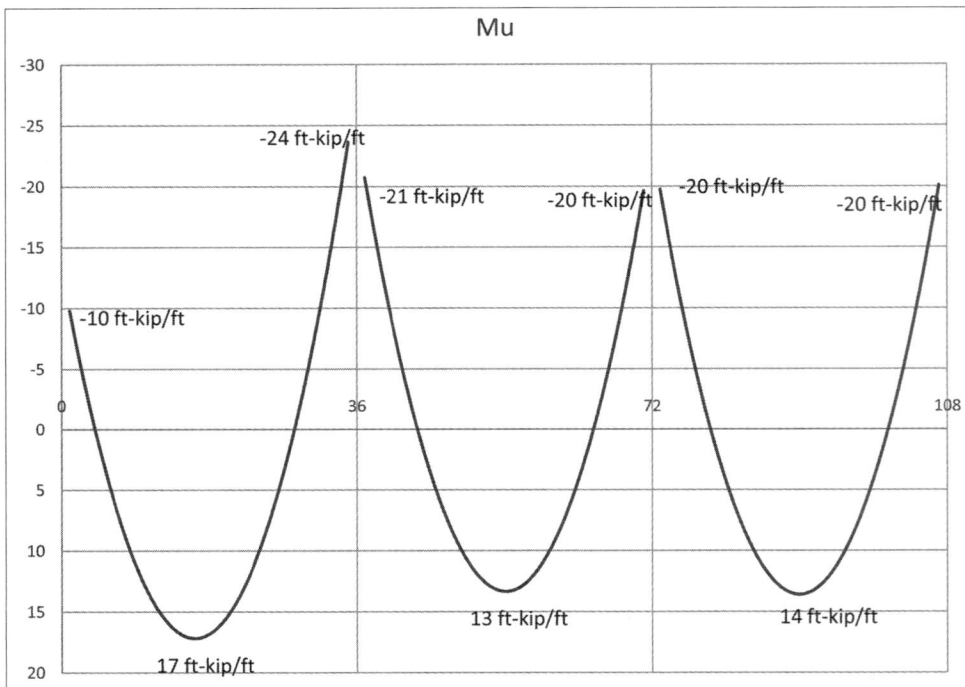

Fig. E3.4—Moment diagram (negative moment at face of column).

| Step 8: Required strength – Calculate required $A_s$ | | |
|---|---|---|
| 8.7.5.2 | Check flexure strength considering PT tendons<br>If the PT tendons alone provide the necessary design strength, $\geq \phi M_n$, then the code permits reinforcement to be detailed with shorter cut-off lengths. If the PT tendons alone do not provide the design strength, then the reinforcement is required to conform to standard lengths.<br>The reinforcing bar and tendons are usually at the same position near the support and midspan. | The depth of the equivalent stress block, $a$, is calculated by<br>$$a = \frac{A_{ps} f_{ps}}{0.85 f'_c (12 \text{ in./ft})}$$<br>where $A_{ps}$ is the tendon area per foot of slab. Section 8.5.2.1 refers to Section 22.3 of the Code for the calculation of $\phi M_n$. Section 22.3 refers to Section 22.2 for calculation of $M_n$. Section 22.2.4 refers to Section 20.3.2.4 to calculate $f_{ps}$. The span-to-depth ratio is $432/10 = 1/43$, so the below equation applies:<br>$$f_{ps} = f_{se} + 10{,}000 + \frac{f'_c}{300 \rho_p}$$ |
| 20.3.2.4 | Each single unbonded tendon is stressed to the value prescribed by the supplier. The value of $f_{se}$ (effective stress in the strand) varies along the tendon length due to friction losses (ACI 423.10R-15), but for design purposes, $f_{se}$ is usually taken as the average value. | The tendon supplier usually calculates $f_{se}$, and 175,000 psi is a common value. The force per strand is therefore 175,000 psi × 0.153 in.² = 26,800 lbs. The required effective force per foot of slab is 30.5 kip/ft, so the spacing of tendons is 26.8 kip/30.5 kip/ft)(12 in./ft) = 10.5 in. The value of $A_{ps}$ is therefore 0.153 in.² × 12/10.5 = 0.175 in.²/ft. The value of $\rho_p$ is $A_{ps}/(b \times d_p)$ = 0.175/108 in.² = 0.00162.<br>$$f_{ps} = 175{,}000 + 10{,}000 + \frac{5000}{0.486} = 195{,}000 \text{ psi}$$<br>This value has upper limits of $f_{se} + 30{,}000$ (= 205,000 psi) and $f_{py}$ (= $0.9 f_{pu}$, or 242,900 psi from commentary), so the design value of $f_{ps}$ is 195,000 psi. |

| 22.2.2.4.1 | Note that the effective depth is 9 in. at critical locations, except at the exterior joint. There, the Code permits a minimum $d$ of $0.8h$, or 8 in. | The compression block depth is therefore: $$a = \frac{A_{ps}f_{ps}}{0.85f_c'(12 \text{ in./ft})} = \frac{0.175 \times 195,000}{0.85 \times 5000 \times 12} = 0.67 \text{ in.}$$ For 3/4 in. cover: $M_n = \phi A_{ps}f_{ps}(d - a/2) = 0.9 \times 0.175 \times 195,000 \times (9 - 0.34) = 266,000$ in.-lb/ft $= 22$ ft-kip/ft. |
|---|---|---|

| | Location from left to right along the span ||||||| 
| | Face of exterior support | First midspan | Face of second support | Second midspan | Face of third support | Third midspan | Face of fourth support |
|---|---|---|---|---|---|---|---|
| $M_n$, only tendons, ft-kip/ft | 20 | 22 | 22 | 22 | 22 | 22 | 22 |
| $M_u$, ft-kip/ft | 10 | 17 | 24 | 13 | 21 | 14 | 20 |

$M_n$ considering the tendons alone are greater than the design moments except for at the face of the second support. The reinforcement required to resist the moments at the face of the second support are required to satisfy the detailing requirements of Section 7.7.3 of the Code while minimum reinforcing bar lengths can be used at all other locations.

### Step 9: Required strength – Minimum area of bonded reinforcement

| 8.6.2.3 | The minimum area of flexural reinforcing bar per foot is a function of the slab's cross sectional area. $A_{cf}$ is based on the greater cross sectional area of the slab-beam strips of the two orthogonal equivalent frames intersecting at a column of a two-way slab. | $A_{s,\,min} = 0.00075 \times A_{cf} = 0.00075 \times 10 \text{ in.} \times 12 \text{ in.} = 0.09 \text{ in.}^2/\text{ft}$. |
|---|---|---|

### Step 10: Required strength – Design moment strength of combined prestressing steel and bonded reinforcement

| 8.5.2 | Determine if supplying the minimum area of reinforcing bar is sufficient to achieve a design strength that exceeds the required strength. | Set the section's concrete compressive strength equal to steel tensile strength, and rearrange for compression block depth $a$: $$a = \frac{A_{ps}f_{ps} + A_s f_y}{0.85 f_c'(12)}$$ $$a = \frac{0.175 \times 195,000 + 0.09 \times 60,000}{0.85 \times 5000 \times 12}$$ $a = 0.78$ in. For 3/4 in. cover: $$M_n = \phi\left[A_{ps}f_{ps} + A_s f_y\right]\left(d - \frac{a}{2}\right)$$ $= 0.9[0.175 \times 195,000 + 0.09 \times 60,000](9 - 0.78)$ $= 292,000$ in.-lb $= 24.3$ ft-kip Comparing this value with the required moment strength $M_u$ indicates that the minimum reinforcement plus the tendons supply enough tensile reinforcement for the slab to resist the factored loads at all locations. |
|---|---|---|

| | | |
|---|---|---|
| **Step 11: Analysis – Distribute moments to column and middle strips** | | |
| 8.7.2.3 | Tests and research have shown that for uniformly loaded structures variations in tendon distribution does not alter the deflection behavior or the capacity for the same total prestressing steel percentage. Section 8.7.2.3 provides specific guidance regarding tendon distribution that allows the use of banded tendon distribution in one direction. | The deflection behavior and capacity differences are not dependent upon the distribution of tendons. It can be extrapolated that distribution of moments to the column and middle strips is unnecessary. |
| **Step 12: Required strength – Factored one-way shear** | | |
| 8.4.3<br>8.4.3.1<br>8.4.3.2 | One-way shear rarely controls thickness design of a two-way slab, but it must be checked. In this section, one-way shear load on the structure is determined. | Fig. E3.5 shows one-way shear diagram with the one-way shear reduced to the face of support. Check maximum factored shear.<br><br>$V_u$ = 58 kip/14 ft = 4.1 kip/ft |

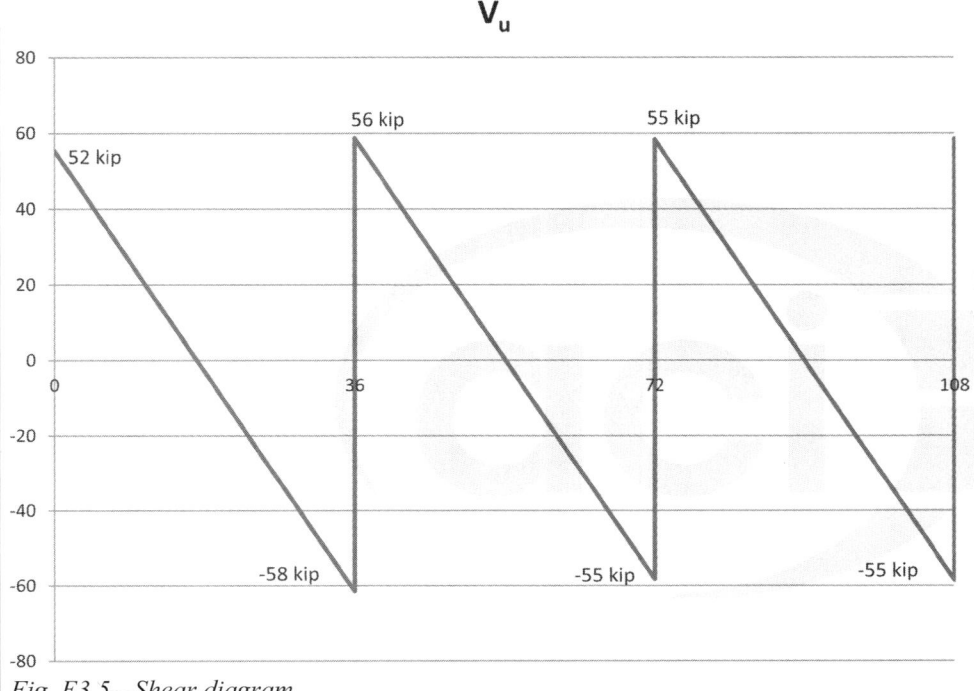

Fig. E3.5—Shear diagram.

| | | |
|---|---|---|
| **Step 13: Required strength – Factored two-way shear** | | |
| 8.3.1.4 | No stirrups are to be used as shear reinforcement. | |
| 8.4.4<br>8.4.4.1<br>22.6.4 | Determine the location and length of the critical section for two-way shear assuming that shear reinforcement is not required. Figure E3.6 shows this examples critical sections. Note that only the exterior and interior columns are calculated in this example. | Exterior columns:<br>$d = 10$ in. $- 1$ in. $= 9$ in.<br><br>$b_o = 2 \times \left( c_1 + \dfrac{d}{2} \right) + (c_2 + d)$<br><br>$b_o = 2 \times \left( 24 \text{ in.} + \dfrac{9 \text{ in.}}{2} \right) + (24 \text{ in.} + 9 \text{ in.})$<br><br>$b_o = 90$ in.<br><br>Interior columns:<br><br>$b_o = \left( c_1 + \dfrac{d}{2} \right) + \left( c_2 + \dfrac{d}{2} \right)$<br><br>$b_o = 2 \times (24 \text{ in.} + 9 \text{ in.}) + 2 \times (24 \text{ in.} + 9 \text{ in.})$<br><br>$b_o = 132$ in. |

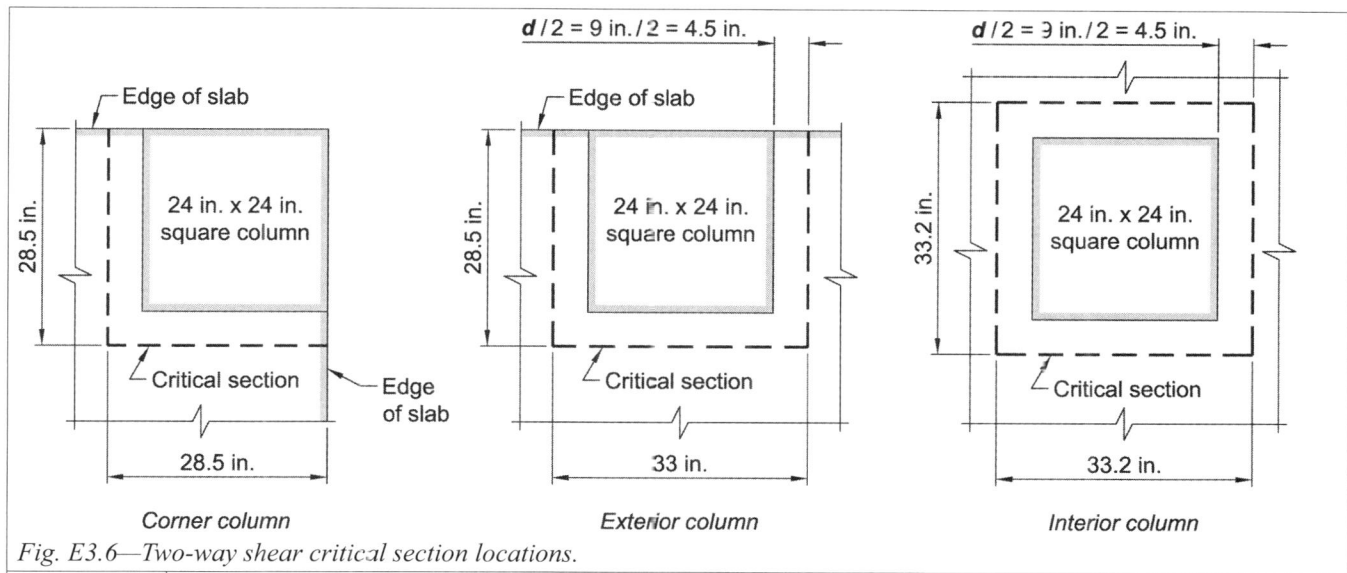

Fig. E3.6—Two-way shear critical section locations.

| 8.4.4.2, 8.4.4.2.1 | Determine factored slab shear stress due to gravity loads $v_{ug}$. | Direct slab shear stress on slab critical section at the exterior columns: $$v_{ug} = \frac{V_u}{b_o d}$$ $$V_u = \left(14\text{ ft} \times 18\text{ ft} - \frac{33\text{ in.} \times 28.5\text{ in.}}{144}\right) \times \frac{232}{1000}\frac{\text{kip}}{\text{ft}^2}$$ $$V_u = 57\text{ ksi}$$ $$v_{ug} = \frac{57\text{ ksi}}{90\text{ in.} \times 9\text{ in.}} = 0.070\text{ ksi}$$ Direct slab shear stress on slab critical section at the interior columns: $$v_{ug} = \frac{V_u}{b_o d}$$ $$V_u = \left(36\text{ ft} \times 14\text{ ft} - \frac{(33\text{ in.})^2}{144}\right) \times \frac{232}{1000}\frac{\text{kip}}{\text{ft}^2}$$ $$V_u = 115\text{ kip}$$ $$v_{ug} = \frac{115\text{ kip}}{132\text{ in.} \times 9\text{ in.}} = 0.097\text{ ksi}$$ |

| | | |
|---|---|---|
| 8.4.4.2.1, 8.4.4.2.2 | Determine the slab shear stress due to factored slab moment resisted by column. | Shear stress on slab due to moments at exterior columns:<br>$\gamma_v = 0.4$<br>$M_{sc} = 10 \text{ ft-kip/ft} \times 14 \text{ ft}$<br>$c_{AB} = 9.02 \text{ in.}$<br>$J_c = 76383 \text{ in.}^4$<br>$\dfrac{\gamma_v M_{sc} c_{AB}}{J_c} = 0.079 \text{ ksi}$<br><br>Shear stress on slab due to moments at interior columns:<br>$\gamma_v = 0.4$<br>$M_{sc} = 24 - 21 \text{ ft-kip/ft} \times 14 \text{ ft}$<br>$c_{AB} = 16.5 \text{ in.}$<br>$J_c = 219633 \text{ in.}^4$<br>$\dfrac{\gamma_v M_{sc} c_{AB}}{J_c} = 0.015 \text{ ksi}$ |
| 8.4.4.2.3 | Determine $v_u$ by combining results from the two-way direct shear and the moment transferred to the column via eccentricity of shear. | $v_u = v_{ug} + \dfrac{\gamma_v M_{sc} c_{AB}}{J_c}$<br><br>Exterior columns:<br>$v_u = 0.070 \text{ ksi} + 0.079 \text{ ksi} = 0.149 \text{ ksi}$<br><br>Corner columns:<br>$v_u = 0.097 \text{ ksi} + 0.015 \text{ ksi} = 0.112 \text{ ksi}$ |
| **Step 14: Design strength – One-way shear** | | |
| 8.5.3.1.1, 22.5 | Nominal one-way shear strength is calculated in accordance with Section 22.5. | $\phi = 0.75$<br>$V_n = 2\sqrt{f_c'}bd$<br>$\phi V_n = 0.75 \times 2 \times \sqrt{5000} \text{ psi} \times 12 \text{ in.} \times 9 \text{ in.} \times \dfrac{1 \text{ kip}}{1000 \text{ lb}}$<br>$\phi V_n = 11.5 \text{ kip/ft}$<br><br>This is greater than the required strength of 4.1 kip/ft from Step 12, therefore, one-way shear is okay. |

# CHAPTER 6—TWO-WAY SLABS

| | Step 15: Design strength – Two-way shear | |
|---|---|---|
| 8.5.3.1.2, 22.6 | Nominal two-way shear strength is calculated in accordance with Section 22.6. Assume shear reinforcement is not required. | Exterior column:<br>$v_n$ = the minimum of the following three equations from Table 22.6.5.2 (ACI 318-14)<br>$v_n = 4\sqrt{f'_c} = \dfrac{4\sqrt{5000}}{1000}$ ksi = 0.283 ksi  **Controls**<br>$v_n = \left(2 + \dfrac{4}{\beta}\right)\sqrt{f'_c} = \dfrac{6\sqrt{5000}}{1000}$ ksi = 0.424 ksi<br>$v_n = \left(2 + \dfrac{\alpha_s \times d}{b_o}\right)\sqrt{f'_c} = \dfrac{4.73\sqrt{5000}}{1000}$ ksi = 0.334 ksi<br>$\phi v_n = 0.75 \times 0.283$ ksi = 0.212 ksi<br><br>This is greater than the required strength for interior columns of 0.149 ksi from Step 6, therefore, two-way shear at interior columns is okay.<br><br>Interior column:<br>$v_n$ = the minimum of the following three equations from Table 22.6.5.2 (ACI 318-14)<br>$v_n = 4\sqrt{f'_c} = \dfrac{4\sqrt{5000}}{1000}$ ksi = 0.283 ksi  **Controls**<br>$v_n = \left(2 + \dfrac{4}{\beta}\right)\sqrt{f'_c} = \dfrac{6\sqrt{5000}}{1000}$ ksi = 0.424 ksi<br>$v_n = \left(2 + \dfrac{\alpha_s \times d}{b_o}\right)\sqrt{f'_c} = \dfrac{4.5\sqrt{5000}}{1000}$ ksi = 0.318 ksi<br>$\phi v_n = 0.75 \times 0.283$ ksi = 0.212 ksi<br><br>This is greater than the required strength for exterior columns of 0.112 ksi from Step 6, therefore, two-way shear at exterior columns is okay.<br><br>The assumption that two-way shear reinforcement is not required at these locations is confirmed. |
| | Step 16: Reinforcement detailing – General requirements | |
| 8.7.1<br>8.7.1.1<br>20.6.1 | Concrete cover, development lengths, and splice lengths are determined in these sections. | Concrete cover is determined using Table 20.6.1.3.2 (ACI 318-14). The bottom of this slab is not exposed to weather or in contact with the ground. The specified cover is 0.75 in. |

| | | |
|---|---|---|
| 8.7.1.2, 25.4 | Development length is used for splice length determination assuming No. 5 bars. | Use Eq. (25.4.2.3(a)) of ACI 318-14 to determine the development length.<br>Using normal weight concrete with No. 6 and smaller uncoated bars and the casting position of less than 12 in. of fresh concrete placed below the horizontal reinforcement, modification factors from Table 25.4.2.4 (ACI 318-14) are:<br><br>$\psi_s = 0.8$<br>$\psi_e = 1.0$<br>$\psi_t = 1.0$<br>$\lambda = 1.0$<br><br>Spacing of the bars is larger than the distance from the center of the bottom bar to the nearest concrete surface.<br><br>$c_b = 0.75 \text{ in.} + \dfrac{0.625 \text{ in.}}{2} = 0.94 \text{ in.}$<br><br>$d_b = 0.625 \text{ in.}$<br><br>$K_{tr}$ is assumed 0 as allowed by 25.4.2.3.<br><br>$\ell_d = \left( \dfrac{3}{40} \dfrac{f_y}{\lambda \sqrt{f_c'}} \dfrac{\psi_t \psi_e \psi_s}{\left( \dfrac{c_b + K_{tr}}{d_b} \right)} \right) d_b$<br><br>$\ell_d = 21.2 \text{ in.}$ |
| 8.7.1.3<br>25.5 | It is likely that splices are required during construction. Allowable locations for splices are shown in ACI 318-14, Fig. 8.7.4.1.3. | Lap splice lengths are determined in accordance with Table 25.5.2.1 (ACI 318-14). The provided $A_s$ is not more than two times larger than the required $A_s$. Therefore, class B splices are required.<br><br>$\ell_{st} = 1.3 \times 21.2 \text{ in.} = 27.5 \text{ in.}$<br><br>use $\ell_{st} = 28$ in. |
| **Step 17: Reinforcement detailing – Spacing requirements** | | |
| 8.7.2<br>8.7.2.1<br>25.2.1<br>8.7.2.3 | Minimum and maximum spacing requirements are determined. The bar spacing for required strength reinforcement are also reviewed. | Minimum spacing is determined in accordance with Section 25.2.1. Minimum spacing is 1 in., $d_b$, and $(4/3)d_{agg}$. Assuming that the maximum nominal aggregate size is 1 in., then the minimum clear spacing is 1.33 in. With a No. 5 bar, this equates to a minimum center-to-center spacing of approximately 2 in.<br><br>Maximum spacing is controlled by Section 8.7.2.3. Assuming that this direction is banded, the maximum spacing requirements of Section 8.7.2.3 are not applicable to this direction. The tendons in the orthogonal direction are limited to a maximum spacing of 5 ft. |

# CHAPTER 6—TWO-WAY SLABS

| | Step 18: Reinforcement detailing – Reinforcement termination | |
|---|---|---|
| 8.7.5.2, 8.7.5.5 | Reinforcement termination is controlled by Section 8.7.5.2. | Bonded nonprestressed reinforcement is required for flexure in one location and Section 8.7.5.2 controls termination of the minimum bonded reinforcement in that location. When the termination location is determined per Section 8.7.5.2, it is approximately 6 in. beyond the face of support. This termination location is within the minimum lengths of Section 8.7.5.5. Therefore, the termination locations indicated in Section 8.7.5.5 satisfy the termination location required by Section 8.7.5.2 for the locations requiring bonded nonprestressed reinforcement for flexural strength. |
| | Step 19: Reinforcement detailing – Structural integrity | |
| 8.7.5.6 | Structural integrity is met using detailing. | Requirement Section 8.7.4.2.2 is met when at least two of the PT tendons pass through the column inside the column reinforcement cage. In this direction, banding of the post-tensioning tendons makes this a simple requirement to satisfy. |
| | Step 20: Slab-column joints | |
| 8.2.7 15.2.1 15.2.2 15.2.3 15.2.5 15.3.1 15.4.2 | Joints are designed to satisfy Chapter 15 of ACI 318-14. | The concrete strength of the slab and columns are identical and therefore, Sections 15.2.1 and 15.3 are met.<br>Requirement Section 15.2.2 is met in Steps 6 and 9 of this example.<br>Requirement Section 15.2.3 is met by satisfying the provisions in Section 15.4.<br>Requirement Section 15.2.5 states that all of the interior columns are restrained because they are laterally supported on four sides by the slab.<br>Requirement Section 15.4.2 applies only to those columns along the exterior of the structure.<br><br>Assuming that No. 4 bars are used as column ties, Eqs. (15.4.2 (a)) and (15.4.2 (b)) are satisfied if the spacing of the column ties in the slab-column joint meets the following:<br><br>(15.4.2(a))<br>$$s = \frac{(0.4 \text{ in.}^2)(60{,}000 \text{ psi})}{(0.75)\sqrt{5000 \text{ psi}}(24 \text{ in.})} = 18.9 \text{ in.}$$<br><br>(15.4.2(b))<br>$$s = \frac{(0.4 \text{ in.}^2)(60{,}000 \text{ psi})}{(50)(24 \text{ in.})} = 20 \text{ in.}$$<br><br>Equation (15.4.2(a)) controls in this case. Because the spacing is larger than the joint depth of 10 in., one tie is required within the slab-column joint in each exterior and corner column. |

| Note: Design post-tensioning is 30.5 kip/ft.<br><br>One strand every 10 in. with this profile. Minimum bonded reinforcement required is 0.09 in.$^2$/ft Figure E3.7 shows the final configuration of the slab. | Assuming (26.5 kip/strand)(1 strand/$x$ in.) = (30.5 kip/ft)/(12 in.)<br><br>No.4 bars: 0.2 in.$^2$/($x$ in.) = 0.09 in.$^2$/0.09 = 26.7 in. → No. 4 bar every 2 ft |
|---|---|

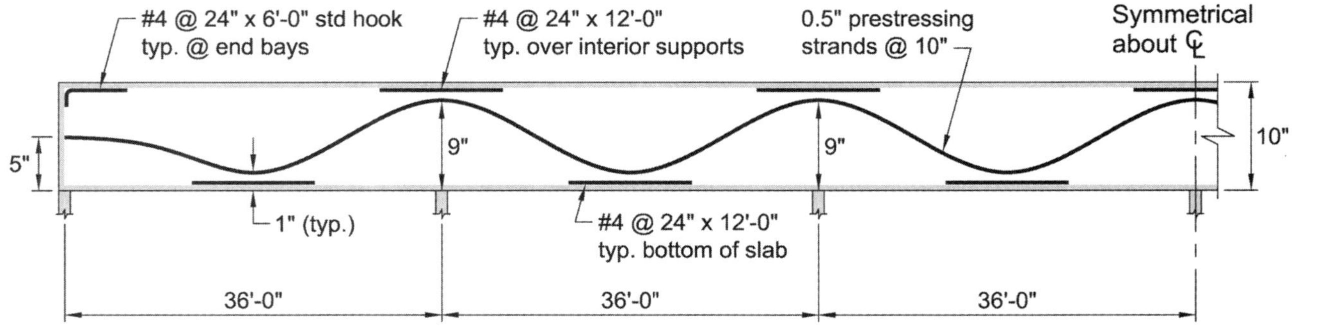

Fig. E3.7—Reinforcement detailing.

Note: a minimum of two unbonded PT strands must be placed in both directions through the column cage.

# CHAPTER 7—BEAMS

## 7.1—Introduction

Structural beams resist gravity and lateral loads, and any combination thereof, and transfer these loads to girders, columns, or walls. They can be nonprestressed or prestressed, cast-in-place, precast, or composite. This chapter discusses cast-in-place, nonprestressed, and post-tensioned (PT) beams. The Code allows for either bonded or unbonded tendons in a PT beam. This chapter does not cover precast, composite, or deep beams. Deep beams are also addressed in ACI 318-14 Chapter 23, Strut-and-Tie.

Beams are designed in accordance with Chapter 9 of ACI 318-14 for strength and serviceability. Beams are assumed to be approximately horizontal, with rectangular or tee-shaped (a stem and a flange) cross sections. The flange width of tee-shaped beams are geometrically limited by Section 6.3.2 and 9.2.4.4, respectively, of ACI 318-14 and the flange is assumed to contribute to the beam's flexural and torsional strength.

Beams, either nonprestressed or prestressed, that are monolithic with the floor framing can be considered laterally braced. For beams that are not monolithic with the floor, ACI 318-14 provides guidance on the spacing of lateral bracing.

## 7.2—Service limits

**7.2.1** *Beam depth*—The engineer determines the beam's concrete strength, steel strength, and other material characteristics to achieve the design performance criteria for strength and service life.

After defining the material properties and the beam's design loads, the engineer chooses the beam's dimensions. These are either provided by architectural constraints, attained from experience, or reached by assuming a depth and width and then adjusting as required through trial and error. Beam depth is addressed in Table 9.3.1.1 of ACI 318-14, which applies if a beam is nonprestressed, not supporting concentrated loads along its span, and not supporting or attached to partitions that may be damaged by deflections. For prestressed or posttensioned beams, the code does not provide a minimum depth-to-span ratio, but for a superimposed live load in the range of 60 to 80 psf a usual span-to-depth ratio is in the range of 20 to 30. Table 9.3 of *The Post-Tensioning Manual* (Post-Tensioning Institute (PTI) 2006) lists span-to-depth ratios for different members that have been found from experience to provide satisfactory structural performance.

The slab thickness is considered as part of the overall beam depth if the beam and slab are monolithic or if the slab is composite with the beam in accordance with Chapter 16 of ACI 318-14.

**7.2.2** *Deflections*—For nonprestressed beams that have depths less than the ACI 318-14. Table 9.3.1.1 minimum, or those that resist a heavy load—usually one- or two-way slabs subjected to above 100 psf—and for PT beams, the designer must calculate deflections. Deflections for calculation, from classical analysis methods or finite element method, can be found in the supplement to this Handbook, ACI Reinforced Concrete Design Handbook Design Aid – Analysis Tables (ACI SP-17DA). The calculated deflections should not exceed the limits in Table 24.2.2 of ACI 318-14, after consideration of time-dependent deflections. Chapter 14, Deflection, of this Handbook includes several design examples on deflection calculations with design aids for T- and L- cross section beams.

For PT beams, Section 24.5.2.1, ACI 318-14 defines three classes of behavior of prestressed flexural members: 1) uncracked, U, 2) cracked, C, and transition between uncracked and cracked, T based on extreme fiber tension stress. Class U in Table 24.5.2.1, ACI 318-14 limits the maximum beam tension stress to less than the cracking stress of concrete, $7.5\sqrt{f_c'}$. Deflection calculations for Class U beams, therefore, can use the gross moment of inertia.

**7.2.3** *Reinforcement strain limits and concrete service stress*

**7.2.3.1** *Strain limits*—For nonprestressed beams with design axial force less than ten percent of gross sectional strength ($< 0.1f_c'A_g$), the minimum strain of the tension reinforcement is 0.004. This limit is to ensure yielding behavior in case of overload.

Nonprestressed beams have no concrete or steel service stress limits, only strength requirements. For pre- and post-tensioned beams, the permissible concrete service stresses are addressed in Section 24.5.3 of ACI 318-14.

For post-tensioned (PT) beams, the analysis of concrete flexural tension stresses is a critical part of the design. In Section 9.3.4.1 of ACI 318-14, beams are classified, presented above, as U (uncracked), C (cracked), or T (transition). Beams with bonded post-tensioning, the combined area of post-tensioning and deformed bars, must be able to resist a factored load that results in moment of at least 1.2 times the section cracking moment.

**7.2.3.2** *Concrete stresses in PT beams*—Before the beam flexural stresses can be calculated, the tendon profile needs to be defined. This profile and the tendon force are directly related to the beam forces and moments created by the post-tensioning. A common approach to calculate PT beam moments is to use the "load balancing" concept, where the profile is usually the maximum practical considering cover requirements, the tendon profile is parabolic, the parabola has an angular "break" at the column centerlines, and that the tendon anchors are at middepth at the exterior (refer to Fig. 7.2.3.2).

The load balancing concept assumes the tendon exerts a uniform upward "load" in the parabolic length, and a point load down at the support. These loads are then combined with the gravity loads, and the analysis is performed with a net service load.

To conform to Code stress limits, the designer can use an iterative approach or a direct approach. In the iterative

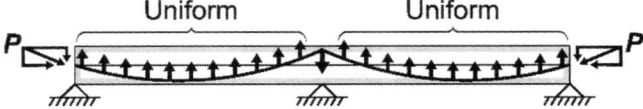

*Fig. 7.2.3.2—Load balancing concept.*

approach, the tendon profile is defined and the tendon force is assumed. The analysis is executed, flexural stresses are calculated, and the designer then adjusts the profile, force, or both, depending on results and design constraints.

In the direct approach, the designer determines the highest tensile stress, then rearranges equations so that the analysis calculates the tendon force needed to achieve the stress limit.

## 7.3—Analysis

Loads and load combinations are obtained from Chapter 5 of ACI 318-14. Beams can be analyzed by any method satisfying equilibrium and geometric compatibility, provided design strength and serviceability requirements are satisfied. Chapter 6 of ACI 318-14 allows for nonprestressed beams satisfying the conditions of Section 6.5.1 to use a simplified approximate method to calculate the design moment and shear forces in beams at the face of support and at midspan. Redistribution of design moments calculated by this method is not permitted.

Beam moments, shear, and deflections along the beams' length are commonly calculated from classic elastic structural analysis. The supplement to this Handbook, ACI Reinforced Concrete Design Handbook Design Aid – Analysis Tables (ACI SP-17DA) has tables that provide moment and shear forces at beam supports and midspan for various boundary and loading conditions. The moment of inertia and modulus of elasticity values used in the classic elastic analysis are addressed in ACI 318. Redistribution of elastic moments calculated by a classical method is permissible.

The engineer can also use finite element software to calculate moments, shear, and deflections along the beams' length. The moment of inertia and modulus of elasticity values used in the finite element model should be carefully considered to obtain realistic deflections and design forces. Redistribution of elastic moments calculated by an elastic finite element method is permissible.

## 7.4—Design strength

Beams resist self-weight and applied loads, which can result in beam flexure, shear, torsion, and axial force. At each section along a beam's length, the design strength is at least equal to the factored load effects; mathematically expressed as $\phi S_n \geq S_u$.

**7.4.1** *Flexure*—Reinforced concrete beam design for flexure typically involves a sectional design that satisfies the conditions of static equilibrium and strain compatibility across the depth of the section.

Following are the assumptions for strength design method listed in Section 22.2 of ACI 318-14; five of these are highlighted as follows:

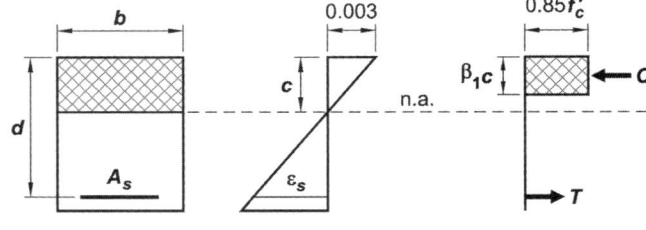

$\beta_1 = 0.85$ for $f'_c \leq 4000$ psi

$\beta_1 = 0.85 - 0.05(f'_c - 4000)/1000 \geq 0.65$ for $f'_c > 4000$ psi

*Fig. 7.4.1—Assumed strain and stress at the nominal condition.*

(a) Strains in reinforcement and concrete are directly proportional to the distance from neutral axis (plane sections remain plane after loading).

(b) Maximum concrete compressive strain in the extreme compression fibers is 0.003 in./in.

(c) Stress in reinforcement varies linearly with strain up to the specified yield strength $f_y$. The stress remains constant beyond this point as strains continue increasing. The strain hardening of steel is ignored.

(d) Tensile strength of concrete is neglected.

(e) Concrete compressive stress distribution is assumed to be rectangular (Fig. 7.4.1).

**7.4.1.1** *Nominal ($M_n$) and design flexural strength ($\phi M_n$)*—A section's $M_n$ is calculated from internal forces assuming the extreme concrete compressive fiber strain reaches 0.003 in./in. Beams exhibit different behaviors depending on the strain level in the extreme tension reinforcement (Section 21.2.2, ACI 318-14); tension controlled, $\varepsilon_s \geq 0.005$; compression controlled, $\varepsilon_s \leq \varepsilon_y = 0.002$; $\varepsilon_y$; and transition, $0.002 < \varepsilon_s < 0.005$.

Reinforced concrete beams behave in a ductile manner by limiting the area of reinforcement such that the tension reinforcement yields before concrete crushes. Tension controlled beam sections produce ductile behavior at the nominal condition, which allows redistribution of stresses and sufficient steel yielding to warn against an imminent failure. The Code requires a beam's extreme tension reinforcement strain (if factored axial compression is less than $0.1f'_c A_g$) to be at least 0.004 (Section 9.3.3.1, ACI 318-14). This strain corresponds to a reinforcement ratio of about $0.75\rho_b$. The Code lowers the strength reduction ($\phi$)-factor for transition-level strains to account for reduced ductility in these sections. Variation of $\phi$-factors with steel tensile strain is shown in Fig. 7.4.1.1 and the corresponding strain profiles at nominal strength (Fig. R21.2.2 (b), ACI 318-14). The basic design inequality is that the factored moment must not exceed the design flexure strength; mathematically expressed as $M_u \leq \phi M_n$.

**7.4.1.2** *Rectangular sections with only tension reinforcement*—Nominal moment strength of a rectangular section with nonprestressed and prestressed tension reinforcement is calculated from the internal force couple shown in Fig. 7.4.1. The area of reinforcement is calculated from the equilibrium of forces. It is assumed that tension steel yields before concrete reaches the assumed compression strain limit of 0.003 in./in. Accordingly, from equilibrium, set steel strength equal to concrete strength:

$$T = C \quad (7.4.1.2a)$$

Substituting the corresponding components for $T$ and $C$:

$$A_s f_y + A_p f_{ps} = 0.85 f_c' \beta_1 c b \quad (7.4.1.2b)$$

where $f_{ps}$ is calculated in Section 20.3 of ACI 318-14.

Assume that $a = \beta_1 c$ and rearrange expressions:

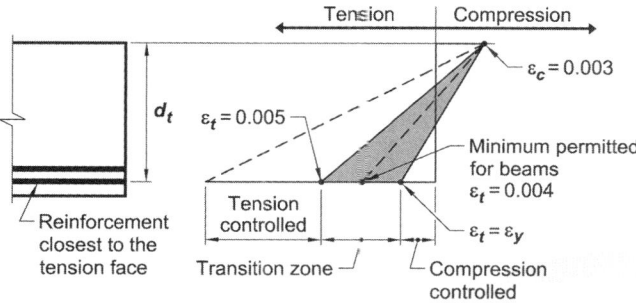

Fig. 7.4.1.1—Strain distribution.

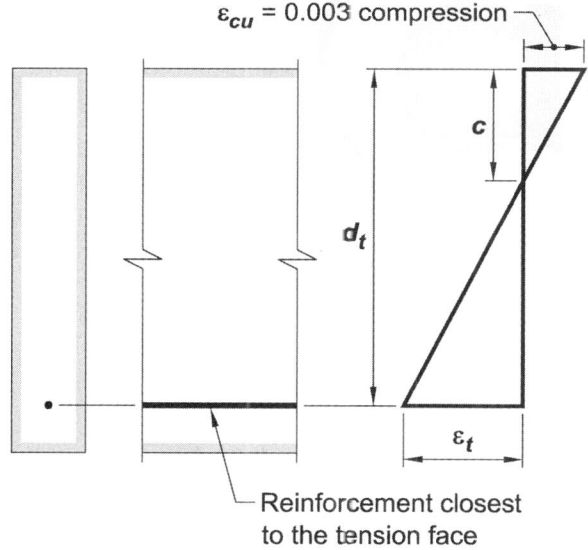

Fig. 7.4.1.2—Strain distribution and net tensile strain in a nonprestressed beam.

$$a = \beta_1 c = \frac{A_s f_y + A_p f_{ps}}{0.85 f_c' b} \quad (7.4.1.2c)$$

Take moments about the concrete resultant, and $M_n$ is calculated as:

$$M_n = T\left(d_t - \frac{\beta_1 c}{2}\right) = \left(A_s f_y\right)\left(d - \frac{a}{2}\right) + \left(A_p f_{ps}\right)\left(d_p - \frac{a}{2}\right) \quad (7.4.1.2d)$$

For reinforced concrete sections with single layer tension reinforcement, $d = d_t$ and $\varepsilon_s = \varepsilon_t$ (7.4.1.3). The stress block geometric parameter $\beta_1$ is between 0.85 and 0.65. For concrete strengths higher than 8000 psi, the value of $\beta_1$ should be reviewed (Ozbakkaloglu and Saatcioglu 2004; Ibrahim and MacGregor 1997). For nonprestressed beams, the $A_p f_{ps}$ term in Eq. (7.4.1.2c) and (7.4.1.2d) of this Handbook is deleted.

**7.4.1.3** *Rectangular sections with tension and compression reinforcement*—Generally, beams are designed with tension reinforcement only. To add moment strength, designers can increase the tension reinforcement area or the beam depth. The cross-sectional dimensions of some applications, however, are limited by architectural or functional considerations, and additional moment strength can be provided by adding an equal area of tension and compression reinforcement. The internal force couple adds to the sectional moment strength without changing the section's ductility. In such cases, the total moment strength consists of adding two components: 1) moment strength from the tension reinforcement-concrete compression couple; and 2) moment strength from the additional tension reinforcement-compression reinforcement couple, assuming both sets of reinforcement yield, as illustrated in Fig. 7.4.1.3.

$$M_n = M_1 + M_2 \quad (7.4.1.3a)$$

$M_1$ is given by Eq. (7.4.1.3a) and $M_2$ is obtained by taking the moment about the tension reinforcement.

$$M_2 = A_s' f_s'(d - d') \quad (7.4.1.3b)$$

Assuming $f_s'$ is equal to $f_y$,

$$M_2 = A_s' f_y(d - d') \quad (7.4.1.3c)$$

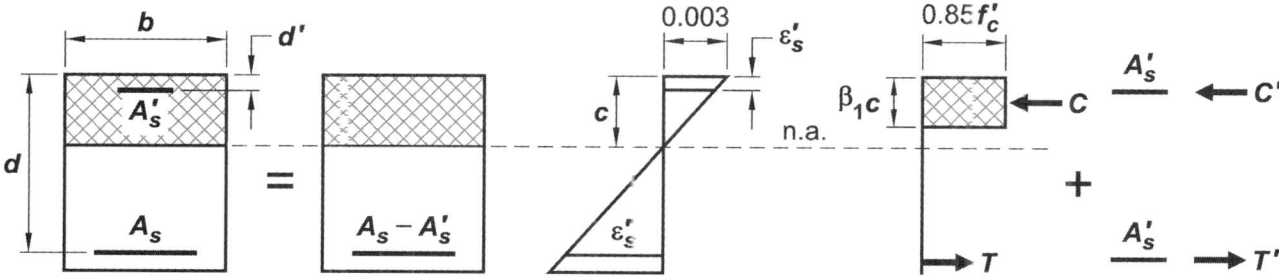

Fig. 7.4.1.3—Forces in double reinforced concrete beam.

Because the steel couple does not require an additional concrete force, adding more tension steel does not create an over-reinforced section as long as an equal area is added in the compression zone. The underlining assumption in calculating the steel force couple is that the steel in compression yields at nominal strength, developing a force equal to the tensile yield strength. This assumption is true in most heavily reinforced sections because the compression Grade 60 steel (0.002 in./in. yield strain) is near the extreme compression fiber, which will strain to 0.003 in./in. at nominal strength.

Depending on the location of the compression reinforcement within the overall strain diagram, it is possible that the compression reinforcement has less strain than 0.002 at nominal strength and, therefore, does not yield. The designer in this case, increases the compression reinforcement area proportional to the ratio of yield strain to compression steel strain. The strain in compression steel, $\varepsilon_s'$, can be computed from Fig. 7.4.1.3 as $\varepsilon_s' = \varepsilon_s(c - d')/(d - c)$, once $\varepsilon_s$ is determined for sections with tension reinforcement to assess if the compression steel yields at nominal strength.

**7.4.1.4** *T-sections*—Cast-in-place and many precast concrete slabs and beams are monolithic, so the slab participates in beam's flexural stiffness, resulting in a T-section. The flange width of a T-section is the effective width of the slab, as defined in Section 6.3.2.1 of ACI 318-14, and the rectangular beam forms the web. Precast double T-sections also benefit from an increase in beam stability during construction.

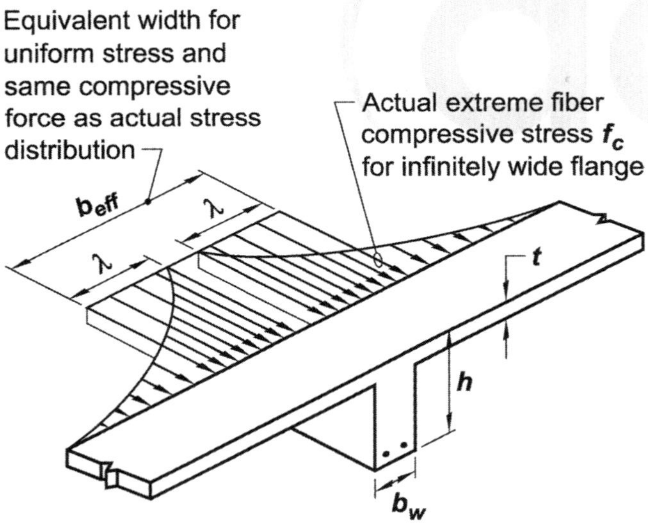

The flange width in most T-sections is significantly wider than the web width (Fig. 7.4.1.4a). For a lightly reinforced section, this often places the neutral axis of the nominal strain diagram within the flange depth. T-sections are analyzed the same as rectangular sections, with section width equal to the effective flange width.

In heavily reinforced T-sections, the area of tension reinforcement in the web (required by the applied moment) brings the neutral axis below the flange, creating a compression zone in the web. In such a case, the total moment strength consists of: 1) tension steel force equal to the flange concrete compression force; and 2) the remaining tension steel force equal to the web concrete compression force. The condition for T-section behavior is expressed mathematically as

$$M_u \leq \phi M_n = \phi(M_{nf} + M_{nw}) \quad (7.4.1.4a)$$

where

$$\phi M_{nf} = \phi \left[ 0.85 f_c' b h_f \left( d - \frac{h_f}{2} \right) \right] \quad (7.4.1.4b)$$

$$\phi M_{nw} = \phi A_s f_y \left( d - \frac{a}{2} \right) + \phi A_p f_{ps} \left( dp - \frac{a}{2} \right) \quad (7.4.1.4c)$$

Many engineers calculate $M_{nf}$ first from equilibrium to find the area of total tension steel needed to balance the flange concrete. The $M_{nw}$ is then calculated assuming a rectangular cross section as shown in Fig. 7.4.1.4b.

For continuous, statically indeterminate, PT beams, effects of reactions induced by prestressing (secondary moments) need to be included per Section 5.3.11 of ACI 318-14. The beam's secondary moments are a result of the column's vertical restraint of the beam against the PT "load" at each support. Because the post-tensioning force and drape are determined during the service stress checks, secondary moments can be quickly calculated by the "load-balancing" analysis concept.

A simple way to calculate the secondary moment is to subtract the tendon force times the tendon eccentricity (distance from the neutral axis) from the total balance moment, expressed mathematically as $M_2 = M_{bal} - P \times e$.

*Fig. 7.4.1.4a—Equivalent stress distribution over flange width.*

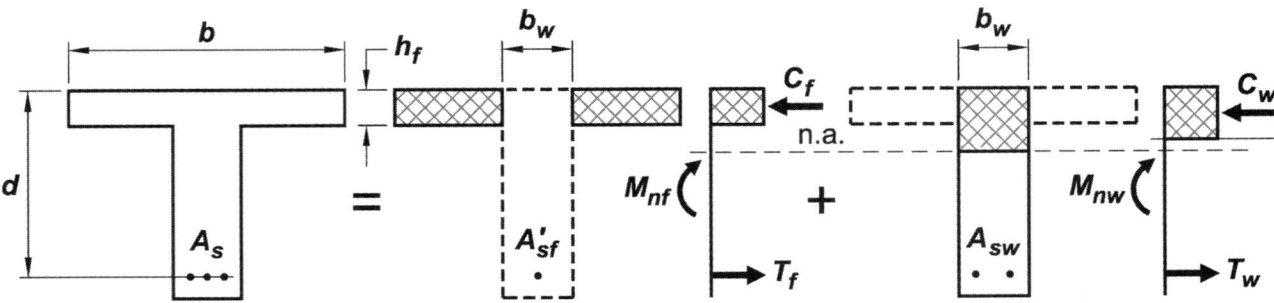

*Fig. 7.4.1.4b—T-section behavior.*

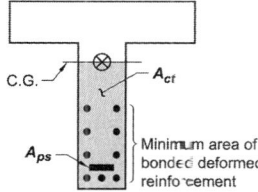

*Fig. 7.4.1.5—Area of minimum bonded deformed longitudinal reinforcement distribution.*

**7.4.1.5** *Minimum flexural reinforcement*—Nonprestressed reinforcement in a section is effective only after concrete has cracked. If the beam's reinforcement area is insufficient to provide a nominal strength larger than the cracking moment, the section cannot sustain its loads upon cracking. This level of reinforcement can be calculated under light loads or beams that are, for architectural and other functional reasons, much larger than required for strength. To protect against potentially brittle behavior immediately after cracking, ACI 318-14 requires a minimum area of tension reinforcement (refer to Section 9.6.1.1 of ACI 318-14).

$$A_{s,min} = \frac{3\sqrt{f_c'}}{f_y} b_w d \quad (7.4.1.5a)$$

but $A_{s,min}$ needs to be at least $200 b_w d/f_y$.

For statically determinate beams, where the T-section flange is in tension, the reinforcement required to provide a nominal strength above the cracking moment is approximately twice that required for rectangular sections. Therefore, $b_w$ in Eq. (7.4.1.5a) is replaced by the smaller of $2b_w$ or the flange width (Section 9.6.1.2 of ACI 318-14). However, when the steel area provided in every section of a member is sufficient to provide flexural strength at least one-third greater than required by analysis, the minimum steel area need not apply (Section 9.6.1.3 of ACI 318-14). This exception prevents requiring excessive reinforcement in overlarge beams.

For prestressed beams with bonded strands, the minimum reinforcement area is that required to develop a design moment at least equal to 1.2 times the cracking moment (Section 9.6.2.1 of ACI 318-14):

$$\phi M_n \geq 1.2 M_{cr} \quad (7.4.1.5b)$$

For prestressed beams with unbonded strands, abrupt flexural failure immediately after cracking does not occur because there is no strain compatibility between the unbonded strands and the surrounding concrete. Therefore, for unbonded tendons, the code only requires a minimum steel area of $0.004 A_{ct}$. These bars should be uniformly distributed over the precompressed tensile zone, close to the extreme tension fibers (Fig. 7.4.1.5).

**7.4.2** *Shear*—Unreinforced concrete shear failure is brittle. This behavior is prevented by providing adequate shear reinforcement that intercepts the assumed inclined cracks. The beam's shear force is usually maximum at a support, and decreases as it moves away from the support, usually at a slope equal to the magnitude of the unit uniform load. In regions of high flexure, cracks form perpendicular to the longitudinal tension reinforcement. Principal tension stresses are approximately horizontal at the longitudinal reinforcement, then change direction gradually to approximately vertical at the location of maximum compression stress (the support). Consequently, cracks in concrete tend to "point" toward the region of maximum compression stress, as indicated by the cracks in Fig. 7.4.2.

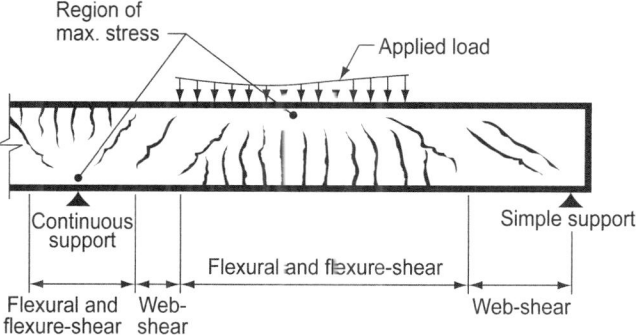

*Fig.7.4.2—Types of cracking in concrete beams (Section R22.5.8.3 of ACI 318-14).*

**7.4.2.1** *Shear strength*—Concrete beams are designed to resist shear and torsion and to ensure ductile behavior at the nominal condition. Shear strength at any location along a beam is calculated as the combination of concrete shear strength, $V_c$, and the steel shear reinforcement, $V_s$ (Section 22.5.1.1 of ACI 318-14). The nominal concrete shear strength $V_c$ is often calculated using the simple ACI 318-14, Eq. (22.5.5.1) $V_c = 2\lambda\sqrt{f_c'} b_w d$ instead of the more detailed equations in Table 22.5.5.1 (ACI 318-14). For beams resisting significant axial force, the concrete shear strength is calculated per ACI 318-14, Table 22.5.6.1 and Eq. (22.5.7.1). Note that for a beam resisting tension, $V_c$ cannot be smaller than zero.

The nominal concrete shear strength, $V_c$, for prestressed beams (defined as $A_{ps} f_{ps} \geq 0.4(A_{ps} f_{pu} + A_s f_y)$) can be calculated using ACI 318-14 simplified equations listed in Table 22.5.8.2, but need not be less than $V_c$ calculated by Eq. (22.5.5.1) of ACI 318-14. The more detailed approach to calculate $V_c$ for prestressed beams is to use the lesser of shear diagonal cracking $V_{ci}$ (Section 22.5.8.3.1 of ACI 318-14) and shear web cracking $V_{cw}$ (Section 22.5.8.3.2).

The factored shear (or required shear strength), $V_u$, can be calculated at a distance $d$ from a support face for the usual support condition (Section 9.4.3.2 of ACI 318-14 and Fig. 7.4.2.1a of this Handbook). For other support conditions, or if a concentrated load is applied within the distance $d$ from a support, the required shear strength is taken at the support face (refer to Fig. 7.4.2.1b and ACI 318-14, Fig. R9.4.3.2 (e) and (f)).

Beam shear reinforcement are usually U-shaped stirrups (Fig. 7.4.2.1c of this Handbook) or closed stirrups. Shear cracking is assumed to occur at 45 degrees from horizontal. ACI 318 uses a truss analog for shear flow, where the stirrups are vertical tension ties in the truss with concrete acting

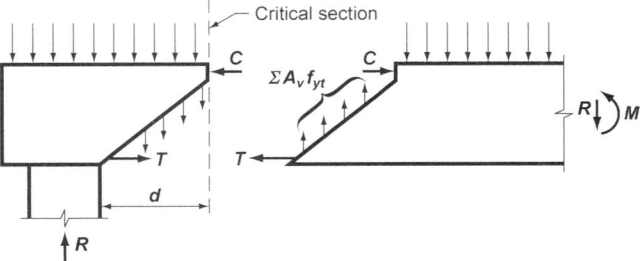

*Fig. 7.4.2.1a—Free body diagrams of the end of a beam.*

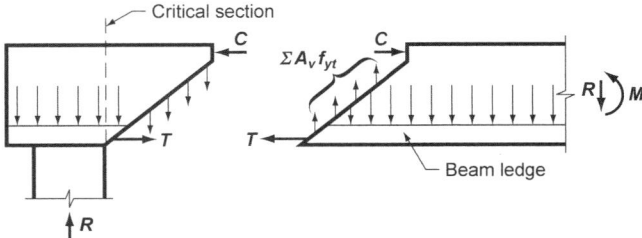

*Fig. 7.4.2.1b—Location of critical section for shear in a beam (R9.4.3.2a, ACI 318-14). Beam loaded near bottom (R9.4.3.2b, ACI 318-14).*

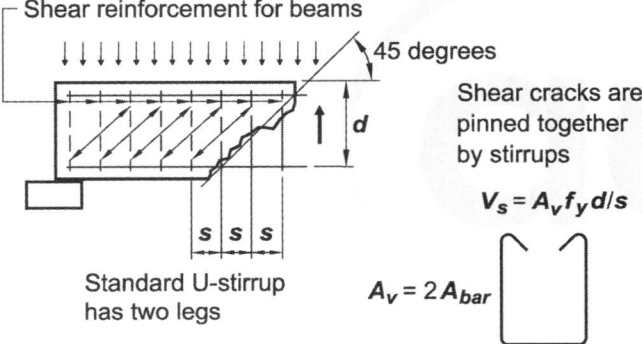

*Fig. 7.4.2.1c—Shear reinforcement.*

as diagonal struts. The longitudinal reinforcement is the tension chord and concrete is the compression chord.

For design, the tension force in each stirrup leg is assumed to be its yield strength times the leg area, and beam stirrups usually have two vertical legs. A U-stirrup has an area $A_v = 2 \times$ (leg area). The beam's nominal steel shear strength is calculated by $V_s = A_v f_{yt} d/s$. Designers usually calculate the required $V_s$ and then determine the stirrup size and spacing, so the equation is rearranged as $A_v/s = V_s/(f_y d)$.

**7.4.2.2** *Designing shear reinforcement*—In general, if a beam's $\phi V_c/2$ is less than $V_u$, shear reinforcement, $A_v$, is needed. The strength reduction factor $\phi$ for shear is 0.75. The required area per unit length is:

$$s \leq (\phi A_v f_{yt} d)/(V_u - \phi V_c) \quad (7.4.2.2)$$

$V_u/\phi - V_c$ represents the nominal shear strength provided by shear reinforcement $V_s$.

There are limited exceptions to the above general rules given in Section 9.6.3.1 of ACI 318-14. For example, a beam having width $b_w$ more than twice the thickness $h$ does not require minimum shear reinforcement as long as the design concrete shear strength exceeds the required shear strength. Note that because the longitudinal beam bars requires support, it is impractical to eliminate beam stirrups, so it is recommended to provide stirrups in all cast-in-place beams, with spacing not exceeding $d/2$.

A type of ribbed floor slab, known as a joist system, is often constructed without shear reinforcement in the joist ribs. A joist system's relative dimensional limits, such as slab thickness, rib width, and rib spacing, are provided in ACI 318-14, Section 9.8.1. If the ribbed floor system does not conform to all the code limits (such as a skip joist system), the system needs to be designed as a beam and slab system.

Section 9.6.3.3 (ACI 318-14) sets lower limits on the $A_v$ to ensure that stirrups do not yield upon shear crack formation. The value of $A_v$ must exceed the larger of $0.75\sqrt{f'_c} b_w s/f_{yt}$ and $50 b_w s/f_{yt}$. The first quantity governs if $f'_c > 4440$ psi. ACI 318-14 has an upper limit of $\sqrt{f'_c}$ to 100 psi, which corresponds to 10,000 psi concrete strength. Section 22.5.3.2 of ACI 318-14 allows the value of $\sqrt{f'_c}$ to be greater than 100 psi if the reinforced and prestressed beam has shear reinforcement per Section 9.6.3.3 and 9.6.4.2 of ACI 318-14. Refer to the Code commentary on this section for more information.

**7.4.3** *Torsion*—Beam torsion (or twisting) creates sectional shear stresses that increase from zero stress at the beam's sectional center to the maximum at the section perimeter. Therefore, ACI 318-14 assumes that significant torsion stress occurs only around the section perimeter. Empirical expressions for torsional strength are provided in Section 9.5.4.1 of ACI 318-14. The torsion shear stress adds to the gravity shear stress on one vertical face, but subtracts from it on the opposite vertical face (Fig. 7.4.3a). Refer also to Fig. 7.4.3a for the definitions of section properties.

When designing for torsion, the engineer needs to distinguish between statically determinate (an uncommon condition) and statically indeterminate torsion (most common condition).

Statically determinate (or equilibrium) torsion is the condition where the equilibrium of the structure requires the beam's torsional resistance, that is the torsional moment cannot be reduced by internal force redistribution to other members. If inadequate torsional reinforcement is provided to resist this type of torsion, the beam cannot resist the applied factored torsion.

Statically indeterminate (or compatibility) torsion exists is the condition where, if the beam loses its ability to resist torsion, the moment is able to be redistributed, equilibrium is maintained, and the torsion load is safely resisted by the rest of the structural system. Torsional moments can be redistributed after beam cracking if the member twisting is resisted by compatibility of deformations with the connected members.

In Fig. 7.4.3b(a), the determinate beam must resist the eccentric load ($w_u e$) on the ledge to columns through beam torsion.

In Fig. 7.4.3b(b), the eccentric load can be resisted by torsion of the beam or by slab flexure. In other words, if the

## Table 7.4.2.2—Shear reinforcement requirements

| Condition | Spacing | | Provision in ACI 318-14 |
|---|---|---|---|
| $V_u \leq \phi V_c/2$ | No shear reinforcement required | | 9.6.3.1 |
| $\phi V_c/2 < V_u \leq \phi V_c$ | Minimum shear reinforcement $s \leq \dfrac{d}{2} \leq 24$ in. | | 9.6.3.1 |
| $\phi V_c < V_u \leq \phi V_c + \phi 4\sqrt{f_c'}b_w d$ | Nonprestressed | $s \leq \dfrac{d}{2} \leq 24$ in. | 9.7.6.2.2 |
| | Prestressed | $s \leq \dfrac{3h}{4} \leq 24$ in. | |
| $\phi V_c + \phi 4\sqrt{f_c'}b_w d < V_u \leq \phi V_c + \phi 8\sqrt{f_c'}b_w d$ | Nonprestressed | $s \leq \dfrac{d}{4} \leq 12$ in. | 9.7.6.2.2 |
| | Prestressed | $s \leq \dfrac{3h}{8} \leq 12$ in. | |
| $V_u > \phi V_c + \phi 8\sqrt{f_c'}b_w d$ | Increase cross section | | 22.5.1.2 |

Definitions:

$A_{cp}$ = area enclosed by outside perimeter of section

$A_c$ = gross area enclosed by shear flow path

$A_{oh}$ = area enclosed by centerline of closed tie

$P_{cp}$ = outside perimeter of concrete section

$P_h$ = perimeter of centerline of closed tie

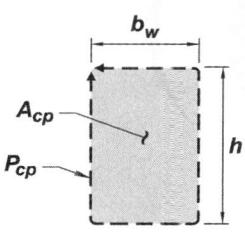

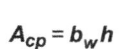

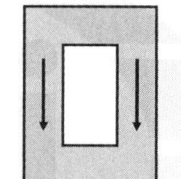

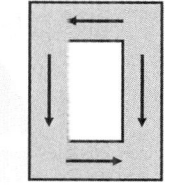

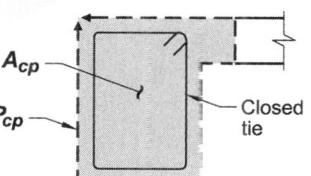

$A_{cp} = b_w h$

$A_{oh} = (h - 3.5)(b_w - 3.5)$

$A_c = 0.85 A_{oh}$

$P_{cp} = 2(b_w + h)$

$P_h = 2[(b_w - 3.5) + (h - 3.5)]$

*Fig. 7.4.3a—Torsion strength definitions of section properties.*

beam loses torsional stiffness, the slab can resist the eccentric loading effects through flexure

A beam's cracking torque, $T_{cr}$, is calculated without consideration of torsion reinforcement.

$$T_{cr} = 4\lambda\sqrt{f_c'}(A_{cp})^2/p_{cp} \quad (7.4.3a)$$

ACI 318-14 assumes that torques less than 1/4 of $T_{cr}$ will not cause a structurally significant reduction in the shear strength and thus is ignored. ACI 318 limits $\sqrt{f_c'}$ to a maximum of 100 psi, which corresponds to 10,000 psi concrete strength. This limit is based on available research. ACI 318-14, Eq. (22.7.7.1a), provides an upper limit to the torque resistance of a concrete beam:

$$T_{max} = 17(A_{oh})^2 \lambda\sqrt{f_c'}/p_h \quad (7.4.3b)$$

where $A_{oh}$ is concrete area enclosed by centerline of the outermost closed transverse torsional reinforcement (Fig. 7.4.3c).

**7.4.3.1** *Torsion reinforcement*—Concrete beams reinforced for torsion per ACI 318-14 are ductile and thus will continue to twist after reinforcement yields. ACI 318-14 specifies beam reinforcement that resists torsion be closed stirrups and longitudinal bars located around the section periphery. Torsion cracks are assumed at angle θ from the member axis, so the torsion strength from closed stirrups is calculated as

$$T_n = \dfrac{2A_o A_t f_{yt}}{s}\cot\theta \quad (7.4.3.1a)$$

where $A_o$ is the gross area enclosed by torsional shear flow path, in.$^2$, $A_t$ is the area of one leg of a closed stirrup, in.$^2$, and

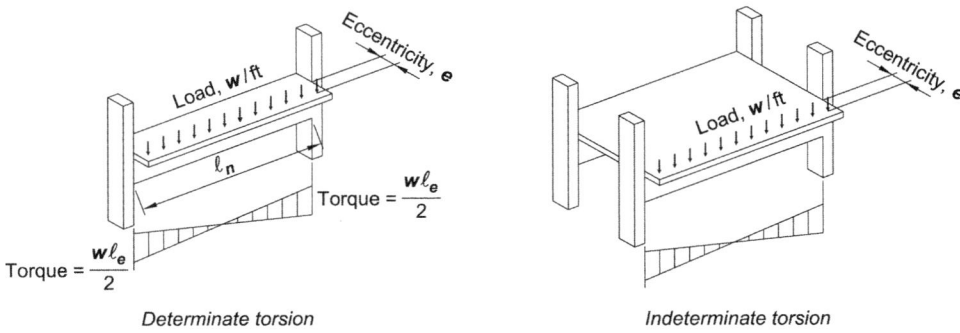

Fig. 7.4.3b—Determinate and indeterminate torsion.

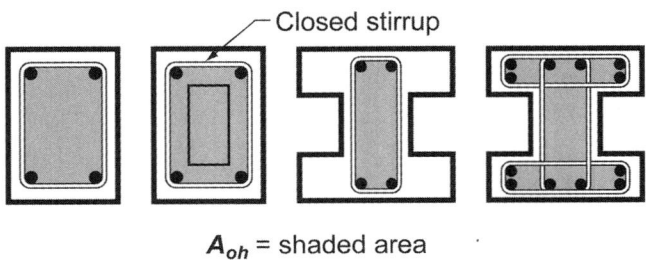

Fig. 7.4.3c—Determining $A_{oh}$.

$f_{yt}$ is the yield strength of transverse reinforcement, psi. ACI 318-14 specifies that angle θ must be greater than 30 degrees and less than 60 degrees; for simplicity in design, use θ = 45 degrees. Solid concrete sections need to be large enough to resist flexural shear $V_u$ and torsion shear $T_u$ within the upper limits given by ACI 318-14, Eq. (22.7.7.1a):

$$\sqrt{\left(\frac{V_u}{b_w d}\right)^2 + \left(\frac{T_u p_h}{1.7 A_{oh}^2}\right)^2} \leq \phi\left(\frac{V_c}{b_w d} + 8\sqrt{f_c'}\right) \quad (7.4.3.1b)$$

Where stirrups are required for torsion in addition to shear, Section 9.6.4.2 of ACI 318-14 requires that the area of two legs of a closed stirrup $(A_v + 2A_t)$ must exceed $0.75(b_w s/f_{yt})$ and $50b_w s/f_y$.

Longitudinal spacing of the closed stirrups must not exceed $p_h/8$, or 12 in. The spacing between the longitudinal bars around the section periphery must not exceed 12 in.

ACI 318, Eq. (22.7.7.1b) requires that the longitudinal bar area, $A_\ell$, be placed around the section periphery. Section 9.6.4.3 of ACI 318-14, requires a minimum area of longitudinal reinforcement $A_{\ell,min}$ be the lesser of (a) and (b):

(a) $\dfrac{5\sqrt{f_c'}A_{cp}}{f_y} - \left(\dfrac{A_t}{s}\right)p_h \dfrac{f_{yt}}{f_y}$

(b) $\dfrac{5\sqrt{f_c'}A_{cp}}{f_y} - \left(\dfrac{25b_w}{f_{yt}}\right)p_h \dfrac{f_{yt}}{f_y}$

The torsion strength from longitudinal bars is calculated as

$$T_n = \frac{2A_o A_\ell f_y}{P_h}\tan\theta \quad (7.4.3.1c)$$

### 7.5—Temperature and shrinkage reinforcement
Refer to Chapter 5, One-way slabs, for information.

### 7.6—Detailing
The longitudinal bar details includes determining the bar size(s), distribution around the perimeter, lengths, and cutoff points. The stirrup details includes determining size, spacing, and configuration.

**7.6.1** *Reinforcement placement*—To limit crack widths, it is preferable to use a larger number of small bars, as opposed to fewer large bars.

**7.6.1.1** *Minimum spacing of longitudinal reinforcement*—Longitudinal reinforcement should be placed at spacing that allows for proper placement of concrete. Table A-3 of ACI Reinforced Concrete Design Handbook Design Aid – Analysis Tables (ACI SP-17DA) shows the 318-14 minimum spacing requirements for beam reinforcement.

**7.6.1.2** *Concrete protection for reinforcement*—The reinforcement should be protected against corrosion and aggressive environments by a sufficiently thick concrete cover (refer to 20.6.1.3.1 of ACI 318-14), as indicated in ACI Reinforced Concrete Design Handbook Design Aid – Analysis Tables (ACI SP-17DA). The engineer should also consider the beam's required fire rating when determining concrete cover (Section 4.11.2 of ACI 318-14). Considering cover, reinforcement should be placed as close to the concrete surface as practicable to maximize the lever arm for internal moment strength and to restrain crack widths.

**7.6.1.3** *Reinforcement in a T-section*—Where a beam's T-section flanges are in tension due to flexure, tension reinforcement needs to be distributed over the flange width or a width equal to 1/10 the span, whichever is smaller (refer to Section 24.3.4 of ACI 318-14). This requirement is intended to limit slab crack widths that can result from widely spaced reinforcement. When 1/10 of the span is smaller than the effective width, additional reinforcement should be provided in the outer portions of the flange to minimize wide cracks in these slab regions.

**7.6.1.4** *Maximum spacing of flexural reinforcement*—Beams reinforced with few large bars could experience cracking between the bars, even when the required tension reinforcement area is provided and the sectional strength is adequate. To limit crack widths to acceptable limits for various exposure conditions, ACI 318-14 Section 24.3.2 specifies a maximum spacing, $s$, for reinforcement closest to the tension face. The spacing limit is the lesser of the two equations that follow:

$$s \le 15\left(\frac{40,000}{f_s}\right) - 2.5c_c$$
$$s \le 12\left(\frac{40,000}{f_s}\right)$$
(7.6.1.4)

In the above equation, $c_c$ is the least distance from the reinforcement surface to the tension face of concrete, and $f_s$ is the service stress in reinforcement. The service stress, $f_s$, can be calculated from strain compatibility analysis under unfactored service loads or may be taken as $2/3 f_y$. Note that Eq. (7.6.1.4) does not provide sufficient crack control for beams subject to very aggressive exposure conditions or designed to be watertight. For such conditions, further investigation is warranted (Section 24.3.5 of ACI 318-14).

**7.6.1.5** *Skin reinforcement*—In deep beams, cracks may develop near the beam's mid-depth, between the neutral axis and the tension face. Therefore, the Code requires beams with a depth $h > 36$ in. to have "skin reinforcement" with a maximum spacing of $s$, as defined in Eq. (7.6.1.4) and illustrated in ACI Reinforced Concrete Design Handbook Design Aid – Analysis Tables (ACI SP-17DA) (refer to Fig. 7.6.1.5 and Section 9.7.2.3 of ACI 318-14). For this case, $c_c$ is the least distance from the skin reinforcement surface to the side face. ACI 318 does not specify a required steel area as skin reinforcement. Research indicates that No. 3 to No. 5 bar sizes or welded wire reinforcement with a minimum area of 0.1 in.$^2$/ft provide sufficient crack control (Frosch 2002).

**7.6.2** *Shear reinforcement*—Stirrup bar size is usually a No.3, No. 4, or No. 5, because larger bar sizes can be difficult to bend. Note that stirrup spacing less than 3 in. can create difficulties in placing concrete. Therefore, some engineers increase the stirrup spacing by doubling the stirrups (refer to Fig. 7.6.2(d)). For wider beams, stirrups should be distributed across the cross section to engage the full beam width and thereby improve shear resistance (Fig. 7.6.2).

**7.6.3** *Torsion reinforcement*—The detailing requirements for beams resisting torsion are listed in ACI 318-14, Sections 9.7.5 and 9.7.6, for longitudinal and transverse reinforcement, respectively. The longitudinal bars are distributed around the stirrup perimeter, with at least one longitudinal bar is placed in each corner (Section 9.7.5.2 of ACI 318-14). To resist torsion, the stirrup ends are closed with 135-degree hooks (Fig. 7.6.2(c) and (e) and 7.6.3). A 135-degree hook may be replaced by a 90 degree hook where the stirrup end is confined and restrained against spalling by a slab or flange of a T-section (refer to Fig. 7.6.2(a), (b), and (d)). Splicing stirrups is not acceptable for torsion reinforcement (Fig. 7.6.3).

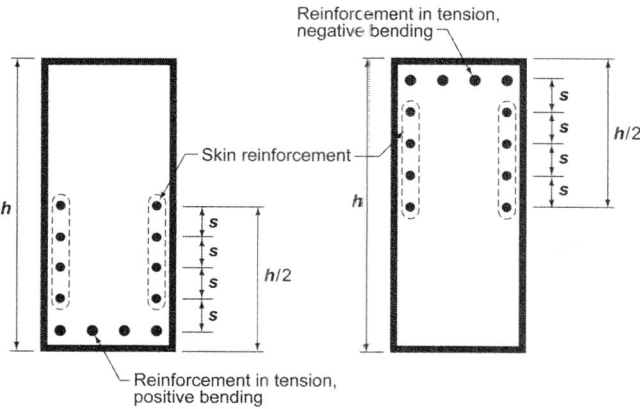

*Fig. 7.6.1.5—Skin reinforcement for beams and joists with* $h > 36$ *in. (R9.7.2.3, ACI 318-14).*

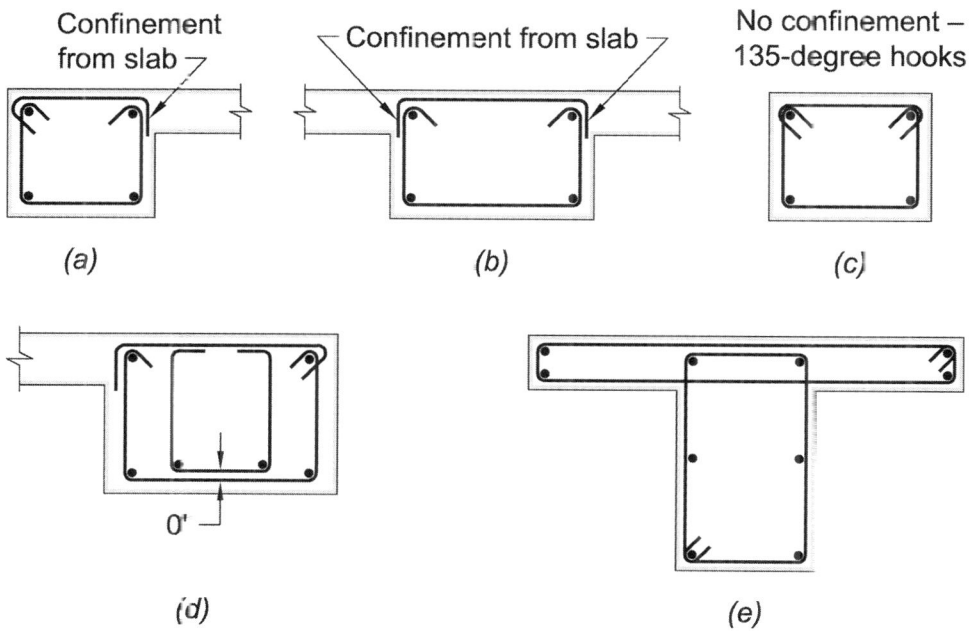

*Fig. 7.6.2a—Longitudinal reinforcement distributed around beam perimeter with closed stirrups.*

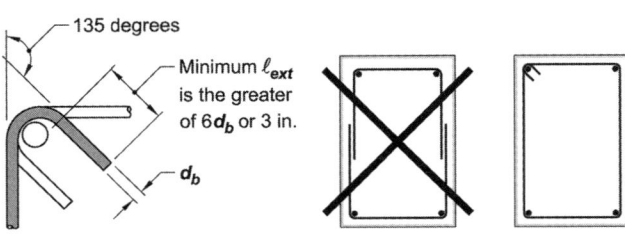

Fig. 7.6.3—Detailing of closed stirrups for torsion.

## REFERENCES

*American Concrete Institute (ACI)*

ACI SP-17DA-14—Reinforced Concrete Design Handbook Design Aid – Analysis Tables; https://www.concrete.org/store/productdetail.aspx?ItemID=SP1714DA

**Authored references**

Frosch, R. J., 2002, "Modeling and Control of Side Face Beam Cracking," *ACI Structural Journal*, V. 99, No. 3, May-June, pp. 376-385.

Ibrahim, H. H. H., and MacGregor, J. G., 1997, "Modification of the ACI Rectangular Stress Block for High-Strength Concrete," *ACI Structural Journal*, V. 94, No. 1, Jan.-Feb., pp. 40-48.

Ozbakkaloglu, T., and Saatcioglu, M., 2004, "Rectangular Stress Block for High-Strength Concrete," *ACI Structural Journal*, V. 101, No. 4, July-Aug., pp. 475-483.

Post-Tensioning Institute (PTI), 2006, *Post-Tensioning Manual*, sixth edition, PTI TAB. 1-06, 354 pp.

## 7.7—Examples

**Beam Example 1**: *Continuous interior beam*

Design and detail an interior, continuous, six-bay beam, built integrally with a 7 in. slab.

Given:

*Load—*
Service additional dead load $D$ = 15 psf
Service live load $L$ = 65 psf
Beam and slab self-weights are given below.

*Material properties—*
$f_c'$ = 5000 psi (normalweight concrete)
$f_y$ = 60,000 psi
$\lambda$ = 1.0 (normalweight concrete)

Span length: 36 ft
Beam width: 18 in.
Column dimensions: 24 in. x 24 in.
Tributary width: 14 ft

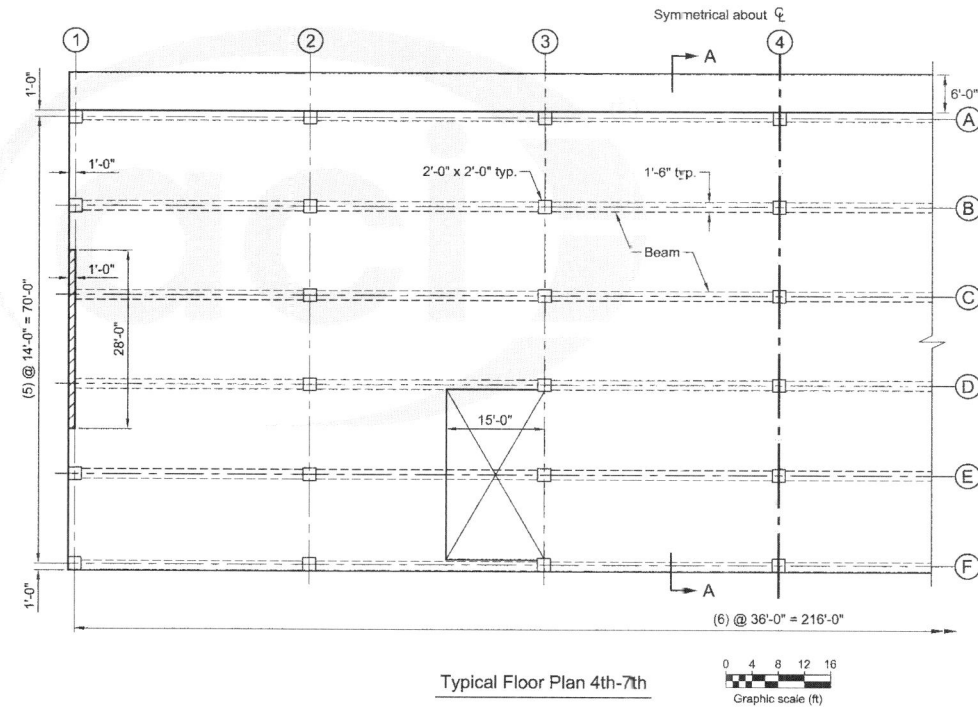

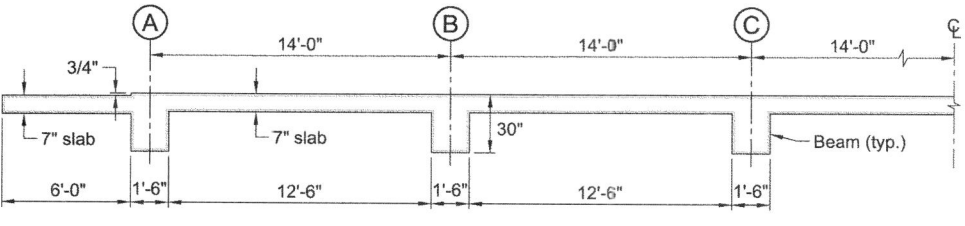

*Fig. E1.1—Plan and partial elevation of 6-span interior beam.*

| ACI 318-14 | Discussion | Calculation |
|---|---|---|
| **Step 1: Material requirements** | | |
| 9.2.1.1 | The mixture proportion must satisfy the durability requirements of Chapter 19 (ACI 318-14) and structural strength requirements. The designer determines the durability classes. Please refer to Chapter 4 of SP-17 for an in-depth discussion of the Categories and Classes.<br><br>ACI 301 is a reference specification that is coordinated with ACI 318. ACI encourages referencing ACI 301 into job specifications.<br><br>There are several mixture options within ACI 301, such as admixtures and pozzolans, which the designer can require, permit, or review if suggested by the contractor. | By specifying that the concrete mixture shall be in accordance with ACI 301-10 and providing the exposure classes, Chapter 19 (ACI 318-14) requirements are satisfied.<br><br>Based on durability and strength requirements, and experience with local mixtures, the compressive strength of concrete is specified at 28 days to be at least 5000 psi. |
| **Step 2: Beam geometry** | | |
| 9.3.1.1 | Beam depth<br>If the depth of a beam satisfies Table 9.3.1.1, ACI 318-14 permits a beam to be designed without having to check deflections, as long as the beam is not supporting or attached to partitions or other construction likely to be damaged by large deflections. Otherwise, beam deflections must be calculated and satisfy the deflection limits in Section 9.3.2 of ACI 318-14. | The beam has four continuous spans, so the controlling condition for beam depth is one end continuous:<br><br>$h = \dfrac{\ell}{18.5} = \dfrac{(36\text{ ft})(12\text{ in./ft})}{18.5} = 23.35$ in.<br><br>Use 30 in. |
| | Self-weight<br>Beam:<br>Slab: | $w_b = [(18\text{ in.})(30\text{ in.})/(144)](0.150\text{ kip/ft}^3) = 0.56$ kip/ft<br>$w_s = (14\text{ ft} - 18\text{ in.}/12)(7\text{ in.}/12)(0.150\text{ kip/ft}^3) = 1.1$ kip/ft |
| 9.2.4.2 | Flange width<br>The beam is placed monolithically with the slab and will behave as a T-beam. The flange width on each side of the beam is obtained from Table 6.3.2.1. | |
| 6.3.2.1 | Each side of web is the least of $\begin{cases} 8h_{slab} \\ s_w/2 \\ \ell_n/8 \end{cases}$ | 8(7 in.) = 56 in.<br>(14 ft)(12)/2 = 84 in.<br>((36 ft)(12 in./ft) − 24 in.)/8 = 51 in.   **Controls** |
| | Flange width:<br>$b_f = \ell_n/8 + b_w + \ell_n/8$ | $b_f = 51$ in. + 18 in. + 51 in. = 120 in. |

| | Step 3: Loads and load patterns | |
|---|---|---|
| 5.3.1 | The service live load is 50 psf in offices and 80 psf in corridors per Table 4-1 in ASCE 7-10. This example will use 65 psf as an average as the actual layout is not provided. A 7 in. slab is a 87.5 psf service dead load. To account for the weight of ceilings, partitions, HVAC systems, etc., add 15 psf as miscellaneous dead load.<br><br>The beam resists gravity only and lateral forces are not considered in this problem.<br>$U = 1.4D$<br><br>$U = 1.2D + 1.6L$ | $U = 1.4(0.56\text{ kip/ft} + 1.1\text{ kip/ft} + (15\text{ psf})(14\text{ft})/1000)$<br>$= 2.6\text{ kip/ft}$<br>$U = 1.2(2.6\text{ kip/ft})/1.4 - 1.6((65\text{ psf})(14\text{ ft})/1000)$<br>$= 3.7\text{ kip/ft}$ **Controls** |
| | Note: Live load is not reduced per ASCE 7-10 in this example. | |
| Step 4: Analysis | | |
| 9.4.3.1 | The beams are built integrally with supports; therefore, the factored moments and shear forces (required strengths) are calculated at the face of the supports. | |
| 9.4.1.2 | Chapter 6 of ACI 318-14 permits several analysis procedures to calculate the required strengths | |
| 6.5.1 | The beam's required strengths can be calculated using approximations per Table 6.5.2 of ACI 318-14, if the conditions in Section 6.5.1 are satisfied:<br>(a) Members are prismatic<br>(b) Loads uniformly distributed<br>(c) $L \leq 3D$<br>(d) Three spans minimum<br><br>Difference between two spans does not exceed 20 percent. | Beams are prismatic<br>Satisfied (no concentrated loads)<br>65 psf < 3(87.5 psf + 15 psf + Beam SW) satisfied<br>Actual 6 spans > 3 spans<br><br>Beams have equal lengths<br><br>All five conditions are satisfied; therefore, the approximate procedure is used. |

| | |
|---|---|
| 6.5.2 | Using $\ell_n = 34$ ft for all bays results in the following and moment and shear forces at face of columns.<br><br>*Shear diagram* 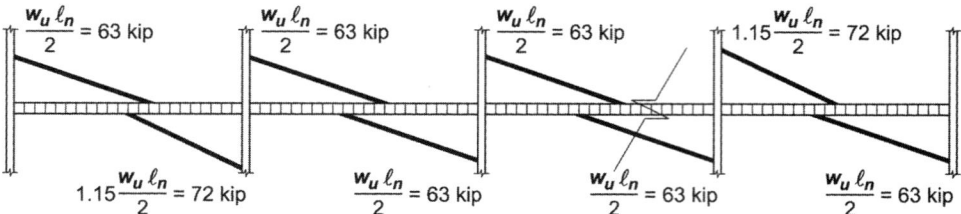 |
| 6.5.4 | *Moment diagram* 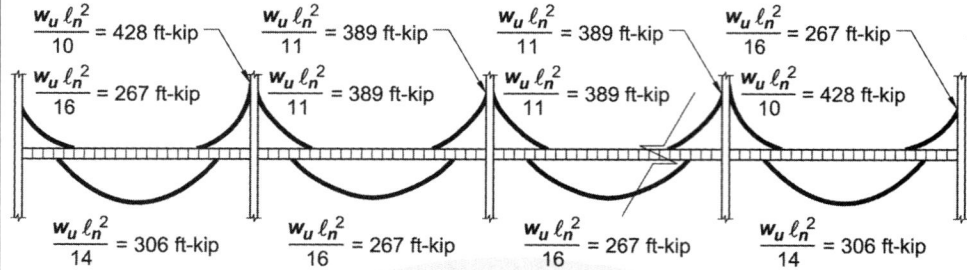<br><br>Fig. E1.2—Shear and moment diagrams. |
| 6.5.3 | Note:<br>The moments calculated using the approximate method cannot be redistributed in accordance with Section 6.6.5.1.<br>Moment diagram drawn on the tension side of the beam. |

## Step 5: Moment design

| | | |
|---|---|---|
| 9.3.3.1 | The code doesn't permit a beam to be designed with steel strain less than 0.004 in./in. at nominal strength. The intent is to ensure ductile behavior.<br><br>In most reinforced concrete beams, such as this example, bar strain is not a controlling issue. | |
| 21.2.1(a) | The design assumption is that beams will be tensioned controlled with a moment reduction factor $\phi = 0.9$. This assumption will be checked later.<br>Determine the effective depth assuming No. 3 stirrups, No. 7 longitudinal bars, and 1.5 in. cover: | |
| 20.6.1.3.1 | One row of reinforcement<br>$d = h - \text{cover} - d_{tie} - d_b/2$ | $d = 30$ in. $- 1.5$ in. $- 0.375$ in. $- 0.875$ in./2 = 27.6 in.<br>use $d = 27.5$ in. |
| 22.2.2.1 | The concrete compressive strain at nominal moment strength is calculated at:<br>$\varepsilon_{cu} = 0.003$ | |
| 22.2.2.2 | The tensile strength of concrete in flexure is a variable property and is approximately 10 to 15 percent of the concrete compressive strength. ACI 318-14 neglects the concrete tensile strength to calculate nominal strength.<br><br>Determine the equivalent concrete compressive stress at nominal strength: | |
| 22.2.2.3<br><br>22.2.2.4.1 | The concrete compressive stress distribution is inelastic at high stress. The Code permits any stress distribution to be assumed in design if shown to result in predictions of ultimate strength in reasonable agreement with the results of comprehensive tests. Rather than tests, the Code allows the use of an equivalent rectangular compressive stress distribution of $0.85f_c'$ with a depth of: | |
| 22.2.2.4.3 | $a = \beta_1 c$, where $\beta_1$ is a function of concrete compressive strength and is obtained from Table 22.2.2.4.3. For $f_c' = 5000$ psi: | $\beta_1 = 0.85 - \dfrac{0.05(5000 \text{ psi} - 4000 \text{ psi})}{1000 \text{ psi}} = 0.8$ |
| 22.2.1.1 | Find the equivalent concrete compressive depth $a$ by equating the compression force to the tension force within the beam cross section:<br>$C = T$<br>$0.85f_c'ba = A_s f_y$ | $0.85(5000 \text{ psi})(b)(a) = A_s(60,000 \text{ psi})$ |
| | For positive moment: $b = b_f = 120$ in. | $a = \dfrac{A_s(60,000 \text{ psi})}{0.85(5000 \text{ psi})(120 \text{ in.})} = 0.118 A_s$ |
| | For negative moment: $b = b_w = 18$ in. | $a = \dfrac{A_s(60,000 \text{ psi})}{0.85(5000 \text{ psi})(18 \text{ in.})} = 0.784 A_s$ |

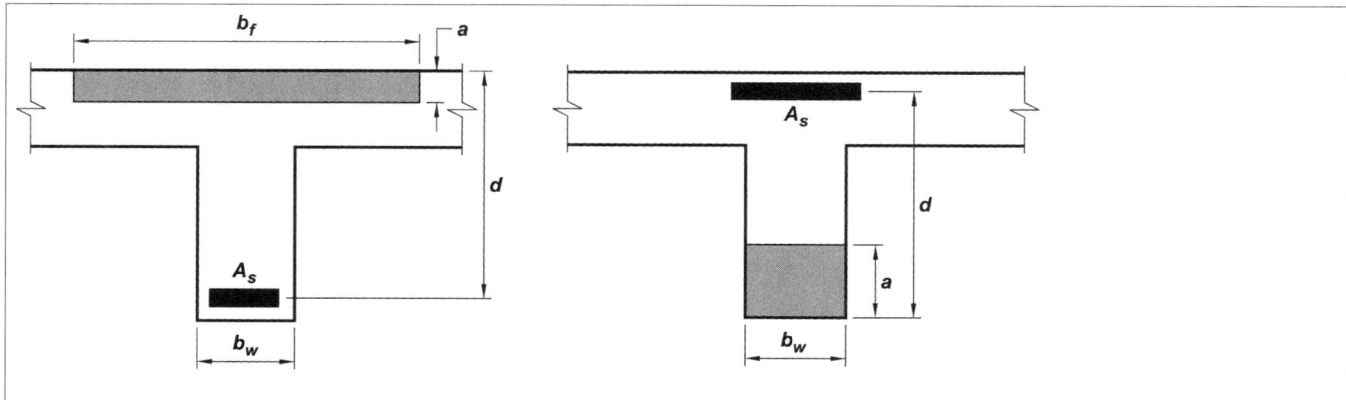

Fig. E1.3— Section compression block and reinforcement locations.

| | |
|---|---|
| | The beam is designed for the maximum flexural moments obtained from the approximate method above. |
| | The first interior support will be designed for the larger of the two moments. |
| 9.5.1.1 | The beam's design strength must be at least equal to the required strength at each section along its length:<br>$\phi M_n \geq M_u$<br>$\phi V_n \geq V_u$ |
| 9.5.2.1 | Beam is not subjected to axial force, therefore, assume $P_u < 0.1 f_c' A_g$ |
| 22.3 | Calculate the required reinforcement area (refer to Fig. E1.2 for design moment values and Fig. E1.4 for moment location).<br>$M_u \leq \phi M_n = \phi A_s f_y \left(d - \dfrac{a}{2}\right)$ |
| 21.2.1a | $\phi = 0.9$<br>A No. 7 bar has a $d_b = 0.875$ in. and an $A_s = 0.6$ in.$^2$<br>$a$ has been calculated above as a function of $A_s$ |
| 21.2.2<br>9.3.3.1 | Check if the calculated strain exceeds 0.005 in./in. (tension controlled (Fig. E1.5), but not less than 0.004 in./in.<br>$a = \dfrac{A_s f_y}{0.85 f_c' b}$ and $c = \dfrac{a}{\beta_1}$<br>where $\beta_1 = 0.8$ (calculated above)<br>Note: $b = 18$ in. for negative moments and 120 in. for positive moments.<br>$\varepsilon_t = \dfrac{\varepsilon_{cu}}{c}(d - c)$ |

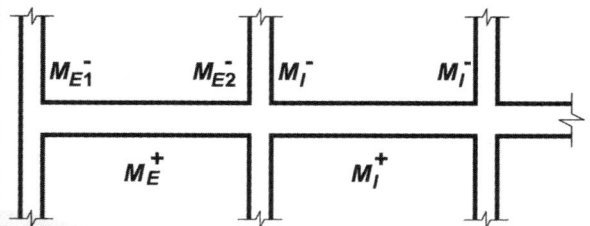

Fig. E1.4—Key to moment; use with table below.

### Table 1.1—Required reinforcement

| Location | $M_u$, ft-kip | $A_{s,req'd}$, in.$^2$ | Number of No. 7 bars | |
|---|---|---|---|---|
| | | | Req'd | Prov. |
| $M_{E1}^-$ | 267 | 2.23 | 3.7 | 4 |
| $M_{E2}^-$ | 428 | 3.65 | 6.1 | 7 |
| $M_I^-$ | 389 | 3.30 | 5.5 | 6 |
| $M_E^+$ | 306 | 2.49 | 4.1 | 5 |
| $M_I^+$ | 267 | 2.17 | 3.6 | 4 |

Note: The beam at the first interior support is designed for the larger of $M_{E2}^-$ and $M_I^-$ refer to Fig. E1.4.

### Table 1.2—Tension strain in reinforcement

| Location | $M_u$, ft-kip | $A_{s,prov}$, in.$^2$ | $a$, in. | $\varepsilon_s$, in./in. | $\varepsilon_s >$ 0.005? |
|---|---|---|---|---|---|
| $M_{E1}^-$ | 267 | 2.40 | 1.88 | 0.0313 | Y |
| $M_{E2}^-$ | 428 | 4.20 | 3.29 | 0.0179 | Y |
| $M_I^-$ | 389 | 3.60 | 2.82 | 0.0208 | Y |
| $M_E^+$ | 306 | 3.00 | 0.35 | 0.166 | Y |
| $M_I^+$ | 267 | 2.40 | 0.28 | 0.208 | Y |

Therefore, assumption of using $\phi = 0.9$ is correct.

# CHAPTER 7—BEAMS

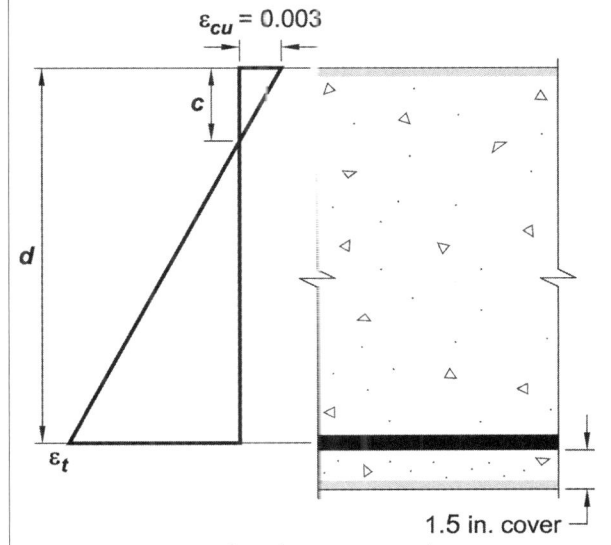

Fig. E1.5—Strain distribution across beam section.

| | | | |
|---|---|---|---|
| 9.6.1.1<br>9.6.1.2 | Minimum reinforcement<br>The provided reinforcement must be at least the minimum required reinforcement at every section along the length of the beam.<br><br>$A_s = \dfrac{3\sqrt{f_c'}}{f_y} b_w d$ (9.6.1.2a)<br><br>$A_s = \dfrac{200}{f_y} b_w d$ (9.6.1.2b)<br><br>Because $f_c' > 4444$ psi, Eq. (9.6.1.2a) controls. | | $A_s = \dfrac{3\sqrt{5000 \text{ psi}}}{60{,}000 \text{ psi}}(18 \text{ in.})(27.5 \text{ in.}) = 1.75 \text{ in.}^2$ **Controls**<br><br>$A_s = \dfrac{200}{60{,}000 \text{ psi}}(18 \text{ in.})(27.5 \text{ in.}) = 1.65 \text{ in.}^2$<br><br>Required reinforcement areas exceed the minimum required reinforcement area at all locations. |

*External spans*

## Step 6: Shear design

| | | |
|---|---|---|
| 9.5.1.1<br>9.5.3.1<br>22.5.1.1 | Shear strength<br>$\phi V_n \geq V_u$<br>$V_n = V_c + V_s$ | |
| 9.4.3.2 | Because conditions (a), (b), and (c) of Section 9.4.3.2 are satisfied, the design shear force is taken at critical section at distance $d$ from the face of the support (Fig. E1.6). | Fig. E1.6—Shear at the critical section.<br><br>$V_{u@d} = (72 \text{ kip}) - (3.7 \text{ kip/ft})(27.5 \text{ in.}/12) = 63.5 \text{ kip}$ |
| 22.5.5.1 | $V_c = 2\sqrt{f_c'} b_w d$ | $V_c = 2\sqrt{5000 \text{ psi}}(18 \text{ in.})(27.5 \text{ in.})/1000 = 70 \text{ kip}$ |
| 21.2.1(b) | Shear strength reduction factor: | $\phi_{shear} = 0.75$<br>$\phi V_c = (0.75)(70 \text{ kip}) = 52.5 \text{ kip}$ |
| 9.5.1.1 | Check if $\phi V_c \geq V_u$ | $\phi V_c = 52.5 \text{ kip} < V_{u@d} = 63.5 \text{ kip}$ **NG**<br>Therefore, shear reinforcement is required. |

| | | |
|---|---|---|
| 22.5.1.2<br><br>21.2.1(b) | Prior to calculating shear reinforcement, check if the cross-sectional dimensions satisfy Eq. (22.5.1.2):<br><br>$V_u \leq \phi(V_c + 8\sqrt{f'_c}b_w d)$<br><br>$\phi = 0.75$<br><br>$V_u = 63.5$ kip | $V_u \leq \phi\left(70 \text{ kip} + \dfrac{8\sqrt{5000 \text{ psi}}(18 \text{ in.})(27.5 \text{ in.})}{1000 \text{ lb/kip}}\right)$<br><br>63.5 kip < 263 kip<br>**OK**, therefore, section dimensions are satisfactory. |
| 22.5.10.1<br><br>22.5.10.5.3<br>22.5.10.5.5<br><br><br><br><br><br><br><br><br><br>9.7.6.2.2<br><br><br>9.6.3.1 | **Shear reinforcement**<br>Transverse reinforcement satisfying Eq. (22.5.10.1) is required at each section where $V_u > \phi V_c$<br><br>$V_s \geq \dfrac{V_u}{\phi} - V_c$<br><br>where $V_s = \dfrac{A_v f_{yt} d}{s}$<br><br>Calculate maximum allowable stirrup spacing:<br><br><br>First, does the beam transverse reinforcement value need to exceed the threshold value of $V_s \leq 4\sqrt{f'_c}b_w d$ ?<br><br><br><br>Because the required shear strength is below the threshold value, the maximum stirrup spacing is the lesser of $d/2$ and 24 in.<br><br>In the region where $V_u \leq \phi V_c/2$, shear reinforcement is not required. In this example, shear reinforcement, however, is provided over the full length. | $\phi V_s \geq (63.5 \text{ kip}) - 52.5 \text{ kip} = 11.0$ kip<br><br>Assume a No. 3 bar, two legged stirrup<br><br>$\dfrac{11.0 \text{ kip}}{\phi} = \dfrac{2(0.11 \text{ in.}^2)(60,000 \text{ psi})(27.5 \text{ in.})}{s}$<br><br>$s = 24.8$ in. This is a very large spacing and must be checked against the maximum allowed.<br><br>$4\sqrt{f'_c}b_w d = 4\left(\sqrt{5000 \text{ psi}}\right)(18 \text{ in.})(27.5 \text{ in.}) = 140.0$ kip<br><br>$V_s = 14.7 \text{ kip} < 4\sqrt{f'_c}b_w d = 140.0 \text{ kip}$ **OK**<br><br><br>$d/2 = 27.5 \text{ in.}/2 = 13.8$ in.<br>Use $s = 12$ in. $< d/2 = 13.8$ in. ∴ **OK** |
| 9.6.3.3 | Specified shear reinforcement must be at least:<br><br>$\dfrac{A_{v,min}}{s} = 0.75\sqrt{f'_c}\dfrac{b_w}{f_{yt}}$<br><br>and<br><br>$\dfrac{A_{v,min}}{s} = 50\dfrac{b_w}{f_{yt}}$ | $\dfrac{A_{v,min}}{s} \geq 0.75\sqrt{5000 \text{ psi}}\dfrac{18 \text{ in.}}{60,000 \text{ psi}} = 0.016 \text{ in.}^2/\text{in.}$<br><br>**Controls**<br><br>$\dfrac{A_{v,min}}{s} \geq 50\dfrac{18 \text{ in.}}{60,000 \text{ psi}} = 0.015 \text{ in.}^2/\text{in.}$<br><br>Provided, No. 3 at 12 in. spacing:<br><br>$\dfrac{A_v}{s} \geq \dfrac{2(0.11 \text{ in.}^2)}{12 \text{ in.}} = 0.018 \text{ in.}^2/\text{in.} > \dfrac{A_{v,min}}{s} = 0.0186 \text{ in.}^2/\text{in.}$<br><br>Spacing satisfies Section 9.6.3.3, therefore, **OK** |

| | Step 7: Reinforcement detailing | |
|---|---|---|
| 9.7.2.1<br>25.2.1 | Minimum bar spacing<br>The clear spacing between the horizontal No. 7 bars must be at least the greatest of:<br><br>Clear spacing greater of $\begin{cases} 1 \text{ in.} \\ d_b \\ 4/3(d_{agg}) \end{cases}$<br><br>Assume maximum aggregate size is 0.75 in. | 1 in.<br>0.875 in.<br>4/3(3/4 in.) = 1 in.<br><br>Therefore, clear spacing between horizontal bars must be at least 1.0 in. |
| 9.7.2.2<br>24.3.4 | Tension reinforcement in flanges must be distributed within the effective flange width, $b_f$ = 120 in. (Step 2), but not wider than: $\ell_n/10$.<br><br>Because effective flange width exceeds $\ell_n/10$, additional bonded reinforcement is required in the outer portion of the flange.<br><br>Use No. 5 for additional bonded reinforcement.<br><br>This requirement is to control cracking in the slab due to wide spacing of bars across the full effective flange width and to protect flange if reinforcement is concentrated within the web width.<br><br>For the first interior support, place tension reinforcement per the higher design moment. For moment locations refer to Fig. E1.4. | $\ell_n/10$ = (34 ft)(12)/10 = 40.8 in. < 120 in., say, 41 in.<br><br>**Table 1.3—Top flange bar distribution**<br><br>| Location | Prov. No. 7 | No. 7 in web | No. 7 in $\ell_n/10^*$ | No. 5 in outer portion$^*$ |<br>|---|---|---|---|---|<br>| $M_{E1}$ | 4 | 4 | — | 5 |<br>| $M_{E2}$ | 7 | 5 | 1 | 3 |<br>| $M_I$ | 6 | 4 | 1 | 3 |<br><br>$^*$Bars on both sides of the web (Refer to Fig. E1.12—Sections). |
| | Exterior span positive moment reinforcement.<br><br>Check if five No. 7 bars (resisting positive moment) can be placed in the beam's web per Reinforced Concrete Design Handbook Design Aid – Analysis Tables, which can be downloaded from: https://www.concrete.org/store/productdetail.aspx?ItemID=SP1714DA. The spacing is calculated below as a demonstration.<br><br>$b_{w,req'd}$ = 2(cover + $d_{stirrup}$ + 0.75 in.) + 4$d_b$<br>　　　　+ 4(1 in.)$_{min,spacing}$　　　(25.2.1)<br><br>where<br>$d_{stirrup}$ = 0.375 in.<br>and $d_b$ = 0.875 in.<br><br>Spacing between longitudinal bars:<br>2.1 in. > 1 in.　**OK** | $b_{w,req'd}$ = 2(1.5 in. + 0.375 in. + 0.75 in.) + 3.5 in. + 4 in.<br>　　　　= 12.75 in. < 18 in.　**OK**<br>Therefore, five No. 7 bars can be placed in one layer in the 18 in. beam web (Fig. E1.7).<br><br>*Fig. E1.7—Bottom reinforcement layout.* |

| | | |
|---|---|---|
| 9.7.2.2<br>24.3.1<br><br>24.3.2 | Maximum bar spacing at the tension face must not exceed the lesser of:<br><br>$$s = 15\left(\frac{40,000 \text{ psi}}{f_s}\right) - 2.5c_c$$<br><br>and<br><br>$$s = 12\left(\frac{40,000 \text{ psi}}{f_s}\right)$$<br><br>where $f_s = 2/3 f_y = 40,000$ psi | $s = 15\left(\dfrac{40,000 \text{ psi}}{40,000 \text{ psi}}\right) - 2.5(2 \text{ in.}) = 10 \text{ in.}$  **Controls**<br><br><br><br>$s = 12\left(\dfrac{40,000 \text{ psi}}{40,000 \text{ psi}}\right) = 12 \text{ in.}$<br><br>Top reinforcement: 10 in. |
| 24.3.2.1 | This limit is intended to control flexural cracking width. Note that $c_c$ is the cover to the No.7 bar, not to the tie. | Bottom reinforcement:<br>If bars are not bundled, 2.3 in. spacing is provided (Fig. E1.7), therefore **OK** |

Bottom bar length along first span
Calculate the inflection points (Fig. E1.8):

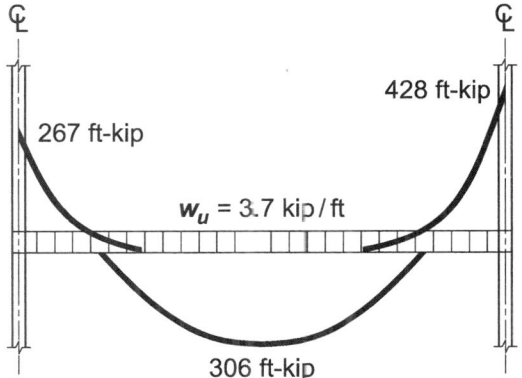

Fig. E1.8—Moment diagram of exterior span.

Inflection point for bottom tension—first span
Assume the maximum positive moment occurs at midspan. From equilibrium, the point of inflection is obtained from the following free body diagram (Fig. E1.9a):

$M_{max} - w_u(x)^2/2 = 0$

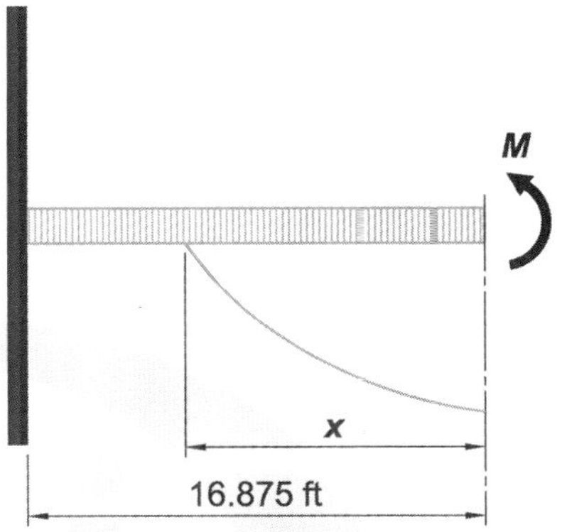

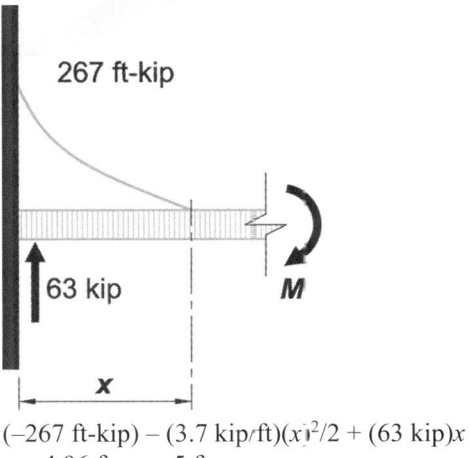

$(306 \text{ ft-kip}) - (3.7 \text{ kip/ft})(x)^2/2 = 0$
$x = 12.86$ ft, say, 13 ft

Fig. E1.9a—Inflection point of maximum positive moment.

Inflection point for top tension—first span
Exterior support:
Calculate the inflection point for negative moment diagram (Fig. E1.9b):

$-M_{max} - w_u(x)^2/2 - V_u x = 0$

$(-267 \text{ ft-kip}) - (3.7 \text{ kip/ft})(x)^2/2 + (63 \text{ kip})x = 0$
$x = 4.96$ ft, say, 5 ft

Fig. E1.9b—Inflection point of exterior negative moment.

| | | | |
|---|---|---|---|
| | | Inflection point for top tension – First interior support<br>Calculate inflection point for the negative moment diagram (Fig. E1.9c):<br><br>$-M_{max} - w_u(x)^2/2 + V_u x = 0$ | (image showing 428 ft-kip moment, 72 kip reaction, distance x)<br><br>$(-428 \text{ ft-kip}) - 3.7 \text{ kip/ft}(x)^2/2 + (72 \text{ kip})x = 0$<br>$x = 7.32$ ft, say, 7 ft 6 in.<br><br>Fig. E1.9c—Inflection point of interior negative moment. |
| | 9.7.1.2<br><br>25.4.2.2<br><br>25.4.2.4<br><br><br><br>25.4.10.1 | Development length of No.7 bar<br>The simplified method is used to calculate the development length of No.7 bars:<br><br>$\ell_d = \dfrac{f_y \psi_t \psi_e}{20\lambda \sqrt{f_c'}} d_b$<br><br>where<br>$\psi_t$ = bar location; $\psi_t = 1.3$ for top bars, because more than 12 in. of fresh concrete is placed below them and $\psi_t = 1.0$ for bottom bars, because not more than 12 in. of fresh concrete is placed below them.<br>$\psi_e$ = coating factor; $\psi_e = 1.0$, because bars are uncoated<br><br>The calculated development lengths could be reduced according to the ratio of:<br>$A_{req'd}/A_{prov.}$ except as required by Section 25.4.10.2. In this example, development reduction is not applied. | Top bars:<br><br>$\ell_d = \left( \dfrac{(60{,}000 \text{ psi})(1.3)(1.0)}{(20)(1.0)\sqrt{5000 \text{ psi}}} \right)(0.875 \text{ in.}) = 48.3$ in.<br><br>Say, 51 in. = 4 ft 3 in.<br><br>Bottom bars:<br><br>$\ell_d = \left( \dfrac{(60{,}000 \text{ psi})(1.0)(1.0)}{(20)(1.0)\sqrt{5000 \text{ psi}}} \right)(0.875 \text{ in.}) = 37.1$ in.<br><br>Say, 39 in. = 3 ft 3 in. |

| | | |
|---|---|---|
| | First span top bars | |
| 9.7.3.2 | Exterior support<br>Bars must be developed at locations of maximum stress and locations along the span where bent or terminated tension bars are no longer required to resist flexure.<br><br>Four No. 7 bars are required to resist the factored negative moment at the exterior column interior face.<br><br>Calculate a distance $x$ from the column face where two No. 7 bars can resist the factored moment. | $-(267 \text{ ft-kip}) - (3.7 \text{ kip/ft})\dfrac{x^2}{2} + (63 \text{ kip})(x)$<br>$= -2(0.5 \text{ in.}^2)(0.9)(60 \text{ ksi})$<br>$\times \left(27.6 \text{ in} - \dfrac{2(0.6 \text{ in.}^2)(60 \text{ ksi})}{2(0.85)(5 \text{ ksi})(18 \text{ in.})}\right)\left(\dfrac{1}{12}\right)$<br><br>$x = 2.05$ ft, say, 24 in.<br><br>At 2 ft 0 in. from the column face, two No. 7 can be cutoff. |
| 9.7.3.3 | Bars must extend beyond the location where they are no longer required to resist flexure for a distance equal to the greater of $d$ or $12d_b$. | For No.7 bars:<br>1. $d = 27.5$ in.  **Controls**<br>2. $12d_b = 12(0.875 \text{ in.}) = 10.5$ in.<br><br>Therefore, extend the middle two No. 7 bars the greater of the development length (51 in.) and the sum of theoretical cutoff point and $d$ from column face (refer to Fig. E1.10): |
| 9.7.3.8.4 | At least one-third of the bars resisting negative moment at a support (two No. 7 > 1/3 of four No. 7) must have an embedment length beyond the inflection point the greatest of $d$, $12d_b$, and $\ell_n/16$. | 24 in. + 27.5 in. = 51.5 in., say, 54 in.  **Controls**<br><br>For No.7 bars:<br>$d = 27.6$ in.  **Controls**<br>$12d_b = 12(0.875 \text{ in.}) = 10.5$ in.<br>$\ell_n/16 = (36 \text{ ft} - 2 \text{ ft})/16 = 2.1$ ft = 26 in.<br><br>Extend the remainder outside two No.7 bars, the greater of the development length (51 in.) beyond the theoretical cutoff point (2 ft 0 in.)<br>51 in. + 24 in. = 75 in. and $d = 27.5$ in. beyond the inflection point (5 ft 0 in.).<br><br>60 in. + 27.5 in. = 87.5 in. > 75 in.  **Controls**<br><br>Therefore, extend bars minimum 87.5 in., increase to, say, 90 in. = 7 ft 6 in. from column face. Refer to Fig. E1.10. |
| | Note: These calculations are performed to present the Code requirements. In practice, the engineer may terminate the two interior No. 7 bars at a distance 4 ft 6 in. from column face and extend the two exterior No. 7 bars the full span length of the beam to support the shear reinforcement stirrups as shown in Fig. E1.11. | |

| | | |
|---|---|---|
| 9.7.3.2<br>9.7.3.3 | First span top bars<br><br>Interior support<br>Following the same steps above, seven No. 7 bars are required to resist the factored moment at the first interior column face.<br><br>Calculate a distance $x$ from the column face where four No. 7 bars can be terminated. | $-(428 \text{ ft} - \text{kip}) - (3.7 \text{ kip/ft})\dfrac{x^2}{2} + (72 \text{ kip})(x) =$<br><br>$\dfrac{-3(0.6 \text{ in.}^2)(0.9)(60 \text{ ksi})}{12}\left(27.5 \text{ in.} - \dfrac{3(0.6 \text{ in.}^2)(60 \text{ ksi})}{2(0.85)(5 \text{ ksi})(18 \text{ in.})}\right)$<br><br>$x = 3.19$ ft, say, 39 in.<br><br>Therefore, extend four No. 7 bars the greater of the development length (51 in.) from column face and $d$ from theoretical cutoff point (39 in.)<br><br>39 in. + 27.5 in. = 66.5 in.; increase to 69 in. (5 ft 9 in.)<br>69 in. > 51 in., therefore, extend four No.7 bars 69 in. |
| 9.7.3.8.4 | $d = 27.5$ in. $> l_n/16 = 25.5$ in. $> 12d_b = 10.5$ in. | Extend the remaining three No. 7 bars the larger of the development length (51 in.) beyond the theoretical cutoff point (38 in.) and $d = 27.5$ in. beyond the inflection point (7 ft 6 in. = 90 in.). The latter controls (Fig. E1.10).<br><br>90 in. + 27.5 in. = 117.5 in. > 39 in. + 51 in. = 90 in.<br>**OK** |
| | Note: These calculations are performed to present the code requirements. In practice, the engineer may terminate the four interior No. 7 bars at a distance $d$ (27.5 in.) beyond the development length from column for a length of (39 in. + 27.5 in. = 66.5 in., say, 5 ft 9 in.). Terminate one No.7 at 10 ft 0 in. from the support and extend the remaining two exterior No.7 bars the full span length of the beam to support the shear reinforcement stirrups as shown in Fig. E1.11. | |
| 9.7.3.2<br>9.7.3.3 | First span bottom bars<br>Following the same steps above, five No. 7 bars are required to resist the factored moment at the midspan of the exterior span.<br><br>Calculate a distance $x$ from the midspan where two No. 7 bars can resist the factored moment. | $(306 \text{ ft-kip}) - (3.7 \text{ kip/ft})\dfrac{x^2}{2} = 2(0.6 \text{ in.}^2)(0.9)(60 \text{ ksi})$<br><br>$\times \left(27.5 \text{ in.} - \dfrac{2(0.6 \text{ in.}^2)(60 \text{ ksi})}{2(0.85)(5 \text{ ksi})(120 \text{ in.})}\right)\left(\dfrac{1}{12}\right)$<br><br>$x = 10.0$ ft<br>Therefore, extend three No. 7 bars the larger of the development length (39 in.) and a distance $d$ beyond the theoretical cutoff point (10 ft = 120 in.) from maximum moment at midspan<br>120 in. + 27.5 in. = 147.5 in., say, 12 ft 6 in. from maximum positive moment at midspan (Fig. E1.10).<br><br>Extend the remaining two No. 7 bars at least the longer of 6 in. into the column or $\ell_d = 39$ in. past the theoretical cutoff point (Fig. E1.10). |

# CHAPTER 7—BEAMS

| | | |
|---|---|---|
| 9.7.3.8.2 | At least one-fourth of the positive tension bars must extend into the column at least 6 in. | Two No. 7 bars out of total five No. 7 will be extended into the column:<br><br>Two No. 7 bars > 1/4 (five No. 7 bars) **OK** |
| 9.7.3.8.3 | At the point of inflection, $d_b$ for positive moment tension bars must be limited such that $\ell_d$ for that bar size satisfies:<br><br>$$\ell_d \leq \frac{M_n}{V_u} + \ell_a$$<br><br>where $M_n$ is calculated assuming all bars at the section are stressed to $f_y$. $V_u$ is calculated at the section. At support, $\ell_a$ is the embedment length beyond the center of the column. The term $\ell_a$ is the embedment length beyond the point of inflection, limited to the greater of $d$ and $12d_b$. | Point of inflection occurs at 4 ft from the column face (Fig. E1.9a).<br><br>$V_u = 63.5$ kip $-$ (3.7 kip/ft)(4 ft) $= 48.2$ kip<br><br>At that location, assume two No. 7 bars are effective:<br><br>$M_n = 2(0.6 \text{ in.}^2)(60 \text{ ksi})$<br>$\times \left( 27.5 \text{ in.} - \frac{2(0.6 \text{ in.}^2)(60 \text{ ksi})}{(2)(0.85)(5 \text{ ksi})(120 \text{ in.})} \right)$ |
| 9.7.3.5 | If bars are cutoff in regions of flexural tension, then a bar stress discontinuity occurs. Therefore, the code requires that flexural tensile bars must not be terminated in a tensile zone unless (a), (b), or (c) is satisfied.<br><br>(a) $V_u \leq (2/3)\phi V_n$ at the cutoff point<br><br>(b) Continuing bars provides double the area required for flexure at the cutoff point and the area required for flexure at the cutoff point and $V_u \leq (3/4)\phi V_n$.<br><br>(c) Stirrup or hoop area in excess of that required for shear and torsion is provided along each terminated bar or wire over a distance $3/4d$ from the termination point. Excess stirrup or hoop area shall be at least $60b_w s/f_{yt}$. Spacing $s$ shall not exceed $d/(8\beta_b)$. | $M_n = 1982$ in.-kip<br><br>$\ell_d \leq \frac{1982 \text{ in.-kip}}{48.2 \text{ kip}} + 27.5 \text{ in.} = 68.6 \text{ in., say, 69 in.}$<br><br>This length exceeds $\ell_d = 39$ in., therefore **OK**<br><br>(a) At 10 ft and $\ell_n/2 = 17$ ft<br>$V_u = 63$ kip $-$ (3.7 kip/ft)(17 ft $-$ 10 ft) $= 37.1$ kip<br><br>$\phi V_n = \phi(V_c + V_s)$; $\phi V_c$ is calculated in Step 6<br><br>$\phi V_n = 0.75\left( 70 \text{ kip} + \frac{2(0.11 \text{ in.}^2)(60 \text{ ksi})(27.5 \text{ in.})}{12 \text{ in.}} \right)$<br><br>$\phi V_n = 75.2$ kip<br>$2/3\phi V_n = 2/3(75.2 \text{ kip}) = 50$ kip<br>50 kip > 37.1 kip, therefore, **OK**.<br><br>Because only one of the three conditions needs to be satisfied, the other two are not checked. |
| **Step 8: Integrity reinforcement** | | |
| 9.7.7.2 | One of the two conditions in 9.7.7.2 must be satisfied: | |
| | At least one-quarter the maximum positive moment bars, but at least two bars, must be continuous. | This condition was satisfied above by extending two No. 7 bars into the support. Also, two bars are more than 1/4 of the provided reinforcement. |
| | Beam longitudinal bars must be enclosed by closed stirrups along the clear span. | Open stirrups are provided, therefore, the second condition will not be satisfied. |
| 9.7.7.3 | Beam structural integrity bars shall pass through the region bounded by the longitudinal column bars. | At least two No. 7 bars are extended through the column longitudinal reinforcement. Therefore, satisfying this condition. |

| | | |
|---|---|---|
| 9.7.7.5 | Splices are necessary for continuous bars. The bars shall be spliced in accordance with (a) and (b):<br><br>(a) Bottom bars (positive moment) shall be spliced at or near the support<br><br>(b) Top bars (negative moment) shall be spliced at or near midspan | splice length = (1.3)(development length)<br><br>$\ell_{st} = 1.3(39$ in.$) = 50.7$ in., say, 4 ft 3 in.<br><br>$\ell_{st} = 1.3(51$ in.$) = 66.3$ in., say, 5 ft 9 in. |

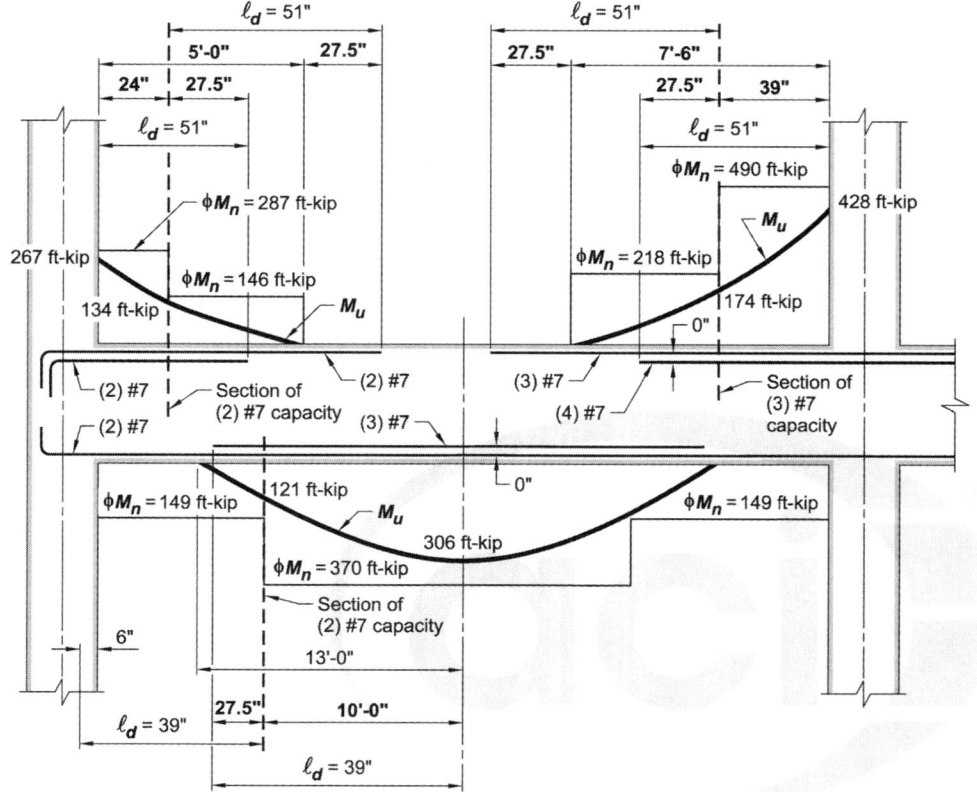

Fig. E1.10—End span bar cutoff locations.

Note: Numbers shown in bold control the bar lengths.

### Step 9: Internal spans

| | | |
|---|---|---|
| | Flexural bars were calculated above in Step 5.<br><br>Six No. 7 top bars are required at supports<br><br>Four No. 7 bottom bars are required at midspan | |
| 9.7.6.2.2 | Stirrup size and spacing were calculated following Step 6: No. 3 at 12 in. are not required over the full length of the beam, it is; however, good practice to maintain stirrups at 12 in. on center. | |

## Step 10: Detailing

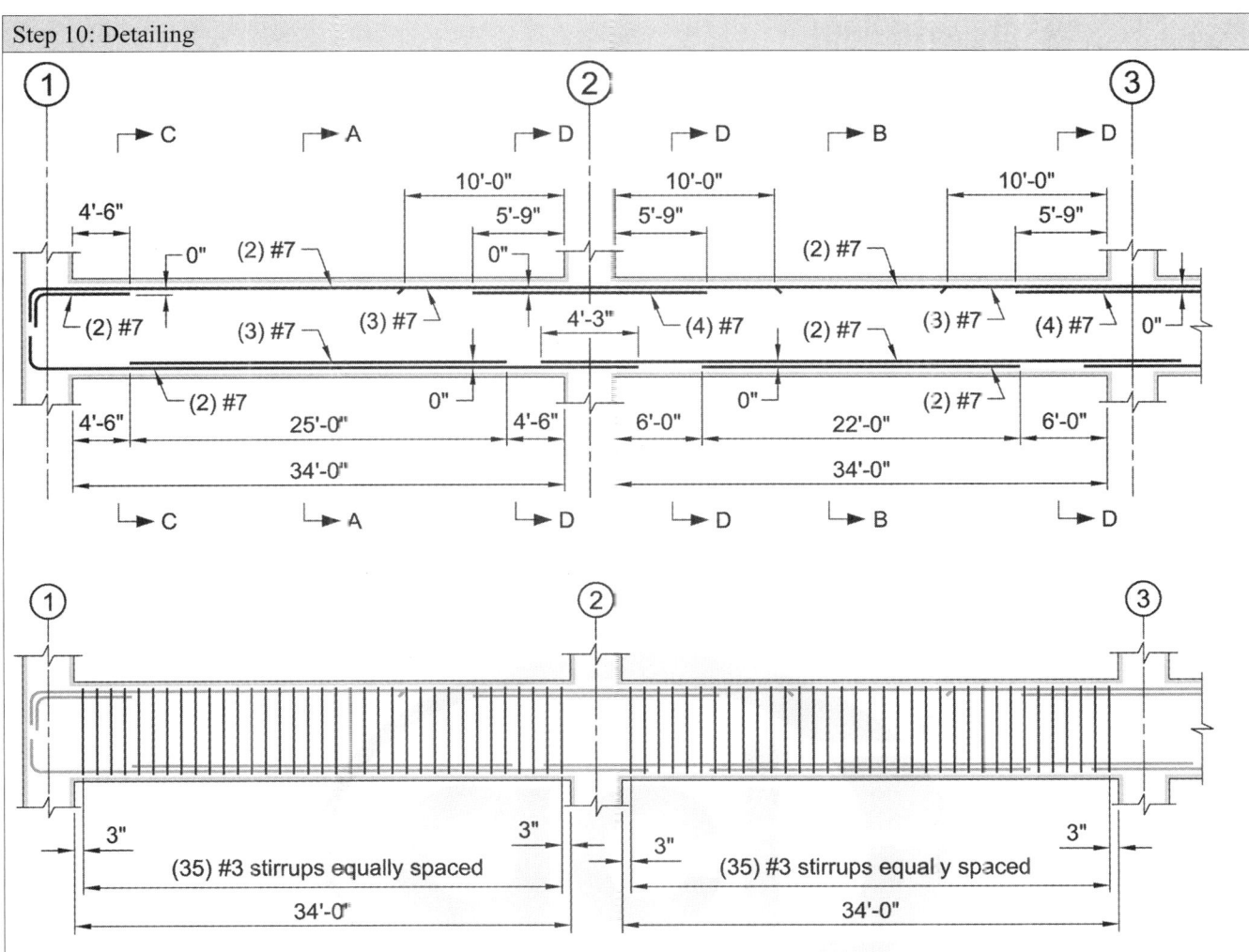

Fig. E1.11—Beam bar details.

Notes:
1. Place first stirrup at 3 in. from the column face.
2. The contractor may prefer to extend two No. 7 top reinforcement over the full beam length rather than adding two No. 5 hanger bars. Bars should be spliced at mid-length.

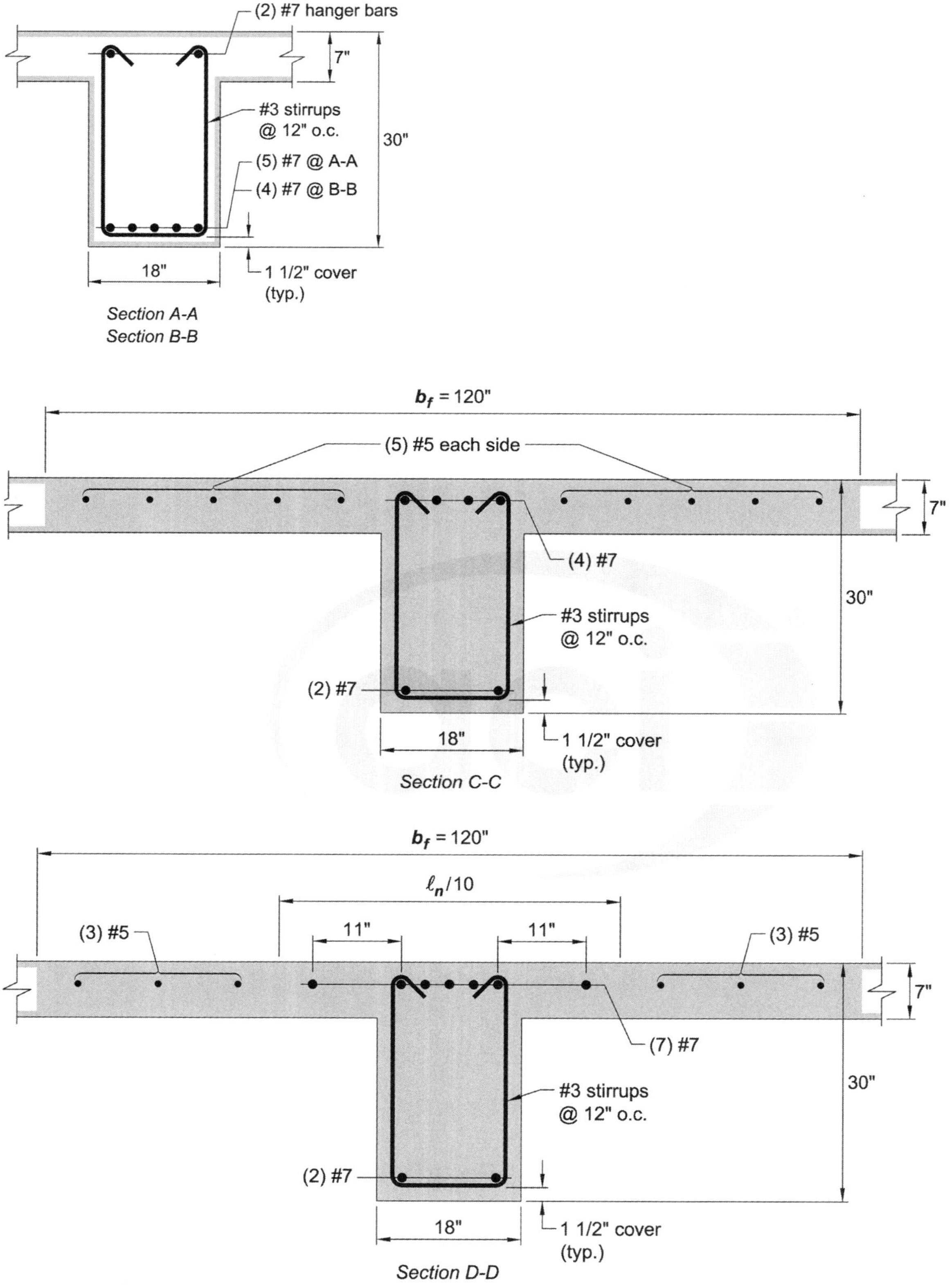

Fig. E1.12—Sections.

Note: Refer to Step 7 Table 1.3 for flange negative moment reinforcement placement.

**Beam Example 2**: *Single interior beam*

Design and detail a one-span Beam B1 built integrally with a 7 in. slab of a seven-story building. The beam frames into girder Beam B2 at each end as shown in Fig. E2.1.

Given:

*Load—*
Service dead load $D$ = 15 psf
Service live load $L$ = 100 psf

*Material properties—*
$f_c'$ = 5000 psi (normalweight concrete)
$f_y$ = 60,000 psi

Span length: 36 ft
Beam width: 18 in.

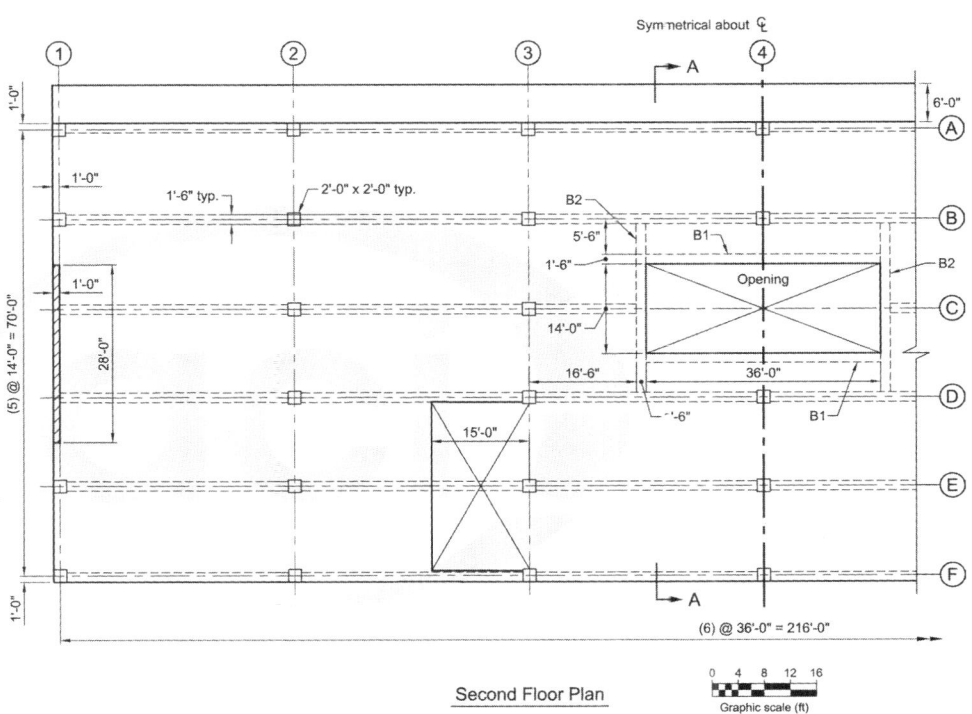

*Fig. E2.1—Plan of one-span interior beam.*

| ACI 318-14 | Discussion | Calculation |
|---|---|---|
| **Step 1: Material requirements** | | |
| 9.2.1.1 | The mixture proportion must satisfy the durability requirements of Chapter 19 and structural strength requirements of ACI 318-14. The designer determines the durability classes. Please refer to Chapter 4 of this Handbook for an in-depth discussion of the categories and classes.<br><br>ACI 301 is a reference specification that is coordinated with ACI 318. ACI encourages referencing ACI 301 into job specifications.<br>There are several mixture options within ACI 301, such as admixtures and pozzolans, which the designer can require, permit, or review if suggested by the contractor. | By specifying that the concrete mixture shall be in accordance with ACI 301-10 and providing the exposure classes, Chapter 19 (ACI 318-14) requirements are satisfied.<br><br>Based on durability and strength requirements, and experience with local mixtures, the compressive strength of concrete is specified at 28 days to be at least 5000 psi. |
| **Step 2: Beam geometry** | | |
| 9.3.1.1 | Beam depth<br>ACI 318-14 permits a beam whose size satisfies Table 9.3.1.1 to be designed without having to check the beam deflection, if the beam is not supporting or attached to partitions or other construction likely to be damaged by large deflections. Otherwise, deflections must be calculated and the deflection limits in Section 9.3.2 must be satisfied.<br><br>The one-span beam is built integrally with the slab and the girders that it frames into. | For a simple supported beam the recommended depth from Table 9.3.1.1:<br>$$h = \frac{\ell}{16} = \frac{(36 \text{ ft})(12 \text{ in./ft})}{16} = 27 \text{ in.}$$<br>Use 28 in. |
| | Self-weight<br>Beam: $b_w = 18$ in.<br>Slab: $t = 7$ in. thick<br><br>Tributary width = 4.75 ft/2 = 2.375 ft (Fig. E2.1) | $w_b = (18 \text{ in.}/12)(28 \text{ in.}/12)(0.150 \text{ kip/ft}^3) = 0.53$ kip/ft<br>$w_s = (2.375 \text{ ft})(7 \text{ in.}/12)(0.150 \text{ kip/ft}^3) = 0.21$ kip/ft |
| 9.2.4.2 | Flange width<br>The beam is poured monolithically with the slab on one side and will behave as an L-beam. The effective flange width on one side of the beam is obtained from Table 6.3.2.1. | |
| 6.3.2.1 | One side of web is the least of: $6h_{slab}$, $s_w/2$, $\ell_n/12$<br><br>Flange width: $b_f = \ell_n/12 + b_w$ | (6)(7 in.) = 42 in.<br>(4.75 ft)(12)/2 = 28.5 in. **Controls**<br>(36 ft(12))/12 = 36 in.<br><br>$b_f = 28.5$ in. + 18 in. = 46.5 in. |

## CHAPTER 7—BEAMS

| | Step 3: Loads and load patterns | |
|---|---|---|
| 5.3.1 | The service live load for public assembly is 100 psf per Table 4-1 in ASCE 7-10. To account for the weight of ceilings, partitions, HVAC systems, etc., add 15 psf as miscellaneous service dead load.<br><br>The beam resists gravity load only and lateral forces are not considered in this example.<br><br>$U = 1.4D$<br><br>$U = 1.2D + 1.6L$ | The superimposed dead load is applied over a tributary width of 4.75 ft/2 + 1.5 ft width of B1 = 3.875 ft (refer to Fig. E2.1).<br><br>$U = 1.4(0.53 \text{ kip/ft} + 0.21 \text{ kip/ft} + (15 \text{ psf})(3.875 \text{ ft})/1000)$<br>    $= 1.12 \text{ kip/ft}$<br><br>$U = 1.2(1.12 \text{ kip/ft})/1.4 + 1.6((100 \text{ psf})(3.875 \text{ ft})/1000)$<br>    $= 1.58 \text{ kip/ft}$ **Controls** |
| | Note: Live load is not reduced per ASCE7-10 in this example. | |
| Step 4: Analysis | | |
| 9.4.3.1 | The beam is built integrally with supports; therefore, the factored moments and shear forces (required strengths) are calculated at the face of the supports. | |
| 9.4.1.2 | Chapter 6 permits several analysis procedures to calculate the required strengths.<br>For this example, assume an elastic analysis results in the beam moment at the face of each support: $w_u\ell^2/16$.<br><br>The total moment is $w_u\ell^2/8$, so the midspan moment is $w_u\ell^2/16$.<br><br>This distribution assumes the girder remains uncracked. If the girder does crack, its stiffness is greatly reduced, which results in a higher moment at midspan. To be conservative, this example assumes the total beam moment, $w_u\ell^2/8$, is resisted by the positive moment reinforcement and the supports resist $w_u\ell^2/16$. | At the face of girders:<br>$M_u = w_u\ell^2/16 = (1.58 \text{ kip/ft})(36 \text{ ft})^2/16 = 128 \text{ ft-kip}$<br><br><br><br><br><br><br><br>At the midspan:<br>$M_u = w_u\ell^2/8 = (1.58 \text{ kip/ft})(36 \text{ ft})^2/8 = 256 \text{ ft-kip}$ |
| | Beam B1 frames into girders on both ends. Because girders are not as rigid as columns or walls and girders supporting concentrated loads may tend to rotate, the end supports may be considered less than fixed end supports. For a single span beam with fixed end supports, the negative moment at the support would be $(1/12)w_u\ell^2$. For this case, the moment achieved would have to be transferred to the girder efficiently. In real terms, the girder may tend to slightly rotate or may endure cracking which would reduce the rigidity and the fixity of the beam to girder joint. To account for this, assume a moment of $(1/16)w_u\ell^2$ to account for a lower capacity to transfer moment at this connection. Furthermore, considering less than fixed end supports, the positive moment at the mid-span approaches the moment of a simple support beam. So conservatively, use a positive midspan moment of $(1/8)w_u\ell^2$. | |

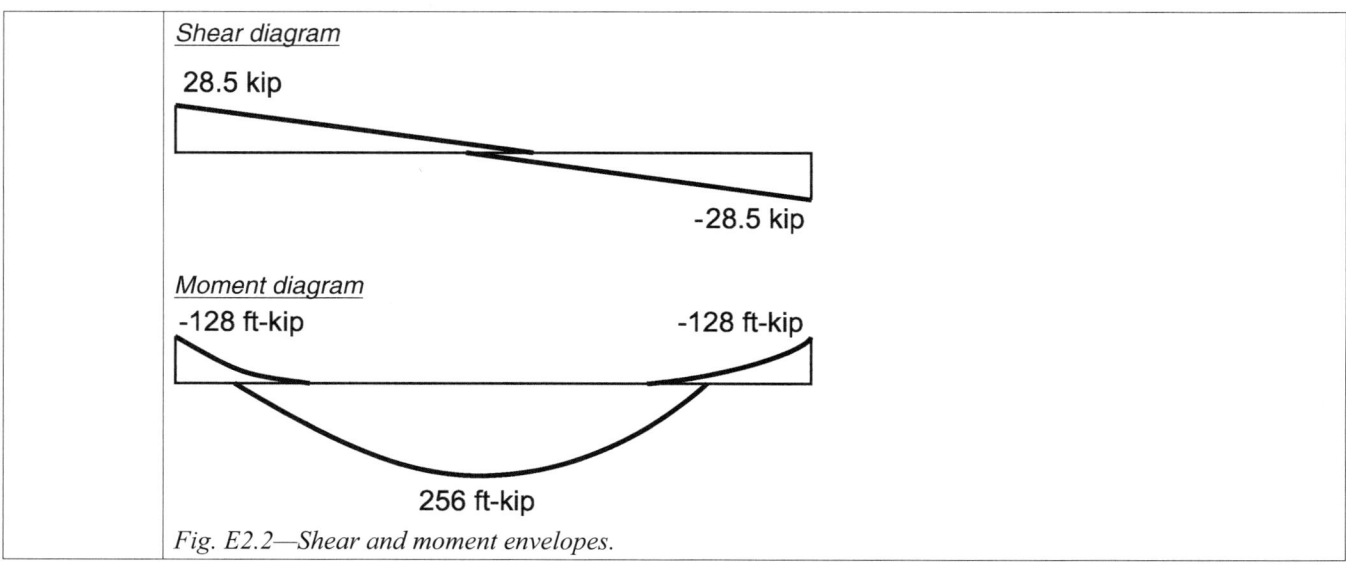

Fig. E2.2—Shear and moment envelopes.

## Step 5: Moment design

| | | |
|---|---|---|
| 9.3.3.1 | The code requires a beam to be designed with steel strain at design strength of at least 0.004 in./in. The intent is to ensure ductile behavior at the nominal strength condition. | |
| | For reinforced beams, such as this example, reinforcing bar strain is usually not a controlling issue. | |
| 21.2.1(a) | Assume the beam will be tensioned controlled with a moment strength reduction factor of $\phi = 0.9$. This assumption will be checked later. | |
| 20.6.1.3.1 | Calculate effective depth assuming No. 3 stirrups, No. 6 longitudinal bars, and 1.5 in. cover: | |
| | The effective depth of one row of longitudinal reinforcement is | |
| | $d = h - \text{cover} - d_{t,stirrup} - d_{b,long}/2$ | $d = 28$ in. $- 1.5$ in. $- 0.375$ in. $- 0.75$ in./2 $= 25.75$ in., say, $d = 25.7$ in. |
| 22.2.2.1 | The concrete compressive strain at which nominal moments are calculated is: $\varepsilon_{cu} = 0.003$ | |
| 22.2.2.2 | The tensile strength of concrete in flexure is a variable property and its value is approximately 10 to 15 percent of the concrete compressive strength. For calculating nominal strength, ACI 318 neglects the concrete tensile strength. Determine the equivalent concrete compressive stress for design: | |
| 22.2.2.3 | The concrete compressive stress distribution is inelastic at high stress. The Code permits any stress distribution to be assumed in design if shown to result in predictions of nominal strength in reasonable agreement with the results of comprehensive tests. Rather than tests, the Code allows the use of an equivalent rectangular compressive stress distri- | |
| 22.2.2.4.1 | bution of $0.85f_c'$ with a depth of: $a = \beta_1 c$, where $\beta_1$ is a function of concrete compressive strength and is obtained from Table 22.2.2.4.3. | |
| 22.2.2.4.3 | For $f_c' = 5000$ psi: | $\beta_1 = 0.85 - \dfrac{0.05(5000 \text{ psi} - 4000 \text{ psi})}{1000 \text{ psi}} = 0.8$ |

| 22.2.1.1 | Find the equivalent concrete compressive depth, $a$, by equating the compression force to the tension force within the beam cross section:<br><br>$C = T$<br>$0.85f_c'ba = A_sf_y$<br><br>For positive moment: $b = b_f = 46.5$ in.<br><br><br><br>For negative moment: $b = b_f = 18$ in. | At midspan<br>$0.85(5000 \text{ psi})(b)(a) = A_s(60,000 \text{ psi})$<br><br>$a = \dfrac{A_s(60,000 \text{ psi})}{0.85(5000 \text{ psi})(46.5 \text{ in.})} = 0.304A_s$<br><br>At support<br>$a = \dfrac{A_s(60,000 \text{ psi})}{0.85(5000 \text{ psi})(18 \text{ in.})} = 0.784A_s$ |
|---|---|---|

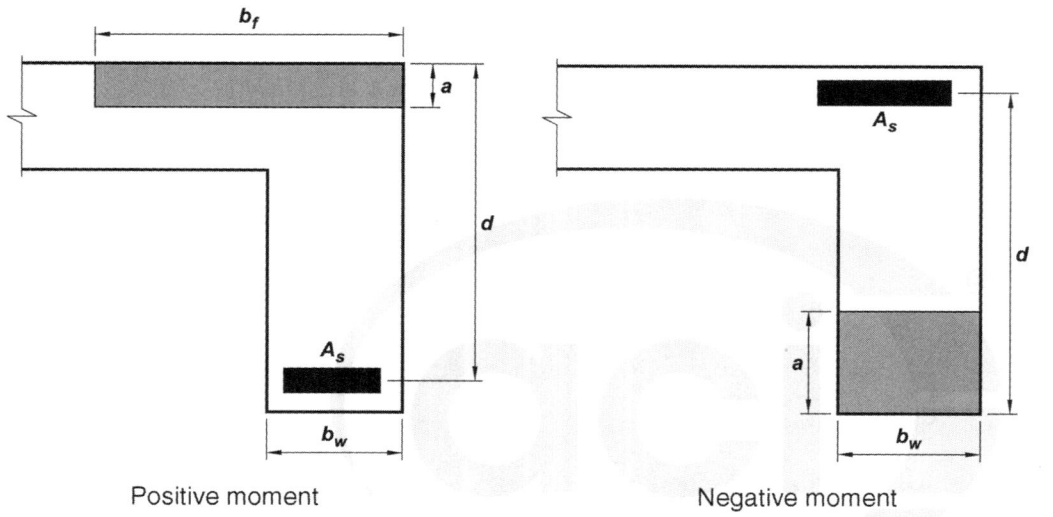

Fig. E2.3—Section reinforcement and compression block at midspan and at support.

| | | | |
|---|---|---|---|
| | | Design the beam for the maximum flexural moment at the midspan and the face of supports. | $M_{u,support} = w_u \ell^2/16 = 128$ ft-kip<br>$M_{u,midspan} = w_u \ell^2/8 = 256$ ft-kip |
| 9.5.1.1 | | The beam strength must satisfy the following equations at each section along its length:<br>$\phi M_n \geq M_u$<br>$\phi V_n \geq V_u$ | |
| | | Calculate required reinforcement area based on the assumptions above: | Midspan |
| | | $M_u \leq \phi M_n = \phi A_s f_y \left(d - \dfrac{a}{2}\right)$ | $256 \text{ ft-kip} \leq \dfrac{(0.9)(60 \text{ ksi})A_s}{12}\left(25.7 \text{ in.} - \dfrac{0.304 A_s}{2}\right)$ |
| | | No. 6 bars $d_b = 3/4$ in. and $A_s = 0.44$ in.² | $A_{s,req'd} = 2.24$ in.²; use six No. 6 |
| | | | Supports |
| | | | $128 \text{ ft-kip} \leq \dfrac{(0.9)(60 \text{ ksi})A_s}{12}\left(25.7 \text{ in.} - \dfrac{0.784 A_s}{2}\right)$ |
| | | | $A_s = 1.13$ in.²; use three No. 6 |
| 21.2.2<br>9.3.3.1 | | Check if calculated strain is greater than 0.005 in./in. (tension controlled), but not less than 0.004 in./in. | |
| | | At midspan. | |
| | | $a = \dfrac{A_s f_y}{0.85 f_c' b}$ and $c = \dfrac{a}{\beta_1}$<br>where $\beta_1 = 0.8$ | $a = 0.304 A_s = (0.304)(6)(0.44 \text{ in.}^2) = 0.80$ in.<br>$c = a/0.8$ in. $= 1.0$ in. |
| | | | $c < h_f$, therefore, beam section behaves as an L-shape. |
| | | $\varepsilon_t = \dfrac{\varepsilon_c}{c}(d - c)$ | $\varepsilon_t = \dfrac{0.003}{1.0 \text{ in.}}(25.7 \text{ in.} - 1.0 \text{ in.}) = 0.074$ |
| | | Note that $b = 18$ in. for negative moments and 54 in. for positive moments. | Therefore, assumption of using $\phi = 0.9$ is correct. |
| | | At support: | |
| | | $a = \dfrac{A_s f_y}{0.85 f_c' b}$ and $c = \dfrac{a}{\beta_1}$<br>where $\beta_1 = 0.8$ | $a = 0.784 A_s = (0.784)(3)(0.44 \text{ in.}^2) = 1.03$ in.<br>$c = a/0.8 = 1.29$ in. |
| | | $\varepsilon_t = \dfrac{\varepsilon_{cu}}{c}(d - c)$ | $\varepsilon_t = \dfrac{0.003}{1.29 \text{ in.}}(25.7 \text{ in.} - 1.29 \text{ in.}) = 0.057$ |
| | | | Therefore, assumption of using $\phi = 0.9$ is correct. |

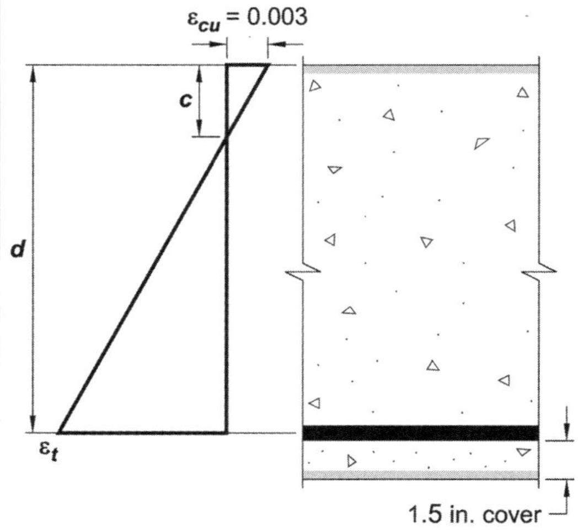

*Fig. E2.4—Strain distribution across beam section.*

| | | | |
|---|---|---|---|
| 9.6.1.1 9.6.1.2 | Minimum reinforcement ratio The provided reinforcement must be at least the minimum required reinforcement at every section along the length of the beams. (a) $A_s = \dfrac{3\sqrt{f'_c}}{f_y} b_w d$ (b) $A_s = \dfrac{200}{f_y} b_w d$ Because $f'_c > 4444$ psi, Eq. (9.6.1.2(a)) controls. | $A_s = \dfrac{3\sqrt{5000 \text{ psi}}}{60,000 \text{ psi}}(18 \text{ in.})(25.7 \text{ in.}) = 1.63 \text{ in.}^2$ **Controls** $A_s = \dfrac{200}{60,000 \text{ psi}}(18 \text{ in.})(25.7 \text{ in.}) = 1.54 \text{ in.}^2$ At midspan: $A_{s,prov.} = (6)(0.44 \text{ in.}^2) = 2.64 \text{ in.}^2 > A_{s,min} = 1.63 \text{ in.}^2$ **OK** At support: $A_s = 1.1 \text{ in.}^2 < A_{s(min)} = 1.63 \text{ in.}^2$ **NG** Therefore, use minimum reinforcement; four No. 6 at support: $A_{s,prov\ (supp)} = 1.76 \text{ in.}^2 > A_{s(min)} = 1.63 \text{ in.}^2$ **OK** | |

# CHAPTER 7—BEAMS

## Step 6: Shear design

| | | |
|---|---|---|
| 21.2.1(b) | **Shear strength**<br>Shear strength reduction factor: | $\phi_{shear} = 0.75$ |
| 9.5.1.1<br>9.5.3.1<br>22.5.1.1 | $\phi V_n \geq V_u$<br>$V_n = V_c + V_s$ | |
| 9.4.3.2 | Design shear force is taken at the face of the support because the vertical reaction causes vertical tension rather than compression (Fig. E2.5). Condition (b), and (c) of Section 9.4.3.2 are satisfied. Condition (c), however, is not satisfied (refer to note at end of this step). | Fig. E2.5—Shear critical section.<br><br>$V_u = 28.5$ kip |
| 22.5.5.1 | $V_c = 2\sqrt{f'_c}b_w d$<br><br>Check if $\phi V_c \geq V_u$ | $V_c = 2(\sqrt{5000 \text{ psi}})(18 \text{ in.})(25.7 \text{ in.})/1000 = 65.4$ kip<br><br>$\phi V_c = (0.75)(65.4 \text{ kip}) = 49$ kip<br>$\phi V_c = 49 \text{ kip} > V_u = 28.5 \text{ kip}$  **OK** |
| 9.6.3.1 | Minimum area of shear reinforcement is required in all regions where $V_u > 0.5\phi V_c$ | $V_u = 28.5 \text{ kip} > 0.5 \phi V_c = 49 \text{ kip}/2 = 24.5$ kip<br><br>However, good engineering practice calls for providing minimum shear reinforcement over the full beam span.<br><br>Provide No. 3 stirrups at 12 in. on center where;<br>12 in. $< d/2 = 25.7$ in. $/2 = 12.8$ in.  **OK** |
| 22.5.1.2 | Cross-sectional dimensions are selected to satisfy Eq. (22.5.1.2):<br><br>$V_u \leq \phi\left(V_c + 8\sqrt{f'_c}b_w d\right)$ | By inspection, this requirement is satisfied. |
| | Note: The shear could be taken at distance $d$ from the face of the girder if hanger reinforcement is provided in the girder as outlined in a paper by A. H. Mattock and J. F. Shen, 1992, "Joints between Reinforced Concrete Members of Similar Depth," *ACI Structural Journal*, V. 89, No. 3, May-June, pp. 290-295. | |

## Step 7: Torsion

|  |  |  |
|---|---|---|
|  |  | Fig. E2.6—Forces transferred from slab to edge beam. |
|  | Calculate the design load at face of slab to beam connection: | $w_u = 1.2(0.21 \text{ kip/ft} + (15 \text{ psf})(2.375 \text{ ft})/1000)$ $+ 1.6(0.1 \text{ ksf})(4.75 \text{ ft})/2$ $w_u = 0.71 \text{ kip/ft}$ |
|  | Calculate the design unit torsion at beam center: | $t_u = (0.76 \text{ kip/ft})(9 \text{ in.}/12) = 0.53 \text{ ft-kip/ft}$ |
|  | Design torsional force: | $T_u = (0.53 \text{ ft-kip/ft})(18 \text{ ft}) = 9.6 \text{ ft-kip}$ |
| 22.7.4.1(a) | Therefore, check threshold torsion $T_{th}$: $$T_{th} = \lambda\sqrt{f_c'}\left(\frac{A_{cp}^2}{p_{cp}}\right)$$ $h_w = 28 \text{ in.} - 7 \text{ in.} = 21 \text{ in.}$ where |  |
|  |  | Fig. E2.7—L-beam geometry to resist torsion. |
|  | $A_{cp} = \sum b_i h_i$ is the area enclosed by outside perimeter of concrete. | $A_{cp} = (18 \text{ in.})(28 \text{ in.}) + (21 \text{ in.})(7 \text{ in.}) = 651 \text{ in.}^2$ |
|  | $p_{cp} = \sum(b_i + h_i)p_c$ is the perimeter of concrete gross area. | $p_{cp} = 2(18 \text{ in.} + 21 \text{ in.} + 7 \text{ in.} + 21 \text{ in.}) = 134 \text{ in.}^2$ |
|  | The overhanging flange dimension is equal to the smaller of the projection of the beam below the slab (21 in.) and four times the slab thickness (28 in.). Therefore, use 21 in. (refer to Fig. E2.7). | $T_{th} = (1.0)\left(\sqrt{5000 \text{ psi}}\right)\left(\frac{(651 \text{ in.}^2)^2}{134 \text{ in.}}\right)$ $T_{th} = 223{,}636 \text{ in.-lb} = 18.6 \text{ ft-kip}$ |
| 21.2.1c | Torsional strength reduction factor: $\phi = 0.75$ | $\phi T_{th} = (0.75)(18.6 \text{ ft-kip}) = 14.0 \text{ ft-kip}$ $T_u = 9.6 \text{ ft-kip} < \phi T_{th} = 14.0 \text{ ft-kip}$ **OK** Torsion reinforcement is not required. |

# CHAPTER 7—BEAMS

## Step 8: Reinforcement detailing

| | | |
|---|---|---|
| 9.7.2.1<br>25.2.1 | **Minimum bar spacing**<br>Minimum clear spacing between the horizontal No. 6 bars must be the greatest of:<br><br>Greatest of $\begin{cases} 1 \text{ in} \\ d_b \\ 4/3(d_{agg}) \end{cases}$<br><br>Assume maximum aggregate size 3/4 in.<br>Check if six No. 6 bars can be placed in the beam's web.<br><br>$b_{w,req'd} = 2(\text{cover} + d_{stirrup} + 0.75 \text{ in.}) + 5d_b$<br>$\qquad + 5(1 \text{ in.})_{min,spacing}$ (25.2.1)<br><br>where<br>$d_{stirrup} = 0.375$ in. and<br>$d_b = 0.75$ in. | 1 in.<br>3/4 in.<br>4/3(3/4 in.) = 1 in.<br><br>Therefore, clear spacing between horizontal bars must not be less than 1 in.<br><br>$b_{w,req'd} = 2(1.5 \text{ in.} + 0.375 \text{ in} + 0.75 \text{ in.}) + 3.75 \text{ in.} + 5 \text{ in.}$<br>$\qquad = 14 \text{ in.} < 18 \text{ in.}$  **OK**<br><br>Therefore, six No. 6 bars can be placed in one layer in the 18 in. beam web with 1.8 in. spacing between bars (Fig. E2.8).<br><br>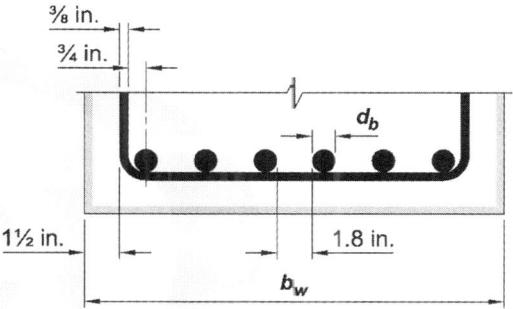<br>Fig. E2.8—Bottom reinforcement layout. |
| 9.7.2.2<br>24.3.1<br>24.3.2 | Maximum bar spacing at the tension face must not exceed the lesser of<br><br>$s = 15\left(\dfrac{40{,}000}{f_s}\right) - 2.5c_c$<br><br>$s = 12\left(\dfrac{40{,}000}{f_s}\right)$<br><br>The maximum spacing concept is intended to limit flexural cracking widths. Note that $c_c$ is the concrete cover to the flexural bars, not the ties. | $s = 15\left(\dfrac{40{,}000 \text{ psi}}{40{,}000 \text{ psi}}\right) - 2.5(2 \text{ in.}) = 10 \text{ in.}$  **Controls**<br><br>$s = 12\left(\dfrac{40{,}000 \text{ psi}}{40{,}000 \text{ psi}}\right) = 12 \text{ in.}$<br><br>18 in. spacing is provided, therefore **OK** |

| | | |
|---|---|---|
| 9.7.3<br>9.7.1.2<br><br>25.4.2.2<br><br><br>25.4.2.4 | **Bar cutoff**<br>_Development length of No. 6 bar_<br>The simplified method is used to calculate the development length of a No. 6 bar:<br>$$\ell_d = \frac{f_y \psi_t \psi_e}{25\lambda\sqrt{f'_c}} d_b$$<br>where<br>$\psi_t$ = bar location; $\psi_t$ = 1.3 for top bars, because more than 12 in. of fresh concrete is placed below them and $\psi_t$ =1.0 for bottom bars, because not more than 12 in. of fresh concrete is placed below them.<br>$\psi_e$ = coating factor; $\psi_e$ = 1.0, because bars are uncoated | Top<br>$$\ell_d = \left(\frac{(60{,}000 \text{ psi})(1.3)(1.0)}{(25)(1.0)\sqrt{5000 \text{ psi}}}\right)(0.75 \text{ in.}) = 33.1 \text{ in.},$$<br>say, 36 in.<br><br>Bottom<br>$$\ell_d = \left(\frac{(60{,}000 \text{ psi})(1.0)(1.0)}{(25)(1.0)\sqrt{5000 \text{ psi}}}\right)(0.75 \text{ in.}) = 25.4 \text{ in.},$$<br>say, 30 in. |
| 9.7.1.3<br>25.5.2.1 | _Splice length of No. 6 reinforcing bar_<br>Per Table 25.5.2.1 splice length is<br>$(\ell_{st}) = 1.3(\ell_d)$ | Top: $1.3\ell_d$ = (1.3)(33.1 in.) = 43.0 in., say, 4 ft 0 in.<br>Bottom: $1.3\ell_d$ = (1.3)(25.4 in.) = 33 in., say, 3 ft 0 in. |
| 9.7.3.8.4 | **Top tension reinforcement**<br>Calculate the inflection point for negative moment diagram:<br>$-M_{max} - w_u(x)^2/2 + V_u x = 0$<br><br><br><br><br>At least one-third of the bars resisting negative moment at a support must have an embedment length beyond the inflection point the greatest of $d$, $12d_b$, and $\ell_n/16$.<br><br>Two of three No. 6 bars are extended over full beam length and one No. 6 bars is terminated beyond the inflection point a distance equal to the embedment length (27 in.). | $(-128 \text{ ft-kip}) - 1.58 \text{ kip/ft}\frac{x^2}{2} + (28.5 \text{ kip})x = 0$<br><br>$x$ = 5.3 ft, say, 5 ft 6 in.<br><br>At 5.5 ft from the girder face two No. 6 can be cutoff and the remainder two No. 6 bars will be extended over the full beam span to support stirrups.<br><br>For No. 6 bars:<br>$d$ = 25.7 in.<br>$12d_b$ = 12(0.75 in.) = 9 in.<br>$\ell_n/16$ = (36 ft)(12)/16 = 27 in.  **Controls**<br><br>66 in. + 27 in. = 93 in.<br>Therefore, the middle two No. 6 bar will be terminated at 7 ft 9 in. (93 in.) from face of support. |

| | | |
|---|---|---|
| 9.7.3.2<br>9.7.3.3 | **Bottom tension reinforcement**<br>Bars must be developed at points of maximum stress and points along the span where bent or terminated tension bars are no longer required to resist flexure. | |
| | Six No. 6 bars are required to resist the factored moment at the midspan.<br><br>Two No. 6 bars can resist a factored moment located at a section $x$ from midspan. | $(256 \text{ ft-kip}) - 1.58 \text{ kip/ft} \dfrac{x^2}{2} = 2(0.44 \text{ in.}^2)(0.9)(60 \text{ ksi})$<br>$\times \left( 25.7 \text{ in.} - \dfrac{2(0.44 \text{ in.}^2)(60 \text{ ksi})}{2(0.85)(5 \text{ ksi})(46.5 \text{ in.})} \right)$<br>$x = 14 \text{ ft} = 168 \text{ in.}$ |
| 9.7.3.3 | A bar must extend beyond the point where it is no longer required to resist flexure for a distance equal to the greater of $d$ or $12d_b$. | For No. 6 bars:<br>1) $d = 25.7$ in. **Controls**<br>2) $12d_b = 12(0.75 \text{ in.}) = 9$ in.<br><br>Therefore, extend four No. 6 bars the greater of the development length (30 in.) from the maximum moment at midspan and a distance $d = 25.7$ in. from the theoretical cutoff point.<br><br>168 in. + 25.7 in. = 193.7 in. say 16 ft-6 in. > $d = 25.7$ in. Therefore, extend four No. 6 bars 16 ft-6 in. from midspan. |
| 9.7.3.8.2 | A minimum of one-fourth positive tension bars must extend into the support minimum 6 in. | Extend the remaining two No. 6 bars > 1/4 six No. 6 a minimum of 6 in. into the support, but not less than $\ell_d = 30$ in. from the theoretical cutoff point (Fig. E2.7). |
| | Note: These calculations are performed to present the code requirements. In practice, all longitudinal bottom bars are extended into the support rather than terminating them 1 ft 6 in. from the support as shown by calculations. | |
| **Step 9: Integrity reinforcement** | | |
| 9.7.7.2 | **Integrity reinforcement**<br>Either one of the two conditions must be satisfied, but not both.<br><br>At least one-fourth the maximum positive moment bars, but not less than two bars must be continuous.<br><br>Longitudinal bars must be enclosed by closed stirrups along the clear span of the beam. | This condition was satisfied above by extending two No. 6 bars into the girders. |

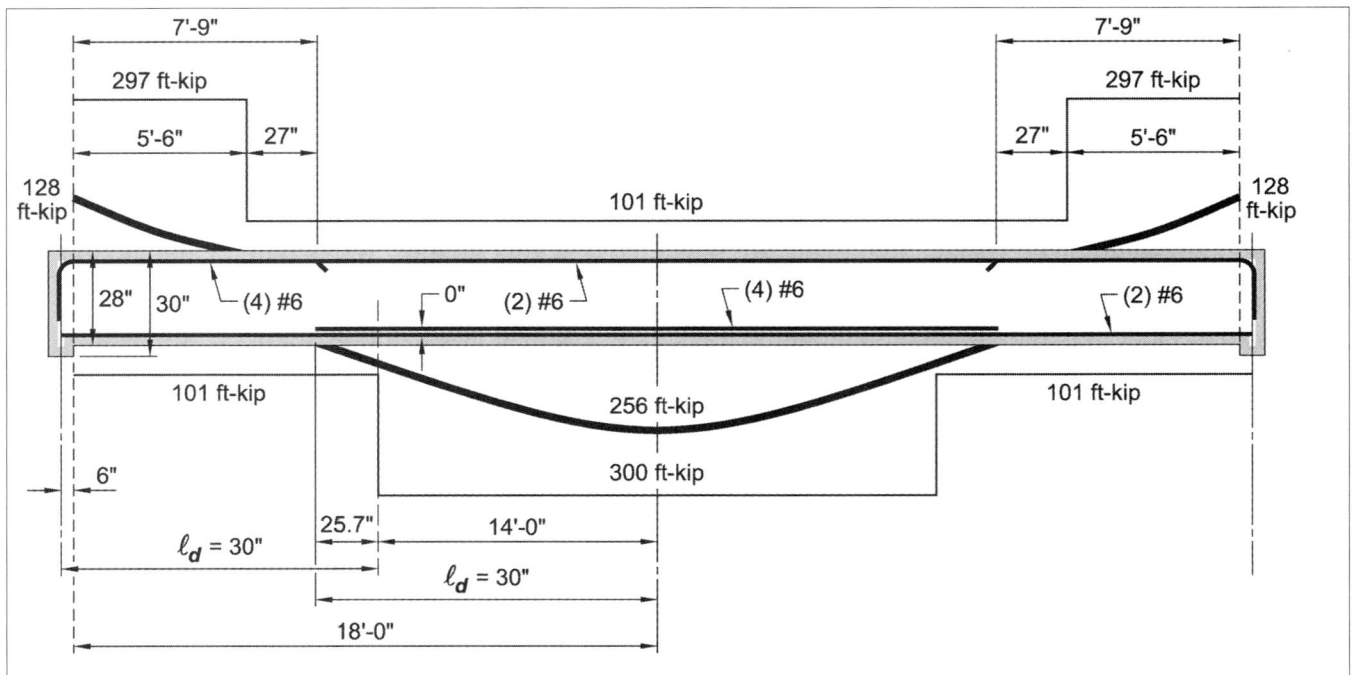

*Fig. E2.9—End span reinforcement cutoff locations.*

## Step 10: Detailing

### Final detailing:

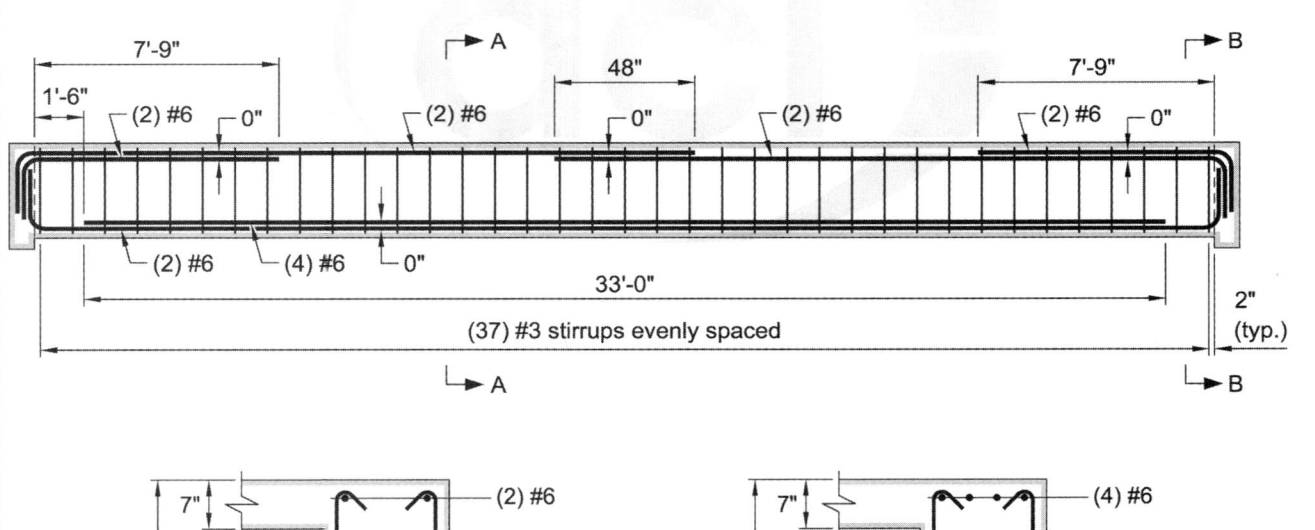

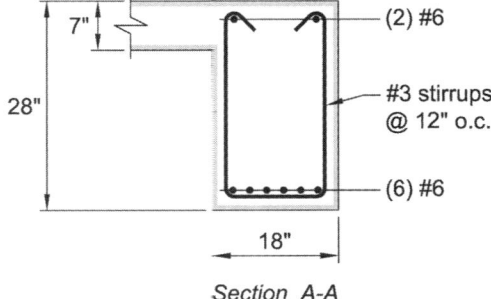

Section A-A

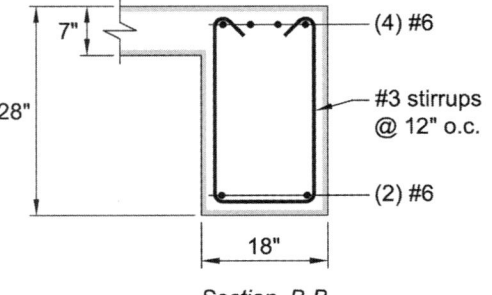

Section B-B

Slab reinforcement not shown for clarity.

*Fig. E2.10—Beam reinforcement details.*

**Beam Example 3**: *Single interior girder beam*

Determine the size of a one-span beam (B2) built integrally with a 7 in. slab of a seven-story building. The beam frames into two girder beams (B3) as shown in Fig. E3.1. Design and detail the beam.

Given:

*Load—*
Service dead load $D = 15$ psf
Service live load $L = 100$ psf
Concentrated loads $P_u = 28.5$ kip ($P_D = 14.4$ kip and $P_L = 7$ kip) located 6 ft 3 in. south and north of Column Lines B and D, respectively. (Refer to Example 2.)

*Material properties—*
$f_c' = 5000$ psi (normalweight concrete)
$f_y = 60,000$ psi

Span length: 28 ft
Beam width: 18 in.

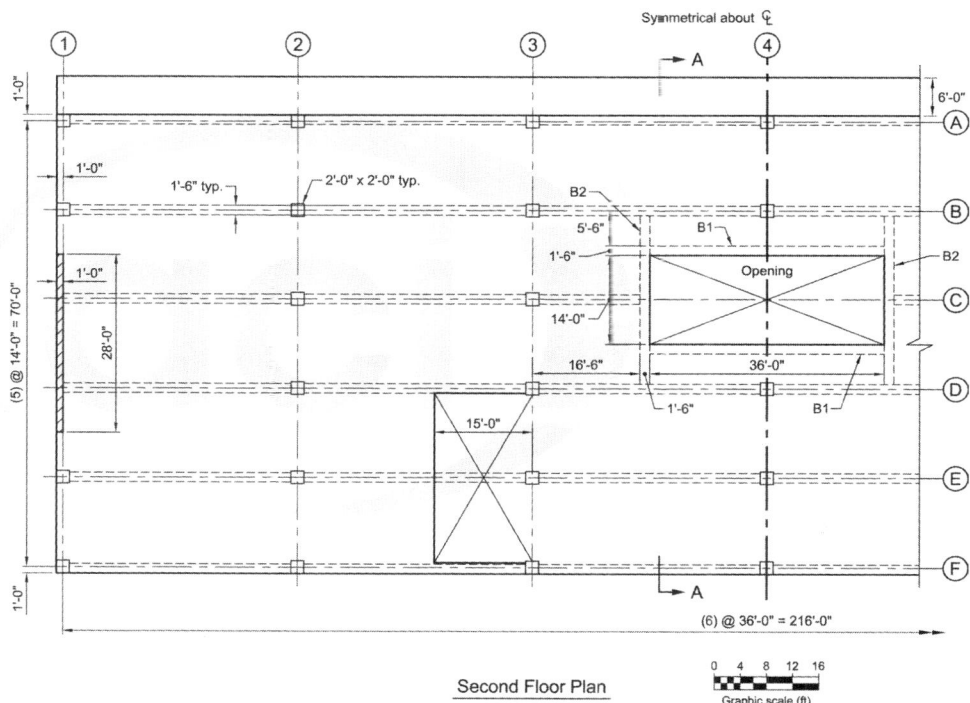

*Fig. E3.1—Plan of Beam B2.*

| ACI 318-14 | Discussion | Calculation |
|---|---|---|
| **Step 1: Material requirements** | | |
| 9.2.1.1 | The mixture proportion must satisfy the durability requirements of Chapter 19 and structural strength requirements of ACI 318-14. The designer determines the durability classes. Please refer to Chapter 4 of SP-17 for an in-depth discussion of the Categories and Classes.<br><br>ACI 301 is a reference specification that is coordinated with ACI 318. ACI encourages referencing ACI 301 into job specifications.<br><br>There are several mixture options within ACI 301, such as admixtures and pozzolans, which the designer can require, permit, or review if suggested by the contractor. | By specifying that the concrete mixture shall be in accordance with ACI 301-10 and providing the exposure classes, Chapter 19 requirements are satisfied.<br><br>Based on durability and strength requirements, and experience with local mixtures, the compressive strength of concrete is specified at 28 days to be at least 5000 psi. |
| **Step 2: Beam geometry** | | |
| 9.3.1.1 | **Beam depth**<br>Beam depth cannot be calculated using Table 9.3.1.1, because two beams frame into it (concentrated loads). For framing simplicity, choose a beam deeper than the beams' depths, framing into it to allow for ease of construction and placement of reinforcement. | Try: $h$ = 30 in. |
| 24.2.2 | The beam deflection will be checked and compared to Table 24.2.2. | |
| 9.2.4.2 | **Flange width**<br>The beam is monolithic with the slab on one side in the middle over a 24 ft long section and slab on both sides for the remainder of the beam. At the maximum positive moment, the beam will behave as an L-beam. Therefore, the effective flange width on one side of the beam is the least from Table 6.3.2.1. | |
| 6.3.2.1 | One side of web is the least of: $6h_{slab}$, $s_w/2$, $\ell_n/2$<br><br>Therefore, flange width: $b_f = \ell_n/12 + b_w$ | (6)(7 in.) = 42 in.<br>(16.5 ft)(12)/2 = 99 in.<br>(28 ft (12) – 18 in.)/12 = 26.5 in  **Controls**<br><br>$b_f$ = 26.5 in. + 18 in. = 44.5 in. |
| | On both sides of the opening, the beam is placed monolithically with the slab and will behave as a T-beam. The flange width on each side of the beam is obtained from Table 6.3.2.1. | |
| 6.3.1 | Each side of web is the least of: $8h_{slab}$, $s_w/2$, $\ell_n/8$<br><br>Therefore, flange width: $b_f = \ell_n/8 + b_w + \ell_n/8$ | 8(7 in.) = 56 in.<br>(16.5 ft)(12)/2 = 99 in.<br>((28 ft)(12) – 18 in.)/8 = 39.75 in.  **Controls**<br><br>$b_f$ = 39.75 in. + 18 in. + 39.75 in. = 97.5 in. |

## CHAPTER 7—BEAMS

| | Step 3: Loads and load patterns | | |
|---|---|---|---|
| | | Self-weights of B2 Beam: beam width $b$ = 18 in. | $w_b$ = [(18 in.)(30 in.)/(144)](0.150 kip/ft$^3$) = 0.56 kip/ft |
| | | Tributary load between Girder B2 and Column Line 3 (refer to Fig. E3.2). The load is transferred to B2 through Beam B4 along Column Line C spanning (15 ft 6 in. clear span). For lobbies and assembly areas, the uniform design live load is 100 psf per Table 4-1 in ASCE 7-10. To account for weights from ceilings, partitions, and HVAC systems, add 15 psf as miscellaneous dead load. | Fig. E3.2—Beams B1 and B4 framing into B2. |
| | | Dead load: Slab self-weight (7 in. thick) supported by Beam B4: | $P_s$ = (7 in./12)(14 ft)(15.5 ft/2)(0.15 kip/ft$^3$) = 10.1 kip |
| | | Beam B4 self-weight: assume beam is: 18 in. wide by 30 in. deep. Note: 7 in. is the slab thickness | $P_B = \left(\dfrac{18 \text{ in.}}{12}\right)\left(\dfrac{30 \text{ in.} - 7 \text{ in.}}{12}\right)\left(\dfrac{16.5 \text{ ft} - 1 \text{ ft}}{2}\right)(0.15 \text{ kip/ft}^3)$ = 3.3 kip |
| | | Superimposed dead load of 15 psf: | $P_{SD}$ = (15 psf/1000)(14 ft)(16.5 ft)/2 = 1.7 kip |
| | | Total dead load at B2 midspan: | $\sum P$ = 10.1 kip + 3.3 kip + 1.7 kip = 15.1 kip |
| | | Live load: Concentrated load between Column Line 3 and girder transferred at midspan | $P_L$ = (0.1 ksf)(14 ft)(16.5 ft)/2 = 11.6 kip |
| | | Beams B1 frame into Beam B2 at 6 ft-3 in. and 21 ft-9 in. from Column Line B (Fig. E3.2). The beams' factored reactions were calculated in Example 2 and were found to be: 28.5 kip in Step 4. | $P_u$ = 28.5 kip |
| | | The beam resists gravity load only. Lateral forces are not considered in this problem. Distributed load: | |
| 5.3.1a | | $w_u$ = 1.4$D$ | $w_u$ = 1.4(0.56 kip/ft + (15 psf)(1.5 ft)/1000) = 0.82 kip/ft |
| 5.3.1b | | $w_u$ = 1.2$D$ + 1.6$L$ | $w_u$ = 1.2(0.82 kip/ft)/1.4 + 1.6((100 psf)(1.5 ft)/1000) = 0.94 kip/ft **Controls** |
| | | Concentrated load | |
| 5.3.1a | | $P_u$ = 1.4$P_D$ | $P_u$ = 1.4(15.1 kip) = 21.1 kip |
| 5.3.1b | | $P_u$ = 1.2$P_D$ + 1.6$P_L$ | $P_u$ = 1.2(15.1 kip) + 1.6(11.6 kip) = 36.7 kip **Controls** |

| Step 4: Analysis | | |
|---|---|---|
| 9.4.1.2 | Beam B2 is monolithic with supports.<br><br>Chapter 6 permits several analysis procedures to calculate the required strengths.<br><br>For this example, calculate beam moment at supports using coefficients from Table B-1 Reinforced Concrete Design Handbook Design Aid – Analysis Tables, which can be downloaded from: https://www.concrete.org/store/productdetail.aspx?ItemID=SP1714DA<br><br>$M_u = w_u \ell^2/12 + P_u \ell/8 + P_u a^2 b/\ell^2 + P_u a b^2/\ell^2$<br><br>and the analysis shows the beam shear is<br>$V_u = w_u \ell/2 + P_u/2$ | $M_u = (0.94 \text{ kip/ft})(28 \text{ ft})^2/12 + (36.7 \text{ kip})(28 \text{ ft})/8$<br>$\quad + (28.5 \text{ kip})(6.25 \text{ ft})^2(21.75 \text{ ft})/(28 \text{ ft})^2$<br>$\quad + (28.5 \text{ kip})(6.25 \text{ ft})(21.75 \text{ ft})^2/(28 \text{ ft})^2$<br>$\quad = 328 \text{ ft-kip}$<br><br>$V_u = (0.94 \text{ kip/ft})(28 \text{ ft})/2 + (36.7 \text{ kip})/2 + 2(28.5 \text{ kip})/2$<br>$\quad = 60 \text{ kip}$<br>Note that the (2)(28.5 kip) shear force represents the two beams framing into the girder beam. |
| | Using coefficients from Appendix B-1 Reinforced Concrete Design Handbook Design Aid – Analysis Tables, which can be downloaded from: https://www.concrete.org/store/productdetail.aspx?ItemID=SP1714DA, assuming maximum moment is at midspan: | $M_u = 255 \text{ ft-kip}$ |
| | This distribution assumes the girder remains uncracked. If the girder, does crack, however, its stiffness is greatly reduced and redistribution of moments occurs.<br><br>Assume that the moments at supports are reduced by 15 percent: | $M_u = (0.85)(328) \text{ ft-kip} = 279 \text{ ft-kip}$ |
| | Accordingly, the moment at midspan must be increased by the same amount: | $M_u = (255 \text{ ft-kip}) + (0.15)(328 \text{ ft-kip}) = 304 \text{ ft-kip}$ |
| | Note:<br>Alan Mattock states that, "…, and it is concluded that redistribution of design bending moments by up to 25% does not result in performance inferior to that of beams designed for the distribution of bending moments predicted by the elastic theory, either at working loads or at failure." (Mattock, A. H., 1959, "Redistribution of Design Bending Moments in Reinforced Concrete Continuous Beams," *Proceedings*, Institution of Civil Engineers (London), V. 13, pp. 35-46.)<br>H. Scholz, however, limits the moment redistribution to 20 percent: "The cut-off point at 20 percent is imposed to avoid excessive cracking at elastic service moments." (Scholz, H., 1993, "Contribution to Redistribution of Moments in Continuous Reinforced Concrete Beams," *ACI Structural Journal*, V. 90, No. 2, Mar.-Apr., pp. 150-155.) | |

Beams B1 and B4 frame into Beam B2; therefore, assume that (B1) reaction of 28.5 kip and 36.7 kip are applied at the face of Beam B2, but in opposite directions. Ignoring the distributed load from the slab, Beam B2 is subjected to torsion:

Refer to the torsion diagram in Fig. E3.3.

From B1:
$$T_u = \frac{(28.5 \text{ kip})(9 \text{ in.})}{12} = 21.4 \text{ ft-kip}$$
and from B4
$$T_u = \frac{(36.7 \text{ kip})(9 \text{ in.})}{12} = 27.6 \text{ ft-kip}$$

*Factored shear diagram*

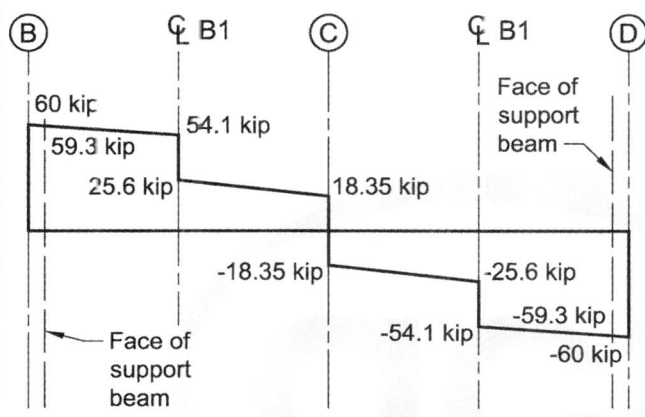

*Factored moment diagram*

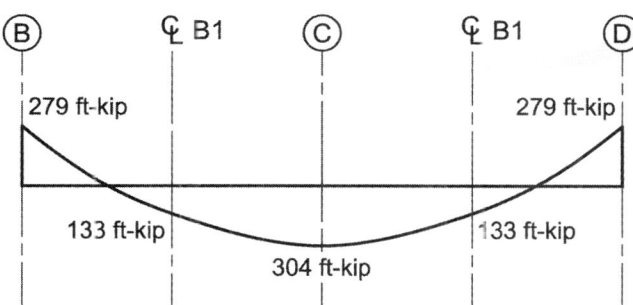

*Factored torsion diagram*

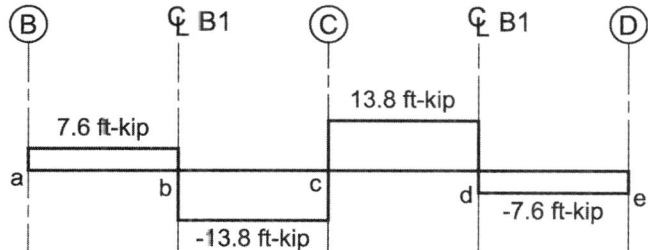

Fig. E3.3—Shear, moment, and torsion diagrams.

## Step 5: Moment design

| | | |
|---|---|---|
| 9.3.3.1 | The Code does not allow a beam to be designed with steel strain at design strength less than 0.004 in./in. The intent is to ensure ductile behavior at the nominal condition.<br><br>For usual reinforced beams, such as this example, reinforcing bar strain is not a controlling issue. | |
| 21.2.1(a) | Assuming the beams will be tension-controlled, $\phi$ = 0.9. This assumption will be checked later. | |
| 9.7.1.1<br>20.6.1.3.1 | Calculate the effective depth assuming No. 3 stirrups, No. 6 bars, and 1.5 in. cover:<br><br>$d = h - \text{cover} - d_{tie} - d_b/2$ | $d$ = 30 in. – 1.5 in. – 0.375 in. – 0.75 in. /2 = 27.75 in., say, $d$ = 27.7 in. |
| 22.2.2.1 | The concrete compressive strain at which nominal moments are calculated is:<br>$\varepsilon_c = 0.003$ | |
| 22.2.2.2 | The tensile strength of concrete in flexure is a variable property and its value is approximately 10 to 15 percent of the concrete compressive strength. For calculating nominal strength, ACI 318 neglects the concrete tensile strength.<br><br>Determine the equivalent concrete compressive stress for design: | |
| 22.2.2.3 | The concrete compressive stress distribution is inelastic at high stress. The actual distribution of concrete compressive stress is complex and usually not known explicitly. The Code permits any stress distribution to be assumed in design if shown to result in predictions of ultimate strength in reasonable agreement with the results of comprehensive tests. Rather than tests, the Code allows the use of an equivalent rectangular compressive stress distribution of $0.85f_c'$ with a depth of:<br>$a = \beta_1 c$, $\beta_1$ is a function of concrete compressive strength and is obtained from Table 22.2.2.4.3. | |
| 22.2.2.4.1 | For $f_c'$ = 5000 psi: | $\beta_1 = 0.85 - \dfrac{0.05(5000 \text{ psi} - 4000 \text{ psi})}{1000 \text{ psi}} = 0.8$ |
| 22.2.2.4.3 | Find the equivalent concrete compressive depth, $a$, by equating the compression force in the section to the tension force (refer to Fig. E3.4):<br>$C = T$<br>$0.85f_c'ba = A_s f_y$ | $0.85(5000 \text{ psi})(b)(a) = A_s(60,000 \text{ psi})$ |

| | | |
|---|---|---|
| 22.2.1.1 | For positive moment: $b = b_f = 44.5$ in. | $a = \dfrac{A_s(60{,}000\text{ psi})}{0.85(5000\text{ psi})(44.5\text{ in.})} = 0.317 A_s$ |
| | For negative moment: $b = b_w = 18$ in. | At support<br>$0.85(5000\text{ psi})(18\text{ in.})(a) = A_s(60{,}000\text{ psi})$<br>$a = \dfrac{A_s(60{,}000\text{ psi})}{0.85(5000\text{ psi})(18\text{ in.})} = 0.784 A_s$ |

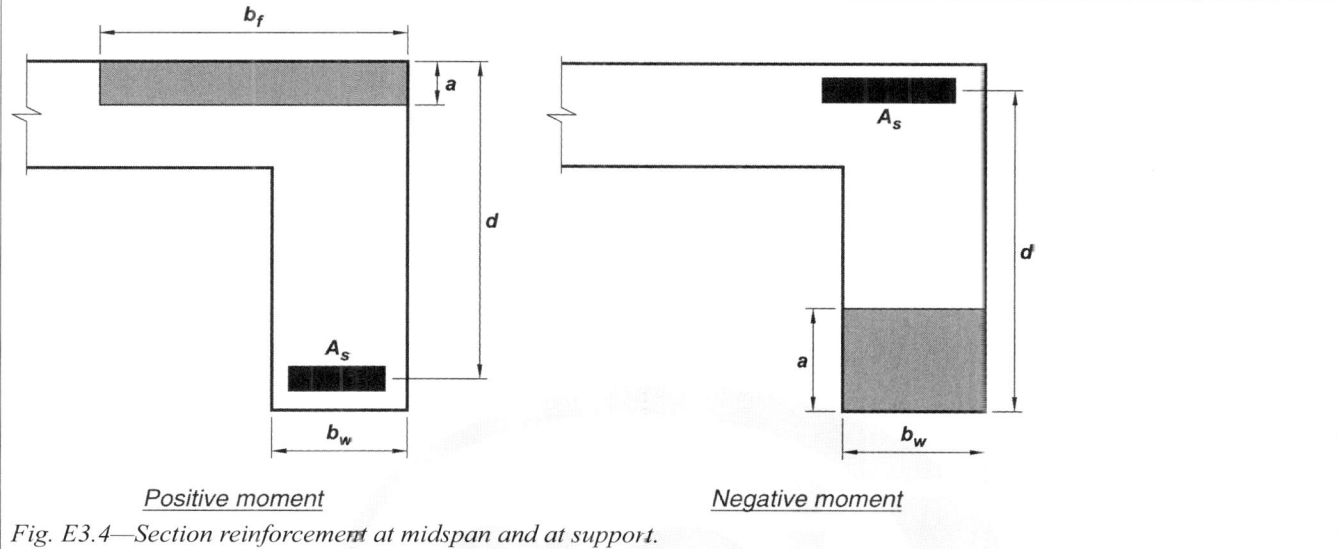

*Fig. E3.4—Section reinforcement at midspan and at support.*

| | | |
|---|---|---|
| 9.5.1.1 | Design the beam for the maximum flexural moment at the midspan and the face of supports.<br><br>The beam strength must satisfy the following inequalities at each section along its length:<br>$\phi M_n \geq M_u$<br>$\phi V_n \geq V_u$<br><br>Calculate required flexural reinforcement area using the following equation:<br><br>$M_u \leq \phi M_n = \phi A_s f_y \left( d - \dfrac{a}{2} \right)$<br><br>No. 6 bars; $d_b = 0.75$ in. and $A_s = 0.44$ in.²<br><br>Note that $b = 18$ in. for negative moments and $b = 44.5$ in. for positive moments. | Midspan<br><br>$304 \text{ ft-kip} \leq (0.9)(60 \text{ ksi}) A_s \left( 27.7 \text{ in.} - \dfrac{0.317 A_s}{2} \right)$<br><br>$A_s = 2.47$ in.²; use six No. 6<br><br>Supports<br><br>$279 \text{ ft-kip} \leq (0.9)(60 \text{ ksi}) A_s \left( 27.7 \text{ in.} - \dfrac{0.784 A_s}{2} \right)$<br><br>$A_s = 2.31$ in.² use 6–No. 6 |
| 21.2.2<br>9.3.3.1 | Check if calculated strain is greater than 0.005 in./in. (tension-controlled), but not less than 0.004 in./in. Refer to Fig. E3.5.<br><br>$a = \dfrac{A_s f_y}{0.85 f'_c b}$ and $c = \dfrac{a}{\beta_1}$<br><br>where $\beta_1 = 0.8$<br><br>$\varepsilon_t = \dfrac{\varepsilon_{cu}}{c}(d - c)$ | Midspan<br>$a = 0.317 A_s = (0.317)(6)(0.44 \text{ in.}^2) = 0.84$ in.<br>$c = a/0.8 = 0.84 \text{ in.}/0.8 = 1.05$ in. < 7 in. slab thickness<br>Therefore, shape assumption is correct<br><br>$\varepsilon_t = \dfrac{0.003}{1.05 \text{ in.}}(27.7 \text{ in.} - 1.05 \text{ in.}) = 0.076 > 0.005$<br><br>Supports<br>$a = 0.784 A_s = (0.784)(6)(0.44 \text{ in.}^2) = 2.07$ in.<br>$c = a/0.8 = 2.07/0.8 = 2.59$ in.<br><br>$\varepsilon_t = \dfrac{0.003}{2.59 \text{ in.}}(27.7 \text{ in.} - 2.59 \text{ in.}) = 0.029 > 0.005$<br><br>Therefore, the assumption of $\phi = 0.9$ is correct. |

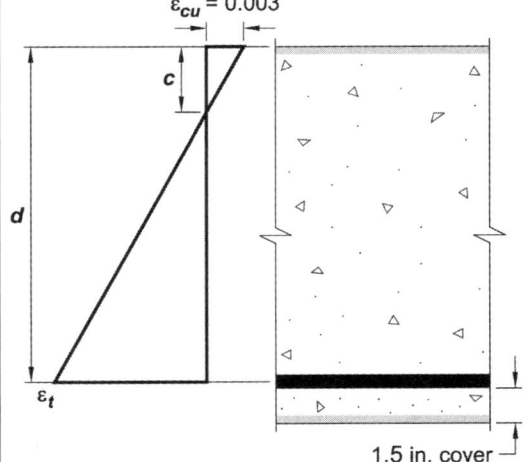

Fig. E3.5—Strain distribution across beam section.

# CHAPTER 7—BEAMS

| | | |
|---|---|---|
| 9.6.1.1<br>9.6.1.2 | **Minimum reinforcement area**<br>The reinforcement area must exceed the minimum required at every section along the length of the beam.<br><br>(a) $A_s = \dfrac{3\sqrt{f_c'}}{f_y} b_w d$<br><br>(b) $A_s = \dfrac{200}{f_y} b_w d$<br><br>Because $f_c' > 4444$ psi, Eq. (9.6.1.2(a)) controls. | $A_s = \dfrac{3\sqrt{5000 \text{ psi}}}{60{,}000 \text{ psi}}(18 \text{ in.})(27.7 \text{ in.}) = 1.76 \text{ in.}^2$<br><br>At midspan: $A_{s(prov.)} = 2.64 \text{ in.}^2 > A_{s(min)} = 1.76 \text{ in.}^2$ **OK**<br>At support: $A_{s(prov.)} = 2.64 \text{ in.}^2 > A_{s(min)} = 1.76 \text{ in.}^2$ **OK** |
| **Step 6: Shear design** | | |
| 21.2.1(b)<br>9.5.1.1 | **Shear strength**<br>Shear strength reduction factor:<br>$\phi V_n \geq V_u$ | $\phi_{shear} = 0.75$ |
| 9.5.3.1<br>22.5.1.1 | $V_n = V_c + V_s$ | 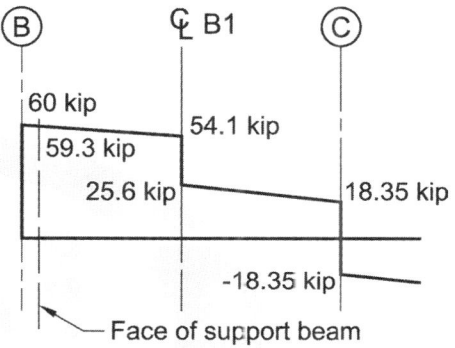<br>Fig. E3.6—Shear-critical section. |
| 9.5.1.2<br>21.2.1b | Shear reduction factor $\phi = 0.75$ | |
| 9.4.3.1<br>9.4.3.2 | Design shear force is taken at the face of the support because the vertical reaction causes vertical tension rather than compression (Fig. E3.5). Condition (b), and (c) of 9.4.3.2 are satisfied. Condition (a), however, is not satisfied. | $V_u = 59.3$ kip |
| 22.5.5.1 | $V_c = 2\sqrt{f_c'} b_w d$<br><br><br>Check if $\phi V_n \geq V_u$ | $V_c = 2\sqrt{5000 \text{ psi}}(18 \text{ in.})(27.7 \text{ in.})/1000 = 70.5$ kip<br><br>$\phi V_c = (0.75)(70.5 \text{ kip}) = 52.9$ kip<br>$\phi V_n = 52.9$ kip $< V_u = 59.3$ kip  **NG**<br><br>Therefore, shear reinforcement is required. |
| | Cross-sectional dimensions are selected to satisfy Eq. (22.5.1.2): | |
| 22.5.1.2 | $V_u \leq \phi(V_c + 8\sqrt{f_c'} b_w d)$ | $V_u \leq \phi\left(70.5 \text{ kip} + 8\sqrt{5000 \text{ psi}}(18 \text{ in.})(27.7 \text{ in.})\right)$<br>$\leq 212$ kip<br><br>Section dimensions are satisfactory. |
| | Note: The shear could be taken at distance $d$ from the face of the girder if hanger reinforcement is provided in the girder, as outlined in a paper by A. H. Mattock and J. F. Shen, "Joints between Reinforced Concrete Members of Similar Depth," *ACI Structural Journal*, V. 89, No.3, May-June 1992, pp. 290-295. | |

| | | |
|---|---|---|
| 22.5.10.1 | **Shear reinforcement**<br>Transverse reinforcement is required at each section where $V_u > \phi V_c$ satisfying Eq. (22.5.10.1): | |
| 22.5.10.5.3<br>22.5.10.5.6 | Section AB<br>$\phi V_s \geq V_u - \phi V_c$<br><br>where $V_s = \dfrac{A_v f_{yt} d}{s}$<br><br>$\dfrac{A_v}{s} \geq \dfrac{V_s}{f_{yt} d}$ | $\phi V_s \geq (59.3 \text{ kip}) - (52.9 \text{ kip}) = 6.4 \text{ kip}$<br><br>$V_s \geq \dfrac{6.4 \text{ kip}}{0.75} = 8.5 \text{ kip}$<br><br>$\dfrac{A_v}{s} \geq \dfrac{(8.5 \text{ kip})}{(60 \text{ ksi})(27.7 \text{ in.})} = 0.0051$ |
| 9.7.6.2.2 | Check maximum allowable stirrup spacing:<br><br>Is $V_s \leq 4\sqrt{f_c'} b_w d$ ? | $4\sqrt{f_c'} b_w d = 4\left(\sqrt{5000 \text{ psi}}\right)(18 \text{ in.})(27.7 \text{ in.}) = 140.5 \text{ kip}$<br><br>say, 141 kip<br><br>$V_s = 8.5 \text{ kip} < 4\sqrt{f_c'} b_w d = 141 \text{ kip}$    **OK**<br><br>Therefore, for stirrup spacing, use the lesser of:<br>$d/2 = 27.7 \text{ in.}/2 = 13.8 \text{ in.}$ or 24 in.<br><br>Try No. 3 stirrups at 12 in. on center.<br><br>$\left(\dfrac{A_v}{s}\right)_{prov} = \dfrac{2(0.11 \text{ in.}^2)}{12 \text{ in.}} = 0.018 \text{ in.}^2/\text{in.} > 0.005 \text{ in.}^2/\text{in.}$<br>**OK** |
| 9.6.3.3 | Specified shear reinforcement must be at least the larger of:<br><br>$A_{s,min}/s = 0.75\sqrt{f_c'} \dfrac{b_w}{f_{yt}}$<br><br>and<br><br>$A_{s,min}/s = 50 \dfrac{b_w}{f_{yt}}$ | $\dfrac{A_{v,min}}{s} \geq 0.75\sqrt{5000 \text{ psi}} \dfrac{18 \text{ in.}}{60,000 \text{ psi}} = 0.016 \text{ in.}^2/\text{in.}$<br>**Controls**<br><br>$\dfrac{A_{v,min}}{s} = 50 \dfrac{18 \text{ in.}}{60,000 \text{ psi}} = 0.015 \text{ in.}^2/\text{in.}$<br><br>Provided:<br><br>$\dfrac{A_v}{s} \geq \dfrac{2(0.11 \text{ in.}^2)}{12 \text{ in.}} = 0.018 \text{ in.}^2/\text{in.} > \dfrac{A_{v,min}}{s} = 0.016 \text{ in.}^2/\text{in.}$<br><br>satisfies 9.6.3.3, therefore **OK** |

# CHAPTER 7—BEAMS

## Step 7: Torsion design

| | | | |
|---|---|---|---|
| 22.7.4.1(a) | Determine if girder torsion can be neglected. Check threshold torsion $T_{th}$: $$T_{th} = \lambda\sqrt{f'_c}\left(\frac{A_{cp}^2}{p_{cp}}\right)$$ | 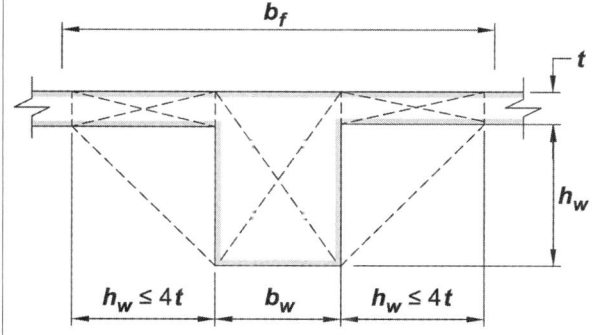 Fig. E3.7—T-beam geometry to resist torsion. | |

Determine portion of slab to be included with the beam for the torsional design:

T-section between $a$ and $b$ and between $d$ and $e$ and

where $A_{cp} = \sum b_i h_i$ is the area enclosed by outside perimeter of concrete

$A_{cp} = (18 \text{ in.})(30 \text{ in.}) + 2(23 \text{ in.})(7 \text{ in.}) = 862 \text{ in.}^2$

$p_{cp} = \sum(b_i + h_i)$ is the perimeter of concrete gross area

$p_{cp} = 2(18 \text{ in.} + 23 \text{ in.} + 23 \text{ in.} + 23 \text{ in.} + 7 \text{ in.}) = 188 \text{ in.}$

$$T_{th} = (1.0)\sqrt{5000 \text{ psi}}\left(\frac{(862 \text{ in.}^2)^2}{188 \text{ in.}}\right)$$

**9.5.1.1**
**9.5.1.2** Refer to Fig. E3.3 for torsional value $T_u$ near supports.

$T_{th} = 279{,}474 \text{ in.-lb} = 23.3 \text{ ft-kip}$

**21.2.1c** Torsional strength reduction factor $\phi = 0.75$

$\phi T_{th} = (0.75)(23.3 \text{ ft-kip}) = 17.5 \text{ ft-kip}$
$\phi T_{th} = 17.5 \text{ ft-kip} > T_u = 7.6 \text{ ft-kip}$ **OK**

Torsion reinforcement is not required between $a$ and $b$ and between $d$ and $e$.

Determine portion of the slab to be included with the beam for torsional design.

L-section between $b$ and $d$

The overhanging flange dimension is equal to the smaller of the projection of the beam below the slab (23 in.) and four times the slab thickness (28 in.). Therefore, use 23 in. (refer to Fig. E3.8).

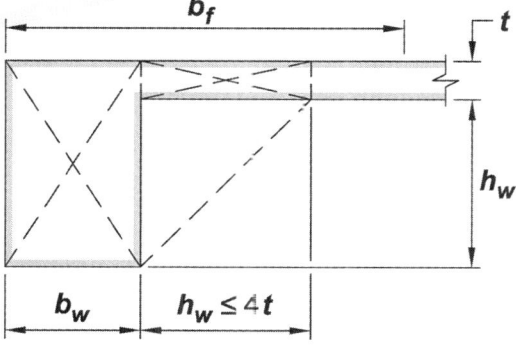
Fig. E3.8—L-beam geometry to resist torsion.

$A_{cp} = (18 \text{ in.})(30 \text{ in.}) + (23 \text{ in.})(7 \text{ in.}) = 701 \text{ in.}^2$
$p_{cp} = 2(18 \text{ in.} + 23 \text{ in.} + 7 \text{ in} + 23 \text{ in.}) = 142 \text{ in.}$

$$T_{th} = (1.0)\left(\sqrt{5000 \text{ psi}}\right)\left(\frac{(701 \text{ in.}^2)^2}{142 \text{ in.}}\right)$$

$T_{th} = 244{,}699 \text{ in.-lb} = 20.4 \text{ ft-kip}$

$\phi T_{th} = (0.75)(20.4 \text{ ft-kip}) = 15.3 \text{ ft-kip}$

Refer to Fig. E3.3 for torsional value $T_u$ at midspan.
$\phi T_{th} = 15.3 \text{ ft-kip} > T_u = 13.8 \text{ ft-kip}$ **OK**

Torsion reinforcement is not required between $b$ and $d$.

### Step 8: Reinforcement detailing

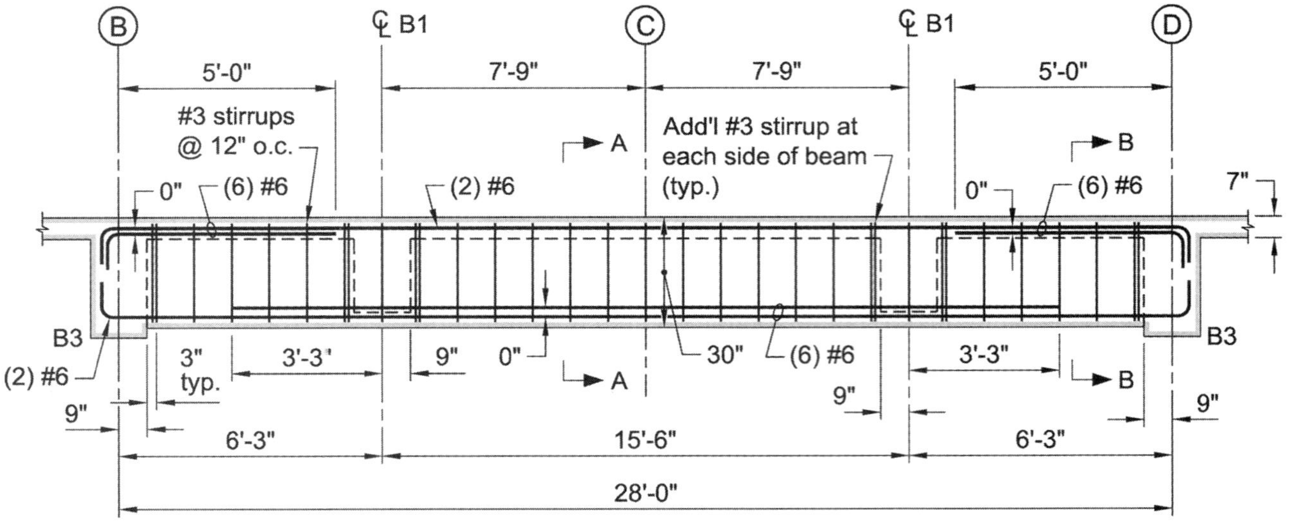

*Fig. E3.9—Longitudinal and transverse reinforcement.*

| | | |
|---|---|---|
| 9.7.2.1 | Minimum top bar spacing<br>From Appendix A of SP-17(14) Reinforced Concrete Design Handbook Design Aid – Analysis Tables, which can be downloaded from: https://www.concrete.org/store/productdetail.aspx?ItemID=SP1714DA, six No. 6 bars can be placed in one layer within an 18 in. wide beam.<br><br>Bar spacing can also be calculated as shown as follows for bottom bars. | |
| 9.7.2.1<br>25.2.1 | Minimum bottom bar spacing<br>Minimum clear spacing between the longitudinal bars is the greatest of:<br><br>Clear spacing $\begin{cases} 1 \text{ in.} \\ d_b \\ 4/3(d_{agg}) \end{cases}$<br><br>Assume 3/4 in. maximum aggregate size.<br><br>Check if six No. 6 bars can be placed in the beam's web.<br><br>$b_{w,req'd} = 2(\text{cover} + d_{stirrup} + 0.75 \text{ in.})$<br>$\qquad + 5d_b + 5(1.0 \text{ in.})_{min,spacing}$ (25.2.1)<br><br>Refer to Fig. E3.10 for reinforcement placement in Beam B2. | 1 in. **Controls**<br>0.75 in.<br>$4/3(3/4 \text{ in.}) = 1 \text{ in.}$ **Controls**<br><br>Therefore, clear spacing between horizontal bars must not be less than 1.0 in.<br><br><br><br>$b_{w,req'd} = 2(1.5 \text{ in.} + 0.375 \text{ in.} + 0.75 \text{ in.}) + 5(0.75 \text{ in.})$<br>$\qquad + 5(1.0 \text{ in.})$<br>$b_{w,req'd} = 14 \text{ in.} < 18 \text{ in.}$ **OK**<br><br>*Fig. E3.10 for reinforcement placement.* |

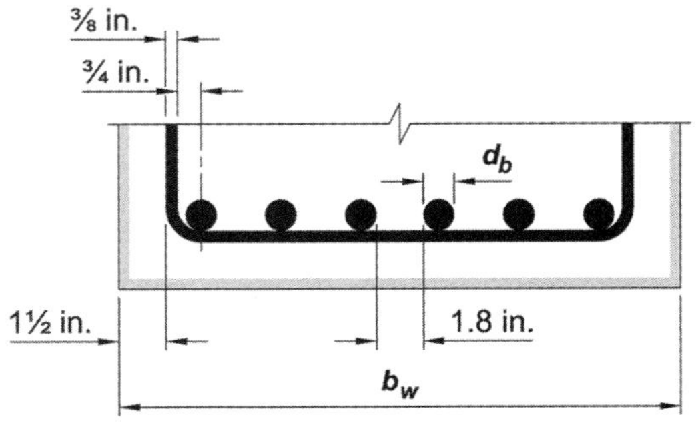

| | | |
|---|---|---|
| 9.7.2.2<br>24.3.1<br>24.3.2 | Maximum bar spacing at the tension face must not exceed the lesser of<br><br>$$s = 15\left(\frac{40{,}000}{f_s}\right) - 2.5c_c$$<br><br>$$s = 12\left(\frac{40{,}000}{f_s}\right)$$<br><br>This spacing is to limit flexural cracking widths, where $c_c = 2$ in. is the least distance from surface of deformed reinforcement to the tension face. | $s = 15\left(\dfrac{40{,}000\text{ psi}}{40{,}000\text{ psi}}\right) - 2.5(2\text{ in.}) = 10$ in.  **Controls**<br><br>$s = 12\left(\dfrac{40{,}000\text{ psi}}{40{,}000\text{ psi}}\right) = 12$ in.<br><br>1.8 in. spacing is provided, therefore **OK** |
| 9.7.1.2<br><br>25.4.2.2<br><br>25.4.2.4 | Development length of No. 6 reinforcing bar<br>The simplified method is used to calculate the development length of No. 6 bars:<br><br>$$\ell_d = \frac{f_y \psi_t \psi_e}{20\lambda\sqrt{f'_c}} d_b$$<br><br>where $\psi_t$ is bar location; $\psi_t = 1.3$ for top horizontal bars, because more than 12 in. of fresh concrete is placed below them, and $\psi_t = 1.0$ for bottom horizontal bars, because not more than 12 in. of fresh concrete is placed below them; $\psi_e$ is coating factor; and $\psi_e = 1.0$ because bars are uncoated | Top bars<br><br>$\ell_d = \left(\dfrac{(60{,}000\text{ psi})(1.3)(1.0)}{(25)(1.0)\sqrt{5000\text{ psi}}}\right)(0.75\text{ in.}) = 33.1$ in.<br><br>say, 36 in.<br><br>Bottom bars<br><br>$\ell_d = \left(\dfrac{(60{,}000\text{ psi})(1.0)(1.0)}{(25)(1.0)\sqrt{5000\text{ psi}}}\right)(0.75\text{ in.}) = 25.5$ in.<br><br>say, 30 in. |
| 9.7.1.3<br>25.5.2.1 | Splice length of No. 6 reinforcing bar<br>Per Table 25.5.2.1, splice length is $1.3(\ell_d)$. | Top: $1.3\ell_d = (1.3)(33.1$ in.$) = 43.0$ in., say, 4 ft 0 in.<br>Bottom: $1.3\ell_d = (1.3)(25.5$ in.$) = 33.2$ in., say, 3 ft 0 in. |
| 9.7.3 | Bar cutoff<br><br>Bottom tension reinforcement<br>Four No. 6 bottom bars are terminated beyond Beam B1 a distance equal to the development length of 30 in.<br><br>Extend two No. 6 bottom bars the full length of the beam and develop into the girder beams along Column Lines B and D at each end with a hook. | Four No. 6 lengths:<br>$\ell = 14$ ft $+ 2(1.5$ ft$) + 2(2.5$ ft$) = 22$ ft |

| | | |
|---|---|---|
| 9.7.3.2 | Top tension reinforcement<br>Reinforcement must be developed at points of maximum stress and points along the span where terminated tension reinforcement is no longer required to resist flexure.<br>Six No. 6 bars are required to resist the factored moment at the support.<br><br>Four No. 6 bars will be terminated at the inflection point. | $-(279 \text{ ft-kip}) - 0.94 \text{ kip/ft} \dfrac{x^2}{2} + 60 \text{ kip}(x) = 0$<br><br>$x = 4.8$ ft, say, 5 ft |
| 9.7.3.8.2 | At least one-third of the negative moment reinforcement at a support must have an embedment length beyond the point of inflection the greatest of $d$, $12d_b$, and $\ell_n/16$. | For No. 6 bars:<br>1) $d = 27.7$ in.   **Controls**<br>2) $12d_b = 12(1.0 \text{ in.}) = 12$ in.<br>3) $\ell_n/16 = 21$ in.<br><br>Therefore, extend four No. 6 bars a distance $d$ beyond the inflection point.<br><br>60 in. + 27.7 in. = 67.7 in. |
| | Place four No. 6 bars within the Beam B2 web and two No. 6 bars on either side of the beam web over 36 in. (refer to Fig. E3.10, Section B) | Extend the remaining two No. 6 bars over the full length of the beam as hanger bars for the stirrups. |
| **Step 9: Integrity reinforcement** | | |
| 9.7.7.2 | Integrity reinforcement<br>Either one of the two conditions must be satisfied, but not both.<br><br>In this example, both are satisfied.<br><br>At least one-fourth the maximum positive moment reinforcement, but not less than two bars must be continuous.<br><br><br>Longitudinal reinforcement must be enclosed by closed stirrups along the clear span of the beam. | <br><br><br><br>This condition was satisfied above by extending two No. 6 bottom reinforcement bars into the support.<br><br>2 No. 6 > (1/4)6 No. 6<br><br>This condition is satisfied by extending stirrups over the full length of the beam. Refer to Fig. E3.9. |

## CHAPTER 7—BEAMS

### Step 10: Deflection

| | | |
|---|---|---|
| 9.3.2 | Calculate deflection limit: | |
| 24.2.3.1 | Immediate deflection is calculated using elastic deflection approach and considering concrete cracking and reinforcement for calculating stiffness.<br><br>Modulus of elasticity: | |
| 19.2.2.1 | $E_c = 57,000\sqrt{f'_c}$ psi     (19.2.2.1b) | $E_c = 57,000\sqrt{5000 \text{ psi}}/1000 = 4030$ ksi |
| 24.2.3.4 | The beam is subjected to a factored distributed force of 0.94 kip/ft or service dead load of 0.58 kip/ft and 0.15 kip/ft service live load.<br><br>The beam is also subjected to concentrated loads from Beam B1 at 5 ft 3 in. and 21 ft 9 in. from Column Line B framing into it having the following reaction; factored 28.5 kip or service dead load of 14.4 kip and 7 kip service live load. Also, Beam B2 is subjected to a concentrated load at midspan from Beam B4 of 36.7 kip factored or 15.1 kip dead service load and 11.6 kip service live load.<br><br>The deflection equation for distributed load with fixity at both ends<br>$\Delta = w\ell^4/384EI$<br><br>For concentrated load at midspan:<br>$\Delta = P\ell^3/192EI$<br><br>Note: $w$ and $P$ are service loads.<br><br>$I_e$ is the effective moment of inertia given by Eq. (25.2.3.5a): | For simplicity, assume that the beam is rectangular for the calculation of the moment of inertia (conservative): |
| 24.2.3.5 | $I_e = \left(\dfrac{M_{cr}}{M_a}\right)^3 I_g + \left[1 - \left(\dfrac{M_{cr}}{M_a}\right)^3\right] I_{cr}$<br><br>where<br><br>$M_{cr} = \dfrac{f_r I_g}{y_t}$     (24.2.3.5b)<br><br>$M_a$ is the moment due to service load.<br><br>For deflection calculation, use moments obtained from elastic analysis. Coefficients from Table B-1 Reinforced Concrete Design Handbook Design Aid – Analysis Tables, which can be downloaded from: https://www.concrete.org/store/productdetail.aspx?ItemID=SP1714DA are used to calculate the moments.<br><br>The beam is assumed cracked; therefore, calculate the moment of inertia of the cracked section, $I_{cr}$. | $I_g = \dfrac{bh^3}{12} = \dfrac{(18 \text{ in.})(30 \text{ in.})^3}{12} = 40,500 \text{ in.}^4$<br><br><br>$M_{cr} = \dfrac{7.5(\sqrt{5000 \text{ psi}})(40,500 \text{ in.}^4)}{(15 \text{ in.})(12,000)} = 120$ ft-kip<br><br>For moment calculation, refer to table below: |

| | | | |
|---|---|---|---|
| | Determine neutral axis of the cracked section: $$nA_s(d-c) = \frac{bc^2}{2} + (n-1)A'_s(c-d')$$ where $c$ is the uncracked remaining concrete depth $n = E_s/E_c$ Cracking moment of inertia, $I_{cr}$: $$I_{cr} = \frac{bc^3}{3} + (n-1)A'_s(c-d')^2 + nA_s(d-c)^2$$ | | For $c$ values, refer to the table below $n$ = 29,000 ksi/4030 ksi = 7.2 For $I_{cr}$ values, refer to the table below |

| | | | |
|---|---|---|---|
| $A'_s$ | 2.64 in.² | 0.88 in.² | 2.64 in.² |
| $A_s$ | 0.88 in.² | 2.64 in.² | 0.88 in.² |
| $c$ | 3.67 in. | 6.5 in. | 3.67 in. |
| $I_{cr}$ | 4016 in.⁴ | 10,314 in.⁴ | 4016 in.⁴ |
| Distributed load $M_{a,dist}$ | | | |
| Dead: | 38 ft-kip | 19 ft-kip | 38 ft-kip |
| Live: | 10 ft-kip | 5 ft-kip | 10 ft-kip |
| Concentrated load $M_{a,conc}$ | | | |
| Dead: | 123 ft-kip | 101 ft-kip | 123 ft-kip |
| Live: | 75 ft-kip | 64 ft-kip | 75 ft-kip |
| Total $M_a$ | 246 ft-kip | 189 ft-kip | 246 ft-kip |
| $I_e$ | 9183 in.⁴ | 15,410 in.⁴ | 9183 in.⁴ |
| Use | 9200 in.⁴ | 15,400 in.⁴ | 200 in.⁴ |

where $M_{a,dist} = \alpha(w_i\text{ kip/ft})(28\text{ ft})^2$; $\alpha = 1/12$ at support and $1/24$ at midspan and $M_{a,conc} = \beta PL$; $\beta = 0.125$ for concentrated load at midspan and $M_a = Pa^2b/\ell^2$, where $a$ is the distance of the concentrated load to the left support

| | | |
|---|---|---|
| 24.2.3.6 | For continuous beams or beams fixed at both ends (positive and negative moments), the Code permits to take $I_e$ as the average of values obtained from Eq.(24.2.3.5a) for the critical positive and negative moments. $$I_{e,avg} = \frac{I_{e,left@supp} + I_{e,right@supp} + I_{e,midspan}}{3}$$ ACI Committee 435 recommends alternate equations to calculate the average equivalent moment of inertia in a beam with two fixed or continuous ends (Eq. (2.15a) of ACI 435R-95). $$I_{e,avg} = 0.7I_{e,midspan} + 0.15(I_{e,left@supp} + I_{e,right@supp})$$ | $$I_{e,avg} = \frac{9200\text{ in.}^4 + 15{,}400\text{ in.}^4 + 9200\text{ in.}^4}{3}$$ $I_{e,avg}$ = 11,267 in.⁴ $I_{e,avg}$ = 0.7(15,400 in.⁴) + 0.15(9200 in.⁴ + 9200 in.⁴) $I_{e,avg}$ = 13,540 in.⁴ |

| | | |
|---|---|---|
| | <u>Immediate deflections</u><br>Deflection due to total distributed load: | $\Delta_{distr} = \dfrac{(0.58 \text{ kip/ft} + 0.15 \text{ kip/ft})(28 \text{ ft})^4 (12)^3}{384(4030 \text{ ksi})(11,267 \text{ in.}^4)} = 0.04$ in. |
| | Deflection due to total concentrated load<br>$P_{D+L} = 15.1 \text{ kip} + 11.6 \text{ kip} = 26.7 \text{ kip}$ | $\Delta_{conc} = \dfrac{(26.7 \text{ kip})(28 \text{ ft})^2 (12)^3}{192(4030 \text{ ksi})(11,267 \text{ in.}^4)} = 0.12$ in. |
| | At midspan:<br>Deflection at midspan due to B1 at 6 ft 3 in. and 21 ft 9 in. from Column Line B.<br>$P_{D+L} = 14.4 \text{ kip} + 7 \text{ kip} = 21.4 \text{ kip}$ | $\Delta_{C.L.} = \dfrac{2(21.4 \text{ kip})(6.25 \text{ ft})^2 (14 \text{ ft})^2 (12)^3}{6(4030 \text{ ksi})(11,267 \text{ in.}^4)(28 \text{ ft})^3}$<br>$\times (3(21.75 \text{ ft})(28 \text{ ft}) - 3(21.75 \text{ ft})(14 \text{ ft})$<br>$- (6.25 \text{ ft})(14 \text{ ft})) = 0.08$ in. |
| | Equation is obtained from Reinforced Concrete Design Handbook Design Aid – Analysis Tables, which can be downloaded from: https://www.concrete.org/store/productdetail.aspx?ItemID=SP1714DA: | |
| 24.2.2 | Total deflection:<br>Check the deflection against the maximum allowable limits in Table 24.2.2: | $\Delta_{D+L} = 0.04$ in. $+ 0.12$ in. $+ 0.08$ in. $= 0.24$ in.<br>$\Delta_{all.} = (28 \text{ ft})(12 \text{ in./ft})/480 = 0.7$ in.<br>$\Delta_{all.} = 0.7$ in. $\gg \Delta = 0.21$ in.  **OK** |
| | For floor supporting or attached to nonstructural elements and not likely to be damaged by large deflection: $\ell/480$ | A shallower beam could have been selected. But because of Beam B1 (28 in. deep) that is framing into it, a 30 in. deep beam was chosen. |

| | | |
|---|---|---|
| | **Long-term deflections**<br>Deflections due to live load are the difference between total deflection and dead load deflection.<br><br>Deflection due to dead load:<br><br>Distributed load: | $\Delta_{distr} = \dfrac{(0.58 \text{ kip/ft})(28 \text{ ft})^4 (12)^3}{384(4030 \text{ ksi})(11,267 \text{ in.}^4)} = 0.04$ in. |
| | Concentrated load at midspan: | $\Delta_{conc} = \dfrac{(15.1 \text{ kip})(28 \text{ ft})^3 (12)^3}{192(4030 \text{ ksi})(11,267 \text{ in.}^4)} = 0.07$ in. |
| | Concentrated load at 6.25 ft and 21.75 ft, respectively. | $\Delta_{C.L.} = \dfrac{2(14.4 \text{ kip})(6.25 \text{ ft})^2 (14 \text{ ft})^2 (12)^3}{6(4030 \text{ ksi})(11,267 \text{ in.}^4)(28 \text{ ft})^3}$<br>$\times (3(21.75 \text{ ft})(28 \text{ ft}) - 3(21.75 \text{ ft})(14 \text{ ft})$<br>$-(6.25 \text{ ft})(14 \text{ ft})) = 0.05$ in. |
| | Total dead load deflection: | $\Delta_D = 0.04$ in. $+ 0.07$ in. $+ 0.05$ in. $= 0.16$ in. |
| | Deflection due to live load: | $\Delta_L = 0.24$ in. $- 0.16$ in. $= 0.08$ in. |
| 24.2.4.1.1 | Calculate long-term deflection:<br>$$\lambda_\Delta = \dfrac{\xi}{1 + 50\rho'}$$ | $\lambda_\Delta = \dfrac{2.0}{1 + 50 \dfrac{0.88 \text{ in.}^2}{(18 \text{ in.})(27.7 \text{ in.})}} = 1.84$ |
| 24.2.4.1.3 | From Table 24.2.4.1.3, the time-dependent factor for sustained load duration of more than 5 years:<br>$\xi = 2.0$<br><br>Long-term deflection due to sustained load is:<br><br>$\Delta_T = (1 + \lambda_\Delta)\Delta_i$ | By inspection, this is satisfied. |

## Step 11: Detailing

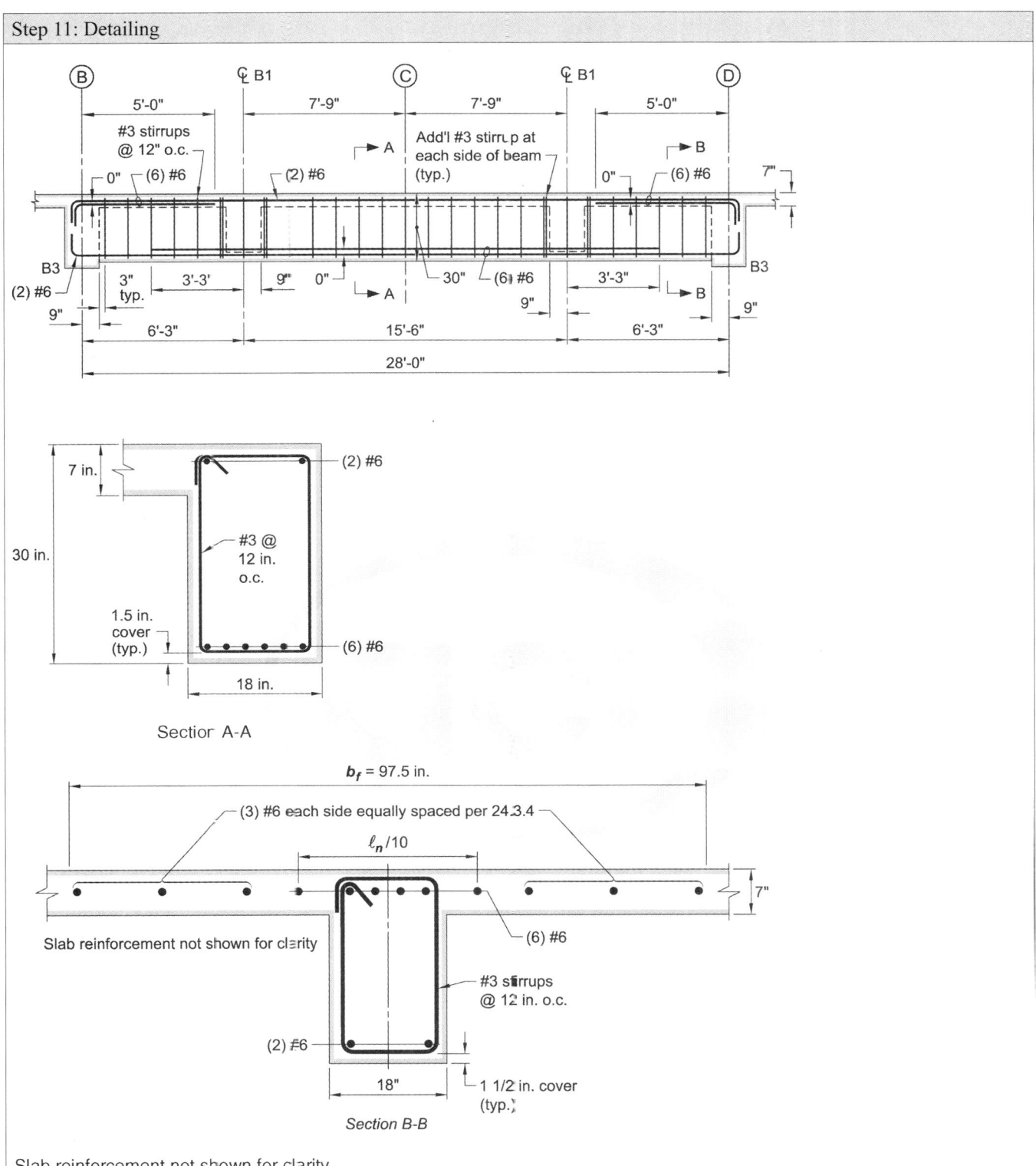

Slab reinforcement not shown for clarity.

*Fig. E3.11—Beam reinforcement details.*

**Beam Example 4**: *Continuous edge beam*

Determine the size of a continuous six-bay edge beam built integrally with a 7 in. slab on the exterior of the building. Design and detail the beam. Ignore openings at Column Lines 3 and 5.

**Given:**

$f_c' = 5000$ psi (normalweight concrete)
$f_y = 60,000$ psi

Beam width: 18 in.
Beam height: 30 in.

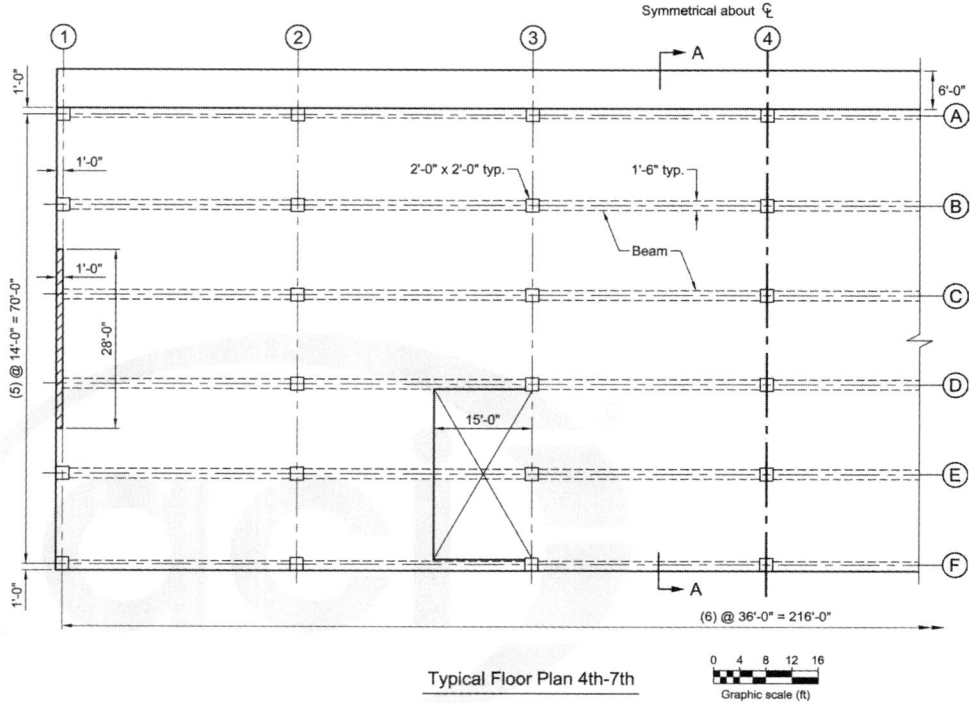

*Fig. E4.1—Plan of a -six-span perimeter beam.*

| ACI 318-14 | Discussion | Calculation |
|---|---|---|
| **Step 1: Material requirements** | | |
| 9.2.1.1 | The mixture proportion must satisfy the durability requirements of Chapter 19 and structural strength requirements of ACI 318-14. The designer determines the durability classes. Please refer to Chapter 4 of SP-17 for an in-depth discussion of the Categories and Classes. ACI 301 is a reference specification that is coordinated with ACI 318. ACI encourages referencing ACI 301 into job specifications. There are several mixture options within ACI 301, such as admixtures and pozzolans, which the designer can require, permit, or review if suggested by the contractor. | By specifying that the concrete mixture shall be in accordance with ACI 301-10 and providing the exposure classes, Chapter 19 requirements are satisfied. Based on durability and strength requirements, and experience with local mixtures, the compressive strength of concrete is specified at 28 days to be at least 5000 psi. |

## Step 2: Beam geometry

| | | |
|---|---|---|
| 9.3.1.1 | **Beam depth**<br>If the depth of a beam satisfies Table 9.3.1.1, ACI 318 permits a beam design without having to check deflections, if the beam is not supporting or attached to partitions or other construction likely to be damaged by large deflections. Otherwise, beam deflections must be calculated and the deflection limits in 9.3.2 must be satisfied. | The beam has six continuous spans. Taking the controlling condition of having one end continuous:<br>$h = \ell = \dfrac{(36\ \text{ft})(12\ \text{in./ft})}{18.5} = 23.4\ \text{in.}$<br>Use 30 in. |
| | **Self-weight**<br>Beam:<br>Slab:<br><br>Façade: assume facade weight is 35 psf spanning 12 ft-0 in. vertically | $w_b = [(18\ \text{in.})(30\ \text{in.})/(144)](0.150\ \text{kip/ft}^3) = 0.56\ \text{kip/ft}$<br>$w_s = [((14\ \text{ft} - 15\ \text{in.}/12)/2)(7\ \text{in.}/12)](0.150\ \text{kip/ft}^3)$<br>$\quad = 0.56\ \text{kip/ft}$<br><br>$w_{cladding} = (35\ \text{psf})(12\ \text{ft})/1000 = 0.42\ \text{kip/ft}$ |
| 9.2.4.2<br><br>6.3.2.1 | **Flange width**<br>The beam is placed monolithically with the slab and will behave as an L-beam. The flange width to one side of the beam is obtained from Table 6.3.2.1.<br><br>One side of web is the least of $\begin{cases} 6h_{slab} \\ s_w/2 \\ \ell_n/12 \end{cases}$<br><br>Flange width: $b_f = \ell_n/12 + b_w$ | <br><br><br><br>$(6)(7\ \text{in.}) = 42\ \text{in.}$<br>$[((14\ \text{ft})(12) - 15\ \text{in.})/2 = 76.5\ \text{in.}$<br>$(34\ \text{ft})(12)/12 = 34\ \text{in.}$ **Controls**<br>$b_f = 34\ \text{in.} + 18\ \text{in.} = 52\ \text{in.}$ |

## Step 3: Loads and load patterns

| | | |
|---|---|---|
| | The service live load is 50 psf in offices and 80 psf in corridors per Table 4-1 in ASCE 7-10. This example will use 65 psf as an average as the actual layout is not provided.<br>To account for the weight of ceilings, partitions, and mechanical (HVAC) systems, add 15 psf as miscellaneous dead load.<br>The beam resists gravity load only and lateral forces are not considered in this problem. | |
| 5.3.1 | $U = 1.4D$<br><br><br><br>$U = 1.2D + 1.6L$ | $w_u = 1.4(0.56\ \text{kip/ft} + 0.56\ \text{kip/ft} + 0.42\ \text{kip/ft}$<br>$\quad + (15\ \text{psf})((14\ \text{ft})/2)/1000))$<br>$\quad = 2.3\ \text{kip/ft}$<br><br>$w_u = 1.2(2.3\ \text{kip/ft})/1.4 + 1.6(65\ \text{psf}/1000)(14\ \text{ft})/2$<br>$\quad = 2.7\ \text{kip/ft}$ **Controls** |

| | Step 4: Analysis | |
|---|---|---|
| 9.4.3.1 | The beams are built integrally with supports; therefore, the factored moments and shear forces (required strengths) are calculated at the face of the supports. | Clear span: $\ell_n = 36 \text{ ft} - 2 \text{ ft} = 34 \text{ ft}$ |
| 9.4.1.2 6.5.1 | Chapter 6 permits several analysis procedures to calculate the required strengths. The beam required strengths can be calculated using approximations per Table 6.5.2, if the conditions in Section 6.5.1 are satisfied: | |
| | Members are prismatic | Beams are prismatic |
| | Loads uniformly distributed $L \leq 3D$ | Satisfied (no concentrated loads) 65 psf < 3(87.5 psf + 15 psf + beam self-weight   **OK** |
| | Three spans minimum | 6 spans > 3 spans |
| | Difference between two spans does not exceed 20 percent. | Beams have equal clear span lengths 34 ft-0 in. All five conditions are satisfied; therefore, the approximate procedure is used. |
| 6.5.4 | *Shear diagram* | |

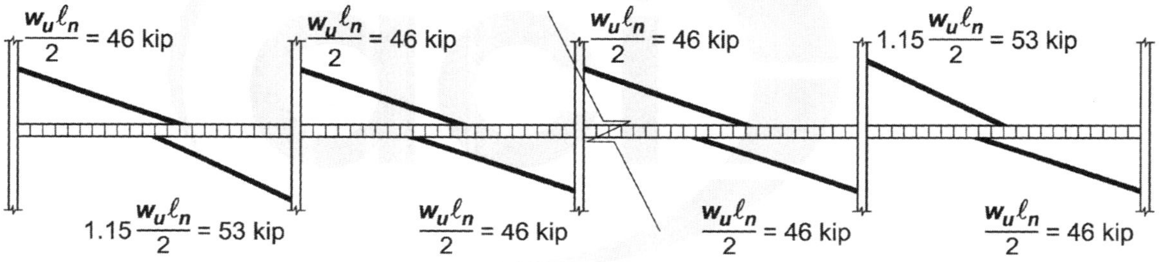

$\dfrac{w_u \ell_n}{2} = 46 \text{ kip}$   $\dfrac{w_u \ell_n}{2} = 46 \text{ kip}$   $\dfrac{w_u \ell_n}{2} = 46 \text{ kip}$   $1.15\dfrac{w_u \ell_n}{2} = 53 \text{ kip}$

$1.15\dfrac{w_u \ell_n}{2} = 53 \text{ kip}$   $\dfrac{w_u \ell_n}{2} = 46 \text{ kip}$   $\dfrac{w_u \ell_n}{2} = 46 \text{ kip}$   $\dfrac{w_u \ell_n}{2} = 46 \text{ kip}$

6.5.2   *Moment diagram*

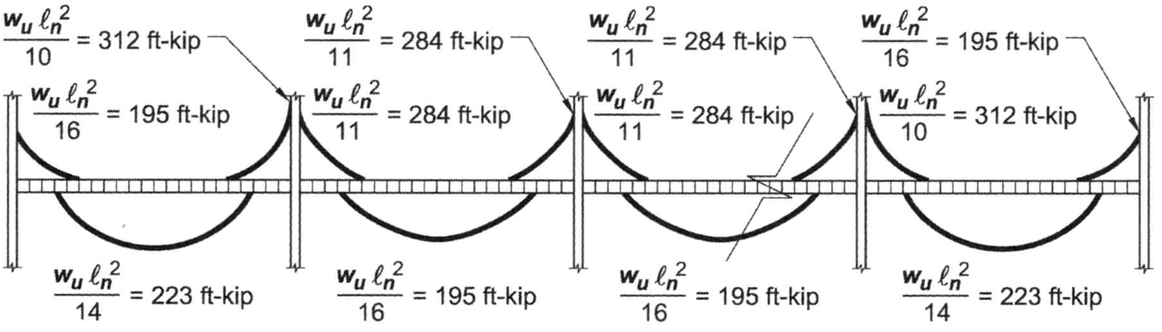

$\dfrac{w_u \ell_n^2}{10} = 312 \text{ ft-kip}$   $\dfrac{w_u \ell_n^2}{11} = 284 \text{ ft-kip}$   $\dfrac{w_u \ell_n^2}{11} = 284 \text{ ft-kip}$   $\dfrac{w_u \ell_n^2}{16} = 195 \text{ ft-kip}$

$\dfrac{w_u \ell_n^2}{16} = 195 \text{ ft-kip}$   $\dfrac{w_u \ell_n^2}{11} = 284 \text{ ft-kip}$   $\dfrac{w_u \ell_n^2}{11} = 284 \text{ ft-kip}$   $\dfrac{w_u \ell_n^2}{10} = 312 \text{ ft-kip}$

$\dfrac{w_u \ell_n^2}{14} = 223 \text{ ft-kip}$   $\dfrac{w_u \ell_n^2}{16} = 195 \text{ ft-kip}$   $\dfrac{w_u \ell_n^2}{16} = 195 \text{ ft-kip}$   $\dfrac{w_u \ell_n^2}{14} = 223 \text{ ft-kip}$

*Fig. E4.2—Shear and moment diagrams.*

6.5.3   Note: The moments calculated using the approximate method cannot be redistributed in accordance with Section 6.6.5.1.

The slab load is eccentric with respect to the edge beam center. Therefore, the beam needs to resist a torsional moment (Fig. E4.3).

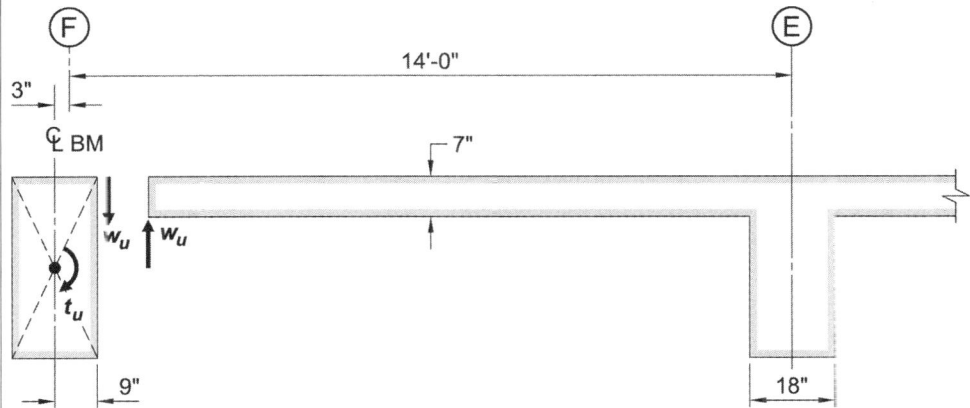

Fig. E4.3—Torsion forces.

Load at slab/beam interface:
$w_u = [((1.2)((7 \text{ in.}/12)(0.15 \text{ kip/ft}^3) + (0.015 \text{ ksf})) + (1.6)(0.065 \text{ ksf}))][(14 \text{ ft} – 1.5 \text{ ft}/2)/2 + 3 \text{ in.}/12]$
$w_u = 1.56 \text{ kip/ft}$

The torsional moment along the beam length is (Fig. E4.4):
$t_u = w_u(b_w/2) = (1.56 \text{ kip/ft})(18 \text{ in.}/2/12) = 1.17 \text{ ft-kip/ft}$

Factored torsional moment at the face of columns: $T_u = (1.17 \text{ ft-kip/ft})(34 \text{ ft})/2 = 20 \text{ ft-kip}$

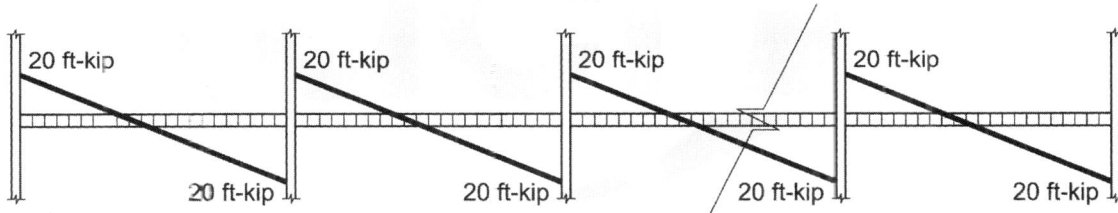

Fig. E4.4—Torsion diagram.

## Step 5: Moment design

| | | |
|---|---|---|
| 9.3.3.1 | The Code does not permit a beam to be designed with steel strain less than 0.004 in./in. at design strength. The intent is to ensure ductile behavior at the factored condition.<br>In most reinforced beams, such as this example, reinforcing bar strain is not a controlling issue. | |
| 21.2.1(a) | The design assumption is that beams will be tension-controlled, $\phi = 0.9$. This assumption will be checked later. | |
| 9.7.1.1<br>20.6.1.3.1 | Determine the effective depth assuming No. 4 stirrups, No. 6 bars, and 1.5 in. cover:<br>One row of reinforcement<br><br>$d = h - \text{cover} - d_{tie} - d_b/2$ | $d = 30 \text{ in.} - 1.5 \text{ in.} - 0.5 \text{ in.} - 0.75 \text{ in.}/2 = 27.6 \text{ in.}$ |
| 22.2.2.1 | The concrete compressive strain at nominal moment strength is:<br>$\varepsilon_{cu} = 0.003$ | |
| 22.2.2.2 | The tensile strength of concrete in flexure is a variable property and is approximately 10 to 15 percent of the concrete compressive strength. ACI 318 neglects the concrete tensile strength to calculating nominal strength. | |
| 22.2.2.3 | Determine the equivalent concrete compressive stress at nominal strength: | |
| 22.2.2.4.1 | The concrete compressive stress distribution is inelastic at high stress. The Code permits any stress distribution to be assumed in design if shown to result in predictions of ultimate strength in reasonable agreement with the results of comprehensive tests. Rather than tests, the Code allows the use of an equivalent rectangular compressive stress distribution of $0.85f_c'$ with a depth of: | |
| 22.2.2.4.3<br>22.2.1.1 | $a = \beta_1 c$, where $\beta_1$ is a function of concrete compressive strength and is obtained from Table 22.2.2.4.3. For $f_c' \leq 5000$ psi: | $\beta_1 = 0.85 - \dfrac{0.05(5000 \text{ psi} - 4000 \text{ psi})}{1000 \text{ psi}} = 0.8$ |
| | Find the equivalent concrete compressive depth, $a$, by equating the compression force to the tension force within the beam cross section (Fig. E4.5):<br>$C = T$<br>$0.85f_c'ba = A_s f_y$ | $0.85(5000 \text{ psi})(b)(a) = A_s(60,000 \text{ psi})$ |
| | For positive moment: $b = b_f = 52$ in. | $a = \dfrac{A_s(60,000 \text{ psi})}{0.85(5000 \text{ psi})(52 \text{ in.})} = 0.271 A_s$ |
| | For negative moment: $b = b_w = 18$ in. | $a = \dfrac{A_s(60,000 \text{ psi})}{0.85(5000 \text{ psi})(18 \text{ in.})} = 0.784 A_s$ |

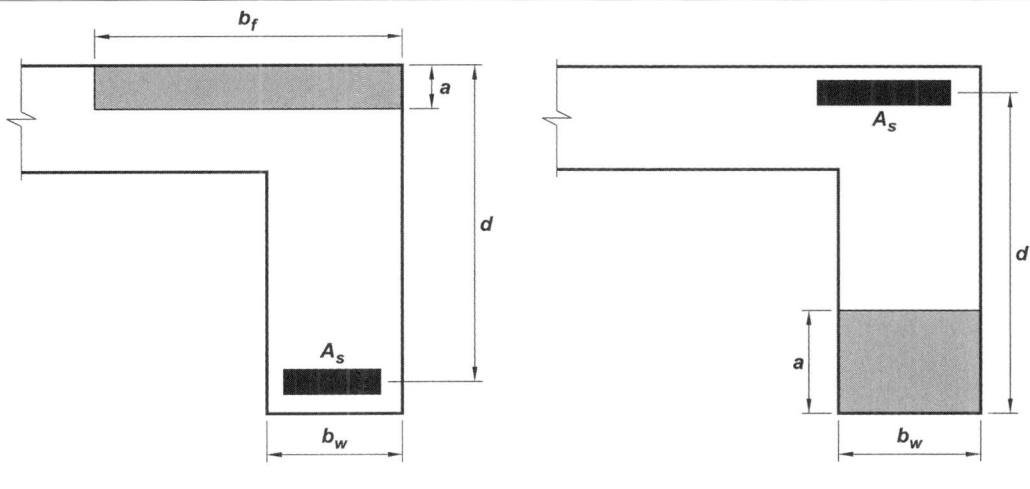

Fig. E4.5—Section compressive block and reinforcement at midspan and at support.

| | | |
|---|---|---|
| | The beam is designed for the maximum flexural moments obtained from the approximate method above. | |
| | The first interior support will be designed for the larger of the moments on either side of the column; | The first interior moment: $M_{max} = 306$ ft-kip |
| 9.5.1.1 | The beam design strength must be at least the required strength at each section along its length (Fig. E 4.6): $\phi M_n \geq M_u$ $\phi V_n \geq V_u$ | 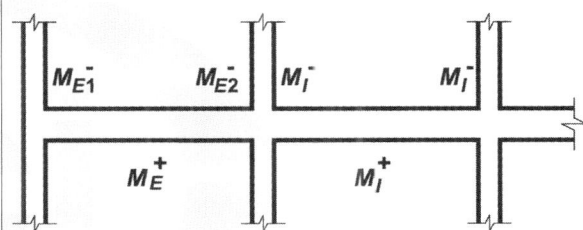<br>Fig. E4.6—Moment key to use with table below. |

Table 4.1—Required reinforcement to resist factored moment

| | | | Number of No. 6 bars | |
|---|---|---|---|---|
| | $M_u$, ft-kip | $A_{s,req'd}$, in.$^2$ | Req'd | Prov. |
| $M_{E1}^-$ | 195 | 1.59 | 3.6 | 4 |
| $M_{E2}^-$ | 312 | 2.6 | 5.91 | 6 |
| $M_I^-$ | 284 | 2.36 | 5.36 | 6 |
| $M_E^+$ | 223 | 1.81 | 4.11 | 5 |
| $M_I^+$ | 195 | 1.57 | 3.57 | 4 |

| | | |
|---|---|---|
| | Calculate required reinforcement area: $\phi M_n \geq M_u = \phi A_s f_y \left( d - \frac{a}{2} \right)$ Each No. 6 bar has a $d_b = 0.75$ in. and an $A_s = 0.44$ in.$^2$ | |
| 21.2.2 9.3.3.1 | Check if the calculated strain exceeds 0.005 in./in. (tension-controlled—Fig. E4.7) but not less than 0.004 in./in. $a = \frac{A_s f_y}{0.85 f_c' b}$ and $c = \frac{a}{\beta_1}$ where $\beta_1 = 0.8$ (calculated above) Note that $b = 18$ in. for negative moments and 57.75 in. for positive moments (refer to Fig. E4.5). $\varepsilon_t = \frac{\varepsilon_{cu}}{c}(d-c)$ | Table 4.2—Strain in tension bars |

| | $M_u$, ft-kip | $A_{s,prov}$, in.$^2$ | $a$, in. | $\varepsilon_s$, in./in | $\varepsilon_s > 0.005$? |
|---|---|---|---|---|---|
| $M_{E1}^-$ | 195 | 1.76 | 1.38 | 0.045 | Y |
| $M_{E2}^-$ | 312 | 2.64 | 2.07 | 0.029 | Y |
| $M_I^-$ | 284 | 2.64 | 2.07 | 0.029 | Y |
| $M_E^+$ | 223 | 2.20 | 0.60 | 0.108 | Y |
| $M_I^+$ | 195 | 1.76 | 0.48 | 0.136 | Y |

Therefore, assumption of using $\phi = 0.9$ is correct.

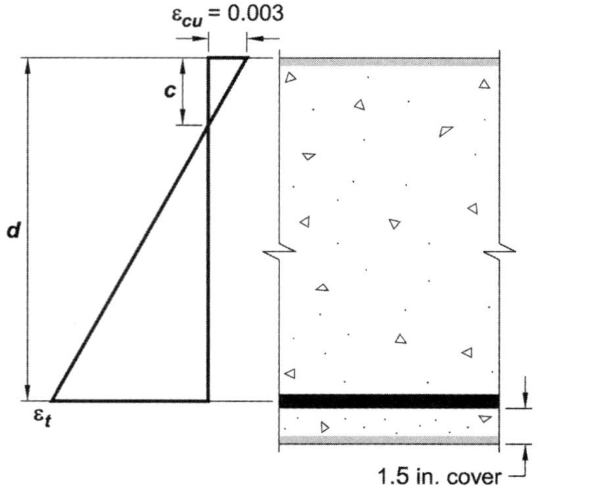

| | | |
|---|---|---|
| | | *Fig. E4.7—Strain distribution across beam section.* |
| 9.6.1.1<br>9.6.1.2 | **Minimum reinforcement**<br>The reinforcement area must be at least the minimum required reinforcement area at every section along the length of the beams.<br><br>$$A_s = \frac{3\sqrt{f'_c}}{f_y} b_w d$$<br><br>Equation (9.6.1.2a) controls because $f'_c > 4444$ psi | $A_s = \dfrac{3\sqrt{5000 \text{ psi}}}{60{,}000 \text{ psi}}(18 \text{ in.})(27.6 \text{ in.}) = 1.76 \text{ in.}^2$ **Controls**<br><br>All calculated reinforcement areas exceed the minimum required reinforcement area. Therefore, **OK** |

# CHAPTER 7—BEAMS

## External spans

### Step 6: Shear design

| | | |
|---|---|---|
| | Shear strength | |
| | The shear forces in the external and internal spans are relatively equal: 53 kip versus 46 kip; therefore, the continuous beam will be designed for 53 kip of shear force | |
| 9.4.3.2 | Because conditions (a), (b), and (c) of 9.4.3.2 are satisfied, the design shear force critical section is taken at a distance $d$ from the face of the support (Fig. E4.8) | Fig. E4.8—Shear critical section. |
| | | $V_{u@d} = (53\text{ kip}) - (2.7\text{ kip/ft})(27.6\text{ in.}/12) = 47\text{ kip}$ |
| | The controlling factored load combination must satisfy: | |
| 9.5.1.1 | $\phi V_n \geq V_u$ | |
| 9.5.3.1 | $V_n = V_c + V_s$ | |
| 22.5.1.1 | | |
| 22.5.5.1 | $V_c = 2\sqrt{f_c'}b_w d$ | $V_c = (2)\sqrt{5000\text{ psi}}(18\text{ in.})(27.6\text{ in.}) = 70.3\text{ kip}$ |
| 21.2.1(b) | Shear strength reduction factor: $\phi V_c = \phi 2\sqrt{f_c'}b_w d$ | $\phi_{shear} = 0.75$ <br> $\phi V_c = (0.75)(70.3\text{ kip}) = 52.7\text{ kip}$ |
| 9.5.1.1b | Check if $\phi V_n \geq V_u$ | $\phi V_c = 52.7\text{ kip} > V_u = 47\text{ kip}$ **OK** <br> Therefore, shear reinforcement is not required. |
| 9.6.3.1 | Code requires that minimum shear reinforcement must be provided over sections where $V_u \geq (1/2)\phi V_c$ | $V_u = 47\text{ kip} > (1/2)\phi V_c = 1/2(52.7\text{ kip}) = 26.4\text{ kip}$ |
| | Check if the cross-sectional dimensions satisfy Eq. (22.5.1.2): | |
| 22.5.1.2 | $V_u \leq \phi(V_c + 8\sqrt{f_c'}b_w d)$ | Therefore, provide minimum shear reinforcement over full beam length. (Refer to torsion calculation Step 7). <br><br> By inspection, this condition is satisfied and section dimensions are satisfactory. |

## Step 7: Torsion design

| | | |
|---|---|---|
| 9.4.4.3 | Torsion design<br>Calculate the torsional moment at $d$ from the face of the support:<br>$t_u$ = 1.17 ft-kip/ft and $T_u$ = 20 ft-kip (Step 4, Fig. E4.3) | $T_{u@d}$ = (20 ft-kip) – (1.17 ft-kip/ft)(27.6 in.)/12 = 17.3 ft-kip |
| 9.2.4.4 | Determine the concrete section resisting torsion.<br><br>The overhanging flange dimension is equal to the smaller of the projection of the beam below the slab (23 in.) and four times the slab thickness (28 in.). Therefore, 23 in. controls (Fig. E4.9). | *Fig. E4.9— L-beam geometry to resist torsion.* |
| 22.7.4.1 | Calculate the threshold torsion value:<br><br>$$T_{th} = \lambda\sqrt{f'_c}\left(\frac{A_{cp}^2}{P_{cp}}\right)$$ | |
| | where $A_{cp}$ is the area enclosed by outside perimeter of concrete cross section; and | $A_{cp}$ = (18 in.)(30 in.) + (23 in.)(7 in.) = 701 in.$^2$ |
| | $p_{cp}$ is the outside perimeter of concrete cross section | $p_{cp}$ = 2(18 in. + 2(23 in.) + 7 in.) = 142 in.<br><br>$T_{th} = (1.0)\sqrt{5000 \text{ psi}}\left(\frac{(701 \text{ in.}^2)^2}{142 \text{ in.}}\right)$ = 20.4 ft-kip |
| 21.2.1(c) | Torsion strength reduction factor $\phi$ = 0.75 | $\phi T_{th}$ = (0.75)(20.4 ft-kip) = 15.3 ft-kip |
| 9.5.4.1 | Check if torsion can be ignored; does $T_{u@d} < \phi T_{th}$? | $T_{u@d}$ = 17.3 kip > $\phi T_{th}$ = 15.3 ft-kip **NG**<br><br>Torsional effects cannot be neglected and reinforcement and detailing requirements for torsion must be considered. |

# CHAPTER 7—BEAMS

| | Torsion reinforcement | |
|---|---|---|
| 9.5.1.1<br>9.5.4.2<br>22.7.5.1 | Calculate cracking torsion:<br><br>$$T_{cr} = 4\lambda\sqrt{f'_c}\left(\dfrac{A_{cp}^2}{P_{cp}}\right)$$ | $T_{cr} = 4(1.0)\sqrt{5000\text{ psi}}\left(\dfrac{(701\text{ in.}^2)^2}{142\text{ in.}}\right) = 81.5$ ft-kip |
| 22.7.3.2 | Check if cross section will crack under the torsional moment. | $T_u = 17.3$ ft-kip $< T_{cr} = 87$ ft-kip   **OK**<br>Reducing $T_u$ to $T_{cr}$ is not required. |

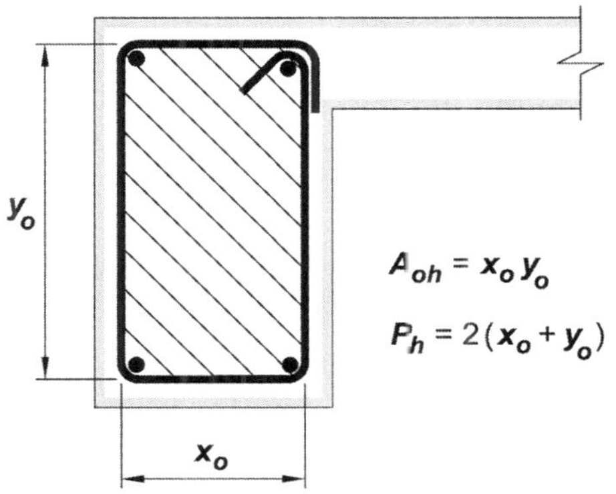

Fig. E4.10—$A_{oh}$ area.

$A_{oh} = x_o y_o$
$P_h = 2(x_o + y_o)$

$p_h = 2[(18\text{ in.} - 2(1.5\text{ in.}) - 0.5\text{ in.})$
$\quad + (30\text{ in.} - 2(1.5\text{ in}) - 0.5\text{ in.})] = 82$ in.

$A_{oh} = (14.5\text{ in.})(26.5\text{ in.}) = 384.25$ in.$^2$

| | Check if cross section is adequate to resist the torsional moment. | |
|---|---|---|
| 22.7.7.1 | $$\sqrt{\left(\dfrac{V_u}{b_w d}\right)^2 + \left(\dfrac{T_u p_h}{1.7 A_{oh}^2}\right)^2} \le \phi\left(\dfrac{V_c}{b_w d} + 8\sqrt{f'_c}\right)$$ | $\sqrt{\left(\dfrac{46{,}000\text{ lb}}{(18\text{ in.})(27.6\text{ in.})}\right)^2 + \left(\dfrac{(17.3\text{ ft-kip})(12\times 10^3)(82\text{ in.})}{1.7(384.25\text{ in.}^2)^2}\right)^2}$<br><br>$\le (0.75)\left(\dfrac{70{,}300\text{ lb}}{(18\text{ in.})(27.6\text{ in.})} + 8\sqrt{5000\text{ psi}}\right)$ |
| | where<br>$p_h$ is the perimeter of centerline of outermost closed transverse torsional reinforcement; and $A_{oh}$ is the area enclosed by centerline of the outermost closed transverse torsional reinforcement | 115 psi $<$ 530 psi   **OK**<br><br>Therefore section is adequate to resist torsion. |
| 9.7.5<br>9.7.6 | Calculate required transverse and longitudinal torsion reinforcement: | |
| 22.7.6.1 | Transverse: $T_n = \dfrac{2A_o A_t f_{yt}}{s}\cot\theta$   (22.7.6.1a) | $\dfrac{T_{u@d}}{\phi} = \dfrac{(17.3\text{ ft-kip})(12)}{0.75} \le T_n = \dfrac{2(327\text{ in.}^2)A_t(60\text{ ksi})}{s}\cot 45$<br>$A_t/s = 0.007$ in.$^2$/in. |
| 22.7.6.1.2 | Longitudinal: $T_n = \dfrac{2A_o A_\ell f_y}{p_h}\tan\theta$   (22.7.6.1b)<br><br>where: $30 \le \theta \le 60$  use $\theta = 45$ degrees | $\dfrac{T_{u@d}}{\phi} = \dfrac{(17.3\text{ ft-kip})(12)}{0.75} \le T_n = \dfrac{2(327\text{ in.}^2)A_\ell(60\text{ ksi})}{82\text{ in.}}\tan 45$<br>$A_\ell \ge 0.58$ in.$^2$ |
| 22.7.6.1.1 | $A_o = 0.85 A_{oh}$ is the gross area enclosed by torsional shear flow path. | $A_o = 0.85(384.25\text{ in.}^2) = 327$ in.$^2$ |

| | | |
|---|---|---|
| 9.5.4.3 | The required area for shear and torsional transverse reinforcement are additive:<br><br>$\dfrac{A_{v+t}}{s} = \dfrac{A_v}{s} + 2\dfrac{A_t}{s}$<br><br>$A_v = 0$ in.$^2$<br>$A_t$ is defined in terms of one leg. Therefore, $A_t$ is multiplied by 2.<br><br>Calculate the maximum spacing of stirrups at $d$ from the column face. | $\dfrac{A_{v+t}}{s} = 0$ in.$^2$/in. $+ 2(0.007$ in.$^2$/in.$) = 0.014$ in.$^2$/in. |
| 9.7.6.3.3 | Maximum spacing of transverse torsional reinforcement must not exceed the lesser of $p_h/8$ and 12 in. | Assume No. 4 stirrup<br>$p_h = 82$ in. calculated above<br>$p_h/8 = 82$ in./8 $= 10$ in. $< 12$ in.; use 10 in. |
| 9.6.4.2 | Check maximum transverse torsional reinforcement: $(A_v + 2A_t)_{min}/s$ must be greater than:<br><br>$0.75\sqrt{f'_c}\dfrac{b_w}{f_{yt}}$<br><br>and<br><br>$50\dfrac{b_w}{f_{yt}}$<br><br>$A_v = 0$ in.$^2$ because calculations showed that shear reinforcement is not required. Minimum shear reinforcement is, however, provided.<br><br>$A_t = (2)(0.20$ in.$^2) = 0.4$ in.$^2$<br><br>Use two legs for torsional reinforcement: | $\dfrac{(A_{v+t})_{min}}{s} \geq 0.75\sqrt{5000\text{ psi}}\,\dfrac{18\text{ in.}}{60{,}000\text{ psi}} = 0.016$ in.<br><br>$\dfrac{(A_{v+t})_{min}}{s} = 50\dfrac{18\text{ in.}}{60{,}000\text{ psi}} = 0.015$ in.<br><br>Provided: $\dfrac{A_{v+t}}{s} = \dfrac{(2)(0.2\text{ in.}^2/\text{in.})}{10} = 0.04$ in.$^2$/in.<br><br>$\dfrac{A_{v+t}}{s} = 0.04$ in.$^2$/in. $> \dfrac{(A_{v+t})_{min}}{s} = 0.016$ in. **OK** |
| 9.6.4.3 | The torsional longitudinal reinforcement $A_{\ell,min}$ must be the lesser of:<br><br>$A_{\ell,min} = \dfrac{5\sqrt{f'_c}A_{cp}}{f_y} - \left(\dfrac{A_t}{s}\right)p_h\dfrac{f_{yt}}{f_y}$<br><br>$A_{\ell,min} = \dfrac{5\sqrt{f'_c}A_{cp}}{f_y} - \left(\dfrac{25b_w}{f_{yt}}\right)p_h\dfrac{f_{yt}}{f_y}$<br><br>$A_p = 701$ in.$^2$ calculated above | $A_{\ell,min} = \dfrac{5\sqrt{5000\text{ psi}}(701\text{ in.}^2)}{60{,}000\text{ psi}} - (0.02\text{ in.}^2/\text{in.})(82\text{ in.})\dfrac{60\text{ ksi}}{60\text{ ksi}}$<br><br>$= 2.49$ in.$^2$  **Controls**<br><br>$A_{\ell,min} = \dfrac{5\sqrt{5000\text{ psi}}(701\text{ in.}^2)}{60{,}000\text{ psi}} - \left(\dfrac{25(18\text{ in.})}{60{,}000\text{ psi}}\right)(82\text{ in.})\dfrac{60\text{ ksi}}{60\text{ ksi}}$<br><br>$= 3.5$ in.$^2$<br><br>$A_{\ell,calc.} = 0.55$ in.$^2 < A_{l,req'd} = 2.49$ in.$^2$  **OK**<br><br>The longitudinal reinforcement must be added to the flexural reinforcement. |

| | | |
|---|---|---|
| 9.5.4.3 | Torsion longitudinal reinforcement, $A_\ell$, must be distributed around the cross section and the portion of $A_\ell$ that needs to be placed where $A_s$ is needed is added to $A_s$ found in Step 5, Table 4.1.<br>Assume that two No. 6 bars will be added at each side face and the remainder will be divided equally between top and bottom of beam with one in each corner.<br>$\Delta A_\ell = A_\ell - 4A_{No.6}$<br><br>Add 0.7 in.$^2$/2 = 0.35 in.$^2$ to $M^-$ and $M^+$ from Step 5, Table 1. | $\Delta A_\ell = (2.49\text{ in.}^2 - 4(0.44\text{ in.}^2)) = 0.7\text{ in.}^2$<br><br>**Table 4.3—Total longitudinal reinforcement at tension side** |

Table 4.3—Total longitudinal reinforcement at tension side

| | $A_{s,req'd}$, in.$^2$ | $\Delta A_\ell/2$, in.$^2$ | $A_s - \Delta A_\ell$, in.$^2$ | Number of No. 6 bars Req'd | Number of No. 6 bars Prov. |
|---|---|---|---|---|---|
| $M_{E1}^-$ | 1.59 | 0.35 | 1.94 | 4.4 | 5 |
| $M_{E2}^-$ | 2.6 | 0.35 | 2.95 | 6.7 | 7 |
| $M_I^-$ | 2.36 | 0.35 | 2.71 | 6.2 | 7 |
| $M_E^+$ | 1.81 | 0.35 | 2.16 | 4.9 | 5 |
| $M_I^+$ | 1.57 | 0.35 | 1.92 | 4.4 | 5 |

| | | |
|---|---|---|
| 9.7.5.1<br>9.7.5.2 | The spacing of longitudinal torsional reinforcement should not exceed 12 in. on center and the minimum diameter must be greater than 0.042 times the transverse reinforcement spacing, but not less than 3/8 in. | 12 in. spacing between longitudinal reinforcement is satisfied; refer to Fig. E4.7<br><br>$d_{b,min} = (0.042)(10\text{ in.}) = 0.42\text{ in.}$<br>$d_{b,No.6} = 0.75\text{ in.} > d_{b,min} = 0.42\text{ in.}$  **OK** |
| | <u>Typical span reinforcement due to torsion moment</u><br>The torsional moment varies from maximum at the face of the support to zero at span mid-length. Theoretically, torsional reinforcement is required over a distance equal to:<br><br>$x = \dfrac{\phi T_{th}}{T_u} \ell_n/2$ | $x = \dfrac{((20\text{ ft-kip}) - (15.3\text{ ft-kip}))(34\text{ ft}/2)}{20\text{ ft-kip}} = 4\text{ ft}$<br><br>from the face of the support |
| 9.7.5.3 | Longitudinal torsional reinforcement must be developed beyond this length a minimum of $x_o + d$ (Fig. E.11). | $x_o + d = 14.5\text{ in.} + 27.6\text{ in.} = 42.1\text{ in.}$ |
| 9.7.5.4 | Develop longitudinal reinforcement at face of support. | Therefore, bars due to torsion moment must extend a minimum distance of:<br>(4 ft)(12) + 42.1 in. = 90.1 in., say, 7 ft 6 in. from the face of the support on both sides of the span. |
| | Note: Bars can be discontinued per the calculations above. Practically, however, bars are extended over the full length of the beam. | |
| | <u>Stirrup spacing:</u> | Extend stirrups No. 4 at 10 in. on center over 7 ft from supports. Remainder of span provide No. 4 stirrups at 12 in. on center. |

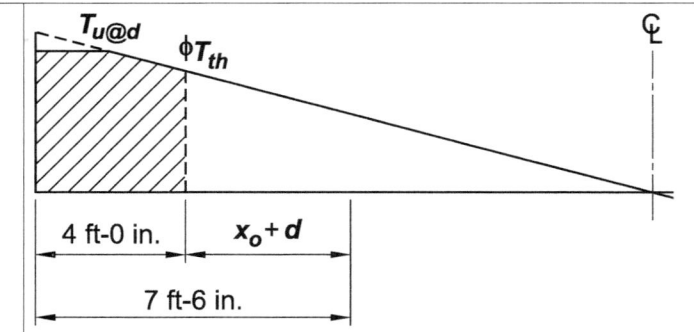

Fig. E4.11—Typical torsion reinforcement in a span applied at both ends of a span.

| Step 8: Reinforcement detailing | | |
|---|---|---|
| 9.7.2.1<br><br>25.2.1 | Minimum top bar spacing<br>Top bars:<br>At maximum and interior negative moments<br><br>The clear spacing between the horizontal bars must be at least the greatest of:<br><br>Clear spacing greater of $\begin{cases} 1 \text{ in.} \\ d_b \\ 4/3(d_{agg}) \end{cases}$<br><br>Assume 3/4 in. maximum aggregate size | 1 in.<br>0.75 in.<br>4/3(3/4 in.) = 1 in.<br><br>Therefore, clear spacing between horizontal bars must be at least 1 in. |
| | Check if seven No. 6 bars can be placed in one layer in the beam's web.<br>$b_{w,req'd} = 2(\text{cover} + d_{stirrup} + 0.75 \text{ in.})$<br>$\quad\quad\quad + 6d_b + 6(1 \text{ in.})_{min,spacing}$  (25.2.1) | $b_{w,req'd} = 2(1.5 \text{ in.} + 0.5 \text{ in.} + 0.75 \text{ in.}) + 4.5 \text{ in.} + 6 \text{ in.}$<br>$= 16 \text{ in.} < 18 \text{ in.}$  **OK**<br><br>Therefore seven No. 6 bars can be placed in one layer in the 18 in. beam web. |

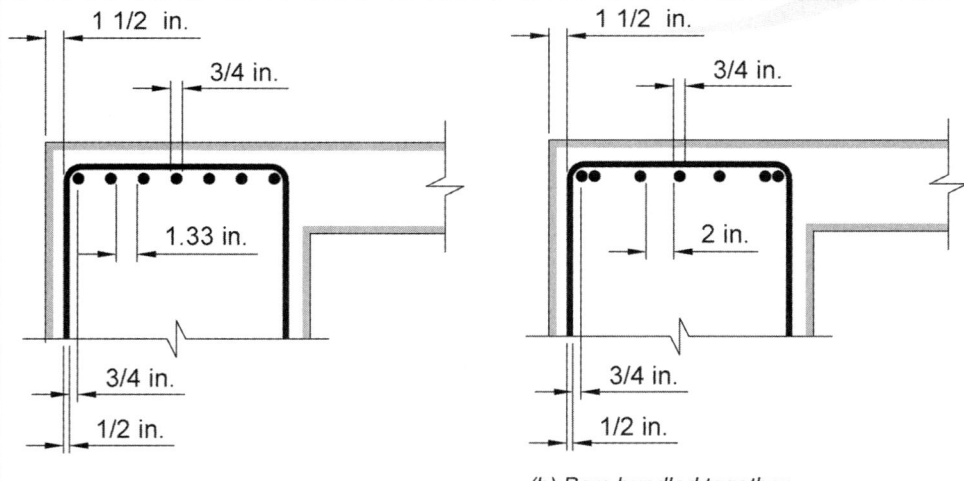

(a) Bars evenly spaced  (b) Bars bundled together

Fig. E4.12—Top bar layout.

Note: a preferred solution would be to bundle a few bars together to provide larger spacing between them and to allow for improved concrete placement (Fig. E4.12b).

| | | |
|---|---|---|
| 25.2.1 | Minimum bottom bar spacing<br>Bottom bars:<br>The clear spacing between the horizontal bars must be at least the greatest of:<br><br>Clear spacing greater of $\begin{cases} 1 \text{ in.} \\ d_b \\ 4/3(d_{agg}) \end{cases}$<br><br>Check if five No. 6 bars can be placed in one layer in the beam's web<br><br>$b_{w,req'd} = 2(\text{cover} - d_{stirrup} + 0.75 \text{ in.})$<br>$\quad + 4d_b + 4(1 \text{ in.})_{min,spacing}$ (25.2.1)<br><br>Refer to Fig. E4.13 for steel placement in beam web. | 1 in.<br>0.75 in.<br>4/3(3/4 in.) = 1 in.<br><br>Therefore, clear spacing between horizontal bars must be at least 1 in.<br><br>$b_{w,req'd} = 2(1.5 \text{ in.} + 0.5 \text{ in.} + 0.75 \text{ in.}) + 3.0 \text{ in.} + 4 \text{ in.}$<br>$\quad = 14.3 \text{ in.} < 18 \text{ in.}$ **OK**<br><br>Therefore five No. 6 bars can be placed in one layer in the 18 in. beam web. |
| | 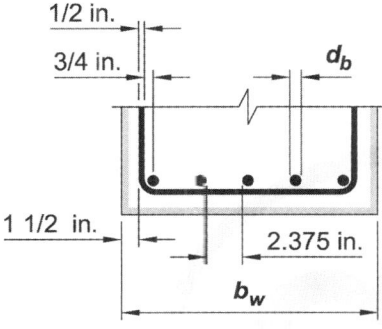<br>Fig. E4.13—Bottom reinforcement layout. | |
| 9.7.2.2<br>24.3.1<br>24.3.2<br><br><br><br><br>24.3.2.1 | Maximum bar spacing at tension face must not exceed<br><br>$s = \text{ the lesser of } \begin{cases} 15\left(\dfrac{40,000}{f_s}\right) - 2.5c_c \\ 12\left(\dfrac{40,000}{f_s}\right) \end{cases}$<br><br>where $f_s = 2/3f_y = 40,000$ psi<br>This limit is intended to control flexural cracking width, where<br>$c_c = 2$ in. is the least distance from the No. 6 bar surface to the tension face. | $s = 15\left(\dfrac{40,000}{40,000}\right) - 2.5(2 \text{ in.}) = 10 \text{ in.}$ **Controls**<br><br>$s = 12\left(\dfrac{40,000}{40,000}\right) = 12 \text{ in.}$<br><br><br><br>2.3 in. spacing is provided; therefore, **OK** |

9.7.3

Bottom reinforcing bar length along first span
Calculate the inflection point for positive moment:

Assume the maximum moment occurs at midspan (Fig. E4.14). From equilibrium, the point of inflection is obtained from a freebody diagram:

$M_{max} - w_u(x)^2/2 = 0$

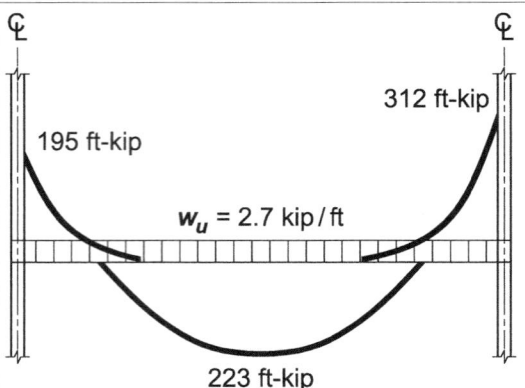

Fig. E4.14—Typical span moment diagram.

Top reinforcing bar length along first span
At the exterior support:
Calculate the inflection point for the negative-moment diagram:
$-M_{max} - w_u(x)^2/2 + V_u x = 0$

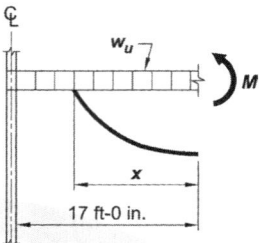

$(223 \text{ ft-kip}) - (2.7 \text{ kip/ft})(x)^2/2 = 0$
$x = 12.86$ ft, say, 13 ft
Inflection point of maximum positive moment

At the first interior support:
Calculate the inflection point for the negative-moment diagram:
$-M_{max} - w_u(x)^2/2 + V_u x = 0$

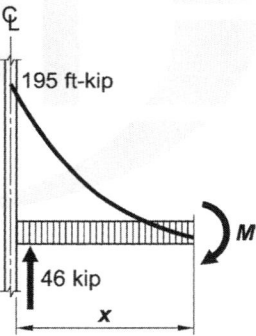

$(-195 \text{ ft-kip}) - (2.7 \text{ kip/ft})(x)^2/2 + 46x = 0$
$x = 4.96$ ft, say, 5 ft 0 in.
Inflection point of exterior negative moment

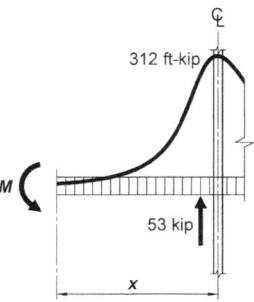

$\sum M = (-312 \text{ ft-kip}) - (2.7 \text{ kip/ft})(x)^2/2 + 53x = 0$
$x = 7.2$ ft, say, 7 ft 3 in.
Inflection point of interior negative moment

Fig. E4.15—Inflection point locations.

| | | |
|---|---|---|
| 25.4.2.2<br><br>25.4.2.4 | **Development length of No. 6 bar**<br>The simplified method is used to calculate the development length of a No. 6 bar:<br><br>$$\ell_d = \frac{f_y \psi_t \psi_e}{20\lambda\sqrt{f'_c}}d_b$$<br><br>where $\psi_t$ is the bar location; $\psi_t = 1.3$, because more than 12 in. of fresh concrete is placed below top horizontal bars, and $\psi_t = 1.0$, because not more than 12 in. of fresh concrete is placed below bottom horizontal bars<br><br>$\psi_e$ is coating factor; and $\psi_e = 1.0$, because bars are uncoated | Top bars:<br>$$\ell_d = \left(\frac{(60{,}000\text{ psi})(1.3)(1.0)}{(25)(1.0)\sqrt{5000\text{ psi}}}\right)(0.75\text{ in.}) = 33.1\text{ in.}$$<br>say, 36 in.<br><br>Bottom bars:<br>$$\ell_d = \left(\frac{(60{,}000\text{ psi})(1.0)(1.0)}{(25)(1.0)\sqrt{5000\text{ psi}}}\right)(0.75\text{ in.}) = 25.4\text{ in.}$$<br>say, 30 in. |
| 9.7.3.2<br><br><br><br><br><br><br><br><br><br><br>9.7.3.3 | **First span top reinforcement**<br><br>Lengths at the exterior support<br>Reinforcement must be developed at sections of maximum stress and at sections along the span where bent or terminated tension reinforcement is no longer required to resist flexure.<br>Four No. 6 bars are required to resist the beam factored negative moment at the exterior column face.<br>Calculate a distance $x$ from the face of the column where two No. 6 bars are sufficient to resist the factored moment.<br><br>Reinforcement must extend beyond the section at which it is no longer required to resist flexure for a distance equal to the greater of $d$ or $12d_b$. | $(-195\text{ ft-kip}) - 2.7\text{ kip/ft}\dfrac{x^2}{2} + 46\text{ kip}(x) = -2(0.44\text{ in.}^2)$<br>$\times (0.9)(60\text{ ksi})\left(27.6\text{ in.} - \dfrac{2(0.44\text{ in.}^2)(60\text{ ksi})}{2(0.85)(5\text{ ksi})(18\text{ in.})}\right)$<br>$x = 2\text{ ft }0\text{ in.}$<br><br>For No. 6 bars:<br>1) $d = 27.6$ in.  **Controls**<br>2) $12d_b = 12(0.75\text{ in.}) = 9$ in<br><br>Therefore, extend two No. 6 bars the longer of the development length (36 in.) and the sum of 24 in. + 27.6 in. = 51.6 in., say, 52 in. or 4 ft 4 in. from the face of the column.<br><br>The sum of the theoretical cutoff point and $d$ controls—extend two No. 6 bars 4 ft 6 in. from the interior face of the exterior support shown bold in Fig. E4.16. |
| 9.7.3.8.4 | At least one-third of the negative moment reinforcement at a support must have an embedment length beyond the point of inflection the greatest of $d$, $12d_b$, and $\ell_n/16$.<br><br>The inflection point is calculated above at 5 ft 0 in. from the face of the column. The remaining two No. 6 bars are extended over the full length of the beam. | For No. 6 bars:<br>1) $d = 27.6$ in.  **Controls**<br>2) $12d_b = 12(0.75\text{ in.}) = 9$ in.<br>3) $\ell_n/16 = (36\text{ ft} - 2\text{ ft})/16 = 2.1\text{ ft} = 25.5$ in.<br>The remaining two No. 6 bars (1/2 of the bars > 1/3) must be extended a minimum of<br>5 ft 0 in. (60 in.) + 27.6 in. = 87.6 in. They are, however, spliced at midspan with the bars from the opposite support to act as hanger bars for stirrups. |

| | | |
|---|---|---|
| 9.7.3.2<br>9.7.3.3<br>9.7.3.8.4 | First span top reinforcement<br><br>Lengths at the interior support<br>Following the same steps above, seven No. 6 bars are required to resist the factored moment at the first interior column face.<br>Calculate a distance $x$ from the face of the column where three No. 6 bars are sufficient to resist the factored moment. (Four No. 6 bars will be discontinued). | $-(312 \text{ ft-kip}) - 2.7 \text{ kip/ft} \dfrac{x^2}{2} + 53 \text{ kip}(x) = -3(0.44 \text{ in.}^2)$<br>$\times (0.9)(60 \text{ ksi})\left(27.6 \text{ in.} - \dfrac{3(0.44 \text{ in.}^2)(60 \text{ ksi})}{2(0.85)(5 \text{ ksi})(18 \text{ in.})}\right)$<br><br>$x = 3.1$ ft, say, 3 ft -3 in.<br><br>Therefore, extend four No. 6 bars the greater of the development length (36 in.) and the sum of theoretical cutoff point (3.25 ft) and $d$.<br>39 in. + 27.6 in. = 56.6 in. The distance of 56.6 in. shown bold in Fig. E4.16 from the exterior face of the exterior support controls. Say 5 ft 0 in. as shown in Fig. E4.16.<br><br>Extend the remaining three No. 6 bars the longer of the development length (36 in.) from where the four No. 6 bars are cut off and $d = 27.6$ in. beyond the inflection point which is 7 ft 3 in. from the interior face of the exterior support.<br><br>The longer length is the distance $d$ beyond the inflection point shown bold in Fig. E4.16. One of the three No. 6 bars will be terminated at 7 ft 3 in. + 27.6 in. $\approx$ 10 ft 0 in. The remaining two No. 6 top bars are extended and spliced at midspan. |

## CHAPTER 7—BEAMS

| | | |
|---|---|---|
| 9.7.3.2<br>9.7.3.3 | First span bottom reinforcement<br><br>Lengths from midspan toward the column<br>Following the same steps above, five No. 6 bars are required to resist the factored moment at the midspan.<br>Calculate a distance $x$ from the face of the column where two No. 6 bars can resist the factored moment. (Three No. 6 bars will be cut off). | $(223 \text{ ft-kip}) - 2.7 \text{ kip/ft} \dfrac{x^2}{2} = 2(0.44 \text{ in.}^2)(0.9)(60 \text{ ksi})$<br>$\times \left(27.6 \text{ in.} - \dfrac{2(0.44 \text{ in.}^2)(60 \text{ ksi})}{2(0.85)(5 \text{ ksi})(52 \text{ in.})}\right)$<br>$x = 9.21$ ft, say, 9 ft 3 in. = 111 in.<br><br>Therefore, extend the three No. 6 bars the longer of the development length (30 in.) and 111 in. + 27.6 in. = 138.6 in. ≅ 11 ft 9 in. from maximum positive moment at midspan.<br>11 ft 9 in. is longer—shown bold in Fig. E4.16 from midspan. |
| 9.7.3.8.2 | A minimum of one-fourth of the positive tension reinforcement must extend into the support minimum 6 in. The 6 in. requirement is superseded by the integrity reinforcement requirement to develop the bar at the column face. | Extend the remaining bars (two No. 6 bars > 1/4 five No. 6) the greater of the development length (30 in.) from the three No. 6 bar cutoff and $d$ = 27.6 in. beyond the inflection point and a minimum of 6 in. into the support.<br><br>The controlling bottom bar length is the distance 6 in. into the support shown in bold in Fig. E4.16. |
| 9.7.3.8.3 | At point of inflection, $d_b$ for positive moment tension reinforcement must be limited such that $\ell_d$ for that reinforcement satisfies:<br>$$\ell_d \leq \dfrac{M_n}{V_u} + \ell_a$$<br>where $M_n$ is calculated assuming all reinforcement at the section is stressed to $f_y$. $V_u$ is calculated at the section. At the support, $\ell_a$ is the embedment length beyond the center of the support. At the point of inflection, $\ell_a$ is the embedment length beyond the point of inflection limited to the greater of $d$ and $12d_b$. | Point of inflection occurs at 4 ft from the face of the column.<br><br>$V_u = 46 \text{ kip} - (2.69 \text{ kip/ft})(4 \text{ ft}) = 35.2 \text{ kip}$<br>At that location assume two No. 6 bars are effective:<br>$M_n = 2(0.44 \text{ in.}^2)(60 \text{ ksi})\left(27.6 \text{ in.} - \dfrac{2(0.44 \text{ in.}^2)(60 \text{ ksi})}{(2)(0.85)(5 \text{ ksi})(52 \text{ in.})}\right)$<br>$M_n = 1451$ in.-kip<br>$\ell_d \leq \dfrac{1451 \text{ in.-kip}}{35.2 \text{ kip}} + 27.6 \text{ in.} = 69 \text{ in.}$<br>This length exceeds $\ell_d = 30$ in., therefore **OK** |
| 9.7.3.5 | If bars are cut off in regions of flexural tension, then stress discontinuity in the continuing bars will occur. Therefore, the Code requires that flexural tensile reinforcement must not be terminated in a tensile zone unless (a), (b), or (c) is satisfied.<br>(a) $V_u \leq (2/3)\phi V_n$ at the cutoff point<br>(b) Continuing reinforcement provides double the area required for flexure at the cutoff point and the area required for flexure at the cutoff point and $V_u \leq (3/4)\phi V_n$.<br>(c) Stirrup or hoop area in excess of that required for shear and torsion is provided along each terminated bar or wire over a distance $3/4d$ from the termination point. Excess stirrup or hoop area shall be at least $60b_w s/f_{yt}$. Spacing $s$ shall not exceed $d/(8\beta_b)$. | (a) At 9 ft 3 in.<br>$V_u = 46 \text{ kip} - (2.69 \text{ kip/ft})(17 \text{ ft} - 9.25 \text{ ft}) = 25.2 \text{ kip}$<br><br>$\phi V_n = \phi(V_c + V_s)$ where $V_c$ is calculated in Step 6<br>$\phi V_n = 0.75(70.3 \text{ kip} + 0) = 52.7 \text{ kip}$<br>$2/3 \phi V_n = 2/3(52.7 \text{ kip}) = 35 \text{ kip}$<br>$V_u = 25.2 \text{ kip} \leq 2/3 \phi V_n = 35 \text{ kip}$     **OK**<br><br>Because only one of the three conditions needs to be satisfied, the other two will not be checked. |

| | Integrity reinforcement for the perimeter beam | |
|---|---|---|
| 9.7.7.1<br>9.7.7.1a | At least one-fourth the maximum positive moment reinforcement, but at least two bars must be continuous. | Two No. 6 bars are extended into the column region two No. 6 > 1/4 (five No. 6)—satisfied (refer to Fig. E4.16). |
| 9.7.7.1b | At least one-sixth the maximum negative moment reinforcement at the support, but at least two bars must be continuous. | Four No. 6 bars are extended into the column region four No. 6 > 1/6 (seven No. 6)—satisfied (refer to Fig. E4.16). |
| 9.7.7.1c | Longitudinal reinforcement must be enclosed by closed stirrups along the clear span of the beam.<br><br>Longitudinal structural reinforcement must pass through the region bounded by the longitudinal reinforcement of the column. | Longitudinal reinforcement is enclosed by No. 4 stirrups at 12 in. on center along the full beam length—satisfied<br><br>This condition is satisfied by extending the two No. 6 top and bottom bars full length and through the column cores. |
| 9.7.7.4<br>25.4.3.1 | Integrity reinforcement must be anchored to develop $f_y$ at the face of the support. Therefore, development length for deformed bars in tension terminating in a standard hook must be the greater of:<br><br>(a) $\ell_{dh} = \left( \dfrac{f_y \Psi_e \Psi_c \Psi_r}{50\lambda \sqrt{f'_c}} \right) d_b$<br><br>(b) $8d_b$<br>(c) 6 in. | At the exterior support, the No. 6 bars must be developed at the face. Calculate if a standard hook will allow a No. 6 bar to develop within the column.<br><br>(a) $\ell_{dh} = \left( \dfrac{(60 \text{ ksi})(1.0)(0.7)(1.0)}{50(1.0)\sqrt{5000 \text{ psi}}} \right)(0.75 \text{ in.}) = 8.9 \text{ in.}$ **Controls**<br><br>(b) $8(0.75 \text{ in.}) = 6$ in.<br>(c) 6 in. |
| 25.4.3.2 | where $\psi_e$ is the coating factor; $\psi_e = 1.0$ because bars are uncoated<br><br>$\psi_c$ is the cover factor; $\psi_t = 0.7$ because bars are smaller than No. 11 and terminate with a 90-degree hook with cover on bar extension beyond hook $\geq 2$ in.<br><br>$\psi_r$ is the confining reinforcement factor; and $\psi_r = 1.0$, because bars are not confined with stirrups spaced at $s \leq 3d_b$. | Therefore, the hook fits within the column: **OK** |
| 9.7.7.5 | Splices are necessary in continuous structural integrity reinforcement. The beam's longitudinal reinforcement shall be spliced in accordance with (a) and (b):<br>  (a) Positive moment reinforcement shall be spliced at or near the support<br>  (b) Negative moment reinforcement shall be spliced at or near midspan | Splice length = (1.3)(development length)<br>$\ell_{dc} = 1.3(27 \text{ in.}) = 35$ in., say, 3 ft 0 in.<br><br>Refer to Fig. E 4.17<br><br>Refer to Fig. E 4.17 |
| 9.7.7.6 | Use Class B tension lap splice | |

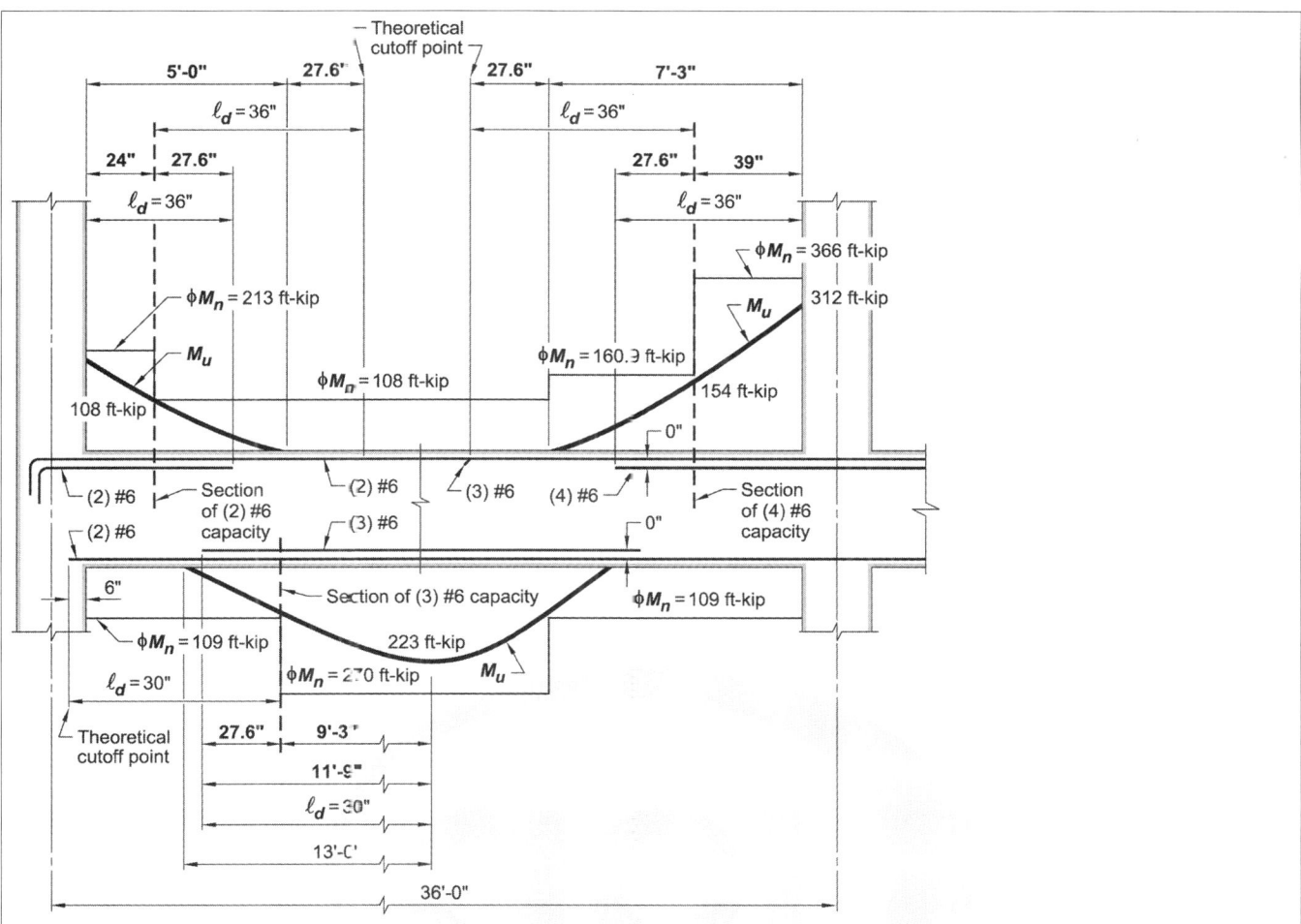

Fig. E4.16—End span reinforcement cutoff locations.

| Step 9: Internal spans | | |
|---|---|---|
| | Flexure reinforcement was calculated above in Step 5. | |
| | Six No. 6 top bars are required at supports | |
| | Five No. 6 bottom bars are required at midspan | |
| 9.7.6.2.2 | Shear and torsion reinforcement following the same calculation in Steps 6 and 7, No. 4 at 10 in. are required for minimum 7 ft 0 in. from face of each column. Space No. 4 stirrups at maximum 12 in. on center ($d/2$) for the remainder of the span. | |

## Step 10: Detailing

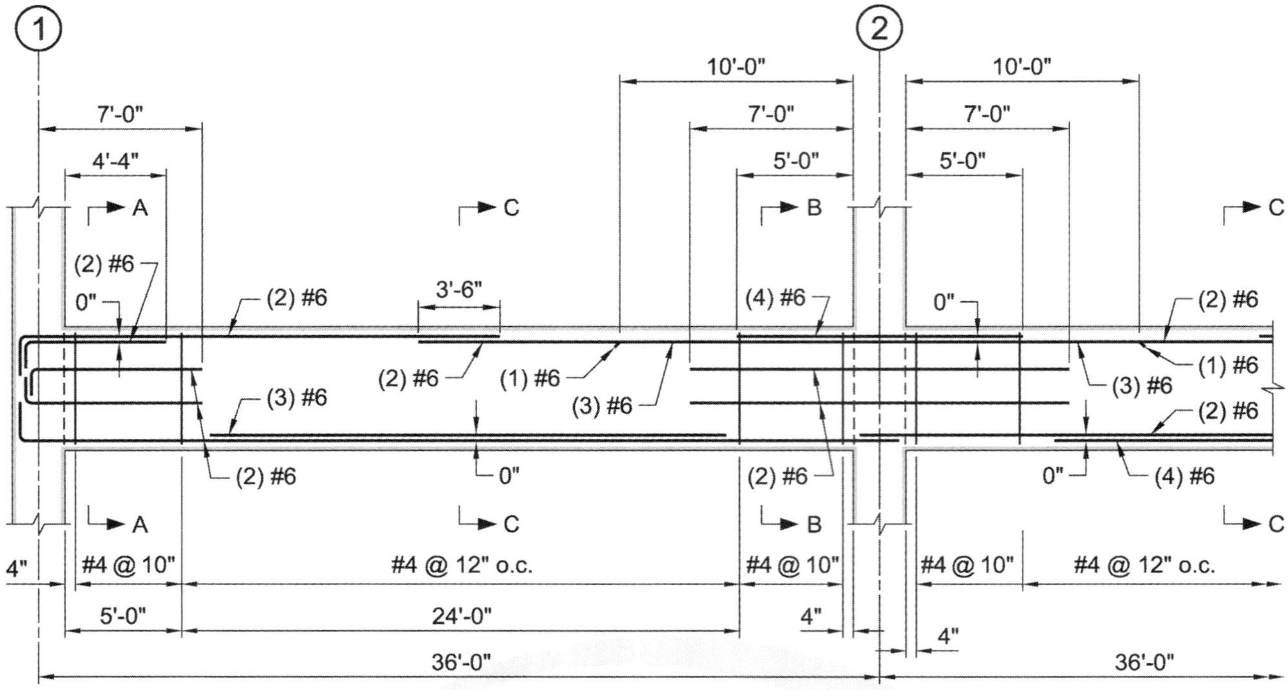

*Fig. E4.17—Beam reinforcement details.*

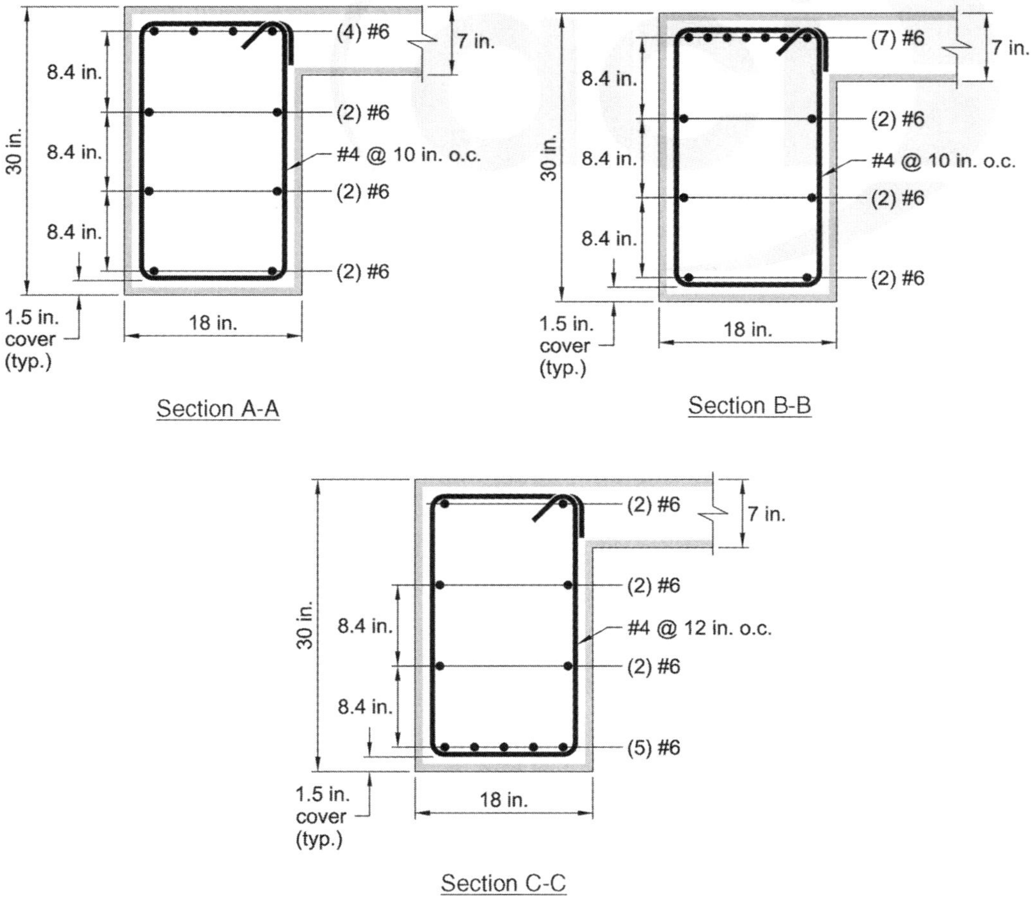

*Fig. E4.18—Sections.*

**Beam Example 5**: *Continuous transfer girder*

Design and detail an interior, continuous, four-bay beam, built integrally with a 7 in. slab. The span between Column Lines B and D is a transfer girder supporting five stories above.

Given:

*Load—*
Service additional dead load $D$ = 15 psf
Service live load $L$ = 65 psf
Girder, beam and slab self-weights are given below.

*Material properties—*
$f_c'$ = 5000 psi (normalweight concrete)
$f_y$ = 60,000 psi
$\lambda$ = 1.0 normal weight concrete

*Span length—*
Typical beam: 14 ft
Girder: 28 ft
Beam and girder width: 24 in.
Column dimensions: 24 in. x 24 in.

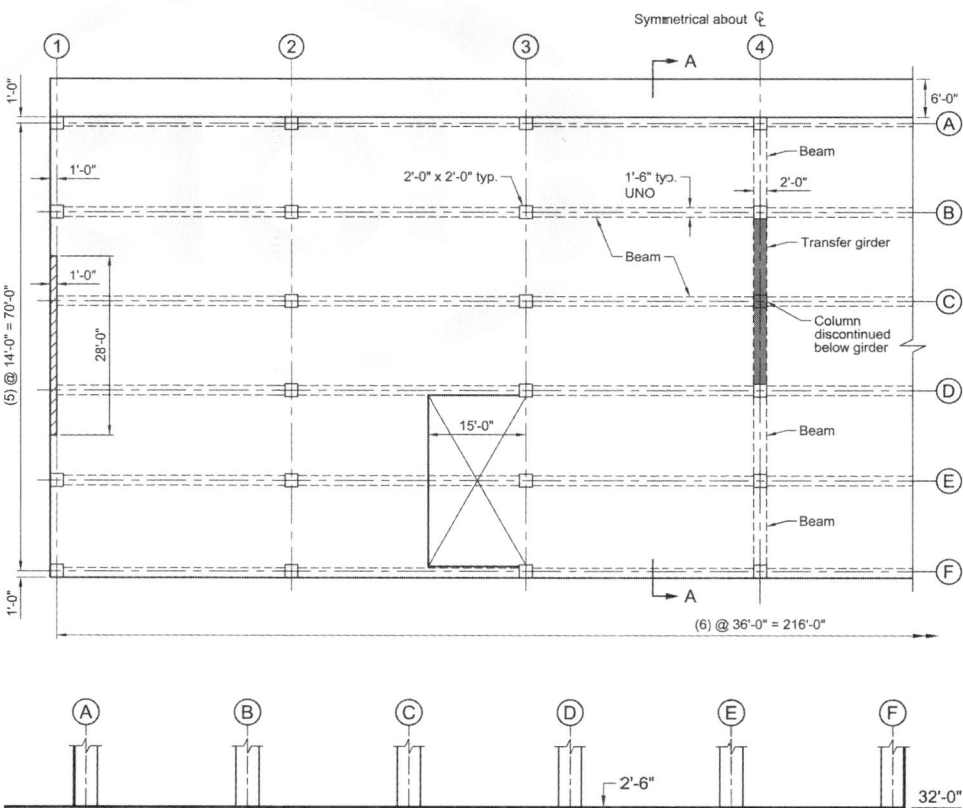

*Fig. E5.1—Plan and elevation of transfer girder and beams.*

| ACI 318-14 | Discussion | Calculation |
|---|---|---|
| **Step 1: Material requirements** | | |
| 9.2.1.1 | The mixture proportion must satisfy the durability requirements of Chapter 19 (ACI 318) and structural strength requirements. The designer determines the durability classes. Please refer to Chapter 4 of SP-17 for an in-depth discussion of the Categories and Classes.<br>ACI 301 is a reference specification that is coordinated with ACI 318. ACI encourages referencing 301 into job specifications.<br>There are several mixture options within ACI 301, such as admixtures and pozzolans, which the designer can require, permit, or review if suggested by the contractor. | By specifying that the concrete mixture shall be in accordance with ACI 301-10 and providing the exposure classes, Chapter 19 requirements are satisfied.<br><br>Based on durability and strength requirements, and experience with local mixtures, the compressive strength of concrete is specified at 28 days to be at least 5000 psi. |
| **Step 2: Beam geometry** | | |
| 9.3.1.1 | <u>Girder depth</u><br>The transfer girder supports a column at midspan with tributary loads from the third level, four stories, and a roof. Therefore, the depth limits in Table 9.3.1.1 cannot be used, and calculated deflections must satisfy the deflection limits in 9.3.2.<br><br>Beams on either side of the girder beam are subjected to uniform load. Therefore, the depth limits in Table 9.3.1.1 are used and the controlling condition for beam depth is one end continuous: | Assume 48 in. deep transfer girder.<br><br>$h = \dfrac{\ell}{18.5} = \dfrac{(14 \text{ ft})(12 \text{ in./ft})}{18.5} = 9.1 \text{ in.}$<br><br>Because the beams are intended to provide continuity to assist the girder, they must have enough stiffness for this purpose.<br>Use a beam depth of 30 in. |
| 9.2.4.2<br><br><br><br>6.3.2.1 | <u>Flange width</u><br>The transfer girder and beams are poured monolithically with the slab and will behave as a T-beam. The effective flange width on each side of the transfer girder is obtained from Table 6.3.2.1.<br><br><u>Transfer girder flange width</u><br>Each side of web is the least of $\begin{cases} 8h_{slab} \\ S_w/2 \\ \ell_n/8 \end{cases}$<br><br>Each side of web is the least of $\begin{cases} 8h_{slab} \\ S_w/2 \\ \ell_n/8 \end{cases}$<br><br>Flange width:<br>$b_f = \ell_n/8 + b_w + \ell_n/8$<br><br><u>Beams flange width</u><br>Each side of web is the least of $\begin{cases} 8h_{slab} \\ S_w/2 \\ \ell_n/8 \end{cases}$<br><br>Flange width:<br>$b_f = \ell_n/8 + b_w + \ell_n/8$ | (8)(7 in.) = 56 in.<br>(does not control)<br>((28 ft)(12) – 24 in.) /8 = 39 in.  **Controls**<br><br><br><br><br><br>$b_f$ = 39 in. + 24 in. + 39 in. = 102 in.<br><br>(8)(7 in.) = 56 in.<br>(does not control)<br>((14 ft)(12) – 24 in.) /8 = 18 in.  **Controls**<br><br>$b_f$ = 18 in. + 24 in. + 18 in. = 60 in. |

## Step 3: Loads and load patterns

**Applied load on transfer girder**
The service live load is 50 psf in offices and 80 psf in corridors per Table 4-1 in ASCE 7-10. This example will use 65 psf as an average live load as the actual layout is not provided. To account for the weight of ceilings, partitions, and HVAC systems, add 15 psf as miscellaneous dead load.

**Dead load:**
Transfer girder self-weight without flanges:
Concentrated load on girder from column above (Fig. E5.2):

$W_{gir} = [(24 \text{ in.})(48 \text{ in.})(0.150 \text{ kip/ft}^3)/144 = 1.2 \text{ kip/ft}$

Slab weight per level:

$P_{sl} = (7\text{in.}/12)(14 \text{ ft})(36 \text{ ft})(0.15 \text{ kip/ft}^3) = 44.1 \text{ kip}$

Column weight per level:

$P_{col} = (2 \text{ ft})(2 \text{ ft})\left(12 \text{ ft} - \dfrac{7 \text{ in.}}{12}\right)(0.15 \text{ kip/ft}^3) = 6.85 \text{ kip}$

Typical beam weight framing into girder at midspan less slab thickness (refer to plan):

$P_{BM} = \left(\dfrac{18 \text{ in.}}{12}\right)\left(\dfrac{30 \text{ in.} - 7 \text{ in.}}{12}\right)(34 \text{ ft})(0.15 \text{ kip/ft}^3)$
$= 14.7 \text{ kip}$

To account for the weight of ceilings, partitions, and mechanical (HVAC) systems, add 15 psf as miscellaneous dead load.

Total dead load applied on girder:

$P_{SDL} = (14 \text{ ft})(36 \text{ ft})(0.015 \text{ kip/ft}^2) = 7.6 \text{ kip}$

Beams self-weights on both ends of the girder:

$P_D = (44.1 \text{ kip} + 14.7 \text{ kip} + 7.6 \text{ kip})(6) + (6.85 \text{ kip})(5)$
$= 433 \text{ kip}$

**Live load:**
The service live load is 50 psf in offices and 80 psf in corridors per Table 4-1 in ASCE 7-10. This example will use 65 psf as an average, as the actual layout is not provided. A 7 in. slab weighs 88 psf service dead load.

$w_s = [(24 \text{ in.})(30 \text{ in.}) + (60 \text{ in.} - 24 \text{ in.})(7 \text{ in.})]$
$\times [(0.150 \text{ kip/ft}^3) / 144] = 1.03 \text{ kip/ft}$

Roof live load 35 psf:
Total live load—per ASCE 7 live load with exception of roof load is permitted to be reduced by:

$P_{Roof} = (14 \text{ ft})(36 \text{ ft})(0.035 \text{ kip/ft}^2) = 18 \text{ kip}$

$L = L_o\left(0.25 + \dfrac{15}{\sqrt{K_{LL}A_T}}\right)$

where $L$ is reduced live load; $L_o$ is unreduced live load; $K_{LL}$ is live load element factor = 4 for internal columns and 2 for interior beam (ASCE 7 Table 4-2); $A_T$—tributary are
and $K_{LL}A_T \geq 400 \text{ ft}^2$

$K_{LL} = 4$; $4(36 \text{ ft})(14 \text{ ft}) = 2016 \text{ ft}^2 > 400 \text{ ft}^2$

Fourth to seventh reduced live load per level:

$L = (0.065 \text{ kip/ft}^2)\left(0.25 + \dfrac{15}{\sqrt{2016 \text{ ft}^2}}\right) = 0.038 \text{ ksf}$

$L = 0.038 \text{ ksf} > 0.4L_o = 0.026 \text{ ksf}$  **OK**

| | | |
|---|---|---|
| | $K_{LL} = 2$; $2(36\text{ ft})(14\text{ ft}) = 1008\text{ ft}^2 > 400\text{ ft}^2$ and $L \geq 0.4L_o = 26$ psf | Reduced third level live load on beam: $$L = (0.065\text{ kip/ft}^2)\left(0.25 + \frac{15}{\sqrt{1008\text{ ft}^2}}\right) = 0.047\text{ ksf}$$ $L = 0.047\text{ ksf} > 0.4L_o = 0.026\text{ ksf}$ **OK** <br><br> Concentrated live load to column per level: $P_{L,col} = (14\text{ ft})(36\text{ ft})(0.38\text{ kip/ft}^2) = 19.2$ kip <br><br> Concentrated live load at third level: $P_{L,3rd} = (14\text{ ft})(36\text{ ft})(0.047\text{ kip/ft}^2) = 23.7$ kip <br><br> Concentrated live load on girder: |
| | $\sum P_L = 4P_{L,col} + P_{L,3rd} + P_{Roof}$ | $\sum P_L = (23.7\text{ kip}) + (19.2\text{ kip})(4\text{ levels}) + (18\text{ kip})$ $= 119$ kip |
| 5.3.1 | The transfer girder resists gravity only and lateral forces are not considered in this problem. <br><br> $U = 1.4D$ <br><br> The superimposed dead load is calculated above and is included in the concentrated load. Live load is applied over the width of the girder (2 ft–0 in.) <br><br> $U = 1.2D + 1.6L$ | Transfer girder <br> Distributed: $w_u = 1.4(1.2\text{ kip/ft}) = 1.68$ kip/ft **Controls** <br> Concentrated: $P_u = 1.4(433\text{ kip}) = 607$ kip <br><br><br><br> Distributed: $w_u = 1.2(1.2\text{ kip/ft}) + 1.6(0.065\text{ ksf})(2\text{ ft})$ $= 1.65$ kip/ft <br> $P_u = 1.2(433\text{ kip}) + 1.6(119\text{ kip}) = 710$ kip **Controls** <br><br> Beams <br> $w_u = 1.4(1.03\text{ kip/ft} + (15\text{ psf})(60\text{ in.}/12)/1000)$ $= 1.55$ kip/ft <br> $w_u = 1.2(1.55\text{ kip/ft})/1.4 + 1.6((65\text{ psf})(60\text{ in.}/12)/(1000)$ $= 1.85$ kip/ft **Controls** |

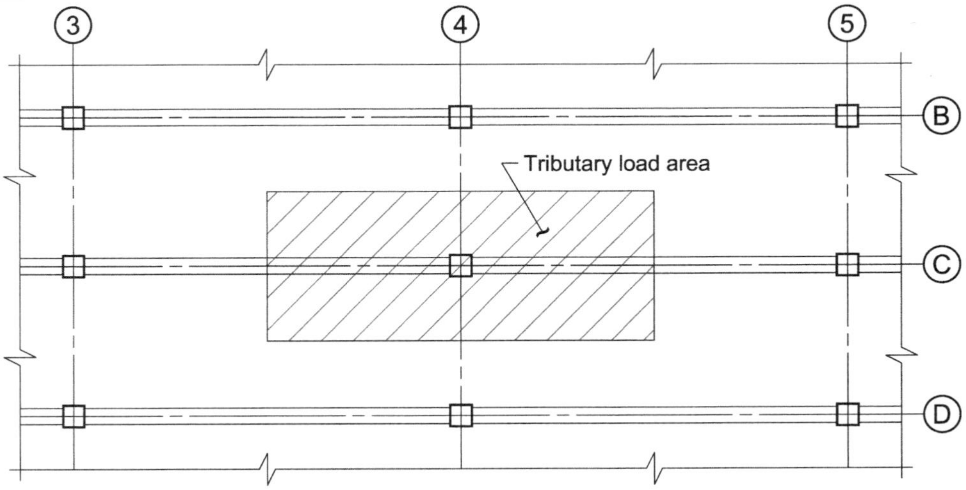

*Fig. E5.2—Column C/4 tributary area.*

| Step 4: Analysis | | |
|---|---|---|
| 9.4.3.1 | The beams are built integrally with supports; therefore, the factored moments and shear forces (required strengths) are calculated at the face of the supports.<br><br>The beams were analyzed as part of a frame. The moment and shear diagram obtained from a software are presented below (Fig. E5.3). | |

### (a) Shear diagram

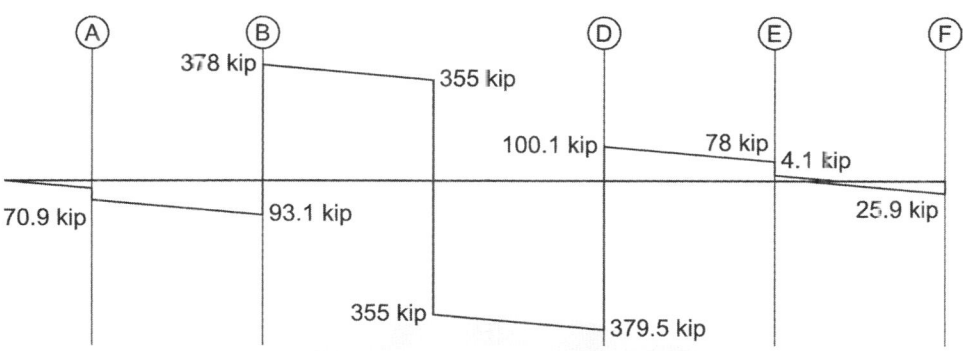

### (b) Moment diagram

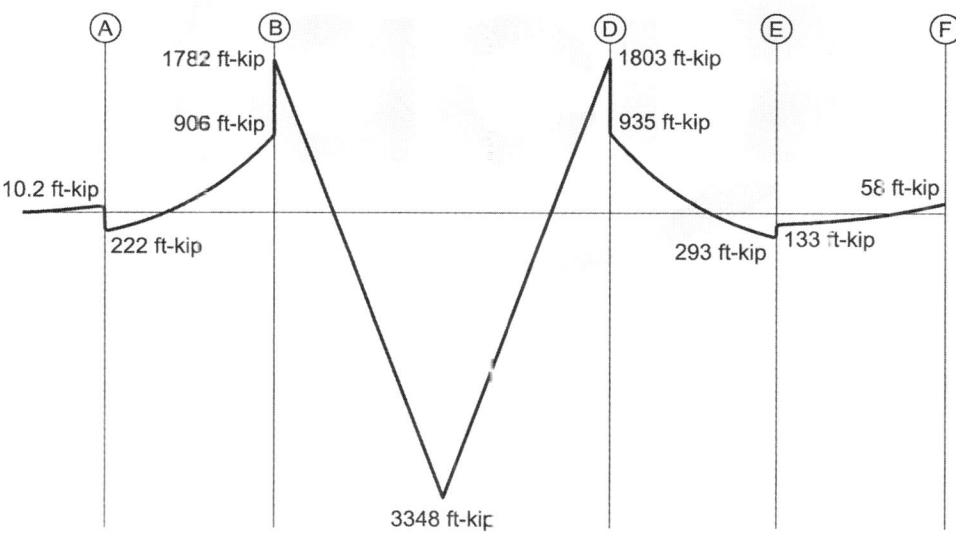

Notes:
1. Factored moments and shear forces are shown at faces of columns.
2. Span BD is subjected to large concentrated load at midspan and relatively small distributed beam self-weight. Therefore, the appearance of a straight line moment diagram.

*Fig. E5.3—Shear and moment diagrams.*

| | Step 5: Moment design | |
|---|---|---|
| 9.3.3.1 | The code does not permit a beam to be designed with steel strain less than 0.004 in./in. at nominal strength. The intent is to ensure ductile behavior.<br><br>In most reinforced beams, such as this example, reinforcing bar strain is not a controlling issue. | |
| 21.2.1(a) | The design assumption is that beams will be tension-controlled; therefore, the moment reduction factor is $\phi = 0.9$. This assumption will be checked later. | |
| 20.6.1.3.1 | Determine the effective depth assuming No. 5 stirrups and No. 11 bars for the transfer girder positive moment and No. 9 bars for the transfer girder negative moment. Assume No. 4 stirrups and No. 6 and No. 9 bars for beams positive and negative moments, respectively. Girder beam and beams will have 1.5 in. cover:<br><br>Assume that the transfer girder requires two rows of reinforcement with one bar spacing between the two rows within the span and one layer of bars at the supports. Beams require one row only<br><br>$d = h - \text{cover} - d_{tie} - 3d_b/2$ | |
| | positive moment | Transfer girder:<br>$d = 48$ in. $- 1.5$ in. $- 0.625$ in. $- 3 (1.41$ in.$)/2 = 43.76$ in.<br>use $d = 43.7$ in. |
| | negative moment | $d = 48$ in. $- 1.5$ in. $- 0.625$ in. $- 1.128$ in.$/2 = 45.3$ in.<br>use $d = 45$ in. |
| | positive moment | Beams:<br>$d = 30$ in. $- 1.5$ in. $- 0.5$ in. $- (0.75$ in.$)/2 = 27.625$ in.<br>use $d = 27.6$ in. |
| | negative moment | $d = 30$ in. $- 1.5$ in. $- 0.5$ in. $- 1.128$ in.$/2 = 27.4$ in. |

| | | |
|---|---|---|
| 22.2.2.1 | The concrete compressive strain at nominal moment strength is calculated at: $\varepsilon_{cu} = 0.003$ in./in. | |
| 22.2.2.2 | The tensile strength of concrete in flexure is a variable property and is approximately 10 to 15 percent of the concrete compressive strength. ACI 318 neglects the concrete tensile strength to calculate nominal strength. | |
| 22.2.2.3 | Determine the equivalent concrete compressive stress at nominal strength:<br><br>The concrete compressive stress distribution is inelastic at high stress. The Code permits any stress distribution to be assumed in design if shown to result in predictions of ultimate strength in reasonable agreement with the results of comprehensive tests. Rather than tests, the Code allows the use of an equivalent rectangular compressive stress distribution of $0.85 f_c'$ with a depth of: $a = \beta_1 c$, where $\beta_1$ is a function of concrete compressive strength and is obtained from Table 22.2.2.4.3. | |
| 22.2.2.4.1 | | |
| 22.2.2.4.3 | For $f_c' = 5000$ psi: | $\beta = 0.85 - \dfrac{0.05(5000 \text{ psi} - 4000 \text{ psi})}{1000 \text{ psi}} = 0.8$ |
| 22.2.1.1 | Find the equivalent concrete compressive depth $a$ by equating the compression force to the tension force within the beam cross section (Fig. E5.4):<br>$C = T$<br>$0.85 f_c' b a = A_s f_y$ | $0.85(5000 \text{ psi})(b)(a) = A_s(60,000 \text{ psi})$ |
| | Transfer girder:<br>For positive moment: $b = b_f = 102$ in. | $a = \dfrac{A_s(60,000 \text{ psi})}{0.85(5000 \text{ psi})(102 \text{ in.})} = 0.138 A_s$ |
| | Beams:<br>For positive moment: $b = b_f = 60$ in. | $a = \dfrac{A_s(60,000 \text{ psi})}{0.85(5000 \text{ psi})(60 \text{ in.})} = 0.235 A_s$ |
| | Transfer girder and beams:<br>For negative moment: $b = b_w = 24$ in. | $a = \dfrac{A_s(60,000 \text{ psi})}{0.85(5000 \text{ psi})(24 \text{ in.})} = 0.588 A_s$ |

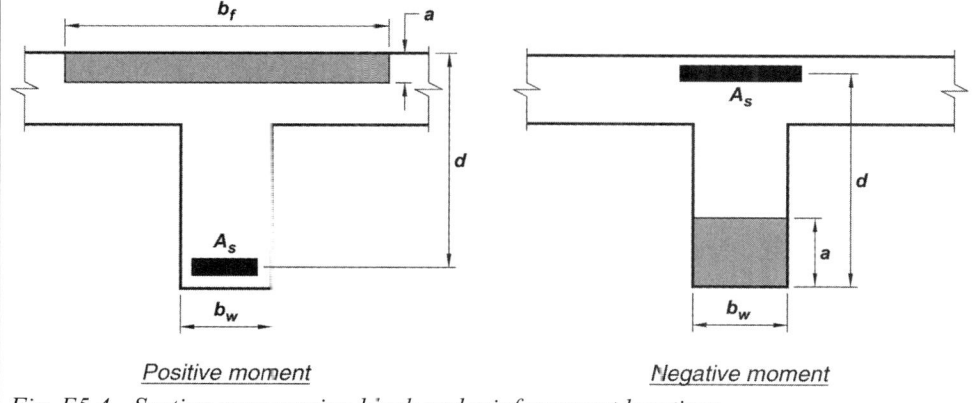

Fig. E5.4—Section compression block and reinforcement locations.

| | | |
|---|---|---|
| 9.5.1.1 | The transfer girder and beams are designed for the maximum flexural moments shown in the above moment diagram (Step 4).<br><br>The beams' design strength must be at least the required strength at each section along their lengths:<br>$\phi M_n \geq M_u$<br>$\phi V_n \geq V_u$<br><br>Calculate the required reinforcement area:<br><br>$\phi M_n \geq M_u = \phi A_s f_y \left(d - \dfrac{a}{2}\right)$<br><br>For top negative reinforcement use: One layer of No. 9 bar; $d_b$ = 1.128 in., $A_s$ = 1.0 in.², and $d$ = 45 in.<br><br>For bottom positive reinforcement use: Two layers of No. 11 bar; $d_b$ = 1.41 in., $A_s$ = 1.56 in.², and $d$ = 43.7 in. | Transfer girder:<br><br>| | $M_u$, ft-kip | $A_{s,req'd}$ in.² | Number of bars | |<br>|---|---|---|---|---|<br>| | | | Req'd | Select |<br>| Max. $M^-$ | 1782 | 9.4 | 9.4 | 10 |<br>| Max. $M^+$ | 3348 | 17.51 | 11.22 | 12 |<br>| Max. $M^-$ | 1803 | 9.5 | 9.5 | 10 | |
| 9.6.1.1<br>9.6.1.2 | <u>Minimum reinforcement</u><br>The provided reinforcement must be at least the minimum required reinforcement at every section along the length of the beam.<br>Use $d_{@support}$ = 45 in. > $d_{@midspan}$ = 43.7 in.<br><br>will yield higher required minimum reinforcement area:<br><br>$A_s = \dfrac{3\sqrt{f_c'}}{f_y} b_w d$<br><br>Because $f_c'$ > 4444 psi, Eq. (9.6.1.2a) only applies. | $A_s = \dfrac{3\sqrt{5000 \text{ psi}}}{60,000 \text{ psi}} (24 \text{ in.})(45 \text{ in.}) = 3.73 \text{ in.}^2$<br><br>Provided reinforcement area exceeds the minimum required. Therefore, **OK** |
| 9.7.2.3<br>24.3.2<br><br><br><br><br><br><br><br><br>24.3.2.1 | The girder is 48 in. deep. To control cracks within the web, ACI 318 requires skin reinforcement to be placed near the vertical faces of the tension zone over a distance of:<br>48 in./2 = 24 in.<br>Spacing of skin reinforcement in girder must not exceed the lesser of:<br><br>$s = 15\left(\dfrac{40,000}{f_s}\right) - 2.5 c_c$ and<br><br>$s = 12\left(\dfrac{40,000}{f_s}\right)$<br><br>where $c_c$ is the clear cover from the skin reinforcement to the side face.<br>$c_c$ = 1.5 in. + 0.625 in. = 2.125 in.<br>and $f_s = 2/3 f_y$ = 40 ksi | $s = 15\left(\dfrac{40,000}{40,000}\right) - 2.5(2.125 \text{ in.}) = 9.7 \text{ in.}$  **Controls**<br><br>$s = 12\left(\dfrac{40,000}{40,000}\right) = 12 \text{ in.}$<br><br>Place two No. 8 skin reinforcement at girder middepth: 24 in. and the second pair of No. 8 bars at 33.5 in. from the top of the girder. |

| | | |
|---|---|---|
| 9.7.2.3 | Skin reinforcement can be used in the strength calculation of the girder.<br><br>Positive reinforcement at midspan:<br><br>Using strain compatibility, try two No. 8 bars in two layers on both sides of the girder. Assume reinforcement is yielding. It will be checked later. | $\phi M_n = (0.9)(60 \text{ ksi})(2)(0.79 \text{ in.}^2)\left(\left(33.5 \text{ in.} - \dfrac{2.58 \text{ in.}}{2}\right) + \left(24 \text{ in.} - \dfrac{2.58 \text{ in.}}{2}\right)\right)$<br><br>$\phi M_n = 4685.8$ in.-kip $= 390.5$ ft-kip, say, 390 ft-kip |
| | Re-evaluating the positive tension reinforcement at girder midspan:<br>Calculated design moment 3348 ft-kip (Step 4).<br><br>Moment to be resisted by No. 11 bars:<br><br>Required reinforcement area by solving the following equation:<br><br>$\phi M_n \geq M_u = \phi A_s f_y \left(d - \dfrac{a}{2}\right)$ | $\phi M_n = (3348 \text{ ft-kip}) - (390 \text{ ft-kip}) = 2958$ ft-kip<br><br><br><br>$A_{s,No.11} = 15.42$ in.$^2$<br><br>Required No. 11 bars = 15.42 in.$^2$/1.56 in.$^2$ = 9.9<br>Choose 10 No. 11 bars. |
| | <u>Negative reinforcement at the supports</u><br>Provide three layers of skin reinforcement on both sides in the top half of the girder. Extend the two No. 8 middepth skin bars over the full length of the girder and use for the top half at the support two layers of No. 6 bars on both sides of the girder. Assume that skin reinforcement reach yielding (will be checked later).<br><br>Calculate the provided moment from the skin reinforcement: | $\phi M_n = (0.9)(60 \text{ ksi})(2)(0.79 \text{ in.}^2)\left(24 \text{ in.} - \dfrac{5.29 \text{ in.}}{2}\right)$<br>$+ (0.9)(60 \text{ ksi})(2)(0.44 \text{ in.}^2)\left(31.25 \text{ in.} - \dfrac{5.29 \text{ in.}}{2}\right)$<br>$+ (0.9)(60 \text{ ksi})(2)(0.44 \text{ in.}^2)\left(38.5 \text{ in.} - \dfrac{5.29 \text{ in.}}{2}\right)$<br><br>$\phi M_n = 4885.1$ in.-kip $= 407.1$ ft-kip, say, 407 ft-kip |
| | Re-evaluating the positive tension reinforcement at girder midspan:<br><br>Calculated design moment 1803 ft-kip (Step 4).<br><br>Moment to be resisted by No. 9 bars:<br><br>Required reinforcement area by solving the following equation:<br><br>$\phi M_n \geq M_u = \phi A_s f_y \left(d - \dfrac{a}{2}\right)$ | $\phi M_n = (1803 \text{ ft-kip}) - (407 \text{ ft-kip}) = 1396$ ft-kip<br><br>$A_{s,No.11} = 7.42$ in.$^2$<br><br>Required No. 9 bars = 7.42 in.$^2$/1.0 in.$^2$ = 7.49<br>Choose 8 No. 9 bars. |

| | | |
|---|---|---|
| 21.2.2<br>9.3.3.1 | Check if the calculated strain exceeds 0.005 in./in. (tension controlled), but not less than 0.004 in./in.<br><br>$$a = \frac{A_s f_y}{0.85 f'_c b} \text{ and } c = \frac{a}{\beta_1}$$<br><br>where $\beta_1 = 0.8$<br><br>$$\varepsilon_s = \frac{0.003}{c}(d-c)$$<br><br>Note that $b = 24$ in. for negative moments and 102 in. for girder positive moments.<br><br>Place eight No. 9 in one layer with $d = 45.3$ in. | Transfer girder:<br><br>| | $M_u$, ft-kip | $A_{s,prov}$, in.$^2$ | $a$, in. | $c$, in. | $\varepsilon_s$, in./in. | $\varepsilon_s >$ 0.005? |<br>|---|---|---|---|---|---|---|<br>| $M^-$ | 1782 | 8 | 4.7 | 5.9 | 0.020 | Y |<br>| $M^+$ | 3348 | 15.6 | 2.15 | 2.7 | 0.0466 | Y |<br>| $M^-$ | 1803 | 8 | 4.7 | 5.9 | 0.020 | Y |<br><br>Therefore, the assumption of using $\phi = 0.9$ is correct. |
| | Check skin reinforcement strain (Fig. E5.5):<br><br>Positive half, strain at middepth<br><br><br><br>Check strain in the upper skin layer at 24 in. from the top: | 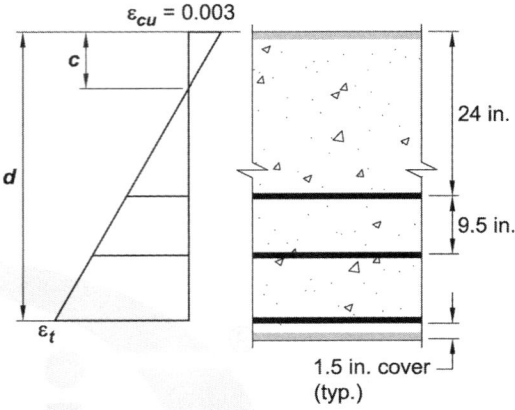<br><br>Fig. E5.5—Strain distribution over beam depth.<br><br>$$\varepsilon_s = \frac{0.003}{2.7 \text{ in.}}(24 \text{ in.} - 2.7 \text{ in.}) = 0.024 > 0.005 \quad \textbf{OK}$$<br><br>By inspection the other skin reinforcement layer has higher strain; therefore, reinforcement in both skin reinforcement layers is yielding and the assumption of using $\phi = 0.9$ is correct. |

Negative half, strain at supports B and D (Fig. E5.6)

Check strain in the lower skin layer at 24 in. from the bottom:

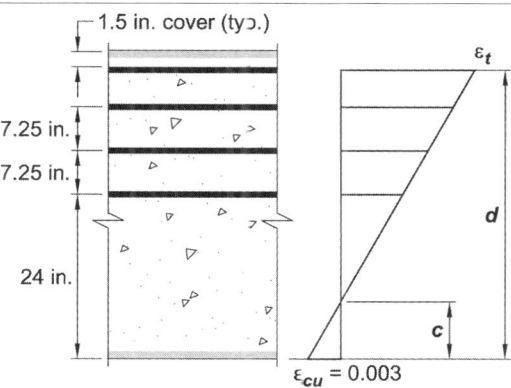

Fig. E5.6—Strain distribution.

$$\varepsilon_s = \frac{0.003}{5.9 \text{ in.}}(24 \text{ in.} - 5.9 \text{ in.}) = 0.009 > 0.005$$

By inspection the other skin reinforcement layers have higher strain; therefore, reinforcement in all skin reinforcement layers is yielding and the assumption made of using $\phi = 0.9$ is correct.

## Beams

Design beams for the maximum load condition. Extend No. 9 top bars from Span BD to resist the 906 ft-kip and 935 ft-kip moments at Column Lines B and D in Spans AB and DE, respectively, and No. 6 bars to resist the rest of the moments.

No. 6 bar; $d_b = 3/4$ in., $A_s = 0.44$ in.$^2$, and $d = 27.6$ in.
No. 9 bar; $d_b = 1.128$ in., $A_s = 1.0$ in.$^2$, and $d = 27.4$ in. for one layer.

Span Column Line AB

|  | $M_u$, ft-kip | $A_{s,req'd}$ in.$^2$ | Number of bars Req'd | Number of bars Select |
|---|---|---|---|---|
| $M^-$ (No. 9) | 906 | 8.0 | 8.0 | 8 |
| $M^+$ (No. 6) | 222 | 1.8 | 4.1 | 5 |

Span Column Line DE

|  | $M_u$, ft-kip | $A_{s,req'd}$ in.$^2$ | Number of bars Req'd | Number of bars Select |
|---|---|---|---|---|
| $M^-$ (No. 9) | 935 | 8.33 | 8.33 | 9 |
| $M^+$ (No. 6) | 293 | 2.38 | 5.42 | 6 |

Span Column Line EF

|  | $M_u$, ft-kip | $A_{s,req'd}$ in.$^2$ | Number of bars Req'd | Number of bars Select |
|---|---|---|---|---|
| $M^-$ (No. 6) | 58 | 0.43 | 0.97 | 2 |
| $M^+$ (No. 6) | 133 | 1.08 | 2.4 | 3 |

| | | |
|---|---|---|
| 9.6.1.1<br>9.6.1.2 | Minimum reinforcement<br>The provided reinforcement must be at least the minimum required reinforcement at every section along the length of the beam.<br><br>$$A_{s,min} = \frac{3\sqrt{f_c'}}{f_y} b_w d$$<br><br>Because $f_c' > 4444$ psi, Eq. (9.6.1.2a) only applies. | Check to see if the required reinforcement areas exceeds the code minimum reinforcement area at all locations.<br>Beams:<br><br>$$A_{s,min} = \frac{3\sqrt{5000 \text{ psi}}}{60,000 \text{ psi}}(24 \text{ in.})(27.6 \text{ in.}) = 2.34 \text{ in.}^2$$<br><br>Provide a minimum six No. 6 bars at all beam tension locations. Except at Column Lines B and D, where transfer girder top reinforcement is extended over adjacent spans to resist the negative moment and Span DE positive moment, where six No. 6 longitudinal bars is required. |
| | Note: Reduce the total number of No. 9 bars to eight No. 9 by extending the two top No. 6 skin bars from the girder into Beam DE to resist part of the negative moment.<br><br>Calculate strain in No. 6 bars (Fig. E5.7). | 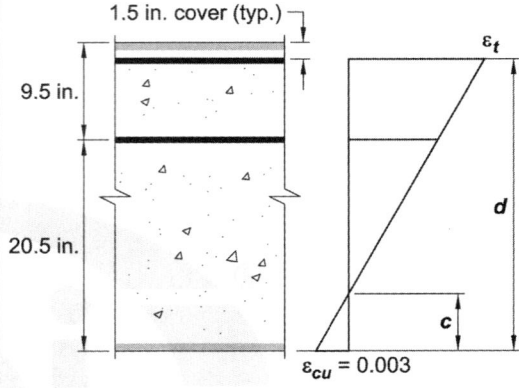<br>Fig. E5.7—Strain diagram.<br><br>$$\varepsilon_s = \frac{0.003}{5.9 \text{ in.}}(20.5 \text{ in.} - 5.9 \text{ in.}) = 0.007 > 0.005$$<br><br>Therefore, reinforcement in the two No. 6 skin bars is yielding and the assumption made of using $\phi = 0.9$ is correct. |
| 21.2.2 | Check if the calculated steel strain exceeds 0.005 in./in. (tension controlled), but not less than 0.004 in./in. (refer to Fig. E5.8):<br><br>$$a = \frac{A_s f_y}{0.85 f_c' b} \text{ and } c = \frac{a}{\beta_1}$$<br><br>where $\beta_1 = 0.8$<br><br>$$\varepsilon_s = \frac{0.003}{c}(d-c)$$<br><br>Note that $b = 24$ in. for negative moments and 102 in. and 60 in. for girder and beam positive moments, respectively.<br><br>$c < h_f$ for both transfer girder and beams; therefore, the T-section members assumption for positive moments is correct. | Beams (only maximum moments are checked)<br><br>| | $M_u$,<br>ft-kip | $A_{s,prov}$,<br>in.² | $a$, in. | $c$, in. | $\varepsilon_s$,<br>in./in. | $\varepsilon_s >$<br>0.005? |<br>|---|---|---|---|---|---|---|<br>| $M^-$ | 935 | 8 | 4.7 | 5.9 | 0.011 | Y |<br>| $M^+$ | 293 | 2.64 | 0.62 | 0.78 | 0.103 | Y |<br><br>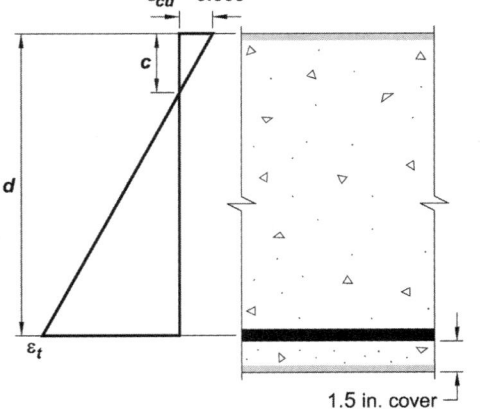<br>Fig. E5.8—Strain distribution across beam section. |

# CHAPTER 7—BEAMS

| Step 6: Shear design | | |
|---|---|---|
| *Transfer girder* | | |
| | Shear strength | |
| 9.4.3.2 | Because conditions a), b), and c) of 9.4.3.2 are satisfied, the design shear force is taken at critical section at distance $d$ from the face of the support (Fig. E5.9). use $d = 45.3$ in. at support<br><br>The controlling factored load combination must satisfy: | *Fig. E5.9—Shear at the critical section.* |
| 9.5.1.1<br>9.5.3.1<br>22.5.1.1<br>22.5.5.1 | $\phi V_n \geq V_u$<br>$V_n = V_c + V_s$<br><br>$V_c = 2\sqrt{f_c'}b_w d$ | $V_{u@d} = (379.5 \text{ kip}) - (1.68 \text{ kip/ft})(45 \text{ in.}/12) = 373.2 \text{ kip}$<br><br>$V_c = \dfrac{(2)(\sqrt{5000 \text{ psi}})(24 \text{ in.})(45 \text{ in.})}{1000 \text{ lb/kip}} = 152.7 \text{ kip}$ |
| 21.2.1(b) | Shear strength reduction factor:<br><br>$\phi V_c = \phi 2\sqrt{f_c'}b_w d$ | $\phi_{shear} = 0.75$<br><br>$\phi V_c = (0.75)(152.7 \text{ kip}) = 114.6 \text{ kip}$ |
| 9.5.1.1b | Check if $\phi V_c \geq V_u$ | $\phi V_c = 114.6 \text{ kip} < V_{u@d} = 373.7 \text{ kip}$ **NG**<br><br>Therefore, shear reinforcement is required. |
| 22.5.1.2 | Prior to calculating shear reinforcement, check if the cross-sectional dimensions satisfy Eq. (22.5.1.2):<br><br>$V_u \leq \phi(V_c + 8\sqrt{f_c'}b_w d)$ | $V_u \leq \phi\left(152.7 \text{ kip} + \dfrac{8\sqrt{5000 \text{ psi}}(24 \text{ in.})(45 \text{ in.})}{1000 \text{ lb/kip}}\right)$<br><br>$\leq 576.6 \text{ kip}$<br><br>**OK**, therefore, section dimensions are satisfactory. |

| | | |
|---|---|---|
| 22.5.10.1 | **Shear reinforcement**<br>Transverse reinforcement satisfying equation 22.5.10.1 is required at each section where $V_u > \phi V_c$<br><br>$$V_s \geq \frac{V_u}{\phi} - V_c$$ | $V_s \geq \dfrac{373.2 \text{ kip}}{0.75} - 152.7 \text{ kip} = 344.9 \text{ kip}$<br><br>Try a No. 5 bar, two legged stirrup |
| 22.5.10.5.3<br>22.5.10.5.6 | Spacing required for No. 5 stirrups:<br><br>where $V_s = \dfrac{A_v f_{yt} d}{s}$ | $344.9 \text{ kip} = \dfrac{(2)(0.31 \text{ in.}^2)(60{,}000 \text{ psi})(45 \text{ in.})}{s(1000 \text{ lb/kip})}$<br><br>$s = 4.85$ in.<br><br>This is a relatively tight spacing.<br><br>Use two No. 5 double stirrups side by side. This will yield a spacing of 9.7 in.; say, 8 in. spacing. |
| 9.7.6.2.2 | Calculate maximum allowable stirrup spacing:<br><br>First, does the beam transverse reinforcement value need to exceed the threshold value?<br><br>$V_s \leq 4\sqrt{f_c'} b_w d$ ? | $4\sqrt{f_c'} b_w d = \dfrac{4\left(\sqrt{5000 \text{ psi}}\right)(24 \text{ in.})(45 \text{ in.})}{1000 \text{ lb/kip}} = 305.5$ kip<br><br>$V_s = 344.9$ kip $> 4\sqrt{f_c'} b_w d = 305.5$ kip **OK** |
| 9.7.6.2.2 | Because the required shear strength is higher than the threshold value, the maximum stirrup spacing is the lesser of $d/4$ and 12 in.<br><br>Because shear force does not vary significantly over the length of the transfer girder (Fig. E5.3a), use two No. 5 stirrups at 8 in. spacing over the full length of the girder. | $d/4 = 45$ in./4 ~ 11 in. < 12 in.<br><br>Use $s = 8$ in. $= d/4 = 11$ in. < 12 in. ∴ **OK** |
| 9.6.3.3 | Specified shear reinforcement must be at least:<br><br>$0.75\sqrt{f_c'} \dfrac{b_w}{f_{yt}}$ and $50 \dfrac{b_w}{f_{yt}}$<br><br>Because $f_c' > 4444$ psi, Eq. (1) controls: | $\dfrac{A_{v,min}}{s} \geq 0.75\sqrt{5000 \text{ psi}} \dfrac{18 \text{ in.}}{60{,}000 \text{ psi}} = 0.016$ in.²/in.<br><br>Provided:<br><br>8 in. spacing: $\dfrac{A_{v,min}}{s} = \dfrac{4(0.31 \text{ in.}^2)}{8 \text{ in.}} = 0.155$ in.²/in.<br><br>Spacing satisfies 9.6.3.3, therefore, **OK** |

## Beams

| | | | |
|---|---|---|---|
| 21.2.1(b) | Shear strength<br>Shear strength reduction factor: | $\phi_{shear} = 0.75$ | |
| 9.5.1.1 | $\phi V_n \geq V_u$ | 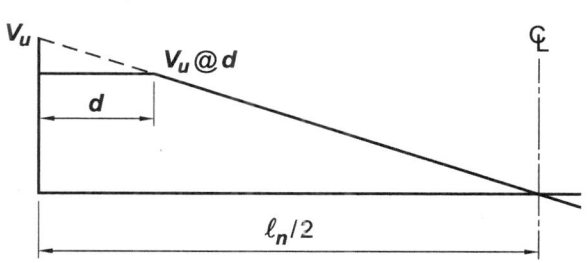<br>Fig. E5.10—Shear at the critical section | |
| 9.5.3.1<br>22.5.1.1 | $V_n = V_c + V_s$ | | |
| 9.4.3.2 | Because conditions a), b), and c) of 9.4.3.2 are satisfied, the design shear force is taken at critical section at distance $d$ from the face of the support (Fig. E5.10). | $V_{u@d} = V_u - (1.85\ \text{kip/ft})(27.4\ \text{in.}/12) = V_u - 4.2\ \text{kip}$ | |
| 22.5.5.1 | $V_c = 2\sqrt{f'_c}b_w d$ | $V_c = (2)\left(\sqrt{5000\ \text{psi}}\right)(24\ \text{in.})(27.4\ \text{in.}) = 93\ \text{kip}$<br>$\phi V_c = (0.75)(93\ \text{kip}) = 69.7\ \text{kip}$ | |
| 9.6.3.1 | Check if $\phi V_c \geq V_u$<br><br>However, ACI 318 requires that minimum shear reinforcement must be provided, where:<br><br>$V_u > 1/2\ \phi V_c = 1/2(69.7\ \text{kip}) = 34.9\ \text{kip}$<br><br>Beam EF does not require shear reinforcement; however, provide minimum shear reinforcement for integrity. | | |

| | | $V_{u@d}$, kip | Is $\phi V_c \geq V_u$? |
|---|---|---|---|
| Beam 1 | Left | 66.7 | Y |
| | Right | 88.9 | N |
| Beam 3 | Left | 95.9 | N |
| | Right | 73.8 | Y |
| Beam 4 | Left | 0 | Y |
| | Right | 21.7 | Y |

Therefore, shear reinforcement is required where concrete strength is less than the factored shear force.

| Beam | | $V_{u@d}$, kip | Is $V_u > 1/2\phi V_c$? |
|---|---|---|---|
| AB | Left | 66.7 | Y |
| | Right | 88.9 | Y |
| DE | Left | 95.9 | Y |
| | Right | 73.8 | Y |
| EF | Left | 0 | N |
| | Right | 21.7 | N |

Provide minimum shear reinforcement over all beams spans.

| | Shear reinforcement<br>$V_s \geq \dfrac{V_u}{\phi} - V_c$<br><br>where $V_s = \dfrac{A_v f_{yt} d}{s}$ and $V_c = 93\ \text{kip}$<br><br>Using No. 3 stirrups | $V_s \geq \dfrac{95.9\ \text{kip}}{0.75} - 93\ \text{kip} = 34.9\ \text{kip}$<br><br>$s = \dfrac{(2)(0.2\ \text{in.}^2)(60{,}000\ \text{psi})(27.4\ \text{in.})}{34{,}900\ \text{lb}} = 18.8\ \text{in.}$<br><br>The spacing exceeds the maximum allowed $d/2 = 13.7\ \text{in.}$; therefore, use 12 in. spacing over the full length of beam.<br><br>By inspection provisions, 9.7.6.2.2 and 9.6.3.3 are satisfied. | |

| | | |
|---|---|---|
| **Step 7: Reinforcement detailing** | | |
| 25.2.1 | Minimum bar spacing<br>Bottom reinforcement—girder:<br>The clear spacing between the horizontal No.11 bars must be at least the greatest of:<br><br>Clear spacing the greater of: $\begin{cases} 1 \text{ in.} \\ d_b \\ 4/3(d_{agg}) \end{cases}$<br><br>Assume 3/4 in. maximum aggregate size. Check if five No.11 bars (resisting positive moment) can be placed in the beam's web; refer to Fig. E5.11.<br><br>$b_{w,req'd} = 2(\text{cover} + d_{stirrup} + 1.0 \text{ in.})$<br>$\quad + 4d_b + 4(1.5 \text{ in.})_{min,spacing}$ (25.2.1)<br><br><br><br>Spacing between longitudinal bars: | 1 in.<br>1.41 in. **Controls**<br>$4/3(3/4 \text{ in.}) = 1$ in.<br><br>Therefore, clear spacing between horizontal bars must be at least 1.41 in., say, 1.5 in.<br><br>$b_{w,req'd} = 2(1.5 \text{ in.} + 0.625 \text{ in.} + 1.0 \text{ in.}) + 5.64 \text{ in.} + 6.0$ in.<br>$\quad = 17.9 \text{ in.} < 24 \text{ in.}$ **OK**<br>Therefore, five No. 11 bars can be placed in one layer in the 24 in. transfer girder web.<br><br>$sp = \dfrac{24 \text{ in.} - [2(1.5 \text{ in.} + 0.625 \text{ in.} + 1.0 \text{ in.}) + 4(1.41 \text{ in.})]}{4}$<br>$= 3.0$ in.<br><br>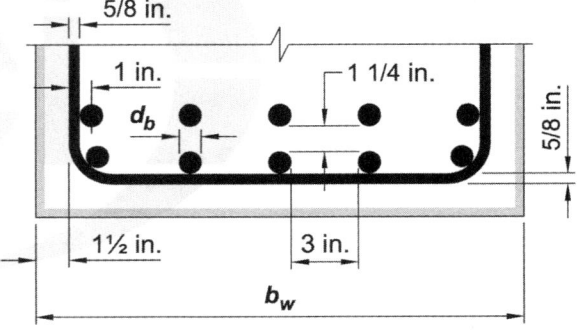<br><br>Fig. E5.11—Bottom reinforcement layout one layer is shown. |
| | Bottom reinforcement—beams:<br>Beams AB, DE, and EF are reinforced with six No. 6 bottom bars uniformly spaced. The calculated spacing is 3 in. Therefore, OK. | |

# CHAPTER 7—BEAMS

| | | | |
|---|---|---|---|
| 24.3.4 | Top reinforcement<br>Tension reinforcement in flanges must be distributed within the effective flange width, $b_f = 102$ in. (Step 2), but not wider than: $\ell_n/10$. | Girder:<br>$\ell_n/10 = (26 \text{ ft})(12)/10 = 31.2$ in. < 102 in. | |
| | Because effective flange width exceeds $\ell_n/10$, additional bonded reinforcement is required in the outer portion of the flange. | Beam:<br>$\ell_n/10 = (12 \text{ ft})(12)/10 = 14.4$ in. < 60 in. | |

| Span | | Prov. No. 9 | No. 9 in web | No. 10 in $\ell_n/10^*$ | No. 6 in outer portion[*] |
|---|---|---|---|---|---|
| AB | L | 4 | 4 | — | 4 |
| | R | 8 | 6 | 2 | 4 |
| BD | L | 8 | 6 | 2 | 6 |
| | R | 8 | 6 | 2 | 6 |
| DE | L | 8 | 6 | 2 | 4 |
| | R | 4 | 4 | — | 4 |
| EF | L | 5[†] | 5[†] | — | 4 |
| | R | 5[†] | 5[†] | — | 4 |

Use No. 6 placed in slab over $b_f$ for additional bonded reinforcement; refer to Fig. E5.12:
This requirement is to control cracking in the slab due to wide spacing of bars across the full effective flange width and to protect flange if reinforcement is concentrated within the web width.

$$\text{Bar spacing} = \frac{24 \text{ in.} - [2(3.125 \text{ in.}) + 5(1.128 \text{ in.})]}{5}$$
$$= 2.4 \text{ in.}$$

[*]Bars to be divided equally on both sides of the web (refer to Fig. E5.12-sections)

[†]Section reinforced with No. 6 bars

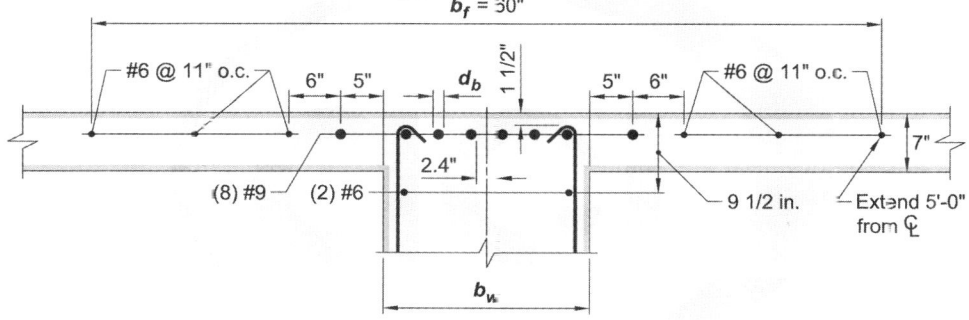

Fig. E5.12—Longitudinal bars distributed with flange.

Note: Slab shrinkage and temperature reinforcement can be used as the additional bonded reinforcement in the outer portion of flange to satisfy ACI 318-14, Section 24.3.4.

## Step 8: Development length

| | | |
|---|---|---|
| 25.4.2.2 | Development length of No. 6, No. 9, and No. 11 bar<br>The simplified method is used to calculate the development length of No. 11 bars:<br>$$\ell_d = \frac{f_y \psi_t \psi_e}{20\lambda\sqrt{f_c'}} d_b$$ | $$\ell_d = \left(\frac{(60{,}000 \text{ psi})(1.0)(1.0)}{(20)(1.0)\sqrt{5000 \text{ psi}}}\right)(d_b) = 42.43 d_b$$ |
| 25.4.2.4 | where $\psi_t$ is the cast position; $\psi_t = 1.3$, if more than 12 in. of fresh concrete is placed below top horizontal bars, and $\psi_t = 1.0$, if not more than 12 in. of fresh concrete is placed below bottom horizontal bars.<br><br>$\psi_e$ is the coating factor; and $\psi_e = 1.0$, because bars are uncoated | Top: $1.3 d_b = 1.3(42.43 d_b) = 55 d_b$ |

| | | No. 6 | No. 9 | No. 8 | No. 11 |
|---|---|---|---|---|---|
| Top | $\ell_d$, in. | 41.3 | 62 | 55 | — |
| | Use $\ell_d$, in. | 42 | 63 | 57 | — |
| Bottom | $\ell_d$, in. | 31.8 | — | — | 59.8 |
| | Use $\ell_d$, in. | 36 | — | — | 60 |

| Step 9: Inflection points | |
|---|---|
| The moment diagram inflection points are calculated at both supports and at midspan (Fig. E5.13(a)). | 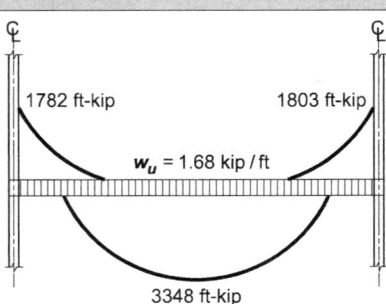<br>(a) Girder moment diagram |
| Bottom bar length along girder<br>Calculate the inflection point for positive moment (Fig. E5.13(b)): | 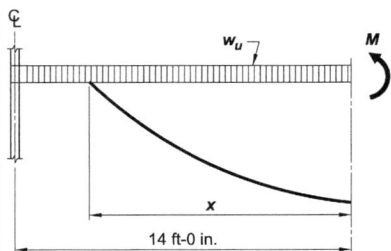<br>(b) inflection point of positive moment<br>$3348 \text{ ft-kip} - (1.68 \text{ kip/ft})(x)^2/2 - 355x = 0$<br>$x = 9.2$ ft, say, 9 ft 3 in. |
| Maximum moment at midspan:<br>3348 ft-kip<br><br>Assume the maximum positive moment occurs at midspan. From equilibrium, the point of inflection is obtained from the following free body diagram:<br>$M_{max} - w_u(x)^2/2 - P/2x = 0$<br><br>Top bar length along transfer span<br>Left support:<br>Calculate the inflection point for negative moment diagram (Fig. E5.13(c)):<br>$-M_{max} - w_u(x)^2/2 + V_u x = 0$ | 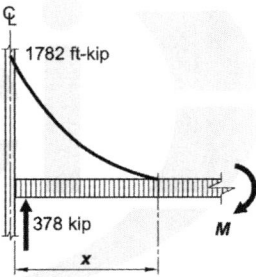<br>(c) Inflection point at Support B<br>$(-1782 \text{ ft-kip}) - (1.68 \text{ kip/ft})(x)^2/2 + (378 \text{ kip})x = 0$<br>$x = 4.76$ ft, say, 4 ft 10 in. |
| Right support:<br><br>Calculate inflection point for the negative moment diagram (Fig. E5.13(d)):<br>$-M_{max} - w_u(x)^2/2 + V_u x = 0$ | 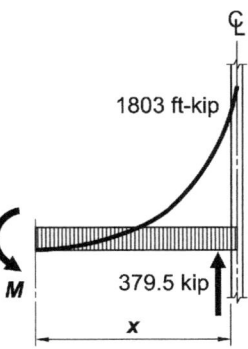<br>(d) Inflection point at Support D<br>$(-1803 \text{ ft-kip}) - (1.68 \text{ kip/ft})(x)^2/2 + (379.5 \text{ kip})x = 0$<br>$x = 4.8$ ft, say, 4 ft 10 in.<br><br>*Fig. E5.13—Girder inflection point locations.* |

Check the inflection point for Spans AB, DE, and EF.

The moment diagram envelop shows that all three spans (AB, DE, and EF) do not have a defined maximum moment at midspan. The moment varies from maximum negative moment at one support and increases to maximum positive moment at the other support.

Span AB:
Calculate inflection point at Support B (Fig. E5.14(a)):
$-M_{max} - w_u(x)^2/2 - V_u x = 0$

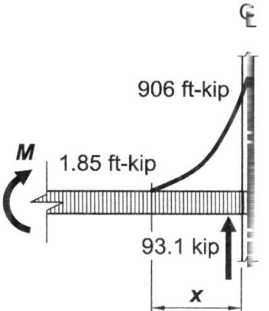

(a) Inflection point location at Support B

$(-906 \text{ ft-kip}) - (1.85 \text{ kip/ft})(x)^2/2 + (93.1 \text{ kip})x = 0$
$x = 10.9$ ft, say, 11 ft 0 in.

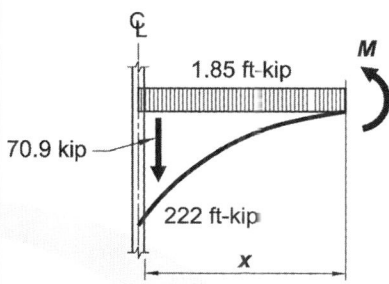

Calculate inflection point at Support A (Fig. E5.14(b)):
$-M_{max} - w_u(x)^2/2 + V_u x = 0$

(b) Inflection point location at Support A
$(222 \text{ ft-kip}) - 1.85 \text{ kip/ft}(x)^2/2 - (70.9 \text{ kip})x = 0$
$x = 3.0$ ft

Fig. E5.14—Beam AB inflection point locations.

Following the same concept for spans three and four the inflection points are calculated at:

| Span | x-left | x-right |
| --- | --- | --- |
| DE | 10.3 ft, say, 10 ft 6 in. | 3.6 ft, say, 3 ft 9 in. |
| EF | 12.2 ft, say, 12 ft 3 in. | 2.44 ft, say, 2 ft 6 in. |

| | | |
|---|---|---|
| **Step 10: Cutoff locations** | | |
| 9.7.3.2<br>9.7.3.3 | Transfer girder<br><br>Support<br>Bars must be developed at locations of maximum stress and locations along the span where bent or terminated tension bars are no longer required to resist flexure.<br><br>Eight No. 9 bars and four No. 6 and two No. 8 skin bars are required to resist the factored moment at Column Lines B and D.<br><br>The two end moments at the supports are close, so the calculation will be applied at one end only. Calculate a distance $x$ from the column face where four No. 9 bars can be discontinued and the contribution from the skin reinforcement is ignored.<br><br>Note: Skin reinforcement are extended over the full girder length and are properly developed or extended into the adjacent spans at the supports. | Cutoff point:<br>$$-1803 \text{ ft-kip} - 1.69 \text{ kip/ft}\frac{x^2}{2} + 379.5 \text{ kip}(x) = -4(1.0 \text{ in.}^2)$$<br>$$(0.9)(60 \text{ ksi})\left(45 \text{ in.} - \frac{4(1.0 \text{ in.}^2)(60 \text{ ksi})}{2(0.85)(5 \text{ ksi})(24 \text{ in.})}\right)$$<br>$x = 2.69$ ft, say, 2 ft 9 in. from column face<br><br>For No. 9 bars:<br>1) $d = 45$ in.  **Controls**<br>2) $12d_b = 12(1.128 \text{ in.}) = 13.5$ in.<br><br>Therefore, extend the four No. 9 bars the greater of the development length (63 in. – Step 8) from the column face and $d$ from theoretical cutoff point (33 in.)<br>33 in. + 45 in. = 78.3 in. > $\ell_d = 63$ in.<br><br>Therefore, 78.3 in. **Controls**<br><br>Four No. 9 bars can be terminated 80 in. from the face of the column, shown bold in Fig. E5.15. |
| 9.7.3.8.4 | At least one-third of the bars resisting negative moment at a support must have an embedment length beyond the inflection point the greatest of $d$, $12d_b$, and $\ell_n/16$. | For No. 9 bars:<br>1) $d = 45$ in.  **Controls**<br>2) $12d_b = 12(1.128 \text{ in.}) = 13.5$ in.<br>3) $\ell_n/16 = (28 \text{ ft} - 2 \text{ ft})/16 = 1.625$ ft = 19.5 in.<br><br>Extend the remaining four No. 9 bars the larger of the development length (63 in.) beyond the theoretical cutoff point (33 in.) and $d = 45$ in. beyond the inflection point (4 ft 10 in.) (Step 9) shown bold in Fig. E5.15).<br><br>63 in. + 33 in. = 96 in.<br>58 in. + 45 in. = 103 in.  **Controls**<br>Extend the remaining four No. 9 bars 8 ft 8 in. from the face of the column.<br><br>The four No. 9 bars will be, however, extended over the full length of the girder. |

| | | |
|---|---|---|
| 9.7.3.2<br>9.7.3.3 | Transfer girder bottom bars<br>Following the same steps above, ten No.11 bars and four No.8 skin bars are required to resist the factored moment at midspan.<br><br>Calculate a distance $x$ from the midspan where four No.11 bars can resist the factored moment.<br><br>Note: Skin bars are extended over full girder length. | $(3348 \text{ ft-kip}) - (1.68 \text{ kip/ft})\dfrac{x^2}{2} - (355 \text{ kip})x = 4(1.56 \text{ in.}^2)$<br>$\times (0.9)(60 \text{ ksi})\left( 43.7 \text{ in.} - \dfrac{4(1.56 \text{ in.}^2)(60 \text{ ksi})}{2(0.85)(5 \text{ ksi})(102 \text{ in.})} \right)$<br><br>$x = 5.92$ ft, say, 6 ft 0 in. from midspan<br><br>Therefore, extend six No.11 bars the larger of the development length (60 in. – Step 8) and a distance $d$ beyond the theoretical cutoff point (6 ft 0 in. **Controls**)<br>72 in. + 43.7 in. = 115.7 in., say, 9.75 ft from maximum positive moment at midspan (Fig. E5.15). Extend the remaining four No. 11 bars at least the longer of 6 in. into the column or $\ell_d = 60$ in. past the theoretical cutoff point (Fig. E5.15);<br>60 in. + 72 in. = 132 in. < (13 ft)(12) + 6 in. = 162 in. The 6 in. into the column controls, however, it is recommended to extend bars to the far face of the column and develop them. |
| 9.7.3.8.2 | At least one-fourth of the positive tension bars must extend into the column at least 6 in. | 4 bars > 1/4(10 bars) = 2.5 bars    **OK** |

| | | |
|---|---|---|
| 9.7.3.8.3 | At the point of inflection, $d_b$ for positive moment tension bars must be limited such that $\ell_d$ for that bar size satisfies:<br><br>$$\ell_d \leq \frac{M_n}{V_u} + \ell_a$$<br><br>where $M_n$ is calculated assuming all bars at the section are stressed to $f_y$. $V_u$ is calculated at the section. The term $l_a$ is the embedment length beyond the point of inflection, limited to the greater of $d$ and $12d_b$. | Point of inflection occurs at 26 ft/2 – 9.25 ft = 3.75 ft from the column face.<br><br>$V_u$ = 379.5 kip – (1.68 kip/ft)(3.75 ft) = 373.2 kip<br><br>At that location, four No. 11 bars are effective:<br><br>$$M_n = 4(1.56 \text{ in.}^2)(60 \text{ ksi})\left(43.7 \text{ in.} - \frac{4(1.56 \text{ in.}^2)(60 \text{ ksi})}{(2)(0.85)(5 \text{ ksi})(24 \text{ in.})}\right)$$<br><br>$M_n$ = 15,674 in.-kip<br><br>$$\ell_d \leq \frac{15{,}674 \text{ in.-kip}}{373.2 \text{ kip}} + 43.7 \text{ in.} = 85.7 \text{ in.}$$<br><br>This length exceeds $\ell_d$ = 60 in., therefore, **OK** |
| 9.7.3.5 | If bars are cut off in regions of flexural tension, then a bar stress discontinuity occurs. Therefore, the code requires that flexural tensile bars must not be terminated in a tensile zone unless (a), (b), or (c) is satisfied.<br>(a) $V_u \leq (2/3)\phi V_n$ at the cutoff point<br>Continuing bars provides double the area required for flexure at the cutoff point and (b) $V_u \leq (3/4)\phi V_n$.<br>(c) Stirrup or hoop area in excess of that required for shear and torsion is provided along each terminated bar or wire over a distance $3/4d$ from the termination point. Excess stirrup or hoop area shall be at least $60b_w s/f_{yt}$. Spacing $s$ shall not exceed $d/(8\beta_b)$.<br><br>$V_{u@7ft}$ = 379.5 kip – (1.68 kip/ft)(7 ft) = 367.7 kip<br><br>$\Delta A_{v,excess} = A_{v,prov} - A_{v,req'd}$ | $V_n = V_c + V_s$ = 153.8 + 344.5 = 498.3 kip<br>$2/3\ \phi V_n$ = 249.2 kip<br>Conditions (a) and (b) are not satisfied.<br>(c) Therefore, over a distance of 3/4(45.3 in.) = 34 in. from the end of the terminated bars space stirrups at 45.3 in./(8(8/12)) = 8.5 in. on center and excess stirrup area must be at least:<br>$A_{v,excess}$ = 60(24 in.)(8 in.)/60,000 psi = 0.192 in.$^2$<br><br>Calculated required stirrup area at this location:<br><br>$$367.7 \text{ kip} - 115.3 \text{ kip} = \frac{(0.75)A_{v,req'd}(60 \text{ ksi})(45.3 \text{ in.})}{8 \text{ in.}}$$<br><br>$A_{v,req'd}$ = 1.0 in.$^2$<br>$\Delta A_{v,excess}$ = 4(0.31 in.$^2$) – 1.0 in.$^2$ = 0.24 in.$^2$<br>0.24 in.$^2$ > 0.192 in.$^2$, therefore, **OK**<br><br>Because only one of the three conditions needs to be satisfied, this requirement is satisfied. |

| | Integrity reinforcement | |
|---|---|---|
| 9.7.7.2 | At least one-fourth of the maximum positive moment bars, but at least two bars, must be continuous and developed at the face of the column | This condition was satisfied above by extending four No.11 bars into the support.<br>4 bars/10 bars = 2/5 > 1/4  **OK** |
| 9.7.7.3 | Beam longitudinal bars must be enclosed by closed stirrups along the clear span. | This condition is satisfied by extending stirrups at 8 in. on center over the full length of the beam. |
| | Beam structural integrity bars shall pass through the region bounded by the longitudinal column bars. | Four No. 11 bars are extended through the column longitudinal reinforcement, therefore satisfying this condition.<br><br>Note: The girder has the same width as the columns' dimensions (24 in.). Therefore, beam longitudinal reinforcement must be offset to clear column reinforcement. |

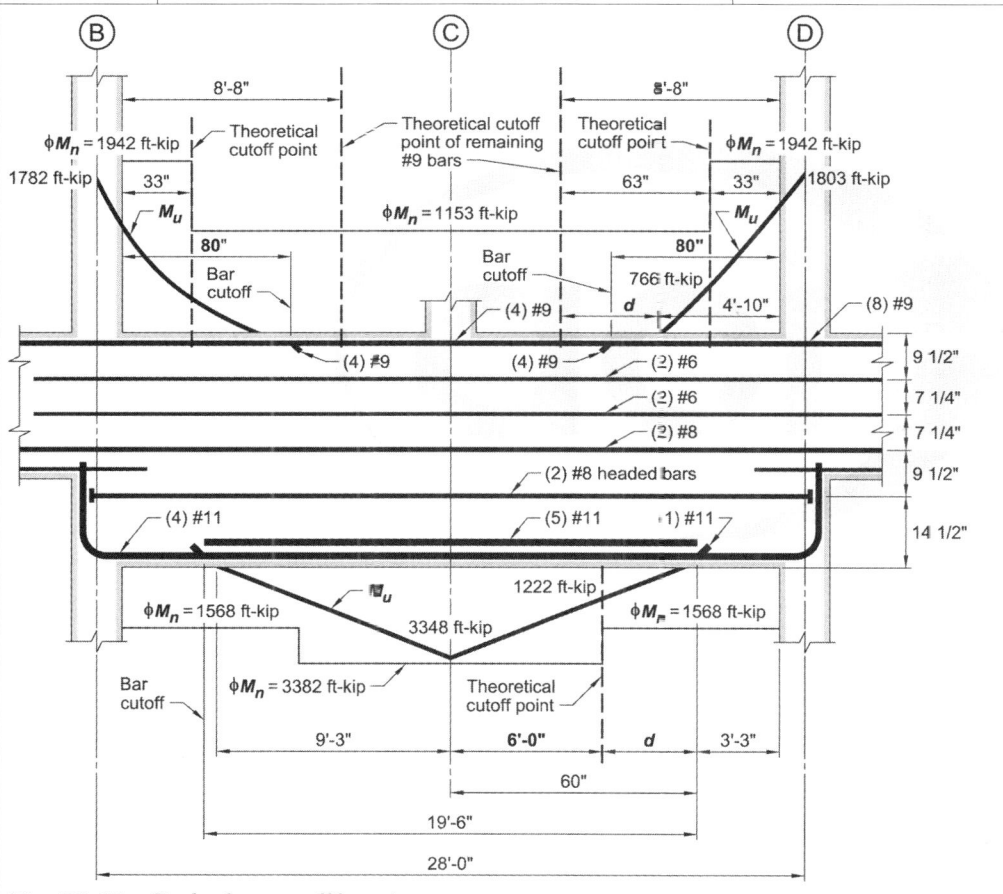

*Fig. E5.15—Girder bar cutoff locations.*

| Step 11: Beams | | |
|---|---|---|
| 9.7.3.2 | Spans AB and DE<br>Beams spanning between column A and B and between D and E are subjected to comparable factored flexure and shear forces. Therefore, the two beams will be designed for the same loads (larger of the two beams).<br><br>To simplify detailing, simply extend eight No. 9 bars and two No. 6 skin bars (span DE only) from the transfer girders to resist the negative moment in the adjacent beams (Step 5).<br><br>Calculate a distance $x$ from the column face where four No. 9 bars can resist the factored moment. | $-(906 \text{ ft-kip}) - (1.85 \text{ kip/ft})\dfrac{x^2}{2} + 93.1 \text{ kip}(x) = -4(1.0 \text{ in.}^2)$<br>$\dfrac{(0.9)(60 \text{ ksi})}{12 \text{ in./ft}}\left(27.4\text{in.} - \dfrac{4(1.0 \text{ in.}^2)(60 \text{ ksi})}{2(0.85)(5 \text{ ksi})(24 \text{ in.})}\right)$<br><br>$x = 4.9$ ft, say, 5 ft<br><br>At 5 ft 0 in. in. from the column face, four No. 9 can be cut off and the remainder four No. 9 bars can resist the factored moment. |
| 9.7.3.3 | The four No. 9 cutoff bars must extend beyond the location where they are no longer required to resist flexure for a distance equal to the greater of $d$ or $12d_b$. | For No. 9 bars:<br>1) $d = 27.4$ in.  **Controls**<br>2) $12d_b = 12(1.128 \text{ in.}) = 13.54$ in. |
| 9.7.3.8.4 | At least one-third of the bars resisting negative moment at a support must have an embedment length beyond the inflection point the greatest of $d$, $12d_b$, and $\ell_n/16$. | Therefore, extend four No. 9 bars the greater of the development length (63 in. – Step 8) from the column face and the sum of theoretical cutoff point and $d$:<br>60 in. + 27.4 in. = 87.4 in. > 63 in.  **Controls**<br>Say, 90 in or 7 ft 6 in.<br>For No. 9 bars:<br>(a) $d = 27.4$ in.  **Controls**<br>(b) $12d_b = 12(1.128 \text{ in.}) = 13.54$ in.<br>(c) $\ell_n/16 = (28 \text{ ft} - 2 \text{ ft})/16 = 1.625 \text{ ft} = 19.5$ in.<br><br>Extend the remainder four No. 9 bars, the greater of the development length (63 in.) beyond the theoretical cutoff point (5 ft) and $d = 27.4$ in. beyond the inflection point (11 ft 0 in.).<br>$\ell = 63$ in./12 + 5 ft = 10.25 ft<br>$\ell = 11.0$ ft + 27.4 in./12 = 13.3 ft  **Controls**<br>Therefore, extend the remainder bars over the full length of the beam; refer to Fig. E5.16 Spans AB and DE. |
| | At the Support E, where there is no negative moment, provide minimum reinforcement of six No. 6 bars extending minimum the development length (42 in.) on both sides of the column. Of the six No. 6 bars, extend two bars over the full length to support the stirrups (hanger bars); refer to Fig. E5.16. | |

| | | |
|---|---|---|
| | For the positive moment region, five No. 6 bottom bars are extended over the full span length for Beams 1, 3, and 4. | |
| **Step 12: Splicing and bar spacing** | | |
| 9.7.7.5 | Splices are necessary for continuous bars. The bars shall be spliced in accordance with (a) and (b):<br>(a) Positive moment bars shall be spliced at or near the support<br>(b) Negative moment bars shall be spliced at or near midspan | Splice length = (1.3)(development length)<br>No. 11: $\ell_{dc}$ = 1.3(60 in.) = 78 in. = 6 ft 6 in.<br>No. 6: $\ell_{dc}$ = 1.3(36 in.) = 46.8 in., say, 48 in. = 4 ft 0 in.<br>No. 9: $\ell_{dc}$ = 1.3(63 in.) = 81.9 in., say, 84 in. = 7 ft 0 in.<br>No. 6: $\ell_{dc}$ = 1.3(42 in.) = 54.6 in., say, 57 in. = 4 ft 9 in. |
| 9.7.2.2<br>24.3.1<br>24.3.2<br><br><br>24.3.2.1 | Maximum bar spacing at the tension face must not exceed the lesser of<br><br>$$s = 15\left(\frac{40,000}{f_s}\right) - 2.5c_c$$<br><br>and<br><br>$s = 12(40,000/f_s)$<br>where $f_s = 2/3\, f_y = 40,000$ psi<br><br>This limit is intended to control flexural cracking width. Note that $c_c$ is the cover to the longitudinal bars, not to the tie. | $s = 15\left(\dfrac{40,000 \text{ psi}}{40,000 \text{ psi}}\right) - 2.5(2 \text{ in.}) = 10 \text{ in.}$  **Controls**<br><br>$s = 12\left(\dfrac{40,000 \text{ psi}}{40,000 \text{ psi}}\right) = 12 \text{ in.}$<br><br>\| \| No. 6 \| No. 9 \| No. 11 \|<br>\|---\|---\|---\|---\|<br>\| Spacing, in. \| 3.75 \| 2.4 \| 3 \|<br><br>All longitudinal bar spacing satisfy the maximum bar spacing requirement; therefore, **OK** |

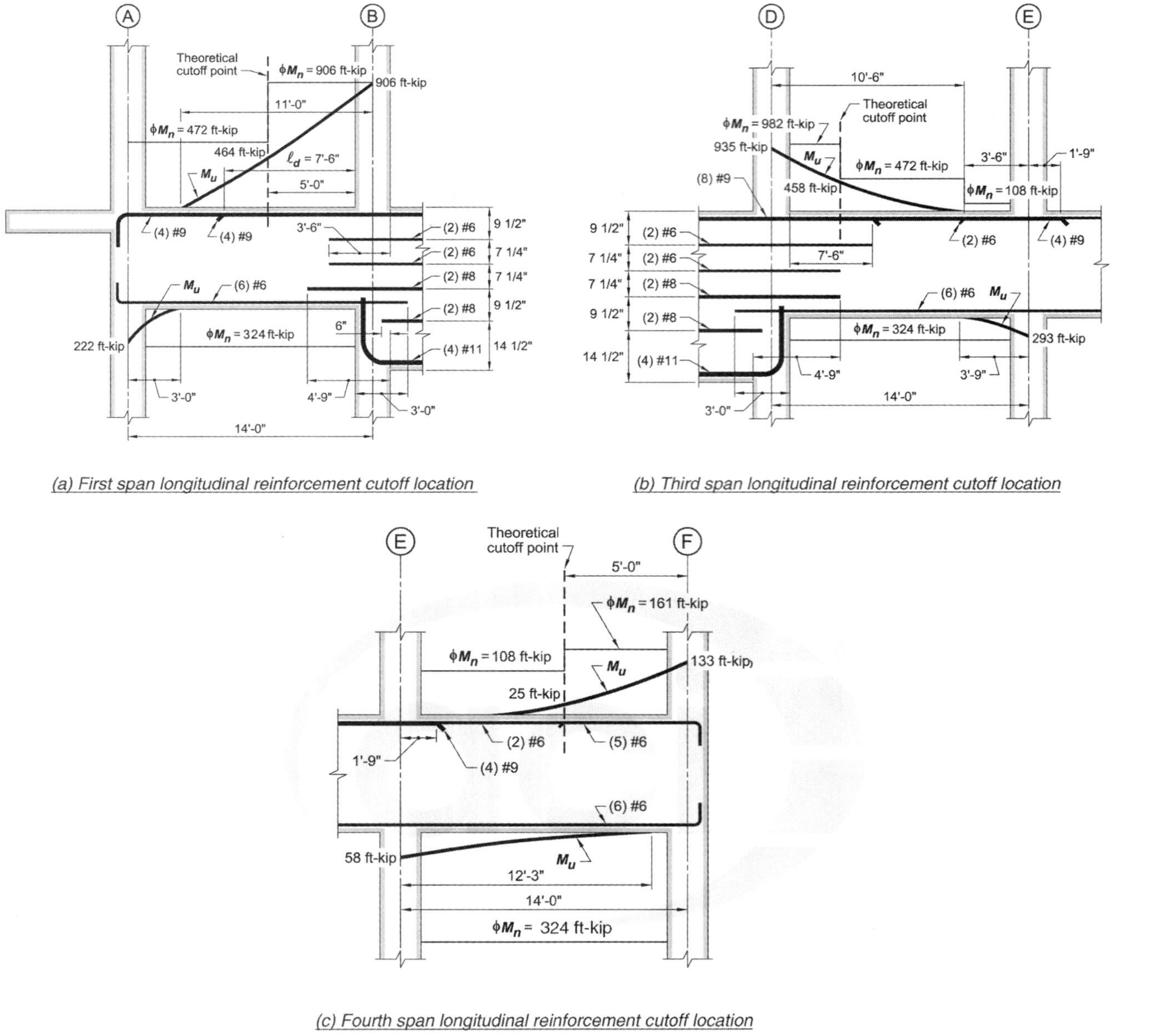

Fig. E5.16—Longitudinal reinforcement cutoff locations.

# CHAPTER 7—BEAMS

## Step 13: Detailing

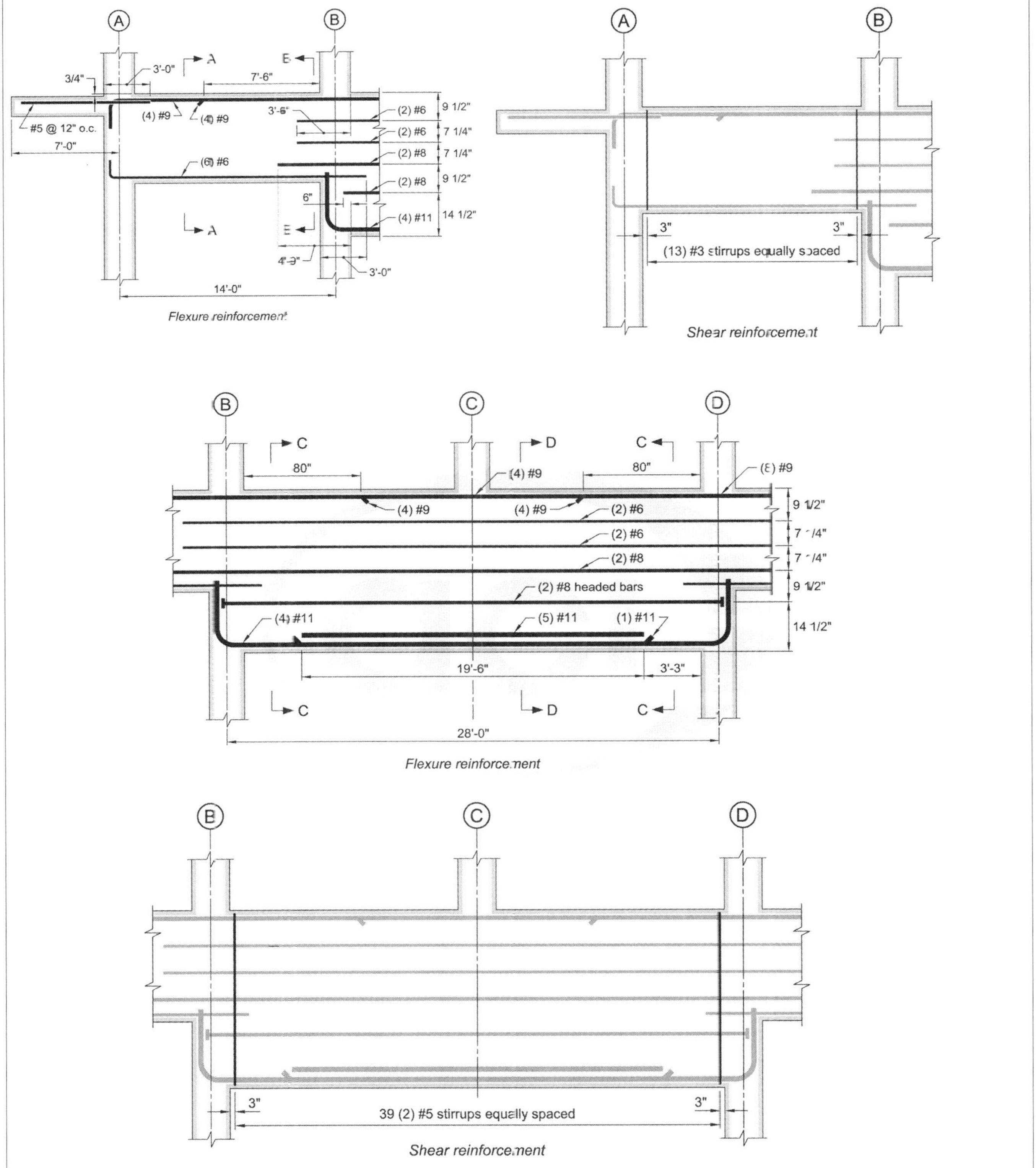

Fig. E5.17—Beam bar details. (Note: Figure continued on next page).

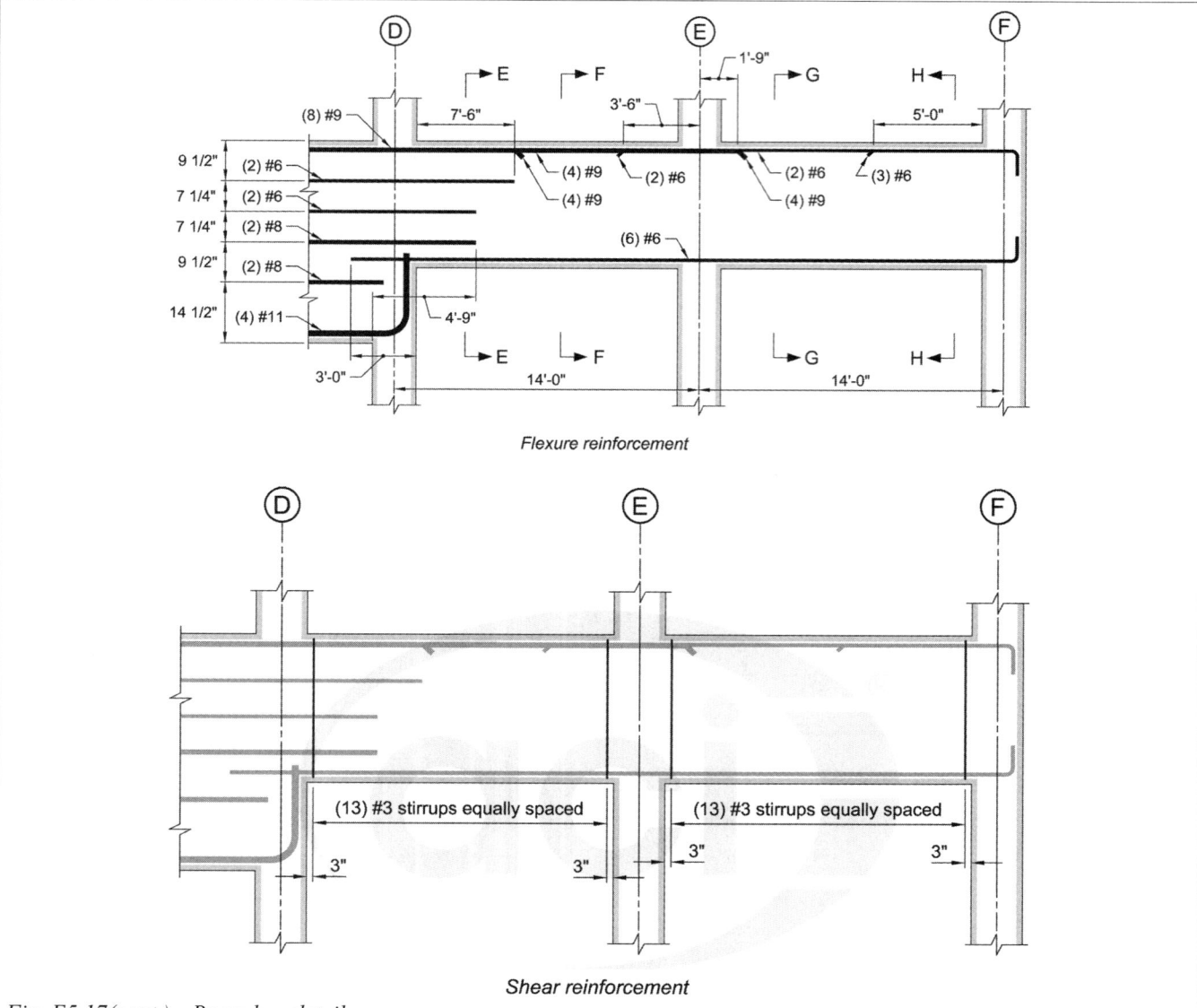

Fig. E5.17(cont.)—Beam bar details.

Notes:
1. Place first stirrup at 3 in. from the column face.
2. The contractor may prefer to extend two No.7 top reinforcement over the full beam length to replace the two No. 5 hanger beams. Bars should be spliced at mid-length

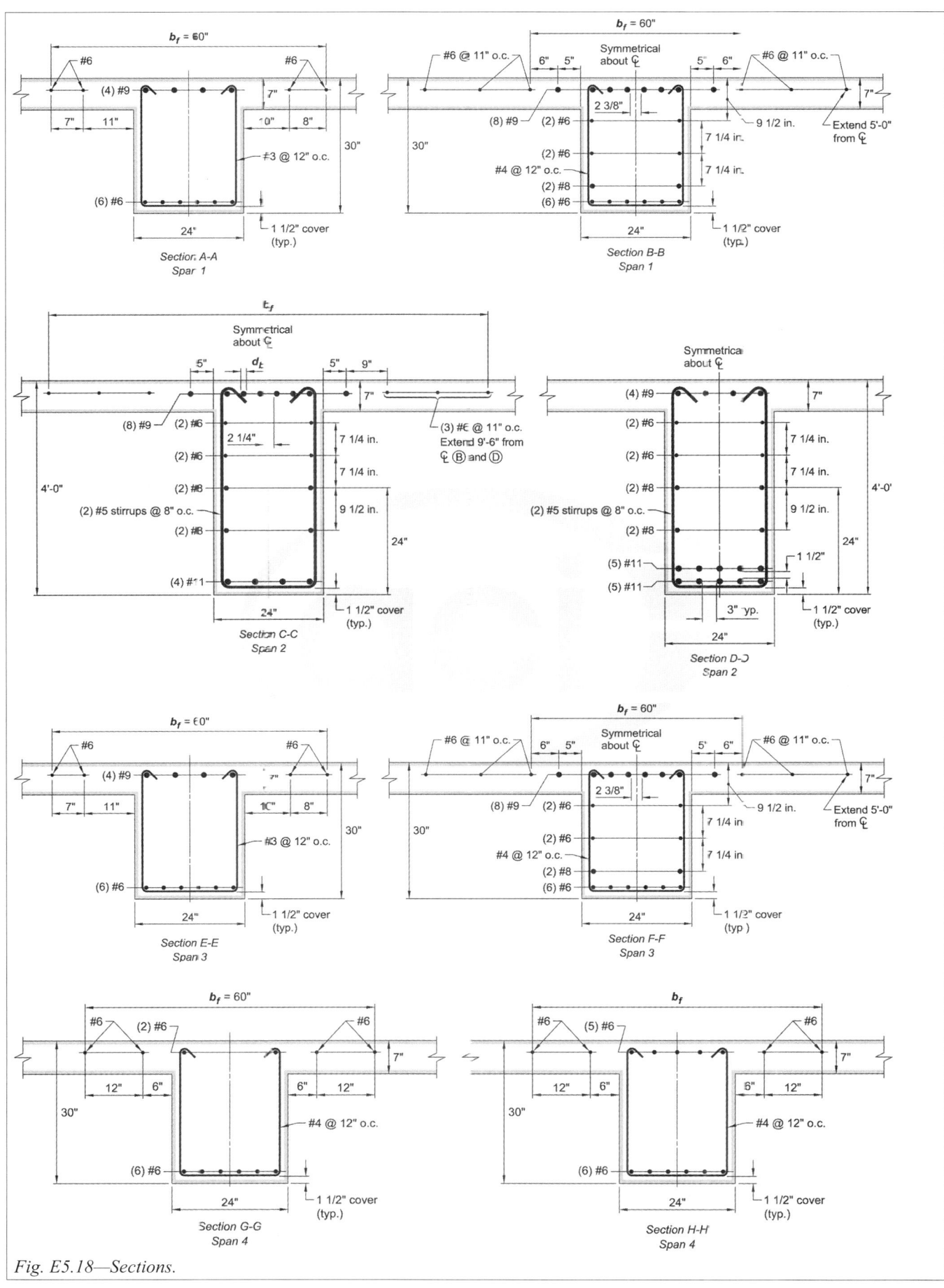

Fig. E5.18—Sections.

**Beam Example 6**: *Post-tensioned transfer girder*

Design and detail an interior, post-tensioned, transfer girder supporting five stories above, built integrally with a 7 in. slab. Girder tendons will not be stressed until concrete compressive strength reaches the specified $f_{ci}$ = 4000 psi. Assume the tendon center of gravity at beam midspan is 4 in. from the bottom and at the beam's center of gravity at the column. Tendon will be composed of 1/2 in. diameter individually coated and sheathed seven-wire prestressing strands.

Given:

*Load*—
Service additional dead load $D$ = 15 psf
Service live load $L$ = 65 psf
Girder, beam and slab self-weights are given below.

*Material properties*—
$f_c'$ = 5000 psi (normalweight concrete)
$f_{ci}$ = 4000 psi
$f_y$ = 60,000 psi
$f_{pu}$ = 270,000 psi
$\lambda$ = 1.0 (normalweight concrete)

*Span length*—
Girder: 28 ft
Beam and girder width: 24 in.
Column dimensions: 24 in. x 24 in.

Area of 1/2 in. diameter strand = 0.153 in.²

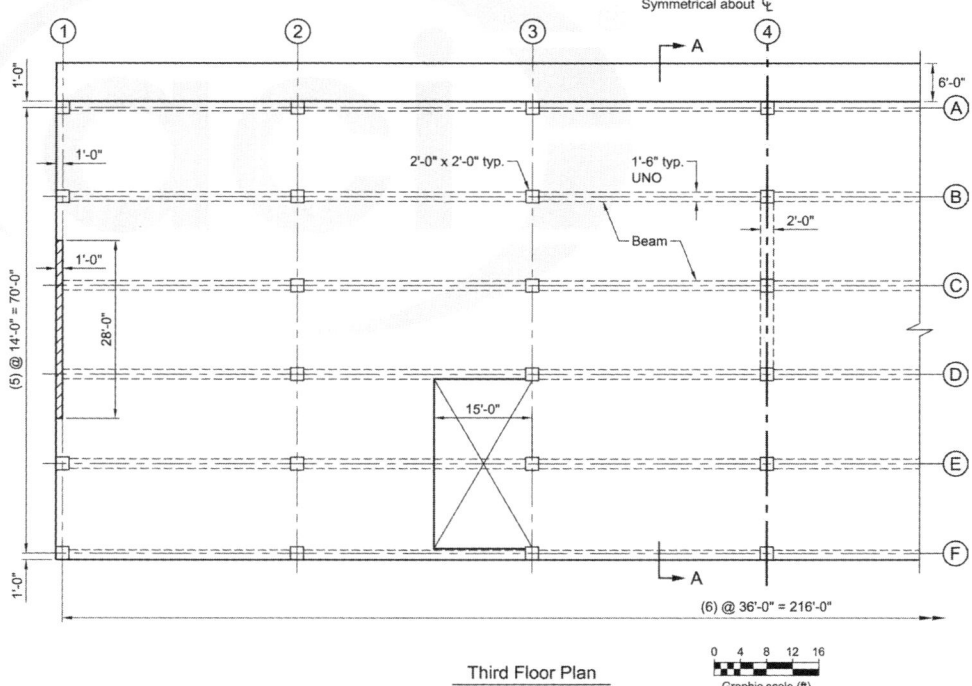

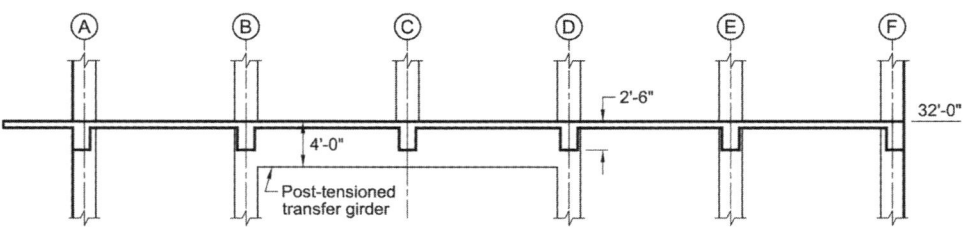

Fig. E6.1—*Plan and partial elevation of third level transfer girder and beams.*

# CHAPTER 7—BEAMS

| ACI 318-14 | Discussion | Calculation |
|---|---|---|
| **Step 1: Material requirements** | | |
| 9.2.1.1 | The mixture proportion must satisfy the durability requirements of Chapter 19 (ACI 318-14) and structural strength requirements. The designer determines the durability classes. Please refer to Chapter 2 of SP-17 for an in-depth discussion of the categories and classes.<br><br>ACI 301 is a reference specification that is coordinated with ACI 318. ACI encourages referencing ACI 301 into job specifications.<br><br>There are several mixture options within ACI 301, such as admixtures and pozzolans, which the designer can require, permit, or review if suggested by the contractor. | By specifying that the concrete mixture shall be in accordance with ACI 301-10 and providing the exposure classes, Chapter 19 requirements are satisfied.<br><br>Based on durability and strength requirements, and experience with local mixtures, the compressive strength of concrete is specified at 28 days to be at least 5000 psi.<br><br>The engineer must specify the transfer stress—in this case, 4000 psi. |
| **Step 2: Beam geometry** | | |
| 9.3.1.1 | <u>Girder depth</u><br>The transfer girder supports a column at midspan. The column load includes self-weight and its tributary loads from the third level, the four stories above it, and the roof. Because of this large concentrated load, the depth limits in Table 9.3.1.1 cannot be used, and calculated deflections must satisfy the deflection limits in 9.3.2. | Assume 48 in. deep transfer girder. |
| 9.2.4.2<br>6.2.3.1<br><br>R6.3.2.3 | <u>Flange width</u><br>The transfer girder is poured monolithically with the slab and will behave as a T-beam.<br><br>It is allowed per ACI 318-14 comment to ignore the flange width requirements based on experience and past performances.<br><br>Determination of an effective flange width for prestressed T-beams is therefore left to the experience and judgment of the licensed design professional. | |
| 6.3.2.1 | Use $b_f = 8h + b_w + 8h$ | $b_f = (8)(7 \text{ in.}) + 24 \text{ in.} + (8)(7 \text{ in.}) = 136 \text{ in.}$ |

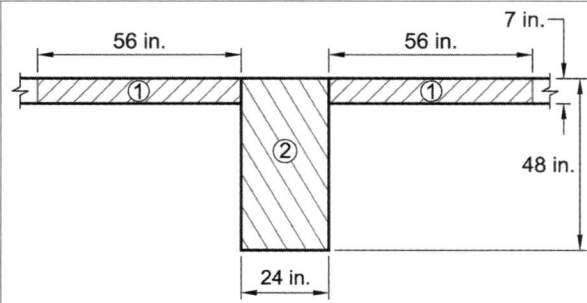

Fig. E6.2—Transfer girder geometry.

| | $A$, in.$^2$ | $y$, in. | $Ay$, in.$^3$ | $A(y - y_t)^2$, in.$^4$ | $I_x$, in.$^4$ | $\sum I_i$, in.$^4$ |
|---|---|---|---|---|---|---|
| 1 | (2)(56 in.)(7 in.) = 784 | 3.5 | 2744 | 116,690 | 3201 | 119,891 |
| 2 | (48 in.)(24 in.) = 1152 | 24 | 27,648 | 79,361 | 221,184 | 300,545 |
| $\sum$ | 1936 | | 30,392 | 196,051 | 224,385 | 420,436 |

Center of gravity: $y_t = \dfrac{30{,}392 \text{ in.}^2}{1936 \text{ in.}^2} = 15.7$ in. from top of girder

$y_b = 48$ in. $- 15.7$ in. $= 32.3$ in. from bottom of girder

Section modulus: $S_{Top} = \dfrac{420{,}436 \text{ in.}^4}{15.7 \text{ in.}} = 26{,}779$ in.

$S_{Top} = \dfrac{420{,}436 \text{ in.}^4}{32.3 \text{ in.}} = 13{,}017$ in.

# CHAPTER 7—BEAMS

## Step 3: Loads

Applied load on transfer girder
The service live load is 50 psf in offices and 80 psf in corridors per Table 4-1 in ASCE 7-10. This example will use 65 psf as an average live load, as the actual layout is not provided. To account for the weight of ceilings, partitions, and HVAC systems, add 15 psf as miscellaneous dead load.

Dead load:
Transfer girder self-weight without flanges:

$w_b = [(24 \text{ in.})(48 \text{ in.})[(0.150 \text{ kip/ft}^3)/144]$
$= 1.2 \text{ kip/ft}$

Slab weight per level:

$P_{s\ell} = (7 \text{ in.}/12)(14 \text{ ft})(36 \text{ ft})(0.15 \text{ kip/ft}^3) = 44.1 \text{ kip}$

Column weight per level:

$P_{col} = (2 \text{ ft})(2 \text{ ft})\left(12 \text{ ft} - \dfrac{7 \text{ in.}}{12}\right)(0.15 \text{ kip/ft}^3) = 6.85 \text{ kip}$

Typical beam weight (refer to plan):
18 in. × (30 in. – 7 in.) × (36 ft – 2 ft)

$P_{bm} = \left(\dfrac{18 \text{ in.}}{12}\right)\left(\dfrac{23 \text{ in.}}{12}\right)(34 \text{ ft})(0.15 \text{ kip/ft}^3) = 14.7 \text{ kip}$

Miscellaneous dead load per level:

$P_{SDL} = (14 \text{ ft})(36 \text{ ft})(0.015 \text{ kip/ft}^2) = 7.6 \text{ kip}$

Total dead load applied on girder:

$P_D = (44.1 \text{ kip} + 14.7 \text{ kip} + 7.6 \text{ kip})(6) + (6.85 \text{ kip})(5)$
$= 433 \text{ kip}$

Live load:
Roof live load: 35 psf:

$P_{L,Roof} = (14 \text{ ft})(36 \text{ ft})(0.035 \text{ kip/ft}^2) = 18 \text{ kip}$

Total live load—per ASCE 7 live load with exception of roof load is permitted to be reduced by:

$L = L_o\left(0.25 + \dfrac{15}{\sqrt{K_{LL}A_T}}\right)$

where $L$ is reduced live load; $L_o$ is unreduced live load; $K_{LL}$ is live load element factor = 4 for internal columns and 2 for interior beam (ASCE 7 Table 4-2) $A_T$ is tributary area
and $K_{LL}A_T \geq 400 \text{ ft}^2$
$4(36 \text{ ft})(14 \text{ ft}) = 2016 \text{ ft}^2 > 400 \text{ ft}^2$
$2(36 \text{ ft})(14 \text{ ft}) = 1008 \text{ ft}^2 > 400 \text{ ft}^2$
and $L \geq 0.4L_o = 26 \text{ psf}$

Fourth to seventh reduced live load per level:

$L = (0.065 \text{ kip/ft}^2)\left(0.25 + \dfrac{15}{\sqrt{2016 \text{ ft}^2}}\right) = 0.038 \text{ ksf}$

$L = 0.038 \text{ ksf} > 0.4L_o = 0.026 \text{ ksf}$ **OK**

Reduced third level live load on beam:

$L = (0.065 \text{ kip/ft}^2)\left(0.25 + \dfrac{15}{\sqrt{1008 \text{ ft}^2}}\right) = 0.047 \text{ ksf}$

$L = 0.047 \text{ ksf} > 0.4L_o = 0.026 \text{ ksf}$ **OK**

Concentrated live load to column per level:
$P_L = (14 \text{ ft})(36 \text{ ft})(0.038 \text{ kip/ft}^2) = 19.2 \text{ kip}$

Concentrated live load at third level:
$P_{L@3rdLevel} = (14 \text{ ft})(36 \text{ ft})(0.047 \text{ kip/ft}^2) = 23.7 \text{ kip}$

Total live load applied on girder (four levels):

$\sum L = (19.2 \text{ kip})(4) + (23.7 \text{ kip}) + (18 \text{ kip}) = 119 \text{ kip}$

| Step 4: Material properties | | |
|---|---|---|
| 20.3.1.1<br>20.3.2.2 | Post-tensioned strands: ASTM A416 | $f_{pu}$ = 270,000 psi |
| 20.3.2.5.1 | Stress in tendon immediately after force transfer: | $0.7 f_{pu}$ = 189,000 psi |
| 20.3.2.6.1 | Considering prestress losses, assume final stress in tendons is: | $0.65 f_{pu}$ = 175,000 psi |
| R20.3.2.1 | Modulus of elasticity is assumed for design and checked against test results after a supplier is contracted.<br><br>For design use:<br>Concrete compressive strength<br>Concrete strength at initial stressing | $E_p$ = 28,500,000 psi<br>$f_c'$ = 5000 psi<br>$f_{ci}'$ = 4000 psi |
| 24.5.2.1 | Assume a maximum concrete flexure stress as $12\sqrt{f_c'}$, Class T, at service.<br><br>Concrete compressive and tensile stresses immediately after transfer: | $7.5\sqrt{f_c'}$ = 530 psi $< f_t \leq 12\sqrt{f_c'}$ = 848 psi |
| 24.5.3.1<br>24.5.3.2 | The transfer girder will be reinforced with nonprestressed bonded reinforcement. Therefore, the $3\sqrt{f_{ci}'}$ can be exceeded. Use $7.5\sqrt{f_{ci}'}$ | <table><tr><th>Location</th><th>Type</th><th>Stress limit, psi</th></tr><tr><td rowspan="2">All</td><td>Compression</td><td>$0.60 f_{ci}'$ = 2400</td></tr><tr><td>Tension</td><td>$7.5\sqrt{f_{ci}'}$ = 474 psi</td></tr></table> |
| 24.5.4.1 | Concrete compressive stress limits at service loads: | <table><tr><th>Load condition</th><th>Concrete compressive stress limits</th></tr><tr><td>Prestress plus sustained load</td><td>$0.45 f_c'$ = 2250 psi</td></tr><tr><td>Prestress plus total load</td><td>$0.60 f_c'$ = 3000 psi</td></tr></table> |
| Step 5: Design assumptions | | |
| | The girder is designed using unbonded single-strand tendons, as it is the preferred construction method in the United States. Bonded tendons may be preferable in other parts of the world. The tendons center of gravity profile is shown in Fig. E6.3. | |
| 9.4.3 | The beam is built integrally with end supports; therefore, the beam is analyzed as part of a frame. The factored moments and shear forces (required strengths) are calculated at the face of the supports. The moment and shear diagrams at different stages are obtained from PTData software.<br>The girder is stressed in two stages. At the first stressing stage, few tendons are stressed after concrete reaches a concrete compressive strength of minimum 4000 psi and subjected only to its self-weight. Before stressing the remaining tendons, the girder will be supporting three levels—dead load only. At the second stage, the remaining tendons are stressed and construction of the upper floors continues. | |

| Step 6: Post-tensioning design | | |
|---|---|---|
| | Find the total number of strands required to resist the total load<br>The required prestress force to support the total dead and live load is calculated at service load condition to satisfy the limit in the selected Class in Table 24.5.2.1.<br><br>The service load is 433 kip concentrated dead load and 119 kip concentrated live load from all levels above the girder—refer to Step 3.<br><br>Assume that approximately 75 percent of the dead load is balanced by the harped post-tensioned tendons.<br><br>From geometry, refer to Fig. E6.3: | $0.75(433 \text{ kip}) = 324.8$ kip, say, 325 kip<br>$325 \text{ kip} = 2F\sin\theta$<br><br>$\theta = \tan^{-1}\left(\dfrac{28.3 \text{ in.}}{(13 \text{ ft})(12 \text{ in./ft})}\right) = 10.3$ degrees<br><br>$F = \dfrac{325 \text{ kip}}{2\sin 10.3} = 908$ kip |
| | Calculate number of strands in the tendon.<br>$f_{pe} = 0.65 f_{pu} = 175$ ksi     (Step 4) | Required tendon area = $(908 \text{ kip})/(175 \text{ ksi}) = 5.19$ in.$^2$<br>Number of strands = $5.19$ in.$^2/(0.153$ in.$^2) = 33.9$<br>Say, 34 strands. Therefore, the design force is:<br><br>$F = (34)(0.153 \text{ in.}^2)(175 \text{ ksi}) = 910 \text{ kip} > 908$ kip   **OK** |
| R20.3.2.6.2 | Note: ACI 318-14 requires a detailed check of losses. Estimation of friction losses in post-tensioned tendons is addressed in PTI TAB.1-06. The lump-sum losses were used herein to determine effective prestress force. ACI 423.3R can be used to calculate refined time-dependent losses. | |
| | 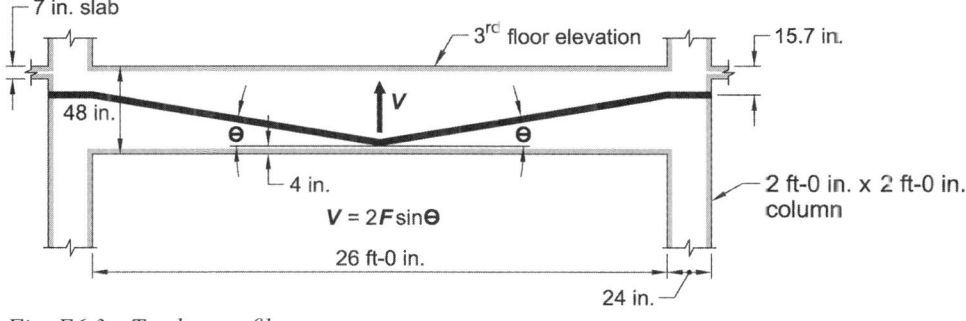<br>Fig. E6.3—Tendon profile. | |
| | Moments at support and midspan for dead load, live load, and post-tensioning are obtained from PTData, Fig. E6.4, 6.5, and 6.6, respectively. | |

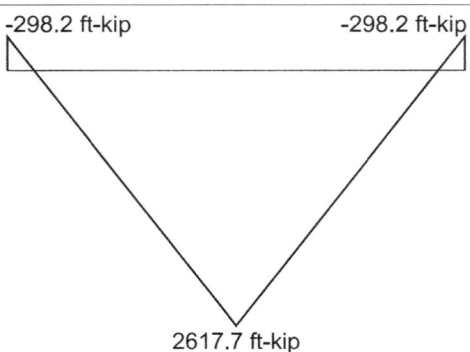

*Fig. E6.4—Service dead load moment diagram.*

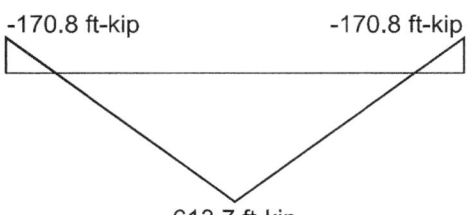

*Fig. E6.5—Service live load moment diagram.*

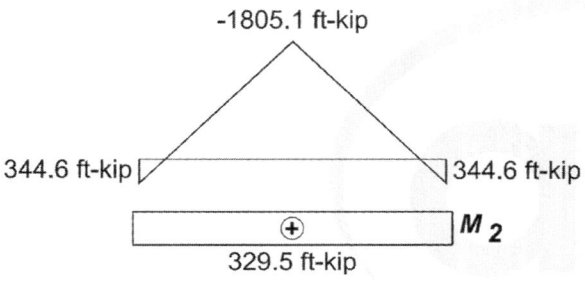

*Fig. E6.6—Post-tensioning moment diagram.*

Stress calculations:
Note: for sign convention:
Compression (−)
Tension (+)

Note: When comparing compression stresses, absolute values are used.

<u>At transfer:</u>
Assume zero dead load present and all 34 diameter strands are stressed. Apply a factor 7/6 to reflect the stress losses in the strands incurred over time.

Midspan stresses:

Top: $f_{top} = -\dfrac{P}{A} + \dfrac{M}{S_{top}}$

Check if actual stress is less than allowable stress (Step 4):

Bottom: $f_{bot} = -\dfrac{P}{A} - \dfrac{M}{S_{bot}}$

Check if actual stress is less than allowable stress (Step 4):

Moment due to balance load at midspan:

$M_{bal} = \dfrac{7}{6}(1805 \text{ ft-kip}) = 2106 \text{ ft-kip}$

$f_{top} = -\dfrac{(7/6)(910 \text{ kip})}{1936 \text{ in.}^2} + \dfrac{(2106 \text{ ft-kip})(12)}{26,779 \text{ in.}^3} = 0.395 \text{ ksi}$

$f_{top} = 395 \text{ psi} < f_{all} = 474 \text{ psi}$  **OK**

$f_{top} = -\dfrac{(7/6)(910 \text{ kip})}{1936 \text{ in.}^2} + \dfrac{(2106 \text{ ft-kip})(12)}{13,017 \text{ in.}^3} = -2.49 \text{ ksi}$

$f_{bot} = 2490 \text{ psi} \approx f_{all} = 2400 \text{ psi}$  **OK**

| | Support stresses: | Moment due to balance at support: $$M_{bal} = \frac{7}{6}(-345 \text{ ft-kip}) = -403 \text{ ft-kip}$$ |
|---|---|---|
| | Top: $f_{top} = -\frac{P}{A} - \frac{M}{S_{top}}$ | $f_{top} = -\frac{(7/6)(910 \text{ kip})}{1936 \text{ in.}^2} - \frac{(403 \text{ ft-kip})(12)}{26,779 \text{ in.}^3} = -0.729 \text{ ksi}$ |
| | Check if actual stress is less than allowable stress (Step 4): | $f_{top} = 729 \text{ psi} < f_{all} = 2400 \text{ psi}$ **OK** |
| | Bottom: $f_{bot} = -\frac{P}{A} + \frac{M}{S_{bot}}$ | $f_{bot} = -\frac{(7/6)(910 \text{ kip})}{1936 \text{ in.}^2} + \frac{(403 \text{ ft-kip})(12)}{13,017 \text{ in.}^3} = -0.177 \text{ ksi}$ |
| | Check if actual stress is less than allowable stress (Step 4): | $f_{bot} = 177 \text{ psi} < f_{all} = 474 \text{ psi}$ **OK** |
| | Conclusion: No stage stressing is required. Stress all tendons after girder beam concrete attained 4000 psi compressive strength. | |
| | At service: Midspan: $M_{TL} = M_D + M_L$ | Service moment at midspan: $M_{TL} = (2618 \text{ ft-kip}) + (614 \text{ ft-kip}) = 3232 \text{ ft-kip}$ $M_{Bal} = -1805 \text{ ft-kip}$ |
| | $\Delta M = M_{TL} - M_{Bal}$ | $\Delta M = (3232 \text{ ft-kip}) - (1805 \text{ ft-kip}) = 1427 \text{ ft-kip}$ |
| | Top: $f_{top} = -\frac{P}{A} - \frac{M}{S_{top}}$ | $f_{top} = -\frac{910 \text{ kip}}{1936 \text{ in.}^2} - \frac{(1427 \text{ ft-kip})(12)}{26,779 \text{ in.}^3} = -1.109 \text{ ksi}$ |
| | Check if actual stress is less than allowable stress (Step 4): | $f_{top} = 1109 \text{ psi} < f_{all} = 2250 \text{ psi}$ **OK** |
| | Bottom: $f_{bot} = -\frac{P}{A} + \frac{M}{S_{bot}}$ | $f_{bot} = -\frac{910 \text{ kip}}{1936 \text{ in.}^2} + \frac{(1427 \text{ ft-kip})(12)}{13,017 \text{ in.}^3} = 0.845 \text{ ksi}$ |
| | Check if actual stress is less than allowable stress – Class T (Step 4): | $f_{bot} = 845 \text{ psi} < f_{all} = 848 \text{ psi}$ **OK** |
| | Support: | Service moment at support: $M_{TL} = (-298 \text{ ft-kip}) - (171 \text{ ft-kip}) = -469 \text{ ft-kip}$ $M_{Bal} = +345 \text{ ft-kip}$ |
| | $\Delta M = M_{TL} - M_{Bal}$ | $\Delta M = (-469 \text{ ft-kip}) + (345 \text{ ft-kip}) = 124 \text{ ft-kip}$ |
| | Top: $f_{top} = -\frac{P}{A} + \frac{M}{S_{top}}$ | $f_{top} = -\frac{910 \text{ kip}}{1936 \text{ in.}^2} + \frac{(124 \text{ ft-kip})(12)}{26,779 \text{ in.}^3} = -0.414 \text{ ksi}$ |
| | Check if actual stress is less than allowable stress (Step 4): | $f_{top} = 414 \text{ psi} < f_{all} = 2250 \text{ psi}$ **OK** |
| | Bottom: $f_{bot} = -\frac{P}{A} - \frac{M}{S_{bot}}$ | $f_{bot} = -\frac{910 \text{ kip}}{1936 \text{ in.}^2} - \frac{(124 \text{ ft-kip})(12)}{13,017 \text{ in.}^3} = -0.548 \text{ ksi}$ |
| | Check if actual stress is less than allowable stress – Class T (Step 4): | $f_{bot} = 548 \text{ psi} < f_{all} = 2250 \text{ psi}$ **OK** |

| | | | | | |
|---|---|---|---|---|---|
| 24.5.2.1<br>24.5.3.1<br>24.5.3.2<br>24.5.3.2.1<br>25.5.4.1 | Conclusions and summary:<br>All service load stresses are acceptable.<br><br>Initial stressing: | | | | |

|  | Location | Stress, psi | Allowable stress, psi | Status |
|---|---|---|---|---|
| Support | Top | −729 | −2400 | OK |
|  | Bottom | −177 | 474 | OK |
| Midspan | Top | 395 | 474 | OK |
|  | Bottom | −2490 | −2400 | ~OK |

At service:

|  | Location | Stress, psi | Allowable stress, psi | Status |
|---|---|---|---|---|
| Support | Top | −414 | −2250 | OK |
|  | Bottom | −584 | −2250 | OK |
| Midspan | Top | −1109 | −2250 | OK |
|  | Bottom | 845 | 848 | OK |

## Step 6: Design strength

### (a) Flexure

| | |
|---|---|
| 5.3.1 | **Factored loads**<br>Shear and moment diagrams are obtained from PTData Fig. E6.7:<br><br>*Moment at midspan:*<br>The beam resists gravity only and lateral forces are not considered in this problem.<br>$U = 1.4D$<br>$U = 1.2D + 1.6L$<br><br>From PTData, secondary moments are: | From Moment diagrams, Fig. E6.4<br><br>$M_u = 1.4(2618$ ft-kip$) = 3665$ ft-kip<br>$M_u = 1.2(2617.7$ ft-kip$) + 1.6(613.7$ ft-kip$) = 4123$ ft-kip **Controls**<br><br>$M_2 = 329.5$ ft-kip, say, 330 ft-kip<br><br>Add secondary moments:<br>$M_u = 4124$ ft-kip $+ 330$ ft-kip $= 4454$ ft-kip |

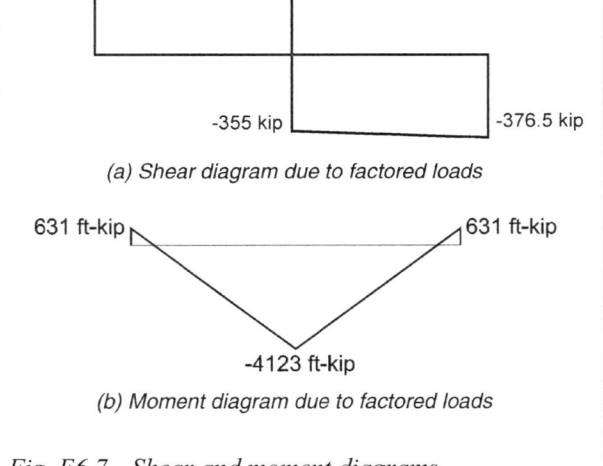

(a) Shear diagram due to factored loads

(b) Moment diagram due to factored loads

*Fig. E6.7—Shear and moment diagrams.*

| | | |
|---|---|---|
| 21.2.1(a) | It is assumed that the girder is tension controlled, $\phi$ = 0.9. This assumption will be checked later. Determine the effective depth assuming No. 10 bars for the girder with 1.5 in. cover: | |
| 20.6.1.3.1 | $d = h - \text{cover} - d_{tie} - d_b/2$ | Transfer girder: $d$ = 48 in. – 1.5 in. – 0.5 in. – (1.128 in.)/2 = 45.4 in |
| 22.2.2.1 | The concrete compressive strain at nominal moment strength is: $\varepsilon_c = 0.003$ | |
| 22.2.2.2 | The tensile strength of concrete in flexure is a variable property and is approximately 10 to 15 percent of the concrete compressive strength. ACI 318 neglects the concrete tensile strength to calculate nominal strength. | |
| 22.2.2.3 | Determine the equivalent concrete compressive stress at nominal strength: The concrete compressive stress distribution is inelastic at high stress. The code permits any stress distribution to be assumed in design if shown to result in predictions of ultimate strength in reasonable agreement with the results of comprehensive tests. Rather than tests, the Code allows the use of an equivalent rectangular compressive stress distribution of $0.85f_c'$ with a depth of: | |
| 22.2.2.4.1 | $a = \beta_1 c$, where $\beta_1$ is a function of concrete compressive strength and is obtained from Table 22.2.2.4.3: | |
| 22.2.2.4.3 | For $f_c'$ = 5000 psi | $\beta_1 = 0.85 - \dfrac{0.05(5000 \text{ psi} - 4000 \text{ psi})}{1000 \text{ psi}} = 0.8$ |

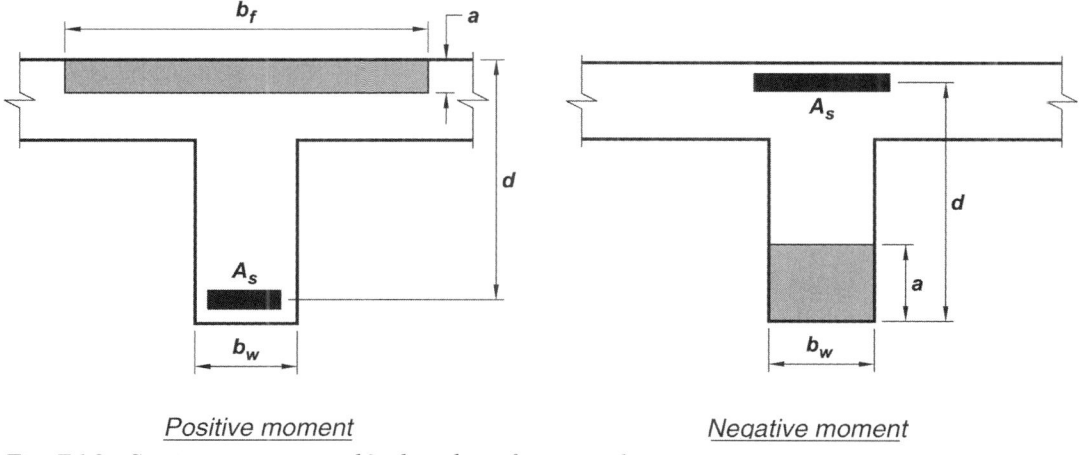

Fig. E6.8—Section compression block and reinforcement locations.

| | | |
|---|---|---|
| 20.3.2.4.1 | For unbonded tendons and as an alternative to a more accurate calculation, the stress in post-tensioned tendons at nominal flexural strength is the least of:<br>(a) $f_{se} + 10{,}000 + f_c'/(100\rho_p)$<br>(b) $f_{se} + 60{,}000$<br>(c) $f_{sy}$<br><br>if<br>$f_{se} = 175$ ksi $> 0.5 f_{pu} = 135$ ksi<br>$\ell_n/h = (26 \text{ ft})(12)/48 \text{ in.} = 6.5 < 35$<br>and where<br>$$\rho_p = \frac{A_{ps}}{bd_p} = \frac{(34)(0.153 \text{ in.}^2)}{(136 \text{ in.})(44 \text{ in.})} = 0.00087$$ | (a) 175 ksi + 10 ksi + 5 ksi/(100(0.00087)) = 242.5 ksi<br>(b) 175 ksi + 60 ksi = 235 ksi  **Controls**<br>(c) $f_{sy} = 0.9 f_{pu} = (0.9)(270 \text{ ksi}) = 243$ ksi<br><br>Therefore, use $f_{ps} = 235$ ksi |
| 9.6.2.3 | The strands are unbonded. A minimum area of deformed reinforcement is required to ensure flexural behavior at nominal girder strength, rather than tied arch behavior. In addition, the reinforcing bar should limit crack width and spacing. To calculate the minimum area:<br>$A_{s,min} = 0.004 A_{ct}$<br>where $A_{ct}$ is the area of that part of the cross section between the flexural tension face and the centroid of the gross section | At midspan:<br>$A_{ct} = (24 \text{ in.})(48 \text{ in.} - 15.7 \text{ in.}) = 775.2 \text{ in.}^2$<br>$A_{s,min} = (0.004)(775.2 \text{ in.}^2) = 3.1 \text{ in.}^2$<br><br>Try four No.8:<br>$A_{s,prov} = (4)(0.79 \text{ in.}^2) = 3.16 \text{ in.}^2$  **OK** |
| | Calculate design moment strength of section at midspan with only PT tendons:<br>$C = T$<br>$0.85 f_c' ba = A_{ps} f_{ps} + A_s f_y$ | $$a = \frac{(34)(0.153 \text{ in.}^2)(235 \text{ ksi}) + (4)(0.79 \text{ in.}^2)(60 \text{ ksi})}{(0.85)(5 \text{ ksi})(136 \text{ in.})}$$<br>$a = 2.44$ in. $< h_f = 7$ in.<br><br>Therefore, compression block in flange.<br>$C = 2.44/0.8 = 3.05$ |
| | Check that section is tension controlled:<br>Is the strain in bars closest to the tension face are greater than 0.005.<br><br>Deformed bars: $\varepsilon_s = \left(\dfrac{d}{c} - 1\right)\varepsilon_{cu}$<br><br>Alternatively:<br>$c/d = (30.5 \text{ in.})/(44 \text{ in.}) = 0.069 < 3/8$ | $\varepsilon_s = \left(\dfrac{45.4 \text{ in.}}{3.05 \text{ in.}} - 1\right)(0.003) = 0.042$ in./in. $> 0.005$<br><br>Therefore, use $\phi = 0.9$. 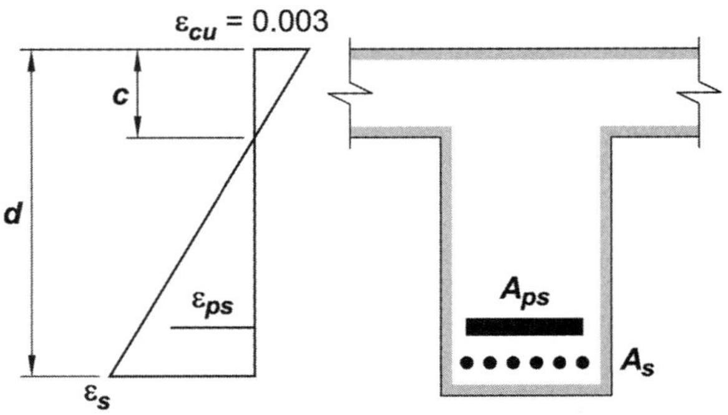 |

Fig. E6.9—Strain distribution over girder depth.

| | | |
|---|---|---|
| | Calculate flexural strength of section: $$M_n = A_{ps}f_{ps}\left(d_p - \frac{a}{2}\right) + A_s f_y \left(d - \frac{a}{2}\right)$$ | $M_n = (5.20\text{ in.}^2)(235\text{ ksi})\left(44\text{ in.} - \frac{2.44\text{ in.}}{2}\right)$ $+ (4)(0.79\text{ in.}^2)(60\text{ ksi})\left(45.5\text{ in.} - \frac{2.44\text{ in.}}{2}\right)$ |
| | $\phi = 0.9$ calculated above | $\phi M_n = (0.9)(52{,}277\text{ in.-kip} + 8395\text{ in.-kip})$ $= 54{,}605\text{ in.-kip}$ $\phi M_n = 4550\text{ ft-kip} > M_u = 4454\text{ ft-kip}$ **OK** |
| 5.3.1 | *Moment at support* The beam resists gravity only and lateral forces are not considered in this problem. $U = 1.4D$ $U = 1.2D + 1.6L$ From PTData, secondary moments are: | From moment diagrams, Fig. E6.4 $M_u = 1.4(298\text{ ft-kip}) = 417\text{ ft-kip}$ $M_u = 1.2(298\text{ ft-kip}) + 1.6(171\text{ ft-kip}) = 631\text{ ft-kip}$ **Controls** $M_2 = 329.5\text{ ft-kip}$, say, 330 ft-kip Add secondary moments: $M_u = -631\text{ ft-kip} + 330\text{ ft-kip} = -301\text{ ft-kip}$ |
| 20.3.2.4.1 | Stress in post-tensioned tendons at nominal flexural strength is the least of: (a) $f_{se} + 10{,}000 + f_c'/(100\rho_p)$ (b) $f_{se} + 60{,}000$ (c) $f_{sy}$ if $f_{se} = 175\text{ ksi} > 0.5 f_{pu} = 135\text{ ksi}$ $\ell/h = (26\text{ ft})(12)/48\text{ in.} = 6.5 < 35$ and where $\rho_p = \frac{A_{ps}}{bd_p} = \frac{(34)(0.153\text{ in.}^2)}{(24\text{ in.})(32.3\text{ in.})} = 0.0067$ | (a) $175\text{ ksi} + 10\text{ ksi} + 5\text{ ksi}/(100(0.0067)) = 192.0\text{ ksi}$ **Controls** (b) $175\text{ ksi} + 60\text{ ksi} = 235\text{ ksi}$ (c) $f_{sy} = 0.9 f_{su} = (0.9)(270\text{ ksi}) = 243\text{ ksi}$ Therefore, use $f_{ps} = 192\text{ ksi}$. |
| 9.6.2.3 | Minimum area of deformed reinforcement at support: $A_{s,min} = 0.004 A_{ct}$ where $A_{ct}$ is the area of that part of the cross section between the flexural tension face and the centroid of the gross section | At supports: $A_{ct} = (136\text{ in.})(7\text{ in.}) + (24\text{ in.})(8.7\text{ in.}) = 1161\text{ in.}^2$ $A_{s,min} = (0.004)(1161\text{ in.}^2) = 4.64\text{ in.}^2$ Try four No.10: $A_{s,prov} = (4)(1.27\text{ in.}^2) = 5.08\text{ in.}^2$ **OK** |
| | Calculate design moment strength of section at midspan with only PT tendons: $C = T$ $0.85 f_c' ba = A_{ps}f_{ps} + A_s f_y$ | $a = \frac{(34)(0.153\text{ in.}^2)(192\text{ ksi}) + (4)(1.27\text{ in.}^2)(60\text{ ksi})}{(0.85)(5\text{ ksi})(24\text{ in.})}$ $a = 12.78\text{ in.}$ Therefore, compression block in flange. $C = 12.78/0.8 = 15.98\text{ in.}$ |

| | | |
|---|---|---|
| | Check that section is tension-controlled:<br>Is the strain in bars closest to the tension face are greater than 0.005?<br><br>Deformed bars: $\varepsilon_s = \left(\dfrac{d}{c} - 1\right)\varepsilon_{cu}$ | $\varepsilon_s = \left(\dfrac{(45.4 \text{ in.})}{15.98 \text{ in.}} - 1\right)(0.003) = 0.0055 \text{ in./in.} > 0.005$<br><br>Therefore, use $\phi = 0.9$.<br><br>*Fig. E6.10—Strain distribution over girder depth.* |
| | Calculate flexural strength of section:<br><br>$M_n = A_{ps} f_{ps}\left(d_p - \dfrac{a}{2}\right) + A_s f_y\left(d - \dfrac{a}{2}\right)$<br><br>$\phi = 0.9$ calculated above | $M_n = (5.20 \text{ in.}^2)(192 \text{ ksi})\left(32.3 \text{ in.} - \dfrac{12.78 \text{ in.}}{2}\right)$<br>$\quad + (4)(1.27 \text{ in.}^2)(60 \text{ ksi})\left(45.5 \text{ in.} - \dfrac{12.78 \text{ in.}}{2}\right)$<br><br>$\phi M_n = (0.9)(25{,}868 \text{ in.-kip} + 11{,}921 \text{ in.-kip})$<br>$\quad\quad = 34{,}010 \text{ in.-kip}$<br>$\phi M_n = 2834 \text{ ft-kip} < M_u = 302 \text{ ft-kip}$ **OK** |

## CHAPTER 7—BEAMS

### (b) Shear design

| | | |
|---|---|---|
| | Shear strength<br>$V_u = w_u \ell/2 + P_u/2$ | $V_u = (1.2)\left((1.2 \text{ kip/ft})(13 \text{ ft}) + \dfrac{433 \text{ kip}}{2}\right)$<br>$\quad + (1.6)\left((0.065 \text{ kip/ft}^2)(2 \text{ ft})(13 \text{ ft}) + \dfrac{119 \text{ kip}}{2}\right)$<br>$V_u = 376.4 \text{ kip}$ |
| 9.4.3.2 | Because conditions a), b), and c) of 9.4.3.2 are satisfied, the design shear force is taken at critical section at distance $h/2$ from the face of the support (Fig. E6.11). | *Fig. E6.11—Factored shear at the critical section.* |
| 21.2.1(b)<br>9.5.1.1 | Shear strength reduction factor:<br>$\phi V_n \geq V_u$ | $V_{u@h/2} = (376.4 \text{ kip}) - (1.65 \text{ kip/ft})(24 \text{ in.}/12) = 373 \text{ kip}$<br>$\phi_{shear} = 0.75$ |
| 9.5.3.1<br>22.5.1.1 | $V_n = V_c + V_s$ | |
| 22.5.5.1<br>22.5.2.1 | $\phi V_c = \phi 2\sqrt{f_c'} b_w d$<br>$d = d_p = 0.8h = 0.8(48 \text{ in.}) = 38.4 \text{ in.}$ | $\phi V_c = 0.75(2)(\sqrt{5000 \text{ psi}})(24 \text{ in.})(38.4 \text{ in.}) = 97.7 \text{ kip}$ |
| 9.5.1.1 | Check if $\phi V_c \geq V_u$ | $\phi V_c = 97.7 \text{ kip} < V_{u@h/2} = 373 \text{ kip}$    **NG**<br>Therefore, shear reinforcement is required. |
| 22.5.1.2 | Before calculating shear reinforcement, check if the cross-sectional dimensions satisfy Eq. (22.5.1.2):<br>$V_u \leq \phi(V_c + 8\sqrt{f_c'} b_w d)$ | $V_u \leq \phi\left(97.7 \text{ kip} - \dfrac{8(\sqrt{5000 \text{ psi}})(24 \text{ in.})(38.4 \text{ in.})}{1000 \text{ lb/kip}}\right)$<br>$= 489 \text{ kip}$<br>$V_u = 373 \text{ kip} \leq \phi(V_c + 8\sqrt{f_c'} b_w d) = 489 \text{ kip}$<br>**OK**, therefore, section dimensions are satisfactory. |

| | | |
|---|---|---|
| 22.5.8.2 | Check if:<br>$A_{ps}f_{se} \geq 0.4(A_{ps}f_{pu} + A_sf_y)$ | $(0.153 \text{ in.}^2)(34)(175 \text{ ksi}) = 910 \text{ kip}$  $0.4((0.153 \text{ in.}^2)(34)(270 \text{ ksi}) + (5.08 \text{ in.}^2)(60 \text{ ksi})) = 684 \text{ kip}$<br><br>910 kip > 684 kip |
| 22.5.8.2 | Therefore, $V_c$ is the least of the following three equations and not 22.5.8.3:<br><br>(a) $\left(0.6\lambda\sqrt{f_c'} + 700\dfrac{V_u d_p}{M_u}\right)b_w d$<br><br>(b) $\left(0.6\lambda\sqrt{f_c'} + 700\right)b_w d$<br><br>(c) $5\lambda\sqrt{f_c'}b_w d$ | |
| 22.5.5.1 | and $V_w \geq 2\lambda\sqrt{f_c'}b_w d$ | $V_w \geq 2(1.0)\sqrt{5000 \text{ psi}}(24 \text{ in.})(38.4 \text{ in.}) = 130.3 \text{ kip}$<br><br>For the calculation of the three equations, refer to the table below: |

| $x$, ft | $V_u$, kip | $M_u$, ft-kip | $d_P$, in. | $V_u d_p / M_u$ | $V_c$ (a), kip | $V_c$ (b), kip | $V_c$ (c), kip |
|---|---|---|---|---|---|---|---|
| 1 | 376.5 | –631 | 38.4 | –1.909 | –1072.7 | 684 | 326 |
| 2 | 374.8 | –255 | 38.4 | –4.694 | –2353.6 | 684 | 326 |
| 3 | 373.2 | 118 | 38.4 | 10.078 | 15,262.9 | 684 | 326 |
| 4 | 371.5 | 491 | 38.4 | 2.422 | 1852.5 | 684 | 326 |
| 5 | 369.9 | 861 | 38.4 | 1.374 | 1001.2 | 684 | 326 |
| 6 | 368.2 | 1231 | 38.4 | 0.958 | 692.9 | 684 | 326 |
| 7 | 366.6 | 1598 | 38.4 | 0.734 | 533.7 | 684 | 326 |
| 8 | 364.9 | 1964 | 38.4 | 0.595 | 436.5 | 684 | 326 |
| 9 | 363.3 | 2328 | 38.4 | 0.499 | 370.9 | 684 | 326 |
| 10 | 361.6 | 2690 | 38.4 | 0.430 | 323.8 | 684 | 326 |
| 11 | 360.0 | 3051 | 38.4 | 0.378 | 288.2 | 684 | 326 |
| 12 | 358.3 | 3410 | 39.6 | 0.347 | 276.3 | 706 | 336 |
| 13 | 356.7 | 3767 | 41.8 | 0.330 | 278.7 | 745 | 355 |
| 14 | 355.0 | 4123 | 44.0 | 0.316 | 282.1 | 784 | 373 |

Equation (22.5.8.2a) controls the middle 8 ft of the girder; the rest of the span is controlled by Eq. (22.5.8.2c), shown shaded in table above. Both equations are greater than Eq. (22.5.5.1).

$V_u > \phi V_c$ at all sections along the girder length. Therefore, shear reinforcement is required. Try No. 4 stirrups; $A_s = 2(0.2 \text{ in.}^2) = 0.4 \text{ in.}^2$ and $d_{p,min} = 38.4$ in.

| $x$, ft | $V_u$, kip | $\phi V_c$ Eq. (22.5.8.2a) and (22.5.8.2c), kip | $\phi V_s = V_u - \phi V_c$, kip | $s = \dfrac{\phi A_v f_y d_p}{V_u - \phi V_c}$, in. | $s_{prov'd.}$, in. |
|---|---|---|---|---|---|
| 1 | 377 | 244 | 132 | 5.23 | 5 |
| 2 | 375 | 244 | 130 | 5.30 | 5 |
| 3 | 373 | 244 | 129 | 5.37 | 5 |
| 4 | 372 | 244 | 127 | 5.44 | 5 |
| 5 | 370 | 244 | 125 | 5.51 | 5 |
| 6 | 368 | 244 | 124 | 5.58 | 5 |
| 7 | 367 | 244 | 122 | 5.66 | 5 |
| 8 | 365 | 244 | 121 | 5.73 | 5 |
| 9 | 363 | 244 | 119 | 5.81 | 5 |
| 10 | 362 | 243 | 119 | 5.82 | 5 |
| 11 | 360 | 216 | 144 | 4.81 | 4 |
| 12 | 358 | 207 | 151 | 4.72 | 4 |
| 13 | 357 | 209 | 148 | 5.10 | 4 |
| 14 | 355 | 212 | 143 | 5.23 | 4 |

| | | | |
|---|---|---|---|
| 9.5.3<br>22.5.10.1 | Shear reinforcement<br>Transverse reinforcement satisfying Eq. (22.5.10.1) is required at each section where $V_u > \phi V_c$<br><br>$V_s \geq \dfrac{V_u}{\phi} - V_c$ | | |
| 22.5.10.5.3<br>22.5.10.5.5 | where $V_s = A_v f_{yt} d/s$ | $V_s = 212$ kip$/0.75 = 282$ kip | |
| 9.7.6.2.2 | Calculate maximum allowable stirrup spacing. First, does the beam transverse reinforcement value need to exceed the threshold value?<br><br>$V_s \leq 4\sqrt{f_c'}b_w d$ ?<br><br>The required shear strength is less than the threshold value; therefore, provide maximum stirrup spacing as the lesser of $3h/8$ and 12 in. | $4\sqrt{f_c'}b_w d = 4(\sqrt{5000 \text{ psi}})(24 \text{ in.})(38.4 \text{ in.}) = 260$ kip<br><br>$V_s = 282$ kip $> 4\sqrt{f_c'}b_w d = 261$ kip  **OK**<br><br>$3h/8 = (3)(48 \text{ in.})/8 = 18$ in. $> d/4 = 12$ in.  **Controls**<br><br>Place first No. 4 stirrup at 3 in. from the face of support. Place No. 4 stirrups at 4 in. on center in the middle 6 ft of the girder. Place No. 4 stirrups at 5 in. on center on both sides of the midsection. | | |

| 9.6.3.3 | Specified shear reinforcement must be the least of the greater of (c) and (d) and (e)<br><br>(c) $0.75\sqrt{f'_c}\dfrac{b_w}{f_{yt}}$, and<br><br>(d) $50\dfrac{b_w}{f_{yt}}$<br><br>(e) $\dfrac{A_{ps}f_{pu}}{80 f_{yt} d}\sqrt{\dfrac{d}{b_w}}$<br><br>to develop ductile behavior. | $\dfrac{A_{v,\min}}{s} \geq 0.75\sqrt{5000\text{ psi}}\dfrac{24\text{ in.}}{60{,}000\text{ psi}} = 0.021\text{ in.}^2/\text{in.}$<br><br>$\dfrac{A_{v,\min}}{s} = 50\dfrac{24\text{ in.}}{60{,}000\text{ psi}} = 0.02\text{ in.}^2/\text{in.}$<br><br>$\dfrac{(0.153\text{ in.}^2)(34)(270\text{ ksi})}{80(60\text{ ksi})(38.4\text{ in.})}\sqrt{\dfrac{38.4\text{ in.}}{24\text{ in.}}} = 0.01\text{ in.}^2/\text{in.}$<br><br>**Controls**<br><br>Provided:<br><br>5 in. spacing: $\dfrac{A_{v,\min}}{s} = \dfrac{2(0.2\text{ in.}^2)}{5\text{ in.}} = 0.08\text{ in.}^2/\text{in.}$<br><br>spacing satisfies 9.6.3.3 ∴ **OK** |
|---|---|---|

## Step 7: Reinforcement detailing

| | | |
|---|---|---|
| 25.2.1 | Minimum longitudinal bar spacing<br>The clear spacing between longitudinal No. 10 bars:<br><br>Clearing spacing greater of $\begin{cases} 1 \text{ in.} \\ d_b \\ 4/3(d_{agg}) \end{cases}$ | 1 in.<br>1.27 in. **Controls**<br>4/3(3/4 in.) = 1 in. assuming a 3/4 in. maximum aggregate size |
| | Check if four No.10 bars (resisting positive moment) can be placed in the beam's web.<br><br>Top bar layout:<br>$b_{w,req'd} = 2(\text{cover} + d_{stirrup} + 1.0 \text{ in.})$<br>$\quad - 3d_b + 3(1.5 \text{ in.})_{min,spacing}$ (25.2.1) | Therefore, clear spacing between horizontal bars must be at least 1.27 in., say, 1.5 in.<br><br>$b_{w,req'd} = 2(1.5 \text{ in.} + 0.5 \text{ in.} + 1.0 \text{ in.}) + 3.81 \text{ in.} + 4.5 \text{ in.}$<br>$= 14.3 \text{ in.} < 24 \text{ in.}$ **OK**<br><br>Therefore, four No.10 bars can be placed in one layer in the 24 in. transfer girder web. |
| 24.3.4 | Tension reinforcement in flanges must be distributed within the effective flange width, $b_f = 136$ in. (Step 2), but not wider than: $\ell_n/10$.<br>Because effective flange width exceeds $\ell_n/10$, additional bonded reinforcement is required in the outer portion of the flange.<br>Use No. 5 placed at slab middepth for additional bonded reinforcement.<br>This requirement is to control cracking in the slab due to wide spacing of bars across the full effective flange width and to protect flange if reinforcement is concentrated within the web width. | 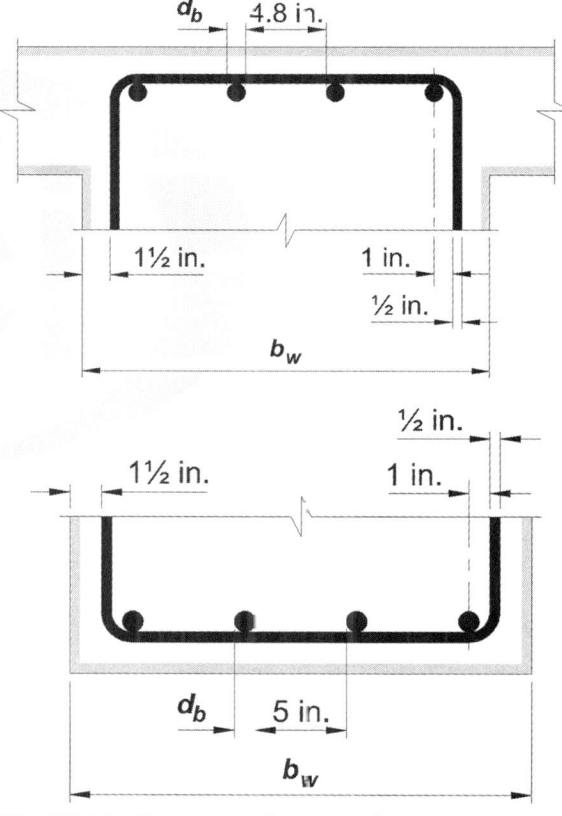<br>Fig. E6.12—Bottom reinforcement layout. |
| | Bottom bar layout:<br>$b_{w,req'd} = 2(\text{cover} + d_{stir} + 1.0 \text{ in.}) + 3d_b$<br>$\quad + 3(1.5 \text{ in.})$ | $b_{w,req'd} = 2(1.5 \text{ in.} + 0.5 \text{ in.} + 1.0 \text{ in.}) + 3 \text{ in.} + 4.5 \text{ in.}$<br>$= 13.5 \text{ in.} < 24 \text{ in.}$ **OK** |

| | | |
|---|---|---|
| 9.7.2.2<br>24.3.1<br>24.3.2 | Maximum bar spacing at the tension face must not exceed the lesser of<br><br>$$s = 15\left(\frac{40,000}{f_s}\right) - 2.5c_c$$<br><br>and<br><br>$s = 12(40,000/f_s)$<br><br>This limit is intended to limit flexural cracking width. Note that $c_c$ is the cover to the longitudinal bars, not to the tie. | $s = 15\left(\frac{40,000 \text{ psi}}{40,000 \text{ psi}}\right) - 2.5(2 \text{ in.}) = 10 \text{ in.}$ **Controls**<br><br>$s = 12\left(\frac{40,000 \text{ psi}}{40,000 \text{ psi}}\right) = 12 \text{ in.}$<br><br>Longitudinal bar spacing satisfy the maximum bar spacing requirement; therefore, **OK** |
| 9.7.2.3<br><br>24.3.2 | Skin reinforcement<br>The transfer girder is 48 in. deep > 36 in. Although the Code does not require skin reinforcement for Class T prestressed beams, many engineers provide it. Skin reinforcement is placed a distance $h/2$ from the tension face. | Use two No. 5 bars each face side as shown in Fig. E6.15 |

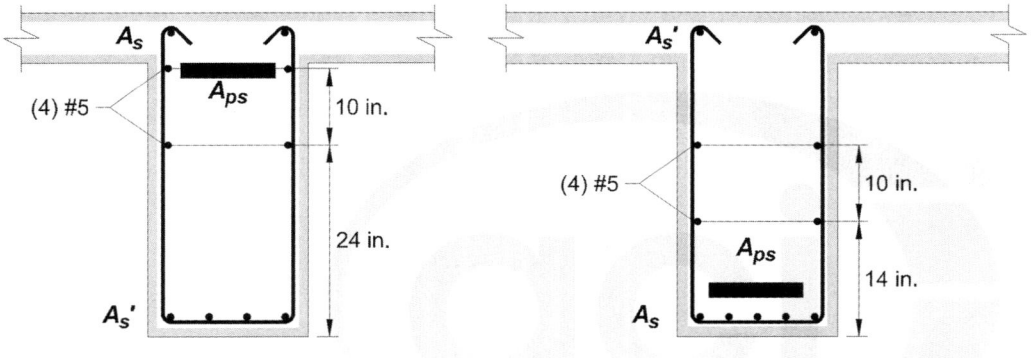

(a) Skin reinforcement at support  (b) Skin reinforcement at midspan
Fig. E6.13—Skin reinforcement in girder.

| | | |
|---|---|---|
| **Step 8: Bar cutoff** | | |
| <br><br><br><br><br><br><br><br><br>25.4.2.2<br><br>25.4.2.4 | Development length<br>Extend top and bottom deformed bars over the full length of the beam. This will ensure better resistance to creep stresses and control of cracks in the girder.<br><br>Therefore, development length calculation is not required into the girder. Bars, however, must be developed within the support at each end.<br>The simplified method is used to calculate the development length of No. 10 bars:<br><br>$$\ell_d = \frac{f_y \psi_t \psi_e}{20\lambda\sqrt{f'_c}} d_b$$<br><br>where $\psi_t = 1.0$ top bar location with not more than 12 in. of fresh concrete below horizontal reinforcement. Otherwise, use 1.3.<br><br>$\psi_t = 1.0$; bars are uncoated | $\ell_d = \left(\frac{(60,000 \text{ psi})(1.0)(1.0)}{(20)(1.0)\sqrt{5000 \text{ psi}}}\right)(d_b) = 42.43d_b$<br><br>For top bars, $\psi_t = 1.3$; $\ell_d = (1.3)(42.43d_b)$<br>No. 8 bars:<br>$\ell_d = 42.43(1.0 \text{ in.}) = 42.43 \text{ in.}$, say, 48 in.<br><br>because No. 10 bars are top bars, development length is:<br>$1.3\ell_d = (1.3)(42.43 \text{ in.})(1.27 \text{ in.}) = 70 \text{ in.}$, say, 6 ft 0 in. |
| 9.7.7.5 | Girder is a single 26 ft long span; therefore, reinforcement splicing is not required. | |

| | | |
|---|---|---|
| 9.7.7.2 | Integrity reinforcement<br>The girder is an internal member; therefore, either a) or b) of 9.7.7.2 must be satisfied. In this example, condition a) is satisfied by having at least one-quarter of the positive moment bars, but not less than two bars continuous. | This condition was satisfied above by extending all four No. 10 bars into the support. **OK** |
| 9.7.7.3 | Beam structural integrity bars must pass through the region bounded by the longitudinal column bars. | Two No. 8 bottom bars are extended and placed between column reinforcement. Because the columns and girder are of the same dimension (24 in.), the two inner girder bars are selected to be the integrity bars. |
| 25.8.1 | Post-tensioning detailing<br>Anchorages for tendons must develop 95 percent of $f_{pu}$ when tested in an unbonded condition. | |
| 25.9 | Post-tensioning anchorage design and detailing is usually provided by the post-tensioning supplier, as well as the detailed tendon layout. | |

## Step 9: Detailing

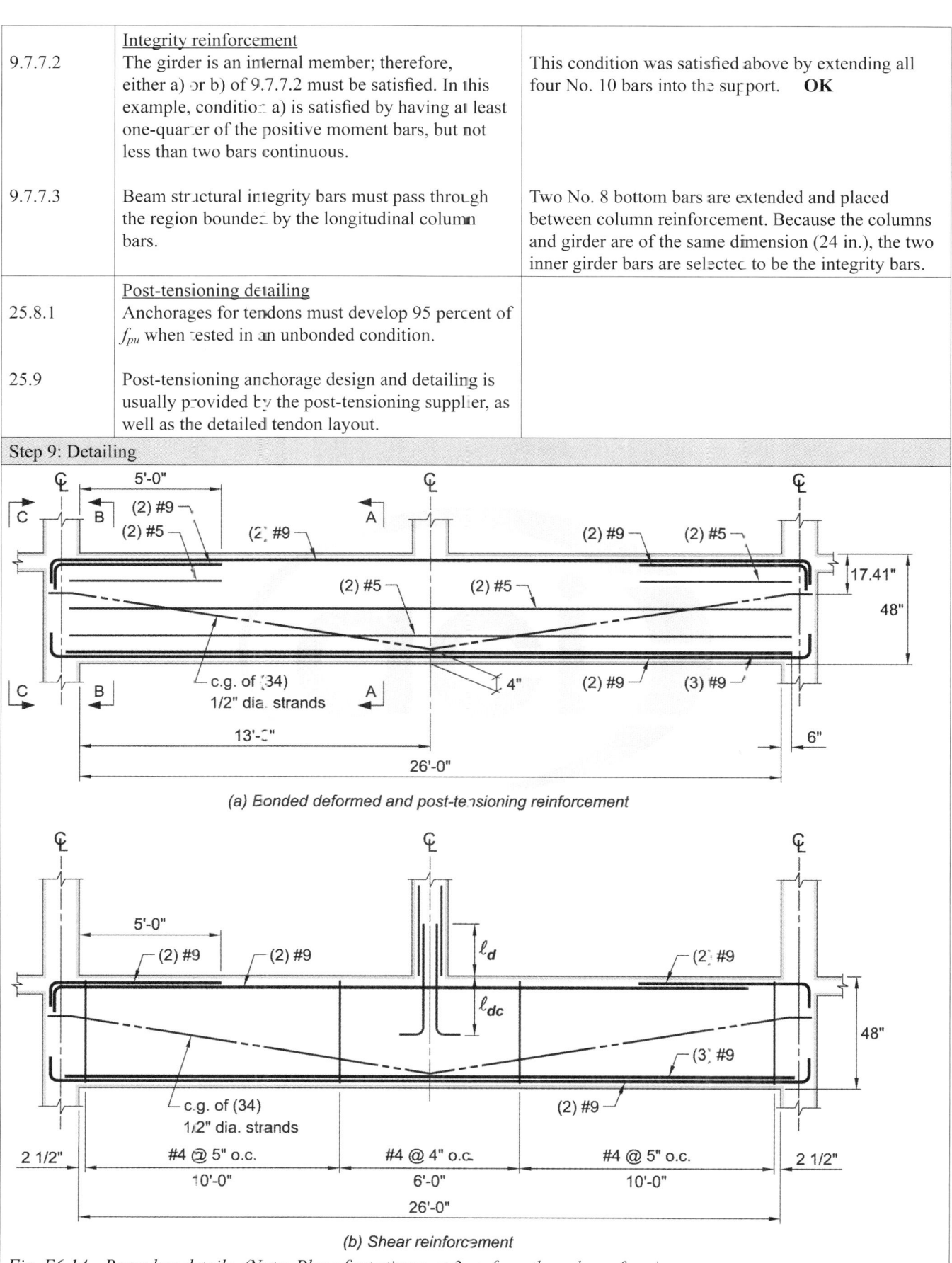

Fig. E6.14—Beam bar details. (Note: Place first stirrup at 3 in. from the column face.)

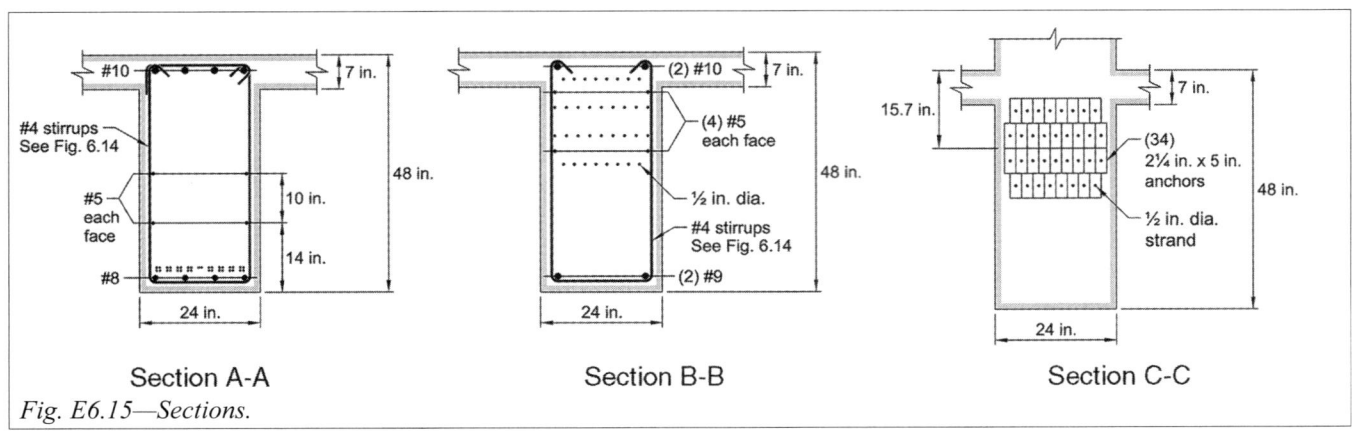

Fig. E6.15—Sections.

## Beam Example 7: *Precast concrete beam*

Design and detail an interior, simply supported precast beam supporting factored concentrated forces of 15 kip located at 4ft 6 in. from each end and a continuously distributed factored force of 4.6 kip/ft. The beam is supported on a 6 in. ledge.

Given:

*Material properties—*
$f_c'$ = 4000 psi (normalweight concrete)
$\lambda$ = 1.0
$f_y$ = 60,000 psi

*Load—*
$P_{u1}$ = 15.0 kip at 4 ft 6 in. from each support
$w_u$ = 4.6 kip/ft
Span length: 18 ft
Beam width: 14 in.
Bearing at support: 6 in.
Bearing at concentrated load: 10 in.

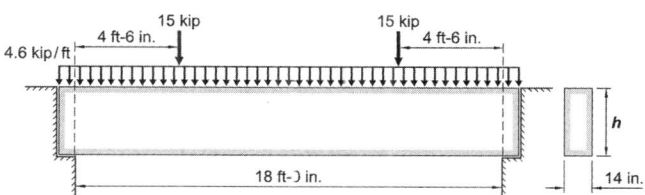

Fig. E7.1—*Simply supported precast concrete beam.*

| ACI 318-14 | Discussion | Calculation |
|---|---|---|
| Step 1: Material requirements | | |
| 9.2.1.1 | The mixture proportion must satisfy the durability requirements of Chapter 19 (ACI 318-14) and structural strength requirements. The designer determines the durability classes. Please refer to Chapter 3 of this Handbook for an in-depth discussion of the categories and classes. ACI 301 is a reference specification that is coordinated with ACI 318. ACI encourages referencing 301 into job specifications. There are several mixture options within ACI 301, such as admixtures and pozzolans, which the designer can require, permit, or review if suggested by the contractor. | By specifying that the concrete mixture shall be in accordance with ACI 301-10 and providing the exposure classes, Chapter 19 (ACI 318-14) requirements are satisfied. Based on durability and strength requirements, and experience with local mixtures, the compressive strength of concrete is specified at 28 days to be at least 4000 psi. |
| Step 2: Beam geometry | | |
| 9.3.1.1 | Beam depth<br>Since beam resists concentrated loads, the beam depth limits in Table 9.3.1.1 of ACI 318-14 cannot be used. | Assume 22 in. deep beam with |

| Step 3: Analysis | | |
|---|---|---|
| | The beam is simply supported and the loads are symmetrical. Therefore, the maximum shear and moment are located at supports and midspan, respectively. | 56.4 kip, 35.7 kip, 20.7 kip, −20.7 kip, −35.7 kip, −56.4 kip

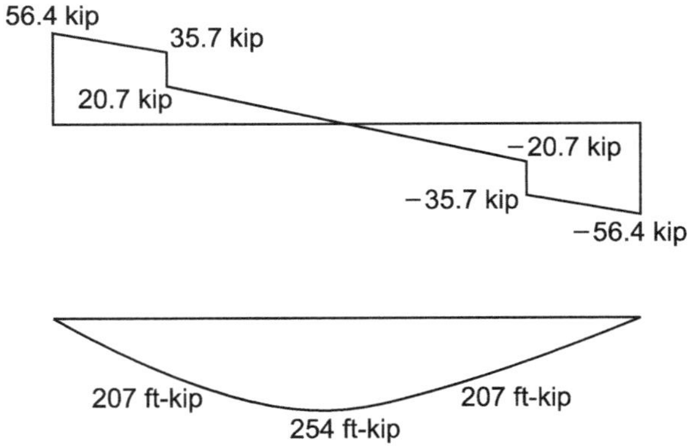

207 ft-kip, 254 ft-kip, 207 ft-kip

*Fig. E7.2—Shear and moment diagrams.* |
| | $V_{u,max} = \dfrac{w_u(\ell)}{2} + (P_u)$ $M_{u,max} = \dfrac{w_u(\ell)^2}{8} + (P_u)(x_1)$ | $V_{u,max} = \dfrac{(4.6 \text{ kip/ft})(18 \text{ ft})}{2} + (15 \text{ kip}) = 56.4 \text{ kip}$ $M_{u,max} = \dfrac{(4.6 \text{ kip/ft})(18 \text{ ft})^2}{8} + (15 \text{ kip})(4.5 \text{ ft}) = 254 \text{ ft-kip}$ |
| **Step 4: Bearing** | | |
| 16.2.6.2 | The minimum seating length of the precast beam on the wall ledge is the greater of: $\ell_n/180$ and 3 in. | $\ell_n/180 = (18 \text{ ft})(12 \text{ in./ft})/180 = 1.2$ in. $< 3$ in. Provided 6 in., therefore **OK**. |
| 22.8.3.2 | Bearing strength Check bearing strength at seat and concentrated load: The supporting surface (ledge) is wider on three of the four sides. Therefore, condition (c) applies: $0.85 f_c' A_1$ A 10 in. wide beam rests on the precast beam: | 0.85(4000 psi)(14 in.)(6 in.)/1000 = 285.6 kip 285.6 kip >> 56.4 kip **OK** 0.85(4000 psi)(14 in.)(10 in.)/1000 = 476 kip >> 15 kip **OK** |

# CHAPTER 7—BEAMS

## Step 5: Moment design

| | | |
|---|---|---|
| 9.3.3.1 | The Code does not permit a beam to be designed with steel strain less than 0.004 in./in at nominal strength. The intent is to ensure ductile behavior. | |
| | In most reinforced beams, such as this example, reinforcing bar strain is not a controlling issue. | |
| 21.2.1(a) | The design assumption is the beams will be tensioned controlled, $\phi = 0.9$. This assumption will be checked later. | |
| 20.6.1.3.3 | Determine the effective depth assuming No. 8 bars and 1.0 in. cover. For precast concrete beam, the minimum cover is the greater of 5/8 in. and $d_b$ and need not exceed 1.5 in. | Use $d_b = 1$ in. cover |
| | One row of reinforcement<br>$d = h - \text{cover} - d_{tie} - d_b/2$ | $d = 22$ in. $- 1.0$ in. $- 0.375$ in. $- 1.0$ in. $/2 = 20.1$ in., say, 20 in. |
| 22.2.2.1 | The concrete compressive strain at nominal moment strength is calculated at:<br>$\varepsilon_{cu} = 0.003$ | |
| 22.2.2.2 | The tensile strength of concrete in flexure is a variable property and is about 10 to 15 percent of the concrete compressive strength. ACI 318 neglects the concrete tensile strength to calculate nominal strength. | |
| | Determine the equivalent concrete compressive stress at nominal strength: | |
| 22.2.2.3 | The concrete compressive stress distribution is inelastic at high stress. The Code permits any stress distribution to be assumed in design if shown to result in predictions of ultimate strength in reasonable agreement with the results of comprehensive tests. Rather than tests, the Code allows the use of an equivalent rectangular compressive | |
| 22.2.2.4.1<br>22.2.2.4.3 | stress distribution of $0.85f_c'$ with a depth of:<br>$a = \beta_1 c$, where $\beta_1$ is a function of concrete compressive strength and is obtained from Table 22.2.2.4.3. | |
| 22.2.1.1 | For $f_c' \leq 4000$ psi: | $\beta_1 = 0.85$ |
| | Find the equivalent concrete compressive depth $a$ by equating the compression force to the tension force within the beam cross section:<br>$C = T$<br>$0.85f_c'ba = A_s f_y$ | $0.85(4000 \text{ psi})(b)(a) = A_s (60,000 \text{ psi})$ |
| | For moment at midspan: $b = 14$ in. | $a = \dfrac{A_s(60,000 \text{ psi})}{0.85(4000 \text{ psi})(14 \text{ in.})} = 1.26 A_s$ |

| | | |
|---|---|---|
| 9.5.1.1 | The beam's design strength must be at least the required strength at each section along its length:<br>$\phi M_n \geq M_u$<br>$\phi V_n \geq V_u$<br><br>Calculate the required reinforcement area:<br>$\phi M_n \geq M_u = \phi A_s f_y \left( d - \dfrac{a}{2} \right)$<br>A No. 8 bar has a $d_b = 1.0$ in. and an $A_s = 0.79$ in.² | $(254 \text{ ft-kip})(12) = (0.9)(60 \text{ ksi})A_s \left( 20 \text{ in.} - \dfrac{1.26 A_s}{2} \right)$<br>$A_{s,\,req'd} = 3.1$ in.²  Use four No. 8<br>$A_{s,\,prov} = 3.16$ in.² > $A_{s,\,req'd} = 3.1$ in.²  **OK**<br><br>Per Reinforced Concrete Design Handbook Design Aid – Analysis Tables, which can be downloaded from: https://www.concrete.org/store/productdetail.aspx?ItemID=SP1714DA, four No. 8 bars require a minimum of 11.5 in. wide beam. Therefore, 14 in. width is adequate. |
| 9.3.3.1 | Check if the calculated strain exceeds 0.004 in./in. (tension controlled).<br>$a = \dfrac{A_s f_y}{0.85 f'_c b}$ and $c = \dfrac{a}{\beta_1}$<br>where $\beta_1 = 0.85$<br>$\varepsilon_t = \dfrac{\varepsilon_{cu}}{c}(d - c)$ | $a = 1.26 A_s = (1.26)(4)(0.79 \text{ in.}^2) = 3.98$ in.<br>$c = a/0.85 = 3.98 \text{ in.}/0.85 = 4.68$ in.<br><br>$\varepsilon_t = \dfrac{0.003}{4.68 \text{ in.}}(20 \text{ in.} - 4.68 \text{ in.}) = 0.01$<br><br>Therefore, assumption of using $\phi = 0.9$ is correct.<br>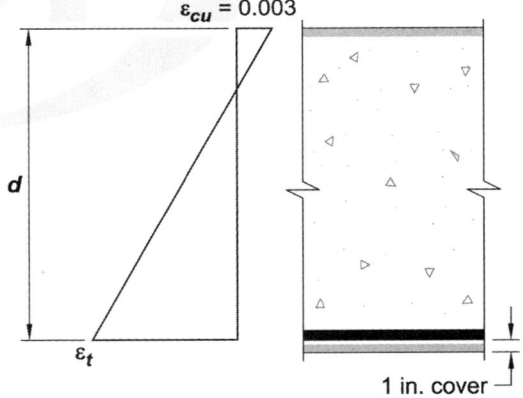<br>*Fig. E7.3—Strain distribution across beam section.* |

# CHAPTER 7—BEAMS

| | | |
|---|---|---|
| 9.6.1.1<br>9.6.1.2 | **Minimum reinforcement**<br>The provided reinforcement must be at least the minimum required reinforcement at every section along the length of the beam.<br>$$A_{s,min} = \frac{3\sqrt{f'_c}}{f_y} b_w d$$<br>$$A_{s,min} = \frac{200}{f_y} b_w d$$ | $A_{s,min} = \frac{3\sqrt{4000 \text{ psi}}}{60,000 \text{ psi}}(14 \text{ in.})(20 \text{ in.}) = 0.89 \text{ in.}^2$<br><br>$A_{s,min} = \frac{200}{60,000 \text{ psi}}(14 \text{ in.})(20 \text{ in.}) = 0.93 \text{ in.}^2$ **Controls**<br><br>$A_{s,prov} = 3.16 \text{ in.}^2 > A_{s,min} = 0.93 \text{ in.}^2$ **OK**<br><br>Required positive moment reinforcement areas exceed the minimum required reinforcement area at all positive moment locations. |
| | **Top reinforcement**<br>While not required by Code, top bars are needed to stabilize the beam's stirrups. Use two No. 5 continuous bars. | |
| **Step 6: Shear design** | | |
| 21.2.1(b) | **Shear strength**<br>Shear strength reduction factor: | $\phi_{shear} = 0.75$ |
| 9.5.1.1 | $\phi V_n \geq V_u$ | $V_u = 56.4$ kip, $V_u @ d = 48.7$ kip, 35.7 kip, 20.7 kip |
| 9.5.3.1<br>22.5.1.1 | $V_n = V_c + V_s$ | |
| 9.4.3.2 | Because conditions (a), (b), and (c) of 9.4.3.2 are satisfied, the design shear force is taken at distance $d$ from the face of the support (Fig. E7.4). | |
| | | *Fig. E7.4—Shear at the critical section.*<br><br>$V_u = (56.4 \text{ kip}) - (4.6 \text{ kip/ft})(20 \text{ in.}/12) = 48.7 \text{ kip}$ |
| 22.5.5.1 | $V_c = 2\sqrt{f'_c} b_w d$<br><br>Check if $\phi V_c \geq V_u$ | $V_c = (2)(\sqrt{4000 \text{ psi}})(14 \text{ in.})(20 \text{ in.}) = 35.4 \text{ kip}$<br><br>$\phi V_c = (0.75)(35.4 \text{ kip}) = 26.6 \text{ kip} < V_u = 48.7 \text{ kip}$ **NG**<br>Therefore, shear reinforcement is required. |
| | Determine required $V_u$ on each side of $P_u$<br>Left of $P_u$: $V_{u,\ell} = V_u - w_u x_1$<br>Right of $P_u$: $V_{u,r} = V_u - w_u x_1 - P_u$ | $V_{u,\ell} = 56.4 \text{ kip} - (4.6 \text{ kip/ft})(4.5 \text{ ft}) = 35.7 \text{ kip}$<br>$V_{u,r} = 56.4 \text{ kip} - (4.6 \text{ kip/ft})(4.5 \text{ ft}) - 15 \text{ kip} = 20.7 \text{ kip}$ |
| 22.5.1.2 | Prior to calculating shear reinforcement, check if the cross-sectional dimensions satisfy Eq. (22.5.1.2):<br><br>$V_u \leq \phi(V_c + 8\sqrt{f'_c} b_w d)$ | $V_u \leq \phi(35.4 \text{ kip} + 8(\sqrt{4000 \text{ psi}})(14 \text{ in.})(20 \text{ in.})) = 132.8 \text{ kip}$<br>**OK**<br>Section dimensions are satisfactory. |

| | | |
|---|---|---|
| 9.5.3<br>22.5.10.1 | **Shear reinforcement**<br>Transverse reinforcement satisfying equation 22.5.10.1 is required at each section where $V_u > \phi V_c$ | |
| 22.5.10.5.3 | $\phi V_s \geq V_u - \phi V_c$<br><br>where $\phi V_s = \dfrac{\phi A_v f_{yt} d}{s}$ | $\phi V_s \geq (48.7 \text{ kip}) - (26.6 \text{ kip}) = 22.1 \text{ kip}$<br><br>Assume a No. 3 bar, two legged stirrup<br><br>$22.1 \text{ kip} = \dfrac{(0.75)(2)(0.11 \text{ in.}^2)(60{,}000 \text{ psi})(20 \text{ in.})}{s}$<br><br>$s = 8.9 \text{ in.}$ |
| 22.5.10.5.6 | Calculate maximum allowable stirrup spacing:<br>First, does the required transverse reinforcement value exceed the threshold value?<br><br>$V_s \leq 4\sqrt{f'_c} b_w d$ ? | $V_s = \dfrac{22.1 \text{ kip}}{0.75} = 29.5 \text{ kip}$<br><br>$4\sqrt{f'_c} b_w d = 4(\sqrt{4000 \text{ psi}})(14 \text{ in.})(20 \text{ in.}) = 71 \text{ kip}$<br><br>$V_s = 29.5 \text{ kip} < 4\sqrt{f'_c} b_w d = 71 \text{ kip}$ **OK** |
| | Because the required shear strength is below the threshold value, the maximum stirrup spacing is the lesser of $d/2$ and 24 in. | $d/2 = 20 \text{ in.}/2 = 10 \text{ in.}$<br>Use $s = 7$ in. $< d/2 = 10$ in., therefore, **OK** |
| 9.7.6.2.2 | It is unnecessary to use No. 3 stirrups at 7 in. on center over the full length of the beam. | |
| | Since the maximum spacing is 10 in., determine the value of:<br><br>$\phi V_n = \phi V_c + \phi V_s$<br>with $s = 10$ in. | $\phi V_n = 26.6 \text{ kip} + \dfrac{(0.75)(2)(0.11 \text{ in.}^2)(60 \text{ ksi})(20 \text{ in.})}{10 \text{ in.}}$<br><br>$\phi V_n = 46.4 \text{ kip}$ |
| | Determine distance x from face of support to point at which<br><br>$V_u = 46.4 \text{ kip}$ | $x = \dfrac{56.4 \text{ kip} - 46.4 \text{ kip}}{4.6 \text{ kip/ft}} = 2.2 \text{ ft}$ |
| | Determine distance $\ell_{v1}$, beyond $x_1$ at which stirrups are not required.<br><br>Find $\ell_{v1} = (V_{u@x1} - \phi V_c/2)/w_u$<br><br>Compute $x_1 + \ell_{v1}$ | $\ell_{v1} = \dfrac{20.7 \text{ kip} - 26.6 \text{ kip}/2}{4.6 \text{ kip/ft}} = 1.6 \text{ ft, say, 2 ft}$<br><br>$x_1 + \ell_{v1} = 4.5 \text{ ft} + 2 \text{ ft} = 6.5 \text{ ft} = 78 \text{ in.}$ |

# CHAPTER 7—BEAMS

| | |
|---|---|
| | Conclude:<br>use $s = 7$ in. until $\phi V_u < 44.5$ kip and<br>use $s = 10$ in. until $\phi V_u < 0.5\phi V_c$ | From face of support use 3 in. space then<br>Five spaces at 7 in. on center (35 in.) and<br>Five spaces at 10 in. on center (3 in.+ 35 in.+ 50 in.) = 88 in. > 78 in.  **OK**<br>The beam middle section of length:<br>(18 ft)(12) − 2(88 in.) = 40 in.<br>does not require shear reinforcement.<br><br>However, extend No. 3 stirrups over the remaining length of 40 in. at 10 in. on center as good practice.<br><br>Note: It is also good practice to add stirrups near a concentrated load. Place six No. 3 stirrups at 4 in. centered on each concentrated load. |

## Step 7: Deflection

**9.3.2.1**
**24.2.3.1**
Immediate deflection is calculated using elastic deflection approach and considering concrete cracking and reinforcement for calculating stiffness.

**24.2.3.4** Modulus of elasticity:

**19.2.2.1**
$$E_c = 57,000\sqrt{f_c'} \text{ psi} \quad (19.2.2.1b)$$

$$E_c = 57,000\sqrt{4000 \text{ psi}} = 3600 \text{ ksi}$$

The beam resists a factored distributed force of 4.6 kip/ft or service dead load of 2.23 kip/ft and 1.2 kip/ft service live load.

The beam also resists a concentrated factored load of 15 kip or service dead load of 7.2 kip and 3.9 kip service live load.

The deflection equation for distributed load with free rotation at both ends:

$$\Delta = \frac{5w\ell^4}{384EI}$$

For concentrated load at thirdspan:

$$\Delta = \frac{P\ell^3}{28.23EI}$$

Note:
· $w$ and $P$ are service loads.
· $I$ is the equivalent moment of inertia given by equation (25.2.3.5a):

**24.2.3.5**
$$I_e = \left(\frac{M_{cr}}{M_a}\right)^3 I_g + \left[1 - \left(\frac{M_{cr}}{M_a}\right)^3\right] I_{cr}$$

where: $M_{cr} = \dfrac{f_r I_g}{y_t}$ (24.2.3.5b)

$$I_g = \frac{bh^3}{12} = \frac{(14 \text{ in.})(22 \text{ in.})^3}{12} = 12,423 \text{ in.}^4$$

$$M_{cr} = \frac{7.5(\sqrt{4000 \text{ psi}})(12,423 \text{ in.}^4)}{11 \text{ in.}} = 535,704 \text{ in.-lb}$$

$M_{cr} = 44.6$ ft-kip

$M_a$ is the moment due to service load.
The beam is assumed cracked, therefore, calculate the moment of inertia of the cracked section, $I_{cr}$.
Calculate the concrete uncracked depth, $c$:

$$n = \frac{29,000 \text{ ksi}}{3600 \text{ ksi}} = 8$$

$$nA_s(d-c) = \frac{bc^2}{2} + (n-1)A_s'(c-a')$$

$$(8)(3.16 \text{ in.}^2)(20 \text{ in.} - c) = \frac{(14 \text{ in.})c^2}{2}$$

where $n = \dfrac{E_s}{E_c}$ and $A_s' = 0$ in.$^2$

Solving for $c$:

$c = 6.9$ in.

Cracking moment of inertia, $I_{cr}$:

$$I_{cr} = \frac{bc^3}{3} + (n-1)A_s'(c-a)^2 + nA_s(d-c)^2$$

$I_{cr} = 5975$ in.$^4$

| | | |
|---|---|---|
| | Deflection due to distributed load: | $\Delta_{distr} = \dfrac{5(2.23 \text{ kip/ft} + 1.2 \text{ kip/ft})(18 \text{ ft})^4(12)^3}{384(3600 \text{ ksi})(5975 \text{ in.}^4)} = 0.38$ in. |
| | Deflection due to concentrated load | $\Delta_{conc} = \dfrac{(7.2 \text{ kip} + 3.9 \text{ kip})(18 \text{ ft})^3(12)^3}{28.3(3600 \text{ ksi})(5975 \text{ in.}^4)} = 0.18$ in. |
| | Total deflection: | $\Delta = 0.38$ in. $+ 0.18$ in. $= 0.56$ in. |
| 24.2.2 | Check the deflection against the maximum allowable limits in Table 24.2.2: | |
| | For floor supporting or attached to nonstructural elements and not likely to be damaged by large deflection: $\ell/240$ | $\Delta_{all.} = (18.0 \text{ ft})(12 \text{ in./ft})/240 = 0.9$ in.<br>$\Delta_{all.} = 0.9$ in. $> \Delta = 0.56$ in. **OK** |
| 24.2.4.1.1 | Calculate long-term deflection:<br>$\lambda_\Delta = \dfrac{\xi}{1+50\rho'}$ | $\lambda_\Delta = \dfrac{2.0}{1+50\dfrac{0.62 \text{ in.}^2}{(14 \text{ in.})(20 \text{ in.})}} = 1.8$ |
| 24.2.4.1.3 | From Table 24.2.4.1.3, the time dependent factor for sustained load duration of more than 5 years:<br>$\xi = 2.0$<br>Therefore, long-term deflection is:<br>$\Delta_T = (1+\lambda_\Delta)\Delta_i$ | $\Delta_T = (1+1.8)(0.56 \text{ in.}) = 1.6$ in.<br><br>Note: The long-term deflection due to sustained loading exceeds $\ell/240$. Therefore, it is recommended to camber the beam 1 in. |

## Step 8: Details

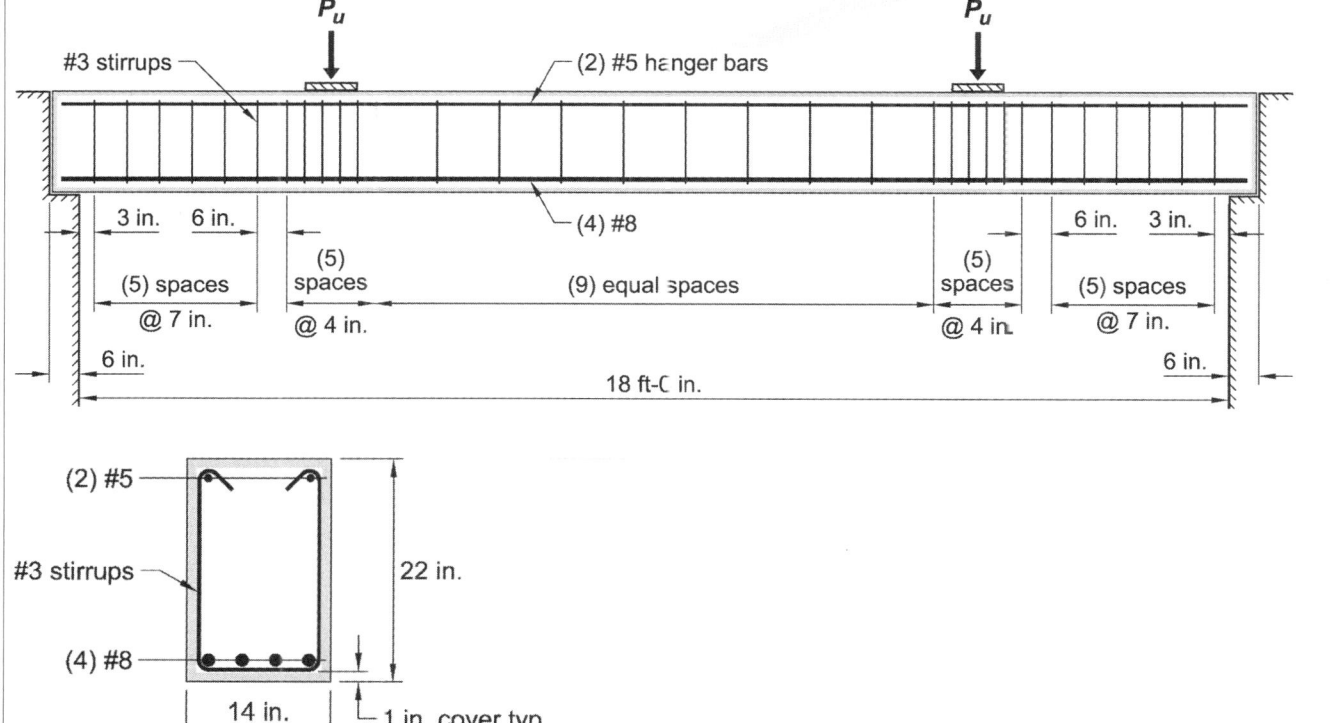

**Beam Example 8**: *Determination of closed ties required for the beam shown to resist shear and torque*

Design and detail a simply supported precast edge beam spanning 29 ft 6in. The beam is subjected to a factored load of 4.72 kip/ft. Structural analysis provided a factored shear and torsion of 61 kip and 53 ft-kip, respectively. The torsional moment is determinate and cannot be redistributed back into the structure.

Given:
$f_c'$ = 5000 psi (normalweight concrete)
$\lambda$ = 1.0
$f_y$ = 60,000 psi
$d$ = 21.5 in.
$V_u$ = 61 kip
$T_u$ = 53 ft-kip
$w_u$ = 4.72 kip/ft

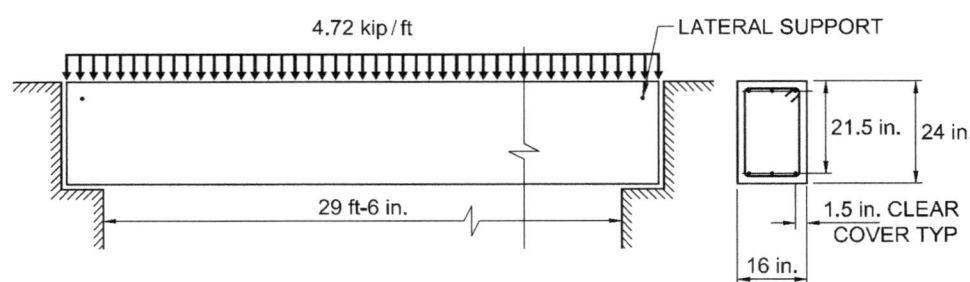

Fig. 8.1—*Beam subjected to determinate torque.*

| ACI 318-14 | Discussion | Calculation |
|---|---|---|
| **Step 1: Section properties** | | |
| 9.2.4.4<br>22.7.6.1<br>22.7.6.1.1<br>9.2.4.4<br>22.7.6.1 | Determine section properties for torsion.<br>$A_{cp} = b_w h$<br>$A_{oh} = (b_w - 3.5 \text{ in.})(h - 3.5 \text{ in.})$<br>$A_o = 0.85 A_{oh}$<br>$p_{cp} = 2(b_w + h)$<br>$p_h = 2(b_w - 3.5 \text{ in.} + h - 3.5 \text{ in.})$ | $A_{cp}$ = 16 in.(24 in.) = 384 in.²<br>$A_{oh}$ = (16 in. – 3.5 in.)(24 in. – 3.5 in.) = 256 in.²<br>$A_o$ = 0.85(256 in.²) = 218 in.²<br>$p_{cp}$ = 2(16 in. + 24 in.) = 80 in.<br>$p_h$ = 2(16 in. – 3.5 in. + 24 in. – 3.5 in.) = 66 in. |
| **Step 2: Cracking torsion** | | |
| 22.7.5.1<br><br><br><br><br>21.2.1c<br><br>9.5.4.1<br>22.7.4.1a | Calculate cracking torsion $T_{cr}$.<br><br>$\phi T_{cr} = 4\phi \lambda \sqrt{f_c'} \left( \dfrac{A_{cp}^2}{p_{cp}} \right)$<br><br>where<br>torsion strength reduction factor $\phi$ = 0.75<br><br>Calculate threshold torsion<br>$T_{th} = 0.25 T_{cr}$ | $\phi T_{cr} = 4(0.75)(1.0)(\sqrt{5000 \text{ psi}}) \left( \dfrac{(384 \text{ in.}^2)^2}{80 \text{ in.}} \right)$<br>= 391,000 in.-lb<br><br>$\phi T_{cr}$ = (391,000 in.-lb)/(12,000 in.-lb/ft-kip)<br>= 32.6 ft-kip<br><br>Threshold torsion = 0.25(32.6 ft–kip) = 8.2 ft-kip<br>Because $T_u$ = 53 ft-kip > 8.2 ft-kip, ties for torsion are therefore required. |
| 22.7.7.1<br><br><br><br><br><br>22.5.1.2 | Is section large enough?<br><br>Calculate $f_v = V_u/(b_w d)$<br>Calculate $f_{vt} = T_u p_h /(1.7 A_{oh}^2)$<br><br><br>Calculate limit = $\phi(2\sqrt{f_c'} + 8\sqrt{f_c'})$<br><br>Is $\sqrt{f_v^2 + f_{vt}^2}$ < limit? | $f_v$ = 61 kip/(16 in. × 21.5 in.) = 0.177 ksi<br>$f_{vt}$ = (53 ft-kip)(12 in./ft)(66 in.)/[1.7(256 in.²)²]<br>= 0.377 ksi<br><br>Limit = $0.75(2+8)(\sqrt{5000 \text{ psi}})$ = 0.53 ksi<br><br>$\sqrt{(0.177 \text{ ksi})^2 + (0.377 \text{ ksi})^2}$ = 0.416 ksi<br>0.416 ksi < limit 0.53 ksi<br><br>Therefore, section is large enough. |

# CHAPTER 7—BEAMS

## Step 3: Calculate shear and torsion reinforcement

| | Required shear tie area/spacing: | |
|---|---|---|
| 9.5.1.1<br>22.5.1.1<br>22.5.5.1<br>22.5.10.5.3<br>21.2.1b | $\dfrac{A_v}{s} = \dfrac{(V_u - 2\phi\lambda\sqrt{f_c'}(b_w d))}{\phi f_y d}$ | $\dfrac{A_v}{s} = \dfrac{(61\text{ kip} - 2(0.75)(1.0)(\sqrt{5000\text{ psi}})(16\text{ in.})(21.5\text{ in.})}{(0.75)(60,000\text{ psi})(21.5\text{ in.})}$<br><br>$\dfrac{A_v}{s} = 0.0253\text{ in.}^2/\text{in.}$ |
| | Required torsional tie area/spacing: | |
| 22.7.6.1a<br>22.7.6.1.2 | $\dfrac{A_t}{s} = \dfrac{T_u}{2\phi A_o f_y \cot\theta}$ | $\dfrac{A_t}{s} = \dfrac{(53\text{ ft-kip})(12\text{ in./ft})}{2(0.75)(218\text{ in.}^2)(50\text{ ksi})\cot 45°} = 0.0324\text{ in.}^2/\text{in.}$ |
| 9.6.4.2 | Calculate total tie area/spacing ($A_v/s + 2A_t/s$) | $\dfrac{A_v}{s} + 2\dfrac{A_t}{s} = 0.0253\text{ in.}^2/\text{in.} + 2(0.0324\text{ in.}^2/\text{in.})$<br><br>$= 0.09\text{ in.}^2/\text{in.}$<br><br>$s = 0.40\text{ in.}^2/(0.09\text{ in.}^2/\text{in.}) = 4.44\text{ in.}$ |
| | Use No. 4 ties for which<br>($A_v + 2A_t$) = 0.40 in<br>Calculate $s = 0.40/(A_v/s + 2A_t/s)$. | Use 4 in. |
| | Check minimum transverse reinforcement.<br><br>Is $\dfrac{0.75\sqrt{f_c'}b_w}{f_{yt}} < \left(\dfrac{A_v}{s} + \dfrac{2A_t}{s}\right)$? | $\dfrac{0.75(\sqrt{5000\text{ psi}})(16\text{ in.})}{60,000\text{ psi}} < \left(\dfrac{0.4\text{ in.}^2}{4\text{ in.}} + \dfrac{2(0.2\text{ in.}^2)}{4\text{ in.}}\right)$<br><br>$0.0141\text{ in.} < 0.2\text{ in.}$ **OK** |

| | Step 4: Longitudinal reinforcement | |
|---|---|---|
| 22.7.6.1<br>22.7.6.1.2 | Calculate torsional longitudinal reinforcement from Eq. (22.7.6.1(b)):<br><br>$$T_n = \frac{2 A_o A_\ell f_y}{p_h} \tan\theta$$<br><br>Set $T_n = \dfrac{T_u}{\phi}$ | $$\frac{(53 \text{ ft-kip})(12 \text{ in./ft})}{0.75} = \frac{2(218 \text{ in.}^2) A_\ell (60 \text{ ksi})}{66 \text{ in.}} \tan 45°$$<br><br>$A_\ell = 2.14$ in.$^2$ |
| 9.6.4.3 | The torsional longitudinal reinforcement $A_{\ell,min}$ must be the lesser of:<br><br>(a) $\dfrac{5\sqrt{f_c'} A_{cp}}{f_y} - \left(\dfrac{A_t}{s}\right) p_h \dfrac{f_{yt}}{f_y}$<br><br>(b) $\dfrac{5\sqrt{f_c'} A_{cp}}{f_y} - \left(\dfrac{25 b_w}{f_{yt}}\right) p_h \dfrac{f_{yt}}{f_y}$ | $\dfrac{5(\sqrt{5000 \text{ psi}})(384 \text{ in.}^2)}{60,000 \text{ psi}} - (0.0324 \text{ in.}^2/\text{in.})(66 \text{ in.})\dfrac{60 \text{ ksi}}{60 \text{ ksi}}$<br><br>$= 0.12$ in.$^2$ **Controls**<br><br>$\dfrac{5(\sqrt{5000 \text{ psi}})(384 \text{ in.}^2)}{60,000 \text{ psi}} - \left(\dfrac{25(16 \text{ in.})}{60,000 \text{ psi}}\right)(66 \text{ in.})\dfrac{60 \text{ ksi}}{60 \text{ ksi}}$<br><br>$= 1.82$ in.$^2$<br>$A_{\ell,prov.} = 2.14$ in.$^2$ > $A_{\ell,req'd} = 0.12$ in.$^2$ **OK** |
| 9.7.5.1 | Distribute torsional longitudinal reinforcement around the perimeter of closed stirrups that satisfy Section 25.7.1.6 (ACI 381-14) (ends of stirrups are terminated with 135-degree standard hooks around a longitudinal bar). | |
| 9.7.6.3.3 | Tie spacing must not exceed the lesser of $p_h/8$ and 12 in.<br><br>Use ten No. 5 longitudinal bars are required, three top and bottom, and two in each vertical face.<br><br>At a distance $d$ from support; $M_u$ decreases by the amount of:<br>$\Delta M_u = V_u d - w_u d^2/2$<br><br>The amount of flexural reinforcement required to resist | $p_h/8 = 66$ in./8 = 8.25 in. < 12 in.<br><br>Use ten No. 5 bars.<br>$A_{\ell,prov.} = (10)(0.31 \text{ in.}^2) = 3.1$ in.$^2$ **OK**<br><br>$\Delta M_u = (61 \text{ kip})(21.5 \text{ in.}/12) - (4.72 \text{ kip/ft})(21.5 \text{ in.}/12)^2/2$<br><br>$\Delta M_u = 102$ ft-kip |
| 9.7.5.2 | $\Delta M_u$: $\Delta M_u = \phi f_y A_s (d - a/2) \sim \phi f_y A_s (0.9 d)$<br><br>No. 5 bars must be at least equal to 0.042 times the transverse reinforcement spacing, but not less than 3/8 in.<br><br>Use No. 5 longitudinal bars. Place five No. 5 in bottom, two No. 5 in each side face, and three No. 5 in top. | $\Delta M_u = (0.9)(60 \text{ ksi}) A_s (21.5 \text{ in.})(0.9) = (102 \text{ ft-kip})(12)$<br>$A_s = 1.17$ in.$^2$<br><br>$(0.042)(4 \text{ in.}) = 0.168$ in. < 0.625 in. = No. 5 **OK**<br>No. 5 > 3/8 in. **OK** |

Step 5: Beam detailing

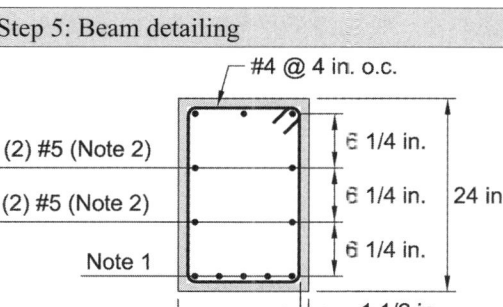

Notes:
1. Bottom reinforcing bars summation of flexure and torsion reinforcement requirements.
2. Side reinforcing bar due to torsional moment

**Beam Example 9:** *Determine closed ties required for the beam of Example 8 to resist shear and torque*

Use the same data as that for Example 8, except that the factored torsion of 53 ft·kip is not an equilibrium requirement, but because the structure is indeterminate, can be redistributed if the beam cracks.

Given:
$f_c'$ = 5000 psi (normalweight concrete)
$\lambda$ = 1.0
$f_y = f_{yt}$ = 60,000 psi
$b_w$ = 16 in.
$h$ = 24 in.
$V_u$ = 61 kip
$T_u$ = 53 ft·kip

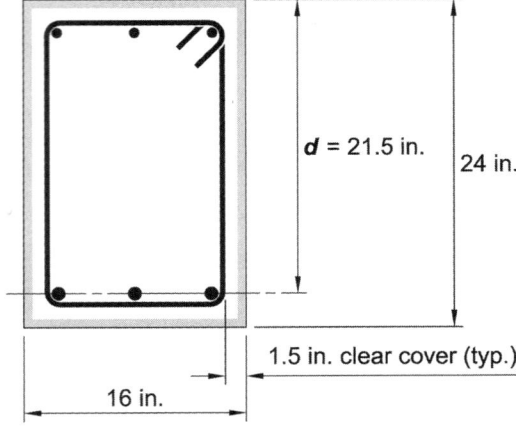

*Fig. E9.1—Beam subjected to torque.*

| ACI 318-14 | Discussion | Calculation |
|---|---|---|
| **Step 1: Section properties** | | |
| 9.2.4.4<br>22.7.6.1<br>22.7.6.1.1<br>9.2.4.4<br>22.7.6.1 | Determine section properties for torsion.<br>$A_{cp} = b_w h$<br>$A_{oh} = (b_w - 3.5 \text{ in.})(h - 3.5 \text{ in.})$<br>$A_o = 0.85 A_{oh}$<br>$p_{cp} = 2(b_w + h)$<br>$p_h = 2(b_w - 3.5 \text{ in.} + h - 3.5 \text{ in.})$ | $A_{cp}$ = (16 in.)(24 in.) = 384 in.²<br>$A_{oh}$ = (16 in. – 3.5 in.)(24 in. – 3.5 in.) = 256 in.²<br>$A_o$ = 0.85(256 in.²) = 218 in.²<br>$p_{cp}$ = 2(16 in. + 24 in.) = 80 in.<br>$p_h$ = 2(16 in. – 3.5 in. + 24 in. – 3.5 in.) = 66 in. |
| **Step 2:** | | |
| 9.5.4.1<br>22.7.4.1a | Calculate threshold torsion:<br>$\phi T_{th} = \phi \lambda \sqrt{f_c'} \left( \dfrac{A_{cp}^2}{p_{cp}} \right)$ | $\phi T_{th} = (0.75)(1.0)(\sqrt{5000 \text{ psi}}) \left( \dfrac{(384 \text{ in.}^2)^2}{80 \text{ in.}} \right) = 97{,}750$ in.-lb |
| 9.5.1.2<br>21.2.1c | Torsion strength reduction factor $\phi$ = 0.75 | |
| 9.5.1.1 | Check if $T_u > \phi T_{th}$ | $\phi T_{th}$ = 8.15 ft–kip<br>$T_u$ = 53 ft–kip > $\phi T_{th}$ = 8.15 ft–kip  **OK**<br>Design section to resist torsional moment. |
| 22.7.5.1 | Calculate cracking torsion<br>$\phi T_{cr} = \phi 4 \lambda \sqrt{f_c'} \left( \dfrac{A_{cp}^2}{p_{cp}} \right)$ | $\phi T_{cr}$ = 32.6 ft-kip |
| 9.5.1.1 | Check if $T_u > \phi T_{cr}$?<br><br>In statically indeterminate structures where $T_u > \phi T_{cr}$, a reduction of $T_u$ in the beam can occur due to redistribution of internal forces after torsion cracking. Therefore, reduce $T_u$ to $\phi T_{cr}$. | $T_u$ = 53 ft-kip > $\phi T_{cr}$ = 32.6 ft-kip<br><br><br><br>Use $T_u$ = 32.6 ft-kip and design for torsional reinforcement. |

# CHAPTER 7—BEAMS

| | | |
|---|---|---|
| 22.7.7.1 | Check if cross-sectional dimensions are large enough. $$\sqrt{\left(\frac{V_u}{b_w d}\right)^2 + \left(\frac{T_u p_h}{1.7 A_{oh}^2}\right)^2} \leq \phi\left(\frac{V_c}{b_w d} + 8\sqrt{f_c'}\right)$$ | $$\sqrt{\left(\frac{61{,}000 \text{ lb}}{(16 \text{ in.})(21.5 \text{ in.})}\right)^2 + \left(\frac{(32.5 \text{ ft-kip})(12{,}000)(66 \text{ in.})}{1.7(256 \text{ in.}^2)^2}\right)^2}$$ $$\leq (0.75)(2\sqrt{5000 \text{ psi}} + 8\sqrt{5000 \text{ psi}})$$ $$\sqrt{(177.3 \text{ psi})^2 + (231.7 \text{ psi})^2} = 292 \text{ psi} \leq 530 \text{ psi} \quad \textbf{OK}$$ Therefore, section is large enough. |
| **Step 3: Torsional reinforcement** | | |
| 9.5.4.1 <br> 22.7.6.1a <br> 22.7.6.1.2 | Find area/spacing of ties due to shear and torsional moment: <br> Calculate torsional tie area/spacing: $$\phi T_n = \phi \frac{2 A_o A_t f_y}{s} \cot \theta \geq T_u = 32.6 \text{ ft-kip}$$ | $$\frac{A_t}{s} = \frac{(32.6 \text{ ft-kip})(12{,}000)}{2(0.75)(218 \text{ in.}^2)(60{,}000 \text{ psi})(\cot 45°)}$$ $$= 0.0199 \text{ in.}^2/\text{in.}$$ |
| 9.5.1.1 <br> 2.5.1.1 <br> 22.5.5.1 <br><br> 22.5.10.5.3 <br> 21.2.1b | Calculate shear tie area/spacing: <br> $\phi V_n = \phi V_c + \phi V_s \geq V_u = 61$ kip <br> and <br><br> $\phi V_s = \phi \dfrac{A_v f_y d}{s}$ | |
| | Calculate total tie area/spacing ($A_v/s + 2A_t/s$) | $$\frac{A_v}{s} = \frac{\dfrac{61{,}000 \text{ lb}}{0.75} - 2\sqrt{5000 \text{ psi}}(16 \text{ in.})(21.5 \text{ in.})}{(60{,}000 \text{ psi})(21.5 \text{ in.})}$$ $$\frac{A_v}{s} = 0.0253 \text{ in.}^2/\text{in.}$$ |
| 9.6.4.2 | Try No. 4 ties and calculate $s$: | $$\frac{A_v}{s} + 2\frac{A_t}{s} = 0.0253 \text{ in.}^2/\text{in.} + 2(0.0199 \text{ in.}^2/\text{in.})$$ $$= 0.065 \text{ in.}^2/\text{in.}$$ $$s = \frac{0.4 \text{ in.}^2}{0.065 \text{ in.}^2/\text{in.}} = 6.1 \text{ in.}$$ Use $s = 6$ in. |
| 22.7.6.1b | Torsional longitudinal reinforcement $$T_n = \frac{2 A_o A_\ell f_y}{p_h} \tan \theta$$ | $$\frac{(32.6 \text{ ft-kip})(12 \text{ in./ft})}{0.75} = \frac{2(218 \text{ in.}^2) A_\ell (60 \text{ ksi})}{66 \text{ in.}} \tan 45°$$ $A_\ell = 1.32$ in.$^2$ |

| | | |
|---|---|---|
| 9.6.4.3 | The minimum torsional longitudinal reinforcement $\ell_{min}$, must be at least the lesser of:<br><br>(a) $\dfrac{5\sqrt{f'_c}A_{cp}}{f_y} - \left(\dfrac{A_t}{s}\right)p_h\dfrac{f_{yt}}{f_y}$<br><br>(b) $\dfrac{5\sqrt{f'_c}A_{cp}}{f_y} - \left(\dfrac{25b_w}{f_{yt}}\right)p_h\dfrac{f_{yt}}{f_y}$ | $\dfrac{5(\sqrt{5000\text{ psi}})(384\text{ in.}^2)}{60,000\text{ psi}} - (0.0324\text{ in.}^2/\text{in.})(66\text{ in.})\dfrac{60\text{ ksi}}{60\text{ ksi}}$<br>$= 0.12\text{ in.}^2$ **Controls**<br><br>$\dfrac{5(\sqrt{5000\text{ psi}})(384\text{ in.}^2)}{60,000\text{ psi}} - \left(\dfrac{25(16\text{ in.})}{60,000\text{ psi}}\right)(66\text{ in.})\dfrac{60\text{ ksi}}{60\text{ ksi}}$<br>$= 1.82\text{ in.}^2$<br>$A_{\ell,prov.} = 1.32\text{ in.}^2 > A_{\ell,req'd} = 0.12\text{ in.}^2$ **OK** |
| 9.7.5.1 | Distribute torsional longitudinal reinforcement around the perimeter of closed stirrups that satisfy Section 25.7.1.6 (ends of stirrups are terminated with 135-degree standard hooks around a longitudinal bar). | |
| 9.7.6.3.3 | Transverse torsional reinforcement spacing must not exceed the lesser of $p_h/8$ and 12 in.<br><br>Use No. 4 longitudinal bars. Place five No. 4 in bottom and two No. 4 in each side face. Excess flexural capacity in top at $d$ from support can serve in place of three No. 4 in top. | $p_h/8 = 66\text{ in.}/8 = 8.25\text{ in.} < 12\text{ in.}$<br><br><br><br>Refer to Beam Example 8. |
| **Step 4: Beam detailing** | | |

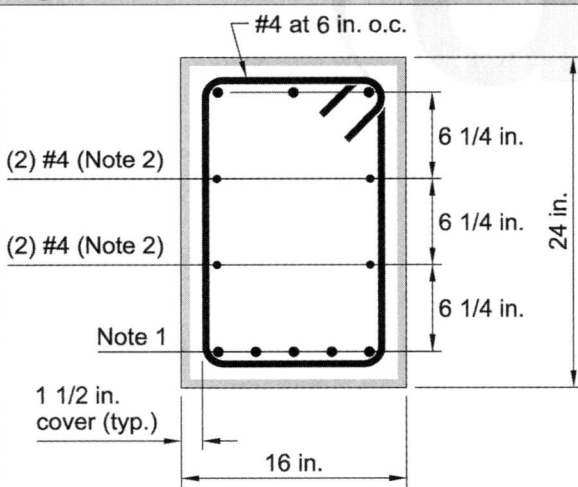

Notes:
1. Bottom bars summation of moment and torsion reinforcement requirements.
2. Side bar due to torsional moments

# CHAPTER 8—DIAPHRAGMS

## 8.1—Introduction

Building diaphragms are usually horizontal, reinforced concrete one-way or two-way slabs spanning between columns or walls, or both columns and walls. They can be built out of cast-in-place (CIP) concrete, precast elements with CIP topping, interconnected precast elements without CIP topping, or precast elements with end strips formed of CIP topping slab or edge beams (Moehle et al. 2010).

Building slabs are designed to resist gravity loads and also to transfer wind, earthquake, fluid, or lateral earth pressure forces to the lateral-force-resisting system, such as moment frames, shear walls, or both (ACI 318-14, Section 12.2). For dual system structures such as shear walls and special moment frames, special moment frames deform in a shear mode, as shown in Fig. 8.1(a), while shear walls deform in a bending mode (cantilever), also as shown in Fig. 8.1(a). Diaphragms maintain compatible deformations between the two systems, thus tying the entire structure together (Fig. 8.1(a) and (b)). Diaphragms also provide lateral support to shear walls and columns. As a rule of thumb, approximately 2 to 5 percent of a column axial force must be resisted by the diaphragm to provide adequate lateral support to the walls and columns. This force is easily achieved in low-rise building diaphragms, but must be checked for columns with high axial force, such as those in high-rise buildings (Moehle et al. 2010). Checks can include:

(a) Slab-bearing force at face of columns
(b) Adequacy of diaphragm slab reinforcement anchored into columns at edge connections
(c) Adequate diaphragm buckling strength to resist the bracing forces

## 8.2—Material

Concrete compressive strength for diaphragms and collectors resisting lateral forces must be at least 3000 psi (ACI 318-14, Section 19.2.1.1). Steel strength for longitudinal and transverse bars is limited to 60,000 psi (ACI 318-14, Section 12.5.1.5).

## 8.3—Service limits

The minimum diaphragm slab thickness must satisfy the requirements of Section 7.3.1 (ACI 318-14) for one-way slabs or Section 8.3.1 (ACI 318-14) for two-way slabs. The diaphragm thickness must also be sufficient to resist in-plane moment, shear, and axial forces (Section 12.5.2.3 of ACI 318-14).

## 8.4—Analysis

Diaphragm slabs must resist gravity loads and lateral in-plane force combinations simultaneously. For concrete slabs, ASCE 7-10 (Section 12.3.1.2) permits the assumption of a rigid diaphragm if the diaphragm aspect ratio, which is the span-to-depth ratio, is 3 or less for seismic design and 2 or less for wind loading (ASCE 7-10, Section 27.5.4) if the structure has no significant horizontal irregularities. While structures are expected to behave inelastically during an earthquake, it is expected that rigid diaphragms will perform elastically under all load conditions. Diaphragm slabs are commonly designed as a deep beam that resists lateral forces at the floor level, with system columns and walls acting as supports for the deep beam.

The diaphragm reinforcement resisting tension due to flexure is placed at the tension edge perpendicular to the

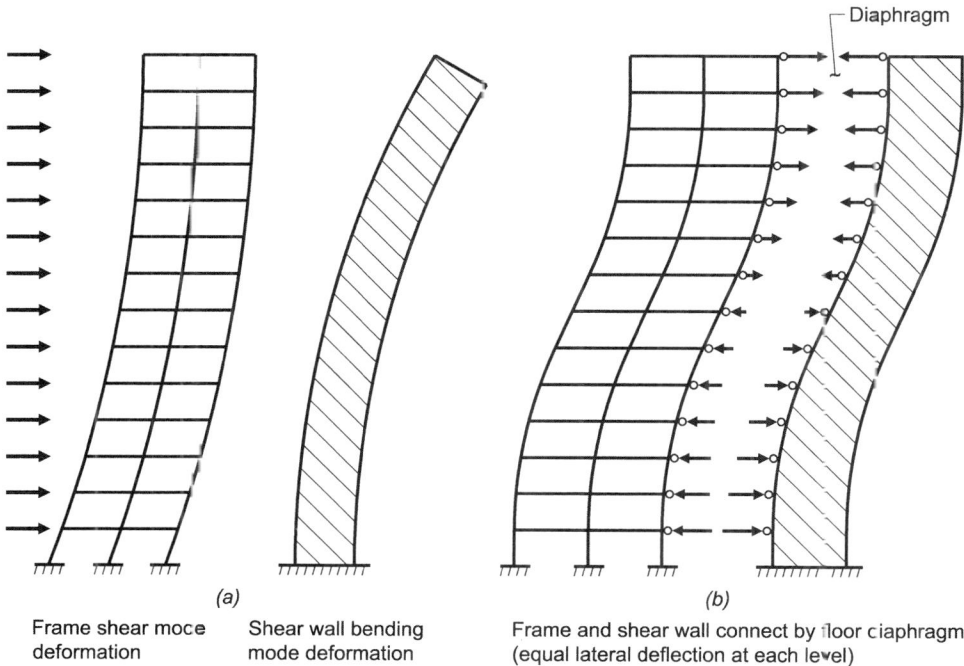

*Fig. 8.1—Shear wall and moment frame dual system deformation.*

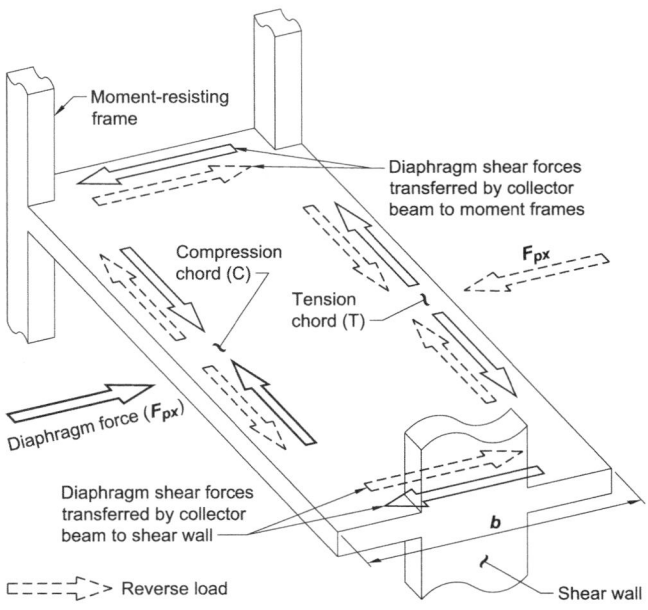

Fig. 8.4a—Diaphragm tension-compression and shear forces due to lateral forces.

applied force. Tension and compression edges are identified as chords. Because earthquake and wind forces are reversible, equal reinforcement should be provided at both chords (Fig. 8.4a). Building edge beams, if provided, are often designed as the diaphragm's chord (Section 8.4.1 of this Handbook). Chords are assumed to resist all the flexural tension from the diaphragm in-plane bending moment resulting from the lateral load. If edge beams are not provided, the slab acts as a deep rectangular beam resisting bending in the plane of the slab, with the chord tension reinforcement placed within $h/4$ of the tension face, where $h$ is the diaphragm width in the direction of analysis (Section 12.5.2.3 of ACI 318-14).

The diaphragm shear forces are resisted by the system columns and walls. The beams or slab sections that transfer shear are identified as collectors. The collector slab or beam connection to the columns and walls must be appropriately designed and detailed to achieve shear transfer.

Rigid diaphragms (Fig. 8.4b(a)) are often modeled as deep beams with spring supports (Fig. 8.4c). Lateral force is distributed to the columns and walls according to their relative lateral stiffness. Flexible diaphragms are modeled with rigid supports (Fig. 8.4b(b)). If all supports have equal

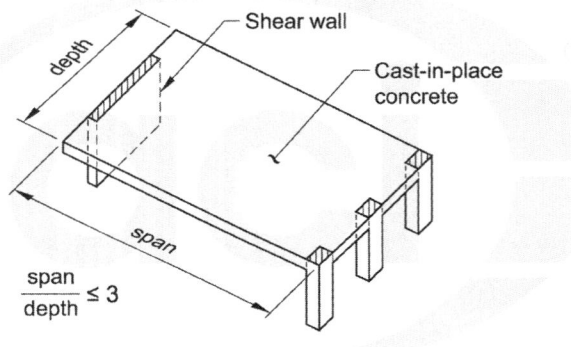

(a) Rigid diaphragm

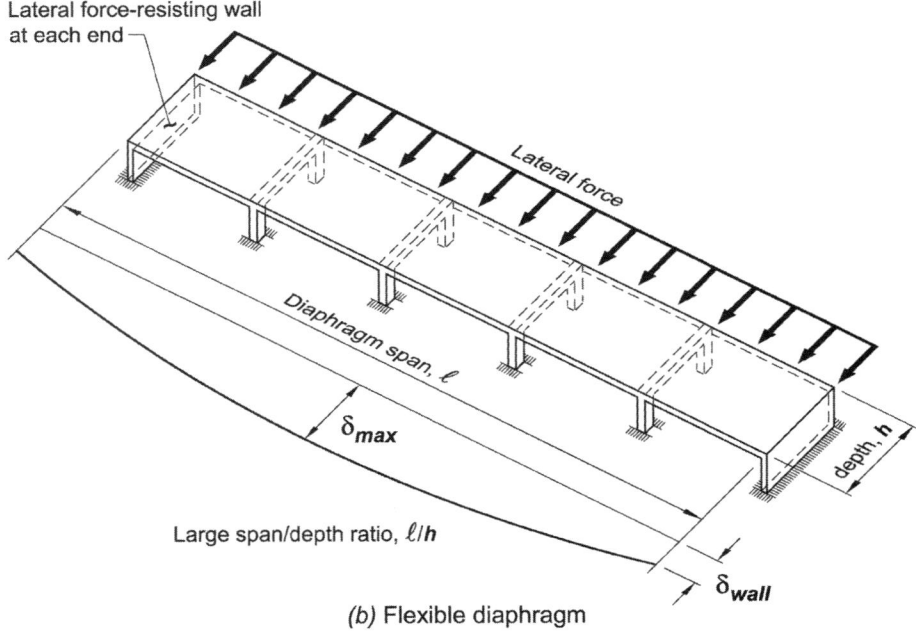

(b) Flexible diaphragm

Fig. 8.4b—Rigid and flexible diaphragm.

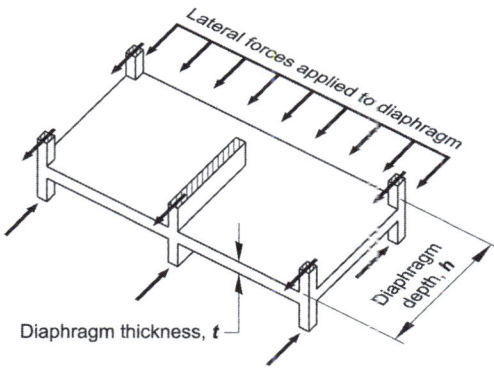

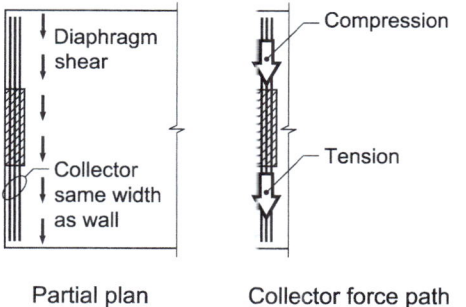

Fig. 8.4.1a—Collector having same width as shear wall—forces are reversible.

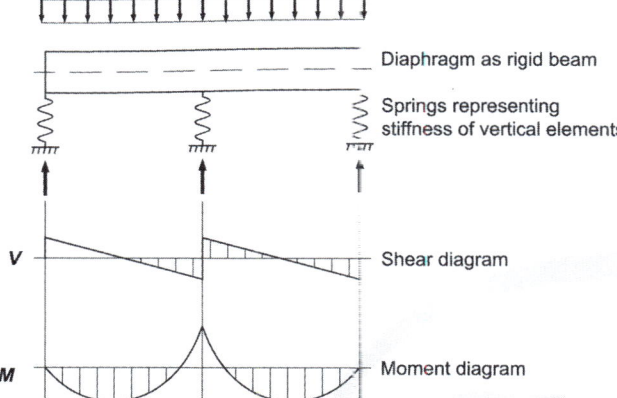

Fig. 8.4c—Rigid diaphragm lateral force distribution.

lateral resistance, the lateral force can be distributed to the columns and walls according to their tributary areas.

Also, finite element and strut-and-tie method can be used to analyze diaphragms. The finite element method should consider diaphragm flexibility (Section 12.4.2.4 of ACI 318-14).

**8.4.1** *Collectors*—Shear walls do not usually extend the full length of a building. Collectors, also called drag members or distributers, are designed to collect lateral forces from the diaphragm and transfer them to the seismic-force-resisting system, or to transfer lateral loads from a shear wall into the diaphragm. Collectors can be the full length of the diaphragm, but not necessarily (ACI 318-14, Section 12.5.4.1). Collectors can be defined as a section within the depth of the diaphragm or as a beam as part of the diaphragm. Collectors, as part of a rigid diaphragm, are expected to perform elastically during an earthquake event.

Collectors parallel to a shear wall can have the same width as a shear wall (Fig. 8.4.1a) or be wider. Collectors eccentric to the wall have an effective width $b_{eff}$, defined as not wider than the thickness of the wall plus one-half the length of the shear wall (Seismology and Structural Standards Committee (SEAOC) 2005; ACI 318-14, Section R12.5.4; Fig. 8.4.1b of this Handbook). Collectors with the same width as the wall will simply transfer the slab lateral forces by axial compression or tension to the shear wall. Collectors having a width wider than the shear wall will transfer part of the diaphragm lateral force by axial compression or tension and the balance will be transferred along the wall length through shear fric-

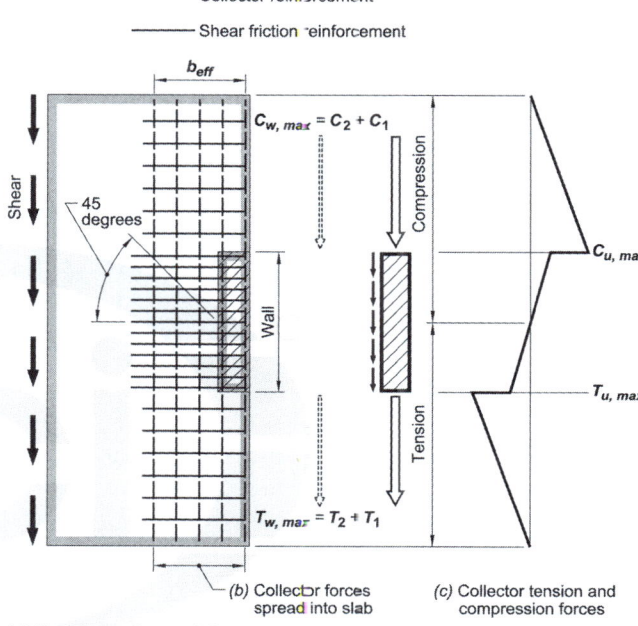

Fig. 8.4.1b—Collector wider than the shear wall—forces are reversible.

tion. An eccentricity results between the resultant force in the collector and the shear wall reaction (Fig. 8.4.1b). This eccentricity creates secondary stresses in the slab transfer region adjacent to the wall. Adequate reinforcement must be provided to resist these stresses (SEAOC 2005).

Collectors, like rigid diaphragms, are expected to behave elastically under axial and compression forces. Reinforcement is usually placed at mid-depth in collectors. Shear reinforcement perpendicular to the walls is needed as shear friction reinforcement for eccentric collectors, and is placed within the slab thickness (SEAOC 2005).

## 8.5—Design strength

Diaphragms in Seismic Design Categories (SDCs) D through F are designed in accordance with Chapter 18 of ACI 318-14.

Diaphragms are designed for stability, strength, and stiffness under factored load combinations; its thickness must be at least that required of that member (ACI 318-14, Section 12.3.1). The shear forces and bending moments resulting from the effects of lateral loads are considered simultaneously.

Diaphragms are designed to resist the design seismic force calculated from the structural analysis, $F_{px}$, which must be at least (ASCE/SEI 7-10, Section 12.10.1.1):

$$F_{px} = \frac{\sum_{i=x}^{n} F_i}{\sum_{i=x}^{n} W_i} W_{px}$$

where $F_{px}$ is the diaphragm design force at level $x$.

The design force applied to level $x_i$ is $F_i$; $W_i$ is the weight tributary to level $x_i$; and $W_{px}$ is the weight tributary to the diaphragm at level $x$. The force calculated from this equation need not exceed $0.45S_{DS}IW_{px}$, but needs to be at least $0.25S_{DS}IW_{px}$ (ASCE/SEI 7-10, Section 12.10.1.1).

Collectors in SDCs C through F are designed for the largest of (a) through (c):

(a) $F_x$ obtained from structural analysis using load combinations with overstrength factor $\Omega_o$ of ASCE/SEI 7-10, Section 12.4.3.2

(b) $F_{px}$ using load combinations with overstrength factor $\Omega_o$ of ASCE/SEI 7-10, Section 12.4.3.2

(c) $F_{px,min} = 0.25S_{DS}IW_{px}$ using load combinations of ASCE/SEI 7-10 (Section 12.4.2.3) forces $F_x$ are applied to all floor levels concurrently.

Forces $F_{px}$ and $F_{px,min}$ "are applied one level at a time to the diaphragm under consideration (Moehle et al. 2010)." The nominal shear strength of a diaphragm is $V_n = A_{cv}(2\lambda\sqrt{f'_c} + \rho_t f_y)$ (ACI 318-14, Section 12.5.3.3) and the cross-sectional dimensions must satisfy $V_n \leq 8\sqrt{f'_c}A_{cv}$ (ACI 318-14, Section 12.5.3.4).

In ACI 318-14, Sections 12.5.3.3 and 12.5.3.4, $A_{cv}$ is the gross area of concrete section bounded by web thickness and section length in the direction of shear force considered, and $\rho_t$ is the ratio of area of distributed transverse reinforcement to gross concrete area positioned perpendicular to the diaphragm flexural reinforcement. The reduction factor $\phi$ for diaphragms is 0.75 (ACI 318-14, Section 12.5.3.2). In SDC D, E, or F, $\phi$ cannot exceed the value corresponding to the seismic-force-resisting system it is part of if the nominal shear strength of the member is less than the shear corresponding to the development of the nominal strength of the member. The least value is 0.6 (ACI 318-14, Section 21.2.4.1).

At diaphragm discontinuities, such as openings and reentrant corners, the design needs to consider the dissipation or transfer of edge (chord) forces. When combined with other forces in the diaphragm, the local design strengths should be within the shear and torsion strength of the diaphragm.

## 8.6—Reinforcement detailing

Generally, chord and collector reinforcement is placed around diaphragm mid-depth. It is common practice (Moehle et al. 2010) to reinforce diaphragm openings smaller than approximately twice the slab thickness with only the displaced reinforcement, but at least one bar on any side. Larger openings require a more rigorous analysis.

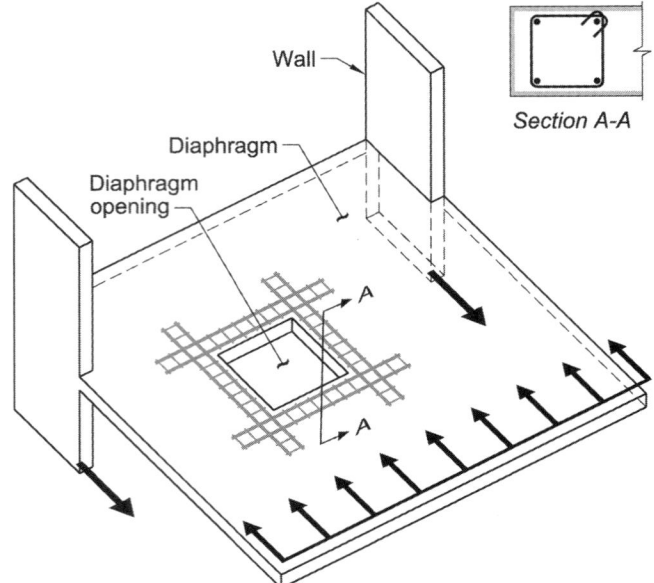

Fig. 8.6a—Reinforcement detail around opening within diaphragm.

### Table 8.6a—Collector and chord reinforcement requirements in SDCs D through F for splice and anchorage zones

| Reinforcement | Splice | Requirement | ACI 318-14 |
|---|---|---|---|
| Longitudinal | Spacing | $3d_b \geq 1.5$ in. | 18.12.7.6(a) |
| | Cover | $2.5d_b \geq 2$ in. | |
| Transverse | Greater of | $0.75\sqrt{f'_c}\dfrac{b_w s}{f_{yt}}$ | 18.12.7.6(b) |
| | | $\dfrac{50b_w s}{f_{yt}}$ | |

Around large openings or other discontinuities, confinement reinforcement (ties) should be placed around the chord bars surrounding the opening (Fig. 8.6a). To properly transfer forces between the diaphragm and columns or walls, chord bar splices should be Type 2 and chord bar spacing should satisfy the requirements of Table 8.6a. Chord bars in higher seismic zones must be confined with closed hoops or spirals per Table 8.6b.

Chords at openings need to be proportioned to resist the sum of the factored axial forces acting in the plane of the diaphragm and the force obtained by dividing the factored moment at the section by the distance between the chords at the section.

A collector parallel to a shear wall has its critical connection at the face of the shear wall. Collector longitudinal bars must extend deep enough into the shear wall to develop and transfer the lateral force to wall reinforcement (Fig. 8.6b). The collector reinforcement is in addition to the horizontal diaphragm reinforcement required to resist the shear force (Moehle et al. 2010). Collector reinforcement must comply with Section 20.2.1 of ACI 318-14 with two exceptions:

(a) Collector or chord reinforcement placed within beams must satisfy ASTM A706/706M, Grade 60. Reinforcement complying with ASTM A615/A615M is permitted if:

## Table 8.6b—Transverse reinforcement requirements for tension and compression collectors and chords in SDCs D through F reinforced with transverse confinement reinforcement (ACI 318-14, Section 18.12.7.6)

| Compressive stress | Transverse reinforcement requirements | Details | | ACI 318-14 |
|---|---|---|---|---|
| $> 0.2f_c'$ ($> 0.5f_c'$ if forces are amplified to account for over-strength) | Yes | Single or overlapping spirals per Sections 25.7.3.5 and 25.7.3.6 of ACI 318-14<br>Circular hoops or rectangular hoops with or without crossties spaced not more than 14 in.<br><br>The dimension $x_i$ from centerline to centerline of laterally supported longitudinal bars is not to exceed 14 inches.<br>The term $h_x$ used in Eq. (18.7.5.3) is taken as the largest value of $x_i$. | | 18.12.7.5 |
| | | Transverse reinforcement spacing along length of the diaphragm is the smallest of:<br>(a) One-fourth minimum member dimension<br>(b) $6d_b$ of smallest longitudinal reinforcement<br>(c) $s_o = 4 + \dfrac{14 - h_x}{3}$ and 4 in. $\geq s_o \leq 6$ in. | | 18.12.7.5 and 18.7.5.3 |
| | | Rectilinear hoop | $A_{sh} = 0.09 \dfrac{sb_c f_c'}{f_{yt}}$ | 18.12.7.5 |
| | | Spiral or circular hoop | Greater of: $\rho_s = 0.45\left(\dfrac{A_g}{A_{ch}} - 1\right)\dfrac{f_c'}{f_{yt}}$<br><br>$\rho_s = 0.12 \dfrac{f_c'}{f_{yt}}$ | |
| $<0.15f_c'$ ($0.4f_c'$ if forces are amplified to account for overstrength) | No | None | | 18.12.7.5 |

(i) Actual yield strength does not exceed $f_y = 60,000$ psi by more than 18,000 psi (ACI 318-14, Section 20.2.2.5)
(ii) Actual tensile strength is at least 1.25 times the actual yield strength (ACI 318-14, Section 20.2.2.5)
(b) If bonded tendons are used to resist collector forces, diaphragm shear, or flexural tension, then the design yield stress for longitudinal and transverse reinforcement is limited to the smaller of the specified yield strength or 60,000 psi (ACI 318-14, Section 12.5.1.5).

For connections to lateral-moment-resisting frames, collector longitudinal bars need to extend at least $\ell_d$ into the frame. Additional reinforcement in the frame's beams could be necessary to transfer the force to other columns of the resisting frame (Fig. 8.6c).

The minimum clear spacing between bars is the greatest of 1 in., one bar diameter, $d_b$, and (4/3) maximum aggregate size $d_{agg}$ (ACI 318-14, Section 25.2). The maximum spacing is the lesser of 18 in. and five times the diaphragm thickness (ACI 318-14, Section 12.7.2.2).

Irregular diaphragms and diaphragms with balconies require special detailing requirements. For a diaphragm with a continuous balcony along one side, chord reinforcement can be placed either along the exterior edge of the balcony or along the exterior frame of a building. It is recommended to place chord reinforcement along the exterior frame of a building and additional crack control reinforcement in the exterior edge of the balcony (refer to Fig. 8.6d). For discontinuous balconies, placing chord reinforcement in the indi-

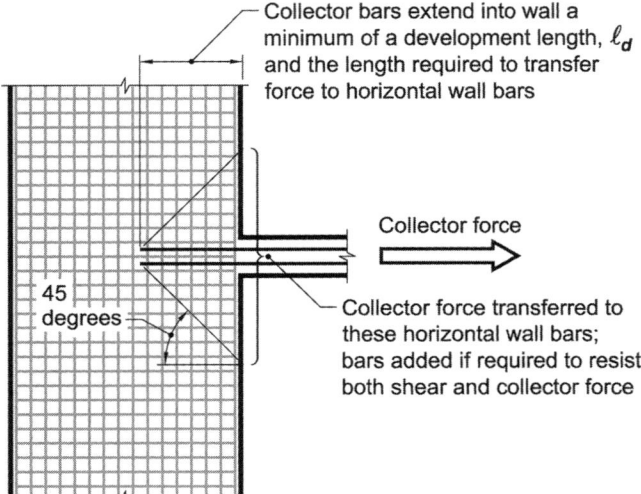

*Fig. 8.6b—Collector reinforcement extended into shear wall.*

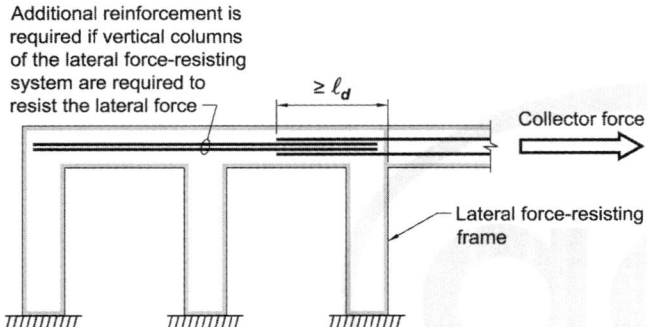

*Fig. 8.6c—Collector reinforcement extended into moment frame.*

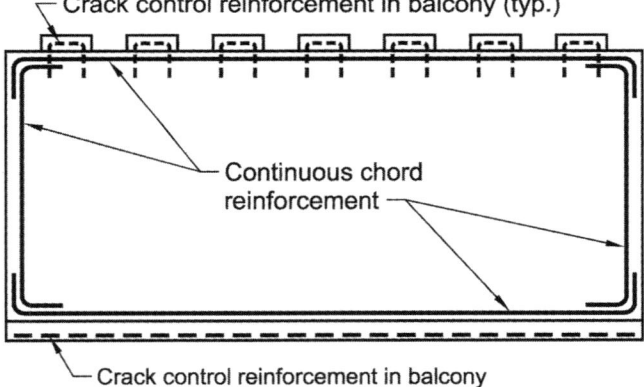

*Fig. 8.6d—Chord reinforcement in a diaphragm with balconies.*

vidual balconies will result in a discontinuous chord that is not structurally integral. It will create a complex load path (refer to Fig. 8.6d). For both cases, especially in cold climates where freezing and thawing could result in concrete cracking and exposing bars to moisture, which may result in deterioration of bars, chord reinforcement is recommended to be placed in line with the exterior frame.

Irregular diaphragms as shown in Fig. 8.6e(a) and (b) can have chord reinforcement placed either around the perimeter of the diaphragm configuration (refer to Fig. 8.6e(a)) or

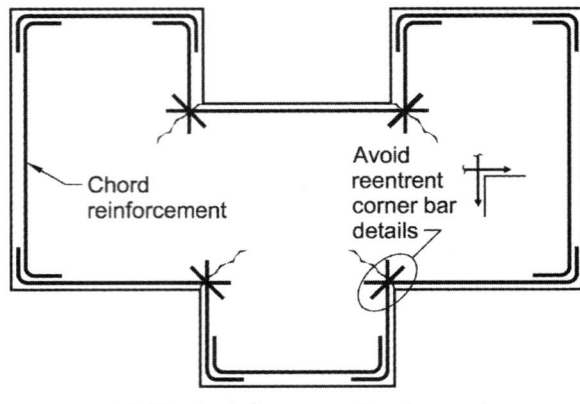

*(a) Chord reinforcement placed around the perimeter of the diaphragm*

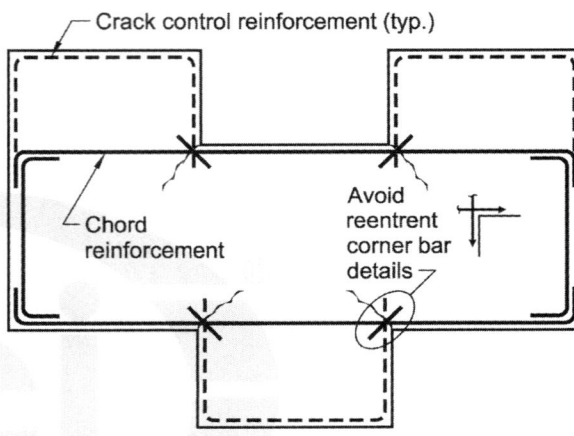

*(b) Chord reinforcement of irregular diaphragms*

*Fig. 8.6e—Chord reinforcement of irregular diaphragms.*

within the core of the diaphragm (refer to Fig. 8.6e(b)). If chord reinforcement is detailed per Fig. 8.6e(a), then special precautions are required to detail reentrant corners. It is recommended to place chord reinforcement within the core of the building and crack control reinforcement in the slab areas extending from the main core, as shown in Fig. 8.6e(b).

## 8.7—Summary steps

1. Determine if diaphragm can be considered as rigid.
2. Calculate $F_x$ at each level from structural analysis.
3. Evaluate the diaphragm inertial force $F_{px}$ at the floor and roof levels:

$$F_{px} = \frac{\sum_{i=x}^{n} F_i}{\sum_{i=x}^{n} W_i} W_{px}$$

4. Check that the diaphragm inertial force is within the maximum limits.
5. Use the larger of $F_x$ and $F_{px}$ to analyze each diaphragm.
6. Add shear forces resulting from the transfer of seismic-load-resisting vertical elements or changes in the relative stiffness to $F_{px}$. The additional forces are multiplied by the

redundancy factor ρ, equal to that used in the design of the structure.

7. Include torsion but ignore its effect if it reduces shear in the lateral-load-resisting vertical elements.

8. Calculate the net shear in the vertical elements due to $F_{px}$, which is the difference in shear forces resisted by the vertical elements immediately above and below the level of the diaphragm being designed.

9. Determine a set of equivalent loads at the diaphragm level that is in equilibrium with the shear forces determined in Step 6 above.

10. Use the equivalent loads to determine the shear and bending moment at critical sections of the diaphragm.

11. Compute the shear per unit length to check the shear strength of the diaphragm.

12. Provide collectors to transfer the shear that is in excess of force transferred directly into the vertical elements.

13. Calculate the shear strength of the diaphragm and compare it to the factored shear force.

14. Check collectors (or equivalent widths of slab assumed to act as a collector) and their connections for diaphragm chord forces.

15. Extend chords at reentrant corners, if any, to develop the forces calculated at the critical sections.

16. Check shear friction at wall-to-slab interface (SEAOC 2005).

# REFERENCES
*American Society of Civil Engineers*
ASCE/SEI 7-10—Minimum Design Loads for Buildings and Other Structures

*ASTM International*
ASTM 615/615M-15—Standard Specification for Deformed and Carbon Steel Bars for Concrete Reinforcement
ASTM A706/706M-14—Standard Specification for Deformed and Plain Low-Alloy Steel Bars for Concrete Reinforcement

## Authored documents
Moehle, J. P.; Hooper, J. D.; Kelly, D. J.; and Meyer, T. R., 2010, "Seismic Design of Cast-in-Place Concrete Diaphragms, Chords, and Collectors: A Guide for Practicing Engineers," *National Earthquake Hazards Reduction Program (NEHRP) Seismic Design Technical Brief No. 3*, The National Institute of Standards and Technology (NIST) GCR 10-917-4, Gaithersburg, MD, 29 pp.

Structural Engineer's Association of California (SEAOC), 2005, "Design of Concrete Slabs as Seismic Collectors," Seismology and Structural Standards Committee, Structural Engineers of California, May, 15 pp.

## 8.8—Examples

**Rigid Diaphragm Example 1**: *Reinforced concrete diaphragm without an opening*—An eight-story structure with 5 x 6 bays in the North-South (N-S) and East-West (E-W) directions, respectively, is located in a low-intensity earthquake region. Design the fourth level diaphragm for the following data:

Given:

*Geometry—*
Bays are 14 ft-0 in. x 36 ft-0 in. (Fig. E1.1(b))
Columns: 24 in. x 24 in.
Story height: Refer to Fig. E1.1(a)
Slab thickness: $h = 7$ in.
Shear wall thickness: $t_w = 12$ in.
Perimeter beams: W x H = 18 in. x 30 in.

*Concrete—*
$f_c' = 5000$ psi
$f_y = 60,000$ psi

*Seismic criteria—*
Site class: D
$S_s = 0.15$ (ASCE 7-10, Fig. 22-1)
$S_1 = 0.08$ (ASCE 7-10, Fig. 22-2)
$T_L = 12$ (ASCE 7-10, Fig. 22-12)
$R = 4$ (ASCE 7-10, Table 12.2-1); ordinary reinforced shear wall along Column Lines 1 and 7
$R = 5$ (ASCE 7-10, Table 12.2-1); intermediate moment frame along Column Lines A and F

Building assigned to: Seismic Design Category (SDC) B

*Wind criteria—*
Building risk category: II ASCE 7-10 Table 1.5-1
Importance, $I_w = 1$ (ASCE 7-10, Table 1.5-2)
Wind speed = 115 mph (ASCE 7-10, Fig. 26.5-1A)
$K_d = 0.85$ (ASCE 7-10, Table 26.6-1)
Exposure category C (ASCE 7-10, Section 26.7)

*Topographic effects—*
$K_{zt} = 1$ (ASCE 7-10, Section 26.8) Flat
$GC_{pi} = 0.18$ (ASCE 7-10, Table 26.11-1) Enclosed building.

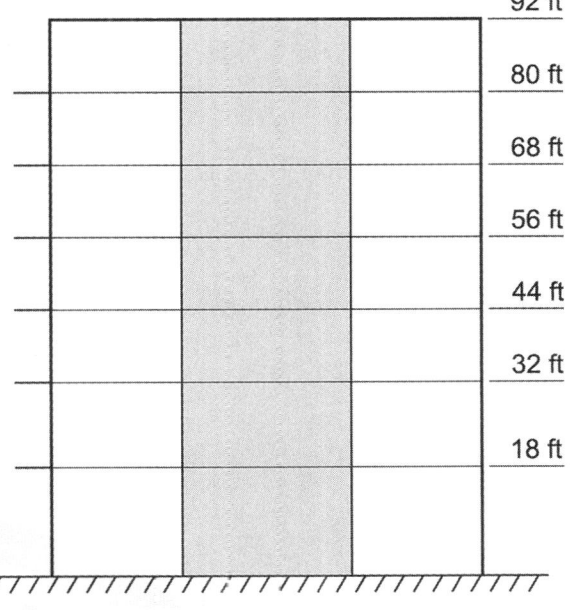

*(a) West elevation*

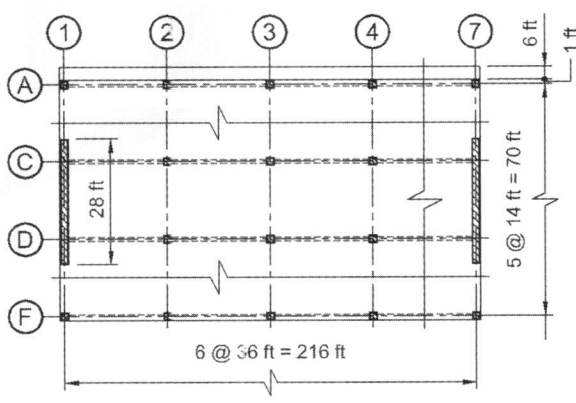

*(b) Fourth level plan*

Fig. E1.1—*Eight-story building.*

| ACI 318-14 | Discussion | Calculation |
|---|---|---|
| **Step 1: Material requirements** | | |
| 7.2.2.1 | The mixture proportion must satisfy the durability requirements of Chapter 19 and structural strength requirements (ACI 318-14).<br><br>The designer determines the durability classes. Please refer to Chapter 4 of this Handbook for an in-depth discussion of the categories and classes.<br><br>ACI 301 is a reference specification that is coordinated with ACI 318-14. ACI encourages referencing ACI 301 into job specifications.<br><br>There are several mixture options within ACI 301, such as admixtures and pozzolans, which the designer can require, permit, or review if suggested by the contractor. | By specifying that the concrete mixture shall be in accordance with ACI 301 and providing the exposure classes, Chapter 19 requirements are satisfied.<br><br>Based on durability and strength requirements, and experience with local mixtures, the compressive strength of concrete is specified at 28 days to be at least 5000 psi. |
| **Step 2: Slab geometry** | | |
| 12.3.1.1 | Diaphragm thickness must satisfy the requirements for stability, strength, and stiffness under factored load combinations. | |
| 12.3.1.2<br>7.3.1.1 | For simplicity, specify floors and roof slab diaphragm thickness that satisfy the minimum one-way slab thickness, spanning in the short direction $\ell_n$ =14 ft with $f_y$ = 60,000 psi and without interior beams.<br>The minimum thickness for exterior panels is $\ell_n/24$, and for interior panels with both ends continuous is $\ell_n/28$:<br>$h_{min} \geq \ell_n/24$  **Controls** | $h_{min} \geq \dfrac{(14 \text{ ft})(12 \text{ in./ft}) - 18 \text{ in.}}{24} = 6.25$ in., say, 7 in. |

## Step 3: Lateral forces

Seismic and wind force calculations are not shown as they are outside the scope of this Handbook
For seismic design, two sets of design forces are usually specified:

1. $F_x$ is the vertical force distribution to the lateral-force-resisting system (LFRS)
2. $F_{px}$ diaphragm design force at Level $x$

The eight story building is analyzed for the seismic and wind effects. Diaphragms are designed for the maximum calculated force. The diaphragm is modeled as rigid in the analysis.

Tables E.1 and E.2 compare the lateral wind, $W$, and seismic, $F_{px}$, lateral forces. The calculations are not shown as it is outside the scope of this Handbook:

### Table E.1—North-South direction

| Story | Height, ft | Seismic $F_{px}$, kip | Wind, kip | | | Controls? |
| --- | --- | --- | --- | --- | --- | --- |
| | | | WW | LW | Combined | |
| Roof | 92 | 118 | 31.83 | 19.89 | 51 | S |
| 7 | 80 | 125 | 61.82 | 39.79 | 102 | S |
| 6 | 68 | 119 | 59.74 | 39.79 | 100 | S |
| 5 | 56 | 108 | 57.34 | 39.79 | 97 | S |
| 4 | 44 | 95 | 54.51 | 39.79 | 94 | S |
| 3 | 32 | 87 | 55.22 | 43.10 | 98 | W |
| 2 | 18 | 81 | 60.21 | 53.05 | 113 | W |

### Table E.2—East-West direction

| Story | Height, ft | Seismic $F_{py}$, kip | Wind, kip | | | Controls? |
| --- | --- | --- | --- | --- | --- | --- |
| | | | WW | LW | Combined | |
| Roof | 92 | 136 | 10.51 | 6.57 | 17 | S |
| 7 | 80 | 157 | 20.42 | 13.14 | 34 | S |
| 6 | 68 | 157 | 19.73 | 13.14 | 33 | S |
| 5 | 56 | 152 | 18.94 | 13.4 | 32 | S |
| 4 | 44 | 140 | 18.00 | 13.14 | 31 | S |
| 3 | 32 | 126 | 18.24 | 14.24 | 32 | S |
| 2 | 18 | 116 | 19.89 | 17.52 | 37 | S |

Note: The controlling force for diaphragm design is represented by S for seismic or W for wind.

From the tables above, seismic forces at the fourth floor control the diaphragm design in both directions.

Where $F_x = C_{vx}V$, $C_{vx} = \dfrac{w_x h_x^k}{\sum w_i h_i^k}$, and $F_{px} = \dfrac{\sum F_i}{\sum w_i} w_{px}$ (ASCE 7-10, Eq. (12.8-11), (12.8-12), and (12.10-1)).

$F_{px}$ is limited to $F_{px,min} = 0.2S_{DS}I_e w_{pi}$, and $F_{px,max} = 0.4S_{DS}I_e w_{pi}$ (ASCE 7-10, Eq. (12.10-2) and (12.10-3)).

### Step 4: Center of mass (COM) and center of rigidity (COR)

Design fourth-level diaphragm:
Take the point of origin at the lower left corner of the building, D1.

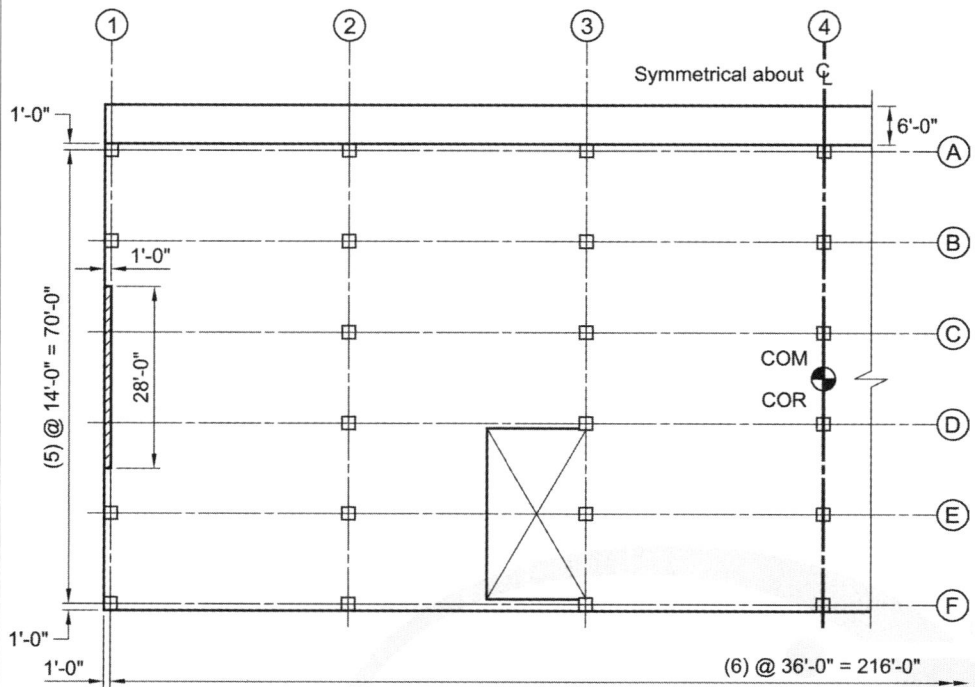

*Fig. E1.2—Mass center and rigidity center location.*

Determine COM
The diaphragm and wall configuration are symmetrical about both x- and y-axes. Therefore, the COM is located at 108 ft-0 in. east of Column Line 1 and 38 ft 0 in. north of Column Line F.

Determine COR
Because of symmetry, the COR and COM coincide.

Accidental torsion
ASCE 7-10 (third edition) commentary Section C12.10.1 considers an additional moment caused by an assumed displacement of COM. A shift of minimum of 5 percent of the building dimension perpendicular to the direction of seismic forces in addition to the actual eccentricity is assumed, referred to as accidental eccentricity.

$e_x = 0$ ft $\pm$ (0.05)(218 ft) $\pm$10.9 ft
$e_y = 0$ ft $\pm$ (0.05)(78 ft) $\pm$3.9 ft

# CHAPTER 8—DIAPHRAGMS

## Step 5: Lateral system stiffness

The diaphragm is idealized as a beam whose depth is equal to the full diaphragm depth spanning between idealized rigid supports (shear walls at Column Line 1 and 7 in the N-S direction) or resisted by the building frame in the E-W direction. Therefore, lateral forces are distributed in proportion to the relative stiffnesses of the resisting walls or frames. Lateral displacement is the sum of flexural and shear displacements.

Wall stiffness in N-S direction

$\Delta = \Delta_{Flexure} + \Delta_{Shear}$

$\Delta = \dfrac{Ph^3}{3EI} + \dfrac{1.2Ph}{AG}$ where $G \cong 0.4E$ and $E = 57,000\sqrt{f'_c} = 4,030,500$ psi

$\Delta_{Flexure} = \Delta_{Fi} = \dfrac{P_i h_i^3}{3EI_i} = \dfrac{P_i h_i^3}{3E \dfrac{L_i^3 t_i}{12}} = \dfrac{4P_i \left(\dfrac{h_i}{L_i}\right)^3}{E t_i}$

$\Delta_{Shear} = \Delta_{Vi} = \dfrac{1.2 P_i h_i}{A_i G} = \dfrac{(1.2) P_i h_i}{(L_i t_i) 0.4 E} = \dfrac{3 P_i \left(\dfrac{h_i}{L_i}\right)}{E t_i}$

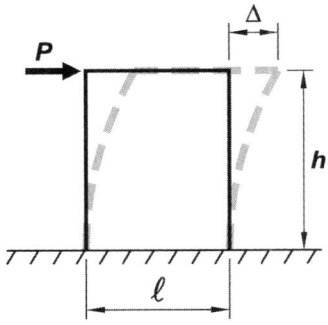

Fig. E1.3—Cantilever wall deflection.

Rigidity $k_i = \sum(1/\Delta_i)$

| Wall at CL* | Height $h$, ft | Length $L$, ft | $h/L$ | $t$, in. | $\Delta_{Fi} \times 10^{-3}$, in. | $\Delta_{Vi} \times 10^{-3}$, in. | $\Delta_i \times 10^{-3}$, in. | $k_i = 1/\Delta_i \times 10^{-3}$, in. |
|---|---|---|---|---|---|---|---|---|
| 1 | 12 | 23 | 0.429 | 12 | $0.0263/E_c$ | $0.107/E_c$ | $0.1335/E_c$ | $7.5E_c$ |
| 7 | 12 | 23 | 0.429 | 12 | $0.0263/E_c$ | $0.107/E_c$ | $0.1335/E_c$ | $7.5E_c$ |

*CL: Column Line.

Equivalent story stiffness of moment frames in the E-W direction

$k_i = \dfrac{12 E_c}{h_i^2 \left(\dfrac{1}{\sum k_c + \sum k_b}\right)}$, where $k_c = I_c/h_i$ column stiffness; and $k_b = I_b/\ell_i$ beam stiffness

$k_c = \dfrac{(24 \text{ in.})(24 \text{ in.})^3}{12(12 \text{ ft})(12 \text{ in./ft})} = 192 \text{ in.}^3$

$k_b = \dfrac{(18 \text{ in.})(30 \text{ in.})^3}{12(36 \text{ ft-2 ft})(12 \text{ in./ft})} = 99 \text{ in.}^3$

In a frame there are seven columns and six beams

$k_i = \dfrac{12 E_c}{((12 \text{ ft})(12 \text{ in./ft}))^2 \left(\dfrac{1}{(7)(192 \text{ in.}^3)+(6)(99 \text{ in.}^3)}\right)} = 1.122 E_c \dfrac{1}{\text{in.}}$

Relative stiffness: Frames = 1 and Walls = 6.7

### Step 6: Lateral resisting system forces

The shear wall reactions resist the direct inertial forces $F_{px}$ and forces from accidental torsion. Forces in walls and moment frames are:

$$F_{uxi} = \frac{k_{ix}}{\sum k_{ix}} F_{px} \pm \frac{k_i d_i}{\sum k_i d_i^2} F_{px} e_y$$

$$F_{uyi} = \frac{k_{iy}}{\sum k_{iy}} F_{py} \pm \frac{k_i d_i}{\sum k_i d_i^2} F_{py} e_x$$

where $d_i$ is the distance ($x_i$ or $y_i$) of each wall from the COR.

$F_{py,y}$ = 95 kip and $F_{px,x}$ = 140 kip are fourth story lateral forces obtained from Step 3 (Tables E.1 and E.2), N-S and E-W directions, respectively. $e_x$ = 10.9 ft and $e_y$ = 3.9 ft are calculated in Step 4.

The torsional moment is calculated by multiplying the lateral inertia force by the corresponding eccentricity:
NS: $T_y = F_{px,y} e_x$ = (95 kip)(±10.9 ft) = ±1036 ft-kip

EW: $T_x = F_{px,x} e_y$ = (140 kip)(±3.9 ft) = ±546 ft-kip

$F_u = F_{vi} + F_{ti}$

Lateral force applied in N-S direction
$F_{py}$ = 95 kip

$$F_{u,wall,max} = \frac{6.7}{6.7+6.7}(95 \text{ kip}) + \frac{6.7(218 \text{ ft}/2)}{2[(1.0)(39 \text{ ft})^2 + (6.7)(109 \text{ ft})^2]}(1036 \text{ ft-kip}) = 52.2 \text{ kip}$$

$$F_{u,wall,min} = \frac{6.7}{6.7+6.7}(95 \text{ kip}) - \frac{6.7(218 \text{ ft}/2)}{2[(1.0)(39 \text{ ft})^2 + (6.7)(109 \text{ ft})^2]}(1036 \text{ ft-kip}) = 42.8 \text{ kip}$$

$$F_{u,MF,max} = \pm \frac{(1.0)(39 \text{ ft})}{2[(1.0)(39 \text{ ft})^2 + (6.7)(109 \text{ ft})^2]}(1036 \text{ ft-kip}) = \pm 0.2 \text{ kip}$$

Lateral force applied in E-W direction
$F_{px}$ = 140 kip

$$F_{u,MF,max} = \frac{1.0}{1.0+1.0}(140 \text{ kip}) + \frac{(1.0)(78 \text{ ft}/2)}{2[(1.0)(39 \text{ ft})^2 + (6.7)(109 \text{ ft})^2]}(546 \text{ ft-kip}) = 70.1 \text{ kip}$$

$$F_{u,MF,min} = \frac{1.0}{1.0+1.0}(140 \text{ kip}) - \frac{(1.0)(78 \text{ ft}/2)}{2[(1.0)(39 \text{ ft})^2 + (6.7)(109 \text{ ft})^2]}(546 \text{ ft-kip}) = 69.9 \text{ kip}$$

$$F_{u,wall,max} = \pm \frac{(6.7)(218 \text{ ft}/2)}{2[(1.0)(39 \text{ ft})^2 + (6.7)(109 \text{ ft})^2]}(546 \text{ ft-kip}) = \pm 2.5 \text{ kip}$$

The difference due to eccentricity in the E-W direction is small. Therefore, use 70 kip.

# CHAPTER 8—DIAPHRAGMS

## Step 7: Diaphragm shear strength

| | | |
|---|---|---|
| 12.5.3.3 | In-plane shear in diaphragm<br>The diaphragm shear force is calculated from<br>$$V_n = A_{cv}\left(2\lambda\sqrt{f'_c} - \rho_t f_y\right) \quad (12.5.5.3)$$<br>Ignoring the strength contribution of reinforcement; $\rho_t = 0$<br>$A_{cv}$ is the diaphragm gross area less the 6.0 ft overhang (refer to Step 9 for further clarification). | Nominal shear strength in E-W direction<br>$V_{n,N} = (218 \text{ ft})(12 \text{ in./ft})(7 \text{ in.})\left(2(1.0)\sqrt{5000 \text{ psi}} + 0\right)$<br>$= 2{,}589{,}708 \text{ lb} \approx 2590 \text{ kip}$<br><br>Nominal shear strength in N-S direction<br>$V_{n,E} = (72 \text{ ft})(12 \text{ in./ft})(7 \text{ in.})\left(2(1.0)\sqrt{5000 \text{ psi}} + 0\right)$<br>$= 855{,}316 \text{ lb} \approx 855 \text{ kip}$ |
| 12.5.3.2<br>21.2.4.2 | Applying the shear strength reduction factor $\phi = 0.75$ at the north and south ends along column lines A and F. | $\phi V_{n,N} = (0.75)(2590 \text{ kip}) = 1940 \text{ kip}$ |
| 12.5.3.2<br>21.4.2.1 | At the east and west ends along column lines 1 and 7, the shear strength reduction factor, $\phi$, must not exceed the least value for shear used for the vertical components of the primary seismic-force resisting system (wall). Therefore, $\phi = 0.6$: | $\phi V_{n,E} = (0.6)(855 \text{ kip}) = 513 \text{ kip}$ |
| 12.5.1.1<br>22.5.1.2 | Check if factored shear force is less than design shear strength calculated in Step 6. | NS: $\phi V_n = 513 \text{ kip} \gg F_u = 52 \text{ kip}$  **OK**<br>EW: $\phi V_n = 1940 \text{ kip} \gg F_u = 70 \text{ kip}$  **OK**<br><br>Therefore, diaphragm has adequate strength to resist the lateral inertia force and shear reinforcement is not required; $\rho_t = 0$. |
| 12.5.3.4 | The nominal shear strength, $V_n$, must not exceed:<br>$$V_n = 8A_{cv}\lambda\sqrt{f'_c}$$<br>$A_{cv}$ is the diaphragm gross area less the 6.0 ft overhang. | $V_n = \dfrac{8(72 \text{ ft})(10 \text{ in.})(12 \text{ in./ft})(1.0)\sqrt{4000 \text{ psi}}}{1000 \text{ lb/kip}} = 4372 \text{ kip}$<br><br>By inspection this is satisfied.  **OK** |

## Step 8: Diaphragm lateral force distribution N-S

| | | |
|---|---|---|
| 12.4.2.4(a) 12.5.1.3(a) | Lateral force is distributed to the walls as follows, refer to Fig. E1.4: Diaphragm is idealized as rigid. Design moments, shear, and axial forces are calculated assuming the diaphragm is a beam supported by idealized rigid supports with a depth equal to full diaphragm depth. The wall and frame forces and the assumed direction of torque due to the eccentricity are shown in Fig. E1.4. | 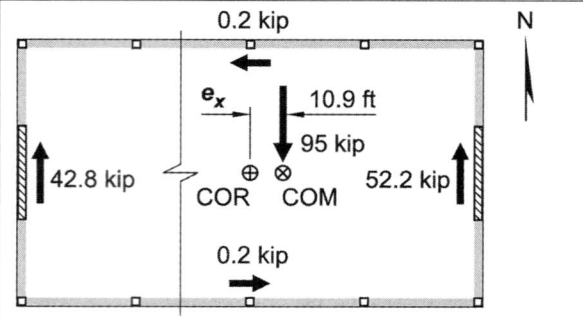 |
| 12.4.2.4 | The diaphragm force distribution is calculated by using $q_L$ and $q_R$ as the left and right diaphragm reactions per unit length (Fig. E1.4): Design force: 95 kip | 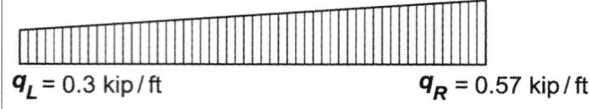 Fig. E1.4— *Seismic forces in the lateral-force-resisting systems in the N-S direction.* |

Force equilibrium

$$q_L\left(\frac{L}{2}\right) + q_R\left(\frac{L}{2}\right) = F_{px,des(NS)}$$

$$q_L\left(\frac{218\text{ ft}}{2}\right) + q_R\left(\frac{218\text{ ft}}{2}\right) = 95\text{ kip} \quad \text{(I)}$$

Moment equilibrium

$$q_L\left(\frac{L}{2}\right)\left(\frac{L}{3}\right) + q_R\left(\frac{L}{2}\right)\left(\frac{2L}{3}\right) = F_{px,des(NS)}\left(\frac{L}{2} + 0.05L\right)$$

$$q_L\frac{(218\text{ ft})^2}{6} + q_R\frac{2(218\text{ ft})^2}{6} = (95\text{ kip})\left(\frac{218\text{ ft}}{2} + 10.9\text{ ft}\right) \quad \text{(II)}$$

From statics solve equations (I) and (II) for $q_L$ and $q_R$:

$q_L + q_R = 0.87$ kip/ft (I)
$q_L + 2q_R = 1.44$ kip/ft (II)
$q_L = 0.3$ kip/ft and $q_L = 0.57$ kip/ft

| | Find the maximum moment by taking the first derivative of the moment equation expressed as a function of $x$ (unknown distance) $dM/dx = 0$ | $x = 114.3$ ft  $M_{max} = 2595$ ft-kip |
|---|---|---|
| | Draw the shear and moment diagrams (Fig. E1.5). Note: In an Aug. 2010 National Institute and Standards Technology (NIST) report, GCR 10-917-4, "Seismic Design of Cast-in-Place Concrete Diaphragms, Chords, and Collectors," by Moehle et al. states that, "This approach leaves any moment due to the frame forces along column lines A and F unresolved. Sometimes this is ignored or, alternatively, it too can be incorporated in the trapezoidal loading." | 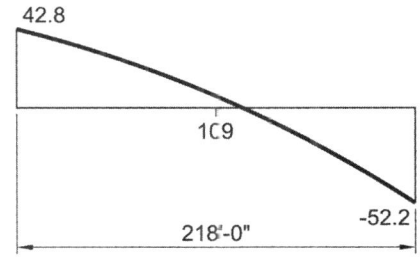 |
| | In this example the small moment due to the frame forces (0.2 kip) is ignored. | 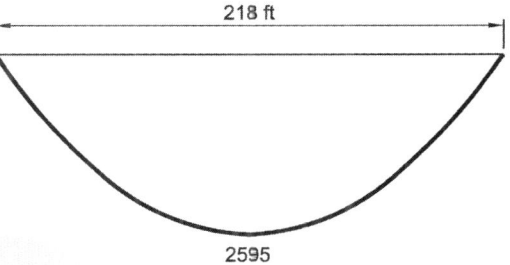  Fig. E1.5—Shear and moment diagrams. |
| | Note: Experienced engineers sometimes simplify the calculations by distributing the load uniformly: | $q = \dfrac{95 \text{ kip}}{218 \text{ ft}} = 0.44$ kip/ft |
| | Resulting in a maximum moment of: | $M_{max} = \dfrac{(0.44 \text{ kip/ft})(218 \text{ ft})^2}{8} = 2614$ ft-kip |
| | | Note: Both approaches, in this example, result in close maximum moments (less than 1 percent), but at different locations (114.3 ft versus 109 ft). |
| | | Shear diagram for the second approach is a straight line with equal shear force at both ends. |
| | | In this example, the detailed approach is presented. |
| **Step 9: Chord reinforcement N-S** | | |
| | Maximum moment is calculated above: | $M_u = 2595$ ft-kip |
| 12.5.2.3 | Chord reinforcement resisting tension must be located within $h/4$ of the tension edge of the diaphragm. | $h/4 = 72.0$ ft/4 = 18 ft |
| | Note: Chord reinforcement can be placed either in the exterior edge of the balcony or it can be placed along the exterior frame of CL A.  Placing chord reinforcement along the exterior frame is a simpler and cleaner load path for the forces in the diaphragm.  Crack control reinforcement should be added in the balcony slab for crack control. | |

| | | |
|---|---|---|
| | Assume tension reinforcement is placed in a 3 ft strip at both north and south sides of the slab edges at CLs A and F. | $3 \text{ ft} < h/4 = 18 \text{ ft}$ **OK** |
| | Chord force<br>The overhang is placed monolithic with the rest of the slab. Chord forces are usually highest furthest from the geometric centroid, in this case, edge of the overhang. To prevent cracking, place chord reinforcement at the outside edge of the overhang. The maximum chord tension force is calculated at 114.3 ft east of CL 1: | |
| | $$T_u = \frac{M_u}{B - 3 \text{ ft}}$$ | $T_u = \dfrac{2595 \text{ ft-kip}}{72 \text{ ft} - 3 \text{ ft}} = 37.6 \text{ kip}$ |
| 12.5.2.2 | Tension due to moment is resisted by deformed bars conforming to Section 20.2.1 of ACI 318-14. | |
| 12.5.1.5 | Steel stress is the lesser of the specified yield strength and 60,000 psi. | $f_y = 60,000$ psi |
| 12.5.1.1 | Required reinforcement<br>$\phi T_n = \phi f_y A_s \geq T_u$ | $A_{s,req'd} = \dfrac{37,600 \text{ lb}}{(0.9)(60,000 \text{ psi})} = 0.70 \text{ in.}^2$ |
| 22.4.3.1 | The building is assigned to SDC B. Therefore, Chapter 18 requirements for chord spacing and transverse reinforcement of Section 18.12.7.6 of ACI 318-14 do not apply. | |
| 18.12.7.5 | Overstrength factor $\Omega_o$ for chord design is not required. Therefore, use the compression stress limit in provision 18.12.7.5 of $0.2f_c'$. | |
| | Required chord width:<br>$$w_{chord} > \frac{C_{Chord}}{0.2 f_c' h_{diaph}}$$ | $w_{chord} > \dfrac{37,600 \text{ lb}}{(0.2)(5000 \text{ psi})(7 \text{ in.})} = 5.4 \text{ in.}$<br>Less than 3 ft assumed. Therefore, **OK** |
| | Choose reinforcement: | Try two No. 6 chord bars.<br>$A_{s,prov.} = 2(0.44 \text{ in.}^2) = 0.88 \text{ in.}^2$ |
| | Check if provided reinforcement area is greater than required reinforcement area: | $A_{s,prov.} = 0.88 \text{ in.}^2 > A_{s,req'd} = 0.70 \text{ in.}^2$ |

The engineer has two options for providing chord reinforcement along the exterior frames:
1. Excess amount of reinforcement used in the beams to resist gravity loads could be used to resist part of the tensile force of the chord. Additional reinforcement is provided to resist the difference.
2. The chord force is resisted with additional reinforcement.

Although the first option is more economical, the second option is detailed in this example. Here again the engineer has several options:
1. Place chord reinforcement outside the web width
2. Place chord reinforcement within the web width

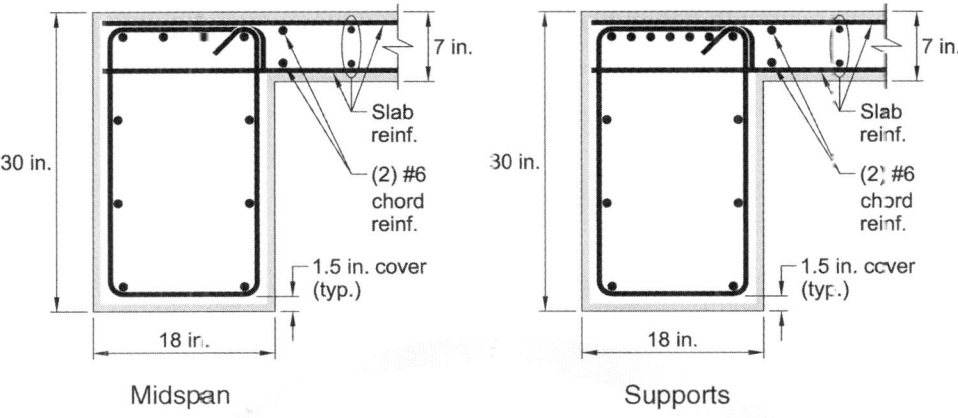

Fig. E1.6—Chord reinforcement along CLs A and F.

| | | |
|---|---|---|
| **Step 10: Collector reinforcement N-S** | | |
| 12.5.4.1 | Collector elements transfer shear forces from the diaphragm to the vertical walls at both east and west ends along column lines 1 and 7 (Fig. E1.2). Collector elements extend over the full width of the diaphragm.<br>Unit shear force:<br><br>$$v_{u@F} = \frac{F_{u@F}}{B}$$<br><br>In diaphragm: $v_{u@F} = \frac{F_{px}}{L_{diaph}}$<br><br>In wall: $v_{u@F} = \frac{F_{px}}{L_{wall}}$<br><br>Force at diaphragm to wall connection<br>East wall south end:<br>$F_{7/D.5} = -(0.72 \text{ kip/ft})(22 \text{ ft}) = -15.8 \text{ kip}$<br><br>East wall north end:<br>$F_{7/B.5} = -15.8 \text{ kip} + (1.14 \text{ kip/ft})(28 \text{ ft}) = 16.0 \text{ kip}$<br><br>At diaphragm end:<br>$F_{7/A} = +16 \text{ kip} - (0.72 \text{ kip/ft})(22 \text{ ft}) = 0.2 \text{ kip}$<br>$\approx 0$ kip due to number rounding.<br><br>Per collector force diagram, the maximum axial force on the collector is $T_u = C_u = 16$ kip. This force must be transferred from the diaphragm to the collector to the shear wall (Fig. E1.7).<br><br>The building is assigned to SDC B. Therefore, the collector force and its connections to the shear wall will not be multiplied by the system overstrength factor, $\Omega_o = 2.5$ (ASCE/SEI 7-10, Table 12.2-1). | From Step 6: $F_u = 52.2$ kip<br><br>$$v_{u@F} = \frac{52.2 \text{ kip}}{72 \text{ ft}} = 0.72 \text{ kip/ft}$$<br><br>$$v_{u@F} = \frac{52.2 \text{ kip}}{28 \text{ ft}} = 1.86 \text{ kip/ft}$$<br><br>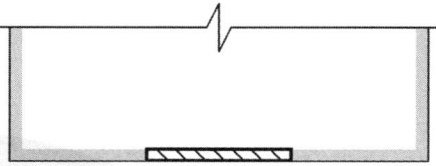<br><br>Unit shear forces:<br>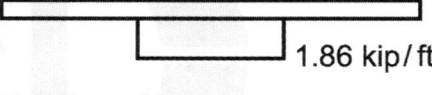<br><br>Net shear forces:<br>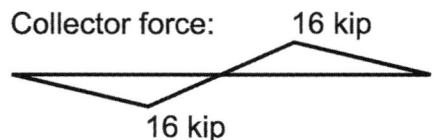<br><br>Collector force: 16 kip<br><br>16 kip<br>*Fig. E1.7—Collector force diagram.* |
| 12.5.4.2 | Collectors are designed as tension members, compression members, or both. | |

| | | |
|---|---|---|
| 12.5.1.1 22.4.3.1 | Tension is resisted by reinforcement as calculated above. Required reinforcement: $\phi T_n = \phi f_y A_s \geq T_u$ | $A_{s,req'd} = \dfrac{T_u}{0.9 f_y} = \dfrac{16{,}000 \text{ lb}}{(0.9)(60{,}000 \text{ psi})} = 0.3 \text{ in.}^2$ Although one No. 5 bar suffices, two No. 5 bars are provided to maintain symmetry of load being transferred from the slab into the wall. |
| 18.12.7.5 | Check if collector compressive force exceeds $0.2 f_c'$. Calculate minimum required collector width using $0.2 f_c'$ $w_{coll.} > \dfrac{C_{Coll.}}{0.2 f_c' t_{diaph}}$ This results in compressive stress on the concrete diaphragm collector being relatively low. The section is adequate to transfer shear stress without additional reinforcement. The collector compression and tension forces are transferred to the lateral force-resisting system within its width (shear wall). Therefore, no eccentricity is present and no-in-plane bending occurs. | $w_{coll} > \dfrac{16{,}000 \text{ lb}}{(0.2)(5000 \text{ psi})(7 \text{ in.})} = 2.3 \text{ in.}$ Use 12 in. wide collector (same width as shear wall). Provide two No. 5 bars at mid-depth of slab to prevent additional out-of-plane bending stresses in the slab. Space the two No. 5 bars at 8 in. on center starting at 2 in. from the edge of the diaphragm within the 12 in. wide collector/shear wall (Fig. E1.8). |
| 12.5.4.1 | ACI 318-14 permits to discontinue the collector along the length of the shear wall where transfer of design collector is not required. | 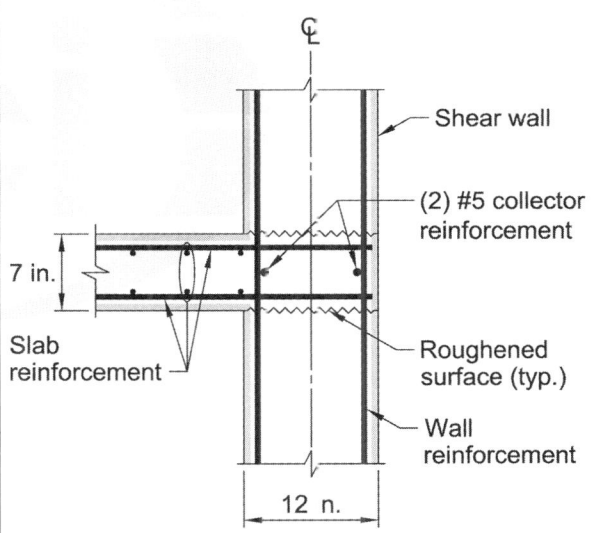 Fig. E1.8—Collector reinforcement. |
| 12.4.2.4 12.5.3.3 21.2.4.2 12.5.1.1 12.5.3.4 | Check slab shear strength along shear walls Slab shear strength along walls: $L = 28$ ft and slab thickness $t = 7$ in. From $\phi V_c = \phi A_{cv} 2\lambda \sqrt{f_c'}$ $\phi = 0.75$ Is the provided shear strength adequate? By inspection, the diaphragm shear design force satisfies the requirement of Section 12.5.3.4 of ACI 318-14. $\phi V_c = \phi A_{cv} 8\lambda \sqrt{f_c'}$ | $\phi V_c = (0.75)(2)(1.0)\left(\sqrt{5000 \text{ psi}}\right)(28 \text{ ft})(12)(7 \text{ in.})$ $\phi V_c = 249{,}467$ lb ~249 kip $\phi V_c = 249$ kip $> V_u = 52$ kip (from Step 7)   **OK** |

| Step 11: Lateral force distribution in diaphragm E-W | | |
|---|---|---|
| 12.4.2.4(a) 12.5.1.3(a) | Design force: 140 kip<br><br>Design moments, shear, and axial forces are calculated assuming a simply supported beam with depth equal to full diaphragm length (refer to Fig. E1.9). | 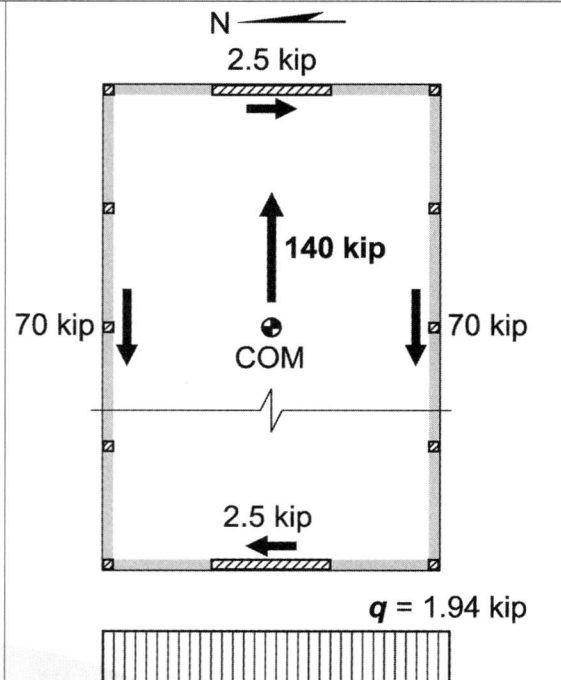<br><br>Fig. E1.9—Seismic forces in the lateral force resisting systems in the E-W direction. |
| | Because of the negligible effect of accidental torsion, the inertial force is uniformly distributed across the diaphragm width. | $q_L = \left(\dfrac{140 \text{ kip}}{72 \text{ ft}}\right) = 1.94 \text{ kip/ft}$<br><br>Shear force: $V = (1.94 \text{ kip/ft})(36 \text{ ft}) = 70 \text{ kip}$ |
| | Maximum moment is located at midlength. | $x = 36 \text{ ft}$ |
| | Draw the shear and moment diagrams (Fig. E1.10). | $M_{max} = \dfrac{w\ell^2}{8} = \dfrac{(1.94 \text{ kip/ft})(72 \text{ ft})^2}{8} = 1260 \text{ ft-kip}$ |
| 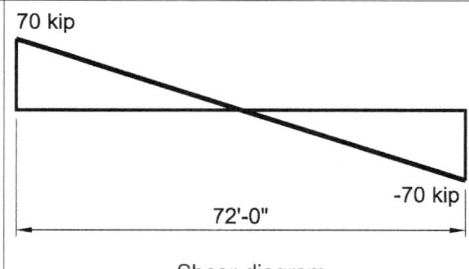<br>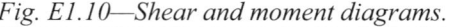<br>Shear diagram<br>Fig. E1.10—Shear and moment diagrams. | | 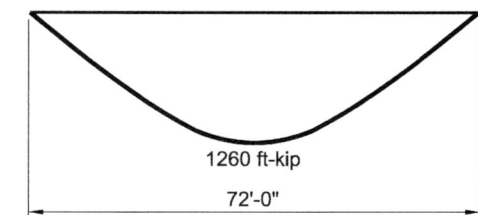<br>Moment diagram |

## Step 12: Chord reinforcement E-W

| | | |
|---|---|---|
| | **Calculate chord reinforcement** Maximum moment is calculated above (Fig. E1.8). | $M_u = 1260$ ft-kip |
| 12.5.2.3 | Chord reinforcement must be located within $h/4$ of the tension edge of the diaphragm. | $h/4 = 218.0$ ft/4 = 54.5 ft |
| | Assume tension reinforcement is placed within a 1 ft strip of the slab edge at both east and west sides of the slab. | 1 ft < $h/4$ = 54.5 ft  **OK** |
| | **Chord force** The maximum chord tension force is at midspan: $$T_u = \frac{M_u}{L - 1 \text{ ft}}$$ | $$T_u = \frac{1260 \text{ ft-kip}}{218 \text{ ft} - 1 \text{ ft}} = 5.8 \text{ kip}$$ |
| 12.5.2.2 | **Chord reinforcement** Tension due to moment is resisted by deformed bars confirming to Section 20.2.1 of ACI 318-14. | |
| 12.5.1.5 | Steel stress is the lesser of the specified yield strength and 60,000 psi. | $f_y = 60,000$ psi |
| 12.5.1.1 22.4.3.1 | **Required reinforcement** $\phi T_n = \phi f_y A_s \geq T_u$ | $$A_{s,req'd} = \frac{T_u}{0.9 f_y} = \frac{5800 \text{ lb}}{(0.9)(60,000 \text{ psi})} = 0.1 \text{ in.}^2$$ |
| | Along column lines 1 and 7, two No. 5 bars collector reinforcement are provided to resist inertia force in the N-S direction. These bars can be used for chord reinforcement in the E-W direction (refer to Fig. E1.8). | $A_{s,prov.} = (2)(0.31 \text{ in.}^2) = 0.62 \text{ in.}^2 > 0.1 \text{ in.}^2$  **OK** |
| | Maximum shear in the E-W direction occurs at CLs 1 and 7: Unit shear force in frame: $$v_{u@1,7} = \frac{F_{u@1,7}}{L}$$ | $$v_{u@1,7} = \frac{70 \text{ kip}}{218 \text{ ft}} = 0.32 \text{ kip/ft}$$ |

## Step 13: Collector reinforcement

| | | |
|---|---|---|
| | Collector along CLs A and F: The continuous reinforced concrete frame over the full length of the building acts as a collector. | |
| | Note: Provide continuous reinforcement with tension splices (Step 15). | |
| 12.5.3.7 | In cast-in-place diaphragms, where shear is transferred from the diaphragm to a collector, or from the diaphragm or collector to a shear wall, temperature and shrinkage reinforcement is usually adequate to transfer that force. | |

| | | |
|---|---|---|
| **Step 14: Shrinkage and temperature reinforcement** | | |
| 12.6.1<br>24.4.3.2 | The minimum area of shrinkage and temperature reinforcement, $A_{S+T}$:<br><br>$A_{S+T} \geq 0.0018 A_g$ | $A_{S+T} = (0.0018)(7 \text{ in.})(12 \text{ in./ft}) = 0.15 \text{ in.}^2/\text{ft}$ |
| 24.4.3.3 | Spacing of S+T reinforcement is the lesser of $5h$ and 18 in.:<br><br>(a) $5h = 5(12 \text{ in.}) = 60 \text{ in.}$<br>(b) 18 in.  **Controls** | Note: Shrinkage and temperature reinforcement may be part of the reinforcing bars resisting diaphragm in-plane forces and gravity loads. If provided reinforcement is not continuous (placing bottom reinforcing bars to resist positive moments at mid-spans and top reinforcing bars to resist negative moments at columns), continuity between top and bottom reinforcing bars can sometimes be achieved by providing adequate splice lengths between them. |
| **Step 15: Reinforcement detailing** | | |
| 12.7.2.1 | Reinforcement spacing<br>Minimum and maximum spacing of chord and collector reinforcement must satisfy 12.7.2.1 and 12.7.2.2. | |
| 25.2.1 | Section 25.2 limits minimum spacing of<br>(a) 1 in.<br>(b) $4/3 d_{agg.}$ ($d_{agg} = 3/4$ in.)<br>(c) $d_b$ (No. 5) | Minimum spacing 1.0 in.  **Controls**<br>$(4/3)(3/4 \text{ in.}) = 1.0 \text{ in.}$<br>0.625 in. |
| 18.12.7.6a | The minimum collector reinforcement spacing at a splice must be at least the largest of:<br>(a) Three longitudinal $d_b$<br>(b) 1.5 in.<br>(c) $c_c \geq \max[2.5 d_b, 2 \text{ in.}]$ | $3(0.625 \text{ in.}) = 1.875 \text{ in.}$<br>1.5 in.<br>2 in.  **Controls** |
| 12.7.2.2 | Maximum spacing is the smaller of $5h$ or 18 in. | 18 in.  **Controls** |

## Step 16: Details

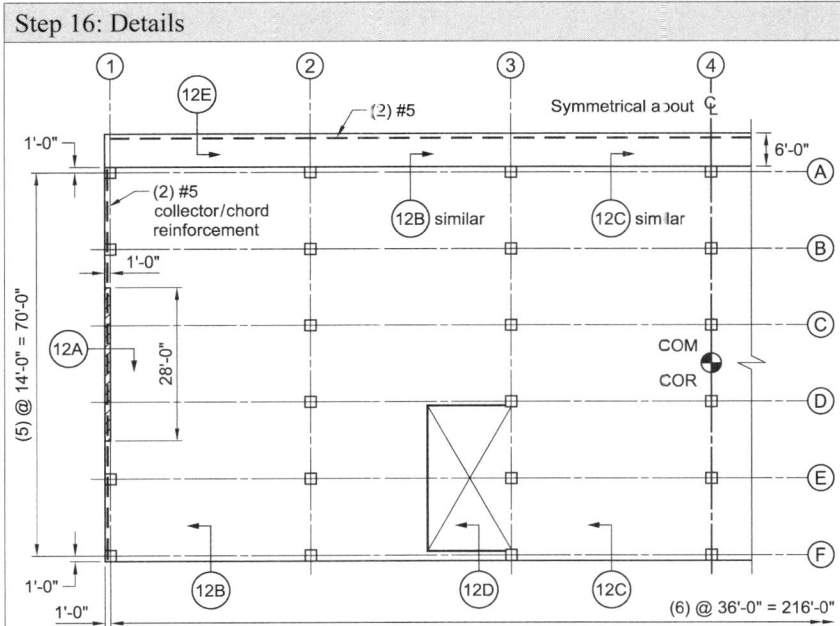

*Fig. E1.11—Diaphragm chord and collector reinforcement.*

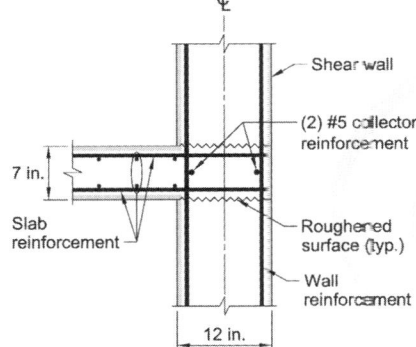

*Section 12A—Chord and collector reinforcement at east and west ends of the diaphragm.*

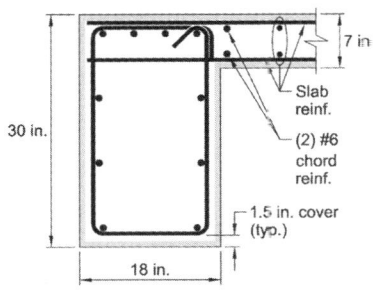

*Section 12B—Chord reinforcement at midspan.    Section 12C—Chord reinforcement at support.*

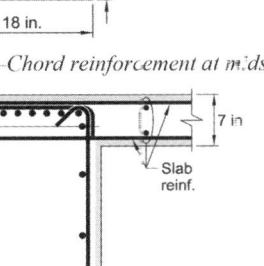   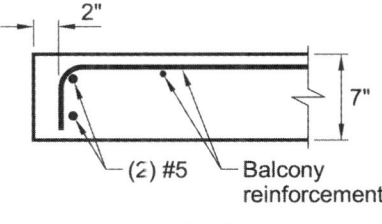

*Section 12D—Chord reinforcement within the beam at midspan and support at opening location.*    *Section 12E—Chord reinforcement at overhang.*

Note: Shrinkage and temperature reinforcement not shown for clarity.

**Rigid Diaphragm Example 2:** *Reinforced concrete diaphragm with opening*—Refer to Diaphragm Example 1 for structure and design data. Analyze and design the second level floor diaphragm with a 14 ft 0 in. x 36 ft 0 in. opening as shown in Fig. E2.1. For diaphragm building elevation, material properties, and design criteria, refer to Diaphragm Example 1.

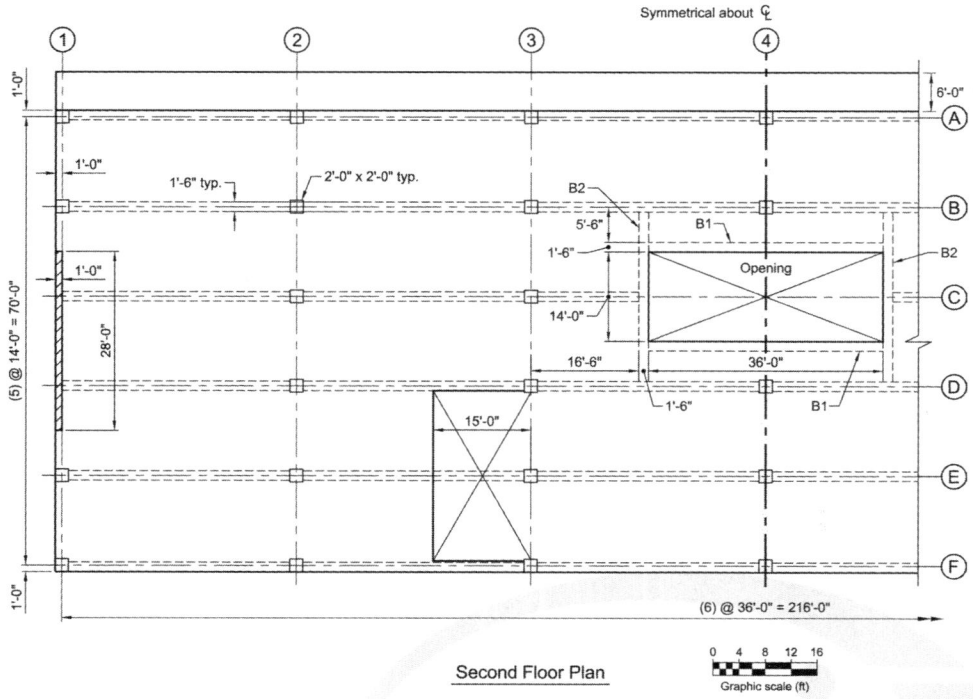

*Fig. E2.1—Eight story building.*

| ACI 318-14 | Discussion | Calculation |
|---|---|---|
| Step 1: Material requirements | | |
| | Refer to Diaphragm Example 1. | |
| Step 2: Slab geometry | | |
| | Satisfied per Diaphragm Example 1. | |
| Step 3: Lateral forces | | |
| For lateral forces and design forces calculations, refer to Diaphragm Example 1, Step 3. North-South (N-S): 81 kip, although wind load controls (113 kip), the diaphragm will be designed for the seismic load in this example. East-West (E-W): 116 kip | | |

## Step 4: Center of mass (COM) and center of rigidity (COR)

Design second-level diaphragm:
Take the point of origin at F1 (Fig. E2.2).

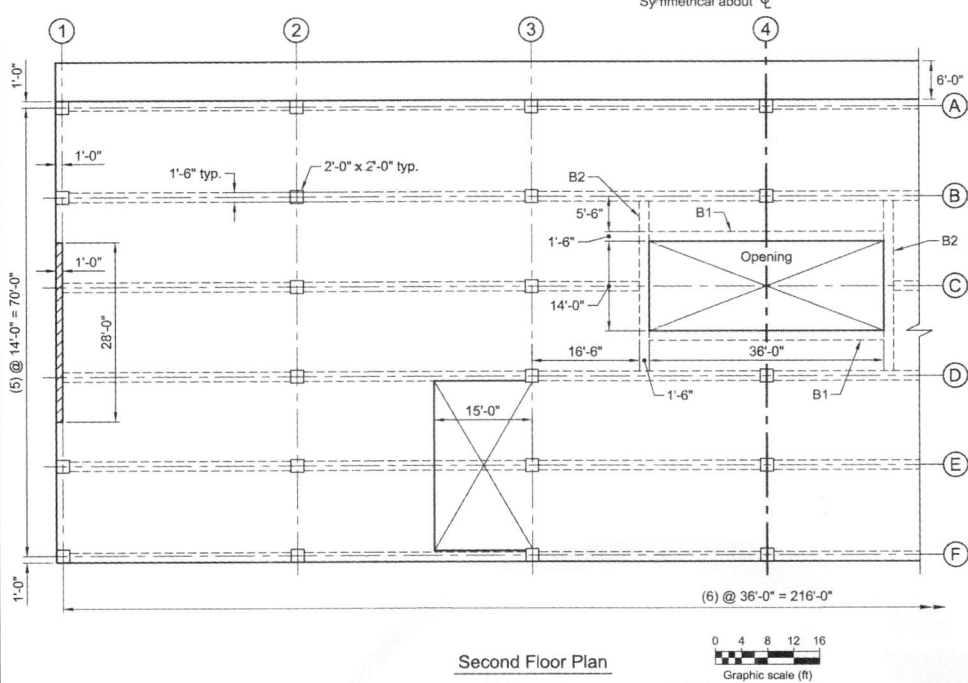

Fig. E2.2—Mass center and rigidity center location excluding accidental torsion.

Determine COM
The COM has shifted to the south because of the opening. Taking the moment area around column line (CL) F:

$$y_{COM} = \frac{(78 \text{ ft})(218 \text{ ft})(37 \text{ ft}) - (36 \text{ ft})(14 \text{ ft})(42 \text{ ft})}{(78 \text{ ft})(218 \text{ ft}) - (36 \text{ ft})(14 \text{ ft})} = 36.8 \text{ ft}$$

Therefore, the COM is located at 108 ft-0 in. east of CL. 1 and 36.8 ft north of CL. F.

Determine COR
Because the lateral resisting systems are symmetrical about both axes, the COR is located at:
$x_{COR} = 218 \text{ ft}/2 - 1 \text{ ft} = 108 \text{ ft}$ and $y_{COR} = 78 \text{ ft}/2 - 1 \text{ ft} = 38 \text{ ft}$ from east of and north of Column Line F1.
$\Delta_y = y_{COR} - y_{COM} = 38 \text{ ft} - 36.8 \text{ ft} = 1.2 \text{ ft}$

Accidental torsion
ASCE 7-10 (third edition), commentary Section C12.10.1 requires an additional moment caused by an assumed displacement of COM. A shift of minimum of five percent of the building dimension perpendicular to the direction of seismic forces in addition to the actual eccentricity is considered, referred to as accidental eccentricity.

$e_x = \pm(0.05)(218 \text{ ft}) = \pm10.9 \text{ ft}$
$e_{y1} = (0.05)(78 \text{ ft}) = 3.9 \text{ ft}$
$e_{y2} = -(0.05)(78 \text{ ft}) = -3.9 \text{ ft}$

## Step 5: Lateral system stiffness calculations

For wall and moment frame stiffness calculations refer to Diaphragm Example 1, Step 6.

## Step 6: Lateral resisting system forces

Force in walls and moment-resisting frames are given by the following equations:

$$F_{uxi} = \frac{k_{ix}}{\sum k_{ix}} F_{px} \pm \frac{k_i d_i}{\sum k_i d_i^2} F_{px} e_y$$

$$F_{uyi} = \frac{k_{iy}}{\sum k_{iy}} F_{py} \pm \frac{k_i d_i}{\sum k_i d_i^2} F_{py} e_x$$

where $d_i$ is the distance ($x_i$ or $y_i$) of each wall from the COR.

$F_{px,y} = 80$ kip and $F_{px,x} = 114$ kip are second-story lateral forces obtained from Example 1, Step 3 (Table), N-S and E-W directions, respectively.

Mass accidental eccentricities are $e_x = 10.9$ ft; $e_{y1} = 3.9$ ft + 1.2 ft = 5.1 ft; and $e_{y2} = -3.9$ ft + 1.2 ft = –2.7 ft are calculated in Step 4 of this example.

The torsional force is calculated by multiplying the lateral inertia force by the corresponding eccentricity:
NS: $T_y = F_{py} e_x = (81 \text{ kip})(\pm 10.9 \text{ ft}) = \pm 883$ ft-kip
EW: $T_{x1} = F_{px} e_{y1} = (116 \text{ kip})(+5.1 \text{ ft}) = +592$ ft-kip
$T_{x2} = F_{px} e_{y2} = (116 \text{ kip})(-2.7 \text{ ft}) = -313$ ft-kip

Lateral force applied in N-S direction
$F_{py} = 80$ kip

$$F_{u,wall,max} = \frac{10.5}{(10.5+10.5)}(81 \text{ kip}) + \frac{(10.5)(218 \text{ ft}/2)}{(2)[(1.0)(39 \text{ ft})^2 + (10.5)(109 \text{ ft})^2]}(883 \text{ ft-kip}) = 44.5 \text{ kip}$$

$$F_{u,wall,min} = \frac{8.1}{(10.5+10.5)}(81 \text{ kip}) - \frac{(10.5)(218 \text{ ft}/2)}{(2)[(1.0)(39 \text{ ft})^2 + (10.5)(109 \text{ ft})^2]}(883 \text{ ft-kip}) = 36.5 \text{ kip}$$

$$F_{u,MF} = \pm \frac{(1.0)(78 \text{ ft}/2)}{(2)[(1.0)(39 \text{ ft})^2 + (10.5)(109 \text{ ft})^2]}(883 \text{ ft-kip}) = \pm 0.2 \text{ kip}$$

Lateral force applied in E-W direction

(a) $F_{px} = 116$ kip and $e_{y1} = 5.1$ ft

$$F_{u,MF,max} = \frac{1.0}{(1.0+1.0)}(116 \text{ kip}) + \frac{(1.0)(78 \text{ ft}/2)}{(2)[(1.0)(39 \text{ ft})^2 + (10.5)(109 \text{ ft})^2]}(592 \text{ ft-kip}) = 58.1 \text{ kip}$$

$$F_{u,MF,min} = \frac{1.0}{(1.0+1.0)}(116 \text{ kip}) - \frac{(1.0)(78 \text{ ft}/2)}{(2)[(1.0)(39 \text{ ft})^2 + (10.5)(109 \text{ ft})^2]}(592 \text{ ft-kip}) = 57.9 \text{ kip}$$

$$F_{u,wall} = \pm \frac{(10.5)(109 \text{ ft})}{(2)[(1.0)(39 \text{ ft})^2 + (10.5)(109 \text{ ft})^2]}(592 \text{ ft-kip}) = \pm 2.7 \text{ kip}$$

(b) $F_{px} = 116$ kip and $e_{y2} = 2.7$ ft

$$F_{u,MF,max} = \frac{1.0}{(1.0+1.0)}(116 \text{ kip}) + \frac{(1.0)(78 \text{ ft}/2)}{(2)[(1.0)(39 \text{ ft})^2 + (10.5)(109 \text{ ft})^2]}(313 \text{ ft-kip}) = 58.1 \text{ kip}$$

$$F_{u,MF,min} = \frac{1.0}{(1.0+1.0)}(116 \text{ kip}) - \frac{(1.0)(78 \text{ ft}/2)}{(2)[(1.0)(39 \text{ ft})^2 + (10.5)(109 \text{ ft})^2]}(313 \text{ ft-kip}) = 57.9 \text{ kip}$$

$$F_{u,wall} = \pm \frac{(10.5)(218 \text{ ft}/2)}{(2)[(1.0)(39 \text{ ft})^2 + (10.5)(109 \text{ ft})^2]}(313 \text{ ft-kip}) = 1.4 \text{ kip}$$

The force distribution in the E-W direction for both calculated eccentricities is small. Therefore, use 58.0 kip.

# CHAPTER 8—DIAPHRAGMS

## Step 7: Check shear force in diaphragm

| | | |
|---|---|---|
| 12.5.3.3 | **In-plane shear in diaphragm** <br> The diaphragm slab is cast-in-place concrete, therefore, shear force is calculated from Eq. (12.5.3.3) <br><br> $V_n = A_{cv}\left(2\lambda\sqrt{f'_c} + \rho_t f_y\right)$ <br><br> Ignoring the strength contribution of reinforcement; $\rho_t = 0$ <br> $A_{cv}$ is the diaphragm gross area less the 6.0 ft overhang (refer to Step 9 for clarification). | **Nominal shear strength in E-W direction** <br><br> $V_{n,N} = (218\text{ ft})(12\text{ in./ft})(7\text{ in.})\left(2(1.0)\sqrt{5000\text{ psi}} + 0\right)$ <br> $= 2{,}589{,}708\text{ lb} \approx 2590\text{ kip}$ <br><br> **Nominal shear strength in N-S direction** <br><br> $V_{n,E} = (72\text{ ft})(12\text{ in./ft})(7\text{ in.})\left(2(1.0)\sqrt{5000\text{ psi}} + 0\right)$ <br> $= 855{,}316\text{ lb} \approx 855\text{ kip}$ |
| 12.5.3.2 <br> 21.2.4.2 | Applying the shear strength reduction factor $\phi = 0.75$ at the north and south ends along column lines A and F. | **Design shear strength in E-W direction** <br> $\phi V_{n,N} = (0.75)(2590\text{ kip}) = 1940\text{ kip}$ |
| 12.5.3.2 <br> 21.4.2.1 | At the east and west ends along Column Lines 1 and 7, the shear strength reduction factor, $\phi$, must not exceed the least value for shear used for the vertical components of the primary seismic-force-resisting system. Therefore, $\phi = 0.75$: | **Design shear strength in N-S direction** <br> $\phi V_{n,E} = (0.75)(855\text{ kip}) = 641\text{ kip}$ |
| 12.5.1.1 <br> 22.5.1.2 | Check if factored shear force is less than design shear strength calculated in Step 7. | NS: $\phi V_n = 641\text{ kip} \gg F_u = 44.5\text{ kip}$  **OK** <br> EW: $\phi V_n = 1940\text{ kip} \gg F_u = 58.0$  **OK** <br><br> Therefore, diaphragm has adequate strength to resist the lateral inertia force and shear reinforcement is not required; $\rho_t = 0$. |
| 12.5.3.4 | The nominal shear strength, $V_n$, must not exceed: <br><br> $V_n = 8 A_{cv} \lambda \sqrt{f'_c}$ <br><br> $A_{cv}$ is the diaphragm gross area less the 6.0 ft overhang. | $V_n = \dfrac{8(72\text{ ft})(10\text{ in.})(12\text{ in./ft})(1.0)\sqrt{4000\text{ psi}}}{1000\text{ lb/kip}} = 4372\text{ kip}$ <br><br> By inspection this is satisfied.  **OK** |

## Step 8: Second-level diaphragm lateral force distribution N-S

| | |
|---|---|
| | Design force: 81 kip |
| 12.4.2.4(a) 12.5.1.3(a) | Diaphragm is idealized as rigid. Design moments and shear axial forces are calculated based on a beam with depth equal to full diaphragm depth satisfying equilibrium requirements. |
| | The wall forces and the assumed direction of torque due to the eccentricity are shown in Fig. E2.3. |
| 12.4.2.4 | The distribution of the applied force on the diaphragm is calculated by using $q_L$ and $q_R$ as the left and right diaphragm reactions per unit length (Fig. E2.3): |

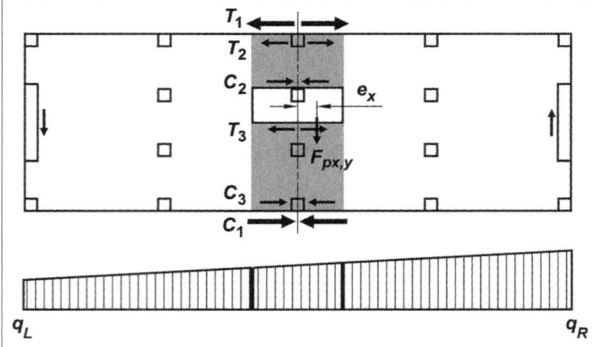

Fig. E2.3—Forces in the structural resisting systems due to a seismic force in the N-S direction.

Force equilibrium

$$q_L\left(\frac{L}{2}\right) + q_R\left(\frac{L}{2}\right) = F_{px,des(NS)}$$

$$q_L\left(\frac{218\text{ ft}}{2}\right) + q_R\left(\frac{218\text{ ft}}{2}\right) = 81\text{ kip} \quad \text{(I)}$$

Moment equilibrium

$$q_L\left(\frac{L}{2}\right)\left(\frac{L}{3}\right) + q_R\left(\frac{L}{2}\right)\left(\frac{2L}{3}\right) = F_{px,des(NS)}\left(\frac{L}{2} + 0.05L\right)$$

$$q_L\frac{(218\text{ ft})^2}{6} + q_R\frac{2(218\text{ ft})^2}{6} = (81\text{ kip})\left(\frac{218\text{ ft}}{2} + 10.9\text{ ft}\right) \quad \text{(II)}$$

From statics solve equations (I) and (II) for $q_L$ and $q_R$:

$q_L + q_R = 0.74$ kip/ft (I)
$q_L + 2q_R = 1.23$ kip/ft (II)
$q_L = 0.25$ kip/ft and $q_L = 0.49$ kip/ft

Find the maximum moment by taking the first derivative of the moment equation expressed as a function of $x$ (unknown distance) $dM/dx = 0$

$x = 114.45$ ft; $M_{max} = 2213$ ft-kip

Draw the shear and moment diagrams and determine the moment and shear forces at opening (Fig. E2.4).

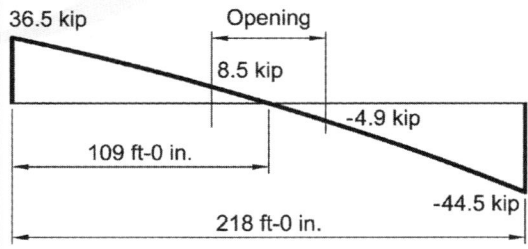

Note: In an Aug. 2010 National Institute and Standards Technology (NIST) report, GCR 10-917-4, "Seismic Design of Cast-in-Place Concrete Diaphragms, Chords, and Collectors," by Moehle et al. states that, "This approach leaves any moment due to the frame forces along column lines A and F unresolved. Sometimes this is ignored or, alternatively, it too can be incorporated in the trapezoidal loading."

In this example the small moment due to the frame forces (0.2 kip) are ignored.

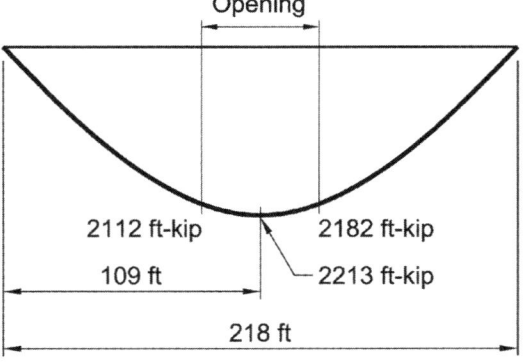

Fig. E2.4—Shear and moment diagrams.

| | | |
|---|---|---|
| | Note: Experienced engineers sometimes simplify the calculations by using uniformly distributed load: | $q = \dfrac{81 \text{ kip}}{218 \text{ ft}} = 0.372 \text{ kip/ft}$ |
| | Resulting in a maximum moment of: | $M_{max} = \dfrac{(0.372 \text{ kip/ft})(218 \text{ ft})^2}{8} = 2210 \text{ ft-kip}$ |
| | Note: Both approaches, in this example, result in similar maximum moment (2213 ft-kip versus 2210 ft-kip), but at different locations (114.5 ft versus 109 ft). In this example the detailed approach is presented. | |
| **Step 9: Chord reinforcement N-S** | | |
| 12.5.2.3 | Maximum moment obtained from moment diagram: | $M_u = 2213$ ft-kip |
| | Chord reinforcement resisting tension must be located within $h/4$ of the tension edge of the diaphragm. | $h/4 = 72.0 \text{ ft}/4 = 18 \text{ ft}$ |
| | Note: Chord reinforcement can be placed either in the exterior edge of the balcony or it can be placed along the exterior frame of CL A.  Placing chord reinforcement along the exterior frame is a simpler and cleaner load path for the forces in the diaphragm.  Crack control reinforcement should be added in the balcony slab for crack control. | |
| | Assume tension reinforcement is placed in a 2 ft strip at both north and south sides of the slab edges at CLs 1 and 5.  Chord force  Maximum chord tension force that must be resisted by the chord at midspan is:  $T_u = \dfrac{M_u}{B - 2 \text{ ft}}$ | 2 ft < $h/4 = 18$ ft   **OK**  $T_u = \dfrac{2213 \text{ ft-kip}}{72 \text{ ft} - 2 \text{ ft}} = 31.6 \text{ kip}$ |

## Chord forces at opening

The opening in the diaphragm results in local bending of the diaphragm segments on either side of the opening (refer to Fig. E2.5).

1. The diaphragm sections above and below the opening are idealized as fixed end beams.
2. The applied loading on the sections above and below the opening are based on the relative mass of each section (1.64:1).
3. The secondary chord forces are calculated based on the internal moment in the diaphragm sections adjacent to the opening.
4. The calculated tension and compression secondary chord forces are added to the primary tension and secondary chord forces.

The opening is located at midlength of the building floor plan in the E-W direction. The total diaphragm forces at left and right edges of the opening are:

The load on the north and south section of the diaphragm bound by the opening is distributed according to the ratio of the masses north and south of the opening. Therefore, 38 percent and 62 percent of the overall applied trapezoidal load will be distributed to the north and south section over this portion of the diaphragm, respectively.

The unit forces magnitude at the east and west ends of the opening are close (0.13 kip/ft versus 0.15 kip/ft north of opening) and (0.22 kip/ft versus 0.24 kip/ft south of opening). Therefore, the average unit force of 0.14 kip/ft and 0.23 kip/ft will be used for calculating the diaphragm moment segments north and south of the opening (Fig. E2.6).

Fixed end moment can be obtained from computer-aided design software programs or from the Reinforced Concrete Design Handbook Design Aid – Analysis Tables, which can be downloaded from: https://www.concrete.org/store/productdetail.aspx?ItemID=SP1714DA

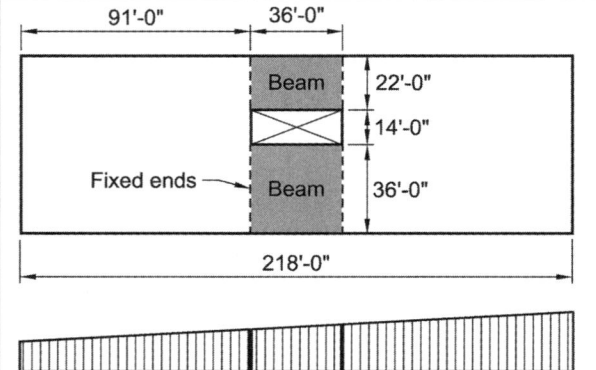

Fig. E2.5—Idealization of sections above and below opening.

Force at east and west ends of opening

$$q'_{u@beg} = 0.25 \text{ kip/ft} + \frac{(0.49 \text{ kip/ft} - 0.25 \text{ kip/ft})(91 \text{ ft})}{218 \text{ ft}}$$
$$= 0.35 \text{ kip/ft}$$

$$q'_{u@end} = 0.25 \text{ kip/ft} + \frac{(0.49 \text{ kip/ft} - 0.25 \text{ kip/ft})(127 \text{ ft})}{218 \text{ ft}}$$
$$= 0.39 \text{ kip/ft}$$

Force north of opening
$q'_{u@bN} = (0.38)(0.35 \text{ kip/ft}) = 0.13 \text{ kip/ft}$
$q'_{u@eN} = (0.38)(0.39 \text{ kip/ft}) = 0.15 \text{ kip/ft}$

Force south of opening
$q'_{u@bS} = (0.62)(0.35 \text{ kip/ft}) = 0.22 \text{ kip/ft}$
$q'_{u@eS} = (0.62)(0.39 \text{ kip/ft}) = 0.24 \text{ kip/ft}$

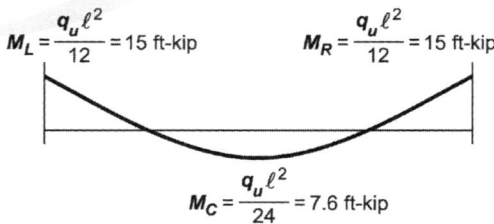

$M_L = \dfrac{q_u \ell^2}{12} = 15 \text{ ft-kip}$    $M_R = \dfrac{q_u \ell^2}{12} = 15 \text{ ft-kip}$

$M_C = \dfrac{q_u \ell^2}{24} = 7.6 \text{ ft-kip}$

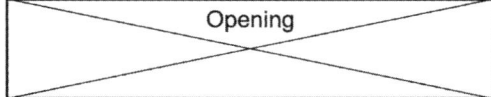

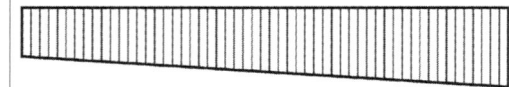

$M_L = \dfrac{q_u \ell^2}{12} = 25 \text{ ft-kip}$    $M_R = \dfrac{q_u \ell^2}{12} = 25 \text{ ft-kip}$

$M_C = \dfrac{q_u \ell^2}{24} = 12.4 \text{ ft-kip}$

Fig. E2.6—Moment diagram of sections at opening.

# CHAPTER 8—DIAPHRAGMS

| Code Ref. | Discussion | Calculation |
|---|---|---|
| 12.5.1.1<br>22.4.3.1 | The secondary chord forces are obtained from the moment diagram. Assuming a 1 ft strip ($< B/4$) at each end of the span between opening and diaphragm edge:<br><br>$$T_{u,opening} = \frac{M_u}{D}$$<br><br>Required reinforcement<br>$\phi T_n = \phi f_y A_s \geq T_u$ | Secondary chord force north of opening:<br><br>$$T^-_{u,op,N} = \frac{15 \text{ ft-kip}}{21 \text{ ft}} = 0.7 \text{ kip}$$<br><br>Secondary chord force south of opening:<br><br>$$T^-_{u,op,S} = \frac{25 \text{ ft-kip}}{35 \text{ ft}} = 0.7 \text{ kip}$$<br><br>$$A_{s,req'd} = \frac{700 \text{ lb}}{(0.9)(60,000 \text{ psi})} = 0.01 \text{ in.}^2$$<br><br>Refer to the following discussion. |
| | The engineer has two options for providing chord reinforcement at the opening:<br>  1. The beams are designed for $1.2D + 1.6L$. For seismic, the governing load combination is $(1.2 + 0.2S_{DS})D + 0.5L + E$. The demand from $1.2D + 1.6L$ is usually higher than the gravity portion of the moments under seismic, $(1.2 + 0.2S_{DS})D + 0.5L$. The two loads are proportioned and then the balance reinforcement is used to carry seismic chord/collector forces.<br>  2. The chord force is resisted with additional reinforcement (conservative).<br><br>In this example the first option is used as the required reinforcement is negligible. Beam top reinforcement is continuous. | |
| | Diaphragm edge<br>Total moment to be resisted is the sum of the main chord force and the secondary chord force:<br>$T_{u,Total} = T_{u1} + T_{u2}$<br><br>$T_{u,Total} = T_{u1} + T_{u3}$ | $T_{u,total,N} = C_{u,total,N} = 31.1 \text{ kip} - 0.7 \text{ kip} = 31.8 \text{ kip}$, say, 32 kip<br><br>$T_{u,total,S} = C_{u,total,S} = 31.1 \text{ kip} + 0.7 \text{ kip} = 31.8 \text{ kip}$, say, 32 kip |
| 12.5.2.2 | Chord reinforcement:<br>Tension due to moment is resisted by deformed bars conforming to Section 20.2.1 of ACI 318-14. | |
| 12.5.1.5 | Steel stress is the lesser of the specified yield strength and 60,000 psi. | $f_y = 60,000$ psi |
| 12.5.1.1<br>22.4.3.1 | Required reinforcement<br>$\phi T_n = \phi f_y A_s \geq T_u$ | $$A_{s,req'd} = \frac{32,000 \text{ lb}}{(0.9)(60,000 \text{ psi})} = 0.59 \text{ in.}^2$$ |
| | The building is assigned to SDC B. therefore, ACI 318-14 Chapter 18 requirements for chord spacing and transverse reinforcement of 18.12.7.6 do not apply. | |
| 18.12.7.5 | Overstrength factor, $\Omega_o$, for chord design is not required. Therefore, use the compression stress limit in Provision 18.12.7.5 of $0.2f_c'$<br>Required chord width:<br><br>$$w_{chord} > \frac{C_{Chord}}{0.2f_c' h_{diaph}}$$<br><br>Choose reinforcement:<br><br>Check if provided reinforcement area is greater than required reinforcement area: | $$w_{chord} > \frac{32,000 \text{ lb}}{(0.2)(5000 \text{ psi})(7 \text{ in.})} = 4.6 \text{ in.}$$<br><br>Less than the assumed 2 ft. **OK**<br>Try two No. 5 chord bars<br>$A_{s,prov.} = 2(0.31 \text{ in.}^2) = 0.62 \text{ in.}^2$<br><br>$A_{s,prov.} = 0.62 \text{ in.}^2 > A_{s,req'd} = 0.59 \text{ in.}^2$ |

| | | |
|---|---|---|
| **Step 10: Collector reinforcement N-S** | | |
| 12.5.4.1 | Collectors transfer shear forces from the diaphragm to the vertical walls at both east and west ends along column lines 1 and 7 (Fig. E2.2). Collectors extend over the entire diaphragm width.<br>Unit shear force:<br><br>$$v_{u@F} = \frac{F_{u@F}}{B}$$<br><br>In diaphragm: $v_{u@F} = \frac{F_{px}}{L_{diaph}}$<br><br>In wall: $v_{u@F} = \frac{F_{px}}{L_{wall}}$<br><br>Force at diaphragm to wall connection<br>Wall west end:<br>$F_{7/D.5} = -(0.62 \text{ kip/ft})(22 \text{ ft}) = -13.6 \text{ kip}$<br>East wall north end:<br>$F_{7/B.5} = -13.6 \text{ kip} + (0.97 \text{ kip/ft})(28 \text{ ft}) = 13.5 \text{ kip}$<br>At diaphragm end:<br>$F_{7/A} = +13.5 \text{ kip} - (0.62 \text{ kip/ft})(22 \text{ ft}) \approx 0 \text{ kip}$<br><br>Per collector force diagram, the maximum axial force on the collector is $T_u = C_u = 13.5$ kip. This force must be transferred from the diaphragm to the shear wall (Fig. E2.7).<br><br>The collector force and its connections to the shear wall will not be multiplied by the system overstrength factor $\Omega_o = 2.5$ (ASCE/SEI 7-10, Table 12.2-1), because this is not a special structural wall. | From Step 6: $F_u = 44.5$ kip<br><br>$$v_{u@F} = \frac{44.5 \text{ kip}}{72 \text{ ft}} = 0.62 \text{ kip/ft}$$<br><br>$$v_{u@F} = \frac{44.5 \text{ kip}}{28 \text{ ft}} = 1.59 \text{ kip/ft}$$<br><br>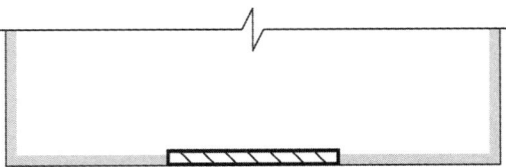<br><br>Unit shear forces:<br>0.62 kip/ft<br><br>1.59 kip/ft<br><br>Net shear forces:<br>0.62 kip/ft<br><br>0.97 kip/ft<br><br>Collector force: 13.6 kip<br>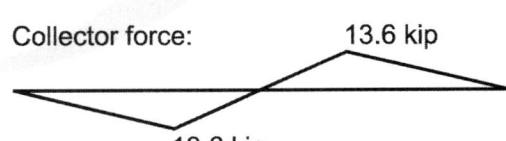<br>13.6 kip<br><br>Fig. E2.7—Collector force diagram. |
| 12.5.4.2 | Collectors are designed as tension members, compression members, or both.<br><br>Tension is resisted by reinforcement as calculated above.<br><br>Required reinforcement: | |
| 12.5.1.1<br>22.4.3.1 | $\phi T_n = \phi f_y A_s \geq T_u$ | $$A_{s,req'd} = \frac{T_u}{0.9 f_y} = \frac{13,600 \text{ lb}}{(0.9)(60,000 \text{ psi})} = 0.25 \text{ in.}^2$$<br><br>Although one No. 5 bar suffices, two No. 5 are provided to maintain symmetry. |

| | | |
|---|---|---|
| 18.12.7.5 | Check if collector compressive force exceeds $0.2f_c'$. Calculate minimum required collector width using $0.2f_c'$<br><br>$$w_{coll.} > \frac{C_{Coll.}}{0.2 f_c' t_{diaph}}$$<br><br>This results in compressive stress on the concrete diaphragm collector being relatively low. The section is adequate to transfer shear stress without additional reinforcement.<br><br>The collector compression and tension forces are transferred to the lateral force-resisting system within its width (shear wall). Therefore, no eccentricity is present and no in-plane bending occurs.<br><br>ACI 318 permits to discontinue the collector along the length of the shear wall where transfer of design collector is not required. | $$w_{coll} > \frac{13{,}600 \text{ lb}}{(0.2)(5000 \text{ psi})(7 \text{ in.})} = 1.9 \text{ in.}$$<br><br>Use 12 in. wide collector (same width as shear wall).<br><br>Provide two No. 5 bars at mid-depth of slab to prevent additional out-of-plane bending stresses in the slab. Space the two No. 5 bars at 8 in. on center starting at 2 in. from the edge of the diaphragm within the 12 in. wide collector/shear wall (Fig. E2.8).<br><br>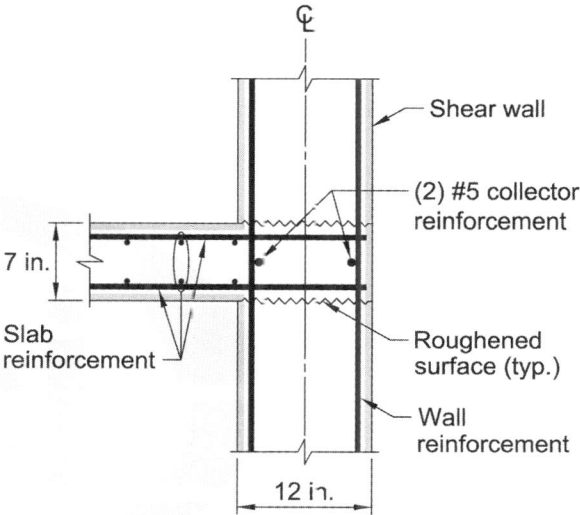<br><br>*Fig. E2.8—Collector reinforcement.* |
| 12.4.2.4<br><br>12.5.3.3<br>21.2.4.2<br><br>12.5.1.1<br><br>12.5.3.4 | <u>Check slab shear strength along shear walls</u><br>Slab shear strength along walls:<br>$L = 28$ ft and slab thickness $t = 7$ in. From<br>$\phi V_c = \phi A_{cv} 2\lambda \sqrt{f_c'}$<br>$\phi = 0.75$<br><br>Is the provided shear strength adequate?<br><br>By inspection, the diaphragm shear design force satisfies the requirement of 12.5.3.4.<br>$\phi V_c = \phi A_{cv} 8\lambda \sqrt{f_c'}$ | $\phi V_c = (0.75)(2)(1.0)(\sqrt{5000})(28 \text{ ft})(12)(7 \text{ in.})$<br>$\phi V_c = 249{,}467 \text{ lb} \sim 249 \text{ kip}$<br><br>$\phi V_c = 249 \text{ kip} > V_u = 44.5 \text{ kip}$ (from Step 7)    **OK** |

| Step 11: Lateral force distribution in diaphragm E-W | | |
|---|---|---|
| 12.4.2.4(a) 12.5.1.3(a) | Design force: 114 kip<br><br>Design moments, shear, and axial forces are calculated based on a beam with depth equal to full diaphragm length satisfying equilibrium requirements.<br><br>The wall forces and the assumed direction of torque due to the eccentricity are shown in Fig. E2.9.<br><br>The distribution of the applied force on the diaphragm is uniform because of the negligible effect of accidental torsion (Fig. E2.9): | 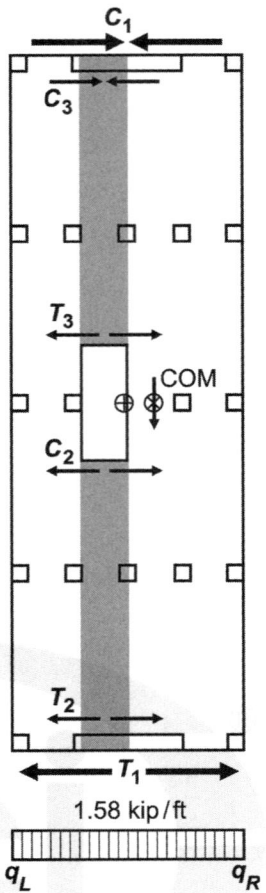<br>Fig. E2.9—Forces in the structural resisting systems due to a seismic force in the E-W direction.<br><br>$q_L = \left(\dfrac{116 \text{ kip}}{72 \text{ ft}}\right) = 1.6$ kip/ft<br><br>Shear force: $V = (1.6 \text{ kip.ft})(36 \text{ ft}) = 58$ kip<br><br>$x = 36$ ft; $M_{max} = 1131$ ft-kip |

| | | |
|---|---|---|
| | Maximum moment is taken at midspan.<br>Draw the shear and moment diagrams (Fig. E2.10). | 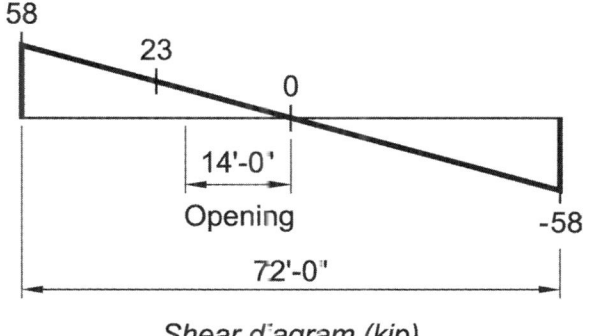<br>*Shear diagram (kip)*<br>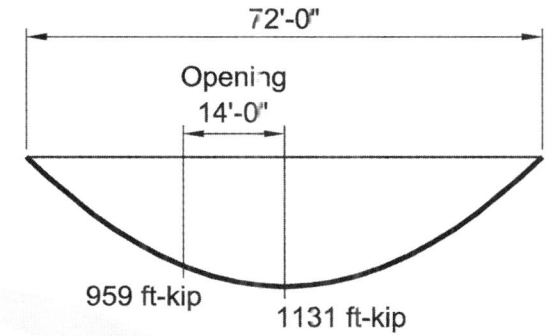<br>*Moment diagram*<br>*Fig. E2.10—Shear and moment diagrams.* |
| **Step 12: Chord reinforcement E-W** | | |
| 12.5.2.3 | Maximum moment is calculated above:<br><br>Chord reinforcement resisting tension must be located within $h/4$ of the tension edge of the diaphragm.<br><br>Assume tension reinforcement is placed in a 1 ft strip at both north and south sides of the slab edges at CLs 1 and 5.<br><br>Chord force<br>Maximum chord tension force that must be resisted by the chord at midspan:<br><br>$$T_u = \frac{M_u}{B - 1 \text{ ft}}$$ | $M_u = 1131$ ft-kip<br><br>$h/4 = 218.0$ ft/4 = 54.5 ft<br><br>1 ft < $h/4 = 54.5$ ft   **OK**<br><br>$$T_{u,1} = \frac{1131 \text{ ft-kip}}{(218 \text{ ft} - 1 \text{ ft})} = 5.2 \text{ kip}$$ |

### Chord forces at opening

The opening in the diaphragm results in local bending of the diaphragm segments on either side of the opening (Fig. E2.11).

1. The diaphragm sections to the east and west of the opening are idealized as fixed end beams.
2. The applied loading on the sections east and west the opening are based on the relative mass of each section (1:1).
3. The secondary chord forces are calculated based on the internal moment in the diaphragm sections adjacent to the opening.
4. The calculated tension and compression secondary chord forces are added to the primary tension and secondary chord forces.

The opening is located at mid-length of the building floor plan in the E-W direction. The load on the north and south sections of the diaphragm bound by the opening are equal to one half of the overall applied trapezoidal load over this portion of the diaphragm (Fig. E2.12).

Because forces at both ends of openings are close, a uniform load is assumed.

Fixed end moment can be obtained from computer-aided aided design software programs or from Reinforced Concrete Design Handbook Design Aid – Analysis Tables, which can be downloaded at: https://www.concrete.org/store/productdetail.aspx?ItemID=SP1714DA

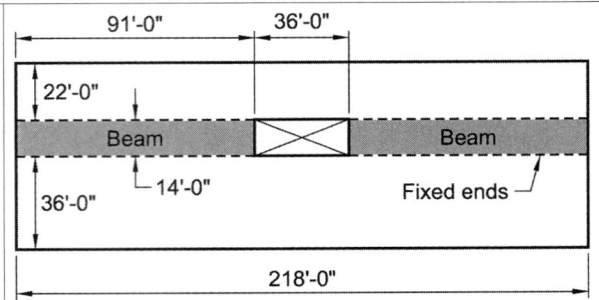

Fig. E2.11—Idealization of sections above and below opening.

Force east and west of opening

$$q'_{u@bE} = q'_{u@bW} = \left(\frac{1.6 \text{ kip/ft}}{2}\right) = 0.8 \text{ kip/ft}$$

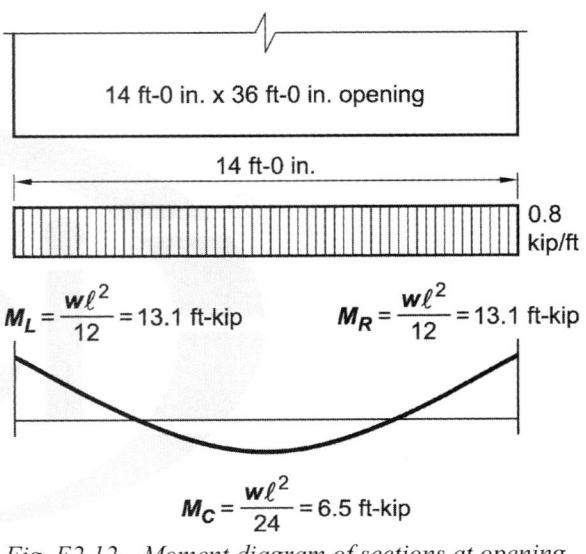

Fig. E2.12—Moment diagram of sections at opening.

---

The secondary chord forces are obtained from the moment diagram. Assuming a 2 ft strip (< B/4) at each end of the span between opening and diaphragm edge:

$$T_{u,opening} = M_u/D$$

$$T^+_{u,op,S} = \frac{13.1 \text{ ft-kip}}{89 \text{ ft}} = 0.15 \text{ kip}$$

Total moment to be resisted is the sum of the main chord force and the secondary chord force:

$$T_{u,total} = T_{u1} + T_{u,O2}$$

$T_{u,total,N} = C_{u,total,N} = 5.2 \text{ kip} + 0.15 \text{ kip} = 5.35 \text{ kip}$
Use 5.4 kip

# CHAPTER 8—DIAPHRAGMS

| | | |
|---|---|---|
| 12.5.2.2 | <u>Chord reinforcement:</u><br>Tension due to moment will be resisted by deformed bars conforming to Section 20.2.1 of ACI 318-14. | |
| 12.5.1.5 | Steel stress is the lesser of the specified yield strength and 60,000 psi. | |
| 12.5.1.1<br>22.4.3.1 | <u>Required reinforcement</u><br>$\phi T_n = \phi f_y A_s \geq T_u$<br><br>The chord forces north of and south of opening are equal. | $A_s = \dfrac{5400 \text{ lb}}{0.9(60,000 \text{ psi})} = 0.1 \text{ in.}^2$<br><br>One No.3 bar satisfies the requirement. The required collector reinforcement in the N-S direction, however, requires two No. 5 bars. Therefore, provided reinforcement is adequate and no additional reinforcement is required. |
| **Step 13: Collector reinforcement E-W** | | |
| | Continuous reinforced concrete frame over the full length of the building will act as a collector.<br><br>Note: Provide continuous reinforcement with tension splices (Step 15). | |
| 12.5.3.7 | In cast-in-place diaphragms, where shear is transferred from the diaphragm to a collector, or from the diaphragm or collector to a shear wall, temperature and shrinkage reinforcement is usually adequate to transfer that force. | |
| **Step 14: Shrinkage and temperature reinforcement** | | |
| 12.6.1<br>24.4.3.2 | The minimum shrinkage and temperature Reinforcement, $A_{S-T}$:<br><br>$A_{S+T} \geq 0.0018 A_g$ | $A_{S+T} = (0.0018)(7 \text{ in.})(12 \text{ in./ft}) = 0.15 \text{ in.}^2$ |
| 24.4.3.3 | Spacing of S+T reinforcement is the lesser of $5h$ and 18 in.<br>$5h = 5(12 \text{ in.}) = 60 \text{ in.}$<br>18 in. **Controls** | Note: Shrinkage and temperature reinforcement may be part of the main reinforcing bars resisting diaphragm in-plane forces and gravity loads. If provided reinforcement is not continuous (placing bottom reinforcing bars to resist positive moments at midspans and top reinforcing bars to resist negative moments at columns), continuity between top and bottom reinforcing bars may be achieved by providing adequate splice lengths between them. |

| | Step 15: Reinforcement detailing | |
|---|---|---|
| 12.7.2.1 | Reinforcement spacing<br>Chord and collector reinforcement minimum and maximum spacing must satisfy 12.7.2.1 and 12.7.2.2. | |
| 25.2.1 | Section 25.2 requires minimum spacing of<br>(a) 1 in.<br>(b) $4/3 d_{agg.}$<br>(c) $d_b$ No. 5 | Minimum spacing 1.0 in. **Controls**<br>$4/3(3/4$ in.$)$ aggregate = 1.0 in.<br>0.625 in. |
| 18.12.7.6a | Collector reinforcement spacing at a splice must be at least the larger of:<br>(a) At least three longitudinal $d_b$<br>(b) 1.5 in.<br>(c) $c_c \geq$ max $[2.5d_b, 2$ in.$]$ | $3(0.625$ in.$) = 1.875$ in.<br>1.5 in.<br>2 in. **Controls** |
| 12.7.2.2 | Maximum spacing is the smaller of $5h$ or 18 in. | 18 in. **Controls** |
| | Edge reinforcement<br>The opening has four beams around its perimeter. Therefore, the beams reinforcement is adequate to resist the tension forces due to inertial forces and additional reinforcement is not required.<br>Note: If beams are not constructed around the opening perimeter a minimum of two No. 5 is recommended around the opening as shown in the Fig. E2.13 and extended a minimum of its development length. | *Fig. E2.13—Two No. 5 reinforcement around opening.* |

## CHAPTER 8—DIAPHRAGMS

### Step 16: Details

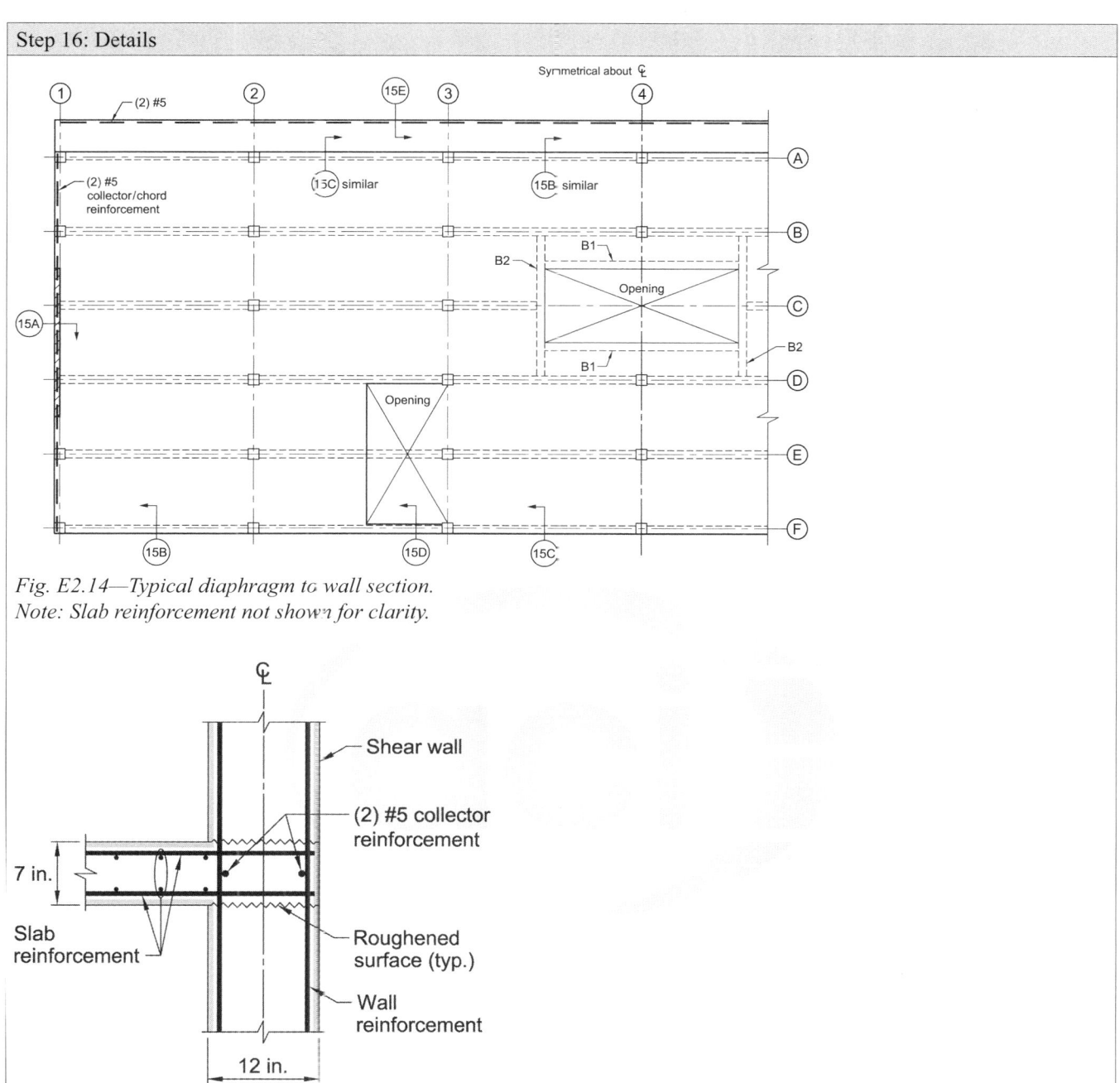

Fig. E2.14—Typical diaphragm to wall section.
Note: Slab reinforcement not shown for clarity.

Fig. 2.15A—Collector reinforcement in shear walls along CL 1 and 7.

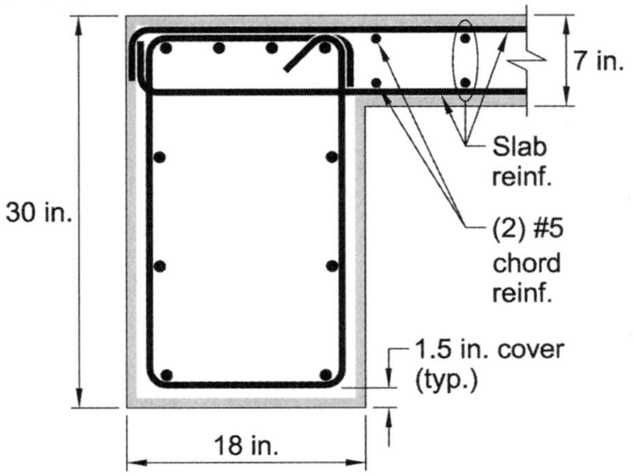

Fig. 2.15B—Chord reinforcement at midspan.

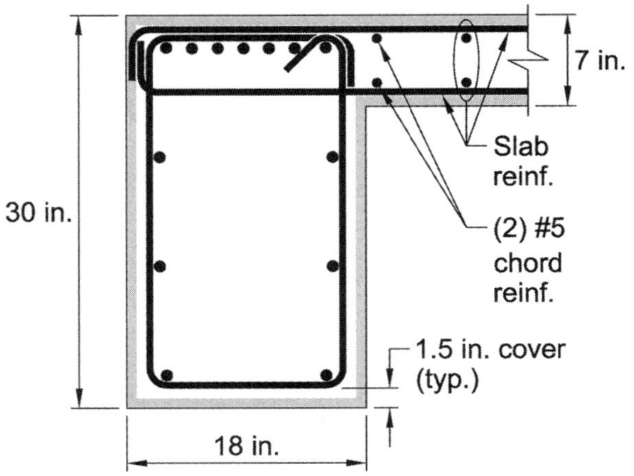

Fig. 2.15C—Chord reinforcement at supports.

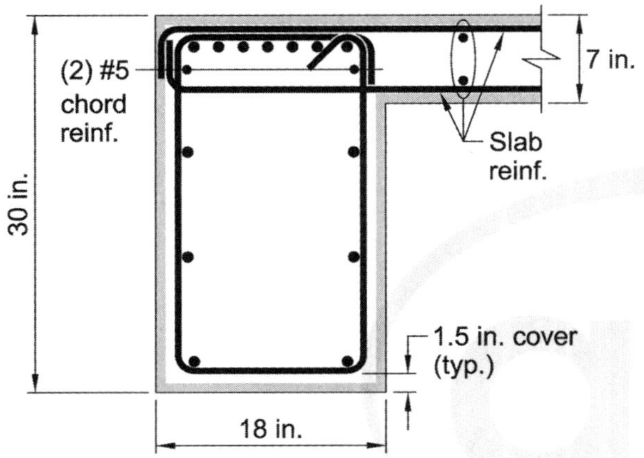

Fig. 2.15D—Chord reinforcement at opening.

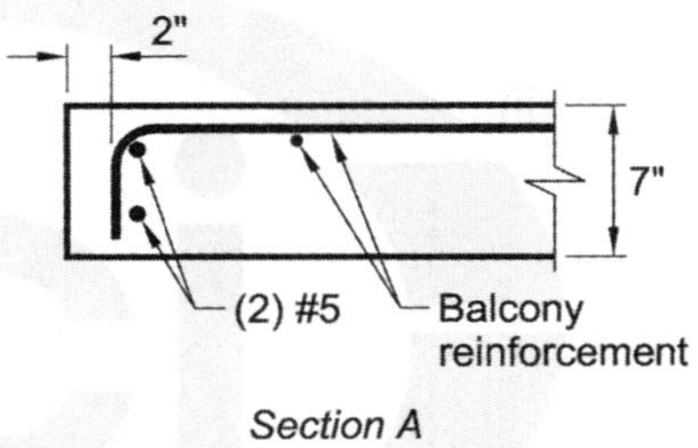

Fig. 2.15E—Crack control reinforcement at balcony edge.

# CHAPTER 8—DIAPHRAGMS

**Rigid Diaphragm Example 3:** *Lateral force distribution of a rigid diaphragm to shear walls*—A three-story wood apartment building is built on a normalweight reinforced concrete one-story slab. The slab is 200 ft x 90 ft with $f_c' = 4000$ psi and $f_y = 60,000$ psi. Assume that the structure is located in an active earthquake region Seismic Design Category (SDC) D and that the seismic analysis of the structural analysis based on ASCE 7-10, resulting in a base shear coefficient of 0.316. The slab supporting the wood structure is 10 in. thick and the wall lengths, height, and thicknesses are shown as follows. Assume the weight of the wood frame building imparts an equivalent uniform dead load of 135 psf to the slab. In addition, add a 10 psf miscellaneous dead load to the slab. Refer to Fig. E3.1 for geometric information.

This example will determine the seismic forces that are resisted by the shear walls, design the diaphragm, chords, and collectors to resist these forces and transmit them to the walls, and then detail the flatwork accordingly.

Given:
*Project data—*
Diaphragm size 200 ft 0 in. x 90 ft 0 in.
Wall 1: 90 ft 0 in. x 8 in.
Wall 2: 30 ft 0 in. x 10 in.
Wall 3: 30 ft 0 in. x 10 in.
Wall 4: 28 ft 0 in. x 10 in.
Wall 5: 40 ft 0 in. x 10 in.
Slab thickness: $t = 10$ in.
Parking structure (top of slab) height is
12 ft above the foundation

*Concrete—*
$f_c' = 4000$ psi
$f_y = 60,000$ psi

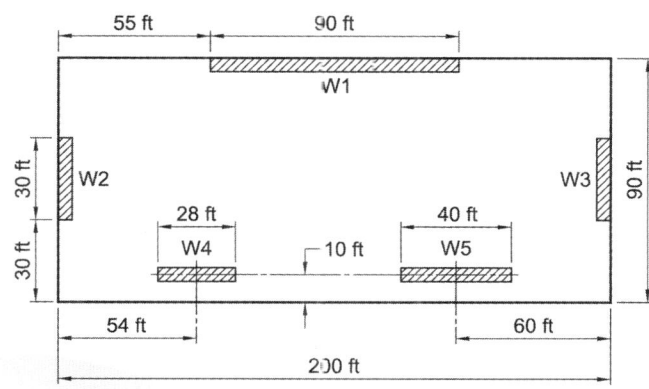

Fig. E3.1—Slab that supports a four-story wood building.

*Seismic criteria—*
SDC D
$C_S = 0.316$

Note: Nonparticipating columns in the lateral-force-resisting system are not shown for clarity.

| ACI 318-14 | Discussion | Calculation |
|---|---|---|
| **Step 1: Material requirements** | | |
| 7.2.2.1 | The mixture proportion must satisfy the durability requirements of Chapter 19 and structural strength requirements (ACI 318-14).<br><br>The designer determines the durability classes. Please refer to Chapter 4 of this Handbook for an in-depth discussion of the categories and classes.<br><br>ACI 301 is a reference specification that is coordinated with ACI 318-14. ACI encourages referencing ACI 301 into job specifications.<br><br>There are several mixture options within ACI 301, such as admixtures and pozzolans, which the designer can require, permit, or review if suggested by the contractor. | By specifying that the concrete mixture shall be in accordance with ACI 301 and providing the exposure classes, Chapter 19 requirements are satisfied.<br><br>Based on durability and strength requirements, and experience with local mixtures, the compressive strength of concrete is specified at 28 days to be at least 4000 psi. |
| **Step 2: Slab geometry** | | |
| 12.3.1.1 | Assume that diaphragm thickness satisfies the requirements for stability, strength, and stiffness under factored load combinations. | Given:<br>$h = 10$ in. |

## Step 3: Lateral forces

The lateral force is obtained by multiplying the self-weight of the reinforced concrete slab, wood frame building dead load, miscellaneous dead load, and the contribution of the shear walls, by the base shear coefficient.

Gravity loads
The reinforced concrete slab self-weight:
$W_{slab} = (L)(B)(h)(\gamma_c)$

$W_{slab} = (200 \text{ ft})(90 \text{ ft})(10 \text{ in.}/12 \text{ in/ft})(150 \text{ lb/ft}^3)$
$= 2{,}250{,}000 \text{ lb} = 2250 \text{ kip}$

Weight of wood frame building dead load and miscellaneous dead load:

$W_D = (135 \text{ psf} + 10 \text{ psf})(200 \text{ ft})(90 \text{ ft}) = 2610 \text{ kip}$

Total gravity dead load:

$W = 2250 \text{ kip} + 2610 \text{ kip} = 4860 \text{ kip}$

Shear wall self-weight contribution to diaphragm lateral force calculation is half the wall height.

N-S direction
$W_i = (L)(H/2)(t_w)(\gamma_c)$

$W_1 = (90 \text{ ft})(12 \text{ ft}/2)(8 \text{ in.})/(12 \text{ in.}/\text{ft})(150 \text{ lb/ft}^3)$
$W_1 = 54{,}000 \text{ lb} = 54 \text{ kip}$

$W_4 = (28 \text{ ft})(12 \text{ ft}/2)(10 \text{ in.})/(12 \text{ in.}/\text{ft})(150 \text{ lb/ft}^3)$
$W_4 = 21{,}000 \text{ lb} = 21 \text{ kip}$

$W_5 = (40 \text{ ft})(12 \text{ ft}/2)(10 \text{ in.})/(12 \text{ in.}/\text{ft})(150 \text{ lb/ft}^3)$
$W_5 = 30{,}000 \text{ lb} = 30 \text{ kip}$

Total gravity dead load in the N-S direction:

$\sum W = 4860 \text{ kip} + 54 \text{ kip} + 21 \text{ kip} + 30 \text{ kip} = 4965 \text{ kip}$

E-W direction
$W_i = (L)(H/2)(t_w)(\gamma_c)$

$W_2 = (30 \text{ ft})(12 \text{ ft}/2)(10 \text{ in.})/(12 \text{ in.}/\text{ft})(150 \text{ lb/ft}^3)$
$W_2 = 22{,}500 \text{ lb} = 22.5 \text{ kip}$

$W_3 = (30 \text{ ft})(12 \text{ ft}/2)(10 \text{ in.})/(12 \text{ in.}/\text{ft})(150 \text{ lb/ft}^3)$
$W_3 = 22{,}500 \text{ lb} = 22.5 \text{ kip}$

Total gravity dead load in the E-W direction:

$\sum W = 4860 \text{ kip} + 22.5 \text{ kip} + 22.5 \text{ kip} = 4905 \text{ kip}$

Lateral loads
Base shear is obtained from ASCE7-10 Section 12.8.1: $V = C_s W$
$C_s$ is calculated using ASCE 7-10 Section 12.8.1.1; not shown here for brevity:

$C_s = 0.316$ given

The equivalent lateral force distribution over the building height is per ASCE 7-10 Eq. (12.8-11).

The diaphragm design forces $F_{px}$ are calculated per ASCE 7-10 Eq. (12.10-1).

$F_{px}$ and $F_{py}$ must be in accordance with ASCE 7-10 Eq. (12.10-2) and (12.10-3).
Calculations not shown here as it is outside the scope of this Handbook.
Equivalent lateral force at the concrete level is:

$F_x = 363.1 \text{ kip}$

| | Diaphragm design forces: N-S: E-W: | $F_{py}$ = 745 kip $F_{px}$ = 726 kip |
|---|---|---|

Note

Conservatively, the weight of all walls–parallel and perpendicular–to the direction of the analysis can be included. In this example, the contribution of wall weights parallel to the applied seismic force is considered in the calculation of diaphragm shears. Walls perpendicular to the applied seismic force are included in determining the lateral force of concrete diaphragms.

### Step 4: Center of mass (COM)

Determine center of mass

Assume that the diaphragm is rigid.

Assume the (0,0) coordinate is located at the bottom left corner of the diaphragm. Center of mass of walls is shown in Table E.1:

#### Table E.1—Determining shear walls center of gravity

| Wall no. | Weight, psf | Length, ft | Area, ft² | Weight, kip | Direction | $x_{cg}$, ft | $Wx_{cg}$, ft-kip | $y_{cg}$, ft | $Wy_{cg}$, ft-kip |
|---|---|---|---|---|---|---|---|---|---|
| 1 (8 in.) | 100 | 90 | 540 | 54 | x | 100 | 5400 | 89.67 | 4842 |
| 2 (10 in.) | 125 | 30 | 180 | 22.5 | y | 0.417 | 9.38 | 45 | 1012.5 |
| 3 (10 in.) | 125 | 30 | 180 | 22.5 | y | 199.583 | 4490.6 | 45 | 1012.5 |
| 4 (10 in.) | 125 | 28 | 168 | 21 | x | 54.0 | 1134 | 10 | 210 |
| 5 (10 in.) | 125 | 40 | 240 | 30 | x | 140.0 | 4200 | 10 | 300 |
| Σ | | | 150 | | | | 15,234 | | 7377.2 |

The values of $x_{cg}$ and $y_{cg}$ are the center of mass of each wall. For example:

Wall 1 has the following coordinates: $x_{cg}$ = 55 ft + 90 ft/2 = 100 ft and y = 90 ft – (8 in./12)/2 = 89.67 ft
Wall 2 has the following coordinates: $x_{cg}$ = 0 ft + (10 in./12)/2 = 0.417 ft and y = 30 ft + 30 ft/2 = 45 ft

Center of mass of all walls:

$$x_1 = \frac{\sum W_i x_{cg,i}}{\sum W_i} = \frac{15,234 \text{ ft-kip}}{150 \text{ kip}} = 101.6 \text{ ft}$$

$$y_1 = \frac{\sum W_i y_{cg,i}}{\sum W_i} = \frac{7377.2 \text{ ft-kip}}{150 \text{ kip}} = 49.2 \text{ ft}$$

Center of mass of the slab is: $x_2$ = 200 ft/2 = 100 ft and $y_2$ = 90 ft/2 = 45 ft
Location of center of mass of the slab and walls combined:

$$x_m = \frac{\sum W_i x_i}{\sum W_i} = \frac{(4860 \text{ kip})(100 \text{ ft}) + (150 \text{ kip})(101.6 \text{ ft})}{4860 \text{ kip} + 150 \text{ kip}} = 100.05 \text{ ft}$$

and

$$y_m = \frac{\sum W_i y_i}{\sum W_i} = \frac{(4860 \text{ kip})(45 \text{ ft}) + (150 \text{ kip})(49.2 \text{ ft})}{4860 \text{ kip} + 150 \text{ kip}} = 45.13 \text{ ft}$$

where 4860 kip and 150 kip are the weight of the slab and walls, respectively.

## Step 5: Center of rigidity (COR) and lateral system stiffness

Determine center of rigidity

From the lateral analysis, the diaphragm is assumed rigid and therefore, diaphragm flexibility is not considered. Therefore, lateral forces are distributed to shear walls in both directions in proportion to their relative stiffnesses. Lateral displacement is the sum of flexural and shear displacements.

Apply a lateral force of 1 kip is applied at the top of a cantilevered wall as shown in Fig. E3.2. The wall's lateral displacement under a unit load, which is related to its stiffness, is the sum of flexural and shear displacements:

$\Delta = \Delta_{Flexure} + \Delta_{Shear}$

$$\Delta = \frac{Ph^3}{3EI} + \frac{1.2Ph}{AG} \text{ where } G \cong 0.4E \text{ and } E = 3{,}605{,}000 \text{ psi}$$

$$\Delta_{Flexure} = \frac{Ph^3}{3EI} = \frac{Ph^3}{3E\frac{L^3t}{12}} = \frac{4P\left(\frac{h}{L}\right)^3}{Et}$$

$$\Delta_{Shear} = \frac{1.2Ph}{AG} = \frac{(1.2)Ph}{(Lt)0.4E} = \frac{3P\left(\frac{h}{L}\right)}{Et}$$

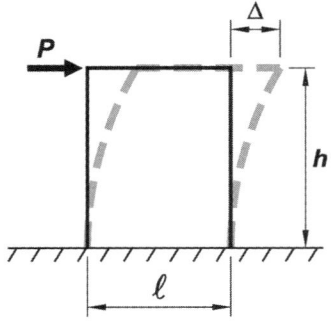

Fig. E3.2—Cantilever wall deflection.

Rigidity $k_i = 1/\Delta_i$ (refer to Table E.2)

### Table E.2—Determining walls' relative stiffnesses

| Wall no. | Height $h$, ft | Length $L$, ft | $h/L$ | $t$, in. | $\Delta_i \times 10^{-4}$, in. | $k_i = 1/\Delta_i \times 10^4$, 1/in. |
|---|---|---|---|---|---|---|
| 1 | 12 | 90 | 0.1333 | 8 | 0.14 | 7.043 |
| 2 | 12 | 30 | 0.4000 | 10 | 0.40 | 2.476 |
| 3 | 12 | 30 | 0.4000 | 10 | 0.40 | 2.476 |
| 4 | 12 | 28 | 0.4286 | 10 | 0.44 | 2.252 |
| 5 | 12 | 40 | 0.3000 | 10 | 0.28 | 3.576 |

### Table E.3—Determining walls' rigidity

| Wall no. | Direction | $x$, ft | $y$, ft | $k_{ix}$ | $k_{iy}$ | $(k_{iy})x$ | $(k_{ix})y$ |
|---|---|---|---|---|---|---|---|
| 1 | x | — | 89.67 | 7.043 | — | — | 631.55 |
| 2 | y | 0.417 | — | — | 2.476 | 1.03 | — |
| 3 | y | 199.58 | — | — | 2.476 | 494.16 | — |
| 4 | x | — | 10.0 | 2.252 | — | — | 22.52 |
| 5 | x | — | 10.0 | 3.576 | — | — | 35.76 |
| Σ | | | | 12.872 | 4.952 | 495.19 | 689.83 |

Calculate the system's center of rigidity:

$$x_r = \frac{\sum k_{iy} x_i}{\sum k_{iy}} = \frac{495.19 \text{ ft/ft}}{4.95/\text{ft}} = 100 \text{ ft}$$

$$y_r = \frac{\sum k_{ix} y_i}{\sum k_{ix}} = \frac{689.83 \text{ ft/ft}}{12.87 \text{ 1/ft}} = 53.6 \text{ ft}$$

Torsional eccentricity
The torsional eccentricity is the difference between the system's center of rigidity and its center of mass (Fig. E3.3):
$e_x = x_r - x_m = 100.05 \text{ ft} - 100.02 \text{ ft} = 0.03 \text{ ft}$, which is negligible
$e_y = y_r - y_m = 53.6 \text{ ft} - 45.1 \text{ ft} = 8.5 \text{ ft}$

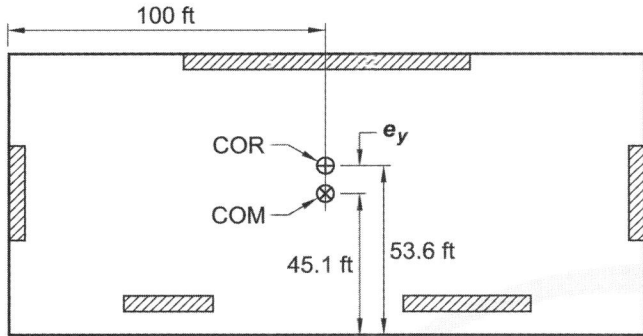

Fig. E3.3—Locations of the system's center of rigidity and center of mass.

ASCE 7-10 requires shifting the center of mass by a minimum of 5 percent of the building dimension, referred to as accidental eccentricity, in addition to the calculated eccentricity.
$e_x = 0 \text{ ft} \pm (0.05)(200 \text{ ft}) = \pm 10 \text{ ft}$
$e_y = 8.5 \text{ ft} \pm (0.05)(90 \text{ ft})$

$$e_{y1} = 8.5 \text{ ft} - 4.5 \text{ ft} = 4 \text{ ft}$$
$$e_{y2} = 8.5 \text{ ft} + 4.5 = 13 \text{ ft}$$

## Step 6: Lateral resisting system forces

In-plane wall forces due to direct lateral shear force are calculated by:

$$F_{vx} = F_{px} \frac{k_{ix}}{\sum k_{ix}}$$

$$F_{vy} = F_{py} \frac{k_{iy}}{\sum k_{iy}}$$

In-plane wall forces due to torsional moment are calculated by:

$$F_{tx} = \frac{k_i x_i}{\sum k_i x_i^2} T_x$$

$$F_{ty} = \frac{k_i y_i}{\sum k_i y_i^2} T_y$$

The torsional moment is the lateral shear force multiplied by the corresponding eccentricity:
NS: $T_y = F_{py} e_x$ = (745 ft)(±10 ft) = ±7450 ft-kip
EW: $T_x = F_{px} e_{y1}$ = (726 lb)(±4 ft) = ±2904 ft-kip
$T_x = F_{px} e_{y2}$ = (726 lb)(±13 ft) = ±9438 ft-kip

The in-plane diaphragm force is the sum of the direct lateral shear force, $F_{vi}$, and the torsional moment, $F_{ti}$ (refer to Tables E.4, E.5. and E.6): $F_u = F_{vi} + F_{ti}$

### Table E.4—Determining wall shear due to seismic forces in the N-S direction

| Wall no | $k_{ix}$ | $k_{iy}$ | $dx_i$, ft | $dy_i$, ft | $k_i d$ | $k_i(d)^2$ | $F_{vi}$, kip | $F_{ti}$, kip | $F_{total}$, kip | $F_{design}$, kip | Use, kip |
|---|---|---|---|---|---|---|---|---|---|---|---|
| 1 | 7.04 | 0 | — | 36.08 | 254.09 | 9166.8 | 0 | 27.3 | –27.3 | 27.3 | –27.3 |
| 2 | 0 | 2.476 | –99.6 | — | –246.56 | 24,553.6 | 372.5 | –26.5 | 26.5 | 346.0 | 399.0 |
| 3 | 0 | 2.476 | 99.6 | — | 246.56 | 24,553.6 | 372.5 | 26.5 | –26.5 | 399.0 | 346.0 |
| 4 | 2.25 | 0 | — | –43.6 | –98.185 | 4280.2 | 0 | –10.5 | 10.5 | –10.5 | 10.5 |
| 5 | 3.57 | 0 | — | –43.6 | –155.91 | 6796.4 | 0 | –16.7 | 16.7 | –16.7 | 16.7 |
| Σ | 12.87 | 4.952 | | | 0 | 69,350.5 | | | | | |

Example on calculating $dy_i$:
Wall 1: $dy_i$ = 90 ft - (8 in./12 in./ft)/2 – 53.6 ft = 36.08 ft
Walls 4 and 5: $dy_i$ = 53.6 ft – 10 ft = 43.6 ft

### Table E.5—Determining wall shear due to seismic forces in the E-W direction $e_{y1}$ = 4 ft

| Wall no | $k_{ix}$ | $k_{iy}$ | $x_i$, ft | $y_i$, ft | $k_i d$ | $k_i(d)^2$ | $F_x$, kip | $F_{x2}^*$, kip | $F_{design}$, kip |
|---|---|---|---|---|---|---|---|---|---|
| 1 | 7.043 | 0 | — | 36.08 | 254.09 | 9166.8 | 397.2 | –10.6 | 386.7 |
| 2 | 0.00 | 2.476 | –99.6 | — | –246.565 | 24,553.7 | 0.0 | 10.2 | 10.2 |
| 3 | 0.00 | 2.476 | 99.6 | — | 246.565 | 24,553.7 | 0.0 | –10.2 | –10.2 |
| 4 | 2.252 | 0 | — | –43.6 | –98.185 | 4280.2 | 127.0 | 4.1 | 131.1 |
| 5 | 3.576 | 0 | — | –43.6 | –155.91 | 6796.4 | 201.7 | 6.5 | 208.2 |
| Σ | 12.87 | 4.952 | | | 0 | 69,350.9 | | | |

### Table E.6—Determining wall shear due to seismic forces in the E-W direction $e_{y2}$ = 13 ft

| Wall no | $k_{ix}$ | $k_{iy}$ | $x_i$, ft | $y_i$, ft | $k_i d$ | $k_i(d)^2$ | $F_x$, kip | $F_{x1}^*$, kip | $F_{design}$, kip |
|---|---|---|---|---|---|---|---|---|---|
| 1 | 7.043 | 0 | — | 36.08 | 254.09 | 9166.8 | 397.2 | –34.5 | 362.8 |
| 2 | 0.00 | 2.476 | –99.6 | — | –246.565 | 24,553.7 | 0.0 | 33.5 | 33.5 |
| 3 | 0.00 | 2.476 | 99.6 | — | 246.565 | 24,553.7 | 0.0 | –33.5 | –33.5 |
| 4 | 2.252 | 0 | — | –43.6 | –98.185 | 4280.2 | 127.0 | 13.3 | 140.4 |
| 5 | 3.576 | 0 | — | –43.6 | –155.91 | 6796.4 | 201.7 | 21.2 | 222.9 |
| Σ | 12.87 | 4.952 | | | 0 | 69,350.9 | | | |

# CHAPTER 8—DIAPHRAGMS

where $d$ is the distance ($dx_i$ or $dy_i$) from the center of each wall to the center of rigidity. $F_{x1}$ is the additional shear force due to eccentricity of 13 ft. $F_{x2}$ is the additional shear force due to eccentricity of 4 ft.

Notes:
- $dx_i$ and $dy_i$ are the distances of a wall from the center of rigidity in the x- and y-direction.
- If torsional moment reduces the magnitude of the direct lateral shear on a wall, then it is ignored.

The wall design shear forces are summarized in Table E.7.

## Table E.7—Summary of wall shear forces due to seismic forces

| Wall no. | Wall length, ft | E-W load, kip | N-S, load | Design, shear, kip |
|---|---|---|---|---|
| 1 | 90.00 | 387 | 27 | 387 |
| 2 | 30.00 | 335 | 399 | 399 |
| 3 | 30.00 | 335 | 399 | 399 |
| 4 | 28.00 | 140 | 11 | 140 |
| 5 | 40.00 | 223 | 17 | 223 |

### Step 7: Diaphragm shear strength

| | | |
|---|---|---|
| 12.5.3.3 | **In-plane shear in diaphragm**<br>The diaphragm nominal shear strength is calculated from Eq. (12.5.3.3)<br>$V_n = A_{cv}\left(2\lambda\sqrt{f_c'} + \rho_t f_y\right)$<br><br>In this example, first check the diaphragm strength without reinforcement; therefore, ignore the strength contribution of reinforcement: $\rho_t = 0$.<br><br>Assume collector length is the full length of diaphragm in both directions. | North and south<br>$V_n = (90\text{ ft})(12\text{ in./ft})(10\text{ in.})\left((2)(1.0)\sqrt{4000\text{ psi}}\right)$<br>$= 1{,}366{,}104\text{ lb} = 1366\text{ kip}$<br><br>East and west<br>$V_n = (200\text{ ft})(12\text{ in./ft})(10\text{ in.})\left((2)(1.0)\sqrt{4000\text{ psi}}\right)$<br>$= 3{,}035{,}787\text{ lb} = 3036\text{ kip}$ |
| 12.5.3.2<br>21.2.4.2 | Applying the reduction factor $\phi = 0.6$. | NS: $\phi V_n = (0.6)(1366\text{ kip}) = 820\text{ kip}$<br>EW: $\phi V_n = (0.6)(3036\text{ kip}) = 1821\text{ kip}$ |
| 18.2.4.1b | Per ACI 318-14, $\phi$ must not exceed the least value for shear used for the vertical components of the primary seismic-force resisting system: | |
| 12.5.1.1<br>22.5.1.2 | Check if design shear strength exceeds the factored shear force. | $\phi V_n = 820\text{ kip} > V_u = 399\text{ kip}$ **OK**<br>$\phi V_n = 1821\text{ kip} > V_u = 387\text{ kip}$ **OK** |
| 18.12.9.2 | The nominal shear strength, $V_n$, must not exceed:<br>$8A_{cv}\sqrt{f_c'}$ | $\dfrac{8(10\text{ in.})(90\text{ ft})(12\text{ in./ft})\left(\sqrt{4000\text{ psi}}\right)}{1000\text{ lb/kip}} = 5464\text{ kip}$<br><br>$V_{n,\,NS} = 1366\text{ kip} < 5464\text{ kip}$ **OK**<br>$V_{n,\,EW} = 3036\text{ kip} < 5464\text{ kip}$ **OK** |

## Step 8: Diaphragm lateral force distribution N-S

| | | |
|---|---|---|
| 12.4.2.4<br>12.5.1.3 | Diaphragm is assumed rigid (ACI 318-14, Section 12.4.2.4(a)). Therefore, the diaphragm design moments, shears, and axial forces are calculated assuming a simply supported beam with depth equal to full diaphragm depth (ACI 318-14, Section (12.5.1.3(a)). | 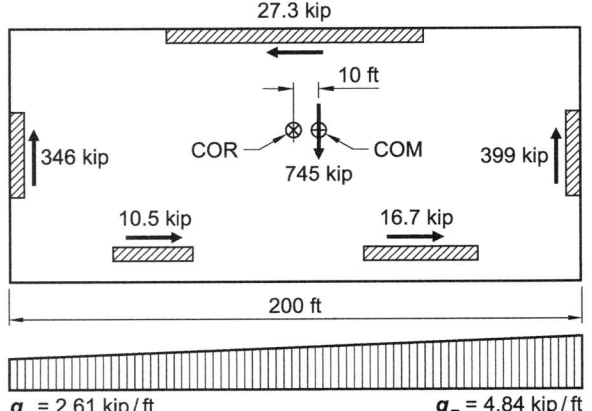 |
| | The wall forces and the assumed direction of the torsional moment are shown in Fig. E3.4. | |
| | Refer to previous Step 5 for calculation of seismic force location. | |
| 6.6.2.3(b) | The seismic force on the diaphragm is distributed within diaphragm as shown in (Fig. E3.4): | Fig. E3.4—Shear wall forces due to seismic force in the N-S direction. |
| | Force equilibrium<br>$$q_L\left(\frac{L}{2}\right) + q_R\left(\frac{L}{2}\right) = F_{px,des(NS)} \quad (I)$$ | $$q_L\left(\frac{200\text{ ft}}{2}\right) + q_R\left(\frac{200\text{ ft}}{2}\right) = 745 \text{ kip}$$ |
| | Moment equilibrium (taken around bottom left corner of the diaphragm)<br>$$q_L\left(\frac{L}{2}\right)\left(\frac{L}{3}\right) + q_R\left(\frac{L}{2}\right)\left(\frac{2L}{3}\right) = F_{px,des(NS)}\left(\frac{L}{2} + 0.05L\right) \quad (II)$$ | $$q_L\frac{(200\text{ ft})^2}{6} + q_R\frac{2(200\text{ ft})^2}{6} = (745\text{ kip})\left(\frac{200\text{ ft}}{2} + 10\text{ ft}\right)$$ |
| | Solve equations (I) and (II) for $q_L$ and $q_R$:<br>Draw the shear and moment diagrams (Fig. E3.5). | $q_L = 2.61$ kip/ft, say, 2.6 kip/ft<br>$q_R = 4.84$ kip/ft, say, 4.8 kip/ft |
| | Note: In an Aug. 2010 National Institute and Standards Technology (NIST) report, GCR 10-917-4, "Seismic Design of Cast-in-Place Concrete Diaphragms, Chords, and Collectors," by Moehle et al. states that, "This approach leaves any moment due to the frame forces along column lines (CL) A and F unresolved. Sometimes this is ignored or, alternatively, it too can be incorporated in the trapezoidal loading." | 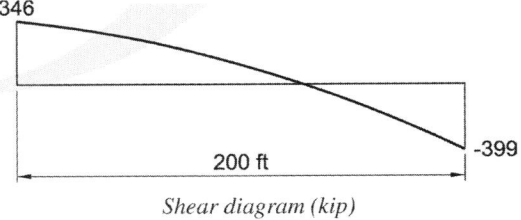<br>Shear diagram (kip) |
| | $V_{max} = 399$ kip<br>$M_{max} = 21{,}106$ ft-kip, say, 21,100 ft-kip | 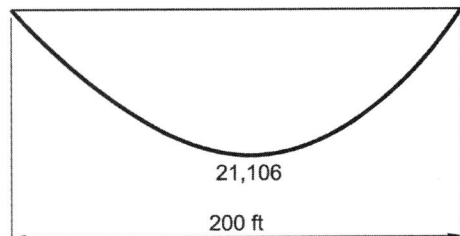<br>Bending moment diagram (ft-kip)<br>Fig. E3.5—Shear and bending moment diagrams due to a lateral seismic force in the N-S direction. |
| | Note: Experienced engineers simplify the calculations by using uniformly distributed load:<br>Calculate a maximum moment: | $q = 745$ kip/200 ft $= 3.723$ kip/ft, say 3.72 kip/ft<br>$$M_{max} = \frac{(3.73 \text{ kip/ft})(200\text{ ft})^2}{8} = 18{,}650 \text{ ft-kip}$$ |

# CHAPTER 8—DIAPHRAGMS

| | Notes:<br>• The difference in maximum moment between the two approaches is 13.3 percent and at different locations (110 ft versus 100 ft).<br>• Shear diagram for the second approach is a straight line with maximum shear force:<br>$V = (3.73 \text{ kip/ft})(200 \text{ ft}/2) = 373 \text{ kip}$ | |
|---|---|---|
| **Step 9: Chord reinforcement N-S** | | |
| R12.1.1 | Assume the slab behaves like a beam with compression and tension forces at the near and far edges, respectively:<br><br>$C_{chord} = T_{chord} = M/d$ | |
| 12.5.2.3 | ACI 318 suggests placement of chord reinforcement within an arbitrary width of $h/4$ of the diaphragm tension edge (Fig. E3.6).<br><br>The maximum chord tension force is:<br><br>Tension due to moment is resisted by deformed bars conforming to Section 20.2.1 of ACI 318-14. Steel stress is the lesser of the specified yield strength and 60,000 psi.<br><br>Required chord reinforcement area:<br>$\phi T_n = \phi f_y A_s \geq T_u$<br><br>It is, however, recommended to place tension reinforcement close to the tension face. Assume tension reinforcement moment arm is approximately $0.95B$ at both north and south sides of the slab edges: The calculated tension force is:<br><br>$T_u = \dfrac{M_u}{0.95B}$<br><br>Required tension reinforcement is:<br>$\phi T_n = \phi f_y A_s \geq T_u$ | $h/4 = 90 \text{ ft}/4 = 22.5 \text{ ft}$<br>$\Rightarrow d = 90 \text{ ft} - 1/2(22.5 \text{ ft}) = 78.75 \text{ ft}$<br><br>$T_u = \dfrac{21,100 \text{ ft-kip}}{78.75 \text{ ft}} = 268 \text{ kip}$<br><br><br><br>$f_y = 60,000 \text{ psi}$<br><br><br>$A_s = \dfrac{268 \text{ kip}}{(0.9)(60 \text{ ksi})} = 5 \text{ in.}^2$<br><br><br><br>$(0.95)(90 \text{ ft}) = 85.5 \text{ ft}$<br><br><br><br>$T_u = \dfrac{21,100 \text{ ft-kip}}{85.5 \text{ ft}} = 247 \text{ kip}$<br><br>$A_s = \dfrac{247 \text{ kip}}{(0.9)60 \text{ ksi}} = 4.6 \text{ in.}^2$ |
| 18.12.7.5 | Per provision 18.12.7.5, the required chord width for the concrete compressive strength limit of $0.2f_c'$.<br><br>$w_{chord} > \dfrac{C_{chord}}{0.2 f_c' h_{diaph}}$<br><br>Note: The chord force does not need to be increased by the overstrength factor. | $w_{chord} > \dfrac{247 \text{ kip}}{(0.2)(4000 \text{ psi})(10 \text{ in.})} = 30.9 \text{ in.}$<br><br>and<br>$w_{chord} = 30.9 \text{ in.} < h/4 = 90 \text{ ft}/4 = 22.5 \text{ ft}$ **OK**<br>Say, 32 in. |
| | Note: Although it is permissible to place bars within 22 ft-6 in. ($h/4$) of diaphragm width, it is recommended to place bars close to the tension end, where it is most effective. Since load is reversible, chord reinforcement is placed at both north and south of diaphragm ends (refer to Fig. E3.6). The final layout of bars will be coordinated with the collector reinforcement due to inertial force in the East–West (E-W) direction. | |

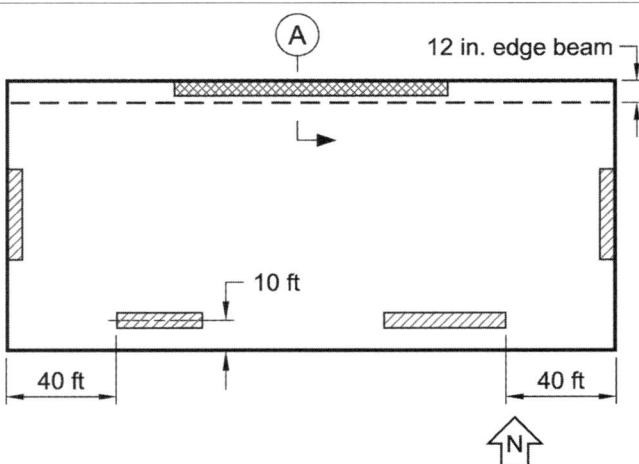

Fig. E3.6—Suggested chord reinforcement at the north and south edges of the diaphragm.

## Step 10: Collectors design N-S

| | | |
|---|---|---|
| | Collectors transfer shear forces from the diaphragm to the vertical walls at both east and west ends (Fig. E3.4). Assume collectors extend over the full width of the diaphragm.<br><br>Partial depth collectors can be considered, but a complete force path should be designed that is capable of transmitting all forces from the diaphragm to the collector and into the vertical elements. | |
| 12.5.4.1 | Unit shear force is the maximum diaphragm shear divided by the diaphragm depth, $B = 90$ ft:<br><br>$$v_{u@F} = \frac{F_{u@F}}{B}$$<br><br>In slab:<br><br>In wall:<br><br>Check if the concrete shear strength excluding reinforcement exceeds the factored shear:<br><br>$$\phi v_c = \phi 2\sqrt{f_c'} b t_{diaph}$$<br><br>Shear reinforcement is, therefore, required. Use the No. 5 at 16 in. on center temperature and shrinkage reinforcement in each direction to increase shear capacity (assuming two-way slab). | From Step 6 (Table E.7): $F_u = 399$ kip<br><br>$$v_{u@F} = \frac{399 \text{ kip}}{90 \text{ ft}} = 4.43 \text{ kip/ft}$$<br><br>$$v_{u@F} = \frac{399 \text{ kip}}{30 \text{ ft}} = 13.3 \text{ kip/ft}$$<br><br>$$\phi v_c = \frac{(0.6)(2\sqrt{4000 \text{ psi}})(12 \text{ in./ft})(10 \text{ in.})}{1000 \text{ lb/kip}} = 9.1 \text{ kip/ft}$$<br><br>$\phi v_c = 9.1$ kip/ft $< v_c = 13.3$ kip/ft  **NG**<br><br>$$\rho_t = \frac{0.31 \text{ in.}^2}{(10 \text{ in.})(16 \text{ in.})} = 0.00194$$<br><br>$\phi v_n = 9.1$ kip/ft $+ (0.6)(0.00194)(10 \text{ in.})(12 \text{ in.})(60 \text{ ksi})$<br>$= 17.5$ kip/ft $> v_u = 13.3$ kip/ft  **OK** |

| | | |
|---|---|---|
| | Force at diaphragm to wall connection<br>The proportional diaphragm force that the collector transfers to walls connection is (Fig. E3.7):<br><br>Wall south end:<br>$-(4.43 \text{ kip/ft})(30 \text{ ft}) = -132.9 \text{ kip}$<br><br>Wall north end:<br>$-132.9 \text{ kip} + (13.3 \text{ kip/ft} - 4.43 \text{ kip/ft}) \times (30 \text{ ft})$<br>$= 133.2 \text{ kip}$<br><br>Slab end:<br>$+133.2 \text{ kip} - (4.43 \text{ kip/ft})(30 \text{ ft}) \approx 0 \text{ kip}$ | |

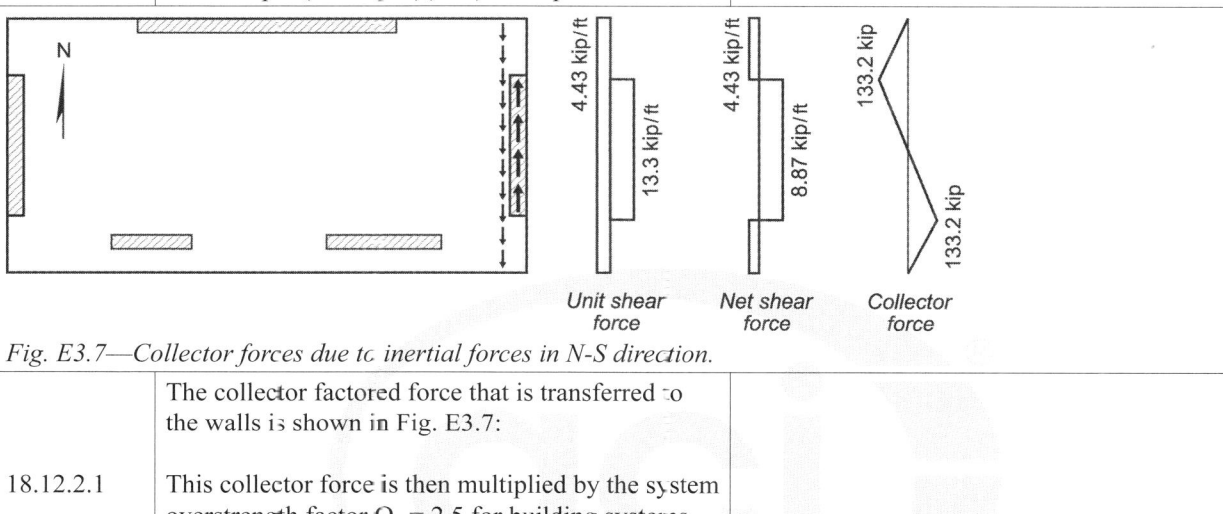

*Fig. E3.7—Collector forces due to inertial forces in N-S direction.*

| | | |
|---|---|---|
| 18.12.2.1 | The collector factored force that is transferred to the walls is shown in Fig. E3.7:<br><br>This collector force is then multiplied by the system overstrength factor $\Omega_o = 2.5$ for building systems with special structural walls in SDC D (ASCE 7-10, Table 12.2-1) | $T_u = \Omega_o T_{Coll} = \Omega_o C_{Coll} = (2.5)(133.2 \text{ kip}) = 333 \text{ kip}$ |
| 12.5.4.2<br><br>R12.5.4<br><br><br><br><br>18.12.7.5 | Collectors are designed as both tension and compression members.<br>There are no beams along the CL 1, so a portion of the slab is used as a collector.<br>The collector width for tension reinforcement is determined by engineering judgement. ACI 318 provides in the commentary that a collector width cannot exceed approximately one-half the contact length between the collector and the vertical element measured from the face of the vertical element.<br><br>The collector width, however, is chosen such that the limiting stresses are not exceeded. When the tension and compression collector forces are increased by the overstrength factor, then the limiting concrete compression stress is $0.5f_c'$. Calculate the required compressive collector width: | $b_{eff} = 30 \text{ ft}/2 + 10 \text{ in.} = 15.8 \text{ ft} = 190 \text{ in.}$<br><br>$w_{collector} = \dfrac{2.5 C_{coll}}{0.5 f_c' t} = \dfrac{333 \text{ kip}}{(2 \text{ ksi})(10 \text{ in.})} = 16.7 \text{ in.}$<br><br>$w_{collector} = 16.7 \text{ in.} < 190 \text{ in.}$<br>but $w_{collector} = 16.7 \text{ in.} > t_v = 10 \text{ in.}$<br><br>Therefore, part of the slab, $b_{eff}$, is needed to resist the collector force. |
| 22.4.3 | Required reinforcement area to resist collector force:<br>$\phi T_n = \phi f_y A_s \geq T_u$ | $A_s = \dfrac{\Omega_o T_{coll}}{0.9 f_y} = \dfrac{333 \text{ kip}}{0.9(60 \text{ ksi})} = 6.2 \text{ in.}^2$ |

Note: Collector reinforcement may be varied along the length of the diaphragm based on required strength and terminated where not required. In this example, the reinforcement is extended over the full length of the diaphragm.

A number of bars are placed in line with the wall. The balance is distributed across the width of the collector element. In this case, for the 10 inch thick wall, two No. 8 bars in line with the wall will result in a reasonable bar spacing of 6 in. Therefore, two No. 8 bars are centered on the wall for

$A_{s,line} = 1.58$ in.$^2$

The balance of the required reinforcement is:

$A_{s,bal} = 6.2$ in.$^2 - 1.58$ in.$^2 = 4.62$ in.$^2$
distributed over the 15.0 ft wide collector;
4.62 in.$^2$/15.0 ft = 0.31 in.$^2$/ft

Try eight No. 5 top and bottom spaced over 15 ft.

$A_{s,prov.} = (2)(8)(0.31$ in.$^2) = 4.96$ in.$^2$
$A_{s,prov.} = 4.96$ in.$^2 > A_{s,req'd} = 4.62$ in.$^2$  **OK**

Design collector region:

The collector geometry results in a moment in the diaphragm section adjacent to the wall because of the eccentricity between the collector and the wall. This moment is solved through shear forces in the diaphragm perpendicular to the collector and bending in the plane of the diaphragm due to eccentric tension and compression forces (Fig. E3.8).

$M_u = T_{dist}e_{ten} + C_{dist}e_{comp} - V\ell_{wall}$

where $T_{dist}$ is the portion of the tension collector force resisted by $A_{s,dist}$; $C_{dist}$ is the portion of the compression collector force resisted by slab outside the wall; and $V$ is the shear strength of the diaphragm.

Where the collector element is in tension, the concrete contribution to $V$ is neglected, $V_c = 0$.
$\phi V_s = \phi f_y \rho_t t_w (w_{comp} - t_{wall})$

Assume No. 5 @ 16 in. on center is provided:

For more in-depth understanding refer to: Structural Engineer Association of California (SEAOC) Seismology Committee (2007) "Concrete Slab Collectors," from the Aug. 2008 *SEAOC Blue Book: Seismic Design Recommendations Compilation*, Structural Engineers Association of California, Sacramento.

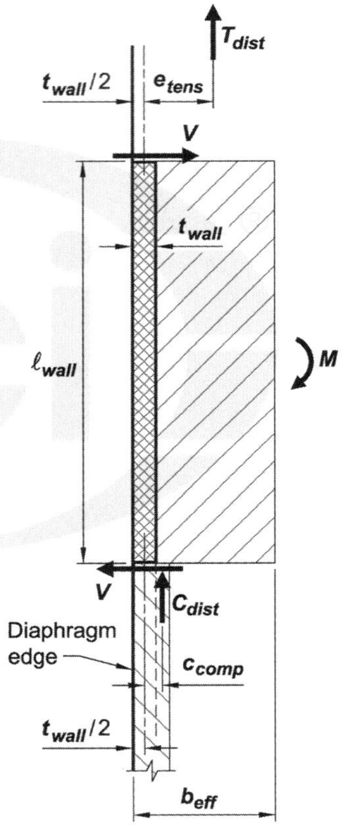

*Fig. E3.8—Diaphragm segment plan.*

$\phi V_s = 0.6(0.00193(60$ ksi$)(10$ in.$)(16.7$ in. $- 10$ in.$)$
$= 4.7$ kip

$\rho_t = \dfrac{0.31 \text{ in.}^2}{(10 \text{ in.})(16 \text{ in.})} = 0.00193 > 0.0018$

| | | |
|---|---|---|
| | Tension force in the slab (outside the wall geometry) is proportional to reinforcement: | $T_{dist} = \left(\dfrac{4.62 \text{ in.}^2}{6.2 \text{ in.}^2}\right)(333 \text{ kip}) = 247 \text{ kip}$ |
| | The moment arm: | $e_{tens} = \dfrac{(15 \text{ ft})(12 \text{ in./ft})}{2} + \dfrac{10 \text{ in.}}{2} = 95 \text{ in.}$ |
| | Compression force in the slab (outside the wall geometry) is proportional to collector width: | $C_{dist} = \left(\dfrac{16.7 \text{ in.} - 10 \text{ in.}}{16.7 \text{ in.}}\right)(333 \text{ kip}) = 137 \text{ kip}$ |
| | Moment arm: | $e_{comp} = \dfrac{10 \text{ in.}}{2} + \dfrac{16.7 \text{ in.} - 10 \text{ in.}}{2} = 8.35 \text{ in.}$ |
| | $M_u = T_{dist}e_{ten} + C_{dist}e_{comp} - V\ell$ | $M_u = (247 \text{ kip})(95 \text{ in.}) + (134 \text{ kip})(8.35 \text{ in.})$ $\quad - (4.7 \text{ kip})(29.5 \text{ ft})(12 \text{ in./ft})$ $M_u = 22{,}920 \text{ in.-kip}$ |
| | Assume $\ell = 30 \text{ ft} - 0.5 \text{ ft} = 29.5 \text{ ft}$ moment arm. Required reinforcement: | $A_{s,req'd} = \dfrac{22{,}920 \text{ in.-kip}}{(0.9)(60 \text{ ksi})(0.9)(29.5 \text{ ft})(12 \text{ in./ft})}$ $A_{s,req'd} = 1.33 \text{ in.}^2$ Use three No. 6 dowels at each end of the wall. Refer to Fig. E3.9. |

| | | |
|---|---|---|
| | Shear transfer design:<br>A number of bars are placed in line with the wall, which results in transferring portion of the diaphragm force in tension and direct bearing of slab against the wall in compression. The diaphragm and shear transfer interface is designed for the balance of the collector element. | |
| | Assuming tension forces are distributed in proportion to collector area, $V_u$ for diaphragm and shear transfer design is then calculated as follows: | $V_u$ = 247 kip + 134 kip + (4.43 kip/ft)(30 ft) = 514 kip |
| | 6.2 in.² is the required reinforcement area to resist the collector force in prior calculation. | |
| | 1.58 in.² is the area of two No. 8 bars placed in-line with the shear wall. | |
| 12.5.3.7 | Shear from the diaphragm is transferred by shear friction to the wall with dowels placed perpendicular to the wall-slab interface: | |
| | $V_n = \mu A_{vf} f_y$ | |
| 22.9.4.2 | Assume that diaphragm slab is placed against hardened wall concrete that is clean, free of laitance, and intentionally roughened to a full amplitude of approximately 1/4 in. From Table 22.9.4.2; $\mu = 1.0$ | |
| 21.2.1(b) | Use a reduction factor of $\phi = 0.75$, because the shear interface is not a member. Otherwise, $\phi = 0.6$. | 514 kip $\leq \phi V_n$ = (0.75)(1.0)$A_{vf}$(60 ksi) |
| | The wall is $\ell_{wall}$ = 30 ft long. | $A_{vf}$ = 11.42 in.² or $A_{vf}/\ell_{wall}$ = 11.42 in.²/(30 ft) = 0.38 in.²/ft<br>Try No. 6 at 12 in. on center.<br>$A_{s,prov.}$ = 0.44 in.²/ft  **OK** |
| | The required diaphragm strength is: | $v_u$ = 514 kip/30 ft = 17.1 kip/ft |
| | The diaphragm shear strength is the contribution of concrete and reinforcement in prior calculation of this step: | |
| | $\phi v_n = \phi(v_c + v_s)$ | $\phi v_n$ = 17.5 kip/ft > $v_u$ = 17.1 kip/ft |
| 22.9.4.4 | The value of $V_n$ across the assumed shear plane must not exceed the lesser of the following limits:<br><br>(a) $0.2 f_c' A_c$<br>(b) $(480 + 0.08 f_c') A_c$<br>(c) $1600 A_c$ | (0.2)(4000 psi) = 800 psi  **OK**<br>(480 + 0.08(4000 psi)) = 800 psi  **OK**<br>1600 psi<br><br>Therefore, $V_n$ must not exceed:<br><br>$\dfrac{(800 \text{ psi})(10 \text{ in.})(30 \text{ ft})(12 \text{ in./ft})}{1000 \text{ lb/kip}}$ = 2880 kip > $V_n$ |

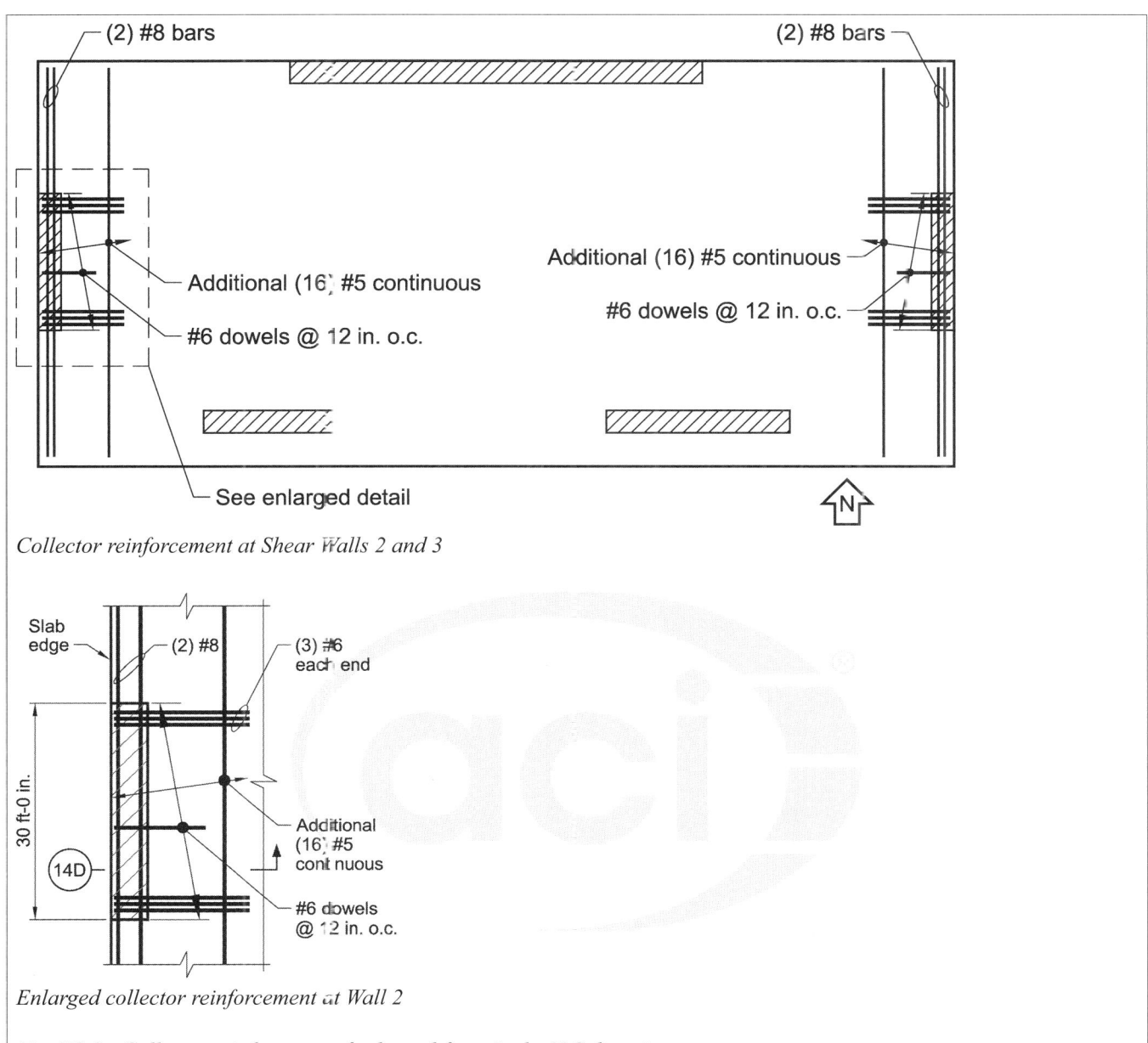

Fig. E3.9—Collector reinforcement for lateral force in the N-S direction.

| Step 11: Diaphragm lateral force distribution E-W |||
|---|---|---|
| | The wall forces and the assumed direction of torque due to accidental eccentricity are shown in Fig. E3.10.<br><br>The distribution of the diaphragm force is calculated by using $q_L$ and $q_R$ as the left and right diaphragm reactions per unit length (Fig. E3.10). | 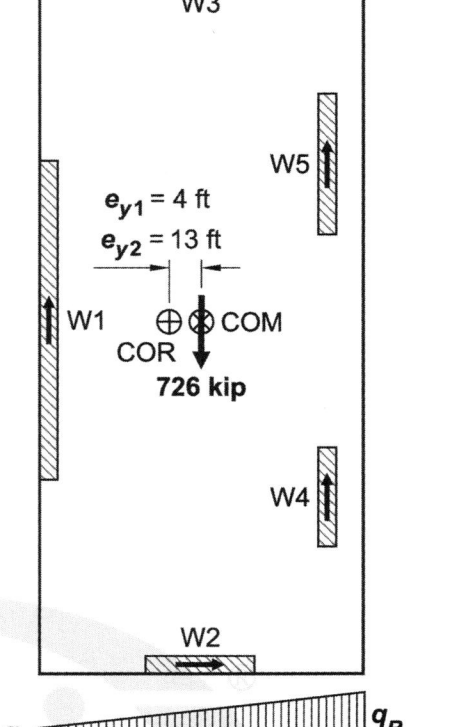<br>Fig. E3.10—Shear wall forces due to a seismic force in the E-W direction at $e_y = 8.5$ ft. |
| | Case I:<br>Eccentricity at $e_y = 4$ ft<br>Refer to Step 5 of this example for calculation of eccentricity.<br><u>Force equilibrium</u><br>$$q_L\left(\frac{L}{2}\right) + q_R\left(\frac{L}{2}\right) = F_{px,des(EW)} \quad \text{(I)}$$ | $$q_L\left(\frac{90 \text{ ft}}{2}\right) + q_R\left(\frac{90 \text{ ft}}{2}\right) = 726 \text{ kip}$$ |
| | <u>Moment equilibrium (taken around CL W1)</u><br>$$q_L\left(\frac{L}{2}\right)\left(\frac{L}{3}\right) + q_R\left(\frac{L}{2}\right)\left(\frac{2L}{3}\right)$$<br>$$= (F_4 + F_5)(80 \text{ ft}) - F_2(200 \text{ ft}) \quad \text{(II)}$$<br><br>$F_2$, $F_4$, and $F_5$ are per Table E.5. | $$q_L\frac{(90 \text{ ft})^2}{6} + q_R\frac{2(90 \text{ ft})^2}{6}$$<br>$$= (208 \text{ kip} + 131 \text{ kip})(80 \text{ ft}) - (10 \text{ kip})(200 \text{ ft})$$ |
| | Solve equations (I) and (II) for $q_L$ and $q_R$: | $q_R = 2.5$ kip/ft<br>$q_L = 13.6$ kip/ft |

| | | |
|---|---|---|
| | Draw the shear and moment diagrams for the diaphragm assuming simply supported beam behavior (Fig. E3.11).<br><br>The maximum moment is located at 40.5 ft from the south end of the diaphragm.<br><br>$V_{max} = 387$ kip<br>$M_{max} = 6542$ ft-kip<br><br>Note: Moehle et al. also state in NIST report number GCR 10-917-4 that, "For a rectangular diaphragm of uniform mass, a trapezoidal distributed force having the same total force and centroid is then applied to the diaphragm. The resulting shears and moments are acceptable for diaphragm design. This approach leaves any moment due to (shear walls perpendicular to the diaphragm inertia lateral force) unresolved; sometimes this is ignored or, alternatively, it too can be incorporated in the trapezoidal loading." | 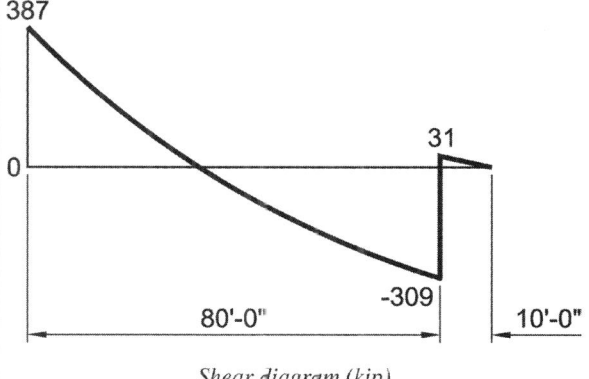<br>Shear diagram (kip)<br><br>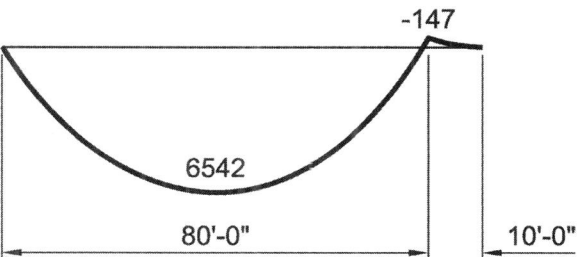<br>Moment diagram (ft-kip)<br><br>Fig. E3.11—Shear and moment diagrams for Case I. |
| | Case II:<br>Eccentricity at $e_y = 13$ ft<br>Refer to previous Step 5 for calculation of eccentricity.<br><br>Force equilibrium<br><br>$q_L\left(\dfrac{L}{2}\right) + q_R\left(\dfrac{L}{2}\right) = F_{px,des(EW)}$ (I)<br><br>Moment equilibrium (taken around CL W1)<br><br>$q_L\left(\dfrac{L}{2}\right)\left(\dfrac{L}{3}\right) + q_R\left(\dfrac{L}{2}\right)\left(\dfrac{2L}{3}\right)$<br>$= (F_4 + F_5)(80\text{ ft}) - F_2(200\text{ ft})$ (II)<br><br>$F_2$, $F_4$, and $F_5$ are per Table E.6.<br><br>Solve equations (I) and (II) for $q_L$ and $q_R$: | $q_L\left(\dfrac{90\text{ ft}}{2}\right) + q_R\left(\dfrac{90\text{ ft}}{2}\right) = 726$ kip<br><br>$q_L\dfrac{(90\text{ ft})^2}{6} + q_R\dfrac{2(90\text{ ft})^2}{6}$<br>$= (223\text{ kip} + 140\text{ kip})(80\text{ ft}) - (33.5\text{ kip})(200\text{ ft})$<br><br>$q_R = 0.4$ kip/ft<br>$q_L = 15.7$ kip/ft |

Draw the shear and moment diagrams for the diaphragm assuming simply supported beam behavior (Fig. E3.12).

The maximum moment is located at 36 ft from the south end of the diaphragm.

$V_{max}$ = 363 kip
$M_{max}$ = 6507 ft-kip, say, 6500 ft-kip

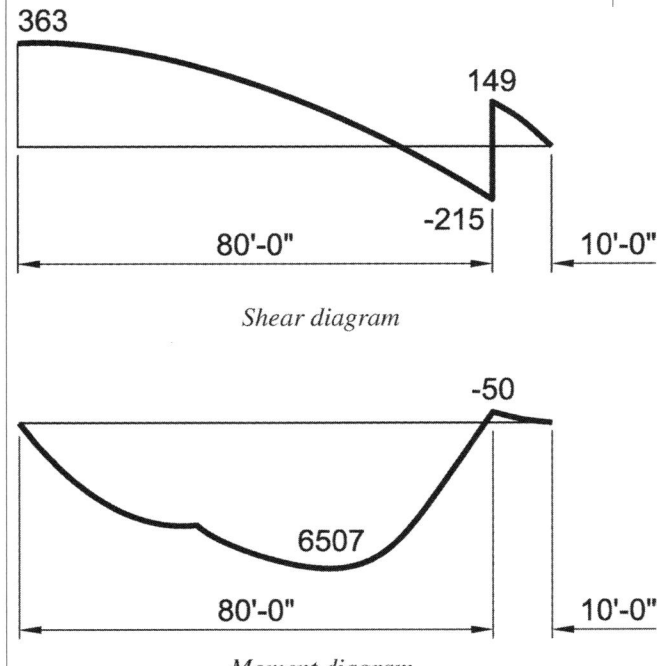

Fig. E3.12—Shear and moment diagrams for Case II.

|  | Case I | Case II |
|---|---|---|
| Shear, kip | 389 | 363 |
| Moment, ft-kip | 6540 | 6500 |

Case I controls. Therefore, design diaphragm in the East-West (E-W) direction for the inertial force obtained from controlling case.

Note: Taking the approach of equivalent uniformly distributed inertial force by ignoring the accidental torsion, the corresponding shear and moment forces are:

Shear: 317.6 kip at wall W1 and 408.4 kip at walls W4 and W5 combined.
Maximum moment: 6252 ft-kip

Comparing the moments of the two approaches, we find that the difference is less than 5 percent - negligible.
Experienced engineers will usually continue the design using the uniformly distributed inertia force. This example, however, will use the detailed approach applying the five percent accidental torsion.

Force summary in the E-W direction

|  | Case I | Case II | Controlling case |
|---|---|---|---|
| Maximum moment, ft-kip | 6540 | 6500 | I |
| $W_1$ shear force, kip | 387 | 363 | I |
| $W_4$ shear force, kip | 131 | 140 | II |
| $W_5$ shear force, kip | 208 | 223 | II |

| Step 12: Chord reinforcement E-W | | |
|---|---|---|
| R12.1.1 | Assume the slab behaves like a beam with compression and tension forces at the near and far edges, respectively: $C_{chord} = T_{chord} = M/d$ | |
| | Chord reinforcement resisting tension must be located within $h/4$ of the tension edge of diaphragm. | $h/4 = 200$ ft/4 $= 50$ ft  $d = 175$ ft |
| | Assume that tension reinforcement will be placed within wall thickness. Therefore, moment arm is approximately 200 ft − 1/2 (10 in./12) = 199.58 ft at both east and west sides of the slab edges: | $d = 199.58$ ft |
| | Chord force<br>The maximum chord tension force is at midspan:<br><br>$$T_u = \frac{M_u}{d}$$ | $$T = \frac{6540 \text{ ft-kip}}{199.58 \text{ ft}} = 32.8 \text{ kip}$$ |
| 18.12.7.5 | Calculate the required chord width for the calculated concrete compressive strength limit of $0.2f_c'$<br><br>$$w_{chord} > \frac{C_{chord}}{0.2 f_c' t}$$ | $$w_{chord} > \frac{32.8 \text{ kip}}{(0.2)(4000 \text{ psi})(10 \text{ in.})} = 4.1 \text{ in.}$$ |
| | Note: Chord force does not need to be increased by the overstrength factor. | |
| 12.5.2.3 | Chord reinforcement resisting tension must be located within $h/4$ of the tension edge of diaphragm. | |
| | Check if required calculated width is less than wall thickness: | $w_{chord} = 4.2$ in. $< t_w = 10$ in.   **OK** |
| | Tension due to moment is resisted by deformed bars conforming to Section 20.2.1 of ACI 318-14. Steel stress is the lesser of the specified yield strength and 60,000 psi. | $f_y = 60,000$ psi |
| | Required reinforcement<br>$\phi T_n = \phi f_y A_s \geq T_u$ | $$A_{s,req'd} = \frac{32.8 \text{ kip}}{(0.9)60 \text{ ksi}} = 0.61 \text{ in.}^2$$ |
| | Note: The chord reinforcement at east and west ends is compared to the partial collector reinforcement placed in the east and west walls due to inertia forces in the North-South (N-S) direction.<br>From Fig. E3.9, two No. 8 bars are placed within the wall thickness that exceed the required chord reinforcement. Therefore, **OK**. | |
| Step 13: Diaphragm shear strength | | |
| | Refer to Step 7 for diaphragm shear strength. | |

| | Step 14: Collector design E-W | |
|---|---|---|
| 12.5.4.1 | Wall 1<br>Collectors transfer shear forces from the diaphragm and transfer them axially to wall W1 (Fig. E3.13). In this example, assume collectors extend over the full length of the diaphragm.<br><br>Unit shear force:<br><br>$$v_{u@F} = \frac{F_{u@F}}{B}$$<br><br>In slab:<br><br><br>In Wall W1: | From Step 6 (Table E.5): $F_u$ = 387 kip<br><br><br><br><br>$v_{u@F} = \dfrac{387 \text{ kip}}{200 \text{ ft}} = 1.94 \text{ kip/ft}$<br><br><br>$v_{u@F} = \dfrac{387 \text{ kip}}{90 \text{ ft}} = 4.3 \text{ kip/ft}$ |
| 12.5.4.1 | Walls 4 and 5<br>Collectors transfer shear forces from the diaphragm and transfer them axially to walls W4 and W5 (Fig. E3.13). Collectors extend over the full width of the diaphragm.<br><br>Unit shear force:<br><br>$$v_{u@F} = \frac{F_{u@F}}{B}$$<br><br>In slab:<br><br><br><br>In wall: | From Step 6 (Table E.6):<br>Slab: $F_u$ = 140 kip + 223 kip = 363 kip<br><br>$v_{u@F} = \dfrac{363 \text{ kip}}{200 \text{ ft}} = 1.82 \text{ kip/ft}$<br><br>Wall 4: $F_u$ = 140 kip<br><br>$v_{u@F} = \dfrac{140 \text{ kip}}{28 \text{ ft}} = 5 \text{ kip/ft}$<br><br>Wall 5: $F_u$ = 223 kip<br><br>$v_{u@F} = \dfrac{223 \text{ kip}}{40 \text{ ft}} = 5.58 \text{ kip/ft}$ |

# CHAPTER 8—DIAPHRAGMS

| Step 15: Collectors design N-S | | |
|---|---|---|
| | Force at diaphragm to wall connection<br>The proportional diaphragm force that the collector transfers to walls connection is (Fig. E3.13):<br><br>Wall 1 west end:<br>$-(1.94 \text{ kip/ft})(55 \text{ ft}) = -106.7 \text{ kip}$<br>Wall 1 east end:<br>$-106.7 \text{ kip} + (2.36 \text{ kip/ft})(90 \text{ ft}) = 106.6 \text{ kip}$<br><br>Diaphragm end:<br>$106.6 \text{ kip} - (1.94 \text{ kip/ft})(55 \text{ ft}) = 0 \text{ kip}$<br><br>Wall 4 west end:<br>$-(1.82 \text{ kip/ft})(40 \text{ ft}) = -72.8 \text{ kip}$<br><br>Wall 4 east end:<br>$-72.8 \text{ kip} + (3.18 \text{ kip/ft})(28 \text{ ft}) = 16.2 \text{ kip}$<br><br>Wall 5 west end:<br>$16.2 \text{ kip} - (1.82 \text{ kip/ft})(52 \text{ ft}) = -78.6 \text{ kip}$<br><br>Wall 5 east end:<br>$-78.6 \text{ kip} + (3.76 \text{ kip/ft})(40 \text{ ft}) = 71.8 \text{ kip}$<br><br>Diaphragm east end:<br>$+71.8 \text{ kip} - (1.82 \text{ kip/ft})(40 \text{ ft}) = 0 \text{ kip}$ | 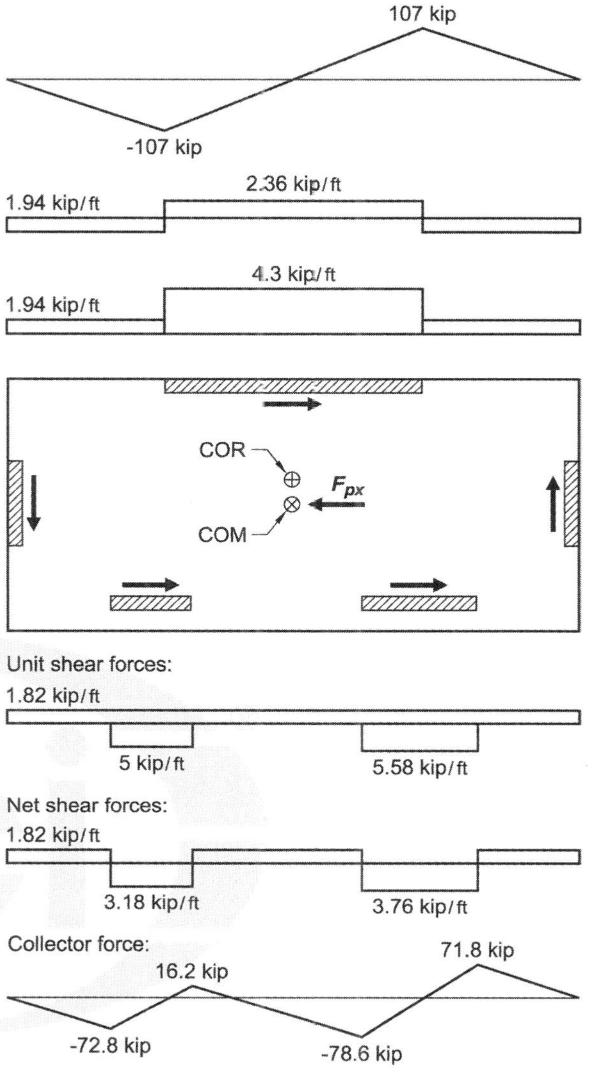 *Fig. E3.13—Collector forces in the E-W direction.* |
| 18.12.2.1 | The collector factored force that is transferred to the walls is shown in Fig. E3.13:<br><br>This collector force is then multiplied by the system overstrength factor, $\Omega_o = 2.5$ for building systems with special structural walls in SDC D (ASCE/SEI 7, Table 12.2-1). | |

| | Wall 1 | | Wall 4 | | Wall 5 | |
|---|---|---|---|---|---|---|
| $2.5T_u$ | 268 | 268 | 182 | 41 | 197 | 180 |
| $2.5C_u$ | −268 | −268 | −182 | −41 | −197 | −180 |

| | | |
|---|---|---|
| 12.5.4.2<br><br>18.12.7.5 | Collectors are designed as tension and compression members.<br>There are no beams along CL 9, so portion of the slab is used as a collector.<br>The collector width is determined by engineering judgement and chosen such that the limiting stresses are not exceeded. When the tension and compression collector forces are increased by the overstrength factor, then the limiting concrete compressive stress is $0.5f_c'$. Calculate the compressive collector width:<br><br>$w_{chord} = 2.5C_{coll}/0.2f_c't$ | |

| | Wall 1 | | Wall 4 | | Wall 5 | |
|---|---|---|---|---|---|---|
| $w_{coll}$, in. | 17 | 17 | 9 | 2 | 9.9 | 9 |
| is $w_{coll} > t_w$? | Y | Y | N | N | N | N |

For Wall 1, the required collector width (17 in.) is wider than the wall thickness (8 in.). Part of the seismic force is resisted by reinforcement placed in-line with the shear wall to transfer the force directly to the end of the shear wall and direct bearing of slab against wall in compression. Therefore, two No. 6 in-line with the wall. The balance of seismic force is resisted by reinforcing bars placed along the sides of the wall and uses the slab shear-friction capacity at the wall-to-slab interface to transfer seismic forces to the wall.
For Walls 4 and 5, the maximum required collectors' widths (9 in. and 9.9 in., respectively) are narrower than the walls widths (10 in.), therefore, place reinforcement within the walls widths.

| | | |
|---|---|---|
| | Wall 1:<br>Required area of collector reinforcement:<br><br>$\phi T_n = \phi f_y A_s \geq T_u$ | $A_s = \dfrac{\Omega_o T_{coll}}{0.9 f_y} = \dfrac{268 \text{ kip}}{0.9(60 \text{ ksi})} = 5 \text{ in.}^2$ |
| | Note: Collector reinforcement along the length of the diaphragm may be varied based on required strength and terminated when not required. In this example, the reinforcement is extended over the full length of the diaphragm. | |
| R12.5.4 | The collector width, as suggested by ACI 318 commentary, is an arbitrary width equal to half the wall width taken from the face of the wall plus the wall width.<br><br>$b_{eff} = \ell_{wall}/2 + t_{wall}$ | $b_{eff} = 90 \text{ ft}/2 + 8 \text{ in.}/(12 \text{ in./ft}) = 45.67 \text{ ft}$ |
| | There are several options to detail the collector reinforcement:<br><br>1. Place bars within the arbitrary width of 45.67 ft. | Spreading bars over more than half the diaphragm width is not practical. |
| | 2. Use the calculated chord reinforcement in the N-S direction to resist the collector inertial forces in the E-W direction over $h/4$. | This option is acceptable as the required chord and collector calculated reinforcement is approximately equal, 5 in.². The collector is wider than the wall, therefore, longitudinal and transverse reinforcement must be provided to transfer forces from the collector into the wall. |
| | 3. Place bars in a 2 ft 0 in. deep by 12 in. wide edge beam | Collector and chord reinforcement are placed in a beam. In this example, this option is used. |

| | | |
|---|---|---|
| | The required reinforcement to resist the gravity dead and live load is calculated to be equal to 3.8 in.$^2$. Required collector reinforcement is 5 in.$^2$ calculated above. Therefore, a total of 8.8 in.$^2$ must be placed within the beam to resist gravity loads combined with either the calculated chord force or collector force due to inertial forces in the N-S and E-W directions, respectively (Fig. E3.14A). Gravity load calculation is not provided in this example. Note: Typically, gravity design is carried out for $1.2D + 1.6L$. For seismic, the gravity loading is $(1.2 + 0.2S_{DS})D + 0.5L$, which is usually less than the previous case. Therefore, it may be possible to count on a portion of provided gravity reinforcement for seismic collectors. | |
| | The beam (12 in.) is wider than the wall (8 in.). Therefore, the force transferred from the beam to the wall is eccentric (2 in.). | $M_u = (268 \text{ kip})(2 \text{ in.}) = 536 \text{ in.-kip}$ |
| | Tension and compression forces at both ends of the wall are: | $T = C = \dfrac{536 \text{ in.-kip}}{(89 \text{ ft})(12 \text{ ft})} = 0.5 \text{ kip}$ |
| | Required reinforcement: | $A_s = \dfrac{(0.5 \text{ kip})(1000 \text{ lb/kip})}{(0.9)(60{,}000 \text{ psi})} = 0.01 \text{ in.}^2$ |
| | Note: The force is very small and the corresponding reinforcement is negligible. Therefore, it is assumed that the result of the eccentric force between beam and wall centerlines is resisted by the diaphragm. In case the force is large (large eccentricity, large force, or shorter wall length), reinforcement is required and placed and properly developed at both ends of the wall and extending into the diaphragm a minimum length equal to the development length of the bars. | |
| | Wall 4: Required collector width is less than the wall thickness. Reinforcement may be placed within the wall width: | $A_s = \dfrac{\Omega_o T_{coll}}{0.9 f_y} = \dfrac{182 \text{ kip}}{0.9(60 \text{ ksi})} = 3.37 \text{ in.}^2$ Try four No. 9 bars: $A_{s,prov.} = 4 \text{ in.}^2 > A_{s, req'd} = 2.94 \text{ in.}^2$ |
| | Shear friction reinforcement is not required as the collector force is already developed into the wall. | |
| 22.9.4.4 | The value of $V_n$ across the assumed shear plane must not exceed the lesser of the following limits: (a) $0.2 f_c' A_c$ (b) $(480 + 0.08 f_c') A_c$ (c) $1600 A_c$ | $(0.2)(4000 \text{ psi}) = 800 \text{ psi}$ **OK** $(480 + 0.08(4000 \text{ psi})) = 800 \text{ psi}$ **OK** $1600 \text{ psi}$ Therefore, $V_n$ must not exceed: $\dfrac{(800 \text{ psi})(10 \text{ in.})(28 \text{ ft})(12 \text{ in.}/\text{ft})}{1000 \text{ lb/kip}} = 2688 \text{ kip} > V_n$ |
| | Wall 5: Required collector width is less than the wall thickness. Reinforcement may be placed within the wall width: | $A_s = \dfrac{\Omega_o T_{coll}}{0.9 f_y} = \dfrac{197 \text{ kip}}{0.9(60 \text{ ksi})} = 3.65 \text{ in.}^2$ Try four No. 9 bars: $A_{s,prov.} = 4.0 \text{ in.}^2 > A_{s, req'd} = 3.65 \text{ in.}^2$ |
| | Shear friction reinforcement is not required as the collector force is already developed into the wall. | |

| Step 16: Shrinkage and temperature reinforcement | | |
|---|---|---|
| 12.6.1<br>24.4.3.2 | Shrinkage and temperature reinforcement:<br><br>$A_{S+T} \geq 0.0018 A_g$ | $A_{S+T} = (0.0018)(10 \text{ in.})(16 \text{ in./ft}) = 0.288 \text{ in.}^2$ |
| 24.4.3.3 | Spacing of S+T reinforcement is the lesser of $5h$ and 18 in.<br>(a) $5h = 5(15 \text{ in.}) = 75$ in.<br>(b) 18 in. **Controls** | Note: Shrinkage and temperature reinforcement may be part of the main reinforcing bars resisting diaphragm in-plane forces and gravity loads. If provided reinforcement is not continuous (placing bottom reinforcing bars to resist positive moments at midspans and top reinforcing bars to resist negative moments at columns), the engineer must ensure continuity between top and bottom reinforcing bars by providing adequate splice lengths between them. |
| **Step 17: Reinforcement detailing** | | |
| 25.4.2.2<br>25.4.10.2<br>12.7.3.2 | Development<br>Chord and collector reinforcement are extended over full length and width of the edges of the diaphragm. Therefore, development length will be calculated only to determine splice lengths.<br>Development length of shear transfer reinforcement<br><br>$\ell_d = \dfrac{f_y \Psi_t \Psi_e}{25 \lambda \sqrt{f_c'}}$ | <table><tr><th>$\ell_d$</th><th>$\ell_d$, in.</th><th>Use $\ell_d$, in.</th></tr><tr><td>No. 5</td><td>23.4</td><td>24</td></tr><tr><td>No. 6</td><td>28.5</td><td>30</td></tr></table> |
| 25.5.2.1 | Splices<br>Because the building lengths are longer than a standard shipping length of the No. 8 longitudinal reinforcement, splices will be needed. Use Class B splice: $1.3(\ell_d)$ | $\ell_d = (1.3)\dfrac{60{,}000 \text{ psi}}{25(1.0)\sqrt{4000 \text{ psi}}}(1.0 \text{ in.}) = 55.6 \text{ in.}$<br>Say, 56 in. (4 ft 8 in.) |
| 18.12.7.6 | The center-to-center spacing of the longitudinal bars for collector and chords at splices and anchorage zones is but not less than 1.5 in. and concrete cover $\geq 2.5 d_b$, but not less than 2 in.<br><br>Therefore, transverse reinforcement is not required. | $3d_b = 3(1.0 \text{ in.}) = 3.0$ in. minimum spacing<br>$(2.5)(1.0 \text{ in.}) = 2.5$ in. cover |
| 12.7.2.1 | Reinforcement spacing<br>Chord and collector reinforcement minimum and maximum spacing must satisfy 12.7.3.2 and 12.7.3.3. | |
| 25.2.1 | Section 25.2 requires minimum spacing of<br>(a) 1 in.<br>(b) $4/3 d_{agg}$.<br>(c) $d_b$       No. 9 | Minimum spacing 1.128 in., say, 1.25 in. **Controls** |
| 18.12.7.6 | Collector reinforcement spacing at splice must be at least the larger of:<br>a. At least three longitudinal $d_b$<br>b. 1.5 in.<br>c. $c_c \geq \max[2.5 d_b, 2 \text{ in.}]$ | $3(1.128 \text{ in.}) = 3.384$ in. say 3.5 in. **Controls** |
| 12.7.2.2 | Maximum spacing is the smaller of $5h$ or 18 in. | 18 in. **Controls** |

# CHAPTER 8—DIAPHRAGMS

**Step 18: Detailing**

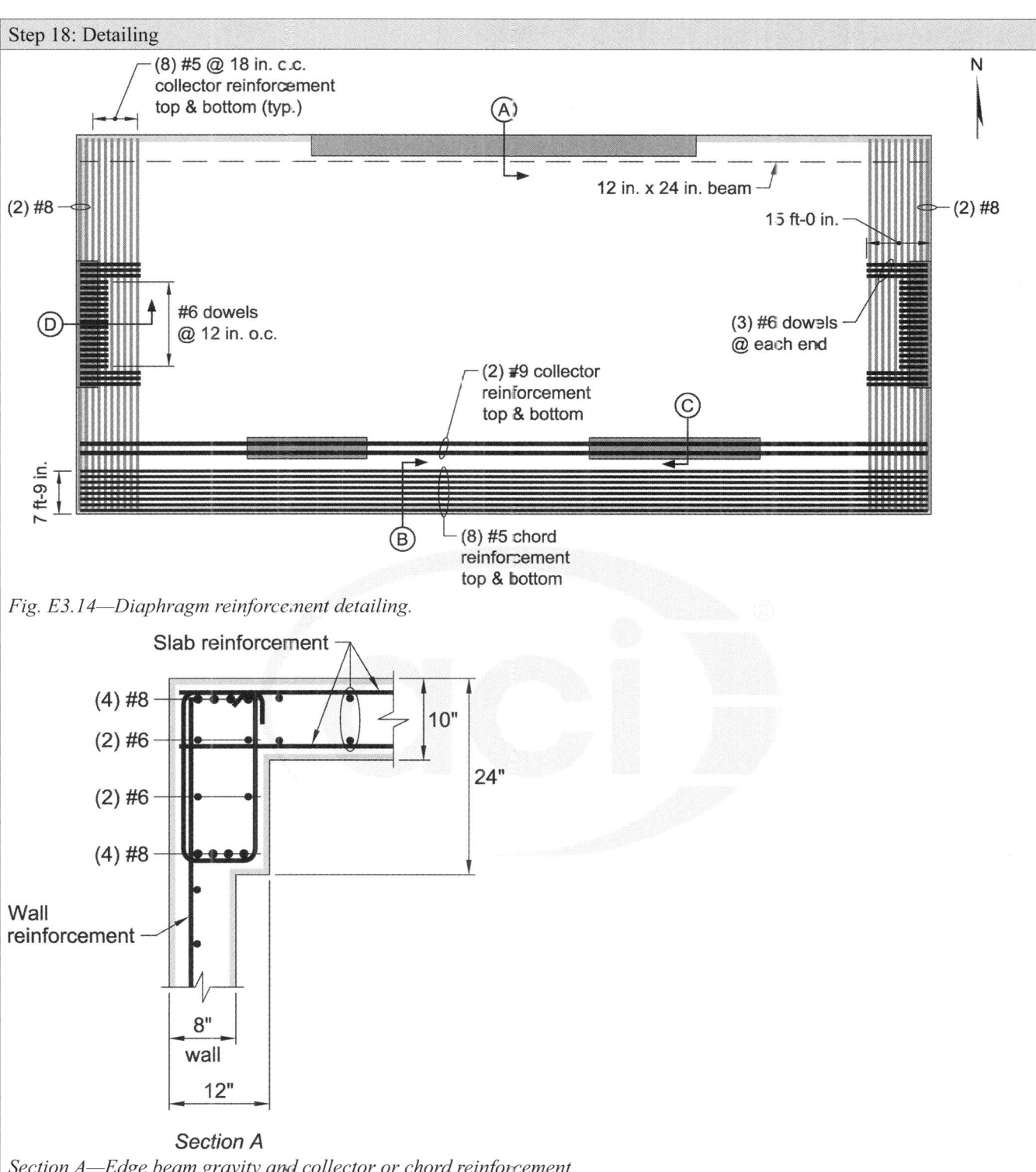

Fig. E3.14—Diaphragm reinforcement detailing.

Section A—Edge beam gravity and collector or chord reinforcement.

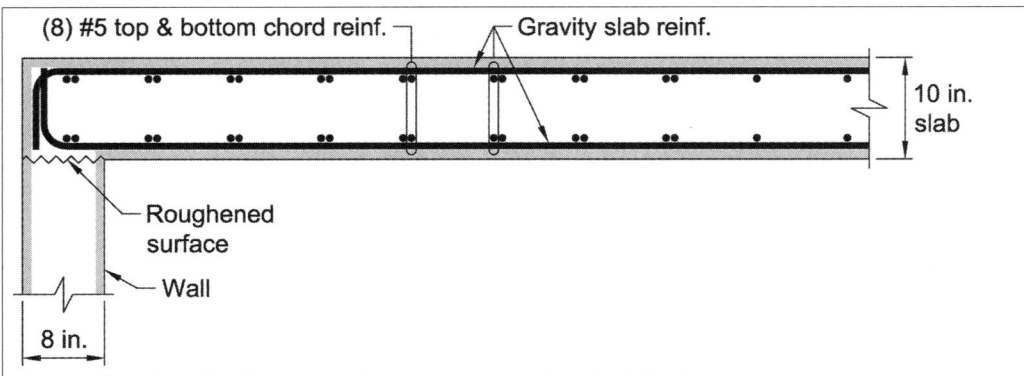

*Section B—Chord/collector reinforcement at south end of diaphragm.*

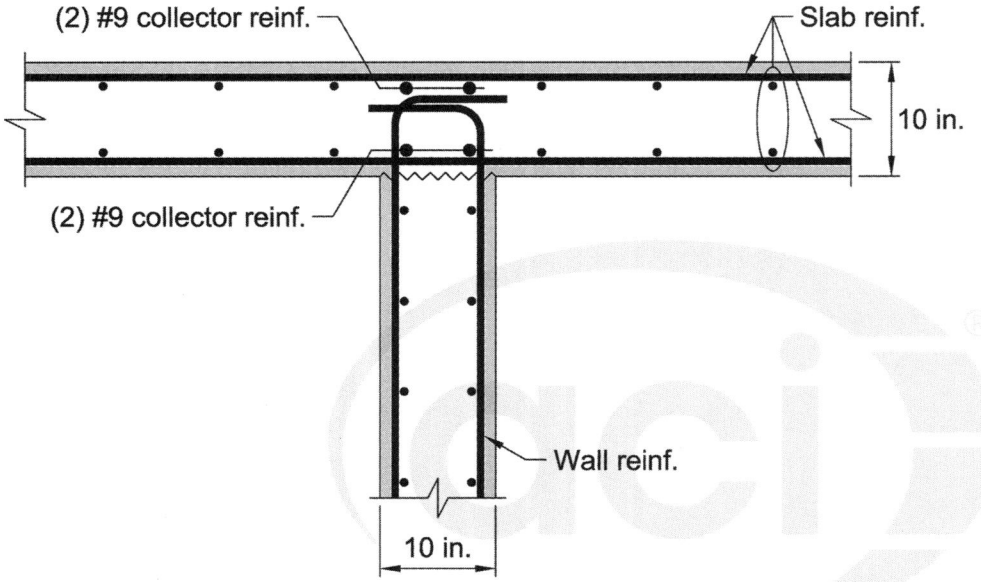

*Section C—Collector reinforcement along Walls 4 and 5.*

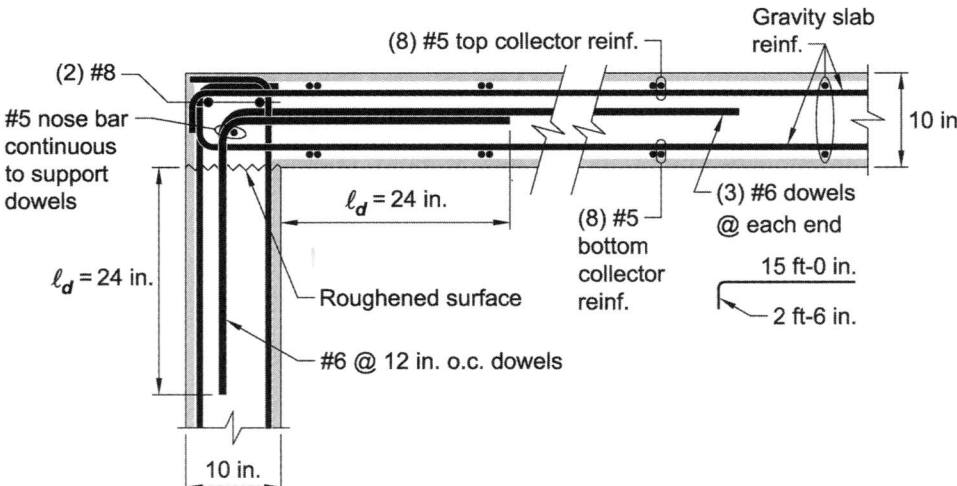

*Section D—Collector reinforcement along Walls 2 and 3.*

Note: Wall reinforcement not shown for clarity for Walls W4 and W5 the detail is similar, however, alternate dowels to either side of the wall.

## Step 19: Discussion

There is no consensus among engineers on how to distribute the diaphragm inertia force. Based on discussions with several respected engineers, the main approaches are as follows:

ASCE 7-10 (third printing) Section 12.10.3.4, recommends shifting the center of mass by a minimum of 5 percent of the building dimension in either direction and perpendicular to the seismic loading, referred to as accidental eccentricity. This five percent torsional eccentricity is applied in addition to the calculated geometric eccentricity. However, the ASCE 7-10 recommendation is located in the commentary, therefore, it is not mandatory.

This example follows the ASCE 7-10 recommendation.

The 5 percent eccentricity is excluded in the analysis. The diaphragm inertia force is uniformly distributed to the diaphragm.

The shear forces in the lateral-force-resisting system due to the equivalent lateral force analysis are compared to the forces due to diaphragm inertia forces. The diaphragm is designed for the larger force.

When center of mass and center of rigidity do not coincide, the lateral-force-is example: (W1 (27.3 kip), W4 (10.5 kip), and W5 (16.7 kip)) for a seismic force acting in the N-S direction and (W2 and W3 (22 kip)) for a seismic force acting in the E-W direction.

Drawing the moment diagram shows a discontinuity at the center of rigidity. Moehle et al. also state in *NIST Report No. GCR 10-917-4* that:

"For a rectangular diaphragm of uniform mass, a trapezoidal distributed force having the same total force and centroid is then applied to the diaphragm. The resulting shears and moments are acceptable for diaphragm design. Note that this approach leaves any moment due to (shear walls perpendicular to the diaphragm inertia lateral force) unresolved; sometimes this is ignored or, alternatively, it too can be incorporated in the trapezoidal loading."

Other engineers incorporate the moment due to shear walls perpendicular to the diaphragm inertia lateral force. This results in discontinuity (jump) in the moment diagram as shown in Fig. E3.15. Equilibrium in the system is obtained by drawing the moment diagram due to the shear forces in the shear walls perpendicular to the direction of the seismic force (Fig. E3.15):

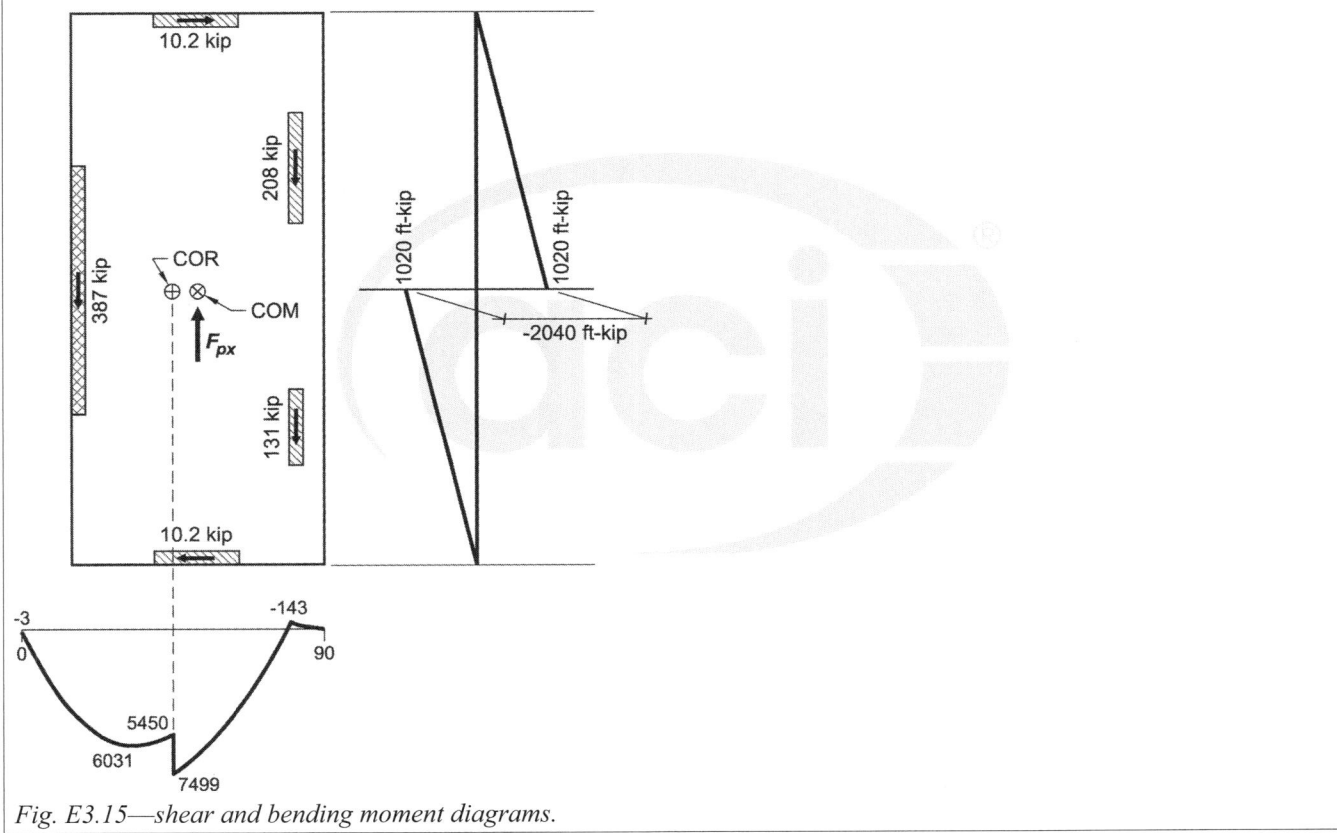

*Fig. E3.15—shear and bending moment diagrams.*

In this example, the moment diagram is constructed by incorporating the shear force in the trapezoidal loading for the construction of the moment diagram. Since Case I resulted in a slightly higher moment, the calculations for that case follow:

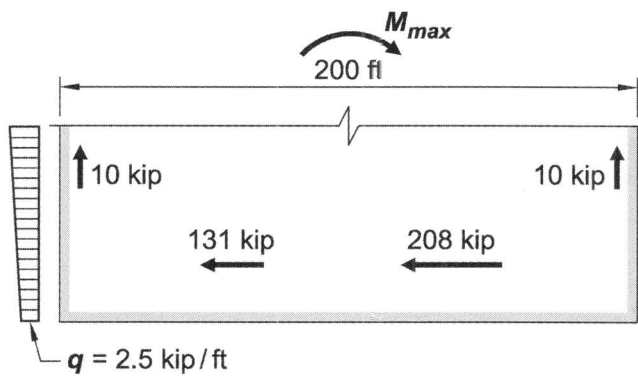

Fig. E3.16—Free body diagram.

$M_x = (131 \text{ kip} + 208 \text{ kip})(x - 10 \text{ ft}) - (2.5 \text{ kip/ft})(x^2/3) - [2.5 \text{ kip/ft} + (13.6 \text{ kip/ft} - 2.5 \text{ kip/ft})/50 \text{ ft} \times x](x^2/6)$

# CHAPTER 9—COLUMNS

## 9.1—Introduction

The column chapter, Chapter 10 in ACI 318-14, follows the organization of the other member chapters: applicability; initial data; analysis to determine the required strength; design area of reinforcement needed to exceed the required strength; check against minimum reinforcement required; and detailing. The analysis must be consistent with ACI 318-14, Chapters 4 through 6. A quick review of Sections 10.1 through 10.4 in ACI 318-14, will remind the designer of limits and rules for columns that should be accounted for in their analysis model. The requirements for column design start in Section 10.5 in ACI 318-14.

A column is always part of the gravity-force-resisting system, and in cast-in-place construction is often part of the lateral force-resisting systems (LFRS). The most common lateral design forces are seismic and wind. For buildings designed to resist seismic forces, the requirements of Chapter 10 in ACI 318-14 apply, along with the additional seismic requirements in Chapter 18 in ACI 318-14 for columns that are part of an ordinary, intermediate, or special moment frame system. There are also seismic requirements for columns that are not part of a LFRS. The seismic requirements are intended to increase column ductility to accommodate the large displacements that are expected during a maximum design earthquake. For buildings that resist only wind forces, columns are designed by Chapter 10 in ACI 318-14 whether they are designated as part of a LFRS or not. There are no additional requirements for wind forces.

This Handbook provides some explanation of ACI 318-14 requirements and how they impact a column's design and detailing. A review of basic engineering principles is provided so that a designer can effectively design columns by hand or computer using only this Handbook.

## 9.2—General

The provisions of Chapter 10 in ACI 318-14 apply for the design of nonprestressed, prestressed, and composite columns. Headings of a section are considered part of the code and care should be taken to notice when a heading limits the following requirements to a particular type of column. The word "nonprestressed" is typically in reference to cast-in-place columns and "prestressed" with precast columns. While ACI 318-14 covers post-tensioned columns, they are not commonly used. Chapter 14 in ACI 318-14 covers the design of plain concrete pedestals.

## 9.3—Design limits

**9.3.1** *General*—Concrete columns offer architects an opportunity to create various cross-sectional shapes for the aesthetic purposes. Unusual cross-sections, however, are more difficult to analyze. Section 10.3 in ACI 318-14 permits the designer to use an effective cross-sectional areas for various situations that allow for a simpler analysis. For

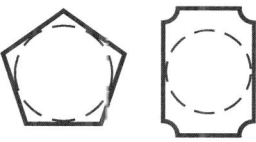

*Fig. 9.3.1—Permitted cross-section for analysis.*

example, Section 10.3.1.1 in ACI 318-14 permits an idealized circular section within the actual section (Fig. 9.3.1). The key point is that a using a portion of an oversized column or wall that is easier to analyze is permitted.

Section 10.3.1.2 in ACI 318-14 allows an oversized column to be designed with a smaller effective area, with a lower limit of one-half the total area. The analysis and design assume the smaller effective area, but the column is detailed considering the actual cross section. Note that the minimum area of steel (refer to 9.6 of this Handbook) is $0.01A_g$ based on the smaller effective area, but $A_g$ cannot be less than half the actual area. Note that the shape of the column can be determined by the building architecture but it must always meet the requirements of ACI 318-14.

**9.3.2** *Initial sizing*—It is not economical to have a unique design for each column in the building. The following guidelines help in economical column construction:

1. Reuse formwork as much as possible. It is common to use only three or four columns sizes for the entire building.
2. Use the same strength of concrete for all columns at a level. Structures under six stories or less commonly use one concrete strength for the full height of the building. Note section 15.3 in ACI 318-14 has additional design requirements at the floor joints if the concrete strength in the columns exceeds 1.4 times the concrete strength of the floor system.
3. Proportion the column cross sections and concrete strengths so that reinforcement ratios are in the range of 1 to 2 percent.
4. If higher axial strength is needed, increasing the strength of concrete is usually more efficient than increasing reinforcement area.

The analysis and design of columns is an iterative process. To begin, the designer assumes sectional properties in order to run an analysis. An initial column area, $A_g$, can be estimated by dividing the maximum factored axial load by $0.4f_c'$ for ordinary columns or $0.3f_c'$ for columns in high seismic areas. Columns are usually rectangular, square, or round. For the first iteration, a 1 percent reinforcement ratio is evenly distributed around the column perimeter. An effective moment of inertia, $I_{eff}$, of concrete members is used in the analysis to account for cracking at the nominal condition. The simple $I_{eff}$ values in Table 6.6.3.1.1(a) in ACI 318-14 are generally used and the cross-section properties are assumed to be uniform for the length of the member. With these

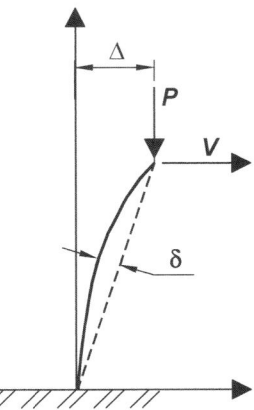

Fig. 9.4.1a—P-Δ effects.

assumptions, an initial analysis is run and, subsequently, section properties or reinforcement area are adjusted as necessary.

## 9.4—Required strength

**9.4.1** *General*—The required strength is calculated using the factored load combinations in Chapter 5 and analysis procedures in Chapter 6 (ACI 318-14). Three methods of analysis: 1) elastic first-order analysis; 2) elastic second-order analysis; and 3) inelastic second-order analysis, are permitted as discussed in Chapter 3 of this Handbook. Regardless of the method chosen, all columns must be checked for slenderness.

Slenderness effects are associated with higher slenderness ratios, $k\ell_u/r$. Higher slenderness ratios occur for long columns, columns with a small cross-sectional dimension, or columns with limited end restraint. As a column becomes more slender, larger lateral deformations and deformation along the length occur due to the applied load. The column must support additional moment created by the column axial load acting on the deformed column, also known as second-order moments. There are two types of second-order moments related to the P-Delta effects shown in Fig 9.4.1a: 1) second-order moments due to translation of the column ends (P-Δ); and 2) second-order moments due to deflection along the member (P-δ). These second-order moments gradually increase due to this geometric nonlinearity until the column stabilizes. If the slenderness ratio is very high, the column becomes unstable and cannot resist the axial load, refer to Fig. 9.4.1b).

**9.4.2** *Slenderness concepts*—A column's degree of slenderness is expressed in terms of its slenderness ratio, $k\ell_u/r$, where $\ell_u$ is unsupported column length; $k$ is effective length factor reflecting end restraint and lateral bracing conditions; and $r$ is the radius of gyration, reflecting the size and shape of a column cross section.

The column's unsupported length $\ell_u$ is the clear distance between the underside of the beam, slab, or column capital above, and the top of the floor below. The unsupported length may be different in two orthogonal directions depending on the building geometry. Figure 9.4.2a shows different framing conditions and corresponding unsupported lengths

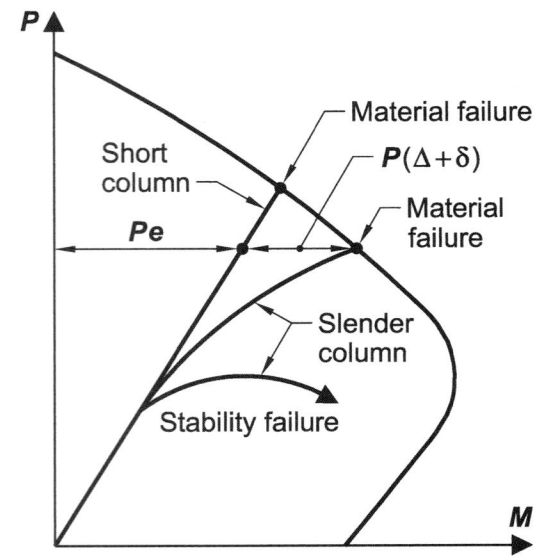

Fig. 9.4.1b—Effects of slenderness on a column.

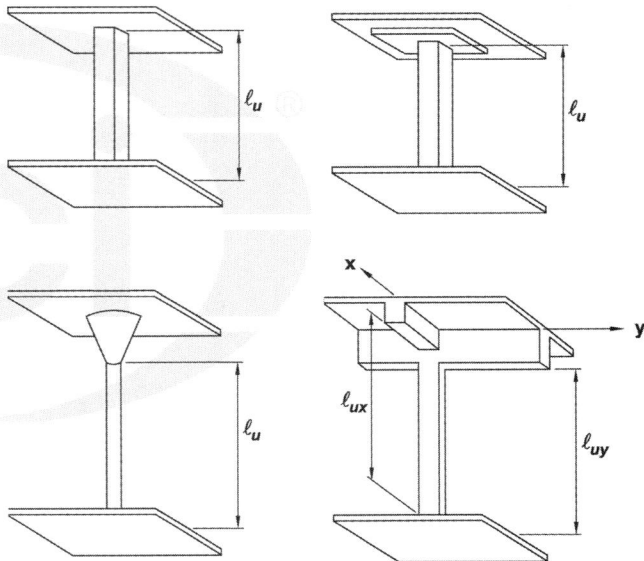

Fig. 9.4.2a—Unsupported column length $\ell_u$.

($\ell_u$). Each coordinate and $x$ and $y$ subscript in the figure indicates the plane of the frame in which the stability of column is investigated.

The effective length factor $k$ reflects the column's end restraint and lateral bracing conditions as shown in Fig. 9.4.2b. The factor $k$ varies between 0.5 and 1.0 for laterally braced columns, and between 1.0 and ∞ for unbraced columns. Most columns have end restraints that are neither perfectly hinged nor fully fixed. The degree of end restraint depends on the floor stiffness relative to the column stiffness. Jackson and Moreland alignment charts, given in Fig. R6.2.5 in ACI 318-14, can be used to determine the factor $k$ for different values of relative stiffness at column ends. The stiffness ratios $\psi_A$ and $\psi_B$ used in the charts should reflect concrete cracking, and the effects of sustained loading. Beams and slabs are flexure dominant members and may

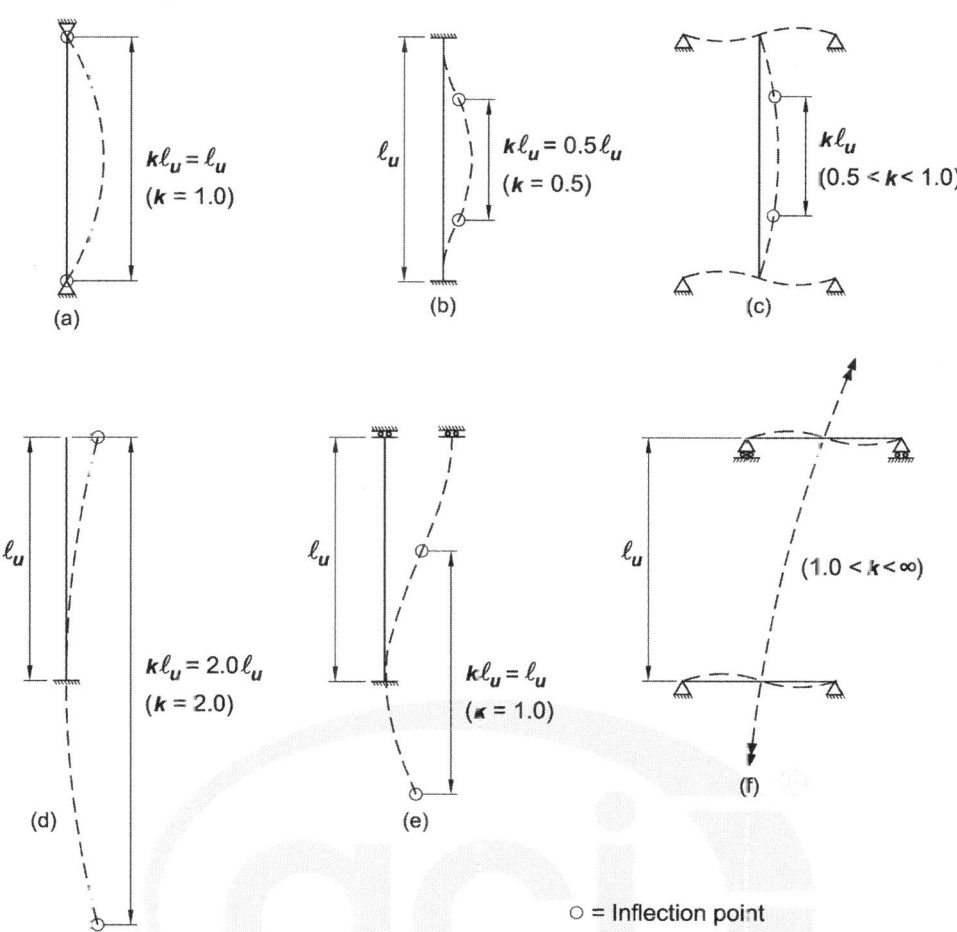

Fig. 9.4.2b—Effective length factor k for columns: (a), (b), and (c) are for nonsway frames; and (d), (e), and (f) are for sway frames.

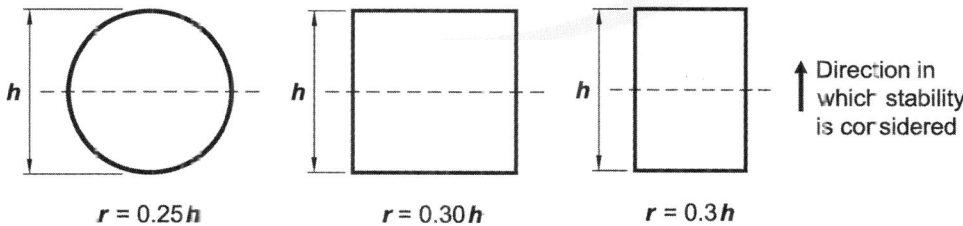

Fig. 9.4.2c—Radius of gyration for circular, square, and rectangular sections.

crack significantly more than columns, which are compression dominant members. The reduced moment of inertia values given in Section 6.6.3.1.1 of ACI 318-14 should be used to determine $k$. Tables D1.1 through D1.5 in the supplement to this Handbook, ACI Reinforced Concrete Design Handbook Design Aid – Analysis Tables (ACI SP-17DA), provide design aids for the calculation of the effective length factor, stiffness, and moment of inertia.

The radius of gyration introduces the effects of cross-sectional size and shape to slenderness. A section with a higher moment of inertia per unit area produces a lower slenderness ratio and thus a more stable column. The radius of gyration, $r$, is defined in Section 6.2.5.1 of ACI 318-14 and shown in the following Eq. (9.4.2)

$$r = \sqrt{\frac{I}{A}} \quad (9.4.2)$$

It is permissible to use $r = 0.3h$ for square and rectangular sections, and $r = 0.25h$ for circular sections, where "$h$" is the dimension in the direction stability is being considered. This is shown in Fig. 9.4.2c.

**9.4.3** *First-order analysis*—For a first-order elastic analysis, Section 6.6.4 in ACI 318-14 provides a moment magnification method which conservatively accounts for slenderness. This method is shown in Example 9.1 of this Handbook. In a first-order elastic analysis, the building frame is analysed once and results are used for input to the moment magnification method.

**9.4.3.1** *Sway or nonsway frames*—The designer needs to determine if the column is in a sway or nonsway frame. A frame is nonsway if it is sufficiently supported by lateral bracing, such as structural walls. Structural walls used for elevator shafts, stairwells, partial building enclosures, or interior stiffening elements provide substantial drift control and lateral bracing. In many cases, even a few structural walls can brace a multi-story, multi-bay building. Sway must be checked for each direction and floor. ACI 318-14 provides three methods to determine if the lateral stiffness is sufficient to designate the frame as nonsway.

1. Section 6.2.6, ACI 318-14: Columns are nonsway if the gross lateral stiffness of the walls (bracing elements) in a story is at least 12 times the gross lateral stiffness of the columns in that story. This is a simple, conservative hand calculation.

2. Section 6.6.4.3(a), ACI 318-14: Columns are nonsway if the increase in column end moments due to second-order effects does not exceed 5 percent of the first-order end moments.

3. Section 6.6.4.3(b), ACI 318-14: Columns are nonsway if the stability index "Q" does not exceed 0.05 as shown in Eq. (9.4.3.1):

$$Q = \frac{\sum P_u \Delta_o}{V_{us} \ell_c} \leq 0.05 \qquad (9.4.3.1)$$

where $\sum P_u$ is total factored axial load acting on all the columns in a story, $V_{us}$ is total factored story shear, $\Delta_o$ is lateral story drift (deflection of the top of the story relative to the bottom of that story) due to $V_{us}$. Story drift $\Delta_o$ should be computed using section properties taking into account the presence of cracked regions along the member, refer to Section 6.6.3.1 in ACI 318-14.

**9.4.3.2** *Column slenderness*—The moment magnification method states that for columns in sway or nonsway frames, secondary effects may be neglected if the slenderness ratios are below the limits given in Section 6.2.5 in ACI 318-14. If these limits are exceeded in a nonsway frame, then second-order effects due to translation ($P$-$\Delta$) may be ignored, but the second-order effects along the member ($P$-$\delta$) given in 6.6.4.5 in ACI 318-14 need be considered. If the slenderness ratio limits are exceeded in a sway frame, column ($P$-$\Delta$) and ($P$-$\delta$) effects are to be calculated. Section 6.6.4.6 in ACI 318-14 is completed first to calculate the amplified end moments due to translation ($P$-$\Delta$). These modified moments are then used in Section 6.6.4.5 in ACI 318-14 to calculate the second-order effects along the member ($P$-$\delta$).

**9.4.4** *Second-order analysis*—Many software analysis programs can directly compute the second-order effects. Such a program must be capable of the iterative calculations to determine:

(a) Second-order moments ($P$-$\Delta$) due to the laterally deflected structure

(b) Second-order moments ($P$-$\delta$) due to deflection along the length of the column

The column needs to be divided into at least two segments to calculate a deflection along its length. A way to check a computer program for this effect is given in the ACI Q&A, Using an Elastic Frame Model for Column Slenderness Calculations (Frosch 2011). Slenderness effects are nonlinear, so the principle of superposition does not apply for calculating the second-order moments. The forces for a load combination must be combined before running the analysis.

Section 6.7.1.1 in ACI 318-14 states, "An elastic second-order analysis shall consider the influence of axial loads, presence of cracked regions along the length of the member, and effects of load duration." The moment magnification method in Section 6.6.4 in ACI 318-14 accounts for these properties by using stiffness reduction factors, $f_k$. The method uses two different factors for the two different types of slenderness effects, $P$-$\Delta$ and $P$-$\delta$, refer to R6.6.3.1.1, R6.6.4.5.2, and R6.6.4.6.2 in ACI 318-14. A lower stiffness reduction factor is required for $P$-$\delta$ compared to $P$-$\Delta$. Computer programs should also account for stiffness reduction in its second-order analysis. These programs should be reviewed for how they account for stiffness reduction.

## 9.5—Design strength

The majority of reinforced concrete columns are designed to resist flexure, axial force, and shear. Figure 9.5 illustrates strains and stresses in a typical column section subjected to combined moment and axial compression. As can be seen, different combinations of moment and accompanying axial force result in different column nominal strengths and corresponding strain profiles, while also affecting the tension or compression controlled behavior. The combination of moment and axial force that result in a column's nominal strength is traditionally presented by "column interaction diagrams." Interaction diagrams are constructed by computing moment and axial force nominal strengths, as shown below, for different strain profiles.

$$P_n = C_c + C_{s1} + C_{s2} - T_s \qquad (9.5a)$$

$$M_n = C_c x_2 + C_{s1} x_1 + C_{s2} + T_s x_3 \qquad (9.5b)$$

As the strains vary using Eqs. (9.5a) and (9.5b) from pure compression to pure bending, a nominal strength curve can be created as shown by the outer curve in Fig. R10.4.2.1 in ACI 318-14. The nominal strength is adjusted to the design strength by multiplying by the appropriate $\phi$ factors. The $\phi$ factor for compression-controlled sections is 0.75 for spirals and 0.65 for other tie configurations. The $\phi$ factor for all tension-controlled sections is 0.9. This factor varies linearly from 0.65 or 0.75 to 0.9 through the transition zone shown in Fig. 9.5. The design strength curve is shown by the inner curve in Fig. R10.4.2.1 in ACI 318-14. The interaction diagram is the plot of both the nominal and design strength curves shown on a graph.

An electronic spreadsheet is provided as a supplement to this Handbook to demonstrate how to make an interaction diagram (ACI SP17DAE). The key points of the diagram

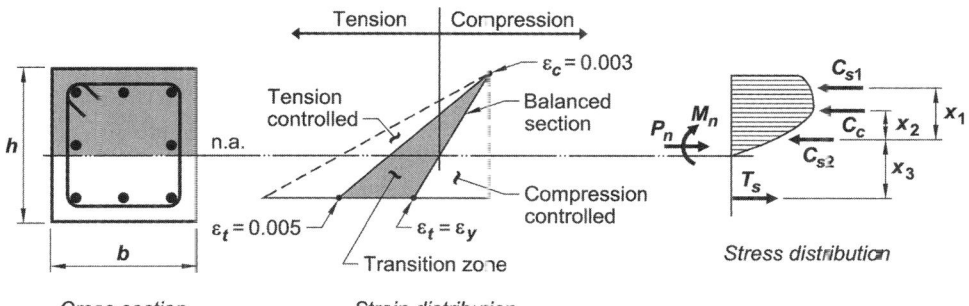

*Fig. 9.5—Column section analysis.*

are: pure compression (zero moment), pure tension (zero moment), pure bending (zero axial force), reinforcement stress at $0.0f_y$, reinforcement stress at $0.5f_y$, and the balanced point where reinforcement stress is at $1.0f_y$ and the concrete reaches it maximum usable strain.

The previous version of this Handbook was a design manual that contained interaction diagrams. These diagrams have been retained in a supplement to this Handbook titled, Reinforced Concrete Design Handbook Design Aid – Analysis Tables (ACI SP-17DA). A brief description of how the diagrams are made is given in the supplement along with information on biaxial moments.

Because shear design in columns is similar to beams, review Beam Chapter 8 of this Handbook for this information. Significant torsion rarely occurs in columns and, therefore, is not specifically addressed in the columns chapter. For column torsion, the beam chapter should be reviewed.

## 9.6—Reinforcement limits

The minimum column vertical reinforcement ratio is $0.01A_g$. This amount is enough to keep the reinforcement from yielding due to concrete creep under sustained service loads (ACI 318-14, Section R10.6.1.1). The maximum reinforcement ratio is $0.08A_g$. This amount is about the maximum that can be realistically provided at the perimeter of a concrete section that would also meet the minimum cover and spacing requirements. This percentage was set in ACI 318-63 when butt splices were common. For present day construction, lap splices are more common and for many projects, bars are spliced at the bottom of the column starting at the floor. In this case, the maximum percent of reinforcement is 4 percent, since the maximum percentage of reinforcement at a section is 8 percent. If the lap splices are staggered, the maximum percent of reinforcement can increase up to 6 percent.

As stated earlier, usual reinforcing ratios are in the range of 1 to 2 percent. There are cases where more reinforcement is necessary, such as to meet a required strength. If the column design routinely requires reinforcement in the 3 percent range, designer should consider a different cross section or higher-strength concrete.

## 9.7—Reinforcement detailing

**9.7.1** *General*—Section 10.5 in ACI 318-14 focuses on the calculation of reinforcement area needed for to resist design forces and moments. Section 10.6 in ACI 318-14 provides minimum column reinforcement area. Section 10.7 in ACI 318-14 provides limitations on the location, spacing, and splicing of longitudinal reinforcement and the location, spacing, geometry, and type of transverse reinforcement. General requirements such as concrete cover, development length, and splice lengths, are covered in Chapters 20 and 25 in ACI 318-14 for all members. Note that many detailing provisions of this chapter are related to how columns are constructed.

**9.7.2** *Longitudinal bars*

**9.7.2.1** *Spacing*—The minimum number of bars in a column is given in Section 10.7.3.1 in ACI 318-14. Square and rectangular columns must have a minimum of four bars. Circular columns must have three or six depending if the tie is triangular or circular. Eight bars are suggested, however, to ensure that the design moment is achieved regardless of the position of reinforcement in the field. The minimum bar spacing is given in Section 25.2 in ACI 318-14. A maximum spacing requirement in Chapter 10 in ACI 318-14 is not explicitly given. Section 18.7.5.2(e) in ACI 318-14, however, states that for columns in a special moment frame, the spacing of longitudinal bars laterally supported by the corner of a crosstie or hoop leg, $a_x$, shall not exceed 14 in. around the perimeter of the column. Note that this spacing is further reduced to 8 in. for conditions given in Section 18.7.5.2(f) in ACI 318-14.

Typically, bars are evenly spaced around the perimeter, as this helps to create a more stable column cage during construction. Designers will often place bars only at the corners of square or rectangular columns to reduce the field work necessary to place ties. Bundled bars are often needed to meet the required area of steel for this arrangement. Note that Section 18.7.5.2 in ACI 318-14 makes this practice impractical for columns in special moment frames due to the 14 in. limitation. If confinement of the concrete core is necessary or desired, evenly spaced longitudinal bars are helpful. If more bars are required to resist flexure at a particular location, some bars can be added to the evenly spaced column bars.

**9.7.2.2** *Splicing*—Splice locations, lengths, and types should be included on the structural drawings. Lap splices are the most common splice type due to their ease of fabrication and construction. Mechanical connectors and butt-welded splices are helpful where bar arrangement becomes congested but they can require additional erection time. End-bearing splices are not common today but are some-

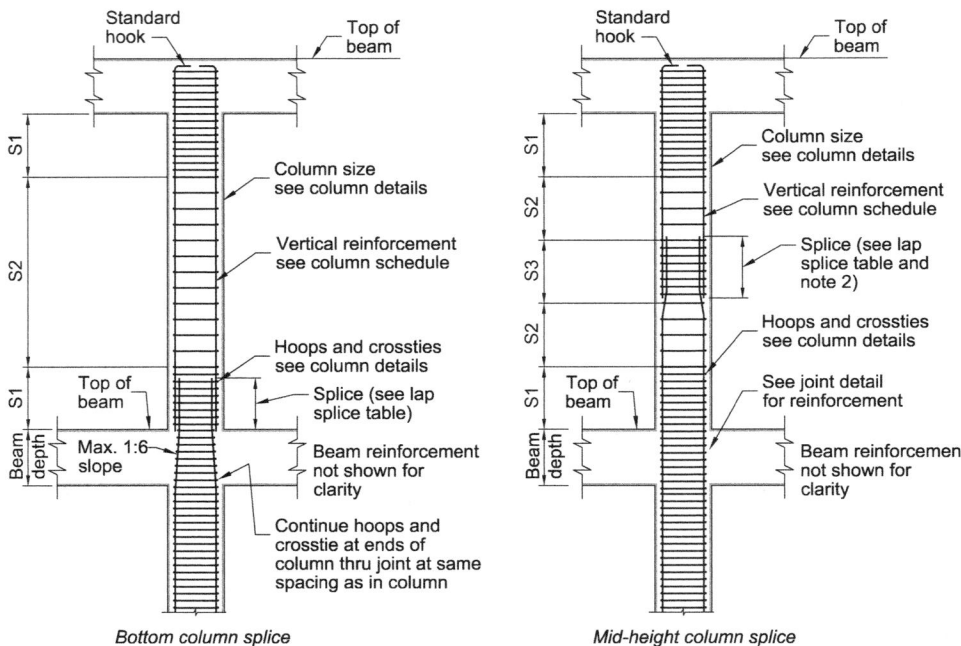

*Fig. 9.7.2.2a—Column splice locations (Fanella 2007).*

times used in bundled bar arrangements that are only in compression under all load combinations. They are not permitted in high seismic applications. Columns splices are usually located at the bottom of columns at each floor, if a mid-height splice is not required. Lap splices for columns in buildings assigned to SDC D, E, or F are required to be located in the center half of the column length according to Sections 18.7.4.3 and 18.14.3.3 in ACI 318-14. Bottom and mid-height column splice locations are illustrated in Fig. 9.7.2.2a.

For lap splices than are about one-third to one-half the story height, it may be more economical to lap-splice the bars every other floor (Concrete Reinforcing Steel Institute (CRSI) 2011). The length of lap splices should be noted on the structural drawings. Lap splices vary with the bar diameter, concrete strength, bar spacing, concrete cover, position of the bar, distance from other bars, and if the bar is tension or compression. Lap splices are not permitted for No. 14 and 18 bars, except for transferring compression (only) to a footing with dowels (ACI 318-14, Section 16.3.5.4).

To maintain bars in the corners of rectangular column, longitudinal bars that are lap spliced are usually offset bent into the column above, whether there is change in column size or not. Circular columns typically need not be offset bent where columns size does not change. The slope of the inclined portion of an offset bent should not exceed one in six (ACI 318-14, Section 10.7.4.1). Additional ties are required at offset bent locations and are placed not more than 6 in. from the point of the bend (ACI 318-14, Section 10.7.6.4). Typically, three closely spaced ties is sufficient to resist the lateral force created by the bend and one of the ties may be part of the regularly spaced ties. Separate splice bars and more ties may be necessary where the column section changes 3 in. or more. Examples of offset bent splices are illustrated in Fig. 9.7.2.2b. Where there is a reduction of reinforcement, longitudinal bars from the column below are typically terminated within 3 in. of the top of the finished floor (ACI 315-99), unless design requires otherwise. Column bar area in the column above must be extended from the column below to lap bars above.

Bundled bars are typically groups of larger bars that span two stories. Lap and end bearing splices in bundled bars require staggering the individual bars. For this reason, bundled bars preassembly is more complicated and can create additional erection time to place longitudinal and transverse bars on the free-standing cage.

**9.7.3** *Transverse bars*

**9.7.3.1** *Column ties*—Standard (nonseismic) arrangements of ties for various numbers of vertical bars are shown in Fig. 9.7.3.1. The one and two-piece tie arrangements shown provide maximum rigidity for column cages preassembled on the site before erection. The spacing of ties depends on the sizes of longitudinal bars, columns, and of ties. The maximum spacing of ties required for shear is shown in Table 10.7.6.5.2 in ACI 318-14.

**9.7.3.2** *Spirals*—Spirals are used primarily for circular columns, piers, and caissons. Spiral reinforcement can be plain or deformed bars or wire. The term "spiral" used in the code is more than a geometric description of a circular tie. It defines the required pitch, reinforcement amount, splicing, and termination, which are listed in Section 25.7.3 in ACI 318-14. A continuously-wound bar or wire not meeting all of the requirements of 25.7.3 is simply a continuous circular tie.

The spiral pitch is between 1 and 3 in., inclusive, and is typically given in 1/4 in. increments. The spiral size and pitch should meet the volumetric reinforcement ratio, $r_s$ (Eq. (25.7.3.3) in ACI 318-14). The continuation of spirals into floor joints is according to Table 10.7.6.3.2. The minimum diameters to which standard spirals can be formed is given in Table 9.7.3.2 (ACI 315-99).

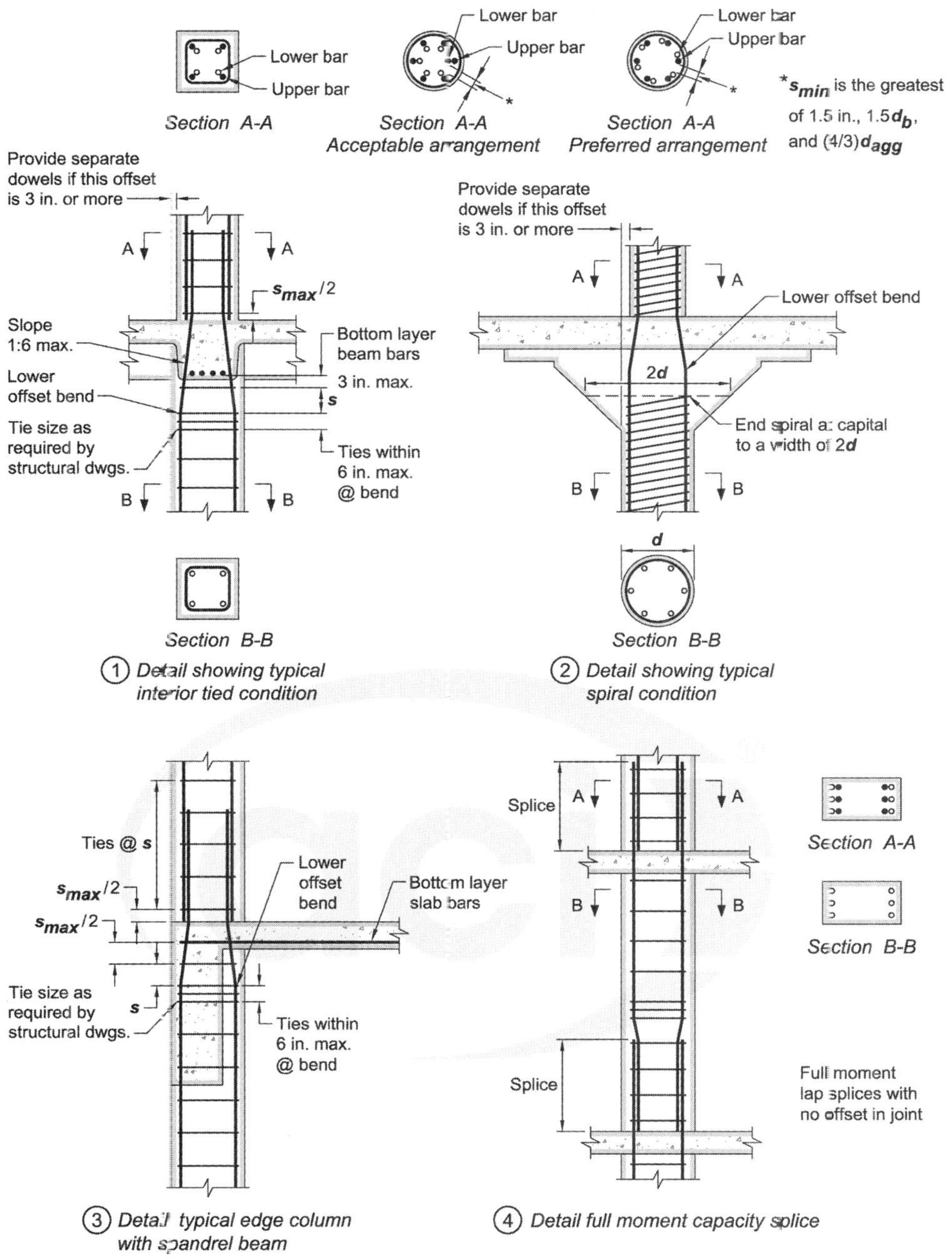

# CHAPTER 9—COLUMNS

Fig. 9.7.2.2b—Column splice details (ACI 315-99).

### Table 9.7.3.2—Minimum diameters of spiral reinforcement

| Spiral bar diameter, in. | Minimum outside diameter that can be formed, in. |
|---|---|
| 3/8 | 9 |
| 1/2 | 12 |
| 5/8 | 15 |
| 3/4 | 30 |

## 9.8—Design steps

1. Determine an initial size of the column and amount of reinforcement.
   (a) Ordinary: $A_g = P_{u,max}/0.4f_c'$; High seismic: $A_g = P_{u,max}/0.3f_c'$
   (b) Shape of column is often dictated by the architect; otherwise, square or round columns are common first estimates
   (c) Reinforcement ratios are typically 1 to 2 percent
2. Run initial analysis to determine column loads.
   (a) If second-order effects are accounted by the computer program, go to Step 3
   (b) If second-order effects are not accounted by the computer program, use the Moment Magnification method in Section 6.6.4 in ACI 318-14 to calculate these effects
3. Create a moment-interaction diagram using an electronic spreadsheet or commercial software. Check the required moment and axial load strengths against the design strength curve.

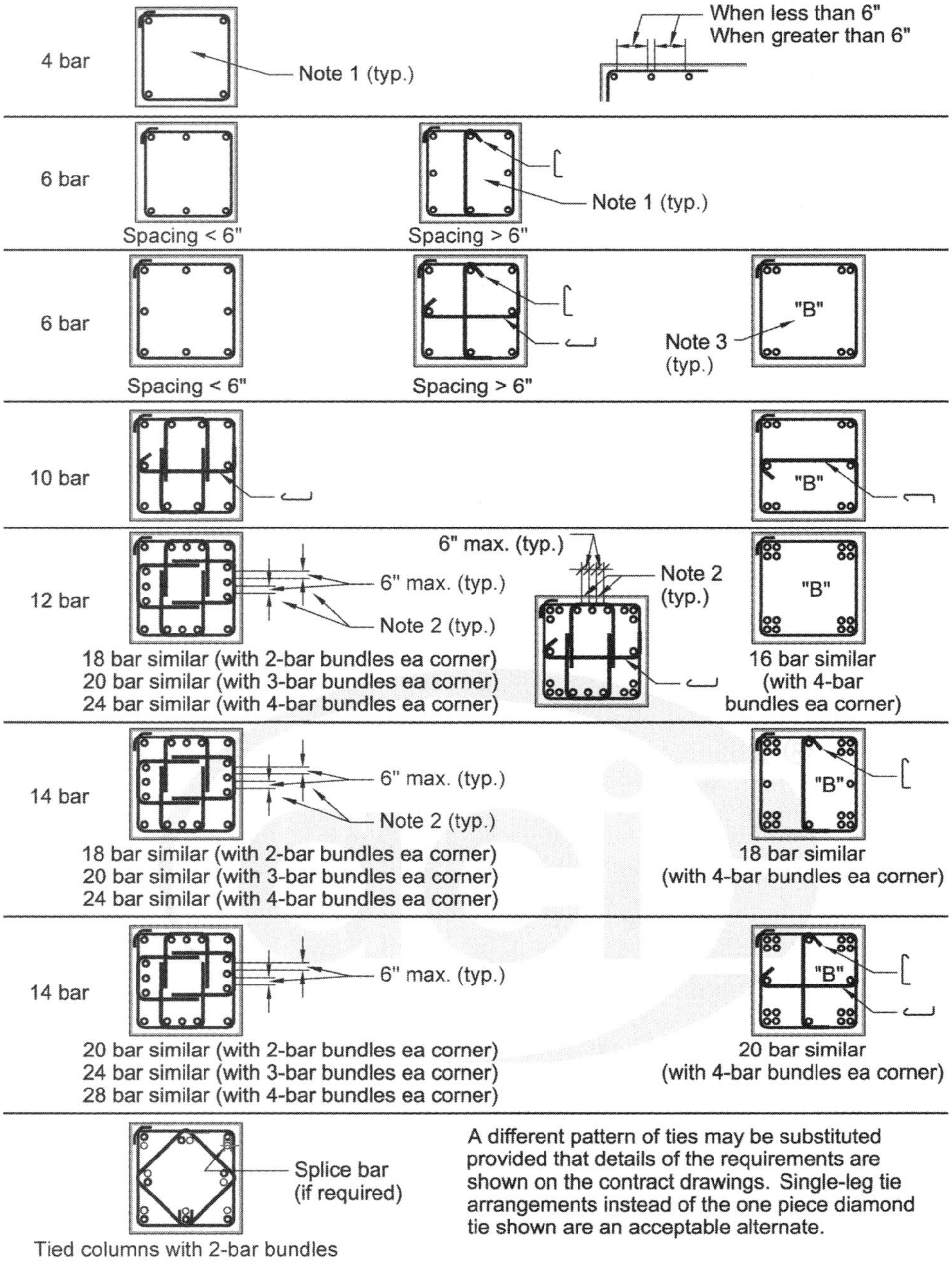

*Fig. 9.7.3.1—Standard column ties (ACI 315-99).*

4. Make adjustments as necessary to the initial column size and reinforcement. Rerun the analysis, if necessary, until design strength is greater than required strength for all load cases.
5. Check shear strength and minimum shear reinforcement requirements.
6. Detail column longitudinal and transverse requirements showing all bar locations, spacing, splices, and bar terminations. It is common to use typical column details and sections along with a column schedule table.

## REFERENCES

*American Concrete Institute (ACI)*

ACI 315-99—Details and Detailing of Concrete Reinforcement

ACI 318-63—Building Code Requirements for Structural Concrete and Commentary

ACI 318-14—Building Code Requirements for Structural Concrete and Commentary

ACI SP-17DA-14—Reinforced Concrete Design Handbook Design Aid – Analysis Tables;
https://www.concrete.org/store/productdetail.aspx?ItemID=SP1714DA

ACI SP-17DAE-14— Interaction Diagram Excel spreadsheet;

https://www.concrete.org/store/productdetail.aspx?ItemID=SP1714DAE

**Authored documents**

Fanella, D., 2007, *Seismic Detailing of Concrete Buildings*, SP382, Portland Cement Association, Skokie, IL, 80 pp.

Frosch, R., 2011, "Using an Elastic Frame Model for Column Slenderness Calculations," *Concrete International*, American Concrete Institute, Farmington Hills, MI, V. 33, No. 6, June, pp. 79-80.

Concrete Reinforcing Steel Institute (CRSI), 2011, "Detailing Concrete Columns," *Concrete International*, American Concrete Institute, Farmington Hills, MI, V. 33, No. 8, Aug., pp. 47-53.

## 9.9—Examples
**Columns Example 1:** *Column analysis*

Analyze first floor interior column in one direction at location E4, from the example building given in Chapter 1 of this Handbook. The moment magnification method is used with a first-order analysis. The column's factored forces and moments are from a first-order frame analysis using hand calculations. The building was also analyzed by first-order and second-order elastic methods using commercial software for comparison purposes.

The factored moments are from an analysis of the moment frame along Grid E. Three common controlling load combinations are considered.

Given:
*Materials—*
Specified concrete compressive strength, $f_c' = 5$ ksi
Specified yield strength, $f_y = 60$ ksi
Modulus of elasticity of concrete, $E_c = 4030$ ksi

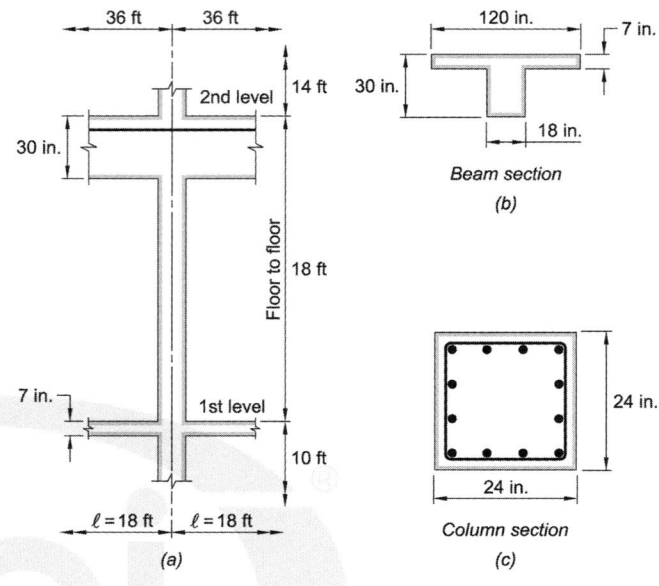

*Loading—*

| Loads considered | Dead + live + snow | Dead + wind + live + snow | | | Dead + EQ + live + snow | |
|---|---|---|---|---|---|---|
| Load Combination | (i) $U = 1.2D+1.6L+0.5S$ | (ii) $U=1.2D+1.0W+1.0L+0.5S$ | | | (iii) $U^*=1.2D+1.0E+1.0L+0.2S$ | |
| Load breakout | $1.2D + 1.6L + 0.5S$ | $1.2D +1.0L + 0.5S$ | | $1.0W$ | $1.2D +1.0L + 0.2S$ | $1.0E$ |
| $V_u$, kip | 0 | 0 | | 6 | 0 | 22 |
| $P_u$, kip | 867 | 777 | | 0 | 789 | 0 |
| $(M_u)_{top}$, kip-in. | 24 | 12 | | ±418 | 12 | ±1740 |
| $(M_u)_{bot}$, kip-in. | −36 | −24 | | ±650 | −24 | ±2328 |

*The load factor on D in Load Combination (iii) is increased as required by ASCE 7-10 by $0.2S_{DS}$. Note that $\rho = 1.0$ for buildings in SDC B.

Reference: SP-17 Supplement, Reinforced Concrete Design Handbook Design Aid – Analysis Tables, found at https://www.concrete.org/store/productdetail.aspx?ItemID=SP1714DA,

| ACI 318-14 | Discussion | Calculation |
|---|---|---|
| | **Step 1—Determine initial column size** | |
| | Estimate the maximum load at the interior column, E4, using Load combination (i): | |
| 5.3.1 | $U = 1.2D + 1.6L + 0.5S$ | $P_u = 867$ kip |
| | Estimate the size of a square column by dividing force by $0.4f_c'$. | area $= \dfrac{867 \text{ kip}}{0.4 \times 5 \text{ ksi}} = 434$ in.$^2$ <br><br> $h = \sqrt{434} = 20.8$ in. |

# CHAPTER 9—COLUMNS

| | | |
|---|---|---|
| 6.2.5<br>6.2.5.1 | Complete a rough check on slenderness: | $r = 0.3h$ and $\dfrac{k\ell_u}{r} = 45$ |
| | Concrete columns become very slender when $k\ell_u/r$ exceeds 45. Since this column is likely to be in a sway frame, assume a $k = 1.5$ and determine a size that satisfies $k\ell_u/r < 45$. | $\ell_u = 18 \times 12 - 30 = 186$ in.<br>Rearranging terms and solve for $h$:<br>$h = \dfrac{1.5 \times 186}{0.3 \times 45} = 20.7$ in. |
| | It is common for an engineer to choose a large enough column size so the column design is permitted to ignore slenderness. For a frame not braced against sidesway, a $k\ell_u/r$ limit of 22 would allow the engineer to ignore slenderness. This example, however, will consider slenderness and show the full set of calculations needed for a slender column. | Try 24 in. x 24 in.<br><br>(Column sizes are often rounded to the nearest increment of 2 in.) |
| **Step 2—Sway or nonsway moment frame** | | |
| Fig. R6.2.6 | A flowchart has been added, Fig. R6.2.6, to the Commentary to ACI 318-14 that helps the engineer determine column analysis options. This example shows the full extent of the provisions when done by hand. These calculations could easily be programmed in a spreadsheet. The first step in the flowchart is to determine if the structure is a sway or nonsway frame. | |
| | A moment frame that is nonsway greatly reduces the required calculations. The Code provides three options to permit the frame to be considered as nonsway. | |
| 6.2.5 | 1) The stiffness of all the bracing elements in a story are at least 12 times the stiffness of all the columns in the direction of evaluation. Bracing elements generally means walls but braces are sometimes used. | There are no bracing elements in the direction of the moment frame. |
| 6.6.4.3(a) | 2) The increase of column end moments due to second-order effects does not exceed 5 percent of the first-order end moments. | Second order moments are calculated in Step 4. The results of the calculation show that the second order effects exceed 5 percent. |
| 6.6.4.3(b)<br>6.6.4.4.1 | 3) $Q$, in accordance with Section 6.6.4.4.1, does not exceed 0.05. $Q$ is determined for a single story and controlling load combination. Different floors of the same moment frame have different $Q$ values. | $Q = \dfrac{\Sigma P_u \Delta_o}{V_{us} \ell_c}$<br>where,<br>$\ell_c = 18 \times 12 - \left(\dfrac{30}{2}\right) + \left(\dfrac{7}{2}\right) = 204.5$ in. |
| | $\ell_c$ is the height of the column from center-to-center of the joints. | |
| | To calculate Q, the load case must consider lateral load and also impose the maximum gravity load. For this example, select Load Combination (iii), which has a larger axial load than Load Combination (ii). | Check Load Combination (iii) |
| | $\Sigma P_u$ is the sum all factored column and wall gravity loads at the floor considered for Load Combination (iii). The value was derived from the loads given in Chapter 1 of this Handbook. | $\Sigma P_u = 25{,}700$ kip |
| | $V_{us}$ is the total factored horizontal story shear for Load Combination (iii). The value was calculated from the lateral forces calculated in Chapter 1 of this Handbook. | $V_{us} = 775$ kip |

| | | |
|---|---|---|
| 6.6.3.1.2 | $\Delta_o$ is the first-order story deflection determined by an elastic analysis. For this hand calculation example, a simple approximation of deflection is provided. The first story of a building is often assumed to have a hinge at 0.67 $\ell_c$. The following equation provides deflection at a distance $\ell$ to the hinge.<br><br>$$\Delta = \frac{V_{us} \times (\ell)^3}{3 \sum EI}$$<br><br>The value for stiffness should be reduced for cracking. A value of 0.5 is commonly used for all members in an analysis calculated by hand. | $$\Delta_o = \frac{V_{us}}{3 \sum EI}\left[\left(\frac{2 \times \ell_c}{3}\right)^3 + \left(\frac{\ell_c}{3}\right)^3\right]$$<br><br>$E_c$ = 4030 ksi<br>$I = 24 \times 24^3/12 = 27{,}650$ in.$^4$<br>$\sum EI$ = 37 columns × 0.5 × 4030 × 27,650<br>   = 2.06 × 10$^9$ kip-in.$^2$<br><br>$\Delta_o = 0.36$ in. |
| | | $$Q = \frac{25{,}700 \times 0.36}{775 \times 204.5} = 0.058 > 0.05$$<br><br>Therefore, the frame is sway. |
| | Note that the deflection from a first-order elastic analysis from software that accounts for the relative stiffness of all the members is 0.64 in. The advantage of a more accurate calculation of lateral deflection is discussed in further detail in Step 4. | A more accurate first-order analysis shows that<br><br>$$Q = \frac{25{,}700 \times 0.64}{775 \times 204.5} = 0.104$$ |

### Step 3—Check to see if slenderness can be neglected

| | | |
|---|---|---|
| | Compute the slenderness ratio, $k\ell_u/r$. The notation $\ell_u$ is the unsupported length. The floor-to-floor distance is 18 ft, and the floor at the second level is 30 in. deep. | $\ell_u$ = 18 ft × 12 in./ft – 30 in. = 186 in. |
| | To calculate the effective length factor $k$ for the column, the member stiffnesses framing at the top and bottom joints need to be calculated.<br><br>For rectangular sections, $I = bh^3/12$.<br><br>For T-sections, find the centroid by<br><br>$$y = \frac{\sum \text{moment of area}}{\sum \text{area}}$$<br><br>then use transformation of sections to calculate $I$. | T-Beam (at top of column):<br>$b_f$ = 120 in. (calculated in Beam Ex. 1)<br>$b_w$ = 18 in.<br>$h_f$ = 7 in.<br>$h$ = 30 in.<br><br>$$y_{beam} = \frac{120 \times 7 \times 3.5 + 18 \times 23 \times 18.5}{120 \times 7 + 18 \times 23} = 8.5 \text{ in.}$$<br><br>$$I_{beam} = \frac{120 \times 7^3}{12} + \left(120 \times 7 \times 5.0^2\right)$$<br>$$+ \frac{18 \times 23^3}{12} + \left(18 \times 23 \times 10.0^2\right) = 84{,}100 \text{ in.}^4$$ |
| | See the plan in Chapter 1 for plan dimensions. | Slab (at bottom of column):<br>$h$ = 7 in.<br>$b$ = 14 ft × 12 in. = 168 in.<br><br>$I_{slab} = 168 \times 7^3/12 = 4800$ in.$^4$ |
| | | Column (all levels):<br>$h$ = 24 in.<br>$b$ = 24 in.<br><br>$I_{col} = 24 \times 24^3/12 = 27{,}650$ in.$^4$ |

| | | |
|---|---|---|
| Table 6.6.3.1.1(a) | Calculate adjusted $EI$ values. For this part, a more detailed values for $EI$ are used. | $E_c = 4030$ ksi<br><br>$0.35(E_c I)_{beam} = 0.35 \times 4030 \times 84{,}100$<br>$= 119 \times 10^5$ kip-in.$^2$<br>$0.25(E_c I)_{slab} = 0.25 \times 4030 \times 4800$<br>$= 4.8 \times 10^6$ kip-in.$^2$<br>$0.70(E_c I)_{col} = 0.70 \times 4030 \times 27{,}650$<br>$= 78 \times 10^6$ kip-in.$^2$ |
| R6.2.5 | Factor $k$ reflects column end restraint conditions, which depends on the relative stiffness of the columns to the floor members at top and bottom joints. At the top joint, the columns frame into beams, and at the bottom joint, the columns frame into a two-way slab. Find the ratio of column stiffness to beam or slab stiffness:<br><br>$\Psi = [(EI/\ell_c)_{col,\,above} - (EI/\ell_c)_{col,\,below}]/$<br>$[(EI/\ell)_{beam,\,left} + (EI/\ell)_{beam,\,right}]$ | Joint at top of column:<br><br>$\Psi_A = \dfrac{\dfrac{78 \times 10^6}{204.5} + \dfrac{78 \times 10^6}{168}}{\dfrac{119 \times 10^6}{432} + \dfrac{119 \times 10^6}{432}}$<br><br>$\Psi_A = 1.5$<br><br>Joint at bottom of column:<br><br>$\Psi_B = \dfrac{\dfrac{78 \times 10^6}{120} + \dfrac{78 \times 10^6}{204.5}}{\dfrac{4.8 \times 10^6}{204.5} + \dfrac{4.8 \times 10^6}{204.5}}$<br><br>$\Psi_B = 23.0$ |
| Fig. R6.2.5 | Read $k$ from the nomograph. | For a sway frame, $k \approx 2.2$<br>(Note: For a nonsway frame, $k \approx 0.9$) |
| 6.2.5.1 | Determine the radius of gyration, $r$<br><br>$r = \sqrt{\dfrac{I_g}{A_g}}$<br><br>Note that Section 6.2.5.1 also allows an approximation of 0.3 times the width of the column in the direction of the frame, which would be $0.3 \times 24 = 7.2$ in this case. | $I_g = 27{,}650$ in.$^4$<br>$A_g = 576$ in.$^2$<br><br>$r = \sqrt{\dfrac{27{,}560}{576}} = 6.9$ |
| 6.2.5 | Check to see if slenderness can be neglected for sway frame. Slenderness for a sway frame can be neglected if $k\ell_u/r \leq 22$. | For a sway frame,<br><br>$\dfrac{2.2 \times 186}{6.9} = 59 > 22$<br><br>Slenderness cannot be neglected. |
| | Note: For a nonsway frame, there are two limits that must be met:<br><br>$\dfrac{k\ell_u}{r} \leq 34 + 12\left(\dfrac{M_1}{M_2}\right)$ and $\dfrac{k\ell_u}{r} \leq 40$<br><br>Use Load Combination (iii) to find $M_1$ and $M_2$. | Note: For a nonsway frame:<br><br>$\dfrac{k\ell_u}{r} = \dfrac{0.9 \times 186}{6.9} = 24$<br><br>$24 \leq 34 + 12\left(\dfrac{1752}{2352}\right) = 40.9$ and $24 \leq 40$<br><br>therefore, slenderness could be neglected if this condition occurred. |

| Step 4—Determine second order effects for $P\Delta$, sway | | |
|---|---|---|
| 6.6.4.6.1 | For a sway frame, the secondary moments at the end of the column due to differential movement of the ends of column must be calculated before calculating deformations along the length. Note that secondary moments are often called "$P\Delta$" (uppercase delta) effects by most reference materials, but are labeled "$P\delta_s$" (lowercase delta) in ACI 318-14.<br><br>The Code provides a conservative method to estimate this effect. The equations are<br><br>$M_1 = M_{1ns} + \delta_s M_{1s}$<br>$M_2 = M_{2ns} + \delta_s M_{2s}$<br><br>Where the first-order moments due to gravity loads for a single load combination ($M_{1ns}$) are added to the first-order moments due to lateral loads ($M_{1s}$) multiplied by the sway moment magnification factor $\delta_s$. | From Load Combination (iii) above:<br><br>|  | $M_{ns}$ | $M_s$ |<br>|---|---|---|<br>|  | 1.2D +1.0L + 0.2S | 1.0E |<br>| $M_1$, kip-in. | 12 | ±1740 |<br>| $M_2$, kip-in. | –24 | ±2328 | |
| 6.6.4.6.2 | The sway moment magnification factor may be determined one of two ways: | |
| Eq. (6.6.4.6.2a) | $\delta_s = \dfrac{1}{1-Q} \geq 1$<br><br>This expression is commonly used if software determines the first order lateral deflection.<br><br>OR | $\delta_s = \dfrac{1}{1-0.058} = 1.06$ |
| Eq. (6.6.4.6.2b) | $\delta_s = \dfrac{1}{1-\dfrac{\sum P_u}{0.75 \sum P_c}} \geq 1$<br><br>This expression is commonly used in hand calculations.<br><br>Note that Section 6.6.4.6.2c also acknowledges that the second order effect may be determined by software that uses a second-order analysis. | $\delta_s = \dfrac{1}{1-\dfrac{\sum P_u}{0.75 \sum P_c}}$<br><br>where $\sum P_u = 25,700$ kip (From Step 2) |

| | | | |
|---|---|---|---|
| 6.6.4.4.2<br>6.6.4.4.4<br><br>R6.6.4.6.2 | Calculate the critical load $P_c$ from Eq. (6.6.4.4.2). $EI_{eff}$ may be calculated from one of three equations in 6.6.4.4.4. Use Eq. (6.6.4.4.4a) since the column reinforcement is not known at this point of the design. Key points about $EI_{eff}$ are<br>(a) Commentary Section R6.6.4.4.4, explains the differences in the equations<br>(b) For sway frames, $\beta_{ds}$ is substituted for $\beta_{dns}$ and is 0.0 for short term lateral loads<br>(c) $I_{se}$ in Eq. (6.6.4.4.4b) may be calculated using Table D.4.5 in the supplement to this Handbook, see reference at the start of this example. | $P_c = \dfrac{\pi^2 (EI)_{eff}}{(k\ell_u)^2}$<br><br>$(EI)_{eff} = \dfrac{0.4 E_c I_g}{1+\beta_{ds}}$<br><br>$= \dfrac{0.4 \times 4030 \times 27{,}650}{1+0}$ kip-in$^2$<br><br>$= 45 \times 10^6$ kip-in.$^2$<br><br>$k = 2.2$ (sway frame)<br>$\ell_u = 186$ in.<br><br>$P_c = \dfrac{\pi^2 \times (45 \times 10^6)}{(2.2 \times 186)^2} = 2630$ kip<br><br>$\delta_s = \dfrac{1}{1 - \dfrac{25{,}700}{0.75(37 \text{ columns} \times 2630)}} = 1.54$<br><br>The large disparity between the two results (1.06 versus 1.54) is unusual. This disparity indicates that the 24 in. column stiffness may be outside normal bounds for stable results, and the engineer should consider increasing the column size. | |
| | Discussion:<br>In the reference by MacGregor and Hage (1977), it is suggested that $\Delta_o$ in Eq. (6.6.4.6.2a) be taken from a first-order analysis using a computer program that accounts for the member stiffnesses. As noted in Step 2, the software lateral deflection is 0.64 in. and the value for $Q$ becomes 0.104. Thus, the revised $\delta_s$ for Eq. (6.6.4.6.2a) is<br><br>$\delta_s = \dfrac{1}{1-0.104} = 1.12$<br><br>The MacGregor and Hage (1977) reference is informative and describes the $Q$ method. A few helpful suggestions from the reference are<br>(a) For $Q$ values between 0.05 and 0.2, the error in second-order moments will be less than 5 percent<br>(b) $\Delta_o/H$ (story height) should be less than 1/500 for nonsway frames and less than 1/200 for sway frames at factored loads<br>(c) $\Sigma P_u/\Sigma P_{cr}$ in Eq. (6.6.4.6.2b) should be less than 0.2<br><br>In this example, $\Sigma P_u/\Sigma P_{cr}$ is 0.26 > 0.2, so the results from Eq. (6.6.4.6.2b) are becoming questionable. Q calculated from deflections by a first-order computer analysis is 0.1 which is in the range suggested by MacGregor and Hage; thus, the second order-moments calculated by the Q method should be within 5 percent of the actual. If the $\Delta_o$ from the computer analysis was not available, it would be prudent to use Eq. (6.6.4.6.2b) and increase the square column size to satisfy $\Sigma P_u/\Sigma P_{cr} < 0.2$. | | |
| | Calculate the magnified moments using the software-based $Q$. Note that only the moment due to the lateral loads are magnified. | $\delta_s M_{1s} = 1.12 \times 1740 = 1950$ kip-in.<br>$\delta_s M_{2s} = 1.12 \times 2328 = 2610$ kip-in. | |
| | Calculate the design moments $M_1$ and $M_2$ | $M_1 = M_{1ns} + \delta_s M_{1s} = 12 + 1950 = 1962$ kip-in.<br>$M_2 = M_{2ns} + \delta_s M_{2s} = 24 + 2610 = 2634$ kip-in. | |

| | | |
|---|---|---|
| 6.2.6 | Check to see if the second-order moments exceed 40 percent of first-order moments. | First order $M_1$ is 1740 + 12 = 1752 kip-in. <br> $1962 \leq 1742 \times 1.4 = 2718$, therefore **OK**. <br><br> First order $M_2$ is 2328 + 24 = 2352 kip-in. <br> $2610 \leq 2352 \times 1.4 = 3292$, therefore **OK**. |
| **Step 5—Determine second order effects for $P\delta$, nonsway** | | |
| 6.6.4.6.4 | Second-order effects along the length of the column must be calculated for a sway or nonsway frame where slenderness cannot be neglected. The magnified moments, $M_1$ and $M_2$, from Section 6.6.4.6 are used in Section 6.6.4.5. | |
| 6.6.4.5.1 | The required moment for design is calculated by multiplying the larger end moment $M_2$ by the moment magnification factor $\delta$ <br> $M_c = \delta M_2$ | |
| 6.6.4.5.2 | The moment magnification factor $\delta$ is calculated by <br> $$\delta = \frac{C_m}{1 - \frac{P_u}{0.75 P_c}} \geq 1.0$$ | $P_u = 789$ kip |
| 6.6.4.5.3 <br><br> 6.6.4.5.4 <br><br> R6.6.4.5.3 | $C_m = 0.6 - 0.4 \dfrac{M_1}{M_2}$ <br><br> Check $M_{2,min}$: <br> $M_{2,min} = P_u(0.6 + 0.03h)$ <br><br> Notice that $M_1/M_2$, has been updated to follow the right hand rule; thus, the sign convention has been changed in ACI 318-14. <br><br> + $M_1/M_2$ is double curvature <br> − $M_1/M_2$ is single curvature | $M_1 = 1962$ kip-in. <br> $M_2 = 2634$ kip-in.  **Controls** <br><br> $M_{2,min} = 789(0.6 + 0.03 \times 24) = 1041$ kip-in., which is less than 2634. <br><br> $C_m = 0.6 - 0.4 \dfrac{1962}{2634} = 0.30$ |
| 6.6.4.4.2 <br><br> R6.6.4.4.4 | The critical buckling load was calculated in Step 4, but $\beta_{ds}$ was substituted for $\beta_{dns}$. It shall now be calculated using $\beta_{dns}$. The commentary states that $\beta_{dns}$ may be assumed to be 0.6. Another common way of calculating $\beta_{dns}$ is to divide the dead load by transient gravity loads for a given load combination. For Load Combination (iii), the calculation is <br><br> $\beta_{dns} = \dfrac{1.23D}{1.23D + 1.0L + 0.2S}$ <br> $= \dfrac{1.2 \times 517}{1.23 \times 517 + 1.0 \times 151 + 0.2 \times 10}$ <br> $= 0.79$ | $P_c = \dfrac{\pi^2 (EI)_{eff}}{(k\ell_u)^2}$ <br><br> $(EI)_{eff} = \dfrac{0.4 E_c I_g}{1 + \beta_{dns}}$ <br><br> $= \dfrac{0.4 \times 4030 \times 27{,}650}{1 + 0.79}$ kip-in.$^2$ <br> $= 25 \times 10^6$ kip-in.$^2$ <br> $k = 2.2$ (sway frame) <br> $\ell_u = 186$ in. <br><br> $P_c = \dfrac{\pi^2 \times (25 \times 10^6)}{(2.2 \times 186)^2} = 1470$ kip |

| | | |
|---|---|---|
| | Magnify the moment for second-order effects along the length of the column. | $\delta = \dfrac{0.3}{1 - \dfrac{789}{0.75 \times 1470}} = 1.06$<br><br>$M_c = \delta M_2 = 1.06 \times 2634 = 2795$ kip-in. |

### Step 6—Summary and discussion

This example calculates the second-order moment for one column, in one-direction, for one load case. It is easy to see that this is a time-consuming process. The following table shows a comparison between the moment and axial loads calculated by hand and by computer software for Load Combination (iii):

| | Hand (first-order) | Hand (second-order) | Computer (first-order) | Computer (second-order) |
|---|---|---|---|---|
| $P_u$, kip | 789 | 789 | 818 | 818 |
| $M_u$, kip-in. | 2352 | 2795 | 2177 | 2401 |

One can see that the hand calculation to find the second-order moment is more conservative than in a computer analysis. The second-order moment increase for the computer is 10 percent which is lower than the 18 percent calculated by the moment magnification method. Notice that the $Q$-method used in Eq. (6.6.4.6.2a) was aided with a $\Delta_c$ computed by a computer analysis. If that deflection was not available, Eq. (6.6.4.6.2b) would have been used and total increase of 63 percent would have been required which is more than the 40 percent allowed. Thus, a larger column would have been selected as suggested in the discussion in Step 4.

**Column Example 2**: *Column for an ordinary moment frame*—Design and detail the first floor interior column at location E4 from the example building given in Chapter 1 of this Handbook. The column is part of an ordinary moment frame. Example 2 is the design of the column analyzed in Example 1. The loads have been modified to match the results of an analysis from commercial software capable of second-order elastic analysis.

Given:
*Materials*—
Specified yield strength, $f_y$ = 60 ksi
Modulus of elasticity of steel, $E_s$ = 29,000 ksi
Specified concrete compressive strength, $f_c'$ = 5 ksi
Modulus of elasticity of concrete, $E_c$ = 4030 ksi
Normalized maximum size of aggregate is 1 in.

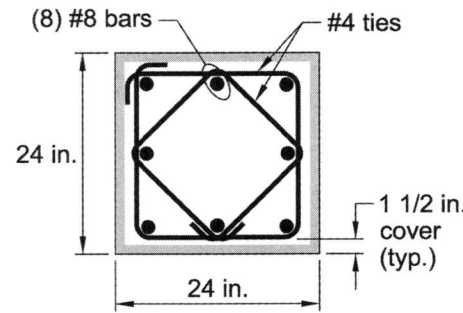

*Loading*—

| Load combinations | $P_u$, kip | $M_u$, kip-in. | $V_u$, kip |
|---|---|---|---|
| (i) $U = 1.2D + 1.6L + 0.5S$ | 890 | 0 | 0 |
| (ii) $U = 1.2D + 1.0W + 1.0L + 0.5S$ | 800 | 651 | 5 |
| (iii) $U^* = 1.2D + 1.0E + 1.0L + 0.2S$ | 818 | 2401 | 18 |
| (iv) $U^* = 0.9D + 1.0E + 1.0L + 0.2S$ | 486 | 2401 | 18 |

*The software adjusts seismic load combinations as required by ASCE 7-10.

Reference: SP-17 Supplement, Interaction Diagram Excel spreadsheet found at https://www.concrete.org/store/productdetail.aspx?ItemID=SP1714DAE

| ACI 318-14 | Discussion | Calculation |
|---|---|---|
| **Step 1—Find the required area of longitudinal reinforcement** | | |
| 10.5.2.1→ 22.4→22.2 | The Code references Section 22.4 for the calculation of $P_n$ and $M_n$. Section 22.4 provides an equation to calculate $P_0$ and references Section 22.2 for strain limits.<br><br>The interaction of $P_n$ and $M_n$ is evaluated by making an interaction diagram. A tutorial spreadsheet is provided with this manual that demonstrates how to make a diagram for a given section and reinforcement ratio ρ. This example used the spreadsheet to create an interactive diagram. The generated interaction diagram shows if the assumed longitudinal reinforcement is satisfactory.<br><br>The section properties and geometry were determined in the analysis from Column Example 1. The next step is to assume a quantity, size, and location of the longitudinal reinforcement. The design of columns is often an iterative process. The final design may show that a larger section or more reinforcement is needed, in which case, the analysis will need to be redone. | |
| 10.6.1.1 | The spreadsheet analyzes rectangular or square columns for combined axial and flexural strength. The designer inputs the number of steel layers in the column cross section. The spreadsheet places the first and last layer as close to the outer face as permitted (concrete cover plus tie bar size).<br><br>The remaining layers are evenly spaced between the outer layers. The spacing of the bars within each layer is similar to the placement of the layers. The first and last bar are as close to the outer face as permitted and the remaining bars are evenly spaced between the outer bars. | |

| Code Ref. | Discussion | Calculation |
|---|---|---|
| 10.7.3.1 | The design moments are not very large, so try the minimum area of reinforcement and assume a uniform distribution of bars around the perimeter. Longitudinal bars are typically larger bars, No. 7 and greater, to make stable column cages for erection and to reduce the number of ties at a section.<br><br>At least four bars are needed in a rectangular column. | $A_{s,min} = 0.01 A_g = 0.01 \times 24^2 = 5.76$ in.$^2$<br><br>Try 8 bars, one in each corner and one on each side (3 layers: 3 bars, 2 bars, 3 bars)<br><br>Area bar needed = 5.76 in.$^2$/8 = 0.72 in.$^2$<br><br>Area of a No. 8 bar is 0.79 in.$^2$; therefore, try (8)-No. 8 bars |
| 10.5.2.1→<br>22.2.2.4.3 | The spreadsheet does a sectional strength analysis using an equivalent rectangular stress distribution according to Section 22.2.2.4. The β is a function of $f_c'$ which is automatically calculated and displayed for the user's information. | **Section properties and geometry**<br><br>No. layers: 3<br>$f_c'$ = 5,000 psi<br>β = 0.80<br>b = 24 in.<br>h = 24 in.<br>$f_y$ = 60 ksi<br>$E_s$ = 29,000 ksi<br>Tie bar size: 4<br>Clear cover to tie = 1.50 in.<br>Long. bar size: 8 |
| 10.7.6.1.2→<br>25.7.2.2 | No. 4 ties are a common starting size since they provide good initial shear strength and are rigid enough to provide column cage stability during erection. | |
| 10.7.1.1→<br>20.6.1.3.1 | Concrete cover protects the reinforcement from corrosion and provides fire protection. The minimum cover is provided in Section 20.6. | |
| | The spreadsheet calculates the distances to the layers of reinforcement, which is needed for later calculations, by using the tie bar size, cover to tie, and longitudinal bar size. | **Number of bars per layer**<br><br>| Layer | $d_i$, in. | No. long. bars | $A_{si}$, in.$^2$ |<br>|---|---|---|---|<br>| 3 | 2.500 | 3 | 2.37 |<br>| 2 | 12.000 | 2 | 1.58 |<br>| 1 | 21.500 | 3 | 2.37 |<br>| | | Σ | 6.32 | |
| 10.7.2.1→<br>25.2.3 | The minimum bar spacing is calculated and displayed for the user's information. | **Bar spacing checks**<br><br>$d_1$ = 21.50 in.<br>c/c bar sp. (h) = 9.50 in. OK<br>c/c bar sp. (b) = 9.50 in. OK<br>Min. clear sp. (in.) = 1.50 in. (25.2.3) |
| 10.5.1.2→<br>21.2<br><br>10.5.2.1→<br>22.4→22.2 | Key variables needed for design are displayed for the user's information. The strain limits and φ factors change as the column interaction diagram transitions from compression-controlled sections to tension-controlled sections. | **Strain definitions**<br><br>$f_y/E$ = 0.00207<br>$\varepsilon_{cu}$ = 0.003<br>Ductile Strain = -0.005<br>Brittle Strain = -0.002<br>$\phi_{tension-controlled}$ = 0.9<br>$\phi_{compression-controlled}$ = 0.65 |
| | The interaction diagram is made by calculating the $P_n$ and $M_n$ for incremental changes in the net tensile strain in the extreme layer of longitudinal tensile reinforcement at nominal strength, $\varepsilon_t$. A random strain of 0.0007 was chosen to illustrate the calculation of $P_n$ and $M_n$ in the following steps. | |
| | 1. Vary $\varepsilon_t$ from pure compression, $\varepsilon_t$ equal to $f_y/E_s$ at all layers, to pure tension, $-f_y/E_s$ at all layers. | Find $P_n$ and $M_n$ for $\varepsilon_t$ = 0.0007 |

| | | |
|---|---|---|
| | 2. Calculate $c$, the distance from the extreme compression fiber to the neutral axis, for the given $\varepsilon_t$ by using similar triangles. $$c = \frac{-\varepsilon_{cu} \times d_1}{\varepsilon_t - \varepsilon_{cu}}$$ | $$c = \frac{-0.003 \times 21.5}{0.0007 - 0.003} = 28.09 \text{ in.}$$ |
| | 3. Calculate $a$, the depth of the equivalent stress block. $$a = \min\begin{bmatrix} \beta_1 \times c \\ h \end{bmatrix}$$ | $$a = \min\begin{bmatrix} 0.80 \times 28.09 = 22.47 \text{ in.} \\ h = 24 \text{ in.} \end{bmatrix}$$ Use $a = 22.5$ in. |
| | 4. Calculate $C_c$, the force at concrete compression block. $C_c = 0.85 \times a \times b \times f_c'$ | $C_c = 0.85 \times 22.47 \times 24 \times 5 = 2292$ kips |
| | 5. Calculate $\varepsilon_{si}$, the strain at each layer of bars. if $c \geq d_i$ then $\varepsilon_{si} = \min\begin{bmatrix} \varepsilon_{cu} \times \dfrac{(c-d_i)}{c} \\ \dfrac{f_y}{E_s} \end{bmatrix}$ if $c < d_i$ then $\varepsilon_{si} = \max\begin{bmatrix} \varepsilon_{cu} \times \dfrac{(c-d_i)}{c} \\ -\dfrac{f_y}{E_s} \end{bmatrix}$ | For $\varepsilon_{s1}$, $\varepsilon_{s2}$, and $\varepsilon_{s3}$, $c \geq d_i$ thus find the minimum of $\varepsilon_{s1} = 0.003 \times \dfrac{(28.1 - 21.5)}{28.1} = 0.000703$ $\varepsilon_{s2} = 0.003 \times \dfrac{(28.1 - 12)}{28.1} = 0.00172$ $\varepsilon_{s3} = 0.003 \times \dfrac{(28.1 - 2.5)}{28.1} = 0.00273$ and $\dfrac{f_y}{E_s} = \dfrac{60}{29,000} = 0.00207$ Use $\varepsilon_{s1} = 0.000703$, $\varepsilon_{s2} = 0.00172$, and $\varepsilon_{s3} = 0.00207$ |
| | 6. Calculate $F_{si}$, the force at each layer of bars. If $\varepsilon_{si} > 0$, then $F_{si} = \varepsilon_{si} \times A_{si} \times E_s - 0.85 \times A_{si} \times f_c'$ Note that positive strain indicates that the bar is in compression. The force is adjusted to account for the displaced concrete. If $\varepsilon_{si} \leq 0$, then $F_{si} = \varepsilon_{si} \times A_{si} \times E_s$ | For $F_{s1}$, $F_{s2}$, and $F_{s3}$, $\varepsilon_s \geq 0$, thus $F_{s1} = 0.000703 \times 2.37 \times 29,000 - 0.85 \times 2.37 \times 5$ $= 38.3$ kip $F_{s2} = 0.00172 \times 2.37 \times 29,000 - 0.85 \times 2.37 \times 5 = 108$ kip $F_{s3} = 0.00207 \times 2.37 \times 29,000 - 0.85 \times 2.37 \times 5 = 132$ kip |
| | 7. Calculate $P_o$. $P_o = C_c + \sum F_{si}$ Note that $P_{n,max}$ is not applied in this spreadsheet until the design strength curve is calculated. | $P_o = 2292 + 38.2 + 180 + 132 = 2570$ kip |
| | 8. Calculate $d_i$, the moment arm to forces $C_c$ and each $F_{si}$. $$d_{Cc} = \frac{h}{2} - \frac{a}{2}$$ $$d_i = \frac{h}{2} - d_i$$ | $d_{Cc} = \dfrac{24}{2} - \dfrac{22.47}{2} = 0.77$ in. $d_1 = 12 - 21.5 = -9.5$ in. $d_2 = 12 - 12 = 0$ in. $d_3 = 12 - 2.5 = 9.5$ in. |

| | | |
|---|---|---|
| | 9. Calculate $M_i$, the moment for each force about the center of the section. $$M_{Cc} = d_{Cc} \times C_c$$ $$M_i = d_i \times F_{si}$$ | $$M_{Cc} = 0.77 \times \frac{2292}{12} = 147 \text{ ft·kip}$$ $$M_1 = -9.5 \times \frac{38.3}{12} = -30.3 \text{ ft·kip}$$ $$M_2 = 0 \times \frac{108}{12} = 0 \text{ ft·kip}$$ $$M_3 = 9.5 \times \frac{132}{12} = 105 \text{ ft·kip}$$ |
| | 10. Calculate $M_n$. $$M_n = M_{Cc} + \sum M_i$$ | $M_n = 147 - 30.3 + 0 + 105 = 222$ ft·kip |
| 10.5.1.2→ 21.2 10.5.2.1→ 22.4.2.1 10.7.5.2.2 | The spreadsheet changes the strain in small increments to create a smooth plot of a nominal strength interaction diagram. The values are then multiplied by $\phi$ and the limits on axial strength are applied to create the design strength interaction diagram. The $\phi$ factor for compression-controlled sections is 0.65 for ties. The $\phi$ factor for all tension-controlled sections is 0.9. This factor varies linearly from 0.65 to 0.9 through the transition zone shown which creates the sharp change in the lower part of the curve. The diagram on the right is for this column example and it has the loads from the four given load combinations plotted. Note that points for load combinations (i), (ii), and (iii) are to the left of the "Stress = $0 f_y$" Line, meaning all the bars remain in compression. Point iv) is for a load combination with lighter gravity loads. It is just to the right of the line, meaning a few bars will be in tension. Thus, a tension splice is required. Another line could be drawn on the diagram showing where the tensile bar stress is $0.5f_y$. That line delineates whether a Class A or B splice is required. For this example, all the bars are going to be spliced at one location so a Class B splice is always required. | Design Strength Interaction Diagram: Use eight No. 8 bars evenly spaced around the perimeter. |
| **Step 2—Find the required area and geometry of transverse reinforcement** | | |
| 10.5.3.1→ 22.5 18.3.3 | The Code references Section 22.5 for the calculation of $V_n$. The $P_u$ given in Load Combination (iii) has the expression $0.2S_{DS}$ is applied in the downward positon as required by ASCE 7-10. The $P_u$ value shown here is changed to reflect $0.2S_{DS}$ in the upward position for the lowest axial load associated with this load combination. Note that $\ell_u$ (186 in.) is greater than $5c_1$ (120 in.); therefore, the additional shear requirement for ordinary moment frame columns does not apply. | Load Combination (iii): $V_u = 18$ kip $P_u = 737$ kip |

| | | |
|---|---|---|
| Eq. (22.5.5.1) | The shear force resisted by the column is minimal in this case. It is conservative to ignore the effect of the compression load on shear resistance. Use Eq. (22.5.5.1) to find $V_c$.<br><br>$V_c = 2\lambda\sqrt{f'_c}b_w d$ | $V_c = 2 \times 1.0\sqrt{5000} \times 24 \times \dfrac{21.5}{1000} = 73.0\,\text{kip} \geq 18\,\text{kip}$ |
| 10.6.2<br>10.5.1.2→<br>21.2 | Minimum shear reinforcement is required when $V_u > 0.5\phi V_c$ | Check for minimum shear reinforcement<br>$18\,\text{kip} \not> 0.5 \times 0.75 \times 73 = 27\,\text{kip}$<br><br>Column ties are not necessary for shear resistance. |
| 10.7.6.1.2→<br>25.7.2<br><br>10.7.6.2 | Ties are not needed for shear but they are necessary for lateral support of longitudinal bars. Thus, tie requirements must meet the geometry requirements of 25.7.2 and location requirements of 10.7.6.2 | |
| 25.7.2.2 | The minimum tie bars size is No. 3 for longitudinal bars No. 10 or smaller; however, a No. 4 tie was chosen as discussed in Step 1. | Use No. 4 ties. |
| 25.7.2.1 | The minimum spacing is $(4/3)d_{agg}$; however, the maximum tie requirements controls for this example.<br><br>The maximum spacing shall not exceed the least of:<br>(a) $16d_b$ of the longitudinal bar<br>(b) $48\,d_b$ of the transverse bar<br>(c) $h$ or $b$ | (a) 16 × 1.00 in. = 16 in.  **Controls**.<br>(b) 48 × 0.50 in. = 24 in.<br>(c) 24 in.<br><br>Use $s$ = 16 in. on center (o.c.) |
| 25.7.2.3<br><br><br><br><br>Fig. R25.7.2.3a | Section 25.7.2.3 requires that lateral support from ties is provided for bars at every column corner and also bars with greater than 6 in. clear on each side. Thus, every vertical bar needs lateral support in this example<br><br>Ties with 90-degree standard hooks are acceptable for this case. A diamond-shaped tie is used to support bars along the sides. This is desirable because it provides a strong column cage for erection; however, it becomes a fabrication problem when the column is rectangular and this tie becomes oblong. It is common to use alternative tie geometry, such as cross ties, for columns that are not square. | Use No. 4 ties @ 16 in. o.c.<br><br>(8) #8 bars, #4 ties, 24 in. × 24 in., 1 1/2 in. cover (typ.) |
| **Step 3—Check the joint** | | |
| 15.2.1→<br>15.3 | The transfer of column axial force through the joint must be checked. It is common to have higher strength concrete in the columns and walls and a lower strength in the floor system. If the difference in concrete strengths is greater than 1.4, then the Code provides three options to account for this difference. This building uses the same $f'_c$ for both the columns and floor system so no additional calculations or adjustments are necessary. | |
| 15.2.2,<br>15.2.3,<br>15.2.4,<br>15.2.5,<br>15.4.1 | Sections 15.2.2, 15.2.3, and 15.4 require that beam-column joints be designed and detailed for shear if moment is being transferred to the column. Section 15.4 does not require minimum shear reinforcement through the joint, if the joint is restrained in accordance with, 15.2.4 or 15.2.5, and is not part of a seismic-force-resisting system. Note that this exception may be superseded by Chapter 18 as noted in Section 15.2.3. This column is part of an ordinary moment frame; therefore, it is part of the seismic-force-resisting system. | |

| | | |
|---|---|---|
| 15.4.2 | Section 15.4.2 applies which requires that minimum shear reinforcement is provided in the joint region even if minimum shear reinforcement is not required in the column. $A_v$ shall be at least the greater of:<br><br>$0.75\sqrt{f'_c}\dfrac{bs}{f_y}$<br><br>$50\dfrac{bs}{f_y}$ | At each tie, $A_v = 0.20$ in.$^2 \times 4$ legs $= 0.80$ in.$^2$ Check s = 16 in.<br><br>$\dfrac{0.75 \times \sqrt{5000} \times 24 \times 16}{60,000} = 0.34$ in.$^2 \le 0.80$ in.$^2$<br><br>$\dfrac{50 \times 24 \times 16}{60,000} = 0.32$ in.$^2 \le 0.80$ in.$^2$ |
| 15.4.2.1<br>15.4.2.2 | The maximum spacing is one-half the depth of the shallowest beam or slab framing into the element and shall be distributed along the column for the full depth of the deepest member. | Find the maximum spacing in joint:<br><br>$s = \dfrac{30}{2} = 15$ in.  **Controls**<br><br>Thus, two ties are required along the depth of the joint at a spacing not greater than 15 in. |
| ACI 352R | Section 15.2.2 of the Code states that, "the shear resulting from moment transfer shall be considered in the design of the joint." The Code does not provide specific requirements to meet this requirement in Ch. 15. If shear strength is a concern, the shear strength requirements of Section 18.18.4 of the Code can be used. Note that joint shear strength is not significantly increased by the increase of shear reinforcement as explained in Section R18.18.4. The shear is so low for this example that a check on shear is not necessary. See Example 4 for a full joint design. For a greater understanding of joint design, reference ACI 562R, "Recommendations for Design of Beam-Column Connections in Monolithic Reinforced Concrete Structures." | |
| **Step 4—Detail the column splice and joint at Level 2** | | |
| | It is common to begin lap splices for the vertical reinforcement at the floor level in ordinary moment frames. Since all the splices are at one location, a Class B tension lap length is provided and checked against the compression lap length as a minimum. | |
| 10.7.5.1.3→<br>25.5→<br>25.5.2<br><br>10.7.5.2.2<br><br>10.7.1.2→<br>25.4→<br>25.4.2 | $\ell_{st}$ is the greater of:<br>(a) $1.3\ell_d$<br>(b) 12 in.<br><br>where,<br><br>$\ell_d = \dfrac{3}{40}\dfrac{f_y}{\lambda\sqrt{f'_c}}\left(\dfrac{\psi_t\psi_e\psi_s}{\dfrac{c_b-K_{tr}}{d_b}}\right)d_b$<br><br>and,<br><br>$K_{tr} = \dfrac{40A_{tr}}{sn}$ | Determine $\ell_{st}$ for a No. 8 bar:<br>$\psi_t = \psi_e = \psi_s = \lambda = 1.0$<br>$d_b = 1.0$ in.<br>$c_b = 1.5 + 0.5 + 1.0/2 = 2.5$ in.<br>$n = 3$, number of longitudinal bars along the splitting plane<br>$s = 16$ in., spacing of ties<br>$A_{tr} = 4$ tie legs $\times 0.2$ in.$^2 = 0.8$ in.$^2$<br><br>$K_{tr} = \dfrac{40 \times 0.8}{16 \times 3} = 0.67$<br><br>$\left(\dfrac{c_b+K_{tr}}{d_b}\right) = \left(\dfrac{2.5+0.67}{1.0}\right) = 3.17 \le 2.5$<br><br>$\ell_d = \dfrac{3}{40}\dfrac{60,000}{1.0\sqrt{5000}}\dfrac{1.0}{2.5}1.0 = 25.4$ in.<br><br>$\ell_{st} = 1.3 \times 25.4 = 33.0$ in. |

| | | |
|---|---|---|
| 10.7.5.1.3→<br>25.5→<br>25.5.5.1 | For $f_y$ equal to 60 ksi, $\ell_{sc}$ is the greater of:<br>(a) $0.0005 f_y d_b = 30 d_b$<br>(b) 12 in. | Check compression lap splice length.<br><br>$\ell_{sc} = 30 \times 1.0 = 30$ in. |
| 10.7.5.2.1 | A common way of expressing this splice on the structural drawings is to make a lap splice table and reference it in the detail. Note that it is common to splice at every other story to save time on labor in the field. | For ease of construction, use $\ell_{st} = 33$ in. for all splices and splice at every level. |
| 10.7.4.1<br>10.7.6.4 | Lateral support of offset bends is provided by column ties at bend. Ties need to resist 1.5 times the horizontal tension component of the computed force in the inclined portion of the offset bar. The horizontal component was determined for the bar strength at maximum incline, 1 in 6 (9.5 degrees).<br><br>The calculations shows that one additional No.4 tie can laterally support one No. 8 bar at the offset. | Nominal vertical bar strength:<br>No. 8 = $(f_y A_{st}) = 60 \times 0.79 = 47.4$ kip<br><br>Horizontal tension at the bend:<br>$P_u = 1.5 \times 47.4 \times \sin(9.5°) = 11.7$ kip<br><br>Nominal tie strength:<br>No. 4 = $(f_y A_{st}) = 60 \times 0.20 = 12$ kips $\geq 11.7$ kip<br><br>Provide one tie leg for each vertical bar at its offset. The current tie detail meets this requirement. |
| 10.7.6.2<br><br><br><br><br><br>15.4.1 | The Code requires the first tie starting on any level to be within, $s/2$, of the top of the slab. It is good construction practice to start the tie at 2 or 3 in. from the top of the floor (beginning of the column cage) and then proceed with the typical tie spacing.<br><br>At the top of the column cage (bottom of the floor above) ties are required through beam / slab joint if the longitudinal bars are not confined by the concrete on all four sides. In this case, only two beams frame into the joint so the beam depth below the slab does not provide joint confinement. The slab is on all four sides so it provides confinement and ties are not required in this joint region. The Code requires the last tie to be within $s/2$ of the lowest horizontal reinforcement in the confined concrete. Again, it is good construction practice to place a tie at 2 or 3 in. below the confining concrete.<br><br>Locate the top offset bend at the tie just below the slab. Provide a tie at the bottom offset bend. This provides two ties in the joint region in required spacing. The tie requirements are not additive in this case. | Ties @ 16" o.c.<br>See lap splice table<br>$30 d_b$ min.<br>Top of slab<br>Tie 3" from top of slab<br>Slope 1:6 max.<br>Lower tie at offset bend<br>Last tie 3" from bottom of slab and at top offset bend<br>Ties @ 16" o.c. |
| **Step 4—Discussion and summary** | | |
| | The 24 x 24 in. column is satisfactory for design. The minimum reinforcement, eight No. 8s, is sufficient to resist the factored loads and moments. Shear is very low and only the minimum tie area and spacing, No 4 ties at 16 in. on center, are required for support of the longitudinal reinforcement. On inspection, this column size and reinforcement will work for the remaining floors, since the columns get shorter and the loads decrease. A smaller column will work for the upper floors but the cost of a change in formwork may not overcome the cost of the small amount of concrete that is saved. For this building, the architect may want to save some space on the upper floors and there may be a compromise between functionality and expense. | |

# CHAPTER 9—COLUMNS

**Column Example 3:** *Column for an intermediate moment frame (IMF)*

Design and detail the first floor interior column at location E4 from the example building given in Chapter 1 of this Handbook. The column is part of an IMF. Example 3 is a continuation of Examples 1 and 2. The loads have been modified to match the results of an analysis from commercial software capable of second-order elastic analysis and for Seismic Design Category C.

## Given:

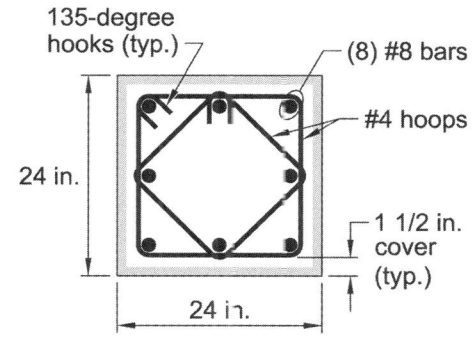

*Materials—*
Specified yield strength, $f_y$ = 60 ksi
Modulus of elasticity of steel, $E_s$ = 29,000 ksi
Specified concrete compressive strength, $f_c'$ = 5 ksi
Modulus of elasticity of concrete, $E_c$ = 4030 ksi
Normalized maximum size of aggregate is 1 in.

*Loading—*

| Load Combinations | $P_u$, kip | $M_u$, kip-in. | $V_u$, kips |
|---|---|---|---|
| (i) $U = 1.2D + 1.6L + 0.5S$ | 890 | 0 | 0 |
| (ii) $U = 1.2D + 1.0W + 1.0L + 0.5S$ | 800 | 651 | 5 |
| (iii) $U^* = 1.2D + 1.0E + 1.0L + 0.2S$ | 848 | 3228 | 25 |
| (iv) $U^* = 0.9D + 1.0E + 1.0L + 0.2S$ | 456 | 3228 | 25 |

\* The software adjusts seismic load combinations as required by ASCE 7-10.

Reference: SP-17 Supplement, Interaction Diagram Excel spreadsheet found at https://www.concrete.org/store/productdetail.aspx?ItemID=SP_714DAE

| ACI 318-14 | Discussion | Calculation |
|---|---|---|
| **Step 1—Discussion on modification of Example 2 for an IMF** | | |
| 10.5.2.1→ 22.4→22.2 | This example demonstrates the column design requirements for an IMF. The design requirements are similar to an ordinary moment frame (OMF) except that hoops are required in the plastic hinge region, $\ell_o$, at a reduced spacing. The additional design requirements for an IMF increases the response modification coefficient $R$ from 3 for an OMF to 5 for an IMF. The increase in R results in a decrease in seismic base shear. An IMF is permitted to be used for structures assigned to SDC B and required for SDC C. For the column designed in Examples 1 and 2, the shear is so low that it is not economical to require the extra detailing to gain the benefit of the reduced shear. For this example, the example building is analyzed for a region assigned to SDC C and Load Combination (iii) are revised. | |
| **Step 2—Find the required area of longitudinal reinforcement** | | |
| | The same column from Example 2 is checked using the interaction diagram spreadsheet referenced at the start of this example. See Example 2 for a detailed discussion on the calculation of nominal and design interaction diagrams.<br><br>The diagram on the right is for this column example and it has the loads from the four given load combinations plotted. | Design Capacity Interaction Diagram:<br>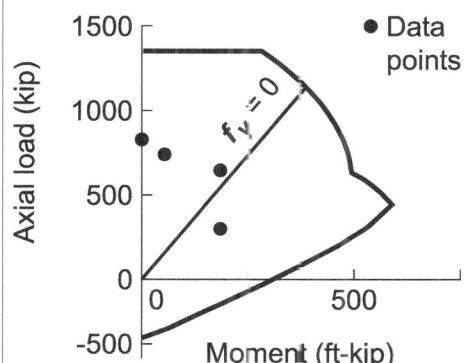<br>Eight No. 8 bars evenly spaced around the perimeter is sufficient. |

| | Step 3—Find the required area of transverse reinforcement | |
|---|---|---|
| 10.5.3.1→ 22.5 10.6.2 | The shear has increased in Load Combination (iii) but it is still very low. The concrete design strength calculated in Example 2 is sufficient to carry the load. | $\dfrac{\phi V_c}{2} = 27$ kip $\geq V_u = 25$ kip |
| 18.4.3.1 | For columns in IMFs, there are additional shear requirements. $\phi V_n$ shall be at least the lesser of: | |
| 18.4.3.1(a) | Shear calculated for reverse curvature from the maximum $M_n$ over the range of factored axial loads with lateral forces. The following figure illustrates how to find $M_n$ (National Institute of Standards and Technology (NIST), "Seismic Design of Reinforced Concrete Special Moment Frames: A Guide for Practicing Engineers," NIST GCR 8-917-1). 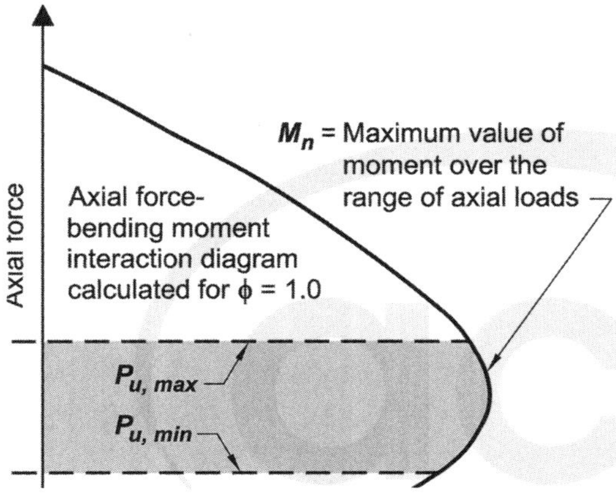 | The range of nominal moments is shown in the following figure. 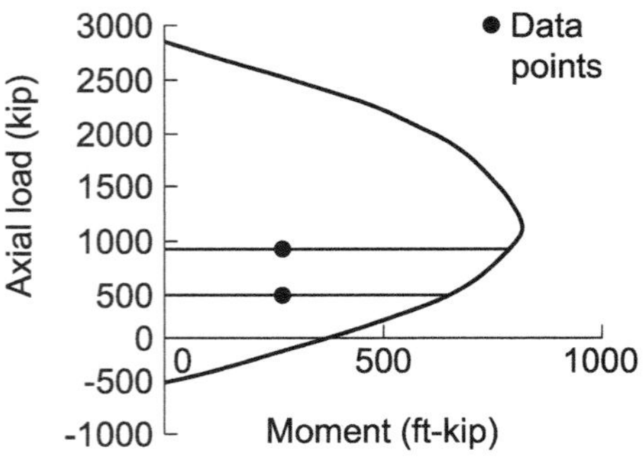 |
| | In this example, the range of $M_n$ does not contain the balance point of the curve. Thus, the maximum value for $M_n$ is at the load combination with the greatest $M_n$.<br><br>OR | The interaction diagram spreadsheet allows the user to input an axial load value and it will calculate the nominal moment, see the "Select Axial Load" tab.<br><br>For $P_u = 848$ kip, $M_n = 9373$ kip-in.<br><br>For $P_u = 456$ kip, $M_n = 7994$ kip-in.<br><br>Use, $M_n = 9373$ kip-in.<br><br>$V_u = \dfrac{M_{nt} + M_{nb}}{\ell_u} = \dfrac{2 \times 9373}{186} = 101$ kips |
| 18.4.3.1(b) | Shear with the overstrength factor $\Omega_o$ applied. | $\Omega_o = 3$ (Table 12.2-1, ASCE 7)<br><br>$V_u = 3 \times 25 = 75$ kip  **Controls** |

| | | |
|---|---|---|
| 10.5.3.1→<br>22.5.10.5.3 | Shear reinforcement is required. The equation for shear reinforcement is<br><br>$$V_s = \frac{A_v f_{yt} d}{s}$$ | $\phi V_c/2 = 27$ kip $\not\geq V_u = 75$ kip  **NG**<br><br>From Example 2<br>Area of a No. 4 bar = 0.20 in.$^2$<br>$A_v$ = 4 legs × 0.20 = 0.80 in.$^2$<br>d = 24 − 2.5 = 21.5 in<br>$f_{yt}$ = 60 ksi |
| 10.7.6.5.2 | When shear is required, there is a limit on spacing. The maximum spacing is the lesser of $d/2$ or 24 in. for<br><br>$$V_s \leq 4\sqrt{f'_c}b_w d = 4\sqrt{5000} \times 24 \times \frac{21.5}{1000} = 145.9 \text{ kip}$$ | Assume maximum spacing of<br><br>$$s_{max} = \frac{d}{2} = \frac{21.5}{2} = 10.75; \text{use } s = 10 \text{ in.}$$<br><br>Calculate $V_s$<br><br>$$V_s = \frac{0.80 \times 60 \times 21.5}{16} = 64.5 \text{ kips}$$<br><br>64.5 ≤ 145.9, thus $s_{max}$ assumption is **OK**. |
| 22.5.1.1 | | $\phi V_n = \phi(V_s + V_c) = 0.75(64.5 + 73) = 103$ kip ≥ 75 kip  **OK** |
| 10.6.2.2 | Check the $A_{v,min}$. $A_{v,min}$ shall be at least the greater of:<br><br>$$0.75\sqrt{f'_c}\frac{bs}{f_y}$$<br><br>$$50\frac{bs}{f_y}$$ | $A_v$ = 4 legs × 0.25 = 0.80 in.$^2$<br><br>$$\frac{0.75 \times \sqrt{5000} \times 24 \times 10}{60,000} = 0.21 \text{ in.}^2 \leq 0.80 \text{ in.}^2 \text{ OK}$$<br><br>$$\frac{50 \times 24 \times 10}{60,000} = 0.20 \text{ in.}^2 \leq 0.80 \text{ in.}^2 \text{ OK}$$ |
| **Step 4—Find the required geometry and spacing of transverse reinforcement** | | |
| 18.4.3.3 | The tie geometry from Example 2 is acceptable for all locations except in the plastic hinge region, $\ell_o$. Section 18.4.3.3 requires that hoops are used in the plastic hinge region, $\ell_o$. | Typical section along $\ell_o$.<br><br>135-degree hooks (typ.)<br>(8) #8 bars<br>#4 hoops<br>24 in.<br>1 1/2 in. cover (typ.)<br>24 in. |
| 10.7.6.1.2→<br>25.7.4 | Hoop geometry is similar to ties but the tie must be closed with seismic hooks at each end. Hoop is further defined in Chapter 2 of the Code. It states that the bend must not be less than a standard 135-degree hook.<br><br>Another common tie arrangement is one exterior hoop with one cross tie for the middle bars in each direction. | |
| 18.4.3.3 | The plastic hinge $\ell_o$ is the greatest of<br>(a) $1/6\ell_u$<br>(b) $h$ or $b$<br>(c) 18 in. | Find $\ell_o$<br>(a) 1/6 × 186 = 31 in.  **Controls.**<br>(b) 24 in.<br>(c) 18 in. |

| | | | |
|---|---|---|---|
| 18.4.3.3 | The maximum spacing, $s_o$, along the plastic hinge region is the smallest of<br>(a) $8d_b$, smallest longitudinal bar<br>(b) $24d_b$, hoop<br>(c) $0.5h$ or $0.5b$<br>(d) 12 in. | | Find the maximum spacing for $s_o$<br><br>(a) $8 \times 1 = 8$ in.  **Controls.**<br>(b) $24 \times 0.5 = 12$ in.<br>(c) $0.5 \times 24 = 12$ in.<br>(d) 12 in. |
| 18.4.3.5 | Outside $\ell_o$ the column transverse reinforcement is provided as shown in Example 2 but with a 10 in. spacing for shear. | | Thus, the transverse reinforcement is No. 4 hoop at 8 in. along $\ell_o$ and No. 4 ties at 10 in. outside of $\ell_o$ |
| **Step 5—Check the joint** | | | |
| 15.2.4<br>18.4.4 | From Example 2, two ties are required along the depth of the joint at a spacing not greater than 15 in. Section 18.4.4 refers the engineer to the Chapter 15 so there are no additional requirements according to the Code for an IMF.<br><br>The shear required for design is larger for the IMF. As discussed in Step 3 of Example 2, the provisions of Section 18.18.4 can be used to check the concrete shear strength through the joint. Note that Section 18.18.4 are requirements for a Type 2 connection as defined in ACI 352R. An IMF could be designed as a Type 1 connection in ACI 352R. | | |
| 18.4.2 | Calculate column shear force $V_u$ associated with the nominal moments and related shears calculated in accordance with Section 18.4.2. The column shear can be approximated with the following equation assuming inflection points occur at the column midheights:<br><br>$$V_{u,col} = \frac{\left[(M_{nl} + M_{nr}) + (V_{ul} + V_{ur})\dfrac{h}{2}\right]}{\ell}$$<br><br>Where $\ell$ is the distance between the mid-height of the column above and below the joint. Note that it is unconservative to ignore the slab for this check; therefore, consider the full effective width, $b_f$, of the T-Beam in the calculations. | | Values adapted from Beam Example 1 in Chapter 7 of this Handbook. Reinforcement is modified to meet the beam requirements for IMFs. Use seven No. 7 bars at the T-beam flange and three No. 7 bars at the T-beam stem.<br><br>$M_{nl} = 6520$ kip-in.<br>$M_{nr} = 2960$ kip-in.<br>$V_{ul} = 77$ kip<br>$V_{ur} = 30$ kip<br><br>Also,<br><br>$\ell = \dfrac{18\,\text{ft} + 14\,\text{ft}}{2} \times 12\,\text{in./ft} = 192\,\text{in.}$<br><br>$h = 24$ in.<br><br>$V_{u,col} = \dfrac{\left[(6520 + 2960) + (77 + 30)\dfrac{24}{2}\right]}{192} = 52$ kip |
| | The shear at the center of the joint is:<br><br>$V_j = T_2 + C_1 - V_u$ | | $T_1 = 3 \times 0.60 \times 60 = 108$ kip<br>$T_2 = 7 \times 0.60 \times 60 = 252$ kip<br><br>$C_1 = T_1 = 108$ kip<br>$C_2 = T_2 = 252$ kip<br><br>$V_j = 252 + 108 - 52 = 308$ kip |

| | | |
|---|---|---|
| 18.8.4.3 | The effective area of the joint, $A_j$, is calculated by the multiplying the column depth, $h$, by the effective width which is the lesser of: | $b_{w,beam} = 18$ in. <br> $h_{col} = b_{col} = 24$ in. <br><br> $x = \dfrac{24 \text{ in.} - 18 \text{ in.}}{2} = 3$ in. |
| Fig. R18.8.4 | (a) $b_{w,beam} + h_{column}$ <br> (b) $b_{w,beam} + 2x$ <br> (c) $b_{column}$ <br><br> where $x$ is the smaller distance between the edge of the beam and edge of the column. | Effective joint width <br> (a) $18 + 24 = 42$ in. <br> (b) $18 + 2 \times 3 = 24$ in. <br> (c) 24 in.  **Controls** <br><br> $A_j = 24 \times 24 = 576$ in.$^2$ |
| 18.8.4.1 <br><br> 21.2.4.3 | Calculate the shear strength of the joint. Table 18.8.4.1 provides $V_n$ for several conditions. This joint is confined by beams on two opposite faces. <br><br> $V_n = 15\lambda\sqrt{f'_c}A_j$ | $V_n = \dfrac{15 \times 1.0 \times \sqrt{5000} \times 576}{1000} = 611$ kips <br><br> $\phi V_n = 0.85 \times 611 = 519$ kip $\geq 308$ kip <br><br> Joint shear is sufficient. |
| | Note that for a Type 1 connection in ACI 352R a higher shear strength is permitted. <br><br> $V_n = 20\lambda\sqrt{f'_c}A_j$ | |
| **Step 6—Detail the column splice and joint at Level 2** | | |
| 18.4.3.4 | The first hoop from the face of the joint must be within $s_o/2$. Similar to Example 2, the first hoop is 3 in. off the top of the slab. For this example the hoops extend into the joint. The last hoop in the joint should be at least $s_o$ from the first hoop at the next level. <br><br> The splice is permitted to start at the top of the slab for an IMF, which is desired for ease of construction. | 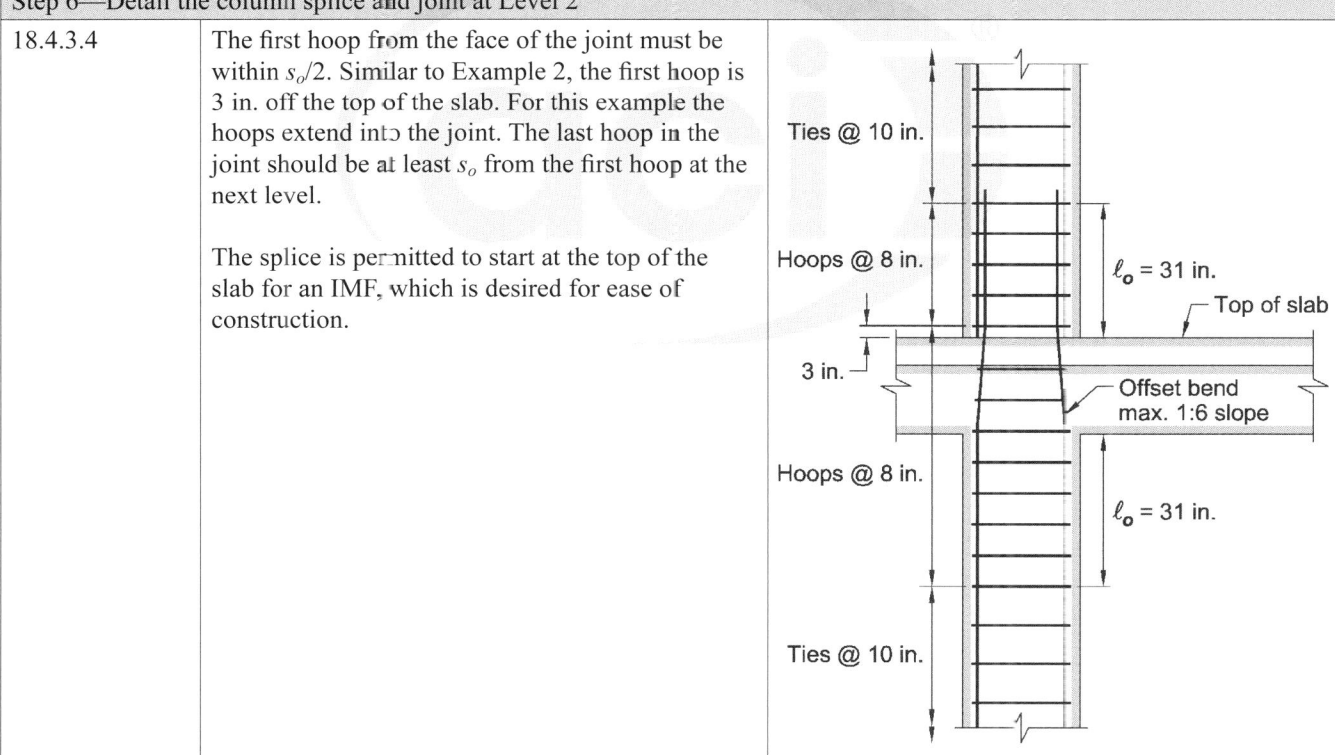 |
| **Step 7—Discussion and summary** | | |
| | This example is an extension of Example 2 but the seismic loads where increased for a Seismic Design Category C. The difference in design resulted in almost twice as much transverse reinforcement. The requirement to use hoops only resulted in a 135-degree hooks instead of 90-degree hooks. The column size and longitudinal reinforcement remained the same. | |

**Column Example 4:** *Column for a special moment frame (SMF)*

Design and detail the first floor interior column at location E4 from the example building given in Chapter 1 of this Handbook. The column is part of an SMF. Example 4 is a continuation of Examples 1, 2, and 3. The loads have been modified to match the results of an analysis from commercial software capable of second-order elastic analysis and for Seismic Design Category D.

**Given:**

*Materials—*
Specified yield strength, $f_y$ = 60 ksi
Modulus of elasticity of steel, $E_s$ = 29,000 ksi
Specified concrete compressive strength, $f_c'$ = 5 ksi
Modulus of elasticity of concrete, $E_c$ = 4030 ksi
Normalized maximum size of aggregate is 1 in.

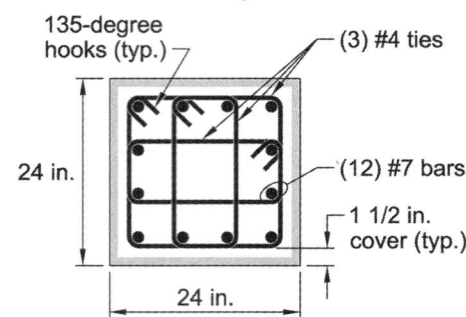

*Loading—*

| Load combinations | $P_u$, kip | $M_u$, kip-in. | $V_u$, kip |
|---|---|---|---|
| (i) $U = 1.2D + 1.6L + 0.5S$ | 890 | 0 | 0 |
| (ii) $U = 1.2D + 1.0W + 1.0L + 0.5S$ | 800 | 651 | 5 |
| (iii) $U = 1.2D + 1.0E + 1.0L + 0.2S$ | 872 | 4491 | 34 |
| (iv) $U = 0.9D + 1.0E + 1.0L + 0.2S$ | 432 | 4491 | 34 |

Reference: SP-17 Supplement, Interaction Diagram Excel spreadsheet found at https://www.concrete.org/store/productdetail.aspx?ItemID=SP1714DAE

| ACI 318-14 | Discussion | Calculation |
|---|---|---|
| **Step 1—Discussion on modification of Example 3 for a special moment frame (SMF)** | | |
| | This example demonstrates the column design requirements for an SMF. The additional design and detailing requirements for a SMF increases the response modification coefficient, R, from 5 for an IMF to 8 for an SMF. A SMF is permitted to be used for structures assigned to SDC B and C and required for SDC D, E, and F. For this example, the example building is analyzed for a region assigned to SDC D and Load Combinations (iii) and (iv) are revised. This example starts with the column designed in Example 3. | The following properties are from the intermediate moment frame (IMF) design and are repeated here for information.<br><br>Column:<br>$h = b = 24$ in.<br>Eight No. 8 longitudinal bars<br>No. 4 hoops<br><br>Beam:<br>$h = 30$ in.<br>$b_f = 120$ in.<br>$b_w = 18$ in.<br>Seven No. 7 long. bars at beam flange<br>Three No. 7 long. bars at beam stem |
| **Step 2—Check dimensional limits and axial load** | | |
| 18.7.2.1 | The column cross section shall satisfy the following:<br><br>a) The least dimension shall be at least 12 in.<br>b) $b/h \geq 0.4$, where $h \geq b$ | $h = 24$ in. $\geq 12$ in.  **OK**<br><br><br>$b/h = 1 \geq 0.4$  **OK** |

# CHAPTER 9—COLUMNS

| | | | |
|---|---|---|---|
| 18.8.2.3 | The beam-column joint must be deep enough to prevent slip of the longitudinal bar from the beam. For normalweight concrete, the joint width parallel to the beam must be 20 times the largest longitudinal reinforcement in the beam. | The largest longitudinal reinforcement in the beam that runs through the joint is a No. 7.<br><br>$20d_b = 20 \times 0.875 = 17.5$ in.<br><br>$h = 24 \geq 17.5$ in.  **OK** | |
| 18.8.2.4 | The depth of the joint shall not be less than one-half the depth of any beam framing into the joint. | $\dfrac{h_{beam}}{2} = \dfrac{30}{2} = 16$ in.<br><br>$h_{column} = 24$ in. $\geq 16$ in.  **OK** | |
| 18.7.5.2(f) | The arrangement of the longitudinal reinforcement is effected if<br><br>$P_u > 0.3 A_g f_c'$ or<br>$f_c' > 10,000$ psi<br><br>The value $h_x$ shall not exceed 8 in. if this occurs; otherwise, it shall not exceed 14 in. | $P_u = 872$ kip<br>$A_g = 576$ in.²<br><br>872 kip > $0.3 \times 576 \times 5 = 864$ kip<br>5000 psi $\not>$ 10,000 psi<br><br>Therefore, $h_x$ shall not exceed 8 in. | |
| | $h_x$ for the cross section in Examples 2 and 3 is<br><br>$h_x = \dfrac{[24 - (1.5 \times 2) - (0.5 \times 2) - 1.0]}{2} = 9.5$ in. $\not\leq$ 8 in.<br><br>Thus, either the cross-section can be increased or the reinforcement should be rearranged to satisfy this requirement.<br><br>This column is at a lower level and the axial load is slightly greater than $0.3 A_g f_c'$. The upper level columns will not exceed this limit. It is therefore recommended to rearrange the reinforcement. The reinforcement for the upper levels can switch back to the previous cross-section if it is found to be more economical. | The area of steel in the current column is<br>$A_{st} = 8 \times 0.79$ in.² $= 6.31$ in.²<br><br>Try using 4 bars at each side using No. 7s<br>$A_{st} = 12 \times 0.60$ in.² $= 7.20$ in.²<br><br>$h_x = [24 - (1.5 \times 2) - (0.5 \times 2) - .875]/3$<br>$= 6.4$ in. $\leq 8$ in.<br><br>Use twelve No. 7 longitudinal bars evenly spaced around the perimeter. | |
| **Step 3—Find the required area of longitudinal reinforcement** | | | |
| 10.5.2.1→<br>22.4→22.2 | Using the Interaction Diagram spreadsheet referenced at the start of this Example, the revised column section is analyzed for the four load combinations given for this example. | Design Capacity Interaction Diagram:<br><br>*[Interaction diagram showing Axial load (kip) vs Moment (ft-kip), with data points and $f_y = 0$ line]*<br><br>Twelve No. 7 bars evenly spaced around the perimeter is sufficient. | |
| 18.7.4.1 | For SMFs, the maximum amount of longitudinal reinforcement, $A_{st}$, is reduced to $0.06 A_g$. The minimum amount of $A_{st}$ stays the same at $0.01 A_g$. | $0.01 A_g = 0.01 \times 24 \times 24 = 5.76$ in.²<br>$0.06 A_g = 0.06 \times 24 \times 24 = 34.56$ in.²<br><br>$A_{st} = 12 \times 0.60$ in.² $= 7.20$ in.²  **OK** | |

| | | |
|---|---|---|
| 18.7.3.2 | The flexural strength of column in a SMF must satisfy:<br><br>$\sum M_{nc} \geq \left(\dfrac{6}{5}\right) \sum M_{nb}$ | From Example 3, the nominal moments in the beam are:<br>$M_{nl}$ = 6520 kip-in.<br>$M_{nr}$ = 2960 kip-in.<br><br>Add one more No. 7 bars to the bottom of the beam so that the positive moment at least one-half the negative moment at the joint face.<br><br>$M_{nl}$ = 6520 kip-in.<br>$M_{nr}$ = 3940 kip-in. |
| 18.6.3.2 | The beam design is not part of this example. The beam used for the ordinary moment frame (OMF) and IMF does not meet all the requirements for a SMF. The beam is modified here as necessary to fully demonstrate the column and joint design. | |
| | $M_n$ is the minimum nominal moment in the range of the interaction curve related to the minimum and maximum axial loads for the seismic load combinations. For more information on how to calculate $M_n$, refer to Step 2 in Example 3. | Column:<br><br>(Interaction diagram: Axial load (kip) vs. Moment (ft-kip), with data points shown) |
| | The Interaction Diagram Spreadsheet allows the user to input an axial load value and it will calculate the nominal moment, see the "Select Axial Load" tab. | For $P_u$ = 872 kip, $M_n$ = 9720 kip-in.<br>For $P_u$ = 432 kip, $M_n$ = 8380 kip-in.<br><br>Use the lowest value, $M_n$ = 8380 kip-in.<br><br>$(2 \times 8380) \geq \dfrac{6}{5}(6520 + 3940)$<br><br>$16{,}750 \geq 12{,}550$  **OK** |
| **Step 4—Determine the geometry of the transverse reinforcement** | | |
| 18.7.4.3<br>18.7.5.1<br>18.7.5.5<br><br>18.7.5.2 | Hoops for rectangular columns or spirals for circular columns are required for the entire column height. The hoops at the plastic hinge, $\ell_o$, and the splice must meet geometry requirements of Section 18.7.5.2. There are six conditions that geometry of the section must satisfy. The cross-section shown on the right meets these conditions. | (Cross-section: 24 in. × 24 in. column with 135-degree hooks (typ.), (3) #4 ties, (12) #7 bars, 1 1/2 in. cover (typ.)) |

# CHAPTER 9—COLUMNS

| | | |
|---|---|---|
| | Note that hoops are closed ties with standard hooks at the end that are at least 135 degrees. Crossties are permitted to support the longitudinal bars between the corners. Where crossties are used they shall be alternated end for end along the longitudinal bar. For this example, every bar needed support. Since there are an even number of bars, overlapping hoops provide the least number of pieces that still provide the necessary support and confinement.<br><br>Notice that Section 18.7.5.2(f) was checked early in this example. The maximum longitudinal spacing requirements of Section 18.7.5.2(e) and (f) can impact the design of the column as it did in this example. | |
| **Step 5—Determine the maximum spacing of the transverse reinforcement** | | |
| 18.7.4.3<br>18.7.5.1<br>18.7.5.3 | The maximum spacing, $s_o$, along the plastic hinge length, $\ell_o$, and splice regions is the smallest of:<br>(a) $0.25h$ or $0.25b$<br>(b) $6d_b$, smallest longitudinal bar<br>(c) $4 + \left(\dfrac{14 - h_x}{3}\right)$<br><br>Also, $s_o$ shall not exceed 6 in. and need not be taken less than 4 in. | Maximum spacing for $s_o$<br>(a) $0.25 \times 24 = 6$ in.<br>(b) $6 \times 0.875 = 5.25$ in. **Controls**<br>(c) $4 + \left(\dfrac{14 - 6.4}{3}\right) = 6.53$ in. |
| 18.7.5.5 | The maximum spacing, $s$, between $\ell_o$ and the column splice regions is the smallest of:<br>(a) $6d_b$, smallest longitudinal bar<br>(b) 6 in. | Maximum spacing for $s$<br>(a) 6 in.<br>(b) $6 \times 0.875 = 5.25$ in. **Controls.** |
| | | Use $s = 5.25$ in. along the column height unless noted otherwise from the following checks. |
| **Step 6—Check the minimum amount of transverse reinforcement required along the plastic hinge length, $\ell_o$** | | |
| 18.7.5.4 | Since $P_u$ is greater than $0.3A_g f_c'$ as shown in Step 2, $A_{sh}/s_o b_c$ shall be greatest of:<br><br>(a) $0.3\left(\dfrac{A_g}{A_{ch}} - 1\right)\dfrac{f_c'}{f_{yt}}$<br><br>(b) $0.09\dfrac{f_c'}{f_{yt}}$<br><br>(c) $0.2 k_f k_n \dfrac{P_u}{f_{yt} A_{ch}}$<br><br>where<br><br>$k_f = \dfrac{f_c'}{25,000} + 0.6 \geq 1.0$ and<br><br>$k_n = \dfrac{n_l}{n_l - 2}$<br><br>$n_l$ is the number of longitudinal bars around the perimeter of the column core that are laterally supported by the corner of hoops or seismic hooks. | $A_{sh} = 0.20$ in.$^2 \times 4 = 0.80$ in.$^2$<br>$b_c = 24 - (2 \times 1.5) = 21$ in.<br>$A_g = 24 \times 24 = 576$ in.$^2$<br>$A_{ch} = 21^2 = 441$ in.$^2$<br><br>$k_f = \dfrac{5000}{25,000} + 0.6 = 0.8 \not\geq 1.0$ ; use 1.0.<br><br>$k_n = \dfrac{12}{12 - 2} = 1.2$<br><br>(a) $0.3\left(\dfrac{576}{441} - 1\right)\dfrac{5}{60} = 0.0077$<br><br>(b) $0.09\dfrac{5}{60} = 0.0075$<br><br>(c) $0.2 \times 1.0 \times 1.2 \dfrac{872}{60 \times 441} = 0.0079$ **Controls** |

| | | |
|---|---|---|
| | | Find the required maximum spacing for this amount of reinforcement $$\frac{A_{sh}}{s_o b_c} \geq 0.0079 \text{ or}$$ $$s_o \leq \frac{A_{sh}}{0.0079 b_c} = \frac{0.80}{0.0079 \times 21} = 4.8 \text{ in.}$$ Use $s = 4.5$ in. along $\ell_o$ |
| **Step 7—Check the minimum amount of transverse reinforcement required for shear** | | |
| 18.7.6.1.1 | Section 18.7.6.1.1 states: "The design shear force $V_e$ shall be calculated from considering the maximum forces that can be generated at the faces of the joints at each end of the column. These joint forces shall be calculated using the maximum probable flexural strengths, $M_{pr}$, at each end of the column associated with the range of factored axial forces, $P_u$, acting on the column. The column shears need not exceed those calculated from joint strengths based on $M_{pr}$ of the beams framing into the joint. In no case shall $V_e$ be less than the factored shear calculated by analysis of the structure." | |
| | The first part of Section 18.7.6.1.1 is very similar to how $M_n$ is calculated in Step 3 above, except that $1.25 f_y$ is used to generate the interaction diagram (National Institute of Standards and Technology (NIST), "Seismic Design of Reinforced Concrete Special Moment Frames: A Guide for Practicing Engineers," NIST GCR 8-917-1). This modified moment is called the probable moment, $M_{pr}$.<br>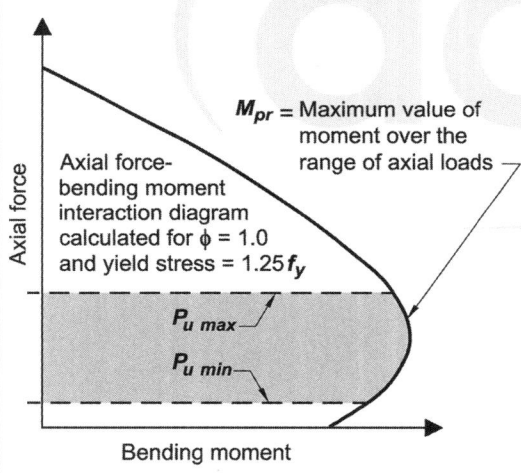 | Using the Interaction Diagram Spreadsheet, generate the curve for $f_y = 75$ ksi.<br>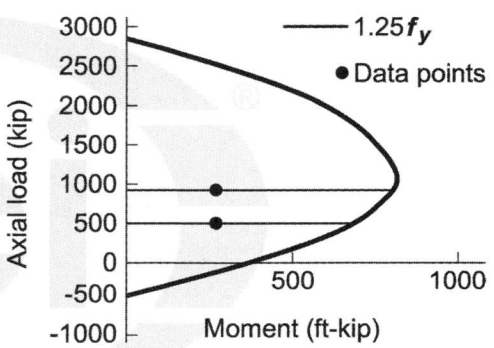<br>Use the "Select Axial Load" sheet in the spreadsheet to find $M_{pr}$.<br><br>For $P_u = 872$ kip, $M_{pr} = 10,284$ kip-in.<br>For $P_u = 432$ kip, $M_{pr} = 9120$ kip-in.<br><br>Use $M_{pr} = 10,284$ kip-in.<br>$$V_e = \frac{M_{prt} + M_{prb}}{\ell_u} = \frac{2 \times 10,284}{186} = 111 \text{ kip}$$ |

| | | |
|---|---|---|
| | The second part of Section 18.7.6.1.1 is similar to how the column shear was calculated for the joint in Step 3 of Example 3. Calculate column shear force $V_e$ associated with the probable moments and related shears of the beam. The column shear can be approximated with the following equation assuming inflection points occur at the column mid-heights: $$V_{e,col} = \frac{\left[(M_{pr1} + M_{pr2}) + (V_{e1} + V_{e2})\frac{h}{2}\right]}{\ell}$$ where $\ell$ is the distance between the mid-height of the column above and below the joint. Note that it is unconservative to ignore the slab for this check; therefore, consider the full effective width, $b_f$, of the T-beam in the calculations. | The probable $M_{pr1} = 8010$ kip-in. $M_{pr2} = 4920$ kip-in. $V_{e1} = 85$ kip $V_{e2} = -22$ kip Also, $$\ell = \frac{18\text{ ft} + 14\text{ ft}}{2} \times 12\text{ in./ft} = 192\text{ in.}$$ $h = 24$ in. $$V_{e,col} = \frac{\left[(8010 + 4920) + (85 - 22)\frac{24}{2}\right]}{192} = 71\text{ kip}$$ |
| | The last part of Section 18.7.6.1.1 states that $V_e$ cannot be less than $V_u$ from the analysis. | The maximum $V_u$ from the load combinations $V_e = 34$ kip |
| | The requirement permits the use of the shear calculated for the second part of Section 18.7.6.1.1, but try to develop the shear related to $M_{pr}$ of the column. | Use $V_e = 111$ kip |
| 10.5.3.1→ 22.5.10.5.3 | Determine the maximum spacing of hoops required for shear | From Example 2: $V_c = 73$ kip $\phi = 0.75$ Area of No. 4 bar = 0.20 in.$^2$ $A_v = 4$ legs × 0.20 = 0.80 in.$^2$ $d = 24 - 2.44 = 21.56$ in. |
| 22.5.1.1 | The equation for shear reinforcement is $$V_s = \frac{A_v f_{yt} d}{s}$$ $\phi V_n = \phi(V_s + V_c) \geq V_e$ | $s = 5.25$ in. from Step 5 $$V_s = \frac{0.80 \times 60 \times 21.56}{5.25} = 197\text{ kips}$$ $\phi V_n = 0.75(197 + 73) = 203$ kip $\geq V_e = 111$ kip |
| **Step 8—Summarize the amount and spacing of transverse reinforcements along the height of the column** | | |
| 18.7.5.1 | The plastic hinge length, $\ell_o$, is the greatest of: (a) $h$ (b) $1/6\ell_u$ (c) 18 in. $\ell_o$ is starts at the top of the slab and bottom of the beam and extends toward the middle of the column. | Find $\ell_o$ (a) 24 in. (b) 1/6 × 186 = 31 in. Controls. (c) 18 in. From Step 6, $s = 4.5$ in. Along $\ell_o$, use No. 4 hoops at 4.5 in. |

| | | |
|---|---|---|
| 18.7.4.3 | Lap splices shall be placed within the center half of the column. The hoop spacing along the splice length must meet Sections 18.7.5.2 and 18.7.5.3 but not 18.7.5.4. In the calculation of hoop spacing for $\ell_o$ regions, 18.7.5.4 controlled but the difference is so small in this case that the same spacing used. | Along the splice, use No. 4 hoops at 4.5 in. |
| 10.7.5.1.3→<br>25.5→<br>25.5.2<br><br>10.7.5.2.2 | $\ell_{st}$ is the greater of:<br>(a) $1.3\ell_d$<br>(b) 12 in.<br><br>where,<br><br>$$\ell_d = \frac{3}{40}\frac{f_y}{\lambda\sqrt{f_c'}}\frac{\psi_t\psi_e\psi_s}{\left(\frac{c_b+K_{tr}}{d_b}\right)}d_b$$<br><br>and<br><br>$$K_{tr} = \frac{40A_{tr}}{sn}$$ | Determine $\ell_{st}$ for a No. 7 bar:<br>$\psi_t = \psi_e = \psi_s = \lambda = 1.0$<br>$d_b = 0.875$ in.<br>$c_b = 1.5 + 0.5 + 0.875/2 = 2.44$ in.<br>$n = 4$, number of bars along the splitting plane<br>$s = 4.5$ in., spacing of ties<br>$A_{tr} = 4$ tie legs × 0.2 in.$^2$ = 0.8 in.$^2$<br><br>$$K_{tr} = \frac{40 \times 0.8}{4.5 \times 4} = 1.78$$<br><br>$$\left(\frac{c_b+K_{tr}}{d_b}\right) = \left(\frac{1.94+1.78}{0.875}\right) = 4.25 \leq 2.5$$<br><br>$$\ell_d = \frac{3}{40}\frac{60,000}{1.0\sqrt{5000}}\frac{1.0}{2.5}0.875 = 22.3 \text{ in.}$$<br><br>$\ell_{st} = 1.3 \times 22.3 = 29.0$ in. |
| 10.7.5.1.3→<br>25.5→<br>25.5.5.1<br><br>10.7.5.2.1 | For $f_y$ equal to 60 ksi, $\ell_{sc}$ is the greater of:<br>(a) $0.0005 f_y d_b = 30d_b$<br>(b) 12 in. | Check compression lap splice length.<br><br>$\ell_{sc} = 30 \times 0.875 = 26.25$ in.<br><br>For ease of construction, use $\ell_{st} = 29$ in. for all splices and splice at mid-height at every level. |
| 18.7.4.3<br>18.2.7<br>18.2.8 | The end result is hoop spacing of 4.5 in. at $\ell_o$ regions and along the splice length. The remainder of the column has a spacing of 5.25 in.<br><br>Note that mechanical and welded splices are permitted. Type 1 mechanical splices and welded splices are not permitted within a distance equal to twice the column depth from the bottom of beam or top of slab. Type 2 mechanical splices do not have a location restriction for cast-in-place SMFs. | Column elevation showing hoop spacing: 4 1/2" spacing at top (2'-7" below top of beam), 5 1/4" spacing, 4 1/2" spacing over 2'-5" center splice at mid-height, 5 1/4" spacing, 2'-7" at top of slab, 4 1/2" spacing at bottom. |

| | | |
|---|---|---|
| **Step 9—Check the joint** | | |
| 18.8<br>18.8.2.1 | The column shear stresses at the joint is calculated in Step 7. The column shear is calculated from the probable moment in the beams where the flexural tensile reinforcement is $1.25f_y$. | $V_{e,col} = 71$ kip |
| | The shear at the center of the joint is:<br>$V_j = T_2 + C_1 - V_u$ | $T_1 = 4 \times 0.6 \times 60 \times 1.25 = 180$ kip<br>$T_2 = 7 \times 0.875 \times 60 \times 1.25 = 315$ kip<br><br>$C_1 = T_1 = 180$ kip<br>$C_2 = T_2 = 315$ kip<br>$V_j = 315 + 180 - 71 = 424$ kip |
| 18.8.4.3<br><br>Fig. R18.8.4 | The effective area of the joint, $A_j$, is calculated by the multiplying the column depth $h$ by the effective width which is the lesser of:<br>(a) $b_{w,beam} + h_{column}$<br>(b) $b_{w,beam} + 2x$<br>(c) $b_{column}$<br><br>where $x$ is the smaller distance between the edge of the beam and edge of the column. | $b_{w,beam} = 18$ in.<br>$h_{col} = b_{col} = 24$ in.<br><br>$x = \dfrac{24\text{ in.} - 18\text{ in.}}{2} = 3$ in.<br><br>Effective joint width<br>(a) $18 + 24 = 42$ in.<br>(b) $18 + (2 \times 3) = 24$ in.<br>(c) 24 in. **Controls**<br><br>$A_j = 24 \times 24 = 576$ in.$^2$ |
| 18.8.4.1<br><br>21.2.4.3 | Calculate the shear strength of the joint. Table 18.8.4.1 provides $V_n$ for several conditions. This joint is confined by beams on two opposite faces. | $V_n = \dfrac{15 \times 1.0 \times \sqrt{5000} \times 576}{1000} = 611$ kips<br><br>$\phi V_n = 0.85 \times 611 = 519$ kip $\geq 424$ kip<br><br>Joint is **OK**. |
| **Step 10—Discussion and summary** | | |
| | This example is an extension of Examples 2 and 3 but the seismic loads where increased for a Seismic Design Category D. The difference in design resulted in a rearrangement of the longitudinal bars and a different hoop arrangement. The amount of transverse reinforcement is almost 4 times that of an OMF and about twice that of an IMF. The column size was sufficient for all three types of moment-resisting frames. | |

# CHAPTER 10—STRUCTURAL REINFORCED CONCRETE WALLS

## 10.1—Introduction

The scope of ACI 318-14, Chapter 11 addresses nonprestressed and prestressed cast-in-place and precast reinforced concrete walls. In addition to Chapter 11, Section 18.10 of ACI 318-14 provides design and detailing requirements for special cast-in-place walls forming part of the seismic-force-resisting system.

Reinforced concrete structural walls are common in buildings, and are almost always part of a building's lateral-force-resisting system (LFRS) due to their high in-plane stiffness. Although structural walls are also part of the gravity-force-resisting system, they are often lightly axially loaded. They are, designed and detailed to resist the combined effects of gravity and lateral forces.

Walls that are part of the LFRS are commonly referred to as shear walls (ASCE 7) because they resist a large portion of the total lateral forces acting on the structure through in-plane shear. In ACI 318, all walls are referred to as structural walls.

ACI 318 and ASCE 7 have coordinated requirements and identify two categories of LFRS for nonprestressed, cast-in-place walls:

1. An ordinary cast-in-place structural wall, permitted in seismic design categories (SDCs) A, B, and C, which is designed and detailed in compliance with Chapter 11 of ACI 318-14.

2. A special structural wall, which is designed and detailed in compliance with Chapters 11 and 18 of ACI 318-14. Special walls are required in SDCs D, E, and F, but can be constructed in all seismic categories

Seismic requirements are intended to increase wall strength and ductility to accommodate the large displacements demands expected during a maximum design earthquake. Walls are laterally connected to diaphragms and vertically to foundations or support elements. In seismic and nonseismic design, the connections to diaphragms are designed to remain elastic, and the energy from the lateral forces is absorbed by the structural wall. In seismic design, connection to the foundation can be the point of maximum wall moment and yielding of vertical reinforcement is expected.

## 10.2—General

**10.2.1** *Distinguishing a column from a wall*—The design of a wall can be so similar to a column that the question of when a rectangular column becomes a wall is often deliberated. For special moment frames, columns are defined as having a minimum aspect ratio of 0.4 in 18.7.2.1(b) of ACI 318-14. Although this limit is necessary to achieve the expected behavior, it might not be the best limit for the consideration of column or wall design. Expected behavior and construction constraints of the member are often the keys to answering the question of wall or column. Columns usually have high axial loads and their shear behavior is similar to beams. Walls usually have low axial loads and their shear behavior is similar to one-way slabs for out-of-plane loads, with unique shear behavior for in-plane loads. For further discussion and information regarding unique shear behavior for in-plane loads in walls, refer to Moehle (2015).

**10.2.1.1** *Longitudinal reinforcement*—In general, longitudinal bars require lateral support to prevent buckling of the bars due to axial compression. In a wall, if longitudinal reinforcement is required for axial strength, or if $A_{st}$ exceeds $0.01A_g$, then Section 11.7.4.1 of ACI 318-14 states the longitudinal reinforcement must be supported by transverse ties. This requirement could be used as a practical limit to determine whether the member should be designed as a column or a wall. If the wall requires heavy reinforcement (exceeding $0.01A_g$), a tie bar would be required at every intersection of longitudinal and transverse reinforcement, which would significantly affect the required amount of construction labor. Designing this same member as a column could be more practical.

**10.2.1.2** *Shear aspect ratios*—The next limit to consider is shear. Most walls have a length-to-thickness ratio of at least 6. For these aspect ratios, it is easy to see how the shear behavior will differ from a column for either in-plane or out-of-plane loads. For smaller aspect ratios of approximately 2.5 to 6, the member is designed either as a wall or column, depending on the shear force applied and the direction of shear, except as limited by 18.7.2.1(b) of the Code. For aspect ratios under 2.5, the member is likely to be designed as a column. Further discussion on this topic is given in Garcia (2003).

**10.2.2** *Wall layout*—Shear walls should be located within a building plan to efficiently resist lateral loading. Locating shear walls in the center half of each building is generally a good location for resisting lateral forces (Fig. 10.2.2(a)). This arrangement, however, can restrict architectural use of space.

Although shear walls are commonly located at the ends of a building, such wall locations will increase slab restraint and shrinkage stresses, especially in long buildings and buildings such as parking structures that are exposed to large temperature changes (Fig. 10.2.2(b)). Symmetrical wall arrangements provide good flexural and torsional stiffness. Walls at the perimeter resist torsional forces most effectively. Walls away from the perimeter, however, could have a higher tributary area and, consequently, larger gravity axial force to resist uplift or overturning. They are, however, less efficient in resisting horizontal torsion.

An unsymmetrical arrangement, however, does not usually provide predictable torsional stiffness due to their eccentricity (Fig. 10.2.2(c)). Such a shear wall layout should be designed explicitly for torsion. A symmetrical arrangement is preferable to avoid designing walls for torsion.

**10.2.3** *Wall configurations*—Shear walls could have several geometric configurations, including plane, flanged, or channel sections. A plane wall section is rectangular with

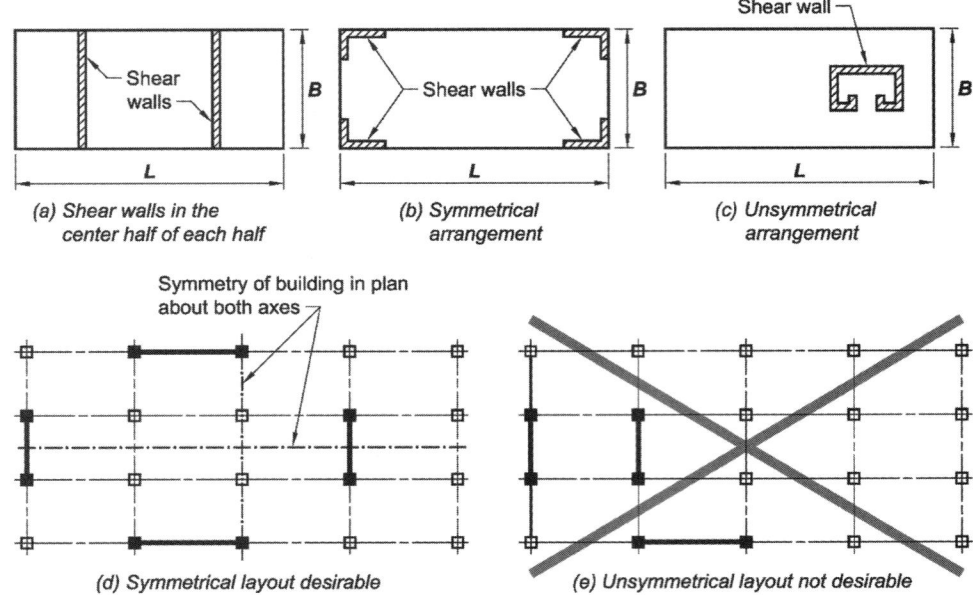

*Fig. 10.2.2—Shear wall layouts.*

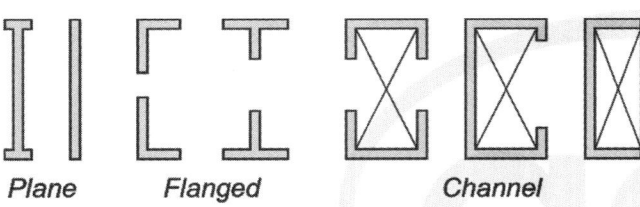

*Fig. 10.2.3a—Shear wall cross sections.*

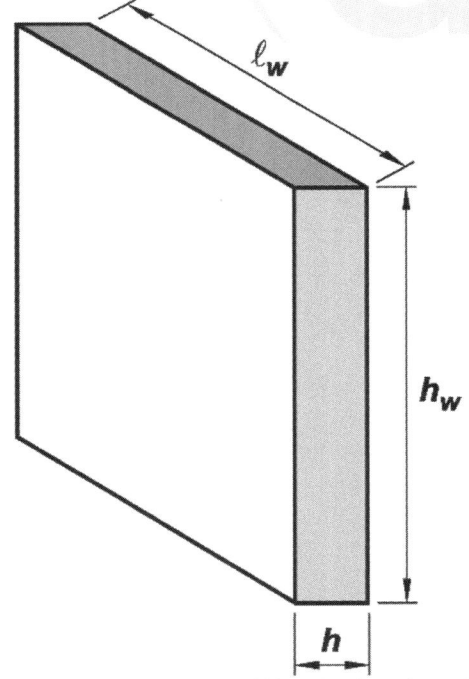

*Fig. 10.2.3b—Defining shear wall height, length, and depth.*

or without enlarged ends (boundary elements). Flanged shear walls are often T- or L-shaped sections. Reinforced concrete walls around building elevator core shafts and stairwells are typically in a C- or U-shape (channel shear wall) (Fig. 10.2.3a).

Notation used to describe the wall dimensions are shown in Fig. 10.2.3b; where $h$ is the wall thickness, $h_w$ is the wall height, and $\ell_w$ in the wall length.

**10.2.4** *Wall type*—The selection of a shear wall type is based on several factors, including functionality, constructability, economy, and seismic performance (Moehle et al. 2011). For low-rise buildings (Fig. 10.2.4(a)), squat, solid walls are predominantly used ($h_w/\ell_w < 2$). As the building increases in height, the wall height to length increases ($h_w/\ell_w \geq 2$), making walls become more slender (Fig. 10.2.4(c)). Perforated walls (Fig. 10.2.4(b)) are acceptable, but depending on a wall's opening percentage, the wall strength could be reduced. A row of vertically aligned openings in a slender wall results in dividing the wall in two sections, termed "coupled walls" because they behave as two individual continuous wall sections connected by coupling beams (Fig. 10.2.4(d)).

**10.2.5** *Design limits*—Minimum wall thicknesses are as shown in Table 10.2.5.

For walls designed by the alternative method for out-of-plane loads in Section 11.8 of ACI 318-14. ACI 551.2R provides the following slenderness limits:

(a) One layer of reinforcement at the center of the wall, $\ell_c/h \leq 50$

(b) Two layers of reinforcement, one at each face of the wall, $\ell_c/h \leq 65$

ACI 318 does not provide separate thickness limits for structural walls resisting in-plane lateral forces. The NEHRP Technical Brief No. 6 (Moehle et al. 2011) suggests the following minimum wall thickness:

(a) Special structural walls: 8 in.
(b) Special structural walls with boundary elements—
  i. Boundary element: 12 in.
  ii. Wall: 10 in.
(c) Coupled shear walls—

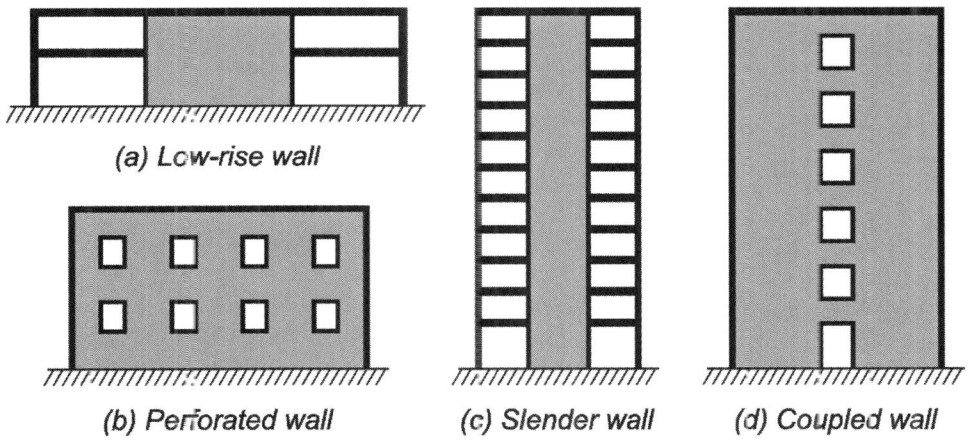

Fig. 10.2.4—Shear wall types.

**Table 10.2.5—Minimum wall thickness h (ACI 318-14, Table 11.3.1.1)**

| Wall type | | Minimum thickness h | |
|---|---|---|---|
| Bearing* | Greater of: | 4 in. | (a) |
| | | 1/25 the lesser of unsupported length and unsupported height | (b) |
| Nonbearing | Greater of: | 4 in. | (c) |
| | | 1/30 the lesser of unsupported length and unsupported height | (d) |
| Exterior basement and foundation* | | 7.5 in. | (e) |

*Only applies to walls designed in accordance with the simplified design method of 11.5.3.

i. Coupling beam designed as a special moment beam: 14 in.

ii. Coupling beam designed with diagonal reinforcement: 16 in.

There is no code limit on the overall building drift. Generally, the relative lateral deflection in any one story should not exceed 1/500 of the story height.

## 10.3—Required strength

Chapter 5 in ACI 318-14 provides the load combinations necessary to design a shear wall for moment, shear, and axial force. Section 12.4 in ASCE 7 has additional seismic load combinations to consider and load effects, such as the overstrength factor $\Omega_o$.

**10.3.1** *Methods of analysis*—ASCE 7 allows for three different types of analysis for determining the lateral seismic forces:

1. Equivalent Lateral Force Analysis (ELF): The equivalent lateral force analysis is the simplest method of analysis and is sufficient for the majority of structures built, using an approximated fundamental period $T_a$, which can be conservative. For long periods greater than 3.5 seconds, this method is not acceptable.

2. Modal Response Spectrum Analysis (MRS): The modal response spectrum analysis accounts for the elastic dynamic behavior of the structure and determines the building period. The calculated base shear can be less than the base shear calculated using the ELF method. The base shear, however, should be scaled to a minimum of 85 percent of the ELF base shear.

3. Seismic Response History Analysis (SRH): The seismic response history method of a three-dimensional model is used to analyze the building.

ASCE 7-10 Table 12.2-1 provides the required response modification coefficient $R$ and deflection amplification factors $C_d$ required in the analysis. The relevant coefficients are listed for shear walls in Tables 10.3.1a through 10.3.1c.

Sections 6.6, 6.7, and 6.8 in ACI 318-14 address first-order elastic analysis, second-order elastic analysis, and second-order inelastic analysis, respectively. Section 18.2.2.1 in ACI 318-14 requires that the interaction of all structural and nonstructural members that affect the linear and nonlinear response of the structure to earthquake motions be considered in the analysis.

For flanged walls, the effective flange width of a wall varies depending on the anticipated deformation level and whether it is in tension or compression. Tests (Wallace 1996) have shown that the effective flange width increases with increasing drift level, and the effectiveness of a flange in compression differs from that for a flange in tension. The value used for the effective compression flange width has little effect on the strength and deformation capacity of the wall; therefore, to simplify design, a single value of effective flange width based on an estimate of the effective tension flange width is used in both tension and compression.

Section 18.10.5.2 in ACI 318-14 defines the effective flange width that extends from the face of the web of L-, T-, C-, or other flanged sections as the lesser of one-half the distance to an adjacent wall web and 25 percent of the total wall height. The full flange width, and not the effec-

### Table 10.3.1a—Design coefficients for shear walls in bearing wall systems

| Seismic-force-resisting system | Response modification coefficient $R$ | Deflection amplification factor $C_d$ |
|---|---|---|
| Special reinforced concrete shear walls | 5 | 5 |
| Ordinary reinforced concrete shear walls | 4 | 4 |
| Intermediate precast shear walls | 4 | 4 |
| Ordinary precast walls | 3 | 3 |

### Table 10.3.1b—Design coefficients for shear walls in building frame systems

| Seismic-force-resisting system | Response modification coefficient $R$ | Deflection amplification factor $C_d$ |
|---|---|---|
| Special reinforced concrete shear walls | 6 | 5 |
| Ordinary reinforced concrete shear walls | 5 | 4.5 |
| Intermediate precast shear walls | 5 | 4.5 |
| Ordinary precast walls | 4 | 4 |

### Table 10.3.1c—Design coefficients for shear walls in dual systems with special moment frames capable of resisting at least 25 percent of prescribed seismic forces

| Seismic-force-resisting system | Response modification coefficient $R$ | Deflection amplification factor $C_d$ |
|---|---|---|
| Special reinforced concrete shear walls | 7 | 5.5 |
| Ordinary reinforced concrete shear walls | 6 | 5 |

tive flange width, may be used in determining the tributary gravity loads that resist uplift (ASCE 7).

## 10.4—Design strength

Walls are a versatile building element used in a variety of ways that determine how the wall is approached for design. A wall is typically very long in one plan dimension, compared to the orthogonal dimension making the wall slender. This slenderness can control the design if there are large loads applied laterally along the smaller wall dimension $h$. Loads applied in this direction are commonly called out-of-plane loads. Loads applied laterally along the larger wall dimension $\ell_w$ are commonly called in-plane loads. Rarely do out-of-plane and in-plane loading have to be considered at the same time, though axial loads are always present. The designer typically designs a wall for the two conditions discussed for axial load with out-of-plane and those with in-plane loads (10.2.1).

**10.4.1** *Design for axial load*—Wall design for axial load is similar to column design. The wall slenderness is considered by using the moment magnification method in Section 6.6.4 of ACI 318-14 for a first-order analysis, or by using a computer program that accounts for $P$-$\Delta$ effects using a second-order analysis. The design is completed according to Section 22.4 of ACI 318-14, which can be quickly evaluated using an interaction diagram generated by software or an electronic spreadsheet. If the resultant of all axial loads is located in the middle third of the wall thickness $h$, then a simplified equation is permitted in 11.5.3 of ACI 318-14, where moment can be ignored and $P_n$ is directly calculated.

The wall section considered effective to resist a concentrated gravity load is the width of the bearing plus four times the wall thickness. The effective width cannot cross a vertical wall joint unless the joint is designed to transfer the load (ACI 318-14, Section 11.2.3.1).

**10.4.2** *Axial and out-of-plane loads*—Walls should be analyzed for combined axial and out-of-plane loads. For walls that are part of a multistory building lateral load system, combined axial and out-of-plane loads rarely control the design of the wall. The design of these walls can be completed according to Section 22.4 of ACI 318-14 and slenderness checked using Section 6.6.4 of ACI 318-14 for a first-order analysis, or by using a computer program that accounts for $P$-$\Delta$ effects using a second-order analysis.

For a tall one-story building with long shear walls, such as a warehouse, combined axial and out-of-plane loads typically control the wall design. There are many one-story commercial buildings that use exterior concrete walls to support roof loads from the adjacent interior bay and resist lateral out-of-plane and in-plane loads. These buildings are typically 40 to 60 ft in height to allow for rack storage or second-story mezzanines. The wall thickness is usually made as slender as possible for economy. The design of these walls is typically completed in accordance with Section 11.8 of ACI 318-14. Alternative methods for out-of-plane slender wall analysis include ACI 318-14, or by using a Finite Element Analysis (FEA) that accounts for $P$-$\Delta$ effects.

The alternative method has several limitations as stated in Section 11.8.1.1 of ACI 318-14. Limits that generally control this method use are:

(a) $P_u$ at midheight cannot exceed $0.06f_c'A_g$

(b) Out-of-plane deflection at service loads cannot exceed $\ell_c/150$

A comprehensive discussion of this method, including its derivation, limitations, use, and worked examples, is given in ACI 551.2R.

**10.4.3** *Axial and in-plane loads, squat walls*—Walls are typically part of the LFRS due to their large in-plane stiffness. In squat walls ($h_w/\ell_w < 2$), the predominant wall failure mode is diagonal shear. Shear applied to the top of the wall is delivered to the base through compressive struts. Diagonal cracks form along the struts at inclined angles of approximately 38 degrees (Barda et al. 1977), as shown in Fig.

10.4.3. The vertical (longitudinal) reinforcement is mostly effective in resisting this type of shear through shear friction. Separation at the crack engages the vertical reinforcement in tension, creating a clamping force and increased resistance to shear. The vertical reinforcement is fully effective for wall height-to-length ratios ($h_w/\ell_w$) of 0.5 or less. As this ratio increases above 0.5, the horizontal (transverse) reinforcement begins to provide a portion of the resistance. At a height-to-length ratios exceeds 2.5, the horizontal reinforcement provides most of the shear strength and the shear behavior is more like a beam.

This change in shear behavior is accounted for in the minimum reinforcement requirement according to Eq. (11.6.2) of ACI 318-14.

$$\rho_\ell \geq 0.0025 + 0.5\left(2.5 - \frac{h_w}{\ell_w}\right)(\rho_t - 0.0025) \quad (11.6.2)$$

where $\rho_t$ shall be at least 0.0025. $\rho_t$ is calculated according to Section 11.5.4.8 of ACI 318-14. If this value is less than a reinforcement ratio of 0.0025, then the minimum reinforcement ratio for both horizontal and vertical reinforcement is 0.0025. If the required shear reinforcement exceeds 0.0025, Eq. (11.6.2) of ACI 318-14 assures that enough longitudinal (vertical) reinforcement is provided for shear-friction resistance in squatter walls. At $h_w/\ell_w \leq 0.5$, $\rho_\ell$ computed from Eq. (11.6.2) could exceed $\rho_t$ required by 11.5.4.8. However, Eq. (11.6.2) does not require that $\rho_\ell$ exceed $\rho_t$ calculated by 11.5.4.8. At $h_w/\ell_w > 0.5$ and $\leq 2.5$, Eq. (11.6.2) provides a minimum amount of longitudinal reinforcement that changes linearly from $\rho_t$ required at an $h_w/\ell_w$ of 0.5 to $h_w/\ell_w$ of 2.5. At $h_w/\ell_w > 2.5$, the transverse (horizontal) reinforcement is fully engaged and a minimum a $\rho_\ell$ of 0.0025 is provided.

For ordinary structural walls, the axial and flexural strength is calculated according to Section 22.4 of ACI 318-14, as discussed in 10.4.2 of this Handbook. Concrete shear strength $V_c$ is calculated according to Table 11.5.4.6 (ACI 318-14), and shear reinforcement strength $V_s$ is calculated according to Section 11.5.4.8 (ACI 318-14). The maximum nominal shear strength $V_n$ is, according to Section 11.5.4.3

$$V_n \leq 10\sqrt{f_c'}hd$$

For special structural walls, the axial and flexural strength may also be calculated according to Section 22.4 of ACI 318-14, which is discussed in 10.4.2 of this Handbook. Section 18.10 in ACI 318-14 implies that a wall is squat when the height-to-length ratio is less than 2.0. For walls with $h_w/\ell_w < 2.0$, the displacement method of Section 18.10.6.2 (ACI 318-14) does not apply, leaving only the stress-based method of Section 18.10.6.3 (ACI 318-14) to determine if boundary elements are necessary. If boundary elements are not necessary, the design of a special structural wall is similar to an ordinary structural wall. A more detailed discussion about boundary elements is given in 10.4.4 of this Handbook. Key differences in squat wall shear design for special structural walls versus ordinary structural walls are

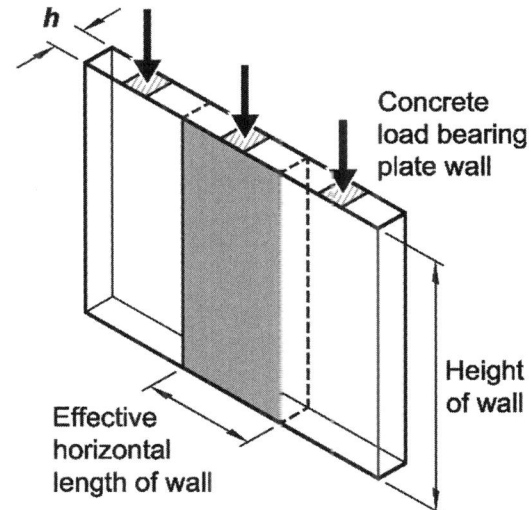

Fig. 10.4.1—Effective wall horizontal length.

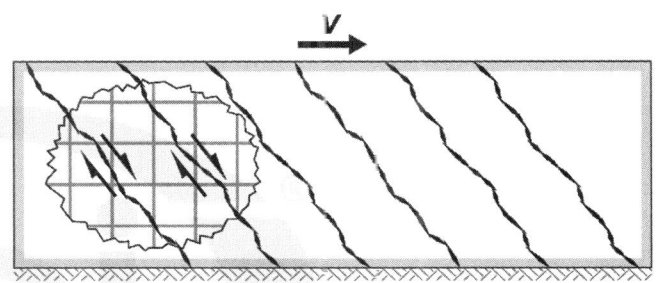

Fig. 10.4.3—Shear in squat walls.

(a) Shear stress is calculated over $A_{cv}$ instead of $hd$. $A_{cv}$ is typically the length of the wall, $\ell_w$, multiplied by the width of the wall web, $h$ (ACI 318-14, Section 18.10.4.1).

(b) Minimum reinforcement is approximately the same:
  i. For ordinary structural walls, the minimum reinforcement is according to Table 11.6.1 (ACI 318-14) for $V_u$ less than $0.5\phi V_c$ or $\phi hd\lambda\sqrt{f_c'}$ (ACI 318-14, Section 11.6.1).
  ii. For special structural walls, the minimum reinforcement is according to Table 11.6.1 for $V_u$ less than $A_{cv}\lambda\sqrt{f_c'}$ (ACI 318-14, Section 18.10.2.1).

Otherwise, the minimum reinforcement ratio for $\rho_t$ and $\rho_\ell$ is 0.0025 (ACI 318-14, Section 18.10.2.1).

(c) For $h_w/\ell_w < 2.0$, $\rho_\ell$ shall be at least $\rho_t$ (ACI 318-14, Section 18.10.4.3).

(d) For $V_u$ greater than $2 A_{cv}\lambda\sqrt{f_c'}$ or $h_w/\ell_w \geq 2.0$, two curtains of reinforcement are required (ACI 318-14, Section 18.10.2.2).

(e) Longitudinal reinforcement must extend $0.8\ell_w$ beyond the point where it is no longer required (ACI 318-14, Section 18.10.2.3(a)).

(f) At locations where yielding of longitudinal reinforcement is expected, development lengths shall be $1.25f_y$ in tension (ACI 318-14, Section 18.10.2.3(b)).

**10.4.4** *Axial and in-plane loads, slender walls*—The term "slender walls" is used if the predominant failure mode is flexure. Boundary elements are often required, which offer increased flexural strength, enhanced curvature capacity,

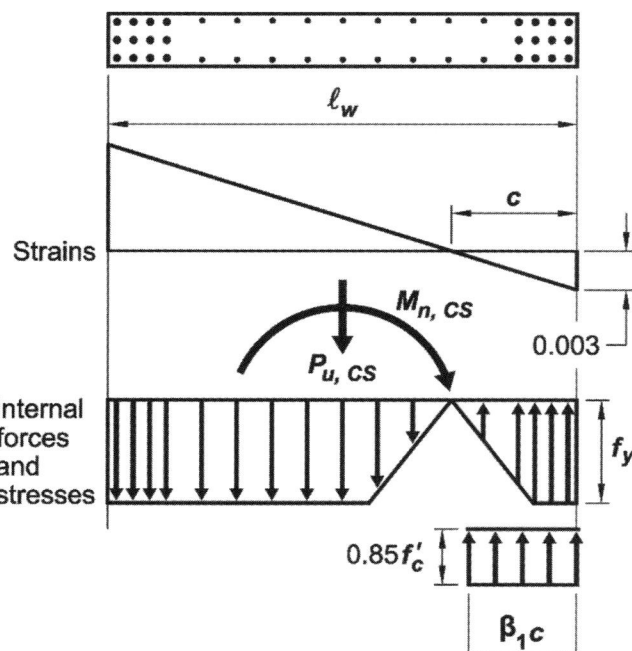

*Fig. 10.4.4—Calculation of neutral axis depth c (Moehle et al. 2011).*

and better distribution of flexural cracks that promote increased displacement capacity (Moehle 2015). For walls designed only by Chapter 11 (ACI 318-14), the design for axial and flexural strength is according to Section 22.4 of ACI 318-14, as discussed in 10.4.2 of this Handbook. For special structural walls, Chapter 18 of ACI 318-14 requires special boundary elements at prescribed limits. There are two methods for determining the need for special boundary elements: the displacement method of Section 18.10.6.2 and the stress-based method of Section 18.10.6.3 (ACI 318-14).

The displacement method is the preferred method of design according to the NEHRP Technical Brief No. 6 (Moehle et al. 2011). This method assumes that:
(a) The longitudinal bars at ends of the wall will yield at a single critical section
(b) The wall has a constant cross section throughout its height
(c) $h_w/\ell_w \geq 2.0$
(d) $\delta_u/h_w \geq 0.005$

A special boundary element is required if the neutral axis depth $c$ calculated exceeds Eq. (18.10.6.2) in ACI 318-14.

$$c \geq \frac{\ell_w}{600(1.5\delta_u/h_w)} \quad (18.10.6.2)$$

This equation was derived from the relationship shown in Fig. 10.4.4 of this Handbook (Moehle et al. 2011). The 1.5 multiplier to $\delta_u$ was added in ACI 318-14 to more accurately evaluate the deflection of the wall at the maximum considered earthquake. The notation $\delta_u$ is the design displacement found through a linear elastic analysis and the section should include the effect of the corresponding axial load for the given lateral seismic force. The limit of $\delta_u/h_w \geq 0.005$ was modified in ACI 318-14. This lower limit provides increased deformation capacity for a range of stiffer walls that did not previously require boundary elements. This method also has an additional requirement for the termination of the transverse reinforcement at special boundary elements, if required. The transverse reinforcement must extend vertically above and below the critical section at least the greater of $\ell_w$ and $M_u/4V_u$, except at the wall base as noted in 18.10.6.4(g) in ACI 318-14. At the foundation, the boundary element ties or hoops must extend into the foundation 12 in., or if the edge of the boundary element is within one-half the foundation depth from an edge of the footing, ties or hoops must extend into the foundation or support a distance equal to the development length of the largest vertical bar in the boundary Section 18.13.2.3 (ACI 318-14). The longitudinal reinforcement of the boundary element must be adequately developed in the foundation.

The stress-based method is used for irregular walls or walls with disturbed regions, for example, around openings, according the NEHRP Technical Brief No. 6 (Moehle et al. 2011). The method requires special boundary elements if the effective compressive stress at the wall ends or around openings exceeds $0.2f_c'$. The boundary elements are discontinued if the effective compressive stress is less than $0.15 f_c'$. The model is based on a linear elastic analysis.

**10.4.5** *Boundary elements*—Special boundary elements are required if the limits in Sections 18.10.6.2 or in 18.10.6.3 of ACI 318-14 are not met. Special boundary elements are defined in Section 18.10.6.4 (ACI 318-14). When the limits in 18.10.6.2 or 18.10.6.3 are met, boundary elements are still required and are defined according to Section 18.10.6.5 (ACI 318-14). If a special boundary element is not required, boundary reinforcement is required if the longitudinal reinforcement ratio at the wall boundary, $\rho$, exceeds $400/f_y$. Boundary reinforcement is also required for the stress-based method where the effective compressive stress is between $0.15f_c'$ and $0.2f_c'$ according to 18.10.6.3. The requirements for size and detailing of these requirements are described in Fig. 10.4.5 of this Handbook. Horizontal reinforcement in structural walls with boundary elements must be anchored into the core of the boundary element with hooks, headed bars, or straight embedment and also has to extend to within 6 in. of the end of wall Section 18.10.6.4.

**10.4.6** *Vertical wall segments and wall piers*—A vertical wall segment is any portion of a wall that is bound by the outer edge of a wall and an edge of an opening, or the portion of a wall bound by the vertical edges of two openings. According to Chapter 11 of ACI 318-14, the design of nonseismic vertical wall segments is the same as that of walls. For special structural walls designed according to Chapter 18 of ACI 318-14, there are additional requirements. The nominal shear strength is reduced for the total cross section of a wall at the vertical wall segments. The calculated $V_n$ may not exceed $8A_{cv}\sqrt{f_c'}$ for the total $A_{cv}$, as shown in Fig. 10.4.6 of this Handbook. For an individual vertical wall segment, $V_n$ may not exceed $10A_{cv}\sqrt{f_c'}$.

Vertical wall segments are designed as walls, columns, or wall piers according to the segment geometry as summarized

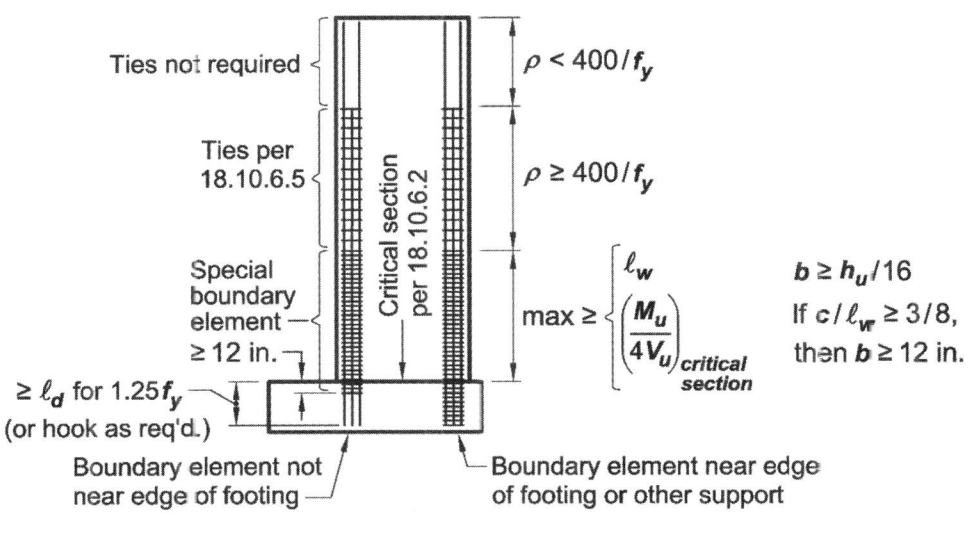

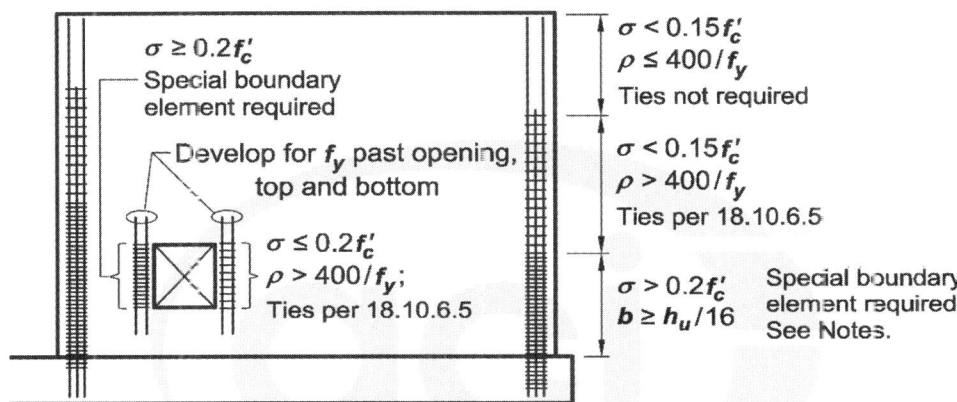

Fig. 10.4.5—Summary of boundary element requirements for special walls (Fig. R18.10.6.4.2 of ACI 318-14).

in Table 10.4.6 of this Handbook. In many cases, the special structural wall requirements apply. If the wall segment is designed as a column, Section 18.10.8.1 requires that the special detailing of 18.7.4, 18.7.5, and 18.7.6 for columns be applied (ACI 318-14). Wall piers are subset of vertical wall segments as defined in Table 10.4.6. They may be designed as special columns or by alternative requirements given in Section 18.10.8 in ACI 318-14. Wall piers designed to these alternative requirements require:

(a) $V_e$ that can develop $M_{pr}$ at the ends of the column or $\Omega_o$ times the factored shear determined by analysis (18.7.6.1 and 18.10.8.1)

(b) Hoops at a spacing not greater than 6 in.

(c) Checking to see if the pier should include a special boundary element

(d) Horizontal reinforcement above and below the wall pier to transfer the design shear into the adjacent wall segments

**10.4.7** *Horizontal wall segments and coupling beams*—A horizontal wall segment is any portion of a wall that is bound by the outer edge of a wall and an edge of an opening, or the portion of a wall bound by the horizontal edges of two openings. According to Chapter 11 of ACI 318-14, the design of nonseismic horizontal wall segments is the same as that of walls. Horizontal wall segments in special structural walls are designed as special structural walls according to Chapter 18 of ACI 318-14. If horizontal wall segments are part of a coupled special structural wall system, the segment is called a coupling beam. The coupling beam is separated into three categories in ACI 318-14.

(a) If $\ell_n/h \geq 4$, the coupling beam is designed as a beam in a special moment frame

(b) If $\ell_n/h < 2$ and $V_u \geq 4\lambda A_{cv}\sqrt{f_c'}$, design the beam with diagonally placed bars for a more effective transfer of shear through the member

(c) For other cases, the beam may be designed either as a special beam or with diagonal placed bars

The design of a coupling beam is beyond the scope of this Handbook. For more information, reference Moehle et al. (2011) and Moehle (2015).

## 10.5—Detailing

Structural walls are, in general, thin, tall, wide members with reinforcement in both the horizontal (transverse) and vertical (longitudinal) directions. Properly designed and detailed shear walls in buildings have resisted seismic forces and sideways effectively in past earthquakes.

If shear walls are the only members in the LFRS, they usually behave as cantilever beams, fixed at the base. They transfer moments, shear, and axial forces to the foundation. If the LFRS includes a stiff frame, the wall could behave more like a column, depending on relative stiffnesses and shear wall locations. In these cases, the shear wall will usually collect the large majority of shear, but their moments are much less due to frame action.

Reinforcement placed in the horizontal and vertical directions resists in-plane shear forces and limits cracking. For taller walls, the vertical reinforcement serves as flexural reinforcement. If significant moment strength is required, additional reinforcement is placed at the ends of a wall or within boundary elements (Fig. 10.5a and 10.5b of this Handbook).

Although one curtain of reinforcement is permitted for ordinary shear walls, two curtains of reinforcement are recommended. It is advantageous to place the transverse reinforcement as the exterior layer to prevent longitudinal reinforcement from buckling and provide better confinement to the concrete. The casting position of transverse reinforcement assumes more than 12 in. of concrete below the bar for the calculation of development and splice lengths ($\psi_t = 1.3$).

The first splice of vertical reinforcement typically occurs immediately above the foundation, where wall longitudinal reinforcement laps with foundation dowel bars. These dowels, lapped with the wall bars, provide the critical mechanism of transferring tension and shear forces from the wall to the foundation. The dowels should extend into the foundation with a depth sufficient enough to be fully developed for tension. For constructability purposes, dowels with 90-degree hooks should extend to the bottom of the foundation where they can be tied firmly to the foundation bottom reinforcement.

Structural walls in SDCs D, E, and F have an additional requirement. Section 18.10.2.3 of ACI 318-14 requires that longitudinal reinforcement be developed or spliced for $1.25f_y$ if the foundation connection is where yielding of wall reinforcement is likely to occur as a result of lateral displacements.

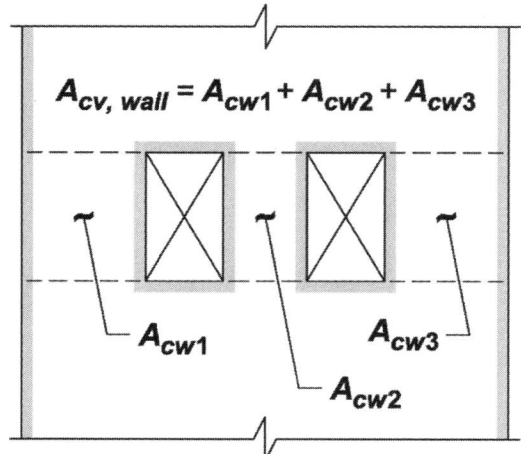

*Fig. 10.4.6—Shear strength for vertical wall segments.*

**Table 10.4.6—Governing design provisions for vertical wall segments* (Table R18.10.1 in ACI 318-14)**

| Clear height of vertical wall segment/length of vertical wall segment ($h_w/\ell_w$) | Length of vertical wall segment/wall thickness ($\ell_w/b_w$) | | |
|---|---|---|---|
| | ($\ell_w/b_w$) ≤ 2.5 | 2.5 < ($\ell_w/b_w$) ≤ 6.0 | ($\ell_w/b_w$) > 6.0 |
| $h_w/\ell_w < 2.0$ | Wall | Wall | Wall |
| $h_w/\ell_w \geq 2.0$ | Wall pier required to satisfy specified column design requirements; refer to Section 18.10.8.1 (ACI 318-14) | Wall pier required to satisfy specified column design requirements or alternative requirements; refer to Section 18.10.8.1 (ACI 318-14) | Wall |

*$h_w$ is the clear height, $\ell_w$ is the horizontal length, and $b_w$ is the width of the web of the wall segment.

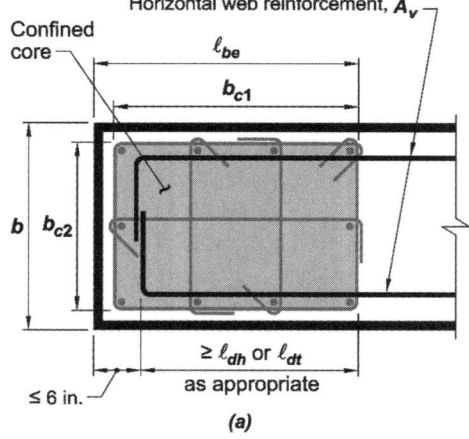

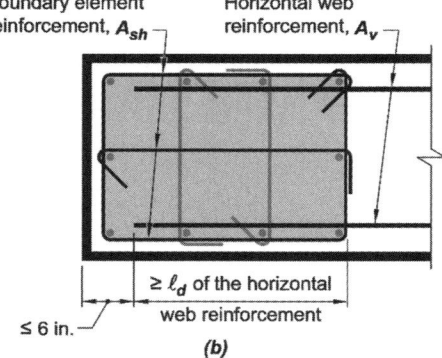

*Fig. 10.5a—Development of wall horizontal reinforcement in confined boundary element.*

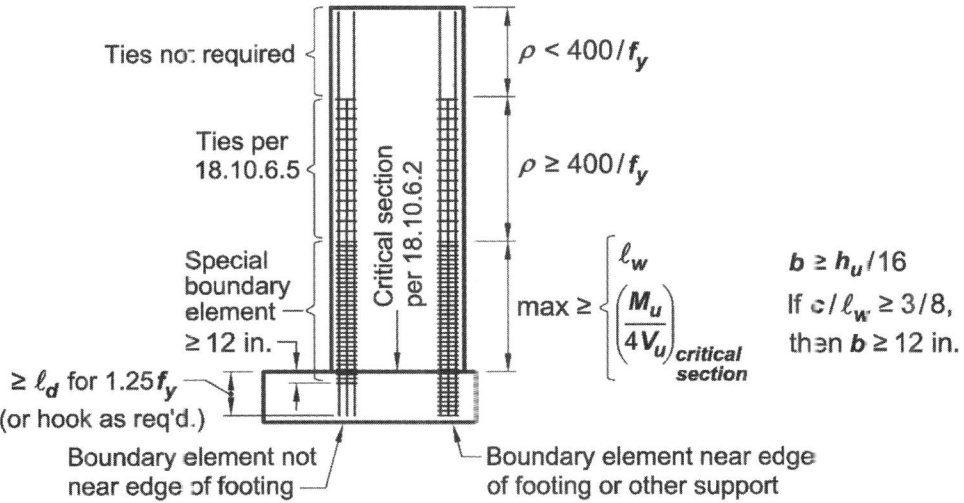

*Fig. 10.5b—Boundary element requirements for special walls.*

The amount of required transverse reinforcement is found in Table 18.10.6.4(f) (ACI 318-14). The spacing of the transverse reinforcement is tighter at the base of the wall for a distance $\ell_w$ or $M_u/4V_u$ at the lesser of 6 in. or $6d_b$. Above this region, the spacing widens to 8 in. or $8d_b$ until the reinforcement ratio drops below $400/f_y$, where transverse reinforcement is not required (Fig. 10.5b).

## 10.6—Summary

Structural walls have two main advantages:

1. They are relatively easy to construct because reinforcement detailing of walls is straightforward.

2. Because of their inherent stiffness, they usually minimize sway and damage in structural and nonstructural elements, such as glass windows and building contents, for buildings exposed to high lateral loads.

Structural walls have disadvantages, including these two:

1. Shear walls can create interior barriers that interfere with architectural and mechanical requirements.

2. Shear walls carry large lateral forces resulting in the possibility of large overturning moments. Attention is required at the wall-foundation interface and foundation design.

## REFERENCES

*American Concrete Institute (ACI)*
ACI 318-14—Building Code Requirements for Structural Concrete and Commentary

ACI 551.2R-10—Design Guide for Tilt-Up Concrete Panels

*American Society of Civil Engineers (ASCE)*
ASCE 7-10—Minimum Design Loads for Buildings and Other Structures

### Authored documents

Barda, F.; Hanson, J. M.; and Corley, W. G., 1977, "Shear Strength of Low-Rise Walls with Boundary Elements," *Reinforced Concrete Structures in Seismic Zones*, SP-53, American Concrete Institute, Farmington Hills, MI, pp. 149-202.

Garcia, L. E., 2003, "Concrete Q&A: 318-02 Questions," *Concrete International*, V. 25, No. 10, Oct., p. 120.

Moehle, J. P., 2015, *Seismic Design of Reinforced Concrete Buildings*, McGraw-Hill Education, New York, 760 pp.

Moehle, J. P.; Ghodsi, T.; Hooper, J. D.; Fields, D. C.; and Gedhada, R., 2011, "Seismic Design of Cast-in-Place Concrete Special Structural Walls and Coupling Beams: A Guide for Practicing Engineers," *NEHRP Seismic Design Technical Brief No.6*, National Institute of Standards and Technology, Gaithersburg, MD, 37 pp.

Wallace, J. W., 1996, "Evaluation of UBC-94 Provisions for Seismic Design of RC Structural Walls," *Earthquake Spectra*, V. 12, No. 2, May, pp. 327-348. doi: 10.1193/1.1585883

## 10.7—Examples

**Shear Wall Example 1:** *Seismic Design Category B/wind*—The reinforced concrete shear wall in this example is nonprestressed. This shear wall is part of the lateral force-resisting-system (a shear wall is at each end of the structure) in the North-South (N-S) direction of the hotel (Fig. E1.1). Material properties are selected based on the code limits and requirements of Chapters 19 and 20 (ACI 318-14), engineering judgment, and locally available materials. The structure is analyzed for all required load combinations by an elastic 3D finite element analysis software model that includes shear wall - frame interaction. The resultant maximum factored moments and shears over the height of the wall are given for the load combination selected. This example provides the shear wall design only at the base.

Given:
$P_u$ = 1015 kip

*In-plane—*
$V_u$ = 235 kip
$M_u$ = 18,600 ft-kip

*Out-of-plane—*
$V_u$ = 16 kip
$M_u$ = 60 ft-kip
*Material properties—*
$f_c'$ = 5000 psi
$f_y$ = 60,000 psi

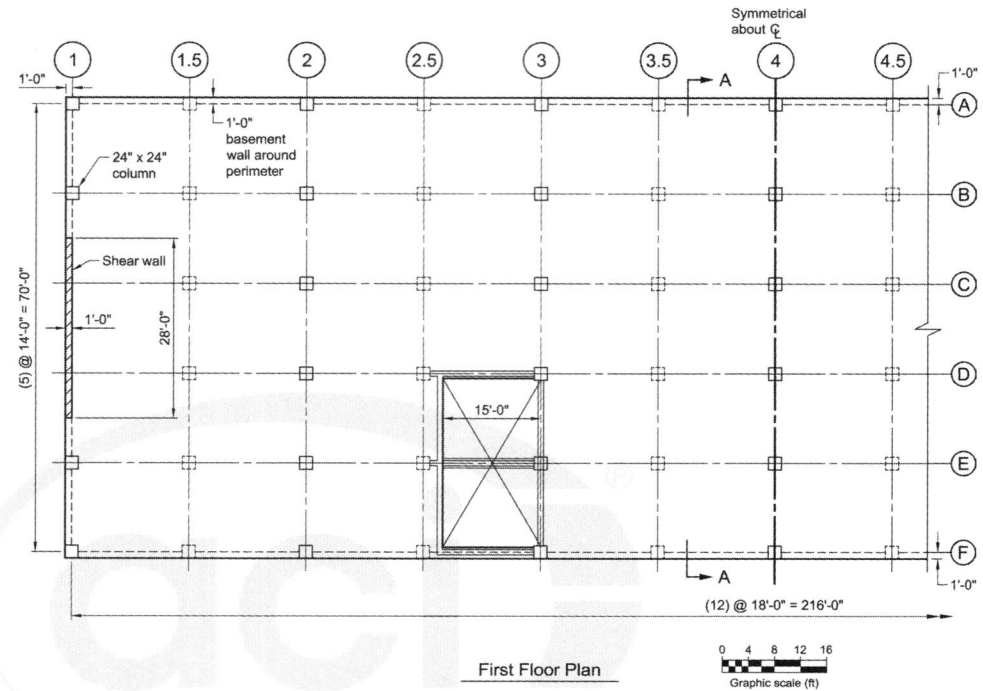

Fig. E 1.1—*Building floor plan, first floor.*

This example uses the Interaction Diagram spreadsheet aid found at https://www.concrete.org/store/productdetail.aspx?ItemID=SP1714DAE.

| ACI 318-14 | Discussion | Calculation |
|---|---|---|
| Step 1: Geometry | | |
| 11.3.1 | This wall design example follows the requirements of Section 11.5.2, and, therefore, the wall thickness does not need to meet the requirements of Table 11.3.1.1 (ACI 318-14). The thickness equations (a) and (b) of Table 11.3.1.1 can, however, provide an indication that the thickness chosen is appropriate. | |
| | From Table 11.3.1.1, the wall thickness must be at least the greater of 4 in. and the lesser of 1/25 the lesser of the unsupported height of the wall (18 ft for the first elevated floor) and the unsupported length of the wall (28 ft from end to end of the wall). | The unsupported height controls; 18 ft < 28 ft $h_{req'd}$ = (18 ft)(12 in./ft)/25 = 8.64 in. Example shear wall $h$ = 12 in. > $h_{req'd}$ = 8.64 in.   **OK** |
| 20.6.1.3.1 | A 12 in. wall is used in this design and the wall is assumed to be exposed to weather on the exterior of the structure. Concrete cover is 1-1/2 in., which is in accordance with Table 20.6.1.3.1 (ACI 318-14). | |

| | Step 2: Loads, load patterns, and analysis of the wall | |
|---|---|---|
| 11.4<br>6.9 | The structure is analyzed using the assumptions and requirements of 11.4.<br><br>The structure was analyzed using 3D elastic Finite Element Analysis (FEA) software that follows the analysis requirements of Section 11.4 and Chapters 5 and 6 of ACI 318-14 for loading and analysis, respectively refer to Fig. E1.2 and E1.3 for in-plane flexure and in-plane shear along the height of the wall, respectively. | The maximum factored axial force, flexural moment, and shear force at the base of the wall are listed in the given section at the start of this example. |

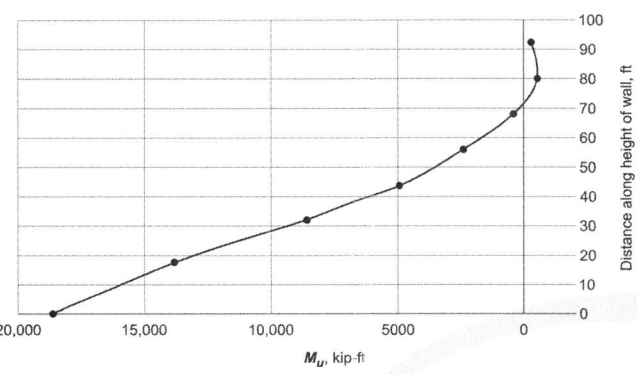
Fig. E1.2—In-plane flexure along the height of the wall.

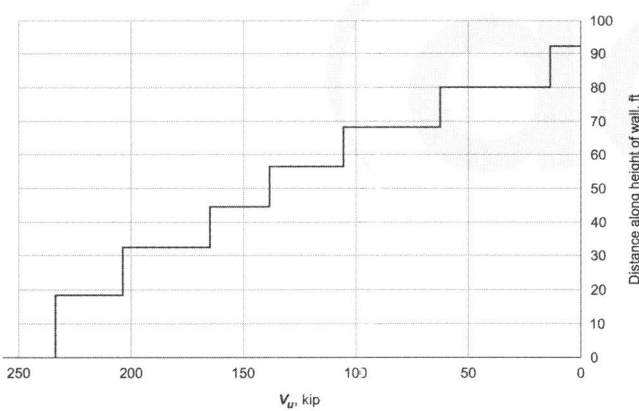
Fig. E1.3—In-plane shear along the height of the wall.

| | Step 3: Concrete and steel material requirements | |
|---|---|---|
| 11.2.1.1 | The mixture proportion must satisfy the durability requirements of Chapter 19 and structural strength requirements (ACI 318-14).<br><br>The designer determines the durability classes. Please refer to Chapter 4 of this Handbook for an in-depth discussion of the categories and classes. ACI 301 is a reference specification that is coordinated with ACI 318-14. ACI encourages referencing ACI 301 into job specifications.<br>There are several mixture options within ACI 301, such as admixtures and pozzolans, which the designer can require, permit, or review if suggested by the contractor. | By specifying that the concrete mixture shall be in accordance with ACI 301 and providing the exposure classes, Chapter 19 requirements are satisfied.<br><br>Based on durability and strength requirements, and experience with local mixtures, the compressive strength of concrete is specified at 28 days to be at least 5000 psi. |

| | | |
|---|---|---|
| 11.2.1.2 | The reinforcement must satisfy Chapter 20 of ACI 318-14.<br>The designer determines the grade of bar and if the bar should be coated by epoxy, galvanized, or both. | By specifying the rebar grade and any coatings, and that the rebar shall be in accordance with ACI 301, Chapter 20 of ACI 318-14 requirements are satisfied. In this case, assume grade 60 bar and no coatings. |
| **Step 4: Axial and flexural design strength** | | |
| 11.5, 11.5.2 | The combined axial and flexural design strength of a shear wall can be determined using an interaction diagram similar to a column interaction diagram.<br><br>The wall interaction diagram is generated using the Interaction Diagram spreadsheet (link in the given section of this example).<br><br>Refer to Column Example 9.2 in this Handbook for additional information about the Interaction Diagram spreadsheet.<br><br>To estimate an initial reinforcement area, the wall is assumed as a cantilever and the amount of flexural reinforcement necessary to resist the moment is calculated. | An initial interaction diagram is made using No. 5 bars at 12 in. spacing throughout the wall (refer to *Note below). It is assumed that all of the longitudinal reinforcement is effective resisting in-plane flexure.<br><br>The first pair of No. 5 bars is assumed to be at 3 in. from the end of the wall, the second pair is placed at 12 in. from the end of the wall, and the remaining pairs at 12 in. spacing. The different spacing at the end of the wall is to allow for cover on the end pairs of bars in the wall and to force the reinforcement to be symmetrical. The reinforcement is symmetrical about the center of the wall and this bar layout is applied to both ends of the wall.<br><br>Fig. E1.4 shows the resulting design strength interaction diagram. The design strength interaction diagram includes the $\phi$-factor.<br><br>The Interaction Diagram spreadsheet contains a sheet named "Select Axial Load." When the user enters a $P_n$, the sheet calculates the associated maximum $M_n$ on the interaction diagram curve and plots a point on the interaction diagram to show that point.<br><br>This point is named the "Input Point" on the interaction diagram. The input point of $P_n$ corresponding to a $P_u$ of 1015 kips calculates a point on the design strength interaction diagram $M_u$ of 24,600 ft-kip. The input point is plotted as a solid triangle.<br><br>The open triangle indicates where the example load resultants are and shows that this iteration does satisfy required strength. Further iterations are unnecessary. |
| | *Note: For constructability, No. 5 bars are selected for the vertical reinforcement. Smaller bars will work for the strength of the wall, but are often too flexible to efficiently work with in a vertical position. | |

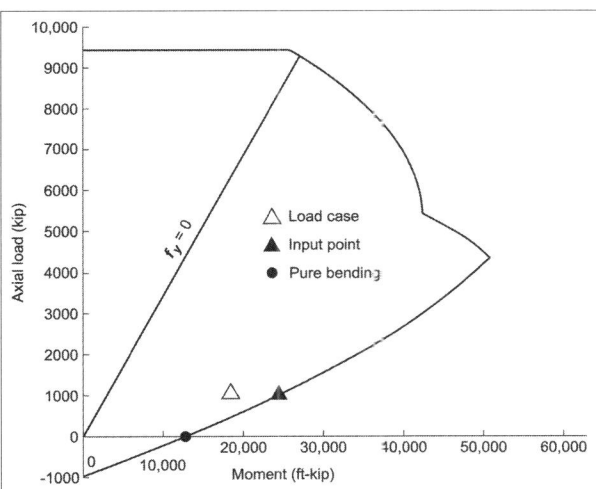

Fig. E1.4—Design strength interaction diagram.

| Step 5: Shear design strength (in-plane) | | |
|---|---|---|
| 11.5.4<br>11.5.4.4<br>11.5.4.5<br>21.2.1<br>11.5.1.1 | Shear walls resist both in-plane and out-of-plane shear. The out-of-plane shear is very small, therefore, by inspection it will not control the wall design.<br>In-plane shear, $V_u$ is given as 235 kips.<br>Equation gives:<br>$V_n = V_c + V_s$<br>$V_c = 2\sqrt{f'_c}(h)(d)$<br>where<br>$h$ = 12 in., wall thickness and $d = 0.8\ell_w = (0.8)(336$ in.$) = 268.8$ in. | $V_c = 2\left(\sqrt{5000 \text{ psi}}\right)(12 \text{ in.})(268.8 \text{ in.}) = 456 \text{ kip}$ |
| | From Table 21.2.1(b) use shear strength reduction factor $\phi = 0.75$ | $\phi V_c = (0.75)(456 \text{ kip}) = 342 \text{ kip}$ |
| | Check if $\phi V_c \geq V_u$ | $\phi V_c = 342 \text{ kip} > V_u = 235 \text{ kip}$ **OK** |
| Step 6: Flexure design strength (out-of-plane) | | |
| 11.5.1<br>11.5.3.1<br>21.2.2 | As shown in Step 4, the layers of No. 5 vertical wall reinforcement satisfies the interaction equation for in-plane bending.<br>The resultant of the out-of-plane moment, $M_u = 60$ ft-kip is within the middle third of the wall. This allows Section 11.5.3.1 to be used to check the out-of-plane strength of the wall. | Eccentricity of the resultant load:<br>$(1015 \text{ kip})(e) = 60$ ft-kip<br>$e = 0.7$ in.<br>$e < 2$ in. |
| | $P_n = 0.55(f'_c)(A_g)\left[1 - \left(\dfrac{k\ell_c}{32h}\right)^2\right]$ | $P_n = 0.55(5 \text{ ksi})(12 \text{ in.})(336 \text{ in.})\left[1 - \left(\dfrac{(0.8)(202 \text{ in.})}{32(12 \text{ in.})}\right)^2\right]$ |
| | From Table 21.2.2(b), use axial strength reduction factor $\phi = 0.65$ | $P_n = 9090$ kip<br>$\phi P_n = (0.65)(9090 \text{ kip}) = 5900$ kip<br>$5900 \text{ kip} \geq 1015 \text{ kip}$ **OK** |

| | | |
|---|---|---|
| **Step 7: Reinforcement limits** | | |
| 11.6 | Because the in-plane shear exceeds $0.5\phi V_c$, 11.6.2 applies. | |
| 11.6.2 | $V_u \geq 0.5\phi V_c$ <br> (a) and (b) must be satisfied: <br> (a) $\rho_\ell$ must be the greater of the three values below: <br> 0.0025 <br> $\rho_\ell \geq 0.0025 + 0.5(2.5 - h_w/\ell_w) \times (\rho_t - 0.0025)$ <br><br> $\rho_\ell$ from equation need not exceed $\rho_t$ in accordance with Eq. (11.5.4.8) <br><br> (b) $\rho = 0.0025$ | Assuming 12 in. for spacing in both directions, the minimum reinforcement results in: <br> $\rho_\ell = 0.0025$ <br><br> $\rho_\ell \geq 0.0025 + 0.5(2.5 - 92\text{ft}/28\text{ ft}) \times \left(\dfrac{0.62 \text{ in.}^2}{(12 \text{ in.})(12 \text{ in.})} - 0.0025\right)$ <br><br> $\rho_\ell \geq 0.0018$ <br> $\rho_t$ (Eq. (11.5.4.8)) = 0.0000* <br><br> $\rho_t$ (11.6.2(b)) = 0.0025 <br><br> Minimum reinforcement in both the longitudinal and horizontal directions is $0.0025sh$. <br> $A_{s,min} = (0.0025)(12 \text{ in.})(12 \text{ in.}) = 0.36 \text{ in.}^2$. <br> Using No. 5 bars in each face and direction at 12 in. gives 0.62 in.$^2$ in both directions. <br> $A_{s,prov.} = 0.62 \text{ in.}^2/\text{ft} > A_{s,req'd} = 0.36 \text{ in.}^2/\text{ft}$ **OK** |
| 11.5.4.8 | | *Equation (11.5.4.8) requires zero reinforcement for strength because the concrete strength is adequate in shear. |
| **Step 8: Reinforcement detailing** | | |
| 11.7 <br> 11.7.2.1 <br> 11.7.3.1 | Placing continuous No. 5 bars in each face and direction at 12 in. meets the detailing requirements of 11.7.2.1 and 11.7.3.1. No. 5 bars were selected in the horizontal direction for ease of construction. | Reinforcement is not required for shear strength; therefore, the maximum spacing for vertical bars cannot exceed $3h$ (36 in.) or 18 in.. The 12 in. spacing of vertical bars are less than these limits. <br> Similarly, the maximum spacing for horizontal bars cannot exceed $3h$ (36 in.) or 18 in. The 12 in. spacing of horizontal bars are less than these limits. |
| 11.7.2.3 | $h > 10$ in., therefore two layers are required. | Two layers are provided having equal reinforcement area. |
| 11.7.4.1 | If the area of vertical reinforcement exceeds $0.01A_g$, or if the reinforcement is needed to resist axial loads, ties are required to confine the vertical reinforcement. The reinforcement ratio for the flexural vertical reinforcement at the wall ends needs to be calculated to determine if ties are required. | The vertical flexural reinforcement used in the design strength interaction diagram is two No. 5 bars at 12 in. on center spacing. The $A_g$ within this length is 144 in.$^2$ <br> $A_{st}$ is 0.62 in.$^2$ <br> The ratio of $A_{st}$ to $A_g$ is 0.0043. This is less than the 0.01 and therefore ties are not required by 11.7.4.1. <br> The maximum factored axial load is 1015 kips or 1,015,000 lb and the maximum factored moment is 18,600 ft-kip or 223,200,000 in-lb. The factored axial stress on the concrete due to the combined loads is: <br> $\sigma = 1{,}015{,}000 \text{ lb}/\{(12 \text{ in.})(28 \text{ ft})(12 \text{ in./ft})\} +$ <br> $223{,}200{,}000 \text{ in.-lb} \times 336 \text{ in.}/37{,}933{,}056 \text{ in.}^4 = 2229$ psi. <br> This is below the design strength of concrete and thus steel is not needed to resist the axial load. Therefore, ties are not required by Section 11.7.4.1. <br> Refer to Fig. E1.5 and E1.6 for wall elevation and section cut at the ends of the wall. |

## Step 9: Detailing

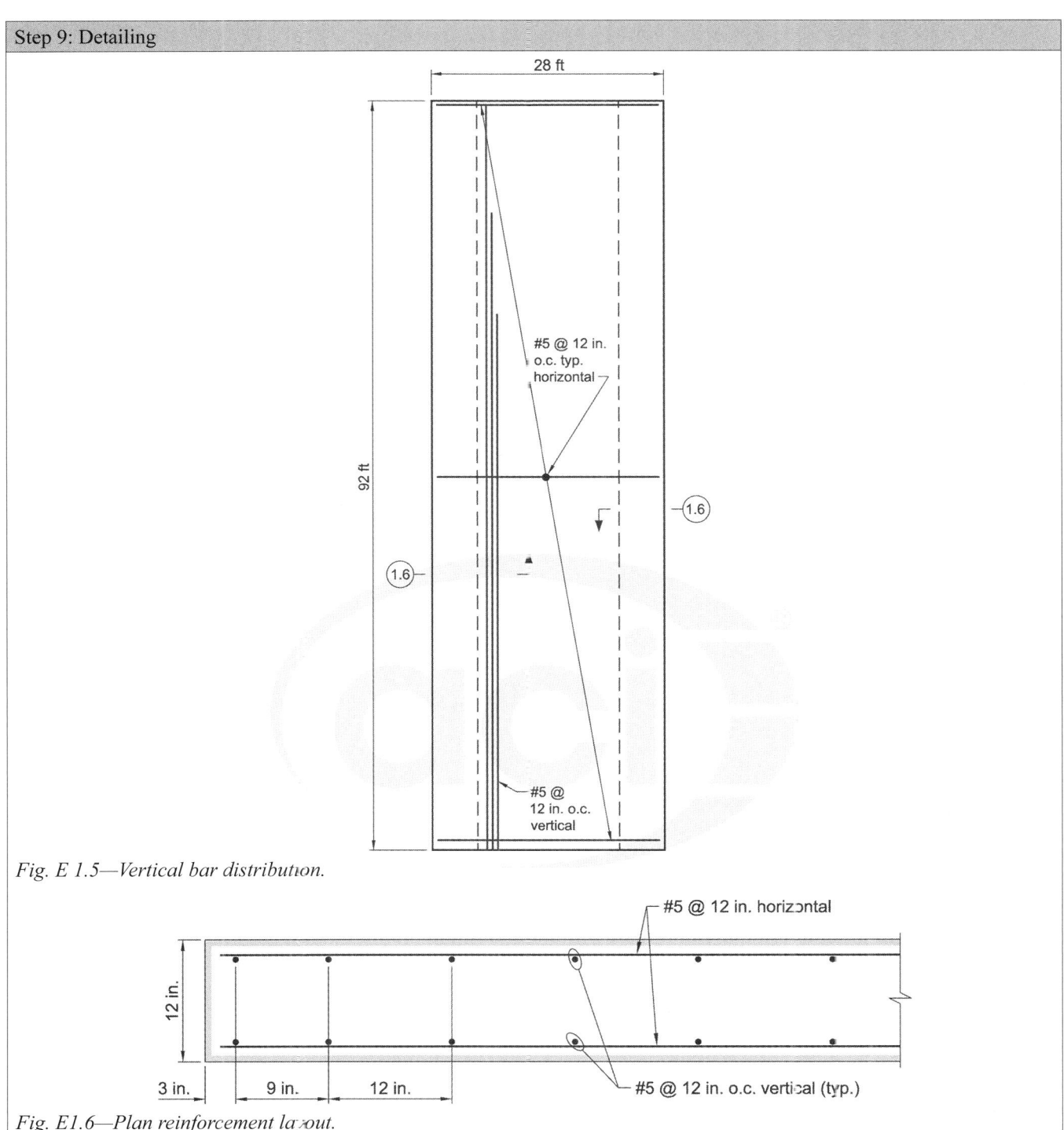

Fig. E1.5—Vertical bar distribution.

Fig. E1.6—Plan reinforcement layout.

**Shear Wall Example 2:** *Seismic Design Category D*

The reinforced concrete shearwall in this example is nonprestressed. This shearwall is part of the lateral force-resisting-system (a shearwall is at each end of the structure) in the North-South (N-S) direction of the hotel (Fig. E2.1). Material properties are selected based on the code limits and requirements of Chapters 19 and 20 (ACI 318-14), engineering judgment, and locally available materials. The structure is analyzed for all required load combinations by and elastic 3D finite element analysis software model that includes shearwall-frame interaction. The resultant maximum factored moments and shears over the height of the wall are given for the load combination selected. This example provides the shearwall design and detailing at the base of the wall.

Given:

*Forces and moments at the wall base—*
$P_u$ = 1015 kip

*In-plane—*
$V_u$ = 470 kip
$M_u$ = 37,200 ft-kip

*Material properties—*
$f_c'$ = 5000 psi
$f_y$ = 60,000 psi

*Out-of-plane—*
$V_u$ = 32 kip
$M_u$ = 120 ft-kip

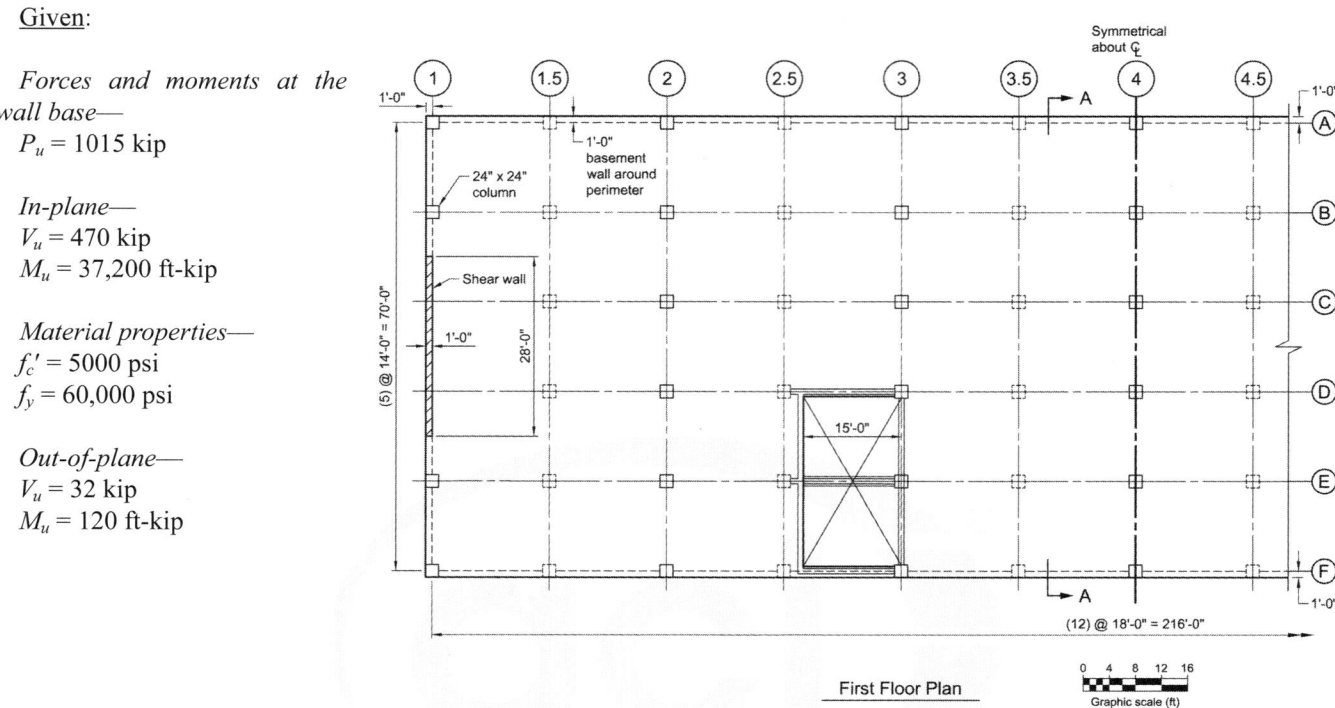

*Fig. E2.1—Building floor plan, first floor.*

This example shows the design and detailing of a special structural shear wall due to in-plane forces, including a seismic boundary element at the wall's edge. In addition, the design strength for the out-of-plane forces is verified. In this example, only one loading condition is checked. In a typical design, several load combinations require checking.

This example uses the Interaction Diagram spreadsheet aid found at https://www.concrete.org/store/productdetail.aspx?ItemID=SP1714DAE.

# CHAPTER 10—STRUCTURAL REINFORCED CONCRETE WALLS

| ACI 318-14 | Discussion | Calculation |
|---|---|---|
| **Step 1: Geometry** | | |
| 11.3.1<br><br><br><br><br><br><br><br><br>20.6.1.3.1 | This wall design example follows the requirements of Chapter 18 and, therefore, does not need to meet the requirements of Table 11.3.1.1 (ACI 318-14). However, the thickness equations (a) and (b) of Table 11.3.1.1 can provide an indication that the thickness chosen is an appropriate design starting point. Note that where special boundary elements are required, the special boundary element will be thicker.<br>From Table 11.3.1.1, the wall thickness must be at least the greater of 4 in. and the lesser of 1/25 the lesser of the unsupported height of the wall (18 ft for the first elevated floor) and the unsupported length of the wall (28 ft from end-to-end of wall). A 12 in. wall is used in this design and the wall is assumed to be exposed to weather on the exterior of the structure. Concrete cover is 1-1/2 in., which is in accordance with Table 20.6.1.3.1 (ACI 318-14). | $h_{req'd} = (18 \text{ ft})(12 \text{ in./ft})/25 = 8.64$ in.<br><br>Example shear wall $h = 12$ in. $> h_{req'd} = 8.64$ in.   **OK** |
| **Step 2: Loads, load patterns, and analysis of the wall** | | |
| 11.4 | The structure is analyzed using the assumptions and requirements of Section 11.4.<br>The structure was analyzed using 3D elastic Finite Element Analysis (FEA) software that follows the analysis requirements of Section 11.4 of ACI 318-14 and Chapter 5 and 6 for loading and analysis, respectively (Fig. E2.2 and E2.3 for in-plane flexure and in-plane shear along the height of the wall, respectively). | The maximum factored axial force, flexural moment, and shear force at the base of the wall are listed in the given section at the start of this example. |

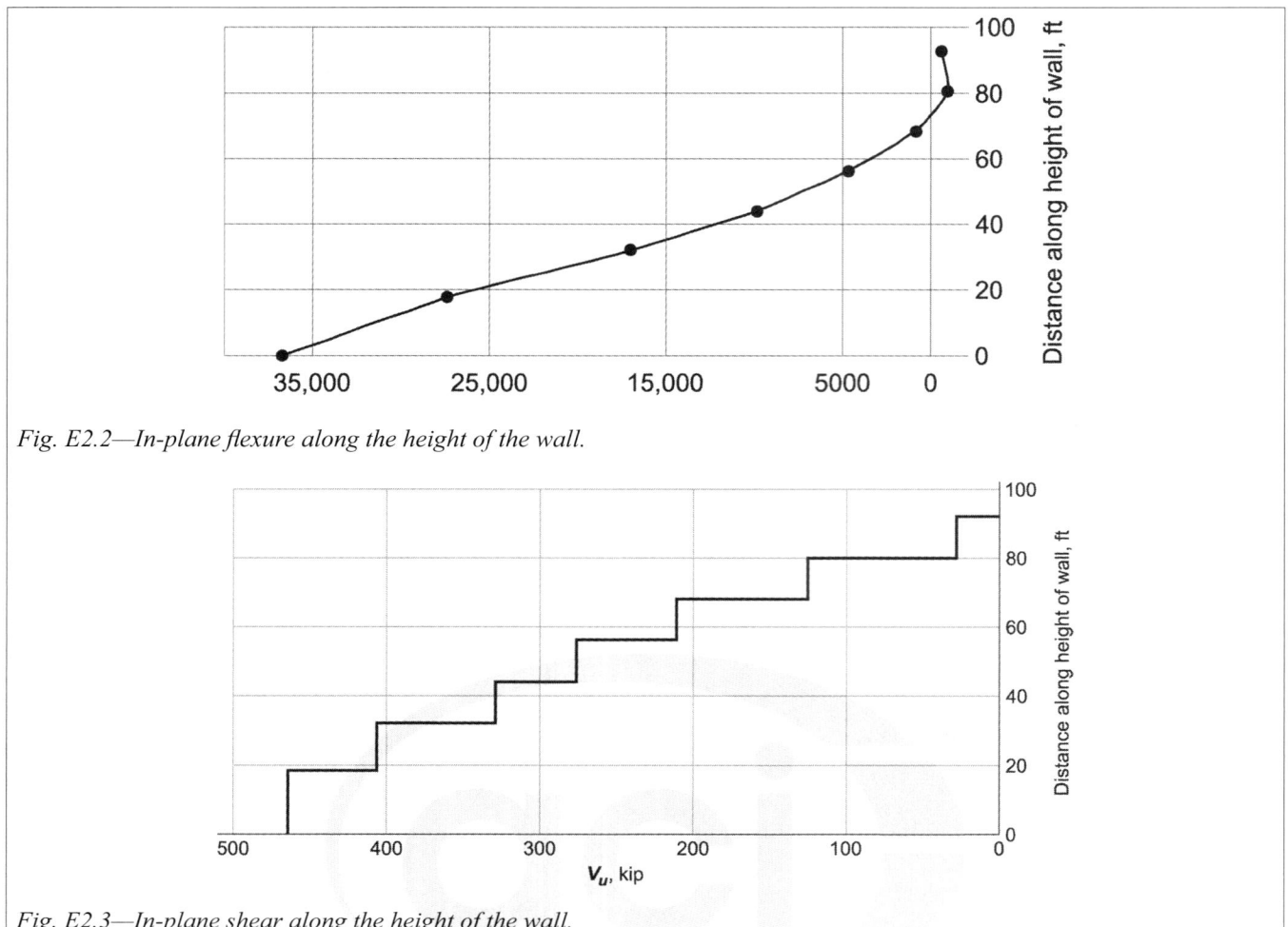

Fig. E2.2—In-plane flexure along the height of the wall.

Fig. E2.3—In-plane shear along the height of the wall.

| | Step 3: Concrete and steel material requirements | |
|---|---|---|
| 11.2.1.1 | The mixture proportion must satisfy the durability requirements of Chapter 19 and structural strength requirements (ACI 318-14). | By specifying that the concrete mixture shall be in accordance with ACI 301 and providing the exposure classes, Chapter 19 requirements are satisfied. |
| | The designer determines the durability classes. Please refer to Chapter 4 of this Handbook for an in-depth discussion of the categories and classes. | Based on durability and strength requirements, and experience with local mixtures, the compressive strength of concrete is specified at 28 days to be at least 5000 psi. |
| | ACI 301 is a reference specification that is coordinated with ACI 318-14. ACI encourages referencing ACI 301 into job specifications. | |
| | There are several mixture options within ACI 301, such as admixtures and pozzolans, which the designer can require, permit, or review if suggested by the contractor. | |
| 11.2.1.2 | The reinforcement must satisfy Chapter 20 of ACI 318-14.<br>The designer determines the grade of bar and if the rebar should be coated by epoxy or galvanized, or both. | By specifying the rebar grade and any coatings, and that the rebar shall be in accordance with ACI 301-10, Chapter 20 (ACI 318-14) requirements are satisfied. In this case, assume Grade 60 bar and no coatings. |
| | Step 4a: Axial and flexural interaction diagram (general) | |
| | The combined axial and flexural design strength of a shearwall is determined using an interaction diagram similar to a column interaction diagram. | |
| | The wall interaction diagram is generated using the Interaction Diagram spreadsheet (link in the given section of this example). | |
| | Refer to Column Example 9.2 in this Handbook for additional information about the Interaction Diagram spreadsheet. | |

| Step 4b: Axial and flexural interaction diagram (in-plane) | | |
|---|---|---|
| 11.1.2, 18.10.5 | Section 11.1.2 requires that special structural walls be designed in accordance with Chapter 18 of ACI 318-14. Chapter 18 covers all requirements necessary to design the wall. | An initial interaction diagram is generated using No. 8 bars at 12 in. spacing throughout the wall. It is assumed that all of the longitudinal reinforcement is effective to resist in-plane flexure. |
| | To estimate an initial reinforcement area, the wall is assumed as a cantilever and the amount of flexural reinforcement necessary to resist the moment is calculated. | The first pair of No. 8 bars is assumed to be at 3 in. from the end of the wall, the second pair is placed at 12 in. from the end of the wall, and the remaining pairs at 12 in. spacing. The wall is symmetrical about the center of the wall and this bar layout is applied at both ends of the wall. |
| | | Fig. E2.4 shows the resulting design strength interaction diagram. The design strength interaction diagram includes the $\phi$-factor. This spreadsheet contains a sheet named "Select Axial Load." When the user enters a $P_u$, the sheet calculates the associated maximum $\phi M_n$ on the design strength interaction diagram curve and plots a point on the design strength interaction diagram. It also generates a corresponding maximum $M_n$ on the nominal strength interaction diagram (not shown). This point is called the "Input Point" on the interaction diagram. The input point of $P_u$ of 1015 kips calculates a maximum $\phi M_n$ point on the interaction diagram of 40,200 ft-kips. The input point is plotted as a solid triangle along the interaction curve. The example has a $P_u$ of 1015 kip and an $M_u$ of 37,200 ft-kip. The open triangle indicates where the example $P_u$ and $M_u$ are and shows that this iteration does satisfy required strength; therefore, further iterations are unnecessary. |

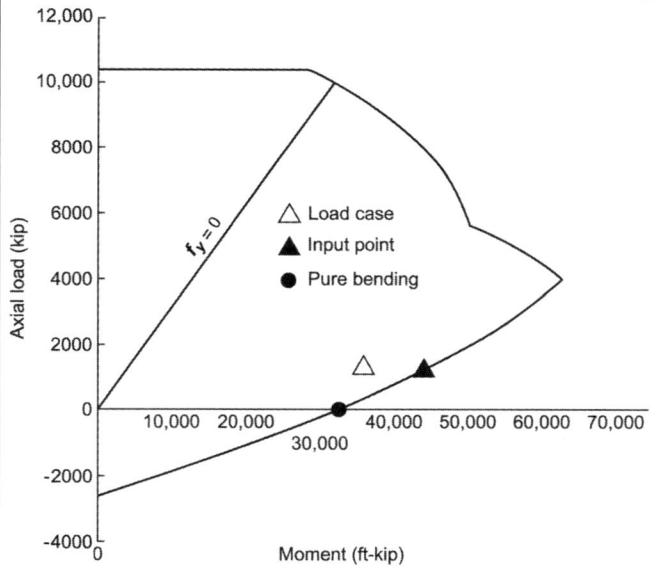

Fig. E2.4—Design strength interaction diagram.

| | Step 4c: Axial, flexural, and shear (out of plane) | |
|---|---|---|
| 11.5.1 | As shown in Step 4b, the layers of No. 8 vertical wall reinforcement satisfies the interaction equation for in-plane bending. | |
| 11.5.3.1 | The resultant of the out-of-plane moment, $M_u = 120$ ft-kip is within the middle third of the wall. This allows Section 11.5.3.1 to be used to check the out-of-plane strength of the wall. $$P_n = 0.55(f_c')(A_g)\left(1-\left(\frac{k\ell_c}{32h}\right)^2\right)$$ | Eccentricity of the resultant load: $(1015\text{ kip})(e) = 120$ ft-kip $e = 1.4$ in. $e < 2$ in. $$P_n = 0.55(5\text{ ksi})(12\text{ in.})(336\text{ in.})\left(1-\left(\frac{(0.8)(202\text{ in.})}{32(12\text{ in.})}\right)^2\right)$$ $P_n = 9090$ kip $\phi P_n = (0.65)(9090\text{ kip}) = 5900$ kip $5900\text{ kip} \geq 1015\text{ kip}$ |
| 21.2.2 | From Table 21.2.2(b) use axial strength reduction factor: $\phi = 0.65$ | **OK** $V_u = 32$ kip The axial and moment design strength over a one foot section of the wall is adequate. |
| 22.5.5.1, 21.2.1 | For out-of-plane shear, the design strength calculation is: $\phi V_c = \phi(2)\sqrt{f_c'}b_w d$ From Table 21.2.1(b) use shear strength reduction factor: $\phi = 0.75$ | $\phi V_c = (0.75)(2)(\sqrt{5000\text{ psi}})(336\text{ in.})(9.25\text{ in.})$ $\phi V_c = 330$ kip  **OK** |

| Step 5: Seismic reinforcement detailing | | |
|---|---|---|
| 18.10.1 | This structure is using a special structural wall to resist the lateral loads applied. | Two curtains of steel are used, the distributed reinforcement ratios are met, and the forces are determined within code allowed analysis methods. Therefore, requirements 18.10.1 through 18.10.3 (ACI 318-14) are met. |
| 18.10.2 | The distributed web reinforcement ratios, $\rho_\ell$ and $\rho_t$, for structural walls must be at least 0.0025, except that if $V_u$ does not exceed $A_{cv}\lambda\sqrt{f_c'}$, then $\rho_\ell$ and $\rho_t$ are permitted to be reduced to the values in Section 11.6 of ACI 318-14. | This example provides No. 4 bars in the horizontal direction in each face at 12 in. This provides 0.40 in.$^2$ per foot in the horizontal direction.<br>The transverse reinforcement ratio is:<br>$\rho_t = 0.4$ in.$^2$/(12 ×12) in.$^2$ = 0.0028 > 0.0025  **OK**<br>This example provides No. 8 bars in the vertical direction in each face at 12 in. This provides 1.58 in.$^2$ per foot in the vertical direction.<br>$\rho_\ell = 1.58$ in.$^2$/(12 × 12) in.$^2$ = 0.0110 > 0.0025  **OK** |
| 18.10.2.2 | The code requires two curtains of distributed reinforcement if;<br>$V_u > 2 A_{cv}\lambda\sqrt{f_c'}$ or $h_w/\ell_w \geq 2.0$. | $h_w/\ell_w \geq$ (92 ft)/(28 ft) = 3.3 > 2<br>Two curtains are required and are provided.  **OK** |
| 18.10.3 | The code allows the $V_u$ calculated from the factored loads to be the design shear. | $V_u = 470$ kip<br>$M_u = 37{,}200$ ft-kip |
| 18.10.4 | The shear strength of special structural walls is affected by the height to length ratio of the wall. The code limits $V_n$ to:<br>$$V_n = A_{cv}\left(\alpha_c \lambda\sqrt{f_c'} + \rho_t f_y\right)$$<br>where $\alpha_c$ is 2.0 for $h_w/l_w \geq 2.0$ and varies between 2.0 and 3 for $h_w/\ell_w < 2.0$. | $V_n = 4032$ in.$^2 (2\sqrt{5000 \text{ psi}} + (0.0028)(60{,}000 \text{ psi})$<br>   $= 1248$ kip<br>where<br>$A_{cv}$ = (12 in.)(28 ft)(12 in./ft) = 4032 in.$^2$ |

# CHAPTER 10—STRUCTURAL REINFORCED CONCRETE WALLS

| | | |
|---|---|---|
| 21.2.4, 21.2.4.1 | The ϕ factor for special structural walls is determined by Sections 21.2.4 and 21.2.4.1. From the analysis of the structure, the maximum axial load under seismic loading combinations for this wall is approximately 1200 kip. | $P_u$ = 1200 kip<br>From the interaction diagram, $M_u$ corresponding to $P_u$ of 1200 kip is:<br>$M_u$ = 41,860 ft-kip<br>$V_u$ from a moment of 41,860 ft-kip is:<br>$V_u$ = 2 × 41,860 ft-kip/18 ft = 4650 kip |
| 18.10.4.4 | In this example, the code limits on shear strength based on concrete strength of<br>$V_n \leq 10 A_{cw}\sqrt{f_c'}$<br>will also limit the ϕ-factor to 0.6. | Max shear:<br>$V_n = 4032 \text{ in.}^2 (10\sqrt{5000 \text{ psi}}) = 2851$ kip<br><br>The $V_u$ calculated from the nominal moment strength of the shear wall is greater than the maximum code allowed shear strength. Therefore, use a ϕ factor of 0.6.<br><br>$\phi V_n = (0.6)(1248 \text{ kip}) = 749 \text{ kip} > 470$ kip OK |
| 18.10.5 | Design for flexure and axial force are the same as for a non-seismic structural wall.<br><br>The code provides geometric limits on wall flanges. | Requirements of Section 18.10.5 are met through the flexural and axial interaction diagram design process in Step 4b.<br><br>This wall is rectangular in plan and does not have end flanges. |
| 18.10.6, 18.10.6.1 | Special boundary elements are often required in special structural walls to resist the large compression forces at the ends of the walls during an earthquake event.<br>Boundary elements are required in accordance with Sections 18.10.6.2 or 18.10.6.3 (ACI 318-14). | |

| 18.10.6.2 | Section 18.10.6.2 applies to walls satisfying the following:<br>18.10.6.2 (a) $h_w/\ell_w \geq 2.0$, are continuous from base of structure to top of wall, and are designed to have a single critical section for flexure and axial loads. Otherwise, Section 18.10.6.3 applies.<br><br>Section 18.10.6.2 requires a special boundary element if the neutral axis depth calculated for the factored axial force and factored moment is greater than the value in Eq. (18.10.6.2)<br>Check if.<br><br>$$c \geq \frac{\ell_w}{600(1.5\delta_u/h_w)}$$<br><br>If so, a special boundary element is needed.<br><br>(b) The special boundary element transverse reinforcement must extend vertically above and below the critical section at least the greater of $\ell_w$ and $M_u/4V_u$ except as permitted by Section 18.10.6.4(g) of ACI 318-14. | $h_w/\ell_w$ = (92 ft)/(28 ft) = 3.3 ft  **OK**<br>Wall is continuous from bottom of structure to top of wall.  **OK**<br>Wall is designed having a single critical section for shear and flexure.  **OK**<br>Therefore, Section 18.10.6.2 will be used.<br><br>The interaction diagram spreadsheet calculates the neutral axis depth $c$, which is 67.85 in.<br>The software that analyzes the structure presents the deflection data for the structure. The value of $\delta_u$ from the software is 2.4 in. and does not include the $C_d$ factor of 5 for a special structural concrete wall from ASCE 7-10.<br><br>$$\delta_u/h_w = \frac{(2.4 \text{ in.})(5)}{(92 \text{ ft})(12 \text{ in./ft})}$$<br><br>$\delta_u/h_w$ = 0.0109<br>Therefore, the value determined from Eq. (18.10.6.2) is:<br><br>$$c \geq \frac{336 \text{ in.}}{600(1.5 \times 0.0109)} = 34.25 \text{ in.}$$<br><br>Because $c$ of 67.85 in. is greater than 34.25 in., Section 18.10.6.2 does require a boundary element. |

| | | |
|---|---|---|
| 18.10.6.4(a), (b), (c), (d) | 18.10.6.4 (a) through (d) impose additional geometric requirements upon the special boundary elements.<br>Section 18.10.6.4 states the following:<br><br>Where special boundary elements are required by 18.10.6.2 or 18.10.6.3, (a) through (h) must be satisfied:<br>a) The boundary element must extend horizontally from the extreme compression fiber a distance at least the greater of $c - 0.1\ell_w$ and $c/2$, where $c$ is the largest neutral axis depth calculated for the factored axial force and nominal moment strength consistent with $\delta_u$.<br><br>(b) Width of the flexural compression zone, $b$, over the horizontal distance calculated by 18.10.6.4(a), including flange if present, shall be at least $h_u/16$.<br><br>(c) For walls or wall piers with $h_w/\ell_w \geq 2.0$ that are effectively continuous from the base of structure to top of wall, designed to have a single critical section for flexure and axial loads, and with $c/\ell_w \geq 3/8$, width of the flexural compression zone $b$ over the length calculated in Section 18.10.6.4(a) shall be greater than or equal to 12 in.<br><br>(d) In flanged sections, the boundary element shall include the effective flange width in compression and shall extend at least 12 in. into the web. | 18.10.6.4(a) requires that the special boundary element extend a minimum from the extreme compression fiber.<br>$c - 0.1\ell_w$ = 67.85 in. – (0.1)(336 in.) = 34.25 in.<br>or<br>$c/2$ = 67.85 in./2 = 33.925 in.<br>The example rounds the length of the boundary element to 34 in.<br><br>18.10.6.4(b) is satisfied by making the wall thicker over the boundary element to meet the requirement of $h_u/16$ = 216 in./16 = 13.5 in. Therefore, the wall thickness will be 14 in. at the boundary elements.<br><br>18.10.6.4(c) does not apply because the example assumes that the wall and boundary element will be reevaluated at different heights to reduce the amount of steel required; therefore, the wall is considered designed for several critical sections for flexure and axial loads.<br><br>Section 18.10.6.4(d) does not apply because this is not a flanged section. |
| 18.10.6.4(e) | Section 18.10.6.4(e) imposes spacing requirements on the transverse reinforcement.<br>The boundary element transverse reinforcement shall satisfy Section 18.7.5.2(a) through (e) and Section 18.7.5.3, except the value $h_x$ in Section 18.7.5.2 must not exceed the lesser of 14 in. and two-thirds of the boundary element thickness, and the transverse reinforcement spacing limit of Section 18.7.5.3(a) must be one-third of the least dimension of the boundary element. | Section 18.10.6.4(e) requires that the geometry and spacing of the vertical and crosstie reinforcement shall meet the requirements of a special structural column. |

| | | |
|---|---|---|
| 18.10.6.4(e)<br>18.7.5.2 | Transverse reinforcement shall be in accordance with (a) through (f):<br>(a) Transverse reinforcement shall comprise either single or overlapping spirals, circular hoops, or rectilinear hoops with or without crossties.<br>(b) Bends of rectilinear hoops and crossties shall engage peripheral longitudinal reinforcing bars. | Section 18.7.5.2 requires that the transverse boundary element reinforcement satisfy essentially the same requirements as those of a special concrete column. However, to satisfy the first part of 18.10.6.4(e), the limit of $h_x$ from Section 18.7.5.2(e) is modified. The permitted $h_x$ is less than 2/3 of the 14 in. width of the boundary element.<br>$2/3 \times 14$ in. = 9-1/3 in.<br>This permitted $h_x$ is for both across the width of the element and along the length of the element. |
| | (c) Crossties of the same or smaller bar size as the hoops shall be permitted, subject to the limitation of Section 25.7.2.2 of ACI 318-14. Consecutive crossties shall be alternated end for end along the longitudinal reinforcement and around the perimeter of the cross section. | The distance between the two curtains of No. 8 bars in a 14 in. thick wall is approximately:<br>14 in. – (2)(1.5 in.) – (2)(0.5 in.) – 1 in. = 9 in.<br>Therefore, for the thickness of the boundary element, there is no need to add vertical reinforcement. |
| | (d) Where rectilinear hoops or crossties are used, they shall provide lateral support to longitudinal reinforcement in accordance with Sections 25.7.2.2 and 25.7.2.3 of ACI 318-14.<br>(e) Reinforcement shall be arranged such that the spacing $h_x$ of longitudinal bars laterally supported by the corner of a crosstie or hoop leg shall not exceed 14 in. around the perimeter of the column. | To satisfy this requirement for $h_x$ along the length of the element, reinforcement is spaced at 7 in. for the first 34 in. to reduce the spacing to below 9 in. Note that for this example, the interaction diagram calculations were modified to match this detailing. However, this added precision may not be necessary in a design office. The designer should use their engineering judgment to determine if this precision is necessary for their specific design. |
| 18.10.6.4(e)<br>18.7.5.3 | 18.7.5.3 Spacing of transverse reinforcement shall not exceed the smallest of (a) through (c):<br>(a) One-fourth of the minimum column dimension<br>(b) Six times the diameter of the smallest longitudinal bar<br>(c) $s_o$, as calculated by:<br>$$s_o = 4 + \left(\frac{14 - h_x}{3}\right) \quad (18.7.5.3)$$<br>The value of $s_o$ from Eq. (18.7.5.3) shall not exceed 6 in. and need not be taken less than 4 in. | The second part of 18.10.6.4(e) and Section 18.7.5.3 determines a maximum spacing of the transverse reinforcement in the special boundary element.<br>18.7.5.3(a) is modified to one-third the minimum column dimension by 18.10.6.4(e):<br>14 in./3 = 4.67 in.<br>18.7.5.3(b):<br>(6)(1 in.) = 6 in.<br>18.7.5.3 (c):<br>$$s_o = 4 + \left(\frac{14 - 9}{3}\right) = 5.67 \text{ in.}$$<br>The example uses a spacing of 4 in. for the transverse reinforcement in the special boundary element and all longitudinal bars in the special boundary element are engaged by a crosstie. |

| | | |
|---|---|---|
| 18.10.6.4(f) | 18.10.6.4(f) sets requirements for the area of reinforcement for the ties in the boundary element.<br>(f) The amount of transverse reinforcement shall be in accordance with Table 18.10.6.4(f).<br><br>Table 18.10.6.4(f)(a):<br><br>$$0.3\left(\frac{A_g}{A_{ch}}-1\right)\frac{f'_c}{f_{yt}}$$<br><br>$A_g = (14 \text{ in.})(34 \text{ in.}) = 476 \text{ in.}^2$<br>$A_{ch} = (14 \text{ in.} - 2 \text{ in.} - 2 \text{ in.})(34 \text{ in.} - 2 \text{ in.} - 2 \text{ in.})$<br>$A_{ch} = 300 \text{ in.}^2$<br><br>Table 18.10.6.4(a):<br><br>$$\frac{A_{sh}}{sb_c} = 0.3\left(\frac{A_g}{A_{ch}}-1\right)\frac{f'_c}{f_{yt}}$$ | For Table 18.10.6.4(f), Eq. (a):<br><br>$$0.3\left(\frac{476}{300}-1\right) = 0.176 > 0.09$$<br><br>Since this is greater than the minimum required by equation (b), use Table 18.10.6.4(a):<br><br>$$A_{sh} = 0.176\frac{5000 \text{ psi}}{60,000 \text{ psi}}(4 \text{ in.})(12 \text{ in.})$$<br><br>$A_{sh} = 0.7 \text{ in.}^2$<br>No. 4 ties at 4 in. vertically satisfy 18.10.6.4(f). |
| 18.10.6.4(g) | Section 18.10.6.4(g) requires that the transverse reinforcement extend into the wall support base when the critical section occurs at the wall base. | This section is satisfied by extending the transverse reinforcement a minimum of 12 in. into the foundation element below the base of the wall. |
| 18.10.6.4(h) | Section 18.10.6.4(h) requires that the horizontal reinforcement in the wall be developed within the core of the boundary element. | The development length of the No.4 bars being used for lateral reinforcement is much less than the 34 in. depth of the boundary element. Extending the No.4 lateral bars through the boundary element to within 6 in. of the end of the wall will satisfy this requirement. |
| 18.10.6.5 | Does not apply. | |
| 18.10.7, 8, and 10 | Do not apply. | |
| 18.10.9 | Section 18.10.9.1 Construction joints in structural walls shall be specified according to Section 25.5.6, and contact surfaces shall be roughened consistent with condition (b) of Table 22.9.4.2 of ACI 318-14. Final sketch of structural wall using the special boundary elements | Section 18.10.9 is satisfied by requiring that all construction joints in the wall be roughened to approximately a 1/4 in. amplitude in the construction documents.<br>Figs. E2.5 and E2.6 show the final configuration of the wall if special boundary elements were required. |

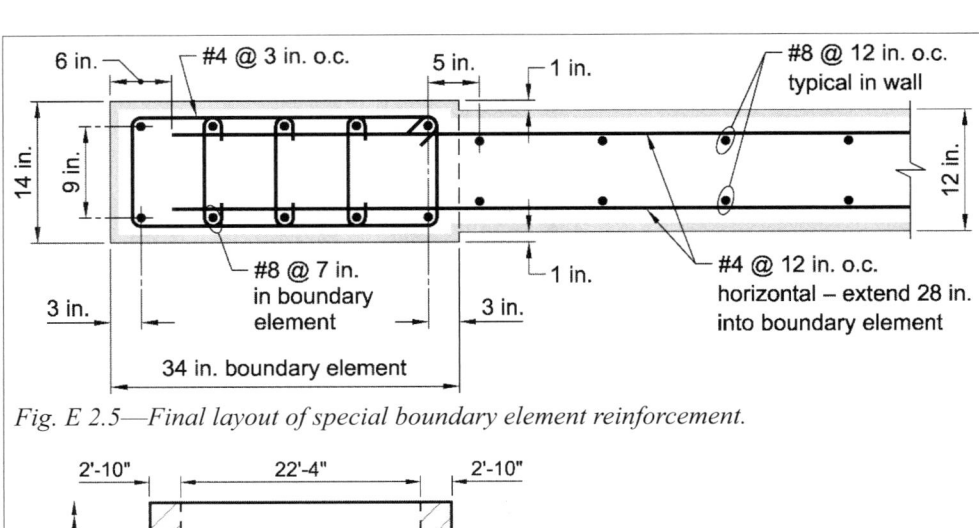

*Fig. E 2.5—Final layout of special boundary element reinforcement.*

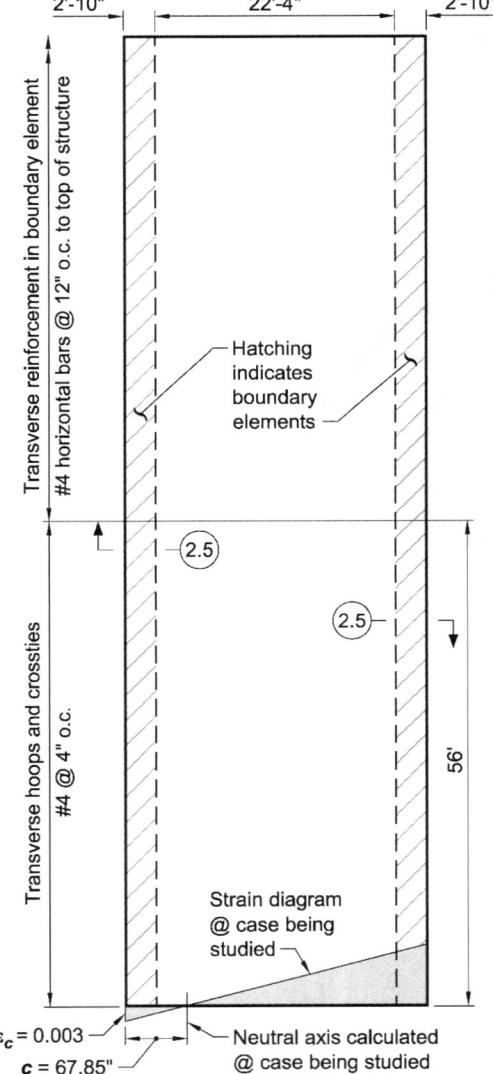

*Fig. E2.6—Elevation of wall with special boundary elements.*

# CHAPTER 11—FOUNDATIONS

## 11.1—Introduction

The foundation is an essential building system that transfers column and wall forces to the supporting soil. Depending on the soil properties and building loads, the engineer may choose to support the structure on a shallow or deep foundation system. Because ACI 318-14 provides design and detailing provisions for shallow foundations and pile caps only, deep foundations are not covered in this Handbook.

Shallow foundation systems include isolated footings that support individual columns (Fig. 11.1(a) and (b), combined footings that support two or more columns (Fig. 11.1(c)), strip footings that support walls (Fig. 11.1(d)), ring footings (Fig. 11.1(e)) that support a tank wall, and mat footings (Fig. 11.1(f)) that support several or all columns or walls.

In this chapter, isolated, combined, and continuous footing examples are presented. A pile cap design is presented in the Strut-and-Tie Chapter of this Handbook.

## 11.2—Footing design

Footing design typically consists of four steps:

1. Determine the necessary soils parameters. This step is often completed by consulting with a geotechnical engineer who furnishes information in a geotechnical report. Important information that a geotechnical report should include are the:

(a) Subsurface profile, which provides physical characteristics of soil, groundwater, rock, and other soil elements

(b) Shear strength parameters to determine the stability of sloped soil

(c) Frost depth to determine the bearing level of footing below frost penetration level

(d) Unit weights, which is the weight of soil and water per unit volume, used to determine the additional load on a footing/structure when backfilled

(e) Bearing capacity, which is the maximum allowable pressure that a footing is permitted to exert on the supporting soil; the size of the footing is based on allowable loads

(f) Predicted settlement, which is the anticipated vertical movement of a footing over time

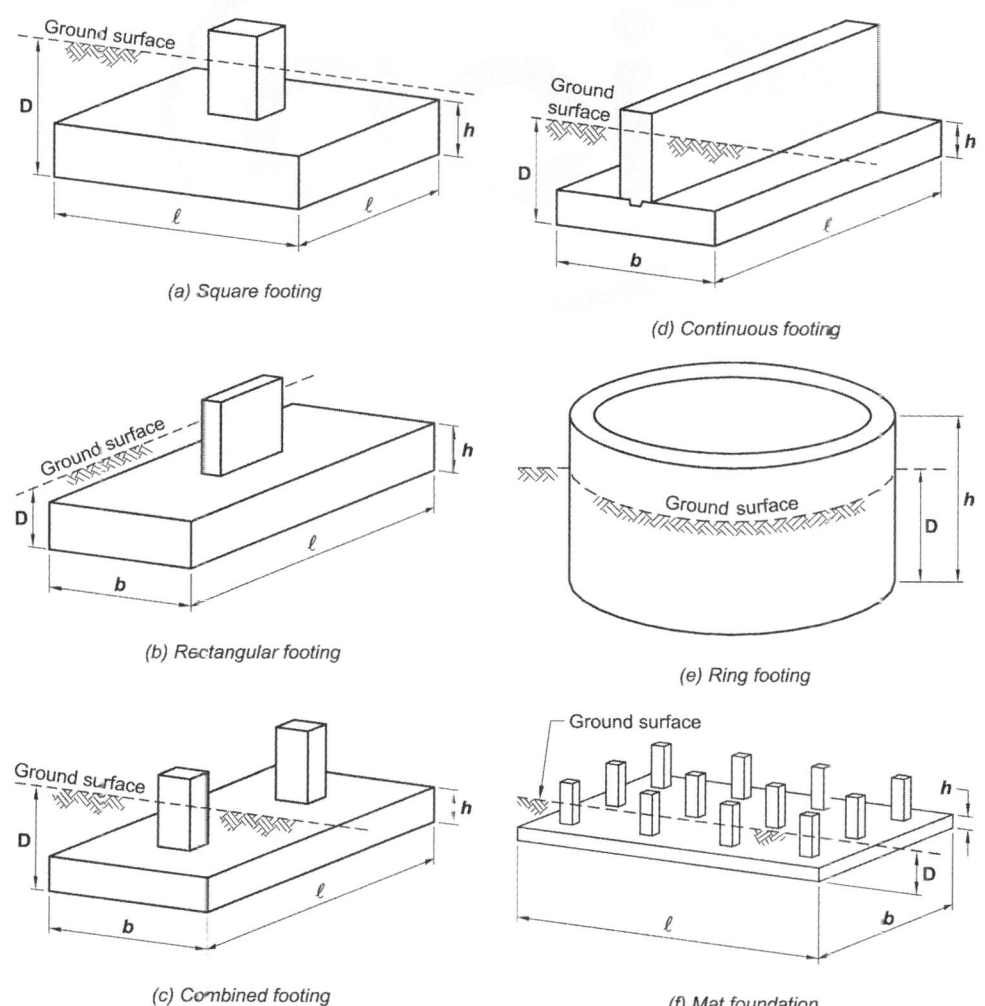

Fig. 11.1—Shallow foundation types.

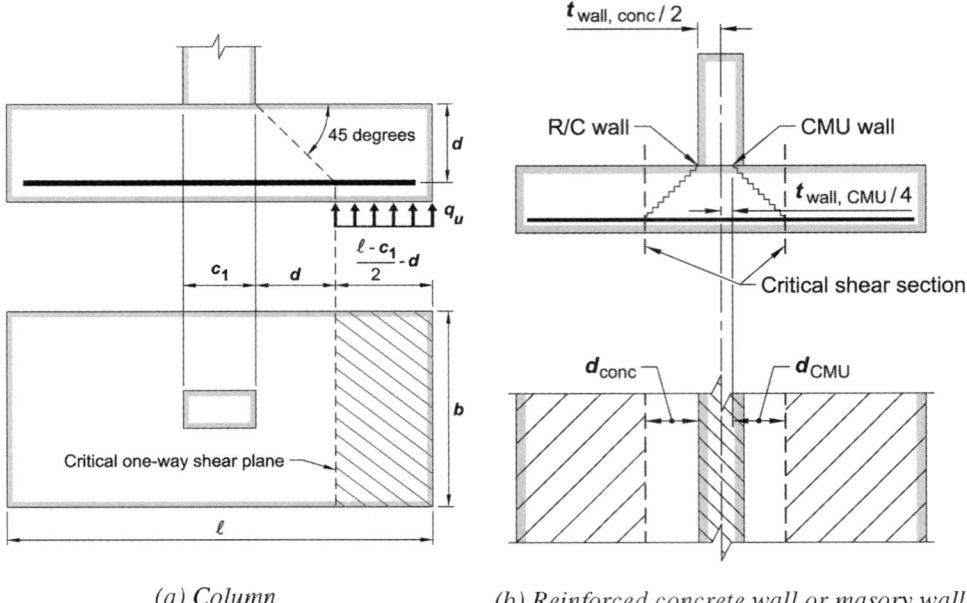

*Fig. 11.3.1a—One-way shear critical section in footings.*

(g) Liquefaction, which is an important soil characteristic if the building is located in an active seismic area.

2. Analyze the building's structure under service loads (ACI 318-14, Section R13.2.6.1) and factored loads (ACI 318-14, Section 5.3.1) to calculate moments and forces on the columns and walls at the footing level; the service load analysis is used to calculate footing bearing areas and the factored load analysis to design the footing.

3. Select the footing geometry so that the soil parameters are not exceeded. The following are typical parameters:

(a) Calculated bearing pressures are assumed to be uniform or to vary linearly; bearing pressure is measured in units of force per unit area, such as pounds per square foot

(b) The effect of anticipated differential vertical settlement between adjacent footings on the superstructure are considered

(c) Footings need to be able to resist sliding caused by any horizontal loads

(d) Shallow footings, assumed not to be able to resist tension, should be able to resist overturning moments from compression reactions only; overturning moments are commonly caused by horizontal loads

(e) Local conditions or site constraints, such as proximity to property lines or utilities, are adequate.

4. Design and detail the footing in accordance with ACI 318-14, Chapter 13. During this step, the previously selected geometry is checked against strength requirements of the reinforced concrete sections. The step-by-step structural design process for concentrically loaded isolated footings follows:

## 11.3—Design steps

1. Find service dead and live column loads: ACI 318-14, Section R13.2.6

The footing geometry is selected using service loads.
$D$ = service dead load from column
$L$ = service live load from column
$P = D + L$
$A_{req} = P/q_{all}$
For square footings, $\ell \geq \sqrt{A_{req}}$
For rectangular footings, choose one of the sides from site constraints and calculate the other such that: $b \times \ell \geq A_{req}$

2. Calculate the design (factored) column load $U$: ACI 318-14, Section 5.3.1

3. Obtain the allowable soil pressure $q_{net}$. Because soil and concrete unit weights are close (120 and 150 pcf, respectively), the footing self-weight may initially be ignored.

4. Calculate the soil pressure based on initial footing base dimensions:
Square footing: $q_u = U/\ell^2$
Rectangular footing: $q_u = U/\ell b$

5. Check one-way (beam) shear:

The critical section for one-way shear extends across the width of the footing and is located at a distance $d$ from the face of a column or wall (Fig. 11.3.1a and Fig. 11.3.1a(b) left side), ACI 318-14, Section 8.4.3.2. The shear is calculated assuming the footing is cantilevered away from the column or wall ACI 318-14, Section 8.5.3.1.1.

For masonry walls the area is halfway between the wall center and the face of the masonry wall (Fig. 11.3.1a(b) right side).

NOTE: If the calculated one-way factored shear exceeds the one-way shear design strength, then increase the footing thickness. Footings are typically not designed with shear reinforcement.

6. Check two-way (slab) shear:
(a) Determine the dimensions of loaded area for:
  i) Rectangular concrete columns, the loaded area coincides with the column area (Fig. 11.3.1b(a))

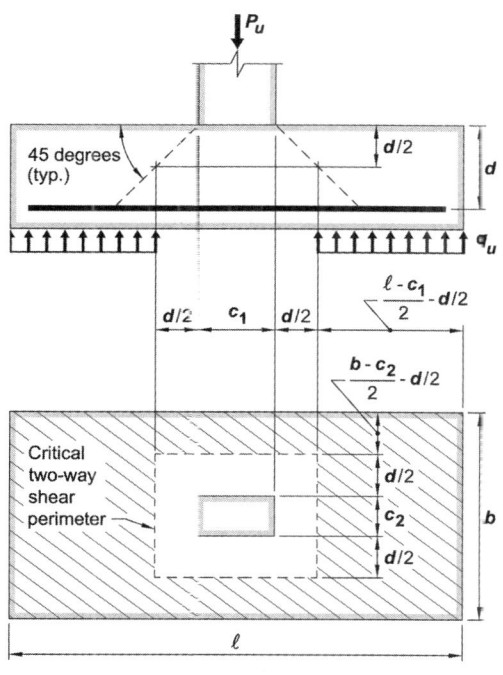
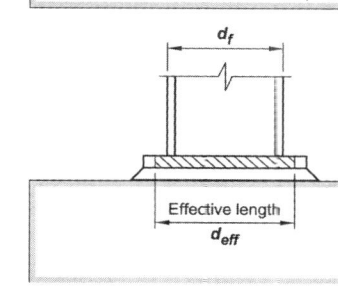

(a) Reinforced concrete column    (b) Steel column with base plate

Fig. 11.3.1b—Two-way shear critical section in footings.

ii) Steel columns, the effective loaded area is assumed to be halfway between the faces of the steel column and the edges of the steel base plates used (Fig. 11.3.1b(b)):

$$b_{eff} = b_f + \frac{b_{bp} - b_f}{2}$$

where $b_f$ is the width of column flange and $b_{bp}$ is base plate side.

$$d_{eff} = d_f + \frac{d_{bp} - d_f}{2}$$

where $d_f$ is the depth of column flange and $d_{bp}$ is base plate side.

(b) Calculate the shear critical section, located at a distance of $d/2$ outside the loaded area (ACI 318-14, Section 13.2.7.2)
(c) Calculate the factored shear force for two-way shear stress, $v_u$
(d) Compare $v_u$ to two-way design stress, $\phi v_n$ (ACI 318-14, Section 22.6.5.2).

NOTE: If the design shear stress is less than factored shear stress, then increase footing thickness and repeat steps starting at (b).

7. Design and detail the footing reinforcement (Fig. 11.3.1c):
Square footings are designed and detailed for moment in one direction and the same reinforcing is placed in the other direction. For rectangular footings the reinforcing must be designed and detailed in each direction. The critical section for moment extends across the width of the footing at the face of the column. ACI 318-14, Sections 13.2.6.4 and 13.2.7.1.

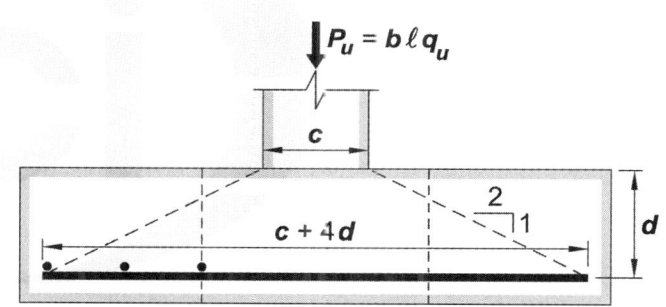

Fig. 11.3.1c—Column load distribution in footing.

(a) Calculate projection, $x$, from the column face (Fig. 11.3.1d):
$x = \ell/2 - c/2$, where $c$ is the smaller dimension of the column for a square footing. For a rectangular footing, $c$ is the dimension perpendicular to the critical section in each direction
(b) Calculate total factored moment, $M_u$, at the critical section
(c) Calculate required $A_s$.

ACI 318-14, Sections 13.3.2.1 and 7.6.1.1, specify a minimum flexural reinforcement $A_s$ must be met, and 7.7.2.3 specifies a maximum bar spacing of 18 in.

8. Check the load transfer from the column to the footing per ACI 318-14, Section 16.3 (Fig. 11.3.1e)
(a) Check the bearing strength of the footing concrete: ACI 318-14, Section 22.8.3.2
(b) Calculate the load to be transferred by reinforcement (usually dowels):

$$\phi B_{dowels} = P_u - \phi B_n$$

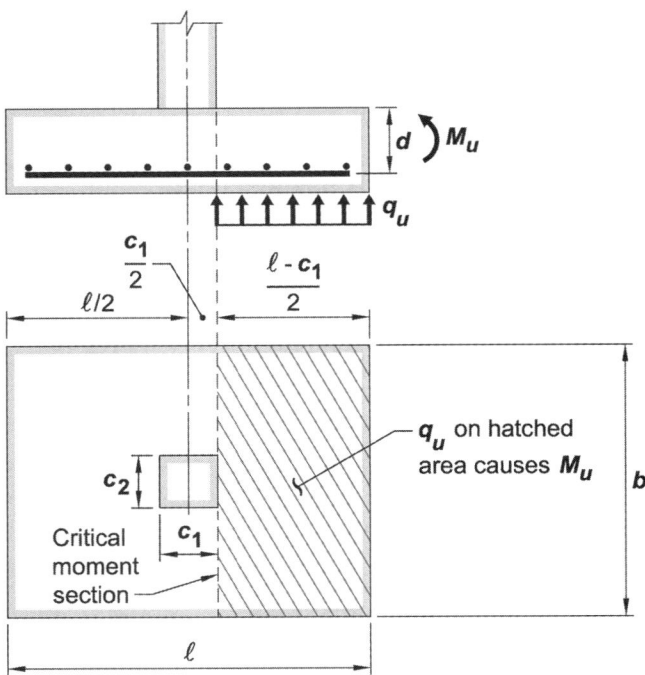

Fig. 11.3.1d—*Moment critical section in footing at reinforced concrete column face.*

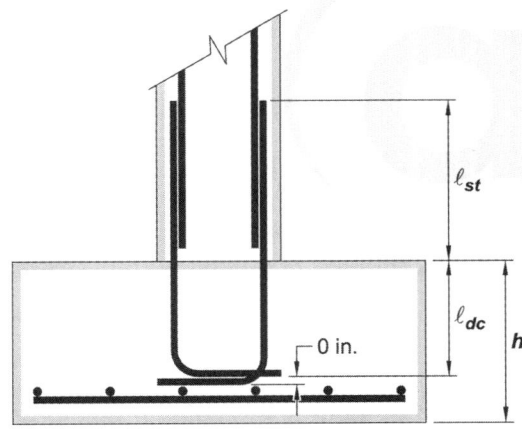

Fig. 11.3.1e–*Column/wall dowels into footing.*

If $\phi B_n \geq P_u$ only a minimum area of reinforcement is required ACI 318-14, Section 16.3.4.1.

(c) Calculate the required reinforcement area and choose bar size and number.

(d) Check dowel embedment into footing for compression: ACI 318-14, Section 25.4.9

NOTE: The footing must be deep enough to develop the dowels in compression, $\ell_{dc}$. Hooks are not considered effective in compression and are used to stabilize the dowels during construction.

(e) Dowels must be long enough to lap with the column bars in compression, $\ell_{sc}$: ACI 318-14, Section 25.5.5

(f) Choose bar size and spacing:

For square footings, $A_s$ must be furnished in each direction. The same size and number of bars should

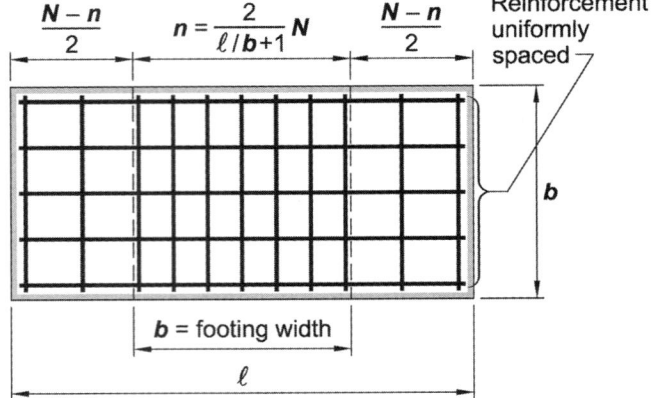

Fig. 11.3.1f—*Bar distribution in short direction.*

be uniformly spaced in each direction (ACI 318-14, Sections 13.3.2.2 and 13.3.3.2).

For rectangular footings, $A_s$ must be furnished in each direction. Bars in long direction should be uniformly spaced. Bars in the short direction should be distributed as follows (ACI 318-14, Section 13.3.3.3):

(i) In a band of width $B_s$ centered on column:

$$\text{\#bars in } B_s = \frac{2}{\ell/b + 1}$$

(total # of bars) (round up to an interger)

(ii) Remaining bars should be uniformly spaced in outer portions of footing (outside the center band width of footing). The remaining bars should satisfy the minimum reinforcement requirements of ACI 318-14, Section 9.6.1 (refer to Fig. 11.3.1f).

(g) Check development length:

Calculate the bar's development length, $\ell_d$, in tension per ACI 318-14, Section 25.4. The development length, $\ell_d$, must be less than ($L_m$ – end cover). If the ($L_m$ – end cover) is shorter than $\ell_d$, use bars of a smaller diameter.

## 11.4—Footings subject to eccentric loading

In addition to vertical loads, footings often resist lateral loads or overturning moments. These loads are typically from seismic or wind forces.

Overturning moments result in a nonuniform soil-bearing pressure under the footing, where soil-bearing pressure is larger on one side of the footing than the other. Nonuniform soil bearing can also be caused by a column located away from the footing's center of gravity.

If overturning moments are small in proportion to vertical loads, that is, the total applied load is located within the kern ($e \leq \ell/6$), then the entire footing bottom is in compression and a $P/A \pm M/S$ analysis is appropriate to calculate the soil pressures, where the parameters are defined as follows:

$P$ = the total vertical service load, including any applied loads along with the weight of all of the foundation components, and also including the weight of the soil located directly above the footing.

$A$ = the area of the footing bottom.

$M$ = the total overturning service moment at the footing bottom.

$S$ = the section modulus of the footing bottom.

If overturning moments are larger, that is, the total applied load falls outside the kern, $e > \ell/6$, then $P/A - M/S$ analysis requires the soil to resist tension (upward movement of the footing), which is not possible.

This soil is only able to transmit compression.

The following are typical steps to calculate footing bearing pressures if nonuniform bearing pressures are present. These steps are based on a footing that is rectangular in plan and assumes that overturning moments are parallel to one of the footing's principal axes. These steps should be completed for as many load combinations as required by the applicable design criteria. For instance, the load combination with the maximum P usually causes the maximum bearing pressure while the load combination with the minimum P usually is critical for overturning.

1. Determine the total service vertical load $P$

2. Calculate the total service overturning moment $M$, measured at the footing bottom

3. Determine whether $P/A$ exceeds $M/S$

4. If $P/A$ exceeds $M/S$, then the maximum bearing pressure equals $P/A + M/S$ and the minimum bearing pressure equals $P/A - M/S$

5. If $P/A$ is less than $M/S$, then the soil bearing pressure is as shown in Fig. 11.4. Such a soil-bearing pressure distribution is structurally inefficient. The maximum bearing pressure, shown in the figure, is calculated as follows: maximum bearing pressure = $2P/[(B)(X)]$ where $X = 3(\ell/2 - e)$ and $e = M/P$.

## 11.5—Combined footing

If a column is near a property line or near a pit or a mechanical equipment in an industrial building, a footing may not be able to support a column concentrically and the eccentricity is very large. In such a case the column footing is extended to include an adjacent column and support both on the same footing, called combined footing (Fig. 11.5). The combined footing is sized to have the resultant force of the two columns within the kern, or preferably to coincide with the center of the footing area. The combined footing can be rectangular, trapezoidal, or having a strap, connecting the two main column footings together (Fig. 11.5(c)).

Design steps:

1. Calculate the total service column loads, $P_1$ and $P_2$ (ACI 318-14, Section R13.2.6)

2. Calculate service column load resultant location
Center for rectangular footing:

$$x_c = \frac{Px_1 + P_2x_2}{\sum P_i}$$

If $P_1$ is much larger than $P_2$, then trapezoidal combined footing may be used.

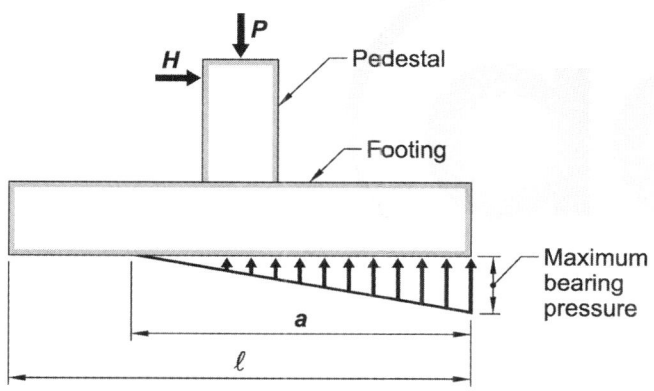

Fig. 11.4—Footing under eccentric loading.

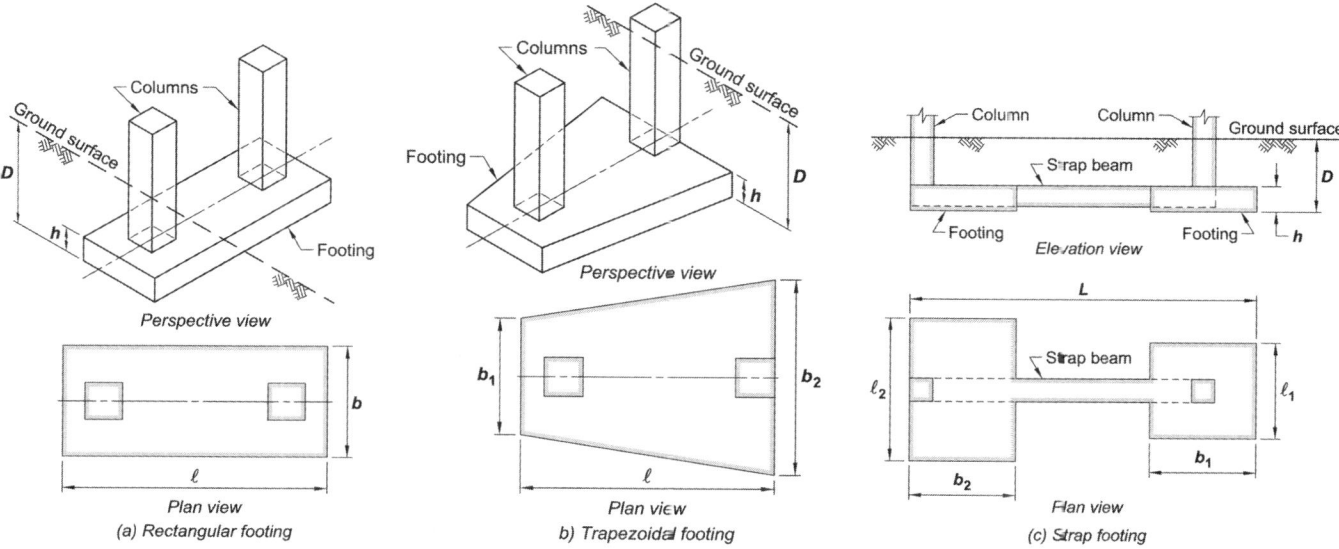

Fig. 11.5—Common types of combined footing geometries.

Determine combined footing length from construction constraints.

Calculate the widths $B_1$ and $B_2$ such that the center of the footing coincides with the force resultant or is at least within $\ell/6$ of the force resultant.

$$\bar{x} = \frac{\ell}{3}\frac{B_1 + 2B_2}{B_1 + B_2} - \frac{c}{2}$$

3. Determine combined footing dimensions assuming uniform bearing

Combined rectangular footing length: $\ell = 2x_c$
Combined rectangular footing width:

$$B = \frac{P_1 + P_2}{\ell q_a}$$

4. Steps to design a combined footing to resist one-way and two-way shear and moment is similar to the isolated footing design steps presented above.

## 11.6—Examples

**Foundation Example 1:** *Design of a square spread footing of a seven-story building*

Design and detail a typical square footing of a six bay by five bay seven-story building, founded on stiff soil, supporting a 24 in. square column. The building has a 10 ft high basement. The bottom of the footing is 3 ft below finished grade (refer to Fig. E1.1). The building is assigned to Seismic Design Category (SDC) B.

Given:
*Column load*—
Service dead load $D = 541$ kip
Service live load $L = 194$ kip
Seismic load $E = \pm 18$ kip

*Material properties*—
Concrete compressive strength $f_c' = 4$ ksi
Steel yield strength $f_y = 60$ ksi
Normalweight concrete $\lambda = 1$
Density of concrete $= 150$ lb/ft$^3$

Allowable soil-bearing pressures—
D only: $q_{all.D} = 4000$ psf
D + L: $q_{all.D+L} = 5600$ psf
D + L + E: $q_{all.Lat} = 6000$ psf

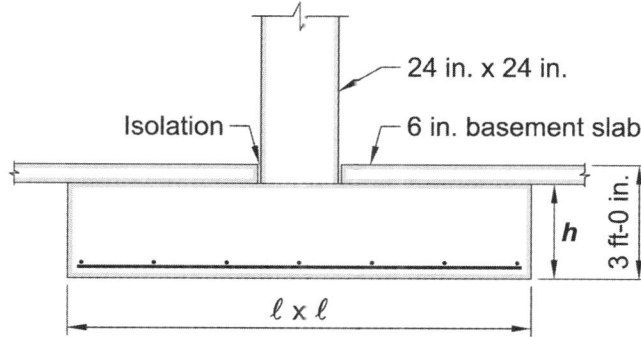

*Fig. E1.1—Rectangular foundation plan.*

| ACI 318-14 | Discussion | Calculation |
|---|---|---|
| **Step 1: Foundation type** | | |
| 13.1.1 | The bottom of the footing is 3 ft below the basement slab. Therefore, it is considered a shallow foundation. | |
| 13.3.3.1 | The footing will be designed and detailed with the applicable provisions of Chapter 7, One-way slabs, and Chapter 8, Two-way slabs of ACI 318-14. | |

| | Step 2: Material requirements | |
|---|---|---|
| 19.2.1.1<br>19.2.1.3 | Concrete compressive strength<br>The value of concrete compressive strength at a given age must be specified in the contract documents. Table 19.2.1.1 provides a lower concrete compressive strength limit of 2500 psi. | Provided: $f_c' = 4000$ psi $> f'_{c,min} = 2500$ psi  **OK** |
| 19.3.1<br><br>19.3.2 | Exposure categories and classes<br>The engineer must either assign exposure classes to the footing with respect to Table 19.3.1.1 (ACI 318-14) so the ready-mix supplier can proportion the concrete mixture, or use the classes to directly specify mixture proportions in the contract documents. Based on the exposure classes, the concrete mixtures must satisfy to the most restrictive requirements of Table 19.3.2.1.<br><br>Concrete exposure categories<br>There are four categories: F, S, W, and C. | |
| 19.3.1.1 | Category F<br>The foundation is placed below the frost line, therefore, it is not exposed to external elements—freezing and thawing cycles. Therefore, class F0 applies. | Class F0<br>Maximum $(w/cm)_{max} = $ N/A<br>Minimum $f_c' = 2500$ psi |
| 19.3.2.1 | Mixture requirements that must be satisfied for F0 are listed in Table 19.3.2.1. | Air content is not required and there are no limits on cementitious materials |
| 19.3.3.1 | Requirements of Table 19.3.3.1 do not apply. | |
| 19.3.1.1<br>19.3.2.1 | Category S<br>Injurious sulfate attack is not a concern. Mixture requirements for S0 are listed in Table 19.3.2.1. | S0 → $(w/cm)_{max} = $ N/A and $f_c' = 2500$ psi |
| 19.3.1.1<br>19.3.2.1 | Category W<br>The footing may be in contact with water and low permeability is not required. | W0 → $(w/cm)_{max} = $ none and $f_c' = 2500$ psi |
| 19.3.1.1<br><br>19.3.2.1 | Category C<br>The concrete is exposed to moisture and there is no external source of chlorides; therefore the class is C1. Mixture requirements for C1 are listed in Table 19.3.2.1. | C1 → $(w/cm)_{max} = $ none and $f_c' = 2500$ psi<br><br>Therefore, there is no restriction on $w/cm$ and $f_c' = 4000$ psi |

Conclusion:
(a) The most restrictive minimum concrete compressive strength is 2500 psi, and no limits on the $w/cm$. Therefore, in the judgment of the licensed design professional, use 4000 psi concrete compressive strength.
(b) Other parameters, such as maximum chloride ion content and air content, are exposure specific, and thus not compared with other exposure limits.
(c) The $f_c'$ utilized in the strength design must be at least what is required for durability.

## Step 3: Determine footing dimensions

| 13.3.1.1 | To calculate the footing base area, divide the service load by the allowable soil pressure. | The unit weights of concrete and soil are 150 pcf and 120 pcf; close. Therefore, footing self-weight will be ignored for initial sizing of footing: |
|---|---|---|
| | $$\text{area of footing} = \frac{\text{total service load } (\sum P)}{\text{allowable soil pressure } q_a}$$ | $$\frac{D}{q_{all.,D}} = \frac{541 \text{ kip}}{4 \text{ ksf}} = 135 \text{ ft}^2 \quad \textbf{Controls}$$ |
| | | $$\frac{(D+L)}{q_{all.,D+L}} = \frac{541 \text{ kip} + 194 \text{ kip}}{5.6 \text{ ksf}} = 131 \text{ ft}^2$$ |
| | | $$\frac{D+L+E}{q_{all.,Lat}} = \frac{541 \text{ kip} + 194 \text{ kip} + (0.7)18 \text{ kip}}{6 \text{ ksf}} = 125 \text{ ft}^2$$ |
| | Assuming a square footing. | $\ell = \sqrt{135 \text{ ft}^2} = 11.6 \text{ ft}$ |
| | The footing thickness is calculated in Step 5, footing design. | Therefore, try a 12 x 12 ft square footing. |

| | Step 4: Soil pressure | |
|---|---|---|
| | Footing stability<br>Because the column doesn't impart a moment to the footing, the soil pressure under the footing is assumed to be uniform and overall footing stability is assumed.<br><br>Calculate factored soil pressure.<br>This value is needed to calculate the footing's required strength.<br><br>$$q_u = \frac{\sum P_u}{\text{area}}$$<br><br>Calculate the soil pressures resulting from the applied factored loads. | |
| 5.3.1(a) | Load Case I: $U = 1.4D$ | $U = 1.4D = 1.4(541 \text{ kip}) = 757 \text{ kip}$<br><br>$q_u = \dfrac{757 \text{ kip}}{144 \text{ ft}^2} = 5.3 \text{ ksf}$ |
| 5.3.1(b) | Load Case II: $U = 1.2D + 1.6L$ | $U = 1.2D + 1.6L = 1.2(541 \text{ kip}) + 1.6(194 \text{ kip}) = 960 \text{ kip}$<br><br>$q_u = \dfrac{960 \text{ kip}}{144 \text{ ft}^2} = 6.7 \text{ ksf}$ **Controls** |
| 5.3.1(d) | Load Case IV: $U = 1.2D + E + L$ | $U = 1.2D + 1.0E + 1.0L = 1.2(541 \text{ kip}) + 18 \text{ kip} + 1.0(194 \text{ kip}) = 861 \text{ kip}$<br><br>$q_u = \dfrac{861 \text{ kip}}{144 \text{ ft}^2} = 6.0 \text{ ksf}$ |
| 5.3.1(e) | Load Case IV: $U = 0.9D + E$ | $U = 0.9D + 1.0E = 0.9(541 \text{ kip}) + 18 \text{ kip} = 505 \text{ kip}$<br><br>$q_u = \dfrac{505 \text{ kip}}{144 \text{ ft}^2} = 3.5 \text{ ksf}$ |
| | The load combinations include the seismic uplift force. In this example, uplift does not occur. | Note: The full definition of $E$ includes not only earthquake loads dues to overturning, but also earthquake loads due to vertical acceleration of ground as per ASCE 7-10, Section 12.4.2. |
| 13.3.2.1 | Because the footing is square, it will only be designed in one direction. | |

## Step 5: One-way shear design

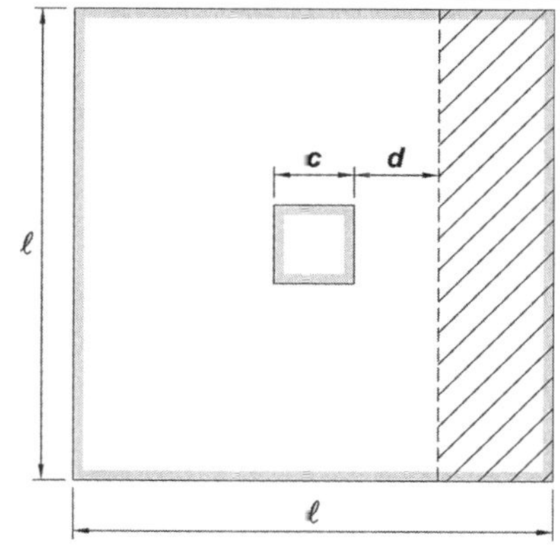

Fig. E1.2—One-way shear in longitudinal direction.

| | | |
|---|---|---|
| 21.2.1(b) 7.5.1.1 | Shear strength reduction factor: $\phi V_n \geq V_u$ | $\phi_{shear} = 0.75$ |
| 7.5.3.1 22.5.1.1 | $V_n = V_c + V_s$ | Assume $V_s = 0$ (no shear reinforcement) $V_n = V_c$ |
| 22.5.5.1 7.4.3.2 | Therefore: $V_c = 2\sqrt{f_c'}b_w d$ and satisfying: $\phi V_c \geq V_u$ | |
| 20.6.1.3.1 | The Code allows the critical section for one-way shear at a distance $d$ from the face of the column (refer to Fig. E1.2). The engineer could either assume a value for $d$ that satisfies the strength Eq. (22.5.5.1) by iteration or equate Eq. (22.5.5.1) to Eq. (7.5.1.1) and solve for $d$. In this example, the first approach is followed: Assume that the footing is 30 in. thick. The cover is 3 in. to bottom of reinforcement. Assume that No. 8 bars are used in the both directions and design for the more critical case (upper layer). Therefore, the effective depth $d$: $d = 30$ in. $- 3$ in. $- 1$ in. $- 1$ in./2 $= 25.5$ in. $\phi V_n \geq V_u = \left(\dfrac{\ell}{2} - \dfrac{c}{2} - d\right)bq_u$ | $V_u = \left(\dfrac{12\text{ ft}}{2} - \dfrac{24\text{ in.}}{2(12\text{ in./ft})} - \dfrac{25.5\text{ in.}}{12\text{ in./ft}}\right)(12\text{ ft})(6.7\text{ ksf})$ $= 231$ kip $\phi V_c = 0.75(2)\sqrt{4000\text{ psi}}\ (12\text{ ft})(25.5\text{ in.})(12\text{ in./ft})$ $= 348$ kip $\phi V_c = 348$ kip $> V_u = 231$ kip    **OK** Therefore, assumed depth is adequate: $h = 30$ in. |

## Step 6: Two-way shear design

| | | |
|---|---|---|
| | The footing will not have shear reinforcement. Therefore, the nominal shear strength for this two-way footing is simply the concrete shear strength: $v_n = v_c$ | |
| 22.6.1.2 22.6.1.4 22.6.4.1 | Under punching shear theory, inclined cracks are assumed to originate and propagate at 45 degrees away and down from the column corners. The area of concrete that resists shear is calculated at an average distance of $d/2$ from column face on all sides (refer to Fig. E1.3). | 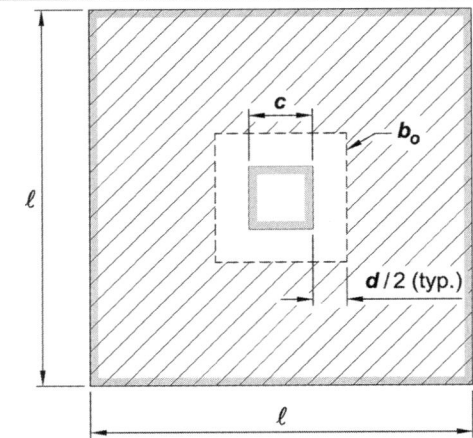 Fig. E1.3—Two-way shear. |

$$b_o = 4(c + d)$$

where $b_o$ is the perimeter of the area of shear resistance.

$b_o = 4(24 + 25.5) = 198$ in.

**22.6.2.1** ACI 318-14 permits the engineer to take the average of the effective depth in the two orthogonal directions when calculating the shear strength of the footing, but in this example the smaller effective depth is used.

**8.4.2.3.4 22.6.5.1 22.6.5.2** The two-way shear strength equations for nonprestressed footings must be satisfied and the least calculated value of (a), (b), and (c) controls:

(a) $v_c = 4\lambda\sqrt{f'_c}$

$v_c = 4(1.0)(\sqrt{4000 \text{ psi}}) = 253$ psi **Controls**

(b) $v_c = \left(2 + \dfrac{4}{\beta}\right)\lambda\sqrt{f'_c}$

$v_c = \left(2 + \dfrac{4}{1}\right)(1.0)(\sqrt{4000 \text{ psi}}) = 379.5$ psi

where $\beta$ is ratio of the long side to short side of column; $\beta = 1$.

(c) $v_c = \left(\dfrac{\alpha_s d}{b_o} + 2\right)\lambda\sqrt{f'_c}$

$v_c = \left(\dfrac{(40)(25.5 \text{ in.})}{198 \text{ in.}} + 2\right)(1.0)(\sqrt{4000 \text{ psi}}) = 452$ psi

**22.6.5.3** $\alpha_s = 40$, considered interior column

Equation (a) controls; $v_c = 253$ psi

$$V_c = 4\lambda\sqrt{f'_c} b_o d$$

$$V_c = \dfrac{4(1.0)(\sqrt{4000 \text{ psi}})(198 \text{ in.})(25.5 \text{ in.})}{1000 \text{ lb/kip}} = 1277 \text{ kip}$$

**21.2.1(b)** Use a shear strength reduction factor of 0.75:

$\phi = 0.75$

$$\phi V_c = (0.75)4\lambda\sqrt{f'_c} b_o d$$

$\phi V_c = 0.75(1277 \text{ kip}) = 958$ kip **OK**

$$V_u = q_u[(a)^2 - (c+d)^2]$$

$$V_u = (6.7 \text{ ksf})\left((12 \text{ ft})(12 \text{ ft}) - \left(\dfrac{24 \text{ in.} + 25.5 \text{ in.}}{12 \text{ in./ft}}\right)^2\right) = 851 \text{ kip}$$

**8.5.1.1** Check if design strength exceeds required strength: $\phi V_c \geq V_u$?

$\phi V_c = 958$ kip $> V_u = 851$ kip **OK**
Two-way shear strength is adequate.

# CHAPTER 11—FOUNDATIONS

| Step 7: Flexure design | | | |
|---|---|---|---|
| 13.2.7.1 | The code permits the critical section to be at the face of the column (refer to Fig. E1.4). | | |
| | | | Fig. E1.4—Flexure in the longitudinal direction. |
| | $M_u = q_u \left(\dfrac{\ell-c}{2}\right)^2 (b)/2$ | $M_u = (6.7 \text{ ksf}) \left(\dfrac{12 \text{ ft} - \dfrac{24 \text{ in.}}{12 \text{ in./ft}}}{2}\right)^2 (12 \text{ ft})/2 = 1005 \text{ ft-kip}$ | |
| 22.2.1.1 | Set concrete compression force equal to the steel tension force at the column face: $C = T$ | | |
| 22.2.2.4.1 | $C = 0.85 f_c' ba$ and $T = A_s f_y$ $a = \dfrac{A_s f_y}{0.85 f_c' b}$ and | $a = \dfrac{A_s (60 \text{ ksi})}{0.85(4 \text{ ksi})(12 \text{ ft})} = 0.15 A_s$ | |
| 7.5.2.1 | $\phi M_n = \phi A_s f_y \left(d - \dfrac{a}{2}\right)$ | | |
| 22.3.1.1 22.2.2.2 | Substitute for $a$ in the equation above. | | |
| 21.2.1(a) | Use moment strength reduction factor from Table 21.2.1. | $\phi = 0.9$ | |
| 8.5.1.1(a) | Setting $\phi M_n \geq M_u = 1005$ ft-kip and solving for $A_s$: | $\phi M_n \geq (0.9) A_s (60 \text{ ksi}) \left(25.5 \text{ in.} - \dfrac{(0.15) A_s}{2}\right)$ $A_s \geq 9.0 \text{ in.}^2$ | |
| 13.3.3.3(a) | Distribute bars uniformly across the entire 12 ft width of footing: | Use 13 No. 8 bars (13 x 0.79 = 10.27 in$^2$) distributed uniformly across the entire 12 ft width of footing. | |
| 8.6.1.1 | Check the minimum reinforcement ratio: $\rho_l = 0.0018$ | $A_{s,min} = 0.0018(12 \text{ ft})(12 \text{ in./ft})(30 \text{ in.}) = 7.8 \text{ in.}^2$ $A_{s,prov} = 10.27 \text{ in.}^2 > A_{s,min} = 7.8 \text{ in.}^2$ | |
| 21.2.1(a) | Check if the assumption of tension controlled behavior and the use of $\phi = 0.9$ is correct. | | |

| | | |
|---|---|---|
| 21.2.2 | To answer the question, the calculated tensile strain in reinforcement is compared to the values in Table 21.2.2. The strain in reinforcement is calculated from similar triangles (refer to Fig. E1.5):<br><br>$\varepsilon_t = \dfrac{\varepsilon_c}{c}(d-c)$ | $a = 0.1(13)(0.79 \text{ in.}^2) = 1.03 \text{ in.}$<br><br>$c = \dfrac{1.03 \text{ in.}}{0.85} = 1.21 \text{ in.}$<br><br>$\varepsilon_t = \dfrac{0.003}{1.21 \text{ in.}}(25.5 \text{ in.} - 1.21 \text{ in.}) = 0.06$<br><br>$\varepsilon_t = 0.06 > 0.005$ |
| 22.2.2.4.1<br>22.2.2.4.3 | where: $c = a/\beta_1$<br>and $a = 0.28 A_s$ | Section is tension controlled and the assumption of $\phi = 0.9$ is correct |
| | | 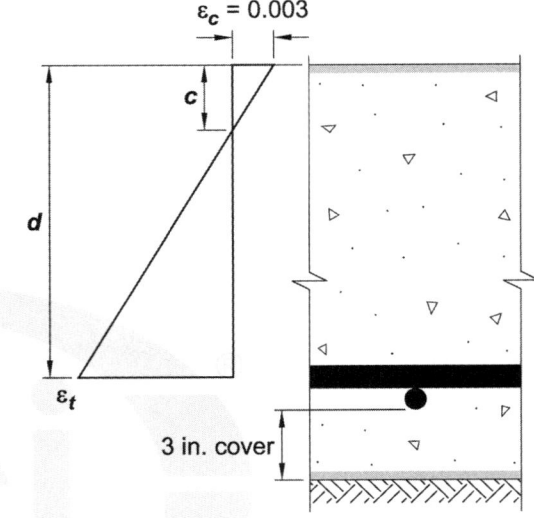<br>*Fig. E1.5—Strain distribution across footing cross section.* |
| | Verify that allowable soil pressure is not exceeded when including footing self-weight and slab self-weight and live load above footing: | |
| | Footing self-weight less soil self-weight: | $W_F = (12 \text{ ft})(12 \text{ ft})\left(\dfrac{30 \text{ in.}}{12}\right)(0.15 \text{ kcf} - 0.12 \text{ kcf}) = 10.8 \text{ kip}$ |
| | Slab self-weight and assume 40 psf live load: | $W_s = (12 \text{ ft})(12 \text{ ft})(0.5 \text{ ft}(0.15 \text{ kcf}) + 0.04 \text{ ksf}) = 16.6 \text{ kip}$ |
| | Total weight on supporting soil: | $W_T = 541 \text{ kip} + 194 \text{ kip} + 10.8 \text{ kip} + 16.6 \text{ kip} = 762.4 \text{ kip}$ |
| | Calculate actual soil pressure: | $q_a = \dfrac{762.4 \text{ kip}}{(12 \text{ ft})(12 \text{ ft})} = 5.3 \text{ ksf} < q_{all} = 5.6 \text{ ksf}$  **OK** |

# CHAPTER 11—FOUNDATIONS

## Step 8: Transfer of column forces to the base

| | | |
|---|---|---|
| 13.2.2.1<br>16.3.1.1 | Factored column forces are transferred to the footing by bearing on concrete and through reinforcement. | |
| 22.8.3.2 | The foundation is wider on all sides than the loaded area. Therefore, the nominal bearing strength, $B_n$, is the smaller of the two equations. | |
| 22.8.3.2(a) | $B_n = \sqrt{\dfrac{A_1}{A_2}}(0.85 f_c' A_1)$ | |
| | and | |
| 22.8.3.2(b) | $B_n = 2(0.85 f_c' A_1)$ | |
| | Check if $\sqrt{\dfrac{A_2}{A_1}} \le 2.0$ | $\sqrt{\dfrac{A_2}{A_1}} = \sqrt{\dfrac{[(12\ \text{ft})(12\ \text{in./ft})]^2}{(24\ \text{in.})^2}} = 6 > 2$ |
| | where $A_1$ is the bearing area of the column and $A_2$ is the area of the part of the supporting footing that is geometrically similar to and concentric with the loaded area. | Therefore, Eq. (22.8.3.2(b)) controls. |
| 21.2.1(d) | The bearing strength reduction factor is 0.65: | $\phi_{bearing} = 0.65$<br>$\phi B_n = (0.65)(2)(0.85)(4000\ \text{psi})(24\ \text{in.})^2$<br>$\phi B_n = 2546\ \text{kip} > 960\ \text{kip}$ (Step 4) **OK** |
| 16.3.4.1 | Column factored forces are transferred to the foundation by bearing and through reinforcement, usually dowels. Provide dowel area of at least $0.005 A_g$ and at least four bars. | $A_{s,dowel} = 0.005(24\ \text{in.})^2 = 2.88\ \text{in.}^2$<br>Use eight No. 6 bars |
| 16.3.5.4 | Bars are in compression for all load combinations. Therefore, the dowels must extend into the footing a compression development length, $\ell_{dc}$, the larger of the two expressions and at least 8 in. (refer to Fig. E1.6): | |
| 25.4.9.2 | $\ell_{dc} = \begin{cases} \dfrac{f_y \psi_r}{50 \lambda \sqrt{f_c'}} d_b \\ \\ (0.0003 f_y \psi_r d_b) \end{cases}$ | $\ell_{dc} = \dfrac{(60{,}000\ \text{psi})(1.0)}{50\sqrt{4000\ \text{psi}}}(0.75\ \text{in.}) = 14.3\ \text{in.}$ **Controls**<br><br>$\ell_{dc} = 0.0003(60{,}000\ \text{psi})(1.0)(0.75\ \text{in.}) = 13.5\ \text{in.}$ |
| | where<br>$\psi_r$ = confining reinforcement factor;<br>$\psi_r = 1.0$, because reinforcement is not confined | |
| | The footing depth must satisfy the following inequality so that the No. 6 dowels can be developed within the provided depth: | |
| 25.3.1 | $h \ge \ell_{dc} + r + d_{b,dwl} + 2 d_{b,bars} + 3\ \text{in.}$ | $h_{req'd} = 14.3\ \text{in.} + 6(0.75\ \text{in.}) + 0.75\ \text{in.} + 2(0.75\ \text{in.})$<br>$\qquad\qquad + 3\ \text{in.} = 24.1\ \text{in.}$ |
| | where $r$ = radius of No. 6 bent = $6 d_b$ | $h_{req'd} = 24.1\ \text{in.} < h_{prov.} = 30\ \text{in.}$ **OK** |

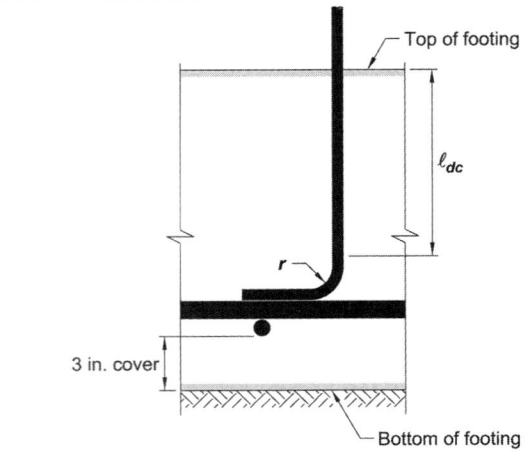

*Fig. E1.6—Reinforcement development length.*

| | | | |
|---|---|---|---|
| **Step 9: Footing details** | | | |
| 13.2.8.3<br>13.2.7.1 | Development length<br>Flexural reinforcement bar development is required at the critical section. This is the point of maximum factored moment, which occurs at the column face. Bars must extend at least a tension development length beyond the critical section. | No. 6: $\dfrac{c_b + K_{tr}}{d_b} = \dfrac{3.5 \text{ in.} + 0}{1.0 \text{ in.}} = 3.5$<br><br>Use maximum $\dfrac{c_b + K_{tr}}{d_b} = 2.5$ | |
| 25.4.2.2<br>25.4.2.4 | $\ell_d = \left( \dfrac{3}{40} \dfrac{f_y}{\lambda\sqrt{f'_c}} \dfrac{\psi_t \psi_e \psi_s}{\dfrac{c_b + K_{tr}}{d_b}} \right) d_b$<br><br>where $\psi_t$ = casting position;<br><br>$\psi_t = 1.0$ because not more than 12 in. of fresh concrete below horizontal reinforcement<br><br>$\psi_e$ = coating factor; $\psi_e = 1.0$, because bars are uncoated<br><br>$\psi_s$ = bar size factor; $\psi_s = 1.0$ for No. 7 and larger<br><br>$c_b$ = spacing or cover dimension to center of bar, whichever is smaller<br><br>$K_{tr}$ = transverse reinforcement index<br>It is permitted to use $K_{tr} = 0$.<br><br>However, the expression: $\dfrac{c_b + K_{tr}}{d_b}$ must not exceed 2.5. | $\ell_d = \left( \dfrac{3}{40} \dfrac{60{,}000 \text{ psi}}{(1.0)\sqrt{4000 \text{ psi}}} \dfrac{(1.0)(1.0)(1.0)}{2.5} \right) d_b = 28.5 d_b$<br><br>No.8 bars: $\ell_d = 28.5(1 \text{ in.}) = 28.5 \text{ in.} > 12 \text{ in.}$<br>Therefore, OK.<br><br>$\ell_d$ in the longitudinal direction:<br>$\ell_{d,prov.} = ((12 \text{ ft})(12 \text{ in./ft}) - 24 \text{ in.})/2 - 3 \text{ in.}$<br>$\ell_{d,prov.} = 57 \text{ in.} > \ell_{d,req'd} = 28.5 \text{ in.}$ **OK**<br><br>Use straight No. 8 bars in both directions. | |

Step 10: Detailing

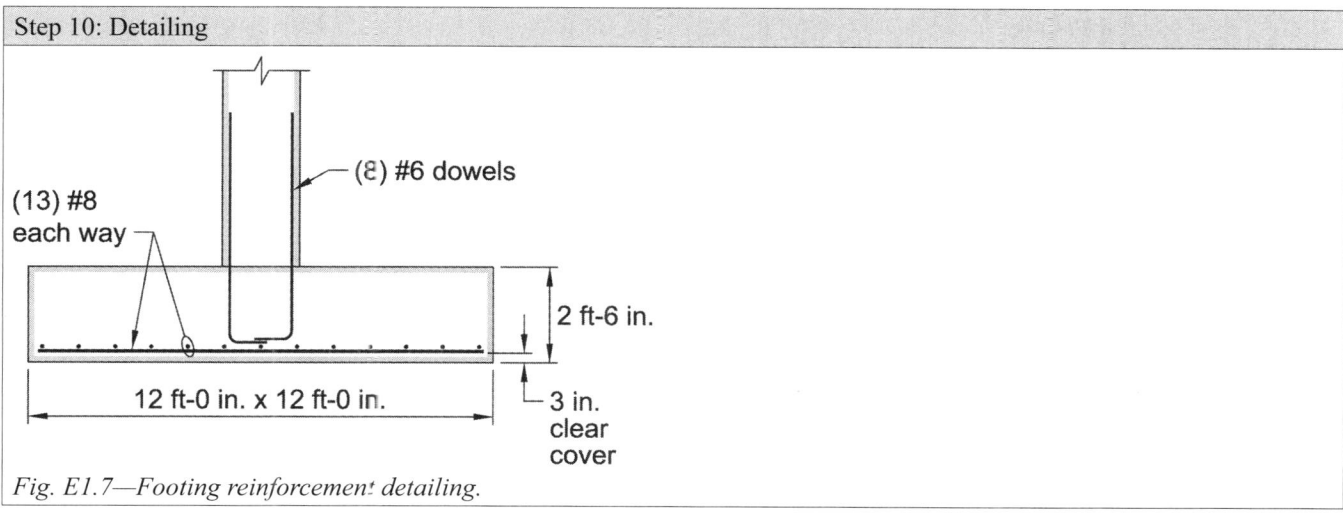

Fig. E1.7—Footing reinforcement detailing.

**Foundation Example 2:** *Design of a continuous footing*

Design and detail of a continuous footing, founded on stiff soil, supporting a 12 in. concrete wall. The footing is located in Seismic Category D and is 3 ft-0 in. below finished grade. Exposure to freezing and thawing is not an issue (refer to Fig. E2.1).

Given:

*Wall load—*
Service dead load $D$ = 25 kip/ft
Service live load $L$ = 12.5 kip/ft
Wind OT $W$ = ±6.4 kip/ft
(Wall vertical force due to resisting wind loads)
Seismic OT $E$ = ±6 kip/ft
(Wall vertical force due to resisting seismic forces)
Note: the wall has no out-of-plane moments or shears.

*Material properties—*
Concrete compressive strength $f_c'$ = 4000 psi
Steel yield strength $f_y$ = 60,000 psi
Normalweight concrete $\lambda$ = 1
Density of concrete = 150 lb/ft$^3$

*Allowable soil-bearing pressures—*
D only: $q_{all,D}$ = 3000 psf
$D + L$: $q_{all,D+L}$ = 4000 psf
$D+L+W$: $q_{all,W}$ = 5000 psf
$D+L+E$: $q_{all,E}$ = 5000 psf

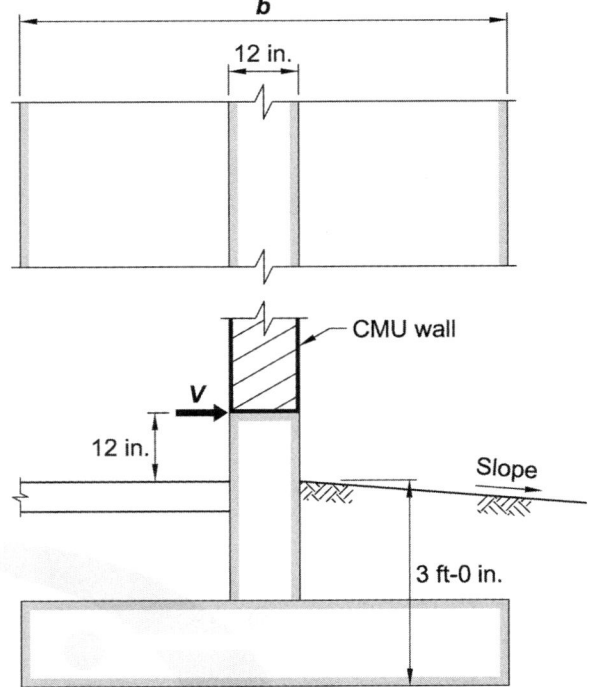

Fig. E2.1—*Plan and elevation of continuous footing.*

| ACI 318-14 | Discussion | Calculation |
|---|---|---|
| **Step 1: Foundation type** | | |
| 13.1.1 | This strip footing is 3 ft below finished grade. Therefore, it is considered a shallow foundation. | |
| 13.3.2.1 | The footing will be designed and detailed with the applicable provisions of Chapter 7, One-way slabs, and Chapter 9, Beams, of ACI 318-14. | |
| 13.2.3.1 | Foundation resisting earthquake forces must comply with Section 18.2.2.3 of ACI 318-14. | |

| | | | |
|---|---|---|---|
| **Step 2: Material requirements** | | | |
| 13.2.1.1 | The mixture proportion must satisfy the durability requirements of Chapter 19 and structural strength requirements (ACI 318-14). | | By specifying that the concrete mixture shall be in accordance with ACI 301 and providing the exposure classes, Chapter 19 requirements are satisfied. |
| | The designer determines the durability classes. Please see Chapter 4 of this Handbook for an in-depth discussion of the categories and classes. | | Based on durability and strength requirements, and experience with local mixtures, the compressive strength of concrete is specified at 28 days to be at least 4000 psi. |
| | ACI 301 is a reference specification that is coordinated with ACI 318-14. ACI encourages referencing ACI 301 into job specifications. | | |
| | There are several mixture options within ACI 301, such as admixtures and pozzolans, which the designer can require, permit, or review if suggested by the contractor. | | |
| | Example 1 of this chapter provides a more detailed breakdown on determining the concrete compressive strength and exposure categories and classes. | | |
| **Step 3: Determine footing dimensions** | | | |
| 13.3.1.1 | To calculate the footing width, divide the service load per foot by the allowable soil pressure. Load combinations are obtained from ASCE7-10, Section 2.4. $$\text{area of footing} = \frac{\text{total service load} \left(\sum P \text{ kip/ft}\right)}{\text{allowable soil pressure } q_a}$$ The footing thickness is calculated in Step 4, footing design. | | Ignoring the footing self-weight: $$\frac{D}{q_{all,D}} = \frac{25 \text{ kip/ft}}{3 \text{ ksf}} = 8.3 \text{ ft}$$ $$\frac{D+L}{q_{all,D+L}} = \frac{25 \text{ kip/ft} + 12.5 \text{ kip/ft}}{4 \text{ ksf}} = 9.4 \text{ ft} \quad \textbf{Controls}$$ $$\frac{D + 0.75L + (0.75)(0.6)W}{q_{all,Lat}}$$ $$= \frac{25 \text{ kip/ft} + (0.75)(12.5 \text{ kip/ft}) - (0.75)(0.6)(6.4 \text{ kip/ft})}{5 \text{ ksf}} = 7.5 \text{ ft}$$ $$\frac{D + 0.75L + (0.75)(0.7)E}{q_{all,Lat}}$$ $$= \frac{25 \text{ kip/ft} + (0.75)(12.5 \text{ kip/ft}) - (0.75)(0.7)(6.0 \text{ kip/ft})}{5 \text{ ksf}} = 7.5 \text{ ft}$$ Use $B = 10$ ft |

| | Step 4: Footing design | |
|---|---|---|
| | Wall stability<br>Because there is no out-of-plane moment, the soil pressure under the footing is assumed to be uniform and overall wall stability is assumed. | |
| 7.4.2.1 | The footing cantilevers on both sides of the wall are designed as one-way slabs. | |
| 5.3.1 | Calculate soil pressure<br>*Factored loads—*<br>Calculate the soil pressures resulting from the applied factored loads. | |
| | Load Case I: $U = 1.4D$ | $U = 1.4D = 1.4(25\text{ kip/ft}) = 35\text{ kip/ft}$ or 3.89 ksf |
| | Load Case II: $U = 1.2D + 1.6L$ | $U = 1.2D + 1.6L = 1.2(25\text{ kip/ft}) + 1.6(12.5\text{ kip/ft})$<br>$= 50\text{ kip/ft}$ or 5.26 ksf **Controls** |
| | Load Case IV: $U = 1.2D + W + L$ | $U = 1.2D + 1.0W + 1.0L$<br>$= 1.2(25\text{ kip/ft}) + 6.4\text{ kip/ft} + 12.5\text{ kip/ft}$<br>$= 48.9\text{ kip/ft}$ or 5.15 ksf |
| | Load Case IV: $U = 0.9D + W$ | $U = 0.9D + 1.0W = 0.9(25\text{ kip/ft}) + 6.4\text{ kip/ft}$<br>$= 28.9\text{ kip/ft}$ or 3.04 ksf |
| | Load Case V: $U = 1.2D + E + L$ | $U = 1.2D + 1.0E + 1.0L$<br>$= 1.2(25\text{ kip/ft}) + 6\text{ kip/ft} + 12.5\text{ kip/ft}$<br>$= 48.5\text{ kip/ft}$ or 5.11 ksf |
| | Load Case VI: $U = 0.9D + E$ | $U = 0.9D + 1.0E = 0.9(25\text{ kip/ft}) + (6\text{ kip/ft})$<br>$= 28.5\text{ kip/ft}$ or 3.0 ksf |

## CHAPTER 11—FOUNDATIONS

| | | |
|---|---|---|
| 21.2.1(b) 7.5.1.1 | *One-way shear design—* Shear strength reduction factor: $\phi V_n \geq V_u$ | $\phi_{shear} = 0.75$ |
| 7.5.3.1 22.5.1.1 | $V_n = V_c + V_s$ | Assume $V_s = 0$ (no shear reinforcement) |
| 22.5.5.1 | Therefore: $V_n = 2\sqrt{f_c'} b_w d$ | $V_n = V_c$ |
| 22.5.1.2 | And satisfying: $V_u \leq \phi V_n$ | 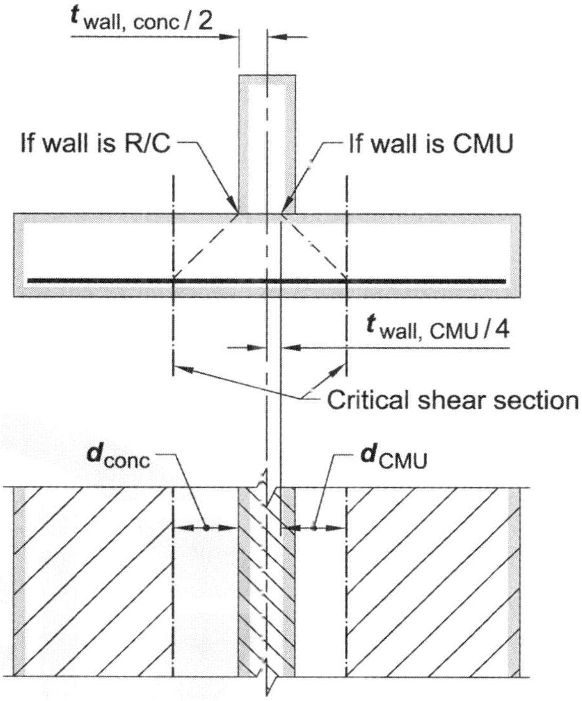 |
| 7.4.3.2 | $V_u$ is calculated at $d$ from the face of the wall. The engineer can either assume a value for $d$ that satisfies the strength Eq. (22.5.5.1) by iteration or equate Eq. (7.5.1.1) to Eq. (22.5.5.1) and solve for $d$. | |
| 13.2.7.1 13.2.7.2 | Note that for a masonry wall, the critical section is located halfway between center and face of masonry wall. | |
| 7.4.3.2 | In this example the second approach is followed: $V_u = \left( \dfrac{B}{2} - \dfrac{t_{wall}}{2} - d \right) q_{all.} \leq \phi V_c$ | $V_u = \left( \dfrac{10 \text{ ft}}{2} - \dfrac{1 \text{ ft}}{2} - d \right)(5.26 \text{ ksf})$  $d$ in ft $\phi V_c = 0.75(2)(\sqrt{4000 \text{ psi}})(d)$, where $d$ is in inches Solving these two equations for $d = 15.0$ in., use 16.5 in. $> 6$ in. |
| 13.3.1.2 | Bottom reinforcement must have an effective depth of more than 6 in. | |
| 20.6.1.3.1 | Concrete cover must satisfy Table 20.6.1.3.1. Use $c = 3$ in. Total footing depth: $h = d + d_b + c$ | Assume No. 8 bars $h = 16.5$ in. $+ 0.5$ in. $+ 3$ in. $= 20$ in. |

Fig. E2.2—Shear critical section.

| | | |
|---|---|---|
| | Flexure design<br>Note: Masonry wall is shown in Fig. E2.2 and E2.3 to indicate that for masonry walls, the critical section for shear or moment are not the face of the masonry wall, but is $t_w/4$ from the face. | 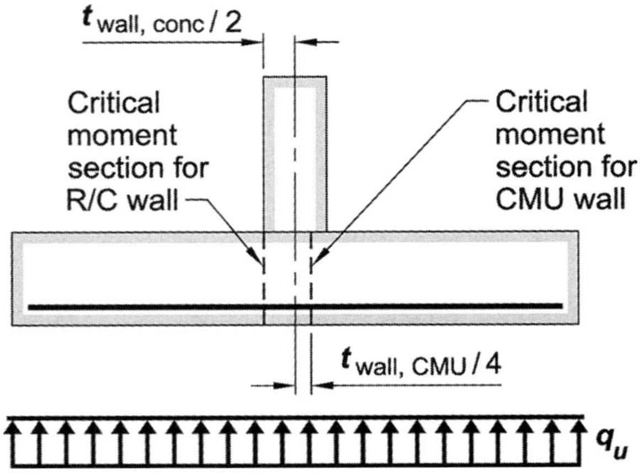 |
| 13.2.7.1 | For concrete walls, the factored moment and moment strength are calculated at the face of the wall. From maximum factored load $M_u$ is: | |
| 22.2.2.4.1 | Set the concrete compression strength equal to the steel tension strength:<br><br>$C = T$; $0.85f_c'ba = A_s f_y$ | |
| 7.5.1.1 | $\phi M_n \geq M_u$ | |

Fig. E2.3—Moment critical sections.

$M_u = (5.26 \text{ ksf})(10 \text{ ft}/2 - 1.0 \text{ ft}/2)^2/2 = 53.3 \text{ ft-kip/ft}$

$C = 0.85(4000 \text{ psi})(12 \text{ in.})a = A_s(60{,}000 \text{ psi})$

$a = 1.47 A_s \text{ in.}$

$(0.9)(60 \text{ ksi})A_s \left(16.5 \text{ in.} - \dfrac{1.47 A_s}{2}\right) = 53.3 \text{ ft} - \text{kip/ft}$

$A_s \geq 0.74 \text{ in.}^2$

Use bottom bars No. 8 at 12 in. on center. If these bars are not hooked, provide calculations to justify the use of straight bars.

| | | |
|---|---|---|
| 7.6.1.1 | Check if $A_s$ exceeds the minimum:<br>$A_{s,min} \geq 0.0018 A_g$ | $A_{s,min} = 0.0018(12 \text{ in.})(20 \text{ in.})$<br>$= 0.43 \text{ in.}^2/\text{ft} < 0.74 \text{ in.}^2/\text{ft}$ **OK** |
| 21.2.1(a) | Check if the section is tension controlled and the use of $\phi = 0.9$ is correct. | |
| 21.2.2 | To answer the question, the tensile strain in reinforcement must be calculated and compared to the values in Table 21.2.2. The strain in reinforcement is assumed to be proportional to the distance from neutral axis calculated from similar triangles (refer to Fig. E2.4): | $a = 1.47 A_s = (1.47)(0.79 \text{ in.}^2) = 1.16 \text{ in.}$<br><br>$c = \dfrac{1.16 \text{ in.}}{0.85} = 1.37 \text{ in.}$ |
| 22.2.1.2 | $\varepsilon_t = \dfrac{\varepsilon_c}{c}(d - c)$<br><br>where: $c = a/\text{and } a = 1.47 A_s$ | $\varepsilon_t = \dfrac{0.003}{1.37 \text{ in.}}(16.5 \text{ in.} - 1.37 \text{ in.}) = 0.033$<br><br>$\varepsilon_t = 0.033 > 0.005$<br>Section is tension controlled and $\phi = 0.9$ |

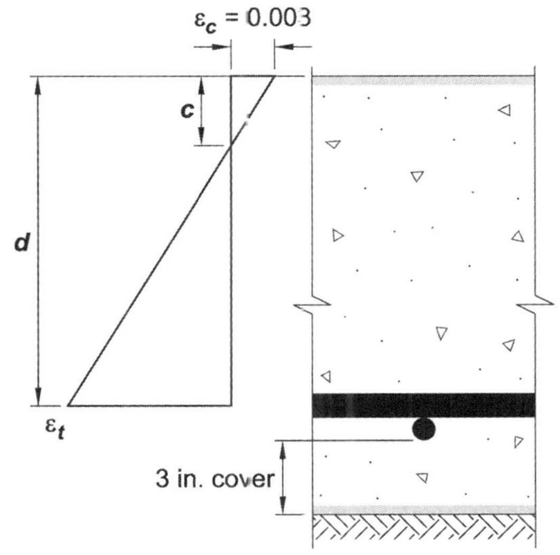

## CHAPTER 11—FOUNDATIONS

*Fig. E2.4—Strain distribution across footing.*

| | | |
|---|---|---|
| | Verify that the allowable soil pressure is not exceeded when including footing self-weight and soil self-weight above footing: | |
| | Footing self-weight: | $W_F = (10 \text{ ft}) \left( \dfrac{20 \text{ in.}}{12} \right) (0.15 \text{ kip/ft}^3 - 0.12 \text{ kip/ft}^3) = 0.5 \text{ kip}$ |
| | Soil weight above footing: | $W_s = (10 \text{ ft}) \left( \dfrac{36 \text{ in.} - 20 \text{ in.}}{12} \right) (0.12 \text{ kip/ft}^3) = 1.6 \text{ kip}$ |
| | Total weight on supporting soil: | $W_T$ = 25 kip/ft + 12.5 kip/ft + 0.5 kip/ft + 1.6 kip/ft <br> = 39.6 ft |
| | Calculate actual soil pressure: | $q_a = \dfrac{39.6 \text{ kip/ft}}{10 \text{ ft}} = 3.96 \text{ ksf} < q_{all} = 4 \text{ ksf}$    **OK** |
| | Note: This is a conservative approach. The footing concrete displaces soil. Therefore, the actual load on soil is the difference between the concrete and soil unit weights multiplied by the footing volume. | |

| | | Step 5: Footing details | |
|---|---|---|---|
| 7.6.4.1<br>7.6.1.1 | | Shrinkage and temperature reinforcement along length of footing<br><br>The area of shrinkage and temperature reinforcement:<br><br>$A_{S+T} \geq 0.0018 A_g$ | $A_{S+T} = (0.0018)(20 \text{ in.})(10.0 \text{ ft})(12 \text{ in./ft}) = 4.3 \text{ in.}^2$<br><br>Ten No. 6 bottom longitudinal bars will satisfy the requirement for shrinkage and temperature reinforcement in the long direction. |
| 25.4.2.3 | | Development length<br>Check if the width of the footing provides adequate length for the bottom tension reinforcement beyond the critical tension section.<br><br>$\ell_d = \left( \dfrac{3}{40} \dfrac{f_y}{\lambda \sqrt{f_c'}} \dfrac{\psi_t \psi_e \psi_s}{\dfrac{c_b + K_{tr}}{d_b}} \right) d_b$ | $\ell_d = \left( \dfrac{3}{40} \dfrac{60{,}000 \text{ psi}}{(1.0)\sqrt{4000 \text{ psi}}} \dfrac{(1.0)(1.0)(1.0)}{2.5} \right) d_b$<br><br>$= 28.5 d_b = 28.5 \text{ in.} > 12 \text{ in.}$ |
| 25.4.2.4 | | where<br>$\psi_t$ = casting position;<br>$\psi_t$ = 1.0 because not more than 12 in. of fresh concrete is placed below horizontal reinforcement<br>$\psi_e$ = coating factor;<br>$\psi_e$ = 1.0, because bars are uncoated<br>$\psi_s$ = bar size factor;<br>$\psi_s$ = 1.0 because bars are larger than No. 7<br>$c_b$ = spacing or cover dimension to center of bar, whichever is smaller | |
| 25.4.2.3 | | $K_{tr}$ = transverse reinforcement index<br><br>It is permitted to use $K_{tr} = 0$.<br><br>However, the expression: $\dfrac{c_b + K_{tr}}{d_b}$ must not exceed 2.5.<br><br>Provided length: $B/2 - t_{wall}/2 - 3 \text{ in.}$ | $\dfrac{c_b + K_{tr}}{d_b} = \dfrac{3.5 \text{ in.} + 0}{1.0 \text{ in.}} = 3.5$<br><br>Use maximum value of 2.5<br><br>$\ell_{avail.} = (10 \text{ ft})(12 \text{ in./ft})/2 - 12 \text{ in.}/2 - 3 \text{ in.} = 51 \text{ in.}$<br>$\ell_{avail.} = 51 \text{ in.} > \ell_d = 28.5 \text{ in.}$  **OK**<br><br>Therefore, the footing is wide enough to use straight bars for development and does not require hooks at both ends. |

## Step 6: Earthquake requirements

| | | |
|---|---|---|
| 13.2.3.2 | **Earthquake load effects**<br>The foundation is in SDC D therefore, ACI 318-14, Section 18.13, must be satisfied. | |
| 18.13.2 | The requirements listed in 18.13.2 for structural walls must be satisfied if calculations show that uplift occurs: | |
| 18.13.2.1 | (a) Vertical reinforcement of structural walls resisting forces induced by earthquake effects must extend into the footing and must be fully developed for tension at the interface. | |
| 18.13.2.3 | (b) Boundary elements of special structural walls that have an edge within one-half the footing depth from an edge of the footing shall have transverse reinforcement in accordance with 18.7.5.2 through 18.7.5.4 provided below the top of the footing. This reinforcement must extend into the footing and be developed a length equal to the development length, calculated for $f_y$ in tension, of the boundary element longitudinal reinforcement. This condition does not apply for this problem. | |
| 18.13.2.4 | (c) Where earthquake effects create uplift forces in boundary elements of special structural walls, flexural reinforcement must be provided in the top of the footing to resist actions resulting from the design factored load combinations, and must be less than required by 7.6.1 or 9.6.1. This condition does not apply for this problem. | |

## Step 7: Detailing

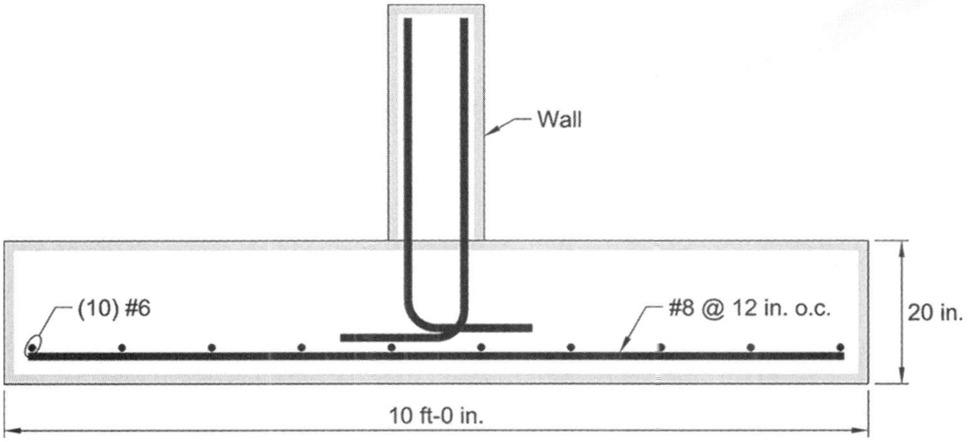

*Fig. E2.5—Continuous footing reinforcement.*

**Foundation Example 3:** *Design of a continuous footing with an out-of-plane moment*

Design and detail a continuous footing, founded on stiff soil, supporting a 12 in. thick bearing wall, founded on stiff soil, and subject to loading that includes an overturning moment. The bottom of the footing is 3 ft below finished grade (refer to Fig. E3.1).

Given:

*Wall load—*
Service dead load = 15 kip/ft (including CMU wall weight)
Horizontal wind shear $V$ = 3.0 kip/ft
(strength level applied at 1 ft above grade)

*Material properties—*
Concrete compressive strength $f_c'$ = 4000 psi
Steel yield strength $f_y$ = 60,000 psi
Normalweight concrete $\lambda$ = 1
Density of concrete = 150 lb/ft$^3$

*Soil data—*
$q_{all}$ = 4000 psf
$q_{u,permitted}$ = 6000 psf
Density of soil = 100 lb/ft$^3$

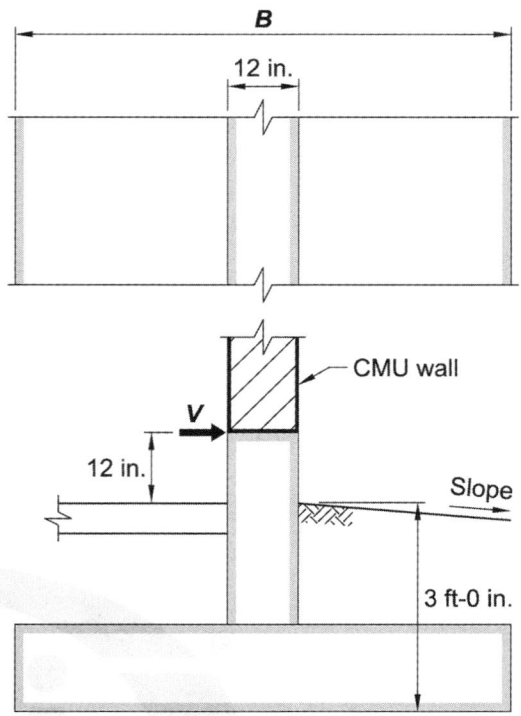

*Fig. E3.1—Plan and elevation of continuous footing.*

| ACI 318-14 | Discussion | Calculation |
|---|---|---|
| **Step 1: Foundation type** | | |
| 13.1.1 | This strip footing is 3 ft below finished grade. Therefore, it is considered a shallow foundation. | |
| 13.3.2.1 | The footing will be designed and detailed with the applicable provisions of Chapter 7, One-way slabs, and Chapter 9, Beams, of ACI 318-14. | |

| | Step 2: Material requirements | |
|---|---|---|
| 13.2.1.1 | The mixture proportion must satisfy the durability requirements of Chapter 19 and structural strength requirements (ACI 318-14).<br><br>The designer determines the durability classes. Please see Chapter 4 of this Handbook for an in-depth discussion of the categories and classes<br><br>ACI 301 is a reference specification that is coordinated with ACI 318-14. ACI encourages referencing ACI 301 into job specifications.<br><br>There are several mixture options within ACI 301, such as admixtures and pozzolans, which the designer can require, permit, or review if suggested by the contractor.<br><br>Example 1 of this chapter provides a more detailed breakdown on determining the concrete compressive strength and exposure categories and classes. | By specifying that the concrete mixture shall be in accordance with ACI 301 and providing the exposure classes, Chapter 19 requirements are satisfied.<br><br>Based on durability and strength requirements, and experience with local mixtures, the compressive strength of concrete is specified at 28 days to be at least 4000 psi. |
| | Step 3: Determine footing dimensions | |
| 13.3.1.1 | Footing width is assumed and then verified through calculations. Iterations may be needed.<br><br>The footing thickness is also assumed and then verified through calculations in Step 4, Footing design. | Try $B = 7$ ft footing width.<br>Area: $A = 1(7) = 7$ ft²/ft<br>Section modulus: $S = 1(7\text{ ft})(7\text{ ft})/6 = 8.167$ ft³/ft |
| 13.3.1.2 | The footing thickness must be such that the bottom reinforcement has an effective depth of at least 6 in. | Try 15 in. footing thickness. |

## Step 4: Footing design

### Wall stability

Because there is an out-of-plane (overturning) lateral force on the stem wall, the overall wall stability must be checked.

To calculate the stability of a footing, the total vertical load is calculated and the resisting moment ($M_R$) is compared to the resulting overturning moment ($M_{OTM}$).

Commonly, engineers require $M_R \geq 1.5 M_{OTM}$ to consider a footing stable.

Weights on bearing soil below footing:

Weight of footing:

$$W_{ftg} = \left(\frac{15 \text{ in.}}{12 \text{ in./ft}}\right)(0.15 \text{ kip/ft}^3) = 0.19 \text{ ksf}$$

Weight of soil above footing:

$$W_{soil} = \left(\frac{36 \text{ in.} - 15 \text{ in.}}{12 \text{ in./ft}}\right)(0.10 \text{ kip/ft}^3) = 0.18 \text{ ksf}$$

Weight of concrete wall pier:

$$W_{conc.pier} = \left(\frac{36 \text{ in.} - 15 \text{ in.}}{12 \text{ in./ft}}\right)(0.15 \text{ kip/ft}^3) = 0.26 \text{ ksf}$$

Total vertical dead load:

$$\sum P = (0.19 \text{ ksf})(7 \text{ ft}) + (0.18 \text{ ksf})(7 \text{ ft} - 1 \text{ ft}) + (0.26 \text{ ksf})(1 \text{ ft}) + 15 \text{ kip/ft}$$
$$= 2.67 \text{ kip/ft} + 15 \text{ kip/ft} = 17.7 \text{ kip/ft}$$

Vertical distance from bottom of footing to location of applied lateral wind shear.

$$H = 3 \text{ ft} + 1 \text{ ft} = 4 \text{ ft}$$

The overturning moment, $M_{OTM}$, is measured at base of footing. The lateral wind force must be multiplied by 0.6 (ASCE7-10 Section 2.4.1) to convert to service load level.

$$M_{OTM} = (0.6)(W)(H) = (0.6)(3.0 \text{ kip/ft})(4 \text{ ft})$$
$$= 7.2 \text{ ft-kip/ft (wind load)}$$

The resisting moment, $M_R$, is calculated as the product of vertical load by distance from the centerline to edge of footing:

$M_R = P(B/2)$

$M_R = (17.7 \text{ kip/ft})(7 \text{ ft}/2) = 61.8 \text{ ft-kip/ft}$
$61.8 \text{ ft-kip/ft} > (1.5)(7.2 \text{ ft-kip/ft}) = 10.8 \text{ ft-kip/ft}$ **OK**

To ensure footing stability, the following inequality must be satisfied:

$M_R > 1.5 M_{OTM}$

| Step 5: Calculate soil pressure | | |
|---|---|---|
| 13.3.1.1 | Service loads<br>The maximum soil pressure is calculated from service forces and moments transmitted by foundation to the soil.<br>To calculate soil pressure, the location of the vertical service resultant force is determined.<br>The distance to the resultant from the front face of stem:<br><br>$$e = \frac{M_{otm}}{\Sigma P}$$<br><br>Check if resultant falls within the middle third (kern) of the footing.<br><br>Because $e \leq B/6$, the footing imposes compression to the soil across the entire width.<br><br>The resulting soil pressure must be less than the allowable bearing pressure provided by the geotechnical report.<br><br>Maximum and minimum soil pressures are calculated by:<br><br>$$q_{1,2} = \frac{\Sigma P}{A} \pm \frac{M_{OTM}}{S}$$ | $$e = \frac{8.4 \text{ ft-kip}}{17.7 \text{ kip}} = 0.47 \text{ ft}$$<br><br>$B/6 = 7 \text{ ft}/6 = 1.17 \text{ ft} > e = 0.47 \text{ ft}$ **OK**<br><br><br><br><br><br><br><br>$$q_{1,2} = \frac{17.7 \text{ kip}}{(7 \text{ ft})(1 \text{ ft})} \pm \frac{8.4 \text{ ft-kip}(6)}{(1 \text{ ft})(7 \text{ ft})^2}$$<br>$q_{max} = 2.53 \text{ ksf} + 1.03 \text{ ksf} = 3.56 \text{ ksf} < q_{all} = 4 \text{ ksf}$<br>$q_{min} = 2.52 \text{ ksf} - 1.03 \text{ ksf} = 1.49 \text{ ksf} > 0 \text{ ksf}$ **OK** |

| Step 6: Factored loads | | | |
|---|---|---|---|
| 13.3.2.1 | | The footing is designed as one-way slab. Calculate the soil pressures resulting from the applied factored loads. | |
| 5.3.1a | | Load Case I: <br> $U = 1.4D$ <br> Use $D = P = 17.7$ kip and <br> $M_{OTM} = W = 12$ ft-kip | $U = 1.4(17.7 \text{ kip/ft}) = 24.7$ kip/ft <br> $q_u = (24.7 \text{ kip/ft}) / (7 \text{ ft}) = 3.53$ ksf $< q_u = 6$ ksf  **OK** |
| 5.3.1d | | Load Case II: <br> $U = 1.2D/A + 1.0W/S + 0.5L/A$ <br> where $S$ is the section modulus (Step 2) <br><br> $e = 1.0(W)/(1.2(P))$ <br><br> Because $e < B/6$, the footing bearing pressure varies as follows (refer to Fig. E3.2): <br><br> $q_u = 1.2(D/A) \pm 1.0(W/S)$ | $1.2D/A = 1.2(17.7 \text{ kip/ft})/(7 \text{ ft}) = 3.03$ ksf <br> $1.0W/S = 1.0(12 \text{ ft-kip})/(8.167 \text{ ft}^3) = 1.47$ ksf <br> $0.5L = 0$ <br><br> $e = 1.0(12 \text{ ft-kip})/(1.2(17.7 \text{ kip})) = 0.56$ ft $< (7 \text{ ft})/6$ <br> $= 1.67$ ft <br><br><br> $q_{u,max} = 3.04$ ksf $+ 1.47$ ksf $= 4.51$ ksf (maximum) <br> $q_{u,min} = 3.04$ ksf $- 1.47$ ksf $= 1.57$ ksf (minimum) <br> $q_{u,max} = 4.51$ ksf $< q_{u,permitted} = 6$ ksf  **OK** |
| 5.3.1f | | Load Case III: <br> $U = 0.9D/A + 1.0W/S$ <br><br> $e = 1.0(W)/(1.2(P))$ <br><br> Because $e < B/6$, <br> bearing pressure $q_u = 0.9(D/A) \pm 1.0(W/S)$ | $0.9D/A = 0.9(17.7 \text{ kip/ft})/(7 \text{ ft}) = 2.28$ ksf <br> $1.0W/S = 1.0(12 \text{ ft-kip})/(8.167 \text{ ft}^3) = 1.47$ ksf <br> $e = 1.0(12 \text{ ft-kip})/(0.9(17.7 \text{ kip})) = 0.75$ ft <br><br><br> $q_{1,2} = 2.27$ ksf $\pm 1.5$ ksf <br> $q_{u,max} = 3.75$ ksf (maximum) $< q_{u,permitted} = 6$ ksf  **OK** <br> $q_{u,min} = 0.81$ ksf (minimum) |

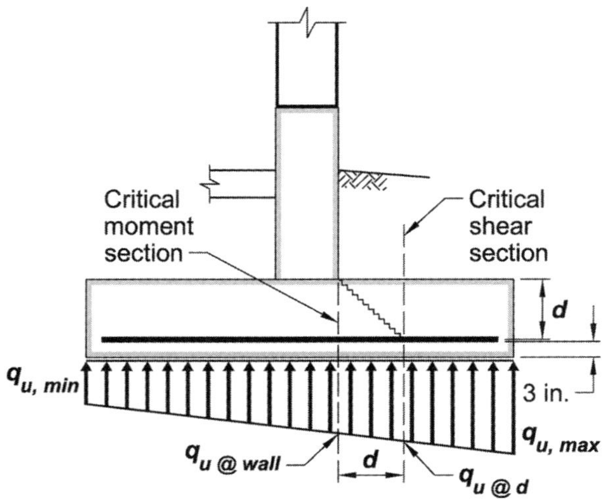

Fig. E3.2—Soil pressure distribution under factored loads, critical shear section, and critical moment section.

# CHAPTER 11—FOUNDATIONS

## Step 7: Shear strength

| | One-way shear design | |
|---|---|---|
| 21.2.1(b) | Shear strength reduction factor: | $\phi_{shear} = 0.75$ |
| 7.5.1.1 | $\phi V_n \geq V_u$ | Assume $V_s = 0$ (no shear reinforcement) |
| 7.5.3.1 | $V_n = V_c + V_s$ | $V_n = V_c$ |
| 22.5.1.1 22.5.5.1 | $V_c = 2\lambda \sqrt{f'_c} b_w d$ $\lambda = 1$ | |
| 20.6.1.3.1a | Effective depth: $d$ = height − cover − $d_b/2$ Assume $d_b$ = 1 in. and $c$ = 3 in. Therefore: $\phi V_c = \phi 2\lambda \sqrt{f'_c} b_w d$ Calculate factored soil pressure at distance $d$ from face of wall: $q_{u,d} = q_{u,min} + \left(\dfrac{q_{u,max} - q_{u,min}}{B}\right)\left(\dfrac{B}{2} + \dfrac{t_{wall}}{2} + d\right)$ | $d$ = 15 in. − 3 in. − 0.5 in. = 11.5 in. $\phi V_c = 0.75(2)\left(\sqrt{4000 \text{ psi}}\right)(12 \text{ in.})(11.5 \text{ in.}) = 13.1$ kip/ft $q_{u,d} = 1.57 \text{ ksf} + \left(\dfrac{4.51 \text{ ksf} - 1.57 \text{ ksf}}{7 \text{ ft}}\right)\left(\dfrac{7 \text{ ft}}{2} + \dfrac{1 \text{ ft}}{2} + \dfrac{11.5 \text{ in.}}{12 \text{ in./ft}}\right)$ = 3.66 ksf Note that $q_{u,max}$ and $q_{u,min}$ are from Step 6, Load Case II. |
| | Calculate factored shear force at $d$ from face of wall: $V_u$ Note: weight of footing and earth above footing are conservatively not deducted from shear force created by $q_u$. | $V_u = \dfrac{4.51 \text{ ksf} + 3.66 \text{ ksf}}{2}(3.5 \text{ ft} - 0.5 \text{ ft} - 0.96 \text{ ft}) = 8.3$ kip/ft |
| 7.4.3.2 22.5.1.2 | Check if: $\phi V_c \geq V_u$ | $\phi V_c = 13.1$ kip/ft $> V_u = 8.3$ kip/ft |

| Step 8: Flexural strength | | |
|---|---|---|
| 13.2.7.1 | Flexure design<br>The footing factored moment is calculated at the face of the wall (refer to Fig. E3.3). | *Fig. E3.3—Moment critical section.* |
| 5.3.1a | Calculate $M_u$ at face of wall from<br>$U = 1.4D/A$.<br>$q_u = 3.53$ ksf    (Step 6) | $M_u = \dfrac{3.53 \text{ ksf}}{2}(3.5 \text{ ft} - 0.5 \text{ ft})^2 = 15.9 \text{ ft-kip/ft}$ |
| 5.3.1d | Calculate $M_u$ at face of wall:<br>$U = 1.2D/A + 1.0W/S + 0.5L/A$<br><br>$q_{u,wall} = q_{u,min} + \dfrac{q_{u,max} - q_{u,min}}{B}\left(\dfrac{B}{2} + \dfrac{t_{wall}}{2}\right)$<br><br>Factored moment:<br><br>$M_u = \dfrac{1}{2}q_{u,wall}\left(\dfrac{B}{2} - \dfrac{t_{wall}}{2}\right)^2$<br>$+ \dfrac{1}{3}(q_{u,max} - q_{u,wall})\left(\dfrac{B}{2} - \dfrac{t_{wall}}{2}\right)^2$ | $q_u = 1.57 \text{ ksf} + \dfrac{4.51 \text{ ksf} - 1.57 \text{ ksf}}{7 \text{ ft}}(3.5 \text{ ft} + 0.5 \text{ ft})$<br>$= 3.25 \text{ ksf}$<br><br>$M_u = \dfrac{3.25 \text{ ksf}}{2}(3.5 \text{ ft} - 0.5 \text{ ft})^2$<br>$+ \dfrac{4.51 \text{ ksf} - 3.25 \text{ ksf}}{3}(3.5 \text{ ft} - 0.5 \text{ ft})^2 = 18.4 \text{ ft-kip/ft}$<br>**Controls** |
| 5.3.1f | By inspection load condition<br>$U = 0.9D/A + 1.0W/S$ does not control.<br><br>Note: Counteracting moments due to footing weight and soil weight are conservatively neglected. | |
| 22.2 | Calculate the required area of flexural reinforcement:<br><br>Set concrete compression strength equal to steel tension strength<br><br>$C = T$ | |
| 22.2.2.4.1<br>22.2.2.3<br>22.2.2.4.3 | $0.85f_c'ba = A_s f_y$ | $0.85(4000 \text{ psi})(12 \text{ in.})a = A_s(60,000 \text{ psi})$<br>$a = 1.47 A_s$ |
| 7.5.1.1<br>22.3.1.1 | $M_u \leq \phi M_n = 0.9 A_s f_y(d - a/2)$ | $\phi M_n = 0.9 A_s(60,000 \text{ psi})(12 \text{ in.} - 1.47 A_s/2)$<br>$\geq 19.38 \text{ ft-kip/ft}$<br><br>$A_s \geq 0.37 \text{ in.}^2/\text{ft}$ |

| | | |
|---|---|---|
| 9.6.1.2 | Check $A_s$ against the minimum: $$A_{s,min} = \frac{200}{f_y}bd$$ | $A_{s,min} = \frac{200}{60,000 \text{ psi}}(12 \text{ in.})(12 \text{ in.}) = 0.475 \text{ in.}^2/\text{ft}$ <br> $A_{s,min} = 0.475 \text{ in.}^2/\text{ft} > A_{s,req'd} = 0.37 \text{ in.}^2/\text{ft}$ <br> Use No. 7 at 15 in. on center bottom bars. <br> $A_{s,prov.} = 0.48 \text{ in.}^2/\text{ft} > A_{s,req'd} = 0.475 \text{ in.}^2/\text{ft}$ **OK** |
| 21.2.1(a) | Check if the section is tension controlled and the use of $\phi = 0.9$ is correct. | |
| 21.2.2 | To answer the question, the tensile strain in reinforcement is calculated and compared to the values in Table 21.2.2. The strain in reinforcement is calculated from similar triangles (refer to Fig. E3.4): | |
| 22.2.1.2 | $$\varepsilon_t = \frac{\varepsilon_c}{c}(d-c)$$ where: $c = a/\beta_1$ and $a = 1.47A_s$ | $c = \frac{1.47(0.48 \text{ in.}^2)}{0.85} = 0.83 \text{ in.}$ <br><br> $\varepsilon_t = \frac{0.003}{0.83 \text{ in.}}(12 \text{ in.} - 0.83 \text{ in.}) = 0.040$ <br><br> $\varepsilon_t = 0.040 > 0.005$ Section is tension controlled and $\phi = 0.9$ |

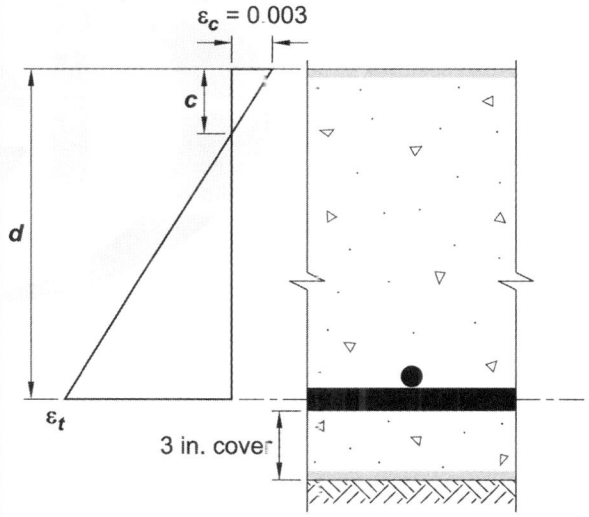

Fig. E3.4—Strain distribution through depth of footing.

## Step 9: Footing details

| | | |
|---|---|---|
| 7.6.4.1<br>24.4.3.2<br>13.2.8.3 | Shrinkage and temperature reinforcement:<br><br>The area of shrinkage and temperature reinforcement:<br>$A_{S+T} \geq 0.0018 A_g$ | $A_{S+T} = (0.0018)(15\text{ in.})(7\text{ ft})(12\text{ in.}) = 2.27\text{ in.}^2$<br><br>Eight No. 5 bottom longitudinal bars (area = 2.48 in.²) satisfies the requirement for shrinkage and temperature reinforcement placed perpendicular to the flexural reinforcement. |
| 13.2.7.1 | Development length<br>Reinforcement development is calculated at the maximum factored moment and the code permits the critical section to be located at the wall face. Bars must extend at least a tension development length beyond the critical section. | |
| 25.4.2.3 | $\ell_d = \left( \dfrac{3}{40} \dfrac{f_y}{\lambda \sqrt{f'_c}} \dfrac{\psi_t \psi_e \psi_s}{\dfrac{c+K_{tr}}{d_b}} \right) d_b$ | $\ell_d = \left( \dfrac{3}{40} \dfrac{60{,}000\text{ psi}}{(1.0)\sqrt{4000\text{ psi}}} \dfrac{(1.0)(1.0)(1.0)}{2.5} \right) d_b$<br>$= 28.5 d_b = 28.5(0.875\text{ in.}) = 25\text{ in.}$ |
| 25.4.2.4 | where<br>$\psi_t$ = casting location;<br>$\psi_t$ = 1.0, because not more than 12 in. of fresh concrete is placed below horizontal reinforcement<br>$\psi_e$ = coating factor;<br>$\psi_e$ = 1.0, because bars are uncoated<br>$\psi_s$ = bar size factor;<br>$\psi_s$ = 1.0, because bars are larger than No. 7<br>$c_b$ = spacing or cover dimension to center of bar, whichever is smaller<br>$K_{tr}$ = transverse reinforcement index | |
| 25.4.2.3 | It is permitted to use $K_{tr} = 0$.<br><br>But the expression: $\dfrac{c_b + K_{tr}}{d_b}$ must not be taken greater than 2.5. | For a No. 7 bar:<br>$\dfrac{c_b + K_{tr}}{d_b} = \dfrac{3.44\text{ in.} + 0}{0.875\text{ in.}} = 3.93$ |
| 25.4.2.1 | The development length is the greater of the calculated value of Eq. (25.4.2.3) and 12 in.<br><br>Check if No. 7 can be developed using straight bars, without hooks. | Use maximum value of 2.5<br>$\ell_d$ = 25 in. 12 in.  **OK**<br><br>$\ell_d$ provided perpendicular to the wall:<br>$\ell_{d,prov.} = ((7\text{ ft})(12\text{ in./ft}) - 12\text{ in.})/2 - 3\text{ in.}$<br>$\ell_{d,prov.} = 33\text{ in.} > \ell_{d,req'd} = 25\text{ in.}$  **OK**<br>use straight No. 7 bars |

## CHAPTER 11—FOUNDATIONS

Step 10: Final design

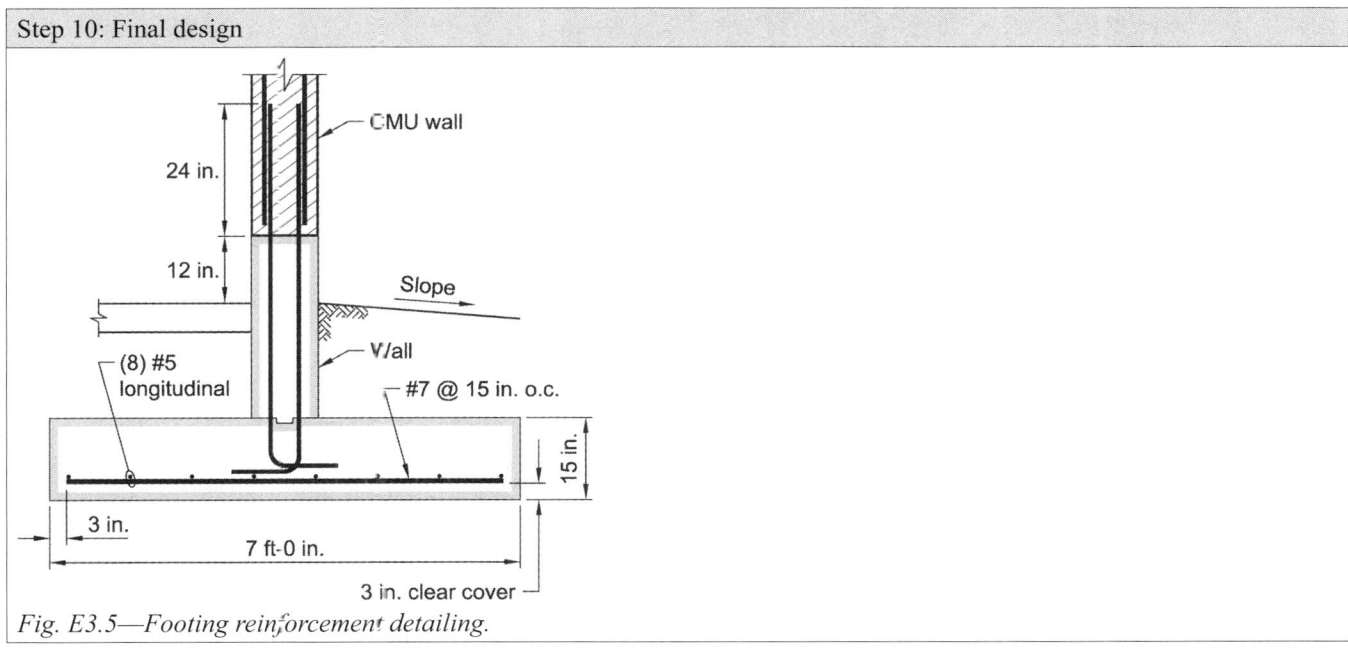

Fig. E3.5—Footing reinforcement detailing.

**Foundation Example 4**—*Design of a rectangular spread footing*

Design and detail a rectangular spread footing founded on stiff soil, supporting an 18 in. square column. The bottom of the footing is 3 ft below finished grade (refer to Fig. E4.1).

Given:

*Column load*—
Service dead load $D$ = 200 kip
Service live load $L$ = 100 kip
Factored wind $W$ = ±175 kip
(Axial column force due to the building frame resisting the wind load)

*Material properties*—
Concrete compressive strength $f_c'$ = 4000 psi
Steel yield strength $f_y$ = 60,000 psi
Normalweight concrete $\lambda$ = 1
Density of concrete = 150 lb/ft$^3$
Density of Soil = 120 lb/ft$^3$

*Allowable service level soil bearing pressures*—
$D$ only: $q_{all,D}$ = 4000 psf
$D + L$: $q_{all,D+L}$ = 5800 psf
$D + L + W$: $q_{all,Lat}$ = 8000 psf

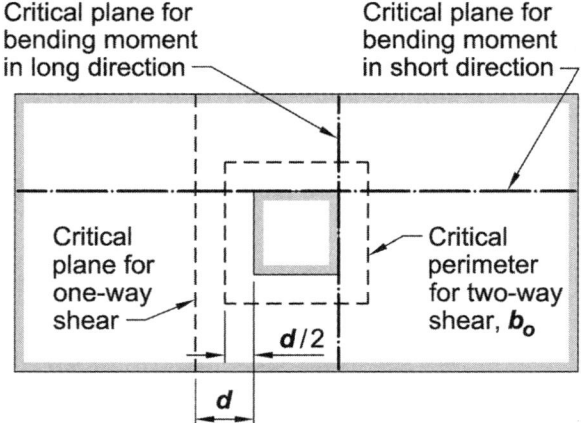

Fig. E4.1—Rectangular footing plan.

| ACI 318-14 | Procedure | Computation |
|---|---|---|
| Step 1: Foundation type | | |
| 13.1.1 | This footing is 3 ft below finished grade. Therefore, it is considered a shallow footing. | |
| 13.3.3.1 | The footing will be designed and detailed with the applicable provisions of Chapter 7, One-way slabs, and Chapter 8, Two-way slabs, of ACI 318-14. | |

# CHAPTER 11—FOUNDATIONS

| | Step 2: Material requirements | |
|---|---|---|
| 13.2.1.1 | The mixture proportion must satisfy the durability requirements of Chapter 19 and structural strength requirements (ACI 318-14). | By specifying that the concrete mixture shall be in accordance with ACI 301-10 and providing the exposure classes, Chapter 19 requirements are satisfied. |
| | The designer determines the durability classes. Please see Chapter 4 of this Handbook for an in-depth discussion of the categories and classes. | Based on durability and strength requirements, and experience with local mixtures, the compressive strength of concrete is specified at 28 days to be at least 4000 psi. |
| | ACI 301 is a reference specification that is coordinated with ACI 318-14. ACI encourages referencing ACI 301 into job specifications. | |
| | There are several mixture options within ACI 301, such as admixtures and pozzolans, which the designer can require, permit, or review if suggested by the contractor. | |
| | Example 1 of this chapter provides a more detailed breakdown on determining the concrete compressive strength and exposure categories and classes. | |
| | **Step 3: Determine footing dimensions** | |
| 13.3.1.1 | To calculate the footing area, divide the service load by the allowable soil pressure. | The unit weights of concrete and soil are 150 pcf and 120 pcf; close. Therefore, footing self-weight will be ignored for initial sizing. Actual soil pressure is checked end of Step 6: |
| | area of footing $\geq \dfrac{\text{total service load }(\Sigma P)}{\text{allowable soil pressure } q_a}$ | $\dfrac{D}{q_{all,D}} = \dfrac{200 \text{ kip}}{4 \text{ ksf}} = 50 \text{ ft}^2$ |
| 5.3.4 | $W_{service} = (0.6)W = (0.6)(175 \text{ kip}) = 105 \text{ kip}$ | $\dfrac{(D+L)}{q_{all,D+L}} = \dfrac{200 \text{ kip} + 100 \text{ kip}}{5.8 \text{ ksf}} = 51.7 \text{ ft}^2$ |
| | | $\dfrac{D+L+W}{q_{all,Lat}} = \dfrac{200 \text{ kip} + 100 \text{ kip} + 105 \text{ kip}}{8 \text{ ksf}} = 50.6 \text{ ft}^2$ **Controls** |
| | The lateral wind force must be multiplied by 0.6 (ASCE 7-10, Section 2.4.1) to convert to service load level. | |
| | Assume that there is a constraint on the width of the footing ($B$) of 5.5 ft. | $(50.6 \text{ ft}^2)/(5.5 \text{ ft}) = 9.2 \text{ ft}$ say 10 ft<br>Use ($\ell \times B$) 10 ft x 5.5 ft |
| | The footing thickness is calculated in Step 4, footing design. | $A_{prov.} = 55 \text{ ft}^2 > A_{req'd} = 50.6 \text{ ft}^2$<br>$B/\ell = (55 \text{ ft})/(10 \text{ ft}) = 0.55$ |

| | Step 4: Factored soil pressure | |
|---|---|---|
| 13.3.3.2 | Footing stability<br>Because there is no out-of-plane moment, the soil pressure under the footing is assumed to be uniform and overall footing stability is assumed.<br>The footing is designed for flexure as one-way slab (Step 5) and checked for two-way punching shear (Step 6).<br>Calculate soil pressure<br>$$q_u = \frac{\sum P_u}{area}$$<br>Factored loads<br>Calculate the soil pressures resulting from the column factored loads. | |
| 5.3.1(a) | Load Case I: $U = 1.4D$ | $U = 1.4(200 \text{ kip}) = 280 \text{ kip}$<br>$$q_u = \frac{280 \text{ kip}}{55 \text{ ft}^2} = 5.1 \text{ ksf}$$ |
| 5.3.1(b) | Load Case II: $U = 1.2D + 1.6L$ | $U = 1.2(200 \text{ kip}) + 1.6(100 \text{ kip}) = 400 \text{ kip}$<br>$$q_u = \frac{400 \text{ kip}}{55 \text{ ft}^2} = 7.3 \text{ ksf}$$ |
| 5.3.1(d) | Load Case IV: $U = 1.2D + W + L$ | $U = 1.2(200 \text{ kip}) + (1.0)(175 \text{ kip}) + 1.0(100 \text{ kip}) = 515 \text{ kip}$<br>$$q_u = \frac{515 \text{ kip}}{55 \text{ ft}^2} = 9.4 \text{ ksf} \quad \textbf{Controls}$$ |
| 5.3.1(e) | Load Case IV: $U = 0.9D + W$ | $U = 0.9(200 \text{ kip}) + 1.0(175 \text{ kip}) = 355 \text{ kip}$<br>$$q_u = \frac{355 \text{ kip}}{55 \text{ ft}^2} = 6.6 \text{ ksf}$$ |
| | The load combinations include the possibility of wind uplift force. In this example, uplift does not occur. | Assume that the calculated $q_u = 9.4$ ksf is acceptable per the geotechnical report. |
| The footing is rectangular in plan. Therefore, it needs to be designed in both directions. Of course, the longer direction will have larger moments and thus is the more critical condition. | | |

# CHAPTER 11—FOUNDATIONS

## Step 5: One-way shear design

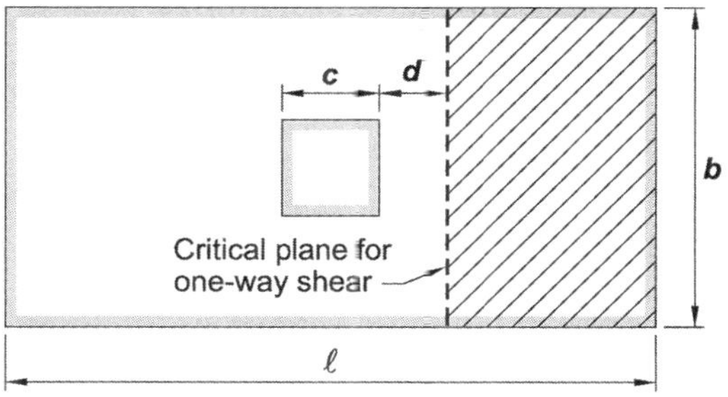

Fig. E4.2—One-way shear in longer direction.

| | | | |
|---|---|---|---|
| 21.2.1(b) | | **Long direction**<br>Shear strength reduction factor: | $\phi_{shear} = 0.75$ |
| 7.5.1.1<br>7.5.3.1<br>22.5.1.1 | | $\phi V_n \geq V_u$<br>$V_n = V_c + V_s$ | Assume $V_s = 0$ (no shear reinforcement)<br>$V_n = V_c$<br>and satisfying: $\phi V_c \geq V_c$ |
| 13.2.7.2<br>7.4.3.2 | | The critical section for one-way shear is permitted at a distance $d$ from the face of the column (refer to Fig. E4.2). | |
| | | The engineer assumes a value of $d$ then checks strength by Eq. (22.5.5.1). If Eq. (7.5.1.1) is not satisfied, a new value of $d$ is selected. | |
| 13.3.1.2 | | Assume that the footing is 28 in. thick. | |
| 20.6.1.3.1 | | The cover is 3 in. to bottom of reinforcement. Assume that No. 6 bars are used in the long direction. Therefore, the effective depth $d$ is: | |
| | | $d = h - \text{cover} - d_b/2$ | $d = 28 \text{ in.} - 3 \text{ in.} - (0.75 \text{ in.})/2 = 24.625 \text{ in.}$<br>say, $d = 24.5$ in. |
| | | $V_u = \left(\dfrac{\ell}{2} - \dfrac{c}{2} - d\right) b q_u$ | $V_u = \left(\dfrac{10 \text{ ft}}{2} - \dfrac{18 \text{ in.}}{2(12 \text{ in./ft})} - \dfrac{24.5 \text{ in.}}{12 \text{ in./ft}}\right)(5.5 \text{ ft})(9.4 \text{ ksf})$<br>$= 114.0$ kip |
| 22.5.5.1 | | and<br>$\phi V_c = 2\sqrt{f_c'} b_w d$ | $\phi V_c = 0.75(2)\sqrt{4000 \text{ psi}} \ (5.5 \text{ ft})(24.5 \text{ in.})(12 \text{ in./ft})$<br>$/(1000 \text{ lb/kip}) = 153.4$ kip |
| 7.5.1.1 | | Is $\phi V_n > V_u$? | $\phi V_c = 153.4$ kip $> V_u = 114$ kip  **OK** |

| | | |
|---|---|---|
| | Short direction<br>Check one-way shear.<br>The effective depth for the short direction is: | 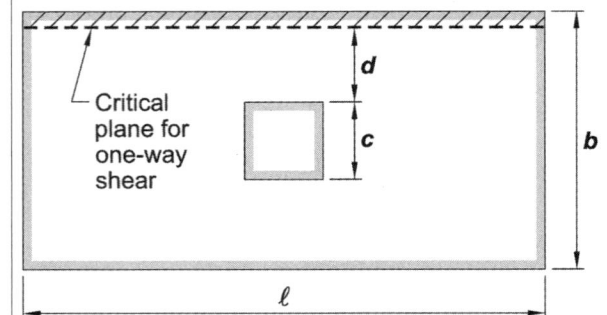 |
| 20.6.1.3.1 | $d = h - \text{cover} - d_b - d_b/2$<br>$d = 28 \text{ in.} - 3 \text{ in.} - 0.75 \text{ in.} - 0.375 \text{ in.}$<br>$d = 23.875 \text{ in.}$, say, $d = 23.5 \text{ in.}$ | |
| 13.2.7.2<br>7.5.1.1 | $\phi V_n \geq V_u = \left(\dfrac{b}{2} - \dfrac{c}{2} - d\right) \ell q_u$ | |
| | | *Fig. E4.3—One-way shear in short direction.* |
| 7.4.3.2 | Distance of critical shear plane from center of footing (refer to Fig. E4.3): | $\dfrac{c}{2} + d = \dfrac{18 \text{ in.}}{2(12 \text{ in./ft})} + \dfrac{23.5 \text{ in.}}{12 \text{ in./ft}} = 2.71 \text{ ft}$ |
| | Half of footing width: | $\dfrac{b}{2} = \dfrac{5.5 \text{ ft}}{2} = 2.75 \text{ ft}$ |
| | | Therefore, one-way shear in the short direction is OK by inspection because the critical shear plane is at the edge of the footing. |

## Step 6: Two-way shear design

**13.3.3.1, 13.2.7.2** — The footing will be without shear reinforcement. Therefore, the nominal shear strength for two-way punching shear is equal to the concrete strength:

**8.5.1.1**
$$V_n = V_c + V_s$$

Footings are usually designed with $V_s = 0$

**22.6.1.2, 22.5.1.1**
$$v_n = v_c$$

**22.6.2.1** — ACI 318 permits the engineer to take the average of the effective depth in the two orthogonal directions when designing the footing.

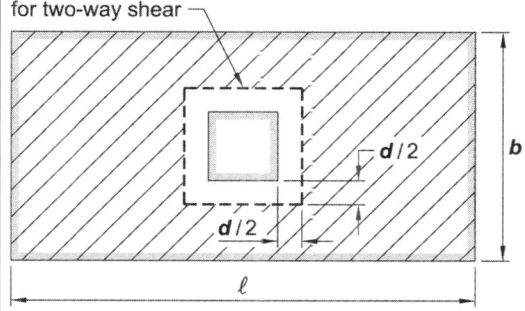

Fig. E4.4—Two-way shear.

$$d = \frac{24.5 \text{ in.} + 23.5 \text{ in.}}{2} = 24 \text{ in.}$$

**13.2.7.1, 22.6.1.4, 22.6.4.1** — Under punching shear theory, inclined cracks are assumed to originate and propagate at 45 degrees away and down from the column corners. The critical section location, $b_o$, is calculated at an average distance of $d/2 = 12$ in. from face of the column on all sides (refer to Fig. E4.4).

$$b_o = 4(c + d)$$

$$b_o = 4(18 + 24) = 168 \text{ in.}$$

**8.4.2.3.4, 22.6.5.1, 22.6.5.2** — Two-way nominal shear strength is the least value of (a), (b), and (c):

(a) $v_c = 4\lambda\sqrt{f_c'}$

$v_c = 4(1.0)(\sqrt{4000 \text{ psi}}) = 253$ psi  **Controls**

(b) $v_c = \left(2 + \dfrac{4}{\beta}\right)\lambda\sqrt{f_c'}$

$v_c = \left(2 + \dfrac{4}{1}\right)(1.0)(\sqrt{4000 \text{ psi}}) = 379.5$ psi

where $\beta$ is ratio of the long side to short side of column; $\beta = 1$

(c) $v_c = \left(\dfrac{\alpha_s d}{b_o} + 2\right)\lambda\sqrt{f_c'}$

$v_c = \left(\dfrac{(40)(24.5 \text{ in.})}{166 \text{ in.}} + 2\right)(1.0)(\sqrt{4000 \text{ psi}}) = 500$ psi

**22.6.5.3** — $\alpha_s = 40$, considered interior column

Equation (a) controls; (a) < (b) < (c); $v_c = 253$ psi

**22.6.5.2(a)**
$$V_c = 4\lambda\sqrt{f_c'}b_o d$$

$$V_c = \frac{(253 \text{ psi})(166 \text{ in.})(24 \text{ in.})}{1000 \text{ lb/kip}} = 1008 \text{ kip}$$

**21.2.1(b)** — Use a shear strength reduction factor of 0.75:

$\phi = 0.75$

$$\phi V_c = (0.75)4\lambda\sqrt{f_c'}b_o d$$

$\phi V_c = 0.75(1008 \text{ kip}) = 756$ kip

$$V_u = q_u((\ell)(B) - (c + d)^2)$$

$$V_u = (9.4 \text{ ksf})\left[(5.5 \text{ ft})(10 \text{ ft}) - \left(\frac{18 \text{ in.} + 24 \text{ in.}}{12 \text{ in./ft}}\right)^2\right]$$

$$= 402 \text{ kip}$$

**8.5.1.1** — Check if design strength exceeds required strength:

$$\phi V_c \geq V_u?$$

$\phi V_c = 756$ kip $> V_u = 402$ kip  **OK**
Two-way shear strength is adequate.

| | |
|---|---|
| Calculate the service-level soil pressure:<br><br>$B = 5.5$ ft, $L = 10$ ft, $h = 28$ in.<br>Weight of displaced soil by the footing.<br><br>Weight of soil above footing:<br><br>The footing weight is added to the dead load: | $W_{ftg} = (5.5\text{ ft})(10\text{ ft})(28\text{in.}/12)(0.15\text{ kcf} - 0.12\text{ kcf})$<br>$= 3.85$ kip, say, 4 kip<br><br>$W_{soil} = \left(\dfrac{36\text{ in.} - 15\text{ in.}}{12\text{ in./ft}}\right)(5.5\text{ ft})(10\text{ ft})(0.120\text{ kip/ft}^3)$<br>$= 4.4$ kip<br><br>$D^* = 200\text{ kip} + 4\text{ kip} + 4.4\text{ kip} = 208.4\text{ kip}$<br><br>$\dfrac{D^* + L + W}{A_{total}} = \dfrac{208.4\text{ kip} + 100\text{ kip} + 105\text{ kip}}{55\text{ ft}^2} = 7.5\text{ ksf}$<br><br>Allowable soil pressure: $q_{all} = 8$ ksf   **OK** |

# CHAPTER 11—FOUNDATIONS

## Step 7: Flexure design

**13.2.7.1** — Long direction

The long direction in the rectangular footing will generate larger moments because of the longer moment arm. The critical section is permitted to be at the face of the column (refer to Fig. E4.5).

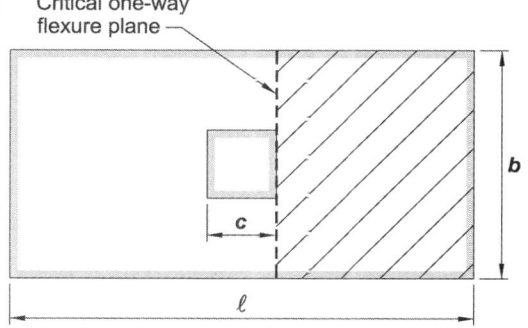

Fig. E4.5—Flexure in the long direction.

$$M_u = q_u \left(\frac{\ell - c}{2}\right)^2 (B)/2$$

$$M_u = (9.4 \text{ ksf}) \left(\frac{10 \text{ ft} - \dfrac{18 \text{ in.}}{12 \text{ in./ft}}}{2}\right)^2 (5.5 \text{ ft})/2 = 467 \text{ ft-kip}$$

**22.2.2.4.1** — Set the concrete compression strength equal to the steel tension strength: $C = T$

$$a = \frac{A_s f_y}{0.85 f_c' b} = 0.28 A_s$$

**22.2.2.4**
**22.2.2.4.3** — $C = 0.85 f_c' b a$ and $T = A_s f_y$
and $f_c' = 4000$ psi

$\beta_1 = 0.85$

**7.5.2.1**
**22.3.1.1**
$$\phi M_n = \phi A_s f_y \left(d - \frac{a}{2}\right)$$

Substitute $0.28 A_s$ for $a$ in the equation above.

**21.2.1(a)** — Use flexural strength reduction factor from Table 21.2.1
$\phi = 0.9$

**8.5.1.1(a)** — Setting $\phi M_n \geq M_u$ and solving for $A_s$, where $M_u = 467$ ft-kip

$$0.9(467 \text{ ft-kip}) = (0.9) A_s (60 \text{ ksi}) \left(24 \text{ in.} - \frac{(0.28) A_s}{2}\right)$$

$A_s \geq 4.0$ in.$^2$

**8.6.1.1** — Check the minimum area:
$A_{s,min} = 0.0018 A_g$

$A_{s,min} = 0.0018(5.5 \text{ ft})(12 \text{ in./ft})(24.5 \text{ in.})$
$= 2.9$ in.$^2 < A_{s,req'd} = 3.96$ in.$^2$ **OK**

Use nine No. 6 bars distributed uniformly across the width of footing.

**13.3.3.3(a)**
**21.2.1(a)**
**21.2.2** — Reinforcement in the longitudinal direction is uniformly distributed.

**22.2.2.4.1**
**22.2.2.4.3** — Check if the section is tension controlled and the use of $\phi = 0.9$ is correct.

To answer the question, the tensile strain in reinforcement is calculated and compared to the values in Table 21.2.2. The strain in reinforcement is calculated from similar triangles (refer to Fig. E4.6):

$$c = \frac{0.28(8)(0.6 \text{ in.}^2)}{0.85} = 1.58 \text{ in.}$$

$$\varepsilon_t = \frac{\varepsilon_c}{c}(d - c)$$

$$\varepsilon_t = \frac{0.003}{1.58 \text{ in.}}(24.5 \text{ in.} - 1.58 \text{ in.}) = 0.044$$

where $c = a/\beta_1$
and $a = 0.28 A_s$

$\varepsilon_t = 0.044 > 0.005$

As assumed, section is tension controlled and $\phi = 0.9$

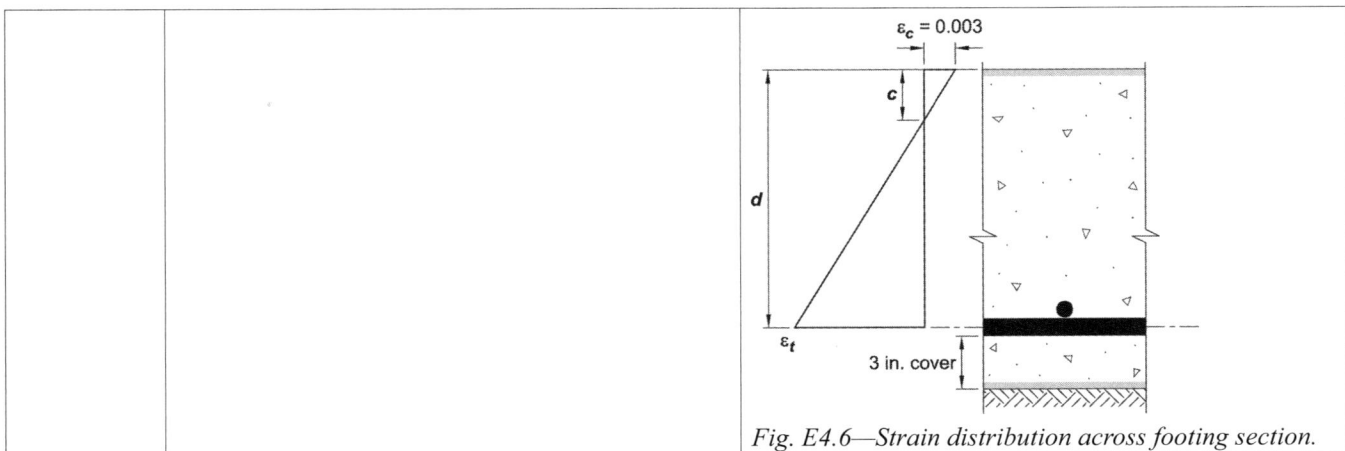

Fig. E4.6—Strain distribution across footing section.

| | Short direction | |
|---|---|---|
| | Calculate moment in the short direction, at the column face. | 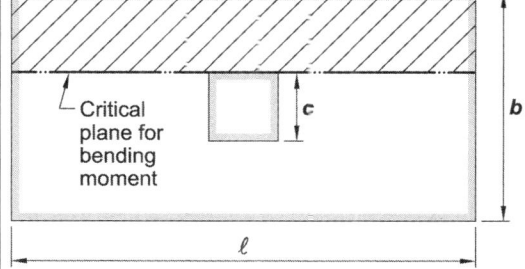 |

Fig. E4.7—Flexure in the short direction.

| | | |
|---|---|---|
| | Note: the effective depth is less than that calculated for the long direction: $d = 23.5$ in. | |
| 13.2.7.1 | $M_u = q_u \dfrac{(b-c)^2}{2}(\ell)/2$ | $M_u = (9.4 \text{ ksf})\left(\dfrac{5.5 \text{ ft} - \dfrac{18 \text{ in.}}{12 \text{ in./ft}}}{2}\right)^2 (10 \text{ ft})/2 = 188$ ft-kip |
| 22.2.2.4 | Set compression force equal to tension force at the column face:<br>$C = T$ | |
| 22.2.2.4.1 | $C = 0.85f_c'ba$ and $T = A_sf_y$ | $a = \dfrac{A_s f_y}{0.85 f_c' b} = 0.147 A_s$ |
| 22.2.2.4.3 | $f_c' = 4000$ psi | $\beta_1 = 0.85$ |
| 7.5.2.1 | $\phi M_n = \phi A_s f_y \left(d - \dfrac{a}{2}\right)$ | |
| 22.3.1.1 | Substitute $0.147A_s$ for $a$ | |
| 21.2.1(a) | Use flexural strength reduction factor from Table 21.2.1. | $\phi = 0.9$ |
| 8.5.1.1(a) | Setting $\phi M_n = M_u$ and solving for $A_s$: | $188 \text{ ft-kip} = (0.9)A_s(60 \text{ ksi})\left(24.0 \text{ in.} - \dfrac{0.147 A_s}{2}\right)$<br>$A_s = 1.75$ in.$^2$ |
| 8.6.1.1 | Check minimum reinforcement area:<br>$A_{s,min} = 0.0018 A_g$ | $A_{s,min} = 0.0018(10 \text{ ft})(12 \text{ in./ft})(28 \text{ in})$<br>$= 6.0$ in.$^2$ $> A_{s,req'd} = 1.75$ in.$^2$<br>Use minimum required reinforcement:<br>$A_s = 6.0$ in.$^2$ |
| 13.3.3.3(b) | In the short direction, a portion of the reinforcement ($\gamma_s A_s$) is distributed within a band width centered on the column. | |
| | $\gamma_s = \dfrac{2}{\beta + 1}$ | $\gamma_s = \dfrac{2}{\dfrac{10 \text{ ft}}{5.5 \text{ ft}} + 1} = 0.71$ |
| | The band width is equal to the length of the short side (5.5 ft). | Reinforcement area in 5.5 ft band width =<br>$(6.0 \text{ in.}^2)(0.71) = 4.26$ in.$^2$<br>Use ten No. 6 bars distributed uniformly across the 5.5 ft band width |
| | The remaining reinforcement $(1 - \gamma_s)A_s$ is distributed equally on both sides outside the band width. The remaining area of reinforcement must be at least the minimum reinforcement with the bars spacing not exceeding the smaller of $3h$ or 18 in. | Reinforcement area outside the central band<br>$= (6.0 \text{ in.}^2) - 10(0.44 \text{ in.}^2) = 1.6$ in.$^2$ |

| | | |
|---|---|---|
| 7.6.1.1 | The area of reinforcement outside the band width must, however, satisfy at least the minimum flexural reinforcement:<br>$A_{s,min} = 0.0018 A_g$ | $A_{s,min} = 0.0018(10 \text{ ft} - 5.5 \text{ ft})/2(12 \text{ in./ft})(28 \text{ in})$<br>$= 1.36 \text{ in.}^2 > A_{s,req'd} = 1.6 \text{ in.}^2/2 = 0.8 \text{ in.}^2$<br>Use three No. 6 bars each side distributed uniformly outside the band width.<br>$A_{s,min} = 1.32 \text{ in.}^2 \approx A_{s,min} = 1.36 \text{ in.}^2$ **OK** |
| **Step 8: Column-to-footing connection** | | |
| 16.3.1.1 | Vertical factored column forces are transferred to the footing by bearing on concrete and the reinforcement, usually dowels. | |
| 22.8.3.2 | The footing is wider on all sides than the loaded area. Therefore, the nominal bearing strength, $B_n$, is the lesser of the two equations. | |
| 22.8.3.2(a) | $B_n = (0.85 f_c' A_1)\sqrt{\dfrac{A_2}{A_1}}$<br>and | |
| 22.8.3.2(b) | $B_n = 2(0.85 f_c' A_1)$<br><br>Check if $\sqrt{\dfrac{A_2}{A_1}} \leq 2.0$ where<br><br>$A_1$ is the area of the column and $A_2$ is the area of the footing that is geometrically similar to and concentric with the column. | $\sqrt{\dfrac{A_2}{A_1}} = \sqrt{\dfrac{[(5.5 \text{ ft})(12 \text{ in./ft})]^2}{(18 \text{ in.})^2}} = 3.6 > 2$<br>Therefore, Eq. (22.8.3.2(b)) controls. |
| 21.2.1(d) | The reduction factor for bearing is 0.65: | $\phi_{bearing} = 0.65$<br>$\phi B_n = (0.65)(2)(0.85)(4000 \text{ psi})(18 \text{ in.})^2$<br>$\phi B_n = 1432 \text{ kip} > 515 \text{ kip (Step 4)}$ **OK** |
| 16.3.4.1 | Provide minimum dowel area of $0.005 A_g$ and at least four bars. This requirement is to ensure ductile behavior between the column and footing. | $A_{s,dowel} = 0.005(18 \text{ in.})^2 = 1.62 \text{ in.}^2$<br>Use four No. 6 bars in each corner of column. |
| 16.3.5.4 | Bars are in compression for all load combinations. Therefore, the bars must extend into the footing a compression development length, $\ell_{dc}$, the larger of the two and at least 8 in.: | |
| 25.4.9.2 | $\ell_{dc} = \begin{cases} \dfrac{f_y \psi_r}{50 \lambda \sqrt{f_c'}} d_b \\ (0.0003 f_y \psi_r d_b) \end{cases}$ | $\ell_{dc} = \dfrac{(60,000 \text{ psi})}{50\sqrt{4000 \text{ psi}}}(0.75 \text{ in.}) = 14.2 \text{ in.}$ **Controls**<br>$\ell_{dc} = (0.0003 \text{ in.}^2/\text{lb})(60,000 \text{ psi})(0.75 \text{ in.}) = 13.5 \text{ in.}$<br>$\ell_{dc} = 14.2 \text{ in. (controls)} > 8 \text{ in.}$ **OK** |
| 25.3.1 | The footing depth $h$ must satisfy the following inequality so that the vertical reinforcement can be developed:<br><br>$h \geq \ell_{dc} + r + d_{b,dwl} + 2 d_{b,bars} + 3 \text{ in.}$<br><br>where<br>$r$ = radius of No. 6 bent = $6 d_b$ | $h_{req'd} = 14.2 \text{ in.} + 6(0.75 \text{ in.}) + 0.75 \text{ in.} + 2(0.75 \text{ in.})$<br>$+ 3 \text{ in.} = 23.95 \text{ in.}$<br>$h_{req'd} = 23.95 \text{ in.} < h_{prov.} = 28 \text{ in.}$ **OK** |

| | | |
|---|---|---|
| | Assume that the column is reinforced with four No.8 bars. | Compression development length for No.8 bars is: |
| 25.5.5.4 25.4.9.1 25.4.9.2 | As mentioned above, bars are in compression for all load cases. Therefore, the compression lap splices is the larger of the two conditions: 1. The development length, $\ell_{dc}$, of the larger bar and | $\ell_{dc} = \dfrac{60,000 \text{ psi}}{50\sqrt{4000 \text{ psi}}}(1.0 \text{ in.}) = 19.0 \text{ in.}$ **Controls** $\ell_{dc} = (0.0003 \text{ in.}^2/\text{lb})(60,000 \text{ psi})(1.0 \text{ in.}) = 18 \text{ in.}$ |
| 25.5.5.1 | 2. The compression lap splice of the smaller bar | $\ell_{sc} = 0.0005(60,000 \text{ psi})(0.75 \text{ in.}) = 22.5 \text{ in.}$ Use $\ell_{sc} = 24$ in. $> 12$ in.  **OK** Therefore extend No.6 bars 24 in. into the column. |

**Step 9: Footing details**

| | | |
|---|---|---|
| 13.2.8.3 13.2.7.1 13.2.8.1 | <u>Development length</u> Reinforcement development is calculated at the maximum factored moment and the code permits the critical section to be located at the column face. Bars must extend a tension development length beyond the critical section. | |
| 25.4.2.3 | $\ell_d = \left( \dfrac{3}{40} \dfrac{f_y}{\lambda \sqrt{f'_c}} \dfrac{\psi_t \psi_e \psi_s}{\dfrac{c+K_{tr}}{d_b}} \right) d_b$ | $\ell_d = \left( \dfrac{3}{40} \dfrac{60,000 \text{ psi}}{(1.0)\sqrt{4000 \text{ psi}}} \dfrac{(1.0)(1.0)(1.0)}{2.5} \right) d_b = 28.5 d_b$ |
| 25.4.2.4 25.4.2.1 | where $\psi_t$ = casting position; $\psi_t = 1.0$, because not more than 12 in. of fresh concrete below horizontal reinforcement $\psi_e$ = coating factor; $\psi_e = 1.0$, because bars are uncoated $\psi_s$ = bar size factor; $\psi_s$ = for No. 6 and larger $c_b$ = spacing or cover dimension to center of bar, whichever is smaller $K_{tr}$ = transverse reinforcement index It is permitted to use $K_{tr} = 0$. But the expression: $\dfrac{c_b + K_{tr}}{d_b}$ must not exceed 2.5. The development length must be the greater of the calculated value of Eq. (25.4.2.2) and 12 in. | No. 6: $\dfrac{c_b + K_{tr}}{d_b} = \dfrac{3.44 \text{ in.} + 0}{0.75 \text{ in.}} = 4.59$ use maximum value of 2.5 No. 6 bars: $28.5(0.75 \text{ in.}) = 22$ in. $> 12$ in. Therefore, OK $\ell_d$ in the long direction: $\ell_{d,prov.} = ((10 \text{ ft})(12 \text{ in./ft}) - 18 \text{ in.})/2 - 3 \text{ in.}$ $\ell_{d,prov.} = 48$ in. $> \ell_{d,req'd} = 22$ in.  **OK** use straight No. 6 bars in long direction $\ell_d$ in the short direction: No. 6: $\ell_{d,prov.} = ((5.5 \text{ ft})(12 \text{ in./ft}) - 18 \text{ in.})/2 - 2 \text{ in.}$ $\ell_{d,prov.} = 22$ in. $\geq \ell_{d,req'd} = 22$ in.  **OK** |

## Step 10: Detailing

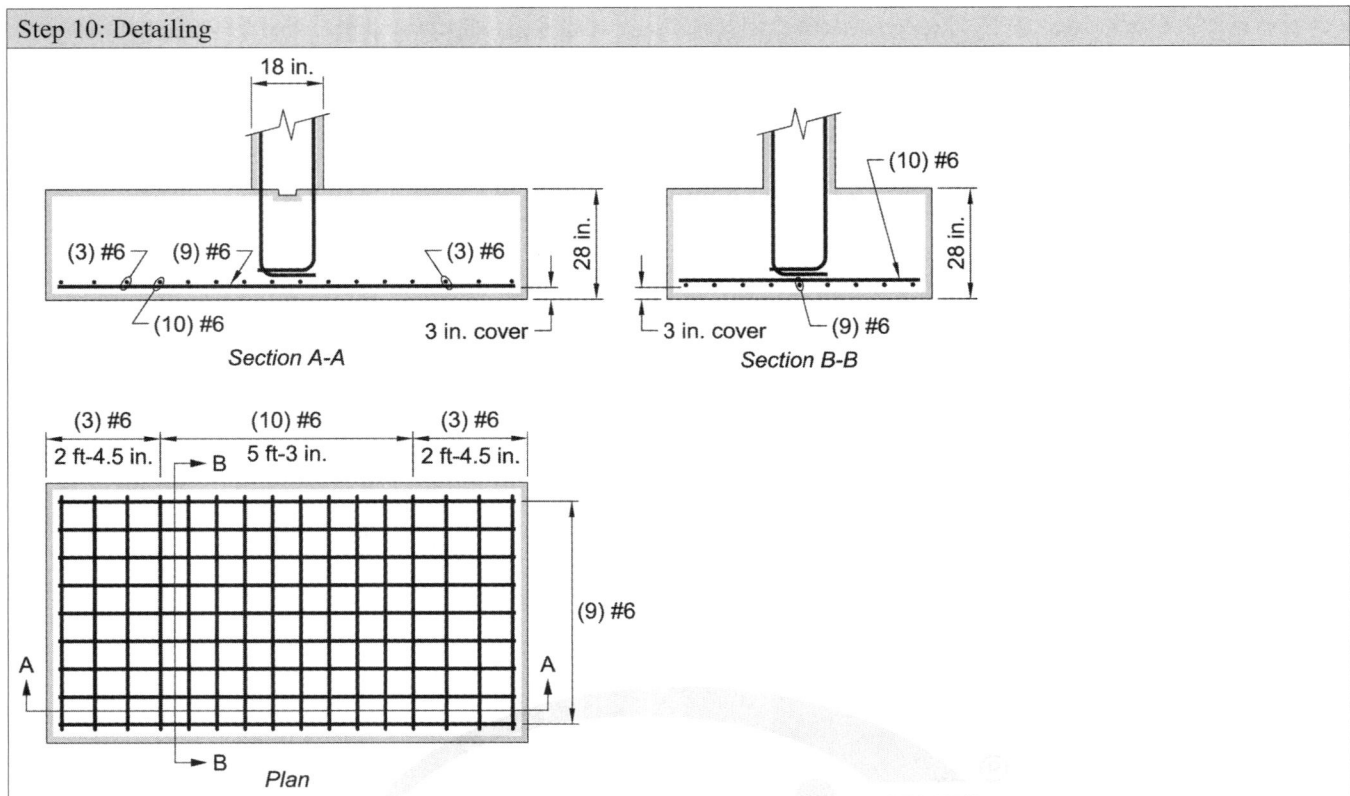

Fig. E4.8—Footing reinforcement detailing.

Square footing

If the problem was solved as square footing, then in Step 3, the following footing dimensions would have been selected: 7 ft 3 in. x 7 ft 3 in.

Following the same calculation procedure, shear strength is satisfied and minimum reinforcement ratio controls the flexure design (10 No.6 each direction). Distribution of reinforcement within a central band does not apply to square footings.

Development lengths and dowel calculations are not affected.

# CHAPTER 11—FOUNDATIONS

**Foundation Example 5:** *Design of a combined footing*

Design and detail a rectangular combined footing, founded on stiff soil, supporting two building columns, oriented as shown in Fig. E5.1. The bottom of the footing is 5 ft below finished grade.

The columns only transmit axial force and neither shear nor moment is transmitted from the frame above into the footing. The soil reaction to column loads is assumed to be uniform across the footing bearing area.

Given:

*Exterior column load—*
Service dead load $D_1$ = 150 kip
Service live load $L_1$ = 100 kip
$c_1 \times c_1$ = 18 in. x 18 in.

*Interior column load—*
Service dead load $D_2$ = 260 kip
Service live load $L_2$ = 160 kip
$c_2 \times c_2$ = 20 in. x 20 in.

*Material properties—*
Concrete compressive strength $f_c'$ = 4000 psi
Steel yield strength $f_y$ = 60,000 psi
Normalweight concrete $\lambda$ = 1
Density of concrete = 150 lb/ft$^3$
Allowable soil bearing pressure
$q_a$ = 5000 psf under all loads

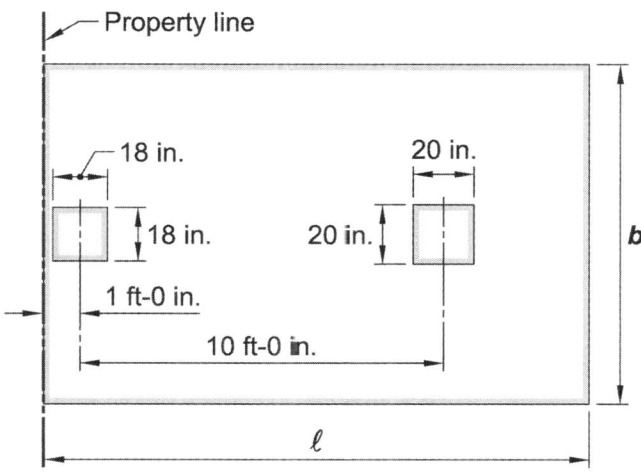

Fig. E5.1—Foundation plan.

| ACI 318-14 | Procedure | Calculation |
|---|---|---|
| **Step 1: Foundation type** | | |
| 13.1.1 | This footing is 5 ft below finished grade. Therefore, it is considered a shallow footing. | |
| **Step 2: Material requirements** | | |
| 13.2.1.1 | The mixture proportion must satisfy the durability requirements of Chapter 19 and structural strength requirements (ACI 318-14). The designer determines the durability classes. Please refer to Chapter 4 of this Handbook for an in-depth discussion of the categories and classes.<br><br>ACI 301 is a reference specification that is coordinated with ACI 318. ACI encourages referencing ACI 301 into job specifications.<br><br>There are several mixture options within ACI 301, such as admixtures and pozzolans, which the designer can require, permit, or review if suggested by the contractor.<br><br>Example 1 of this chapter provides a more detailed breakdown on determining the concrete compressive strength and exposure categories and classes. | By specifying that the concrete mixture shall be in accordance with ACI 301 and providing the exposure classes, Chapter 19 requirements are satisfied. Based on durability and strength requirements, and experience with local mixtures, the compressive strength of concrete is specified at 28 days to be at least 4000 psi. |

| | Step 3: Determine footing dimensions | |
|---|---|---|
| 13.1.1<br>13.3.1.1 | Service loads<br>To calculate the footing area, assume the columns are supported on isolated square footings. Divide the column service loads by the allowable soil pressure. | The unit weights of concrete and soil are 150 pcf and 120 pcf; close. Therefore, footing self–weight will be checked later: |
| | Exterior column:<br>$A_{req'd} = (D_1 + L_1)/q_a$ | $A_{req'd} = (100\text{ kip} + 150\text{ kip})/5\text{ ksf} = 50\text{ ft}^2$<br>Use 7 ft 3 in. x 7 ft 3in. |
| | Interior column:<br>$A_{req'd} = (D_2 + L_2)/q_a$ | $A_{req'd} = (260\text{ kip} + 160\text{ kip})/5\text{ ksf} = 84\text{ ft}^2$<br>Use 9 ft 3 in. x 9 ft 3in. |
| | Because the column is in close proximity to the property line, the exterior column footing cannot be concentric with the column, and the footing needs external bracing to remain stable. This can be supplied by a moment connection between the exterior and the interior footing, but in this case, the two footings are simply combined.<br><br>The footing thickness is calculated in Step 5, Two-way shear design. | |
| 13.2.6.2 | Determine the location of the resultant of the two service column loads by taking the moments about the center of the exterior column.<br><br>$x_c = \dfrac{P_1 x_1 + P_2 x_2}{\Sigma P_i}$ | $x_c = \dfrac{(260\text{ kip} + 160\text{ kip})(10\text{ ft})}{(150\text{ kip} + 100\text{ kip}) + (260\text{ kip} + 160\text{ kip})} = 6.3\text{ ft}$ |
| | The distance of the resultant from the property line is: | $x = 6.3\text{ ft} + 1\text{ ft} = 7.3\text{ ft}$ |
| 13.3.1.1 | The footing length, $L$, is taken equal to $2x$ so the soil pressure can be assumed as uniform under the two column loads: | $2(7.3\text{ ft}) = 14.6\text{ ft}$, |
| 13.3.4.3 | Distribution of bearing pressure under combined footing must be consistent with the soils properties and structure. The footing width is calculated (refer to Fig. E5.2):<br><br>$B = \dfrac{P}{q_a L}$ | $B = \dfrac{(150\text{ kip} + 100\text{ kip}) + (260\text{ kip} + 160\text{ kip})}{(5\text{ ksf})(14.6\text{ ft})} = 9.2\text{ ft}$<br>Use 9 ft 6 in. x 15 ft 0 in. |

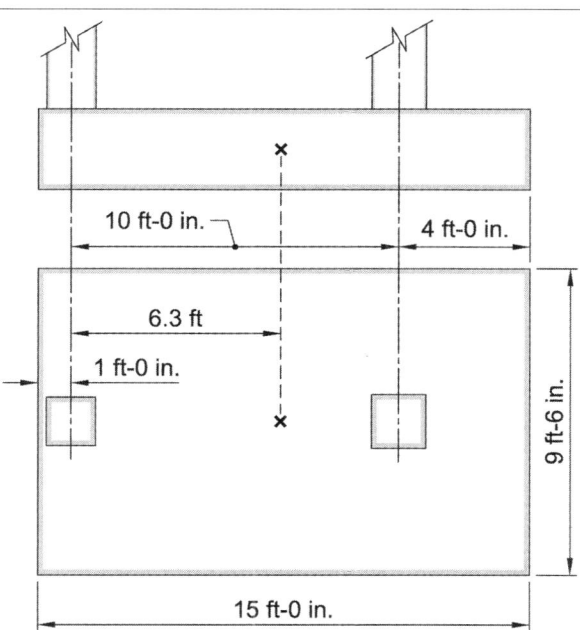

Fig. E5.2—Combined footing dimensions.

| Step 4: Design forces | | |
|---|---|---|
| 13.3.4.3<br>13.2.6.1 | Calculate soil pressure<br>Factored loads<br>Calculate the soil pressures resulting from the applied factored loads including footing self-weight. Assume 2 ft 6 in. thick footing.<br>Footing self-weight: | $W = (0.15 \text{ kip/ft}^3)(15 \text{ ft})(9.5 \text{ ft})(2.5 \text{ ft}) = 53.5 \text{ kip}$<br>See Note that follows. |
| | Soil self-weight above combined footing: | $W = (0.12 \text{ kip/ft}^3)(15 \text{ ft})(9.5 \text{ ft})(2.5 \text{ ft}) = 42.8 \text{ kip}$ |
| 13.2.6.1<br>13.3.4.3 | Total dead load: | dead load = 150 kip + 260 kip + 53.5 kip + 42.8 kip<br>= 506.3 kip |
| | Total live load: | live load = 100 kip + 160 kip = 260 kip |
| 5.3.1a | Load Case: $U = 1.4D$ | $q_u = \dfrac{1.4(506.3 \text{ kip})}{(15 \text{ ft})} = 47.3 \text{ kip/ft}$ |
| 5.3.1b | Load Case: $U = 1.2D + 1.6L$ | $q_u = \dfrac{1.2(506.3 \text{ kip}) + 1.6(260 \text{ kip})}{(15 \text{ ft})} = 68.2 \text{ kip/ft}$<br>**Controls** |
| | Distributed soil pressure per square area below combined footing (refer to Fig. E5.3): | $q_u = \dfrac{1024 \text{ kip}}{(15 \text{ ft})(9.5 \text{ ft})} = 7.2 \text{ kip/ft}^2$ |
| | Note: This is a conservative approach. The footing concrete displaces soil. Therefore, the actual load on soil is the difference between the concrete and soil unit weights multiplied by the footing volume. | |

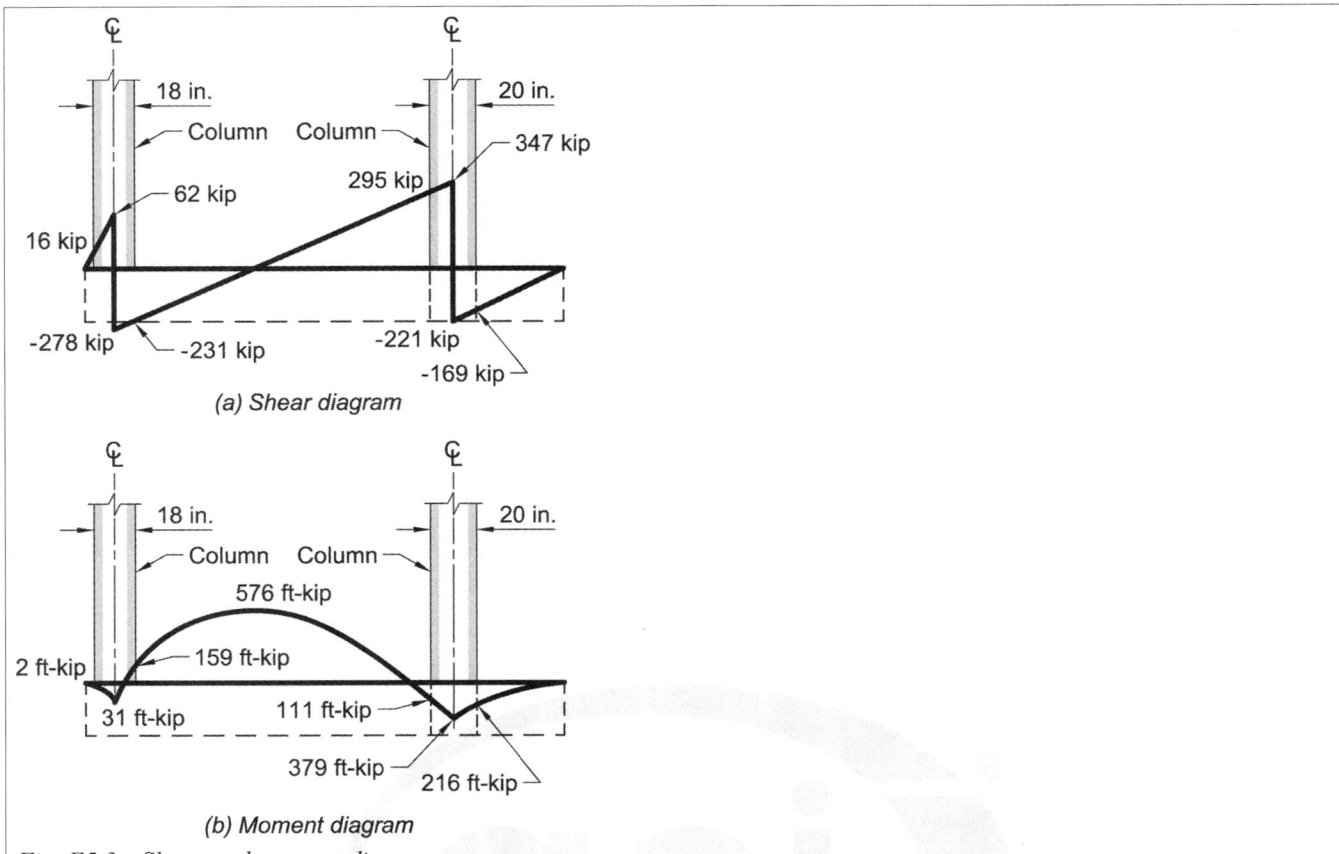

Fig. E5.3—Shear and moment diagrams.

Note: –231 kip and 295 kip are shear forces taken at the exterior and interior column faces, respectively. 159 ft-kip and 111 ft-kip and 216 ft-kip are flexure moment taken at the exterior column face and both interior column faces.

# CHAPTER 11—FOUNDATIONS

## Step 5: Two-way shear design

| | | |
|---|---|---|
| 13.3.4.1 | Design of combined footing must satisfy the requirements of Chapter 8 for two-way slab of ACI 318-14. | |
| | Check footing two-way shear strength at both columns. | |
| 21.2.1b | Shear strength reduction factor: | $\phi_{shear} = 0.75$ |
| 22.6.1.4 | Exterior column<br>The footing critical shear perimeter, $b_o$, at the exterior column is three sided. From the free body diagram, the direct shear force, $V_{ug}$, is the result of the factored column force less the factored soil pressure force within the critical shear perimeter (refer to Fig. E5.4 and Fig. E5.5). Therefore, $V_{ug} = P_{u1} - q_u A_1$ | Fig. E5.4—Determining centroidal axis of shear perimeter. |
| 22.6.4.1 | where $A_1 = (c_2 + d)((c_1 + d)/2 + e)$<br>$d = 30$ in. $- 3$ in. $- 1.128$ in./2 $= 26.4$ in. | $A_1 = (18 \text{ in.} + 26.4 \text{ in.})\left(\dfrac{18 \text{ in.} + 26.4 \text{ in.}}{2} + 12 \text{ in.}\right) = 1518.5 \text{ in.}^2$ |
| | Assume No. 9 bars and $e = 1$ ft edge distance from the column centerline (refer to Fig. E5.4) | |
| 5.3.1 | Solving for $V_{ug}$, where | |
| | $P_{u1} = 1.2 D_1 + 1.6 L_1$ | $P_{u1} = (1.2)(150 \text{ kip}) + (1.6)(100 \text{ kip}) = 340 \text{ kip}$ |
| | $F_u = q_u A_1$ | $F_u = \dfrac{(7.2 \text{ ksf})}{144 \text{ in.}^2/\text{ft}^2}(1518.5 \text{ in.}^2) = 76 \text{ kip}$ |
| | Substituting into:<br>$V_{ug} = P_{u1} - q_u A_1$ | $V_{ug} = 340 \text{ kip} - 76 \text{ kip} = 264 \text{ kip}$ |

| | | |
|---|---|---|
| | The Code requires the footing moment at the critical shear centroid is transferred into the column by direct flexure and by eccentricity of shear. | |
| | Calculate the centroid axis of shear perimeter (refer to Fig. E5.4 and Fig. E5.5): where | |
| | $$\bar{x} = \frac{2(b_1)^2/2}{2b_1 + b_2}$$ where | $$\bar{x} = \frac{2(34.2 \text{ in.})^2/2}{2(34.2 \text{ in.}) + 44.4 \text{ in.}} = 10.4 \text{ in.}$$ |
| | $$b_1 = e + \frac{c}{2} + \frac{d}{2}$$ and | $$b_1 = 12 \text{ in.} + \frac{18 \text{ in.}}{2} + \frac{26.4 \text{ in.}}{2} = 34.2 \text{ in.}$$ |
| | $b_2 = c + d$ | $b_2 = 18 \text{ in.} + 26.4 \text{ in.} = 44.4 \text{ in.}$ |
| | From the free body diagram (refer to Fig. E5.4 and Fig. E5.5), summing the factored column load and soil pressure force about the critical section centroid: | |
| 8.4.4.2.1 | $$M_u^* = P_{u1}(b_1 - e - \bar{x}) - F_u\left(\frac{b_1}{2} - \bar{x}\right)$$ | $M_u^* = (340 \text{ kip})(34.2 \text{ in.} - 12 \text{ in.} - 10.4 \text{ in.})$ $- 76 \text{ kip}\left(\frac{34.2 \text{ in.}}{2} - 10.4 \text{ in.}\right)$ |
| 8.4.4.2.3 | The maximum shear stress due to direct shear and shear due to moment transfer is: | $M_u^* = 3503$ in.-kip |
| | $$v_u = \frac{V_{ug}}{A_c} + \frac{\gamma_v M_u^* c}{J_c}$$ (Eq. (R8.4.4.2.3)) | |
| | $A_c = b_o d$; is the area of concrete within the critical section $b_o$; (refer to Fig. E5.4 and Fig. E5.5). | $A_c = (44.4 \text{ in.} + 2(34.2 \text{ in.}))(26.4 \text{ in.}) = 2978$ in.$^2$ |
| 8.4.4.2.3 | The shear perimeter moment of inertia $J_c$ is: | |
| | $$J_c = 2\left[b_1\frac{d^3}{12} + d\frac{b_1^3}{12} + (b_1 d)\left(\frac{b_1}{2} - \bar{x}\right)^2\right] + b_2 d\bar{x}^2$$ | $J_c = 2\left[(34.2 \text{ in.})\frac{(26.4 \text{ in.})^3}{12} + 26.4 \text{ in.}\frac{(34.2 \text{ in.})^3}{12}\right.$ $+ (34.2 \text{ in.})(26.4 \text{ in.})\left(\frac{34.2 \text{ in.}}{2} - 10.4 \text{ in.}\right)^2\right]$ $+ (44.4 \text{ in.})(26.4 \text{ in.})(10.4 \text{ in.})^2$ $= 488{,}727$ in.$^4$ |
| 8.4.2.3.2 | The portion of the moment is transferred by flexure is $\gamma_f M_u^*$, where $\gamma_f$ is: | |
| | $$\gamma_f = \frac{1}{1 + \frac{2}{3}\sqrt{\frac{b_1}{b_2}}}$$ | $$\gamma_f = \frac{1}{1 + \frac{2}{3}\sqrt{\frac{34.2 \text{ in.}}{44.4 \text{ in.}}}} = 0.63$$ |
| 8.4.4.2.2 | The moment fraction transferred by shear, $\gamma_v M_u^*$, where $\gamma_v$ is: $\gamma_v = 1 - \gamma_f$ | $\gamma_v = 1 - 0.63 = 0.37$ |

# CHAPTER 11—FOUNDATIONS

| | | |
|---|---|---|
| 8.4.4.2.1 | Solving for $v_u$ from Eq. (R8.4.4.2.3) above: <br><br> where $c = b_1 - \bar{x}$ <br> $c = 34.2$ in. $- 10.4$ in. $= 23.8$ in. | $v_u = \dfrac{264 \text{ kip}}{2978 \text{ in.}^2} + \dfrac{(0.37)(3503 \text{ in.}-\text{kip})(23.8 \text{ in.})}{488,727 \text{ in.}^4}$ <br> $= 0.089$ ksi $+ 0.063$ ksi $= 0.152$ ksi |
| 22.6.5.1 <br> 22.6.5.2 | Two-way shear strength equations must be satisfied and the least value controls: <br><br> $v_u \leq \phi v_c = \phi \begin{Bmatrix} \left(\dfrac{\alpha_s d}{b_o} + 2\right) \\ \left(2 + \dfrac{4}{\beta}\right) \\ 4 \end{Bmatrix} \lambda \sqrt{f'_c}$ | $\dfrac{\alpha_s d}{b_o} + 2 = \dfrac{(30)(26.4 \text{ in})}{(2)(34.2 \text{ in.}) + 44.4 \text{ in.}} + 2 = 9 > 4$  **NG** <br><br> $2 + \dfrac{4}{\beta} = 2 + \dfrac{4}{1} = 6 > 4$  **NG** <br><br> Therefore, 4  **Controls** |
| 22.6.5.3 | $\alpha_s = 30$, edge column | $\phi v_c = (0.75)(4)(1)\sqrt{4000}$ psi $= 190$ psi |
| 22.6.5.2 | $\phi v_c = \phi 4 \lambda \sqrt{f'_c}$ | $\phi v_c = 190$ psi $> v_u = 156$ psi  **OK** |
| 21.2.1 | Shear strength reduction factor: 0.75 | The factored stress exceeds the footing design shear stress. |

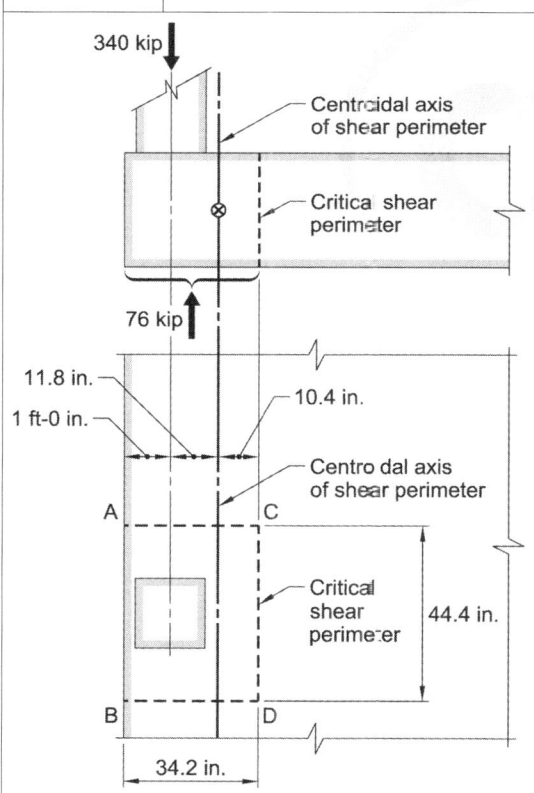

*Fig. E5.5—Two-way shear at the exterior column.*

| | Interior column: | |
|---|---|---|
| 8.4.4.1.1 | The maximum factored shear force at the critical section is equal to the factored column load less the factored soil pressure within the critical section (refer to Fig. E5.6): | |
| 22.6.1.4 | $V_u = P_{u2} - q_u(c_2 + d)(c_2 + d)$ | $V_u = (1.2)(260 \text{ kip}) + (1.6)(160 \text{ kip})$ |
| 5.3.1 | where $P_{u2} = 1.2D_2 + 1.6L_2$ | $- \dfrac{(7.2 \text{ ksf})(20 \text{ in.} + 26.4 \text{ in.})(20 \text{ in.} + 26.4 \text{ in.})}{(12 \text{ in./ft})^2}$ |
| | | $V_u = 568 \text{ kip} - 107.6 \text{ kip} = 459.4 \text{ kip, say, } 460 \text{ kip}$ |
| 22.6.4.1 | Critical section, $b_o$: | $b_o = (4)(20 \text{ in.} + 26.4 \text{ in.}) = 185.6 \text{ in.}$ |
| 22.6.5.1 | Two-way shear is the least of: | |
| 22.6.5.2 | | Check if the $\sqrt{f_c'}$ factors are less than 4. |
| | | 4 is used if the other factors are larger than 4. |
| | $v_u \le \phi v_c = \phi \begin{pmatrix} 4 \\ \left(2 + \dfrac{4}{\beta}\right) \\ \left(\dfrac{\alpha_s d}{b_o} + 2\right) \end{pmatrix} \lambda \sqrt{f_c'}$ | $(2 + 4/\beta) = 6 > 4$; with $\beta = 1$<br>Eq. (22.6.5.2(b)) does not control. |
| | | $(\alpha_s d/b_o + 2) = (40)(26.4 \text{ in.})/185.6 \text{ in.} = 5.7 > 4$.<br>Eq. (22.6.5.2(c)) does not control. |
| 22.6.5.3 | Use $\alpha_s = 40$ (interior column) | Therefore, use the factor 4. |
| | Shear strength Eq. (22.6.5.2(a)) controls: | $\phi v_c = (0.75)(4)(1.0)\left(\sqrt{4000 \text{ psi}}\right) = 189.7 \text{ psi}$ |
| | | $\phi V_c = (189.7 \text{ psi})(4)(20 \text{ in.} + 26.4 \text{ in.})(26.4 \text{ in.})$<br>$= 929.5 \text{ kip, say, } 930 \text{ kip}$ |
| 8.5.1.1 | Check if $\phi V_c > V_u$ | $\phi V_c = 930 \text{ kip} \gg V_u = 460 \text{ kip}$   **OK** |

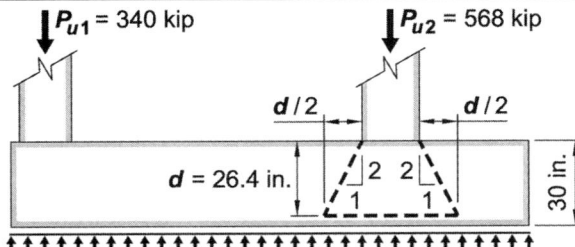

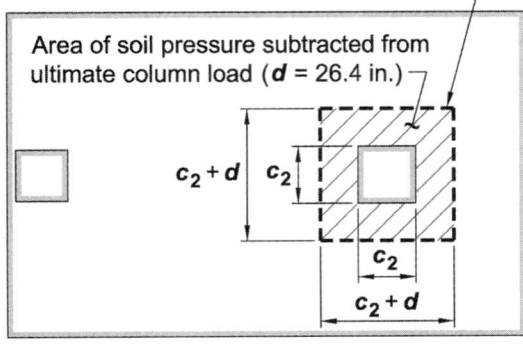

Fig. E5.6—Two-way shear at interior column.

## CHAPTER 11—FOUNDATIONS

| Step 6: One-way shear design | | |
|---|---|---|
| 13.3.2.1<br>13.2.7.2<br>7.4.3.2 | One-way shear design:<br>One-way shear strength is calculated at a distance $d$ from the interior column face (refer to Fig. E5.7) where the maximum shear force is permitted to be calculated.<br><br>Check if $\phi V_c$ with $d = 26.4$ in. exceeds $V_u = 295$ kip (refer to Fig. E5.3(a)). | |
| 7.4.1.1<br>7.4.3.2 | Calculate required strength from column factored loads less soil pressure<br><br>$V_u \leq V_{u@face} - q_u(c_2/2 + d)$<br><br>Calculate shear strength and verify that it exceeds the calculated required strength: | $V_u = 295 \text{ kip} - 68.2 \text{ kip/ft}\left(\dfrac{26.4 \text{ in.}}{12 \text{ in./ft}}\right) = 145 \text{ kip}$ |
| 7.5.1.1<br>7.5.3.1<br>22.5.1.1 | $\phi V_n \geq V_u$<br>$V_n = V_c + V_s$ | $V_s = 0$ (foundations are usually sized such that shear reinforced is not required). Therefore,<br>$V_n = V_c$ |
| 7.5.3.1<br>22.5.5.1<br>21.2.1 | Shear strength is calculated from:<br>$\phi V_c = \phi 2\sqrt{f'_c} b_w d$ and $\phi = 0.75$ | $\phi V_c = (0.75)(2)\left(\sqrt{4000 \text{ psi}}\right)(9.5 \text{ ft})(12 \text{ in./ft})(26.4 \text{ in.})$<br>$= 285.5$ kip |
| 7.5.1.1 | Check if $V_u < \phi V_c$ | $V_u = 145$ kip ~ $1/2\phi V_c = 285.5$ kip/2 = 143 kip<br><br>Therefore, shear reinforcement is not required. |

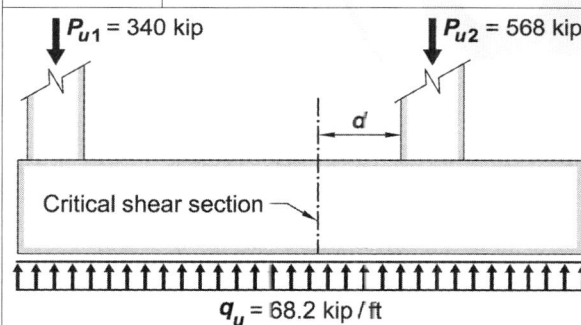

Fig. E5.7—One-way shear.

Summary: The combined footing thickness of 30 in. satisfies both one-way and two-way shear reinforcement.

## Step 7: Flexure design

Calculate the flexural reinforcement in the combined footing.

<u>Longitudinal direction</u>
Note that flexural tension occurs at the top of the footing between the two columns and at the bottom of the footing at both interior and exterior columns (Fig. E5.3(b)).

<u>Top reinforcement between columns</u>
At the section of maximum moment, set the internal compression force equal to internal tension force to calculate the reinforcement area:

| Code ref. | | |
|---|---|---|
| 22.2.2.4 | $C = T$ | |
| 22.2.2.4.1 22.2.3.1 | $0.85 f_c' ba = A_s f_y$ | $0.85(4000 \text{ psi})(9.5 \text{ ft})(12 \text{ in./ft})a = A_s 60{,}000 \text{ psi}$ <br> $a = 0.155 A_s$ |
| 22.3.1.1 | From moment diagram (Fig. E5.3(b)) <br> $M_u \leq \phi M_n = \phi A_s f_y (d - a/2)$ | $M_u = 576$ ft-kip |
| 21.2.1a | Use flexural strength reduction factor from Table 21.2.1 | $\phi = 0.9$ |
| 9.5.1.1a | Setting $\phi M_n \geq M_u$ and substitute for $a$ in the equation above. | $(576 \text{ ft-kip})\left(12 \dfrac{\text{in.-kip}}{\text{ft-kip}}\right) \geq 0.9 A_s (60 \text{ ksi})\left(26.4 \text{ in.} - \dfrac{0.155 A_s}{2}\right)$ |
| | Solve for $A_s$ | $A_s \geq 4.9$ in.$^2$ |
| 9.6.1.2 | Check if the minimum reinforcement area controls: | |
| | $A_{s,min} = \dfrac{3\sqrt{f_c'}}{f_y} bd$   (9.6.1.2a) | $A_{s,min} = \dfrac{3\sqrt{4000 \text{ psi}}}{60{,}000 \text{ psi}} (9.5 \text{ ft})(12 \text{ in./ft})(26.4 \text{ in.}) = 9.5$ in.$^2$ |
| | $A_{s,min} = \dfrac{200}{f_y} bd$   (9.6.1.2b) | $A_{s,min} = \dfrac{200}{60{,}000 \text{ psi}} (9.5 \text{ ft})(12 \text{ in./ft})(26.4 \text{ in.}) = 10.0$ in.$^2$ |
| | Eq. (9.6.1.2b) controls because concrete compressive strength $f_c'$ is less than 4444 psi. | $A_{s,min} = 10.0 \text{ in.}^2 > A_{s,req'd} = 4.9 \text{ in.}^2$ <br><br> Therefore, minimum reinforcement controls. <br><br> Use 10 No. 9 top continuous and evenly distributed over the width of the footing. <br><br> $A_{s,prov.} = (10)(1.0 \text{ in.}^2) = 10$ in.$^2$ |
| 22.2.2.4.1 22.2.2.4.3 | Check if section is tension controlled and the use of $\phi = 0.9$ is correct: | $a = \left(0.155 \dfrac{1}{\text{in.}}\right)(10 \text{ in.}^2) = 1.55$ in.$^2$ |
| 21.2.2 | $c = \dfrac{a}{\beta_1}$ | $c = \dfrac{1.55 \text{ in.}}{0.85} = 1.82$ in. |

| | | |
|---|---|---|
| 7.7.2.3 | Calculate the strain in the tension reinforcement and compare to the minimum strain required for tension-controlled section: $$\varepsilon_t = 0.003\left(\frac{d-c}{c}\right)$$ Place bars such that the spacing between them does not exceed $3h$ or 18 in. | $$\varepsilon_t = 0.003\left(\frac{26.4 \text{ in.} - 1.82 \text{ in.}}{1.82 \text{ in.}}\right) = 0.0405 > 0.005$$ Therefore, section is tension controlled. Use No.9 at 12 in. on center < $3h$ = 90 in. and 18 in. Place first bar placed at 3 in. from the edge. |

| | | |
|---|---|---|
| 13.2.7.1 | **Reinforcement at interior column**<br>The moment is taken at the interior face of the interior column: | $M_u = 216$ ft-kip (Fig. E5.3(b)) |
| 22.2.2.4 | At the section of maximum moment, set the internal compression force equal to internal tension force to calculate the reinforcement area: | |
| 22.2.2.4.1 | $C = T$ | |
| 22.2.3.1 | $0.85 f_c' b a = A_s f_y$ | $0.85(4000 \text{ psi})(9.5 \text{ ft})(12 \text{ in./ft})a = A_s 60{,}000 \text{ psi}$<br>$a = 0.155 A_s$ |
| 22.3.1.1<br>22.2.2.1<br>22.2.2.2 | $\phi M_n = \phi f_y A_s (d - a/2)$<br>Substitute for $a$ in the equation above. | |
| 21.2.1a | Use flexural strength reduction factor from Table 21.2.1: | $\phi = 0.9$ |
| 9.5.1.1a | Setting $\phi M_n \geq M_u$ and solving | $(216 \text{ ft-kip})(12 \text{ in./ft}) \geq 0.9(60 \text{ ksi}) A_s \left( 26.4 \text{ in.} - \dfrac{0.155 A_s}{2} \right)$ |
| | Solve for $A_s$ | $A_s \geq 1.83 \text{ in.}^2$ |
| 9.6.1.2 | This is less than the minimum reinforcement area calculated above. | Therefore, use $A_{s,min} = 10.0 \text{ in.}^2 > A_{s,req,d} = 1.83 \text{ in.}^2$<br><br>Use ten No. 9 continuous bottom bars evenly distributed over the width of the footing.<br><br>$A_s = (10)(1.0 \text{ in.}^2) = 10 \text{ in.}^2$<br><br>$a = \left( 0.155 \dfrac{1}{\text{in.}} \right)(10 \text{ in.}^2) = 1.55 \text{ in.}^2$ |
| | Check if section is tension controlled and the use of $\phi = 0.9$ is correct: | |
| 22.2.2.4.1<br>22.2.2.4.3 | $c = \dfrac{a}{\beta_1}$ | $c = \dfrac{1.55 \text{ in.}}{0.85} = 1.82 \text{ in.}$ |
| | Calculate the strain in the tension reinforcement and compare to the minimum strain required for tension-controlled section: | |
| 21.2.2 | $\varepsilon_t = 0.003 \left( \dfrac{d - c}{c} \right)$ | $\varepsilon_t = 0.003 \left( \dfrac{26.4 \text{ in.} - 1.82 \text{ in.}}{1.82 \text{ in.}} \right) = 0.0405 > 0.005$<br><br>Therefore, section is tension controlled. |
| | Note: The calculated factored moment at the exterior column face (159 ft-kip) and the exterior moment at the interior column (111 ft-kip) is smaller than the calculated factored interior moment of the interior column face (216 ft-kip). Therefore, minimum reinforcement area controls. Provide 10 No. 9 bottom bars over full length of combined footing and spaced at 12 in. on center < $3h$ = 90 in. and 18 in. | |

## CHAPTER 11—FOUNDATIONS

| | | |
|---|---|---|
| | Transverse reinforcement<br>In a combined footing, transverse moment distribution may be addressed similar to an isolated spread footing. A strip over the width of the footing is considered to resist the column load. This strip is, however, not independent of the footing itself. | |
| | Darwin et al. (2015) and Fanella (2011) recommend the width of the strip to be half the effective depth ($d/2$) on either side of the footing from the face of columns. | Calculate $d$ to the center of the second layer.<br>$d = 30$ in. $- 3$ in. $- 1.128$ in. $- 1.0$ in. $= 25.37$ in., say, $d = 25.3$ in.<br><br>$w = 20$ in. $+ 2(25.3$ in./2$) = 45.3$ in. |
| | Interior column<br>The factored column load distributed over the width of the footing is used to determine the transverse bending moment. | |
| 5.3.1 | Factored distributed soil reaction is:<br><br>$q^*_{u,TI} = \dfrac{P_{u,int}}{B}$ | $q^*_{u,TI} = \dfrac{1.2(260 \text{ kip}) + 1.6(160 \text{ kip})}{9.5 \text{ ft}} = 59.8$ kip/ft<br><br>say, 60 kip/ft |
| 13.2.7.1 | The factored moment at the column face is:<br><br>$M_u = \dfrac{1}{2} q^*_u \left( \dfrac{b}{2} - \dfrac{c}{2} \right)^2$<br><br>where<br><br>$\dfrac{b}{2} - \dfrac{c}{2} = \dfrac{9.5 \text{ ft}}{2} - \dfrac{20 \text{ in.}}{2(12 \text{ in./ft})} = 3.92$ ft<br><br>Refer to Fig. E5.8 | $M_u = \dfrac{1}{2}(60 \text{ kip/ft})(3.92 \text{ ft})^2 = 460$ ft-kip |
| 8.5.1.1 | Calculate required reinforcement:<br>$\phi M_n = \phi f_y A_s j d \geq M_u$<br><br>Coefficient on $d$: $j = 0.9$ | 460 ft-kip $= 0.9 A_s (60,000$ psi$)(0.9)(25.3$ in.$)$<br>$A_s = 4.5$ in.$^2$ |
| 21.2.1 | Flexural strength reduction factor: | $\phi = 0.9$ |
| 9.6.1.2 | Check if the minimum reinforcement area controls:<br><br>$A_{s,min} = \dfrac{200}{f_y} bd$ | $A_{s,min} = \dfrac{200}{60,000 \text{ psi}}(45.3 \text{ in.})(25.3 \text{ in.}) = 3.8$ in.$^2$/ft |
| | Eq. (9.6.1.2b) controls because concrete compressive strength $f_c'$ is less than 4444 psi.<br><br>Check if section is tension controlled and the use of $\phi = 0.9$ is correct: | Required reinforcement is greater than the minimum required. Therefore use eight No. 7 spaced at 6 in. on center and placed within the calculated width 46.4 in.<br>$A_{s,prov} = (8)(0.6 \text{ in.}^2) = 4.8$ in.$^2 > A_{s,req'd} = 4.5$ in.$^2$<br><br>$a = \dfrac{(0.9)(60,000 \text{ psi})(4.8 \text{ in.}^2)}{0.85(4000 \text{ psi})(45.3 \text{ in.})} = 1.68$ in. |
| 22.2.2.4.1<br>22.2.2.4.3 | $c = \dfrac{a}{\beta_1}$ | $c = \dfrac{1.68 \text{ in.}}{0.85} = 1.98$ in. |

| Section | | | |
|---|---|---|---|
| 21.2.2 | Calculate the strain in the tension reinforcement and compare to the minimum strain required for tension-controlled section: $$\varepsilon_t = 0.003\left(\frac{d-c}{c}\right)$$ | | $\varepsilon_t = 0.003\left(\dfrac{26.4 \text{ in.} - 1.98 \text{ in.}}{1.98 \text{ in.}}\right) = 0.035 > 0.005$ |
| | | | Therefore, section is tension controlled. |
| | **Exterior column** The factored column load distributed over the width of the footing is used to determine the transverse bending moment. | | $d = 25.3$ in. $w = 12$ in. $+ 18$ in./2 $+ 25.3$ in./2 $= 33.7$ in. |
| 5.3.1 | Factored distributed soil reaction is: $$q^*_{u,TI} = \frac{P_{u,int}}{B}$$ | | $q^*_{u,TI} = \dfrac{1.2(150 \text{ kip}) + 1.6(100 \text{ kip})}{9.5 \text{ ft}} = 35.8$ kip/ft |
| 13.2.7.1 | The factored moment at the column face is: $$M_u = \frac{1}{2}q^*_u\left(\frac{b}{2} - \frac{c}{2}\right)^2$$ $$\frac{b}{2} - \frac{c}{2} = \frac{9.5 \text{ ft}}{2} - \frac{18 \text{ in.}}{2(12 \text{ in./ft})} = 4 \text{ ft}$$ Refer to Fig. E5.8. | | $M_u = \dfrac{1}{2}(35.8 \text{ kip/ft})(4 \text{ ft})^2 = 286$ ft-kip |
| 8.5.1.1 | Calculate required reinforcement: $\phi M_n \geq M_u = \phi f_y A_s jd$ Coefficient on $d$: $j = 0.9$ | | 286 ft-kip $= 0.9 A_s (60,000 \text{ psi})(0.9)(25.3 \text{ in.})$ $A_s = 2.8$ in.$^2$ |
| 21.2.1 | Flexural strength reduction factor: | | $\phi = 0.9$ |
| 9.6.1.2 | Check if the minimum reinforcement area controls: $$A_{s,min} = \frac{200}{f_y}bd$$ | | $A_{s,min} = \dfrac{200}{60,000 \text{ psi}}(33.7 \text{ in.})(25.3 \text{ in.}) = 2.8$ in.$^2$/ft |
| | | | Minimum reinforcement is equal to the calculated required reinforcement. Therefore use five No. 7 spaced at 10 in. on center. $A_{s,prov} = (5)(0.6 \text{ in.}^2) = 3.0$ in.$^2 > A_{s,req'd} = 2.8$ in.$^2$ |
| | Check if section is tension controlled and the use of $\phi = 0.9$ is correct: | | $a = \dfrac{(0.9)(60,000 \text{ psi})(3.0 \text{ in.}^2)}{0.85(4000 \text{ psi})(33.7 \text{ in.})} = 1.41$ in. |
| 22.2.2.4.1 | $$c = \frac{a}{\beta_1}$$ | | $c = \dfrac{1.41 \text{ in.}}{0.85} = 1.66$ in. |
| 22.2.2.4.3 | Calculate the strain in the tension reinforcement and compare to the minimum strain required for tension-controlled section: | | |
| 21.2.2 | $$\varepsilon_t = 0.003\left(\frac{d-c}{c}\right)$$ | | $\varepsilon_t = 0.003\left(\dfrac{25.3 \text{ in.} - 1.66 \text{ in.}}{1.66 \text{ in.}}\right) = 0.043 > 0.005$ |
| | | | Therefore, section is tension controlled. |

| | For sections outside the effective width at the exterior and interior columns, provide minimum reinforcement area. | $A_{s,min} = \dfrac{200}{60,000 \text{ psi}} (12 \text{ in.})(25.3 \text{ in.}) = 1 \text{ in.}^2/\text{ft}$ <br><br> Use No. 7 at 7 in. on center < $3h$ = 90 in. or 18 in. **OK** |
|---|---|---|
| | \<diagram showing footing dimensions: 3.92 ft / 20 in. / 3.92 ft (Interior); 4.0 ft / 18 in. / 4.0 ft (Exterior)\> | |

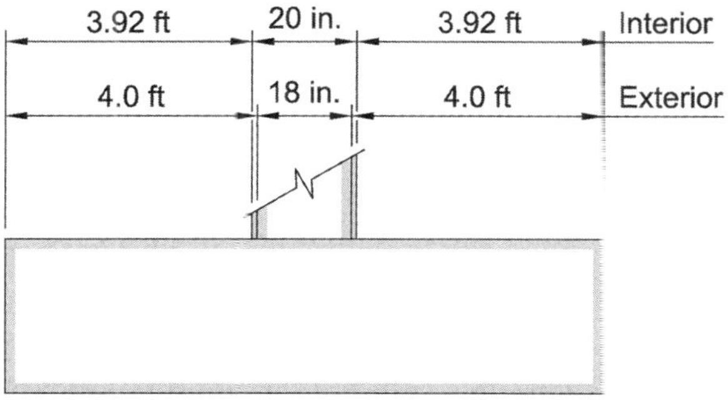

Fig. E5.8—*Footing width at columns for transverse reinforcement calculations.*

## Step 8: Footing details

### Development length of No. 9 top bars

From the moment diagram in Fig. E5.3(b), the positive moment inflection points at the exterior and interior columns occur at 0.12 ft and 0.45 ft from the respective column centerlines (Fig. E5.9). Therefore, extend top bars to the edge of the footing.

Check if the available distance is sufficient to develop the top bars at midspan in tension.

The development length for No. 9 bar is calculated using a simplified equation as allowed by ACI 318-14 code rather than the more detailed Eq. (25.4.2.3a):

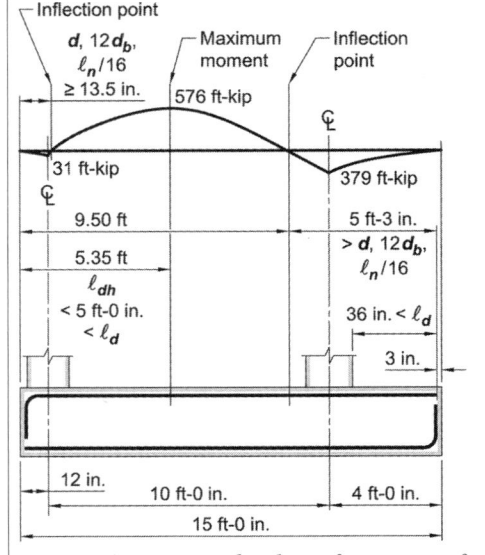

Fig. E5.9—Longitudinal reinforcement of combined footing.

**25.4.2.2**
$$\ell_d = \left( \frac{f_y \psi_t \psi_e}{20 \lambda \sqrt{f'_c}} \right) d_b$$

$$\ell_d = \left( \frac{(60,000 \text{ psi})(1.3)(1.0)}{(20)(1.0)\sqrt{4000 \text{ psi}}} \right) d_b = 61.7 d_b = 69.6 \text{ in.}$$

where

**25.4.2.4**
$\psi_t$ (casting position) = 1.3 for top bars because more than 12 in. of fresh concrete is placed below horizontal bars and
$\psi_e$ (coating factor) = 1.0 because bars are uncoated

Use 70 in. = 5 ft 10 in.

**25.4.1.4** Check if $\sqrt{f'_c}$ is less than 100 psi

$\sqrt{4000 \text{ psi}} = 63.2 \text{ psi} < 100 \text{ psi}$ **OK**

**25.4.2.1** Check if development is less than 12 in.

$\ell_d = 69.6 \text{ in.} = 5.8 \text{ ft} > 12 \text{ in.}$ **OK**

$15 \text{ ft} - 5.35 \text{ ft} = 9.65 \text{ ft} > 5.8 \text{ ft}$ **OK**

The available length from maximum moment at midspan to the interior column is greater than the calculated development length (Refer to Fig. E5.9):

Enough distance is available to develop No. 9 bars.

The available length from maximum moment at midspan to the exterior column is less than the calculated development length, $\ell_d = 5.8$ ft.

**25.4.3.1** Therefore, a hook is required at the exterior support and must be the greater of:

a. $\ell_{dh} = \left( \frac{f_y \psi_e \psi_c \psi_r}{50 \lambda \sqrt{f'_c}} \right) d_b$

$\ell_{dh} = \left( \frac{(60,000 \text{ psi})(1.0)(0.7)(1.0)}{50 \lambda \sqrt{4000 \text{ psi}}} \right)(1.128 \text{ in.})$

$\ell_{dh} = 14.98$ in., say, 15 in. **Controls**

b. $8 d_b$
c. 6 in.

$8(1.128 \text{ in.}) = 9 \text{ in.}$
6 in.

where:
$\psi_e$ (coating factor) = 1.0 because bars are uncoated
$\psi_c$ (cover) = 0.7 for No.9 bars with 3 in. side cover and 3 in. cover for the 90-degree hook
$\psi_r$ (confinement) = 1.0 bars are not confined.

Therefore, No. 9 top straight bars can be placed full length and will be developed at the point of maximum moment at the interior column and 90-degree hook at the exterior column.

| | | |
|---|---|---|
| 25.4.2.2<br><br>25.4.2.4<br><br><br><br><br>25.4.2.1 | Development of bottom bars<br>Longitudinal bars No. 9<br>The factored moment at the exterior column is negligible, 2 ft-kip at face of column, 32 ft-kip at column centerline (refer to Fig. E5.3(b)).<br><br>Calculate the development length at the interior column $M_u = 279$ ft-kip at exterior face (Fig. E5.3(b)):<br><br>$$\ell_d = \left(\frac{f_y \psi_t \psi_e}{20\lambda\sqrt{f'_c}}\right) d_b$$<br><br>where<br>$\psi_t$ = casting location; $\psi_t = 1.0$, because not more than 12 in. of fresh concrete is placed below horizontal reinforcement<br>$\psi_e$ = coating factor; $\psi_e = 1.0$, because bars are uncoated<br><br>Check if development is less than 12 in.<br><br>Interior column:<br>Column is located 4 ft from footing edge, which is less than the required calculated development length of 54 in. = 4 ft 6 in. Therefore, provide a hook for the bottom No. 9 bars at the interior column. Refer to prior calculations for top bars. | $\ell_d = \left(\frac{(60,000 \text{ psi})(1.0)(1.0)}{(20)(1.0)\sqrt{4000 \text{ psi}}}\right) d_b = 47.4 d_b$<br><br>\| Bar size \| $\ell_d$, in. \| Use, in. \|<br>\|---\|---\|---\|<br>\| No. 9 \| 53.5 \| 54 \|<br>\| No. 7 \| 41.5 \| 42 \|<br><br><br><br>Both required development length exceeds 12 in. Therefore, **OK**<br><br><br><br><br>$\ell_{dh}$ = 15 in. < 15 ft – 11.83 ft = 3.17 ft   **OK** |
| | Note: That if the more detailed development length Eq. (25.4.2.3a) is used, then adequate distance is available to place the No. 9 bars without having to bend them. | |
| | Transverse reinforcement:<br>From Fig. E5.8, the overhang at the interior column is 3.92 ft = 47 in., which is greater than the required calculated development length = 42 in. Therefore, No.7 bars are placed straight. | |

## Step 9: Column-to-footing connection

| | | |
|---|---|---|
| 16.3.1.1 | Interior column<br>Factored column forces are transferred to the footing by bearing and through reinforcement, usually dowels. | |
| 22.8.3.2 | The footing is wider on all sides than the loaded area. Therefore, the nominal bearing strength, $B_n$, is the smaller of the two equations.<br><br>(a) $B_n = (0.85 f_c' A_1)\sqrt{\dfrac{A_2}{A_1}}$<br><br>and<br><br>(b) $B_n = 2(0.85 f_c' A_1)$<br><br>$A_1$ is the bearing area of the column and $A_2$ is the area of the part of the supporting footing that is geometrically similar to and concentric with the loaded area.<br><br>The sides of the pyramid tapered wedges are sloped 1 vertical to 2 horizontal.<br><br>Check if $\sqrt{\dfrac{A_2}{A_1}} \leq 2.0$ where | Center of column is located 3 ft 2 in. from the end of footing and 3 ft 11 in. of the combined footing long sides.<br><br>3.16 ft /2 = 1.58 ft = 19 in. < 30 in. footing thickness<br>$A_2 = [2(10 \text{ in.}) + (3.16 \text{ ft})(12 \text{ in./ft})]^2 = 9063 \text{ in.}^2$<br><br>$\sqrt{\dfrac{A_2}{A_1}} = \sqrt{\dfrac{9063 \text{ in.}^2}{(20 \text{ in.})^2}} = 4.76 > 2$<br><br>Therefore, Eq. (22.8.3.2(b)) controls.<br>$B_n = 2(0.85 f_c' A_1)$ |
| 21.2.1 | The bearing strength reduction factor is 0.65: | $\phi_{bearing} = 0.65$<br><br>$\phi B_n = (0.65)(2)(0.85)(4000 \text{ psi})(20 \text{ in.})^2$<br>$\phi B_n = 1768 \text{ kip} > 1.2D + 1.6L = 600 \text{ kip}$ |
| 16.3.4.1 | Column factored forces are transferred to the footing by bearing and through dowels. The minimum dowel area is $0.005 A_g$ and at least four bars across the interface between interior column and combined footing. | $A_{s,dowel} = 0.005(20 \text{ in.})^2 = 2.0 \text{ in.}^2$<br><br>Use one No. 7 bar in each corner of column. |
| 16.3.5.4 | The four No. 7 dowels must be developed in the footing depth.<br><br>Bars are in compression for all load combinations. Therefore, the bars must extend into the footing at least a compression development length $\ell_{dc}$, which is the larger of the following two expressions: | $A_s = (4)(0.6 \text{ in.}^2) = 2.4 \text{ in.}^2 > A_{s,dowel} = 2.0 \text{ in.}^2$ |
| 25.4.9.2 | $\ell_{dc} = \begin{cases} \dfrac{f_y \psi_r}{50 \lambda \sqrt{f_c'}} d_b \\ (0.0003 f_y \psi_r d_b) \end{cases}$ | $\ell_{dc} = \dfrac{(60{,}000 \text{ psi})(1.0)}{(50)\sqrt{4000 \text{ psi}}}(0.875 \text{ in.}) = 16.6 \text{ in.}$ **Controls**<br><br>$\ell_{dc} = (0.0003)(60{,}000 \text{ psi})(0.875 \text{ in.}) = 15.75 \text{ in.}$ |

| | | |
|---|---|---|
| 25.4.9.3 | $\psi_r$ = confining reinforcement factor;<br>$\psi_r$ = 1.0, because reinforcement is not confined<br>The footing depth must satisfy the following inequality:<br>$h \geq \ell_{dc} + r + d_{b,dwl} + d_{b,\#7} + d_{b,\#9} + 3$ in. | $h_{req'd}$ = 16.6 in. + 6(0.875 in.) + 0.875 in. + 0.875 in.<br>+ 1.128 in. + 3 in. = 27.728 in., say, 28 in.<br>$h_{req'd}$ = 28.0 in. < $h_{prov.}$ = 30 in.   OK |
| 20.6.1.3.2 | 3 in. cover (refer to Fig. E5.10)<br><br>Check development length of dowel reinforcement into the column.<br><br>The length of dowels in the column is the greater of the development length and lap splice length. Assume that the column is reinforced with six No. 8 bars.<br>$d_{b,dowel} < d_{b,column}$ | 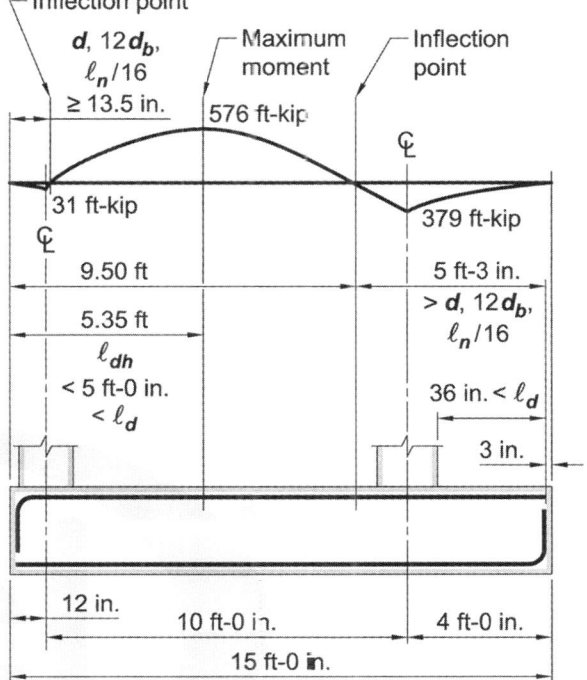<br>Fig. E5.10—Reinforcement development length. |
| 25.4.9.2 | Therefore, the lap splice length must be the greater of a) and b):<br><br>a. No. 8 bars is the larger of:<br><br>$\ell_{sc}$ = larger of $\begin{cases} \dfrac{f_y \psi_r}{50\lambda\sqrt{f_c'}} d_b \\ 0.0003 f_y d_b \end{cases}$<br><br>where<br>$\psi_r$ = confining reinforcement factor;<br>$\psi_r$ = 1.0, because stirrup spacing is greater than 4 in. (condition 3) | $\dfrac{(60{,}000 \text{ psi})(1.0)}{(50)(1.0)\sqrt{4000 \text{ psi}}}(1.0 \text{ in.}) = 19.0$ in.<br><br>$0.0003(60{,}000 \text{ psi})(1.0 \text{ in.}) = 18.0$ in. |
| 25.5.5.1 | b. The compression lap splice length of No. 7 is the larger of:<br><br>$\ell_{sc}$ = larger of $\begin{cases} 0.0005 f_y d_b \\ 12 \text{ in.} \end{cases}$ | $0.0005(60{,}000 \text{ psi})(0.875 \text{ in.}) = 26.25$ in.<br><br>12 in.<br><br>Use 27 in. = 2 ft 3 in. long lap splice. |

| | | |
|---|---|---|
| 16.3.1.1 | <u>Exterior column</u><br>The column factored forces are transferred to the footing by bearing and through dowels. | |
| 22.8.3.2 | The footing is wider on all sides than the loaded area. Therefore, the nominal bearing strength, $B_n$, is the smaller of the following two equations.<br><br>(a) $B_n = (0.85 f_c' A_1)\sqrt{\dfrac{A_2}{A_1}}$<br><br>and<br><br>(b) $B_n = 2(0.85 f_c' A_1)$<br><br>$A_1$ is the bearing area of the column and $A_2$ is the area of the part of the supporting footing that is geometrically similar to and concentric with the loaded area.<br><br>The sides of the pyramid tapered wedges are sloped 1 vertical to 2 horizontal.<br><br>Check if $\sqrt{\dfrac{A_2}{A_1}} \leq 2.0$, where | Center of column is located 1 ft-0 in. from the end of footing and 3 ft-11 in. of the combined footing long sides.<br><br>1 ft /2 = 0.5 ft = 6 in. < 30 in. footing thickness<br>$A_2 = [2(9 \text{ in.} + 3 \text{ in.})]^2 = 576 \text{ in.}^2$<br><br>$\sqrt{\dfrac{A_2}{A_1}} = \sqrt{\dfrac{576 \text{ in.}^2}{(18 \text{ in.})^2}} = 1.33 < 2$<br><br>Therefore, Eq. (22.8.3.2(a)) controls. |
| 21.2.1 | The bearing strength reduction factor for bearing is 0.65: | $\phi_{bearing} = 0.65$ |
| 16.3.4.1 | Column factored forces are transferred to the footing by bearing and dowels. The minimum dowel area is $0.005 A_g$ and at least four bars are needed across the interface between column and footing.<br><br>The four No.6 dowels must be developed in the footing. | $\phi B_n = (0.65)(0.85)(4000 \text{ psi})(18 \text{ in.})^2 \sqrt{\dfrac{576 \text{ in.}^2}{(18 \text{ in.})^2}}$<br><br>$\phi B_n = 955 \text{ kip} > 340 \text{ kip}$ **OK**<br><br>$A_{s,dowel} = 0.005(18 \text{ in.})^2 = 1.62 \text{ in.}^2$ |
| 16.3.5.4 | The bars are in compression for all load combinations. Therefore, the bars must extend into the footing a compression development length $\ell_{dc}$, which is the larger of the two following expressions: | Use one No. 6 bars in each corner of the column.<br><br>$A_s = (4)(0.44 \text{ in.}^2) = 1.76 \text{ in.}^2 > A_{s,dowel} = 1.62 \text{ in.}^2$ **OK** |
| 25.4.9.2 | $\ell_{dc} = \begin{cases} \dfrac{f_y \psi_r}{50 \lambda \sqrt{f_c'}} d_b \\ (0.0003 f_y \psi_r d_b) \end{cases}$<br><br>where<br>$\psi_r$ = confining reinforcement factor;<br>$\psi_r = 1.0$, because reinforcement is not confined | $\ell_{dc} = \dfrac{(60,000 \text{ psi})(1.0)}{(50)\sqrt{4000 \text{ psi}}}(0.75 \text{ in.}) = 14.2 \text{ in.}$ **Controls**<br><br>$\ell_{dc} = 0.0003(60,000 \text{ psi})(1.0)(0.75 \text{ in.}) = 13.5 \text{ in.}$ |
| | The footing depth $h$ must satisfy the following inequality:<br>$h \geq \ell_{dc} + r + d_{b,dwl} + d_{b,\#7} + d_{b,\#9} + 3 \text{ in.}$ | $h_{req'd} = 14.2 \text{ in.} + 6(0.75 \text{ in.}) + 0.75 \text{ in.} + 0.875 \text{ in.}$<br>$\quad + 1.128 \text{ in.} + 3 \text{ in.} = 24.5 \text{ in.}, \text{ say, } 25 \text{ in.}$<br>$h_{req'd} = 25 \text{ in.} < h_{ftg.prov.} = 30 \text{ in.}$ **OK** |
| 20.6.1.3 | 3 in. cover (refer to Fig. E5.10) | |

| | | |
|---|---|---|
| | Check development length of dowel reinforcement into the column.<br><br>The length of dowels in the column is the greater of the development length and lap splice length. Assume that the column is reinforced with six No. 8 bars.<br>$d_{b,dowel} < d_{b,column}$ | |
| 25.4.9.2 | Therefore, the lap splice length must be at least equal to the larger of (a) and (b):<br><br>(a) For column bars, at least the larger of: | |
| 25.5.5.1 | $\ell_{dc} = \begin{cases} \dfrac{f_y \psi_r}{50 \lambda \sqrt{f'_c}} d_b \\ (0.0003 f_y \psi_r d_b) \end{cases}$<br><br>where<br>$\psi_r$ = confining reinforcement factor;<br>$\psi_r = 1.0$, because reinforcement is not confined | $\dfrac{(60,000 \text{ psi})(1.0)}{(50)(1.0)\sqrt{4000 \text{ psi}}}(0.75 \text{ in.}) = 14.3$ in. **Controls**<br><br>$0.0003(60,000 \text{ psi})(0.75 \text{ in.}) = 13.5$ in. |
| | (b) The compression lap splice length of dowel must be at least the larger of:<br><br>$\ell_{sc}$ = larger of $\begin{cases} 0.0005 f_y d_b \\ 12 \text{ in.} \end{cases}$ | $0.0005(60,000 \text{ psi})(0.75 \text{ in.}) = 22.5$ in. **Controls**<br><br>12 in.<br><br>Use 24 in. long lap splice. |

## Step 10: Details

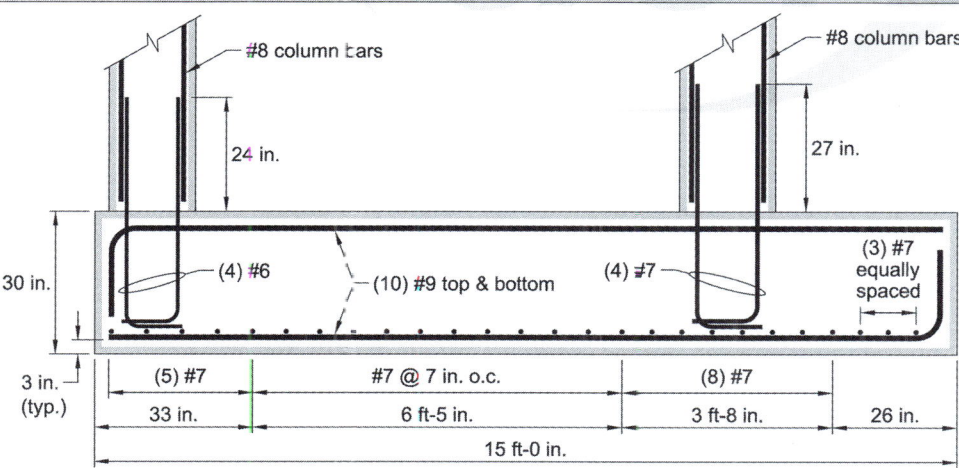

Fig. E5.11—Combined footing dimensions and reinforcement.

## References

Darwin, D.; Dolan, C., Nilson, A., eds., 2015, *Design of Concrete Structures*, McGraw-Hill Professional Publishing, 15th edition, New York, 786 pp.

Fanella, D., ed., 2011, *Reinforced Concrete Structures: Analysis and Design*, McGraw-Hill Professional Publishing, first edition, New York, 615 pp.